Progress in Mathematics

Volume 181

Resolution of Singularities

A research textbook in tribute to Oscar Zariski

Based on the courses given at the Working Week in Obergurgl, Austria, September 7–14, 1997

H. Hauser
J. Lipman
F. Oort
A. Quirós
Editors

Springer Basel AG

Editors:

Herwig Hauser
Institut für Mathematik
Universität Innsbruck
A-6020 Innsbruck

Joseph Lipman
Department of Mathematics
Purdue University
West Lafayette, IN 47907
USA

Frans Oort
Department of Mathematics
Universiteit Utrecht
3508 TA Utrecht
The Netherlands

Adolfo Quirós
Departamento de Matemáticas
Universidad Autónoma de Madrid
28049 Madrid
Spain

1991 Mathematics Subject Classification 14B05, 32Sxx, 58C27

A CIP catalogue record for this book is available from the Library of Congress, Washington D.C., USA

Deutsche Bibliothek Cataloging-in-Publication Data

Resolution of singularities : a research textbook in tribute to Oscar Zariski ; based on the courses given at the working week in Obergurgel, Austria, September 7 - 14, 1997 / H. Hauser . . . - Boston ; Basel ; Berlin : Birkhäuser, 2000
(Progress in mathematics ; Vol. 181)
ISBN 978-3-0348-9550-7 ISBN 978-3-0348-8399-3 (eBook)
DOI 10.1007/978-3-0348-8399-3

ISBN 978-3-0348-9550-7

Originally published by Birkhäuser Verlag in 2000
Softcover reprint of the hardcover 1st edition 2000
Printed on acid-free paper produced of chlorine-free pulp. TCF ∞

ISBN 978-3-0348-9550-7

9 8 7 6 5 4 3 2 1

Table of Contents

Part 1: Classes of the Working Week

Part 2: Contributions

Preface

"Given a variety X, determine a subvariety Z of X such that blowing up X in Z resolves the singularities of X."

This is – in its simplest form – the ultimate goal of resolution of singularities. The contributions in this volume show how much and how little we know about this problem. The intentions are twofold.

One: The articles have been written to introduce the reader in an instructive and accessible way to a selection of the most important techniques and results. In this sense the volume is a textbook, written by several authors. Special care has been taken to explain the underlying ideas and concepts and to illustrate them by striking examples.

Two: The articles have been collected to state the main questions and open problems. There are few research articles, but many research problems. In this sense, the volume is a research book, pointing at many mysteries, ranging from the level of exercises to hard unresolved problems.

Oscar Zariski's work on resolution of singularities, continued among others by his students S. Abhyankar and H. Hironaka, has decisively advanced the subject. The volume starts with a biographical sketch of Zariski's life and career. It is followed by an account on the development and achievements of the theory of resolution of singularities, including a dictionary of the standard notions and an assorted bibliography.

Part I is formed by the expanded manuscripts of the classes held at the Working Week on Resolution of Singularities at Obergurgl in Tirol, Austria, September 7–17, 1997. They cover: Alterations and weak resolution (D. Abramovich and F. Oort); Desingularization of differential equations (J.-M. Aroca); Constructive resolution in characteristic zero (S. Encinas and O. Villamayor). We regret that the contribution 'Valuation theory and local uniformization' by B. Teissier, though refereed and accepted for publication in this volume, could not be included, as a final version was not produced by the last possible deadline before printing.

Part II contains articles written especially for this volume. The topics were chosen so as to introduce and discuss aspects complementary or auxiliary to part I. The articles treat: Resolution of surfaces (V. Cossart, H. Hauser, D.T. Lê, G. Müller); Toric varieties and toric resolution (D. Cox, R. Goldin and B. Teissier); Valuation theory and local uniformization (F.-V. Kuhlmann, M. Vaquié); Equisingularity and equiresolution (J. Lipman); Moduli spaces (B. van Geemen and F. Oort); Applications of alterations (T. Geisser, A.J. de Jong, F. Pop); Computer implementation of resolution algorithms (G. Bodnár and J. Schicho); False proofs and their reviews (H. Reitberger). All articles have been refereed by two independent experts.

Some important contributions to resolution of singularities could not be represented by an article in this volume. The historical account and the various bibliographies of the papers of this book give credit and reference to these. We mention here only some of the most recent ones.

Macaulayfication and Cohen-Macaulay properties of Rees rings (Kawasaki, Brodmann, Herrmann, Korb, Goto, Faltings, ...)

Complete ideals and integral closure (Lipman, Huneke, Cutkosky, Spivakovsky, Teissier, ...)

Factorization of morphisms into blowups (Zariski, Abhyankar, Spivakovsky, Cutkosky, ...)

Hilbert-Samuel functions, initial ideals and Newton diagrams (Hironaka, Bennett, Oka, Bierstone and Milman, Galligo, Hauser, ...)

Weak resolution by blowups (Bierstone and Milman, Abhyankar, Urabe, ...)

Approximate roots, Puiseux expansions (Abhyankar, Moh, Spivakovsky, ...)

Proximity relations (Gonzalez-Sprinberg, Campillo, Lejeune, Oda, Sánchez-Giralda, Granja, ...)

Nash manifold of arcs on a variety (Nash, Lejeune, Gonzalez-Sprinberg, Denef and Loeser, Grinberg and Kazhdan, ...)

Minimal model program (Mori, Kollár, Reid, Kawamata, ...)

Essential for the preparation of this volume were the efforts of the authors to write on a suggested topic and to try to explain it in a comprehensible way. This attempt was supported by the numerous referees who carefully read the manuscripts and proposed many improvements. The people from Birkhäuser Verlag did a fine professional job in editing the volume.

We are very indebted to all for their kind collaboration.

Shortly after the Working Week, Manfred Herrmann from Cologne died unexpectedly. He was a dedicated, energetic mathematician, and an affable colleague. He took great interest in the mathematical and social program of the conference, participating actively in it. We will remember him as he was with us in what turned out to be the last months of his life.

The Working Week on Resolution of Singularities and this volume were supported by a grant from the Austrian Ministery of Science, the Land Tirol, by Gemeinde Sölden and Obergurgl, the University of Innsbruck, Hypobank Tirol, Tyrolean Airways, Tiwag, Tirol Werbung and Reisebüro Schenker.

Tirol, October 1, 1999
The editors

Program of the Working Week on Resolution of Singularities, Obergurgl 1997

Sunday 7/9: Arrival.

8:30 p.m.	Opening, presentation of speakers, formation of working groups.

Monday 8/9 – Saturday 13/9 except Wednesday 10/9:

8:30–9:30	Villamayor: Constructive resolution.
9:45–10:45	Abramovich: Alterations and resolution.
11:30–12:30	Aroca: Singularities of differential equations.
2:30–3:30	Oort: Stable curves and moduli spaces.
4:00–5:00	Teissier: Toric resolution and valuations.
5:15–6:45	Working groups Encinas-Villamayor and Abramovich-Oort
8:30–9:30	Working groups Teissier and Aroca (not on Friday, Saturday)

Tuesday 9/9

after dinner	Lipman, Hironaka: History of resolution of singularities.

Wednesday 10/9:

8:30–9:30	Hironaka: Resolution problems in characteristic p.
10:00–11:00	Lipman: Equisingularity and equiresolution.
11:30–12:30	Abhyankar: Local resolution in characteristic 0.
1:30–7:00	Excursion to glacier.

Thursday 11/9

after dinner	Abhyankar: Life and work of O. Zariski.

Friday 12/9:

8:30 p.m.	Questions and Problem Session.

Saturday 13/9:

5:30–7:30	Plenary Session.
7:30	Invited Dinner.

Sunday 14/9: Departure.

Participants of the Working Week

Abhyankar Shreeram
Dep. of Mathematics
Purdue University
West Lafayette, IN 47907 USA
abhyankar@math.purdue.edu

Abramovich Dan
Dep. of Mathematics
Boston University
111 Cummington
Boston, MA 02215 USA
abrmovic@math.bu.edu

Acquistapace Francesca
Dip. di Matematicà
Università di Pisa
Via F. Buonarotti 2
Pisa, 56127 Italy
acquistf@dm.unipi.it

Alberich Carraminyana Maria
Dep. d'Algebra i Geometria
Universitàt de Barcelona
Gran Via de les Corts Catalanes, 585
Barcelona, 08071 Spain
alberich@cerber.mat.ub.es

Alonso Garcia Mariemi
Fac. de Matemáticas
Universidad Complutense
E-28040 Madrid
alonsog@eucmax.sim.ucm.es

Alonso Gonzalez Clementa
Sección de Matemáticas
Universidad Valladolid
E-47005 Spain

Amnon Yekutieli
Dep. of Theoretical Mathematics
Weizmann Institute
Rehovot, 76100 Israel
amnon@wisdom.weizmann.ac.il

Anta de la Iglesia Florencia
Sección de Matemáticas
Universidad Valladolid
E-47005 Spain

Araujo Antonio
Centro de Matemática
Universidade de Lisboa
Av. Gama Pinto 2
Lisbon 1699 Portugal

Aroca Bisquert Fuensanta
Sección de Matemáticas
Universidad Valladolid
E-47005 Spain
fuen@cpd.uva.es

Aroca José Manuel
Dep. de Algebra y Geometría
Facultad de Ciencias
E-47005 Valladolid, Spain
aroca@cpd.uva.es

Audoubert Benoit
Université de Nantes
Dép. de Mathématiques
2 rue de la Houssinière
Nantes cedex 03, 44072 France
audouber@math.univ-nantes.fr

Ban Chunsheng
Dep. of Mathematics
Ohio State Univ.
Columbus 231 West 18th Ave
Columbus, OH 43210 USA
cban@math.ohio-state.edu

Begue Aguado Alvaro
Sección de Matemáticas
Universidad Valladolid
E-47005 Spain
alvaro@wamba.cpd.uva.es

Belenkiy Ari
Mathematics & Comp. Sci. Dep.
Bar-Ilan University
Ramat Gan Israel
abelen@macs.biu.ac.il

Bermejo Isabel
Dep. Matemática Fundamental
Facultad de Matemáticas
Universidad La Laguna
La Laguna, 38271 Tenerife Spain
ibermejo@ull.es

Blanloeil Vincent
Institut de Recherche
Mathématique Avancée
Université Louis Pasteur et CNRS
7 rue Descartes
F-67084 Strasbourg France
blanloei@math.u-strasbg.fr

Bravo Zarza Ana
Dep. de Matemáticas
Universidad Autónoma de Madrid
E-28049 Spain
ana.bravo@uam.es

Cabral Joao
Centro de Matemática
Universidade de Lisboa
Av. Gama Pinto 2
Lisbon 1699 Portugal

Alberto Calabri
Dip. di Matematica
G. Castelnuovo
Università di Roma La Sapienza
00185 Roma Italy
calabri@mercurio.mat.uniroma1.it

Carlyle Gabriel
The Mathematical Institute
24-29 St. Giles
Oxford, OX1 3LB England
carlyle@maths.ox.ac.uk

Cioffi Francesca
Dip. di Matematica e Applicazioni
Università di Napoli Federico II
Napoli, Italia
cioffi@matna2.dma.unina.it

Corral Perez Nuria
Sección de Matemáticas
Universidad Valladolid
E-47005 Spain

D'Souza Harry
Dep. of Mathematics
UM-Flint,
MI 48502-2186 USA
dsouza_h@msb.flint.umich.edu

Dais Dimitrios
Mathematisches Institut
der Universität Tübingen
Auf der Morgenstelle 10
D-72076 Tübingen
ddais@aol.com

D'Anna Marco
Dip. di Matematica
Università di Roma, La Sapienza
Piazzale A. Moro, 2
Roma, 00185 Italy
danna@mat.uniroma1.it

de Bobadilla de Olazabal Javier
Avd/ Reina Victoria 72, 8D
Madrid, 28003 Spain
j.debobadilla@math.ruu.nl

Elzein Fouad
Université de Nantes
Nantes, 44072 France
elzein@math.univ-nantes.fr

Encinas Santiago
Dep. Matemática Aplicada
E.T.S. Arquitectura
Avda. Salamanca s/n
Valladolid, 47014 Spain
sencinas@cpd.uva.es

Fortuna Elisabetta
Dip. di Matematica
Università di Pisa
Via F. Buonarotti, 2
Pisa, I-56127 Italy
fortuna@dm.unipi.it

Franco Nuno
Centro de Matemática
Universidade de Lisboa
Av. Gama Pinto 2
Lisbon 1699 Portugal

Galbiati Margherita
Dipartimento di Matematica
Via Buonarotti 2
Pisa, I-56127 Italy
galbiati@dm.unipi.it

Garcia Barroso Evelia
Dep. de Matemática Fundamental
Facultad de Matemáticas
Universidad de La Laguna
La Laguna, Tenerife, 38271 Spain
ergarcia@vilaflor.ccti.ull.es

Garcia Zamora Alexis Miguel
Instituto de Matemáticas, Unam
Unidad Morelia
Nicolas Romero 150, Colonia Centro
Morelia, Michoacan, Mexico
amgarcia@jupiter.ccu.umich.mx

Gardener Tim
The Queen's College
Oxford OX1 4AW,
United Kingdom
gardener@maths.ox.ac.uk

Gimenez Philippe
Dep. Algebra, Geometria y Topologia
Universidad Valladolid
Valladolid, E 47005 Spain
pgimenez@cpd.uva.es

Gonzalez Diez Carlos
Sección de Matemáticas
Universidad Valladolid
Valladolid, 47005 Spain

Gonzalez Perez Pedro
Dep. Matemática Fundamental
Facultad de Matemáticas
La Laguna, Tenerife, 38271 Spain
pedro.gonzalez@ens.fr

Hamm Helmut
Fachbereich Mathematik
Universität Münster
Einsteinstr. 62
Münster, 48149 Deutschland
hamm@escher.uni-muenster.de

Hauser Herwig
Mathematik
Universität Innsbruck
A-6020 Austria
herwig.hauser@uibk.ac.at

Heiss Werner
Mathematik
Universität Innsbruck
A-6020 Austria
werner.heiss@uibk.ac.at

Herrmann Manfred
Mathematisches Institut
Universität Köln

Hertling Claus
Mathematisches Institut
Universität Bonn
Beringstrasse 3
Bonn, 53115 Deutschland
hertling@rhein.iam.uni-bonn.de

Hironaka Heisuke
Dep. Mathematics
Yamaguchi University
1677-1 Yoshida
Yamaguchi, 753 Japan
as6h-hrnk@j.asahi-net.or.jp

Huang I-Chiau
Institute of Mathematics
Academia Sinica
Nankang, Taipei
Taiwan 11529, R.O.C.,
ichuang@math.sinica.edu.tw

Ito Hiroyuki
Dep. of Mathematics
Harvard University
1 Oxford St.
Cambridge MA, 02138 USA
hiroito@abel.math.harvard.edu

Ito Yukari
Dep. of Mathematics
Tokyo Metroplitan University
Hachioji
192-03 Tokyo, Japan
yukari@comp.metro-u.ac.jp

Jaworski Piotr
Instytut Matematyki
Uniwersytet Warszawski
ul. Banacha 2
Warszawa, 02097 Poland
jwptxa@mimuw.edu.pl

Jeddi Ahmed
Université Nancy 1
Institut Elie Cartan
Vandoeuvre-Les-Nancy, 54506 France
ahmed.jeddi@antares.iecn.u-nancy.fr

Kim Minhyong
Dep. of Mathematics
University of Arizona USA
kim@shire.math.columbia.edu

Kuhlmann Franz-Viktor
Dept. of Mathematics
Univ. of Saskatchewan
106 Wiggins Rd
Saskatoon, S7N 5E6 Canada
fkuhlman@fields.utoronto.ca

Lavila Vidal Olga
Dep. d'Algebra i Geometria
Universitàt de Barcelona
Gran Via de les Corts Catalanes 585
Barcelona, 08071 Spain
lavila@cerber.mat.ub.es

Lejeune-Jalabert Monique
Institut Fourier
Université de Grenoble 1
Grenoble, France
monique.lejeune-jalabert@ujf-grenoble.fr

Lipman Joseph
Dep. of Mathematics
Purdue University
West Lafayette, IN 47907 USA
lipman@math.purdue.edu

Liu Qing
Université Bordeaux 1
351, Cours de la Libération
Talence, 33405 France
liu@math.u-bordeaux.fr

Lönne Michael
Institut für Mathematik
Universität Bonn
Bonn, 53115 Deutschland
loenne@math.uni-bonn.de

Luengo Ignacio
Fac. de Matemáticas
Universidad Complutense
E-28040 Madrid, Spain
iluengo@eucmos.sim.ucm.es

Maeda Hironobu
Dep. of Mathematics
Tokyo University of
Agriculture and Technology
3-5-8 Saiwai-cho, Fuchu-shi
Tokyo 183, Japan
maeda@cc.tuat.ac.jp

McEwan Lee J.
Dep. of Mathematics
Ohio State Univ.
Columbus 231 West 18th Ave
Columbus, OH 43210 USA
mcewan@math.ohio-state.edu

Marin Perez David
Dep. de Matematiques
Universitàt Autonoma de Barcelona
Bellaterra Barcelona, 08193 Spain
david@manwe.mat.uab.es

Melkersson Leif
Dept. of Mathematics
University of Lund
Lund, S-221 00 Sweden
leif.melkersson@math.lu.se

Melle Hernandez Alejandro
Fac. de Matemáticas
Universidad Complutense
Madrid, 28040 Spain
amelle@eucmos.sim.ucm.es

Michler Ruth
Dep. of Mathematics
University of North Texas
Denton, TX 76201 USA
michler@mickey.math.unt.edu

Neto Orlando
Centro de Matemática
Universidade de Lisboa
Av. Gama Pinto 2
Lisbon 1699 Portugal
orlando@ptmat.lmc.fc.ul.pt

Notari Roberto
Dip. di Matematica
Politecnico di Torino
Corso Duca degli Abruzzi, 24
Torino, 10129 Italy
notari@polito.it

Olalla Miguel
Dep. de Algebra
Universidad de Sevilla
Sevilla, 41012 Spain
olalla@atlas.us.es

Oort Frans
Dep. Mathematics
Universiteit Utrecht
Budapest Laan 6
NL-3508 TA Utrecht, Netherlands
oort@math.ruu.nl

Peraire Rosa
Dep. d'Algebra i Geometria
Universitàt de Barcelona
Gran Via, 585
Barcelona, 08007 Spain
peraire@cerber.mat.ub.es

Pernazza Ludovico
V. Agri, 1
Roma, 00198 Italy
pernazza@cibs.sns.it

Pfeifle Julian
calle Doce 3b
Los Arroyos
E-28280 El Escovial, Spain
jpfeifle@escet.urjc.es

Pichon Anne
Section de Mathématiques
de l'Université de Genève
2-4, rue du Lievre
Genève 24, 1211 Switzerland
anne.pichon@math.unige.ch

Polishchuk Alexander
Dep. of Mathematics
Harvard University
1 Oxford Street
Cambridge MA, 02138 USA
apolish@abel.math.harvard.edu

Quirós Adolfo
Dep. de Matemáticas
Universidad Autónoma de Madrid
Madrid, 28049 Spain
adolfo.quiros@uam.es

Reguera Ana
Sección de Matemáticas
Universidad Valladolid
E-47005 Spain
areguera@cpd.uva.es

Reitberger Heinrich
Mathematik
Universität Innsbruck
A-6020 Austria
heinrich.reitberger@uibk.ac.at

Resina Gil Debora
Dep. de Matematiques
Universitàt Autonoma de Barcelona
Bellaterra Barcelona, 08193 Spain
debora@manwe.mat.uab.es

Ribon Herguedas Javier
Sección de Matemáticas
Universidad Valladolid
E-47005 Spain

Riemenschneider Oswald
Mathematisches Seminar
Universität Hamburg
Hamburg, 20146 Germany
riemenschneider@math.uni-hamburg.de

Rieseneder Christian
Mathematik
Universität Innsbruck
A-6020 Austria
christian.rieseneder@uibk.ac.at

Roczen Marko
Institut für Mathematik
Humboldt-Universität
Berlin, 10099 Germany
roczen@mathematik.hu-berlin.de

Roe i Vellve Joaquim
Universitàt de Barcelona
Dep. d'Algebra i Geometria
Gran Via de les Corts Catalanes 585
Barcelona, 08007 Spain
jroevell@cerber.mat.ub.es

Romero Gamarra Alfonso
Sección de Matemáticas
Universidad Valladolid
E-47005 Spain

Rosenberg Joel
Dep. of Mathematics
Massachusetts Institute of Technology
Cambridge, MA 02139 USA
joelr@math.lsa.umich.edu

Sageaux Thierry
Université de Bordeaux I
351 cours de la Libération
Talence, 33405 France
sageaux@math.u-bordeaux.fr

Sanz Sanchez Fernando
Sección de Matemáticas
Univ. Valladolid
E-47005 Spain

Schicho Josef
RISC-Linz
Johannes Kepler Universität
Linz, 4040 Austria
josef.schicho@risc.uni-linz.ac.at

Schmitt Alexander
Institut für Mathematik
Universität Zürich
Winterthurerstrasse 190
Zürich, 8057 Switzerland
schmitt@math.unizh.ch

Seiler Wolfgang
Fakultät für Mathematik und Informatik
Universität Mannheim
Mannheim, 68131 Germany
seiler@math.uni-mannheim.de

Silva Pedro
Centro de Matemática
Universidade de Lisboa
Av. Gama Pinto 2
Lisbon 1699 Portugal

Slembek Silke
UFR de Mathématique et d'Informatique,
Université Louis Pasteur
7 rue René Descartes
F-67084 Strasbourg France
slembek@irma.u-strasbg.fr

Smith Karen
Dep. of Mathematics
University of Michigan
Ann Arbor, MI 48109 USA
kesmith@math.lsa.umich.edu

Soto Manuel J.
Dep. de Algebra
Universidad de Sevilla
Sevilla, 41012 Spain
soto@atlas.us.es

Spreafico Maria Luisa
Dipartimento di Matematica
Politecnico di Torino
Corso Duca degli Abruzzi, 24
Torino, 10129 Italy
spreafico@polito.it

Stadelmeyer Peter
RISC-Linz
Johannes Kepler Universität
Linz, A-4040 Austria
peter.stadelmeyer@risc.uni-linz.ac.at

Teissier Bernard
Dép. de Mathématiques
Ecole Normale Supérieure
45, rue d'Ulm
F-75005 Paris, France
teissier@dmi.ens.fr

Thalhammer Mechthild
Mathematik
Universität Innsbruck
A-6020 Austria
mechthild.thalhammer@uibk.ac.at

Tomaru Tadashi
School of Health Sciences
Gunma University
Showa-machi, Maebashi
Gunma, 371 Japan
ttomaru@sb.gunma-u.ac.jp

Tornero José Maria
Dep. de Algebra
Facultad de Matemáticas
Universidad de Sevilla
Sevilla, 41012 Spain
tornero@atlas.us.es

Ueno Kenji
Dep. of Mathematics
Faculty of Science
Kyoto University
Kyoto, 606-01 Japan
ueno@kusm.kyoto-u.ac.jp

Vaquié Michel
Dép. de Mathématiques
Ecole Normale Supérieure
45 rue d'Ulm
Paris, F-75230 France
michel.vaquie@ens.fr

Velez Melon Pilar
Dep. de Ingenieria Informatica
Universidad Antonio de Nebrija
c/Pirineos, 55
28040 Madrid Spain
pvelez@dii.unnet.es

Veys Wim
Dep. Wiskunde
University of Leuven
Celestijnenlaan 200B
Leuven, 3001 Belgium
wim.veys@wis.kuleuven.ac.be

Villamayor Orlando
Departamento de Matemáticas
Universidad Autónoma de Madrid
E-28049 Madrid, Spain
villamayor@uam.es

Vosegaard Henrik
Dep. of Mathematics
Utrecht University
TA Utrecht, NL-3508 Netherlands
vosegaard@math.ruu.nl

Progress in Mathematics, Vol. 181, © 2000 Birkhäuser Verlag Basel/Switzerland

Oscar Zariski 1899–1986

Oscar Zariski was born in Kobrin, Belarus, in 1899. After attending, in 1918–1920, the University of Kiev, where he was largely self-educated (of necessity, in those unsettled times), he went to Rome and studied under the preeminent Italian geometers G. Castelnuovo, F. Enriques, and F. Severi. He received his doctorate at the University of Rome in 1924, and shortly thereafter came to the U.S.A. There ensued, first at Johns Hopkins and then at Harvard, sixty years of remarkable mathematical activity, lasting well into his eighties – an inspiring counterexample to the dictum that the domain of creative mathematics is reserved for the young.

Except for some early publications on set theory, which reflected the historical and philosophical interests of his teacher Enriques, he worked entirely in the field of algebraic geometry. His work was characterized by its concern with matters of fundamental importance; its profound effect on the field is universally recognized. His pioneering applications of the methods of abstract commutative algebra to questions in geometry set the stage for massive developments in both subjects.

Zariski's far-reaching influence was due in no small part to his qualities as a teacher. He projected the power which he exerted over mathematics, and which mathematics held over him – he was a vibrant embodiment of the ideal of a scholar-teacher, a model of devotion to the pursuit of the good mathematical life. Through the force of his personality, his interest in people, and his ability to communicate with and motivate others, he attracted and helped to develop the talents of many young mathematicians.

Among his students were a good proportion of the algebraic geometers educated in the United States during the 1950's and 60's. (See the volume dedicated to Zariski on the occasion of his seventieth birthday, Publ. Math. IHES. **36** (1969).) Of them, Abhyankar, Hironaka, and Lipman lectured at the Obergurgl working week on Resolution of Singularities which the present book commemorates, as did Teissier, who was closely associated with Zariski after 1970.

Zariski was honored with a number of high recognitions accorded by the American Mathematical Society: he was awarded the Cole Prize in Algebra in 1944, was invited to deliver the Colloquium Research Lectures at the summer meeting of the A.M.S. at Yale in 1947, served as President of the Society in 1969 and 1970, and was awarded the Steele prize for cumulative influence (1981). He received honorary doctorates from Holy Cross College (1959), Brandeis University (1965), Purdue University (1974), and Harvard University (1981). At the national level, he was awarded the Medal of Science by President Lyndon Johnson (1965), and the Wolf prize for lifetime achievement from the government of Israel (1982).

Zariski's Collected Papers were published in four volumes by the M.I.T. Press. In his preface to those volumes (see [Z3] for the last version) he gives his view of

the major landmarks of his career. There is more to be found in David Mumford's appreciation of Zariski as a scientist and a person [M], and in Carol Parikh's full-length biography [P].

Here is a brief summary of Zariski's scientific work.

In his earlier papers, he dealt with topological questions on algebraic varieties, especially problems having to do with the fundamental group. At that time, he was much influenced by Solomon Lefschetz at Princeton. (Castelnuovo himself had advised Zariski that the methods of the Italian school, though they had had many successes, had reached their limit, and that the topological work of Lefschetz would be of great importance for the future of algebraic geometry.)

In his 1935 monograph *Algebraic Surfaces* he expounded masterfully the achievements of the Italian geometers. This book provided a most important link between the classical and modern theories; its value was such that a second updated edition was put out by Springer-Verlag in 1971 (reprinted 1995). The book was a turning point in Zariski's career – and hence in the development of algebraic geometry – as it brought him to the realization that the Italian theory, for all its exciting beauty, had no adequate mathematical foundation: few of the main results had been properly proved.

It had been recognized by some mathematicians, notably Krull and van der Waerden, that a suitable basis for geometry might be found in what was then called modern algebra, a subject which had been developed under the leadership of Emmy Noether and Emil Artin. Zariski began, ab initio, an intensive study of algebra. (He had the good fortune at that time of being able to attend a seminar held weekly at Princeton by Emmy Noether, who had recently emigrated from Germany.) His efforts bore fruit when he foresaw the geometric potentialities of Krull's valuation theory. He tested his ideas against the basic problem of resolution of singularities. This led him to the formulation of new and solid foundations for algebraic geometry, especially in its birational aspects, and to rigorous algebraic proofs for the resolution of singularities on three-dimensional varieties and for local uniformization on varieties of any dimension (over fields of characteristic zero). It was in these latter, previously unexplored, areas that his new methods revealed their scope, opening the way for the subsequent successes of Abhyankar and Hironaka.

Around the same time Zariski found important applications for Krull's theory of local rings, in the study of local properties of algebraic varieties. This work produced substantial additions to the theory, culminating in his 1951 memoir on holomorphic functions [Z2]. Emphasis on local rings became a basic feature of the theory of schemes; and that memoir is the birthplace of formal schemes.

While Zariski did a great deal of algebra, he always considered geometric intuition to be the prime source of his ideas; and indeed, powerful geometric insight is an essential feature of his work. So, although Krull was instrumental in the development of crucial algebraic topics such as valuation theory and local algebra, he nevertheless made no direct contribution to algebraic geometry as such. Krull had anticipated the implications of algebra for geometry, but it remained for Zariski to realize them. Also van der Waerden had begun to write on geometry in abstract algebraic terms before Zariski, but he did not reach, as Zariski did, to the deeper questions which lie at the heart of the subject.

The interaction between geometry and commutative algebra benefited both fields. Some of the more immediate happy results of their merging were described in Zariski's 1950 address to the International Congress of Mathematicians [Z1]. He continued during the 1950's to contribute from the new point of view to classical geometrical theories – algebraic surfaces, linear systems on algebraic varieties, etc. Commutative algebra grew rapidly too; in the late 1950's, Zariski collaborated with Pierre Samuel on a two-volume treatise, *Commutative Algebra,* which remains today a useful reference for the subject.

From 1962 on, Zariski worked mainly (not exclusively) on the algebraic theory of equisingularity. His multiple papers on the subject take up most of the 650-page volume [Z3]. Several young researchers, especially in France, joined him in this enterprise. Unfortunately, with his passing from the scene, Zariski's particular vision of equisingularity has been rather neglected in the past fifteen years. The wider study of stratified spaces is active, and much progress has been made e.g., in the algebra and geometry associated with Whitney stratifications. But with his unique geometric point of view, Zariski bequeathed us a number of open and apparently deep problems which still promise to lead to significant enhancements of our understanding of algebraic and analytic varieties. In this, as in his other work, he extended the horizons of our knowledge.

Oscar Zariski died at his home in Brookline, Massachusetts, on July 4, 1986.

References

[M] Mumford, D.: Oscar Zariski, 1899–1986, Notices Amer. Math. Soc. bf 33 (1986), 891–894. (See also the video "Oscar Zariski and his work," Amer. Math. Soc., Providence, 1988.)

[P] Parikh, C. A.: The Unreal Life of Oscar Zariski, Academic Press, San Diego, London, 1991. (See also review by R. Hartshorne, American Math. Monthly **99** (1992), 482–486.)

[Z1] Zariski, O.: The fundamental ideas of abstract algebraic geometry, Proc. Internat. Congress of Mathematicians, Cambridge, Mass., 1950.

[Z2] Zariski, O.: Theory and applications of holomorphic functions on algebraic vasrieties over arbitraty ground fields, Memoirs Amer. Math. Soc. **69** (1951), 1–90.

[Z3] Zariski, O.: Collected Papers, vol. IV, MIT Press, Cambridge, Mass., 1979.

Joseph Lipman
Purdue University
West Lafayette, IN 47907, USA
`lipman@math.purdue.edu`
`www.math.purdue.edu/~lipman`

Progress in Mathematics, Vol. 181,

Resolution of Singularities 1860–1999

Herwig Hauser

This account shall serve as a quick guide to the historical development, the main contributors and the basic notions in the context of resolution of singularities. For detailed information we give references to the literature for each item. The presentation, which is necessarily subjective, does not claim completeness or utmost rigor. It shall merely help the reader to access the field and to find further sources.
We are very indebted to A. Quirós and the referees for a careful reading of earlier drafts and many suggestions and improvements. The photograph of Zariski was kindly made available by the Archive of the Mathematisches Forschungsinstitut Oberwolfach. The drawings of Abhyankar, Hironaka and Lipman were realized by Maria Alberich and Joaquim Roé during the Working Week at Obergurgl.

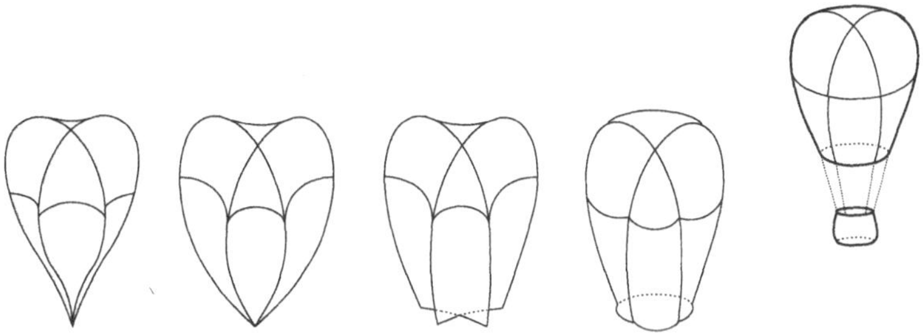

FIGURE 1. Blowing up singularities

Main achievements

We list a selection of important stages in the development of resolution of singularities, without evaluating the results. The period considered lasts from 1860 till 1999. The definitions used are explained in the dictionary. Dates are sometimes approximate. Results or generalizations which have been announced but whose proofs do not exist in written form are not included. Scheme over R shall mean scheme of finite type over the ring R [or their localizations or completions]. The letter K is reserved for fields. Characteristic 0 signifies that all residue fields are of characteristic 0. By surface we understand a reduced two dimensional variety or scheme over a ring, by hypersurface a scheme which can be embedded into a regular scheme as a closed codimension 1 subscheme. Some notions like local resolution may vary with the context.

Curves over **C**, resolution by quadratic transformations: Several authors between 1860 and 1929, varying level of rigor (see section on contributors).
Surfaces over **C**, hypersurface, local resolution, non-embedded: Jung 1908 [Ju, Lp 4].
Surfaces over **C**, hypersurface, global resolution, non-embedded: Walker 1935 [Wa 1, Lp 4].
Surfaces over K, hypersurface, characteristic 0, K algebraically closed, global resolution, non-embedded: Zariski 1939 [Za 1], simplified in [Za 3]. The assumption K algebraically closed is dispensable, Cossart 1999 [Co 3].
Normalization: Zariski 1939 [Za 1, Za 2, Za 11, MZ].
Local uniformization, varieties of arbitrary dimension over K, characteristic 0, K algebraically closed: Zariski 1940 [Za 2, Co 3]. For arbitrary characteristic, but admitting finite field extensions, Kuhlmann 1998 [Ku 1, Ku 2]. For formal germs in arbitrary characteristic and valuations of maximal rank, but admitting centers outside the singular locus, Urabe 1999 [Ur]. See also [Cu 3, BM 2, BM 3, BM 4].
Threefolds over K, hypersurface, characteristic 0, K algebraically closed, global resolution, non-embedded: Zariski 1944 [Za 4, Lp 4]
Surfaces over K, hypersurface, positive characteristic, K perfect, local uniformization, non-embedded: Abhyankar 1956 [Ab 1].
Excellent schemes of arbitrary dimension over K, characteristic 0, global resolution, embedded: Hironaka 1964 [Hi 1, Gr 2, Gi 1, Kz].
Arithmetic surfaces, hypersurface, global resolution, non-embedded: Abhyankar 1965 [Ab 2].
Surfaces over K, hypersurface, arbitrary characteristic, K perfect, global resolution, embedded: Abhyankar 1966 [Ab 3, Ab 4, Lp 4, CGO].
Threefolds over K, hypersurface, characteristic > 5, global resolution, non-embedded: Abhyankar 1966. Special cases by Cossart 1987, Moh 1996 [Ab 3, Ab 5, Co 2, Mo 3].
Excellent surfaces over K, hypersurface, arbitrary characteristic, K algebraically closed, local resolution, non-embedded: Hironaka 1967 [Hi 4]. Extended by Cossart 1981 to the case of non algebraically closed fields [Co 1]. Extended by Hauser to global embedded resolution for algebraically closed fields [Ha 4].
Normal quasi-homogeneous surfaces over **C**, global resolution by weighted blowups: Orlik-Wagreich 1971 [OW, La 1, Mü].
Toric and toroidal varieties over **C**, toric resolution by subdivision of fans, nonembedded, not equivariant: Kempf, Knudson, Mumford, Saint-Donat 1973 [KKMS, Od 4, Fu 2, Cx, Kt 1, Kt 2]. Resolution of curves by toric modification: Oka 1996, Goldin-Teissier, this volume [Ok, LO, GT].

Arbitrary excellent two dimensional schemes, global resolution, non-embedded: Lipman 1978 [Lp 1, At]. Announced by Hironaka 1967 in [Hi 4] and by Abhyankar 1968 in [Ab 13], both without giving the details of a proof valid for the general situation.
Macaulayfication: Special cases by Faltings 1978, Brodmann 1983, general case by Kawasaki 1998 [Fa, Bm 3, Ka].
Combinatorial resolution, winning strategy for polyhedral game: Zariski 1944, Spivakovsky 1983 [Za 2, p. 861-863, Sp 1, Sp 2, Ha 3].
Sandwich singularities, surfaces over **C**, resolution by Nash blowup: Rebassoo 1977 for certain surfaces in $\mathbf{C}^3$, Spivakovsky 1990, expanding work of Gonzalez-Sprinberg and combining Nash blowups with normalization [Rb, GS 1, GS 2, GS 3, Sp 5, Lê].
Schemes of arbitrary dimension over K, characteristic 0, global constructive and canonical resolution, equivariant, embedded: Villamayor 1989 and 1992, Bierstone and Milman 1991 and 1997 (treat also analytic case), Encinas and Villamayor 1998 [Vi 1, Vi 2, BM 3, BM 1, EV 1, EV 2, EV 3].
Alteration of varieties of arbitrary dimension over a field of arbitrary characteristic to a regular one: de Jong 1996 [dJ, AO, Be, AW, BP].
Weak resolution: de Jong, Abramovich, Bogomolov, Pantev, Wang 1996 [AJ, AW, BP, AO].
Equisingularity and equiresolution: Zariski 1962, Teissier 1980, Laufer 1987, Encinas 1996, Villamayor 1998 [Za 6, Za 13, Lp 3, Te 2, E 1, Vi 4].
Factorization of birational morphisms as a composition of blowups and blowdowns: For dimension 2, Zariski 1944 [Za 4], Abhyankar 1956 [Ab 12]. Counterexamples in dimension 3 by Sally 1972 and Shannon 1973 [Sa, Sh]. Positive results: Christensen 1984, Morelli 1996, Cutkosky 1997 for dimension 3, 1998 for arbitrary dimension, Włodarczyk 1997 and 1999, Abramovich, Karu, Matsuki, Rashid, Włodarczyk 1998 and 1999 [Cr, Mr, Cu 1, Cu 2, W 1, W 2, AMR, AMKW].
Reduction of vector fields and differential equations: Seidenberg 1968, F. Cano, J. Cano, D. Cerveau since 1990 [Sd 1, CF, CJ, CC, Ar 1, Ar 2].

Some research problems

We mention key phrases but do not give details or precise statements. As such, the list below shall merely manifest that the subject is far from being closed or thoroughly understood. For accurate formulations of questions and problems the reader is referred to the articles of this volume.

(1) Global embedded resolution of arbitrary excellent surfaces and threefolds in arbitrary characteristic.
(2) Resolution of higher dimensional schemes over a field of positive characteristic.
(3) Simple proof of strong resolution in characteristic 0.
(4) Investigation of schemes over **Z**. Description of equimultiple locus.

(5) Resolution in mixed characteristic.
(6) Description of a (canonical) ideal which, when taken as center of blowup, resolves in one step the singularities of the scheme. Computation of explicit examples.
(7) Extension of the notion of idealistic exponents to positive characteristic.
(8) Investigation of coefficient ideals in arbitrary characteristic.
(9) Uniform treatment of characteristic 0 and p.
(10) Independence of embedded resolution of embedding.
(11) Effective description of composition of blowups as one blowup of a non-reduced ideal.
(12) Factorization of modifications as composition of blowups.
(13) Refinement of normalization, so as to give e.g. resolution of surfaces in one step.
(14) Equivariant toric resolution.
(15) Equisingularity in higher dimensions.
(16) Equiresolution in higher dimensions.
(17) Factorization of ideals in higher dimensions.
(18) Search for new invariants measuring the complexity of singularities.
(19) Refinement of the notion of blowup (weighted blowup, non-reduced centers) as to make inductions easier.
(20) Economic computer implementation of resolution algorithm.
(21) Bounding complexity of resolution algorithms.
(22) Search for applications of resolution of singularities in other fields, e.g., mathematical physics, numerical analysis, computer graphics.

Contributors

The following list is far from being complete, both as to names and subjects. For technical and human reasons not all relevant contributions could be included. Neither, credits could be splitted and distributed up to the smallest detail. The reader will understand that such a list cannot avoid inaccurrancies and injusticies, and that it must reflect the personal view of the chronicler. Often, the citations refer to articles with extensive literature on the subject. Works of authors in other fields are not mentioned.

O. Zariski's contribution to the field is predominant and represents the start of a systematic treatment of techniques and problems. This impact has been continued and reinforced by his students, among them S. Abhyankar, H. Hironaka and J. Lipman, as well as many other researchers.

Resolution of curves (over **C**): B. Riemann (1865), M. Noether (1870–1888), L. Kronecker (1882), R. Dedekind and W. Weber (1882), G. Halphen (1884), E. Bertini (1888), P. del Pezzo (1888), M. Noether and A. Brill (1892), C. Segre (1897), E. Picard and G. Simart (1897), F. Enriques (1918), G.A. Bliss (1923). See the extensive bibliographies in [Za 9, Sg, Ab 10].

Resolution of surfaces (over **C**, mostly not rigorous): P. del Pezzo (1892), G. Kobb (1892), K. Hensel (1900), F. Severi (1914–1926), G. Albanese (1924), H.W. Jung (1908), O. Chisini (1921), G. Castelnuovo and F. Enriques (1901–1914), B. Levi (1889–1899), B. Segre (1947), P. du Val (1944–1948), L. Derwidué (1951–1955). See again the bibliographies in [Za 9, Sg, Ab 10].

W. Krull 1931: General theory of valuations [Kr].

R.J. Walker 1935: Global resolution of algebraic surfaces over **C**, patching Jung's local arguments; first rigorous proof for surfaces [Wa 1, Wa 2, Lp 4].

O. Zariski (1899–1986, Ph. D. 1924 at Rome with G. Castelnuovo, see [Lp 7] for more details): Influenced by E. Noether's treatment of algebra, Zariski develops between 1935 and 1944 an algebraic and systematic approach to resolution of singularities for varieties over an algebraically closed field of characteristic zero. After establishing a first proof of resolution of surfaces in 1939 [Za 1, Za 3], he investigates the equimultiple locus of a surface and its behaviour under permissible blowup, showing that it can be transformed by point blowups into a union of points and smooth curves intersecting transversally. Once this is achieved, a sequence of blowups in maximal permissible centers lowers the highest multiplicity occurring on the variety (Theorem of Beppo Levi). This gives by induction embedded resolution for surfaces and, via a detailed study of linear systems, also resolution of threefolds in characteristic 0 [Za 4].

Oscar Zariski

At about the same time, Zariski exploits the theory of valuations to prove local uniformization in any dimension over fields of characteristic zero [Za 2], see also [Co 3]. The case of arbitrary characteristic, but allowing specific finite field extensions, has been established by Kuhlmann through valuation theoretic methods [Ku 1, Ku 2]. This can also be proven as a consequence of de Jong's theorem on alterations [dJ]. In [Za 5], Zariski proves the compactness of the Zariski-Riemann surface.

Around 1939, Zariski introduces the concept of normalization [Za 1, Za 2, Za 11, MZ]. This provides a direct method to simplify the singularities of a variety and to eliminate singular loci of codimension 1. In particular, normalization suffices to resolve curves. For surfaces, normalization does not suffice but has turned out to be useful when combined with point blowups or Nash modifications [GS 3, Sp 5, Lê]. Related to this is the theory of complete ideals and their factorization [Za 8], see also [Lp 2, Lp 6, Sp 4].

Zariski studied the notion of equisingularity of a hypersurface along a subvariety and, related to it, equiresolution [Za 6, Za 13, Te 2, La 2, Lp 3, E]. Here, the dimensionality type of a hypersurface is defined so as to give an invariant which measures in a certain way the complexity of a singularity.

L. Derwidué, W. Gröbner, B. Segre 1951–1955: Produce (among others) false but published proofs of resolution, commented on by Zariski, Abellanas and Semple in the corresponding Math. Reviews [Gb, Re].

F. Hirzebruch 1953: Analytic version of Jung's proof [Hz].

S. Abhyankar (Ph. D. 1955 with Zariski): He is the first one to establish results in positive characteristic. In his thesis he proves local uniformization of surfaces over a field of characteristic p [Ab 1] . This is extended to embedded resolution of surfaces over perfect fields of any characteristic and non-embedded resolution of threefolds over algebraically closed fields of characteristic $\neq 2, 3, 5$ in 1966 [Ab 3, RS], as well as to the case of arithmetic surfaces [Ab 2]. Resolution of arbitrary excellent two dimensional schemes is announced in [Ab 13]. There, only a lemma is proven; details of a proof seem never to have appeared. In [Ab 3], also principalization, dominance and uniformization of surfaces and threefolds are treated. The case of characteristic $2, 3, 5$ for threefolds has been announced [Ab 10, p. 253], but not published [Ab 5].
Abhyankar introduced and exploited the concept of Tschirnhausen transformation in order to produce coordinates which exhibit significant local invariants of the singularity [AM]. This is the algebraic analogue of Hironaka's hypersurface of maximal contact [Hi 6, AHV 1]. Both are known to fail in positive characteristic [Hi 9, Ha 2, ex. 8]. The idea of considering coefficient ideals as a means to apply induction on the embedding dimension, already implicit in Jung's approach [Ju, Lp 4], is extended further in [Ab 4]. Due to Abhyankar is the theory of good points, taken up by Encinas and Villamayor in [EV 1], and, together with Moh, of approximate roots [AM], revived by Spivakovsky recently. Abhyankar also proved several

results on valuation theory [Ab 11] and the factorization of birational morphisms in dimension 2 into a composition of blowups [Ab 12].

H. Hironaka (Ph. D. 1960 with Zariski): First to consider the non-hypersurface case. Proves in 1964 embedded resolution of varieties of arbitrary dimension over any field of characteristic zero [Hi 1], see also [Gi 1, Kz]. Introduces for this the concept of standard basis [Hi 1, p. 208] and the local invariant ν^* generalizing the order of a hypersurface at a point. To control its behaviour under blowup, the existence of reduced standard bases is proven via the Division Theorem for ideals of power series [Hi 6, Hi 11]. The induction invariant ν^* is replaced later by the Hilbert-Samuel function, investigated by his student Bennett w.r.t. localization, semicontinuity and permissible blowup [Bn]. The constancy of both along a subvariety is equivalent and known as normal flatness. Hironaka's induction argument is very involved. It has been modified and simplified since 1989 by Villamayor, Bierstone-Milman and Encinas-Villamayor. Their proofs rely on invariants which take into account the history of prior blowups of the resolution process [Vi 1, Vi 2, BM 1, BM 2, BM 3, BM 4, EV 1, EV 2, EV 3]. This allows them to establish equivariant resolution in characteristic zero by an algorithmic procedure. See [Lp 8, Ha 6] for featured reviews on these works. For a proof in the analytic category, see also [AHV 1, AHV 2, AHV 3].

In order to handle the Hilbert-Samuel function and to apply induction on the local embedding dimension, Hironaka introduces the notion of idealistic exponents [Hi 6]. These describe the equimultiple locus as a certain singularity locus of a

hypersurface, compare with the basic objects in [EV 2] and the regular presentations in [BM 1].

As to positive characteristic, Hironaka gives in [Hi 9] an example showing that the equimultiple locus need not have smaller local embedding dimension than the scheme itself, and in [Hi 5] an example where his invariant $n - \tau$ increases under blowup, see also [Ha 2, ex. 7]. However, Hironaka proposes in [Hi 4] different induction invariants in order to prove local resolution for surfaces of arbitrary characteristic. See [Ha 4] for a conceptual treatment of these invariants and a complete proof of global resolution of surfaces over algebraically closed fields.

J. Lipman (Ph. D. 1965 with Zariski): He establishes resolution of arbitrary excellent surfaces, using cohomology and duality to first reduce the singularities to rational ones [Lp 1, Lp 2, At]. Though the resolution is not embedded, since it depends on normalization and other non-embedded transformations, this is the only (completely proven) result in the field which works for excellent schemes (of dimension 2) in all generality. Applies principalization and resolution to investigate complete ideals and their factorization [Lp 2, Lp 6]. Pursues Jung's approach through the study of quasi-ordinary singularities in higher dimensions [Lp 5, BE]. Several contributions on equisingularity and equiresolution [Lp 3, Lp 5].

A. Grothendieck (1965): Introduces the notion of excellent schemes and shows that this is a necessary condition on schemes for their resolution [Gr 1, IV.7.8].

J.-P. Serre (1965); Detailed description of Hilbert-Samuel function [Se].

B. Bennett, B. Singh (1970–1980): Study the behaviour of the Hilbert-Samuel function under blowup and localization and prove the corresponding upper-semi-continuity properties. Describe its compatibility under passage to completion and characterize normal flatness [Bn, Si 1, Si 2].

J.-M. Aroca, J.-L. Vicente (1975): Prove together with Hironaka resolution in the complex analytic category, treat extensively division theorems, maximal contact and the weighted tangent cone [AHV 1, AHV 2].

G. Faltings, M. Herrmann, U. Orbanz, B. Ulrich, M. Brodmann, W. Vasconcelos, C. Huneke, T. Kawasaki (1978–1999): Attempt to reduce the singularities of a scheme to Cohen-Macaulay-singularities (Macaulayfication), accomplished recently in all generality by Kawasaki [Fa, Bm 3, Ka]. Study properties of Rees algebras [HIO, Vs].

J. Giraud (1974): Results on normal flatness, study of maximal contact and of the invariant τ in positive characteristic [Gi 2, Gi 3].

T. Oda (1973–1986): Study of additive group schemes related to Hironaka's invariant τ, and of infinitely near singularities [Od 1, Od 2, Od 3]. Treatment of toric varieties [Od 4].

O. Villamayor, E. Bierstone, P. Milman, S. Encinas (1989–1999): Develop constructive and equivariant resolution algorithms for varieties of characteristic zero (the first two authors apply the reasoning also to analytic spaces) [Vi 1, Vi 2, BM 1, BM 2, BM 3, BM 4, EV 1, EV 2, EV 3, E 1, E 2]. Define invariants which take into account the history of what has occurred in earlier blowups. Extend and simplify Hironaka's notion of idealistic exponents. Provide explicit examples of resolutions. The algorithm of Encinas and Villamayor has been implemented by Bodnár and Schicho [BS 1, BS 2].

H. Hironaka, E. Bierstone, P. Milman, H. Sussmann (1988–1990): Applications of real analytic desingularization to subanalytic sets [BM 4, Su].

M. Spivakovsky (1983–1999): Classifies valuations and factorizes ideals in dimension 2, continuing the theory and results developed by Krull in the 30', by MacLane in the 40' and by Zariski, Abhyankar and Lipman in the 60' [Sp 4]. Finds winning strategy for Hironaka's simple polyhedral game, and a counterexample to the hard game, as well as to the Theorem of Beppo Levi in dimension three [Sp 1, Sp 2, Sp 3]. Extends and investigates the notions of approximate roots and Newton-Puiseux expansions.

V. Rebassoo, G. Gonzalez-Sprinberg, M. Spivakovsky (1977–1990): Study behaviour and improvement of surfaces under Nash blowup [Rb, GS 1, GS 2, GS 3, Sp 5, Lê]. Spivakovsky introduces notion of Sandwich singularities and shows, following a suggestion of Hironaka, that surfaces over **C** can be resolved by normalizations and Nash modifications.

T. Sánchez-Giralda, A. Campillo, M. Lejeune, G. Gonzalez-Sprinberg (1976–1999): Study of infinitely near points and clusters [SG 1, SG 2, Ca, CGL, Cs].

T.T. Moh, V. Cossart (1981–1996): Resolution of special cases of surfaces and threefolds in arbitrary characteristic. Construct examples where induction invariants (slopes) increase under blowup in positive characteristic [Mo 1, Mo 3, Ha 2, ex. 16].

H. Hauser (1992–1999): Global embedded resolution of surfaces over an algebraically closed field of arbitrary characteristic [Ha 4], extending and completing arguments of Hironaka in [Hi 4]. Establishes Gauss-Bruhat decomposition for automorphisms of complete local rings, and compatibility of flag varieties with blowup, as to control invariants under blowup. Introduces minimal and maximal initial ideals as a general concept to define invariants, containing as special cases the classical resolution invariants [Ha 1].

H. Hauser, J. Pfeifle, J. Rosenberg (1997–1999): The first two authors construct non-reduced structures on normal crossings centers in $\mathbf{A}^3$ and $\mathbf{A}^4$ so as to give regular blowups [Ha 4, Pf]. Extended to arbitrary dimension by J. Rosenberg [Ro].

P. Orlik, P. Wagreich, H. Pinkham, H. Flenner, E. Brieskorn, H.B. Laufer (1971–1983): Resolution of normal, quasihomogeneous surfaces. Study resolution graphs, Dynkin diagrams, minimal and good resolutions [OW, La 1, Mü].

G. Kempf, F. Knudson, D. Mumford, B. Saint-Donat, T. Oda, V.I. Danilov (1973–1988): Toroidal varieties and toric resolution by subdivision of fans [KKMS, Od 4, Fu 2, Cx].

M. Oka, R. Goldin, B. Teissier (1996–1999): Resolution of curves by toric modification [Ok, LO, GT].

M. Lejeune, B. Teissier, B. Youssin (1973, 1990): Extract coordinate free invariants from the Newton polyhedra of a singularity [LT, Yo, Hi 2, Hi 3].

A. Seidenberg (1968), F. Cano, J. Cano, D. Cerveau, C. Camacho, A. Lins Neto, P. Sad, O. Neto (1984–1999): Reduction of singularities of vector fields, foliations and differential equations in dimension 2 and 3 [CC, CF, CJ, CLS, CS, Sd, Ar 1, Ar 2. Ne]. Application to the existence of separatrices.

A.J. de Jong (1996): Introduction and use of alterations. Resolution of varieties in arbitrary characteristic up to a finite map, fibering the scheme in curves and applying semi-stable reduction [dJ, Be, AO].

D. Abramovich, A.J. de Jong, J. Wang, K. Karu, F. Bogomolov, T. Pantev (1996–1999): Weak equivariant resolution in characteristic 0 [AO, AJ, BP, AW].

P. Berthelot, O. Gabber, A.J. de Jong, F. Oort, T. Geisser, F. Pop (1996–1999): Relations and applications of de Jong's theorem to arithmetic geometry [Be, dJ, Ge, Po].

F.-V. Kuhlmann (1998): Local uniformization in arbitrary characteristic up to well controlled finite field extensions [Ku 1, Ku 2].

F.-V. Kuhlmann, B. Teissier (1998): Study of valuation theory as a tool for resolution and uniformization problems [Ku 1, Ku 2, Te 1].

T. Urabe (1999): Local uniformization in arbitrary characteristic for formal germs and valuations of maximal rank, admitting centers outside the singular locus [Ur].

J. Sally, D. Shannon, C. Christensen, H. Pinkham, R. Morelli, D. Cutkosky, J. Włodarczyk, D. Abramovich, K. Karu, K. Matsuki, S. Rashid (1972, 1996–1999): Factorization of local birational morphisms as a composition of blowups and blowdowns [Sa, Sh, Cr, Mr, Cu 1, Cu 2, Cu 3, W 1, W 2, AMR, AKMW].

G. Bodnár, J. Schicho (1999): Implementation of resolution algorithm of Encinas and Villamayor through Maple [BS 1, BS 2]. Computes simple examples automatically, complexity problems for more difficult examples.

Dictionary

The definitions below shall give a general idea of the concepts. They are not rigorous. Some notions may vary with the context. For precise information please consult the references. Throughout, W denotes the (regular) ambient scheme, and X the singular scheme to be resolved.

Albanese's method: Argument to resolve surfaces locally [Al, Lp 4, sec. 1.5].

Alteration: A proper, surjective and generically finite morphism [dJ, AO, I.1.2, Be].

Approximate root: A polynomial whose d-th power coincides in terms of degree $\geq n - \frac{n}{d}$ with a given polynomial of degree n in Weierstrass form [AM, part 2, p. 42].

Arithmetic case: The schemes are defined over $\mathbf{Z}$ or over the ring of integers of an algebraic number field [Ab 2, Lp 1].

Bad point: See good point.

Blowup (blowing-up, quadratic, monoidal transformation, sigma-process, dilatation, Hopf map): Birational proper map $\pi : W' \to W$ associated to a given (not necessarily regular) scheme W and a chosen center Z. The scheme W' is obtained by glueing for each i the graphs of the maps $f_1/f_i, \dots, f_k/f_i : W \to \mathbf{A}^k$, where $f_1, \dots, f_k$ range over a generator system of the ideal I_Z defining Z in an affine chart of W. Equivalently, $W' = \text{Proj} \oplus_{j \geq 0} I_Z^j$. [Hi 1, III.2, Bn, Bm 1, Bb, Ha 5, Hs, II.7 and V.3, Sf, VI.2.2, Dd].

Combinatorial blowup: See monomial blowup.

Local blowup: Map given by the ring homomorphism between the local rings of a point in X and a point above it [Ha 1].

Monomial blowup (combinatorial blowup): Local blowup described by a ring homomorphism which is given by a monomial substitution of the chosen regular system of parameters [Ha 1, Sp 1]. See also toric maps.

Nash blowup (Nash modification): Associating to a complex analytic variety X the closure of the tangent bundle of the regular locus $Y = \text{Reg}\, X$ in the correct Grassmannian [GS 1, GS 2, GS 3, Lê, Nb, Rb, Sp 5]. For hypersurfaces, it corresponds to blowing up the Jacobian ideal generated by the first order partial derivatives of the defining function.

Blowup-algebra (Rees algebra): Homogeneous coordinate ring of blowup of an ideal, equals $\oplus_{k \geq 0} I^k$ where I is the ideal of the center [HIO, Ha 5, Vs]. For an extensive bibliography on properties of the Rees algebra, see [Vs].

Taut blowup: Blowup of surfaces with center the singular points of the equimultiple locus, respectively the whole equimultiple locus (with suitable non-reduced structure) in case the equimultiple locus has at most normal crossings [Ha 4].

Center of blowup: Ideal or closed subscheme which has been blown up [Hi, III.2, Hs, II.7]. It is called permissible, if it is smooth and contained in the equimultiple locus (or in a Hilbert-Samuel stratum) of the variety.

Characteristic: For a scheme over a field, it can be zero or positive, equal to a prime number p. Over a ring, one considers all residue fields, distinguishing then between the case of equal or mixed characteristic.

Characteristic pair: See slope.

Coefficient ideal: Ideal associated to a formal power series and generated by powers of its coefficients when expanded w.r.t. one variable. For $f(x_1, \dots, x_n) = x^o + \sum_{i=0}^{o-1} a_i(x_2, \dots, x_n) x_1^i \mod x_1^{o+1}$ with $o = \text{ord}_0 f$, the coefficient ideal can be defined as the ideal $(a_i(x_2, \dots, x_n)^{\frac{o!}{o-i}})$ [Ab 3, BM 1, EV 1, EV 2, Hi 6]. Serves to define slopes and to apply induction on the embedding dimension. Depends on the coordinates.

Complete ideal: Ideal I of a ring which is integrally closed in the sense that each ring element satisfying a monic polynomial equation with coefficients in certain powers of the ideal belongs to the ideal [Lp 6, Sp 5, ZS, appendix]. Mostly studied in rings of dimension two.

Completion: Passage from a local ring to its completion so as to achieve more flexibility for the definition of the invariants [Na, ZS]. The Hilbert-Samuel function and stratification behave well when passing to completions of local rings [Si 2].

Contact exponent: See slope.

Cremona transformation: Birational map from $\mathbf{P}^2$ to $\mathbf{P}^2$ [Hu, Hs].

Desingularization: See resolution.

Dilatation: See blowup.

Division theorem: Describes normal form of a polynomial or power series modulo an ideal. Used to construct a unique generator system of an ideal in a local ring, the reduced standard basis, by dividing the inital monomials through the ideal [Hi 6, Hi 11]. Works for generic coordinates also in the Henselization of a polynomial ring and thus allows to stay within finite algebra extensions.

Domination: Given a projective variety X and a birationally equivalent variety X^*, find by a permissible sequence of blowups of X a projective variety X' with dominant morphism $X' \to X^*$ (i.e., with dense image) [Ab 3].

Equimultiple locus (ν-fold curve): Reduced subscheme of points of X along which the defining ideal of X in W has constant maximal multiplicity ν or constant Hilbert-Samuel function [Za 4, Ha 4, BM 1, EV 2]. For hypersurfaces, equal to the subscheme where the order of the defining equation is maximal.

Equiresolution: See under resolution.

Equisingularity: Same kind of singular complexity of a scheme along a subscheme. May refer to equimultiplicity along the subscheme, topological constancy, simultaneous resolution, etc. [Za 6, Lp 3, La 2, Te 2].

Examples: Explicit resolution of certain singularities can be found in [BK, Ok, La 1, Mu, BM 1, sec. 2, EV 1, EV 2, KKMS]. See also the implementation of resolution by Bodnár and Schicho [BS]. Counterexamples to various generalizations of results are assembled in [Ha 2].

Excellent schemes: Class of schemes introduced by Grothendieck as a necessary condition to be resolvable [Gr 1, IV.7.8]. Excellent schemes allow passage to completion without loss of substantial local information [Ab 6].

Exceptional divisor (exceptional components): Inverse image of the center under blowup. Locus where the blowup is not an isomorphism (if the center is not a hypersurface).

Factorization of ideals: Decomposition of an ideal into a product of ideals. The theory is rather complete in dimension two [Lp 6, Za 8, Sp 5].

Factorization of morphisms: Description of birational morphisms as a composition of blowups and blowdowns [Za 4, Ab 12, Hs, p. 166, Sa, Sh, Cr, Mr, Cu 1, Cu 2, Cu 3, W 1, W 2, AMR, AKMW].

Flag varieties: Flag of local regular subschemes of a regular ambient scheme at a point. Needed to define Hironaka's invariant β in [Hi 4], or other asymmetric invariants extracted from the Newton polyhedron, see [Ha 4]. Flags are compatible with blowup [Ha 1].

Gauss-Bruhat decomposition: Decomposition of the group of automorphisms of a formal power series ring into a product BUS with B a generalized Borel subgroup, U a generalized unipotent subgroup, and S the group of permutations [Ha 1]. Needed to realize resolution invariants in coordinates in which the blowup is monomial.

Good point: Class of points of surfaces where the resolution algorithm requires only curve blowups, in contrast to bad points. Can be described by arithmetic conditions on the order of the defining function and certain coefficient ideals [Ab 6, EV 1, RS].

Henselization: Intermediate ring between polynomial ring and power series ring. Introduced by Artin so as to be able to work with algebras of finite type instead of completions [Hi 6, Ra, Na].

Hilbert-Samuel function at a point: Local invariant of a scheme x given by the integer function $h(k) = \dim_K M_a^{k+1}/M_a^k$ with $K = \mathcal{O}_a/M_a$ the residue field of the local ring $\mathcal{O}_a$ of x at a. There is a unique polynomial H, the Hilbert-Samuel polynomial of X at a, which coincides with h for large k [Bn, Se, Si 2].

Hilbert-Samuel stratification of a scheme: Each stratum consists of the points of X with the same Hilbert-Samuel function, and is locally closed, by the upper-semicontinuity of the Hilbert-Samuel function. Compare with notion of normal flatness [Bn].

Hopf map: See blowup.

Hypersurface of maximal contact: Any smooth hypersurface which contains the equimultiple locus of a scheme and whose strict transforms contain the equimultiple loci of the strict transforms of the scheme as long as the multiplicity remains constant. Exists always in characteristic zero [AHV 1, Hi 6, EV 2, Bm 1, Gi 2, Gi 3, Co 4], and forms the basis for induction on the embedding dimension and the concept of idealistic exponents. Need not exist in positive characteristic [Hi 9, Ha 2, ex. 8].

Idealistic exponent: Description of equimultiple locus of a scheme in characteristic zero as an intersection of hypersurfaces [Hi 6, BM 1, EV 2].

Infinitely near point: Points of the exceptional divisors of a sequence of blowups mapping to the given point [CGL, SG 2, Lp 6, Cs]. The point is called infinitely very near, if the (local) multiplicity of the singular scheme has remained constant.

Infinitesimal semicontinuity: Signifies that a certain invariant does not increase when passing from a point of the scheme to a point above in the blown up scheme, provided the center is permissible.

Initial ideal: Monomial ideal associated to an ideal of power series with respect to a monomial order as the ideal of initial monomials. Can be made coordinate free by taking the minimal or maximal initial ideal [Ha 1]. If the monomial order is graded, it refines the Hilbert-Samuel function.

Invariants: Mostly local numerical data associated to a singularity so as to measure its complexity. Often defined through coordinates and then made coordinate independent by specifying certain sets of coordinates (e.g., maximizing coordinates). The most current invariants are the order or multiplicity, Hironaka's ν^*, the Hilbert-Samuel function, minimal and maximal initial ideals, slopes of Newton polyhedra or orders of coefficient ideals, dimensions and local embedding dimensions, number of exceptional components passing through the point. These are put together in form of a vector considered with the lexicographic order and then used as induction invariant [Za 4, Hi 1, Hi 2, Hi 3, Hi 4, Bn, HIO, Ha 1, Ha 4, BM 1, EV 2, Ab 8, BK, Yo].

History invariants: Invariants which take into account the events of prior blowups, in particular the accumulated exceptional exponents since the last drop of multiplicity. Needed to distinguish between components of the equimultiple locus which have no apparent difference. History allows to equip the components with their age and selects the candidate for the next blowup [BM 1, EV 1, EV 2]. Can be dispensed with in low dimensions by admitting normal crossing centers [Ha 4, Pf, Ro].

Jung's method: Project a surface in three space to two space, consider the resulting discriminant curve and transform it into a normal crossing curve by point blowups. This yields quasi-ordinary singularities whose resolution is relatively simple [Ju, Wa 1, Lp 4, Lp 5, BE].

Local uniformization: See uniformization.

Macaulayfication: Birational proper map $X' \to X$ with X' a Cohen-Macaulay scheme [Fa, Bm 3, Ka].

Maximal contact: See hypersurface of maximal contact.

Model, smooth, of projective scheme X: Smooth projective scheme with function field isomorphic to that of X [ZS II, p. 116, Ab 3, p. 25].

Modification: A proper birational morphism between varieties [AO, I.1.1].

Moduli space (or scheme): Scheme which classifies a class of objects up to isomorphism. There are coarse and fine moduli spaces, e.g. the coarse moduli space of curves of genus g, or the fine moduli space of curves of genus g with a level-m-structure [AO, GO].

Monomialization: Transformation of a morphism into a monomial or toroidal map by blowups or toric modifications of the source and target space [Cu 2, Mr, AK, Kt 1, Kt 2].

Monoidal transformation: See blowup.

Multiplicity: Leading coefficient of the Hilbert-Samuel polynomial of X at a point a multiplied by $\dim_a X!$ [Se, HIO]. Independent of embedding. Coincides for hypersurfaces with the order of the Taylor expansion of the defining equation.

Nash blowup: See under blowup.

Newton-Puiseux expansion: See Puiseux expansion.

Newton polyhedron: Positive convex hull in $\mathbf{N}^n$ of the set of exponents of the nonzero monomials of the Taylor expansion at a given point of the defining function f of a hypersurface in n-space [LT, Yo, Hi 2, Hi 3, Ha 3, CP]. Depends on the choice of local coordinates. Allows to read off many invariants such as the order or multiplicity or characteristic pairs of the hypersurface.

Newton polygon: Name for the Newton polyhedron in the case of dimension 2.

Normal crossings divisor (or scheme): Scheme which can be embedded in a regular scheme such that it is given there locally by a monomial ideal with respect to some choice of coordinates [Or, Ab 3, I.1.5].

Normal flatness: A scheme X is normally flat along a subscheme Z if Z is contained in a Hilbert-Samuel stratum of X. Equivalently, if the local Hilbert-Samuel function of X or the invariant ν^* is constant along Z, or if the graded rings associated to the local rings of X form a flat family over Z [Bn, Si 1, Si 2, Hi 1, chap. II, HSV].

Normalization: Associating to a scheme X a scheme X' with finite birational morphism $X' \to X$ whose local rings are given by the integral closure of the local rings of X [ZS II, p. 93, Hs, p. 23 and p. 91]. Resolves the singularities of curves, and reduces singularities of surfaces to isolated singularities. Non-embeddable process. Construction of normalization by Stolzenberg, Seidenberg and Traverso [St, Sd 2, Tr]. Implemented in the computer algebra system Singular.

ν^*: Local invariant defined by Hironaka to generalize the order of a power series to the case of ideals. Equals the sequence of orders of a standard basis of the ideal [Hi 1, chap. III.1].

Order of an ideal I at a point: Maximal integer k such that the stalk I_a of I at a is contained in M_a^k, where M_a denotes the maximal ideal at a of the ambient space of I. Depends on embedding.

Patching (glueing, globalization): Centers are mostly defined as strata where certain local invariants achieve their maximal value. Thus they are often defined on neighborhoods of points and the blowup can only be globalized if the local definitions patch so as to give as center a global closed subscheme [Wa 1, Vi 2, BM 1, Lp 4, Ab 3].

Permissible center: See center of blowup.

Polygon: See Newton polygon.

Polyhedral game (weak form): Given are fixed coordinates and the Newton polyhedron of the defining function of a hypersurface in affine space at the origin. Player

A chooses a coordinate subspace as center of blowup. Player B chooses one of the origins of the affine charts of the blowup. The new polyhedron is the one associated to the strict transform of the defining function at the point chosen by B. Then the moves are repeated. Player A wins, if the Newton Polyhedron becomes an orthant (i.e., a quadrant, octant, etc.) after finitely many moves, else B wins. There exists a winning strategy for A [Za 2, p. 861-863, Sp 1, Ha 3]. In the strong version of the game, the polyhedron is in addition modified after each move of B, reflecting certain changes of coordinates. There, a winning strategy for B does not exist [Sp 2]. The strong version would have been a key result to advance in positive characteristic, though its formulation is definitely too strong.

Polyhedron: See Newton polyhedron.

Principalization: Making an ideal sheaf locally principal by taking its inverse image under a suitable composition of blowups [Ab 3].

Proper morphism: Algebraic analogue of maps where inverse images of compacta are compact [Hs, II.4].

Puiseux expansion: Parametrization of a curve by means of a power series with rational exponents of bounded denominator [Ca].

Quadratic transformation: See blowup.

Quasi-ordinary singularities: Surfaces whose discriminant under projection to the plane has at most normal crossings [Lp 4, Lp 5, BE].

Reduction of singularities of differential equations: Attempt to simplify the singularities of differential equations to a stable form (i.e., unchangeable by further blowups) and such that the resulting differential equation can be solved [Ar 1, Ar 2, Sd, CF, CJ, CC, CLS, CS, Ne].

Rees algebra: See blowup-algebra.

Resolution: Overall notion of simplifying a system of algebraic or differential equations, a variety or a scheme to a simplest and accesible form, with different meanings according to the context.

Resolution of singularities (desingularization): Given a scheme (or variety) X, a resolution of X is a proper birational map $\pi : X' \to X$ with X' a regular scheme.

Resolution of excellent schemes: Excellence is a necessary hypothesis for the existence of resolutions [Gr 1, IV.7.8, Ab 3]. The conjecture is that it is also sufficient.

Resolution by induction: The map π is constructed as a composition of blowups with prescribed centers, and the improvement of the scheme X under each blowup is exhibited by the decrease of an invariant belonging to a well ordered set, whose minimal element corresponds to regular schemes (or normal crossings schemes, in case of embedded resolution) [Hi 1, BM 1, EV 2, Ha 2]. Such invariants are mostly vectors whose components can be the multiplicity, the Hilbert-Samuel function, initial ideals, embedding dimensions, or slopes of Newton polyhedra [Ha 1]. The invariant usually takes into account the exceptional divisors occurred so far.

Resolution by point blowup and normalization: Works for surfaces as shown by Zariski [Za 2, Hi 1, p. 150], see [Lp 4, sec. 4] for an explicit example.

Resolution over a field: The scheme X to be resolved is of finite type over a field, respectively a localization or completion thereof. The field is often assumed perfect, or algebraically closed, with different degree of difficulties according to positive or zero characteristic. Over the complex numbers, resolution over **C** sometimes refers to complex analytic spaces [AHV 1, AHV 2, BM 1].

Resolution graph: Graph whose vertices are the exceptional components of the minimal resolution of a surface, and whose edges indicate intersections of the components [La 1, Mü].

Algorithmic resolution: See constructive resolution.

Canonical resolution: The centers of blowups are chosen in a canonical way, i.e., not depending on any artificial choice, for instance as the smallest stratum of a stratification of the scheme by local invariants [BM 1]. See also equivariant resolution.

Combinatorial resolution (monomial resolution, polyhedral game): Local coordinates are fixed at a chosen point of X. The blowup is only considered at the origins of the induced charts, and with the canonically induced coordinates. Thus, all associated ring homomorphisms are described by monomial substitutions of the coordinates [Sp 1, Sp 2, Ha 1, Ha 3].

Constructive resolution (algorithmic resolution): The map π is described as a sequence of explicitly given blowups in (mostly regular) centers, often defined as the smallest stratum of X with respect to a stratification given by the constancy of certain local invariants (equimultiple locus, Hilbert-Samuel stratification) [BM 1, EV 2]. Allows sometimes computer implementation [BS].

Embedded resolution: Given a scheme X embedded in a regular excellent ambient scheme W, find a proper birational map $\pi : W' \to W$ such that the strict transform X' of X is a regular scheme having normal crossings with the exceptional divisor [Hi 1, Ab 3, Lp 4]. If this last condition of transversality is not imposed, the resolution is called non-embedded. Under embedded resolution, the map $X' \to X$ should not depend on the chosen embedding of X in W [E 2].

Equiresolution (simultaneous resolution): Given a scheme X over a base scheme S, find a resolution $\pi : X' \to X$ such that all fibers of X' over S resolve the fibers of X over S [La 2, Lp 3, E 1, Te 2].

Equivariant resolution: The map π is compatible with group actions on X (i.e. the action lifts to X') and (sometimes) also smooth morphisms $Y \to X$, e.g. étale morphisms [EV 2, AW, BM 1, RY].

Good resolution: Resolution of surfaces such that the exceptional curves are smooth intersecting at most transversally and no three meet in a point [La 1, Mü].

Local resolution: Ambiguous notion used with different meanings [Lp 4]. See also local uniformization.

Minimal resolution: Resolution of surfaces $X' \to X$ such that any other resolution $Y \to X$ factors through X' via a morphism $Y \to X'$. It exists and is unique up to isomorphism [La 1, Lp 2, p. 277, Lp 4, sec. 4, Mü, Hi 1, p. 150, Br 1, Lemma 1.6].

Monomial resolution: See combinatorial resolution.

Simultaneous resolution: Resolving a family of schemes simultaneously [Lp 3, La 2, Te 2, E, Cu 3].

Strong resolution (mostly called simply resolution): The map π is an isomorphism over the regular locus of X [Hi 1, Ab 3]. Else, the resolution is called weak [dJ, AW, I.1.4, Be].

Taut resolution: Resolution of surfaces obtained by transforming first through point blowups the equimultiple locus to normal crossings and then blowing up the whole equimultiple locus with non-reduced structure at the intersection points of the equimultiple locus [Ha 4].

Toric resolution: The resolution of a toric variety is given by subdivisions of the associated fans [KKMS, Fu 2, Cx, Od]. In general, there is no canonical subdivision.

Ridge: Additive scheme reflecting the translations under which the tangent cone of a scheme is invariant [Hi 5, Od 2, Od 3].

Riemann surface: See Zariski-Riemann surface.

Sandwich singularity: Normal complex surfaces whose local rings dominate two dimensional regular local rings. Can be obtained by blowing up a complete ideal in $\mathbf{C}^2$ [Sp 5, GS 3, Lê].

Semicontinuity with respect to deformations and localizations: In order to use invariants for induction purposes, it is convenient and sometimes necessary that they be upper semicontinuous with respect to deformations and localizations. Then the strata of constancy are locally closed, and the invariants cannot increase when passing to localizations (thus often reducing the induction argument to the consideration of residually rational points) [Hi 1, III.3, Bn, BM 1]. See also infinitesimal semicontinuity.

Semi-stable reduction: Let R be a discrete valuation ring with fraction field K. A smooth scheme X over K has semistable reduction if there exists a scheme Y over R whose generic fiber is X and whose special fiber is a reduced normal crossings divisor [DM, AtW, KKMS].

Sigma-process: See blowup.

Singularity (singular point): Point of a variety or scheme where the local ring is not regular [Za 10].

Singular locus: Reduced subscheme of singular points of X. For embedded schemes over a perfect field it is described by certain minors of the Jacobian matrix of the defining functions [Za 10].

Slope of Newton polygon (characteristic pair, contact exponent): Coordinate free invariants extracted from the Newton polyhedron of a hypersurface. For plane curves, given by an equation of form $f = x^o + \sum_{i=0}^{o-1} a_i(y)x^i$, it equals the maximum over all coordinate choices of the slope $\min\{\frac{i}{o-i},\ a_i \neq 0\}$ of the first segment of the Newton polygon. Can be interpreted as the order of certain coefficient ideals [Ab 8, AHV 1, EV 1, Ha 1, Ca, BK, LT].

Standard basis: In the definition of Hironaka, a generator system of an ideal of a local ring whose initial forms of lowest degree generate the tangent cone of the ideal, i.e., the ideal of all initial forms [Hi 1, III.1]. Used to define the invariant ν^* as the increasing sequence of the orders of a standard basis, who served as induction invariant until it was commonly replaced by the Hilbert-Samuel function [Bn]. Nowadays, a standard basis is defined with respect to a monomial order and has initial monomials (with respect to this order) who generate the ideal of all initial monomials [Hi 11]. Allows to define maximal initial ideals as a simpler and finer invariant replacing ν^* and the Hilbert-Samuel function [Ha 1].

Stratification: Decomposition of a scheme into locally closed regular subschemes, mostly defined through the constancy of some local invariant.

Tangent cone: Scheme defined by the homogeneous polynomials of minimal degree of the Taylor expansions of the elements of the ideal defining a scheme at a point [Hi 1, chap. III, AHV 2, Bn].

Weighted tangent cone of a hypersurface: Scheme defined by the weighted homogeneous polynomial approximating best the defining equation of a hypersurface [Ab 8, AHV 1].

τ: Local invariant defined by Hironaka and used implicitly already by Zariski and Abhyankar. Denotes the minimal number of variables necessary to express the tangent cone of an ideal or variety. Behaves well under blowup in characteristic zero, but not over imperfect fields [Hi 1, III.4, Hi 5, Od 2, Od 3].

Toric map: Morphism given by a monomial substitution of a regular system of parameters [GT, Cx].

Transform: Variety, scheme, ideal, vector field, foliation, differential equation etc. obtained by blowup, modification or alteration.

Proper transform: See strict transform.

Total transform: Inverse image $\pi^{-1}X$ of a subscheme X of W under a blowup $\pi : W' \to W$ [Hi 1, II.2, Bn]. Consists of the exceptional divisor $\pi^{-1}(X \cap Z)$ and the strict transform of X'.

Strict transform (proper transform): Closure X' of $\pi^{-1}(X \setminus Z)$ in W' [Hi 1, III.2, Bn, Sf, p. 74]. Deletes all exceptional components from the total transform. If I_X is the defining ideal of X in W, X' is defined in W by the ideal generated by all $E^{-\mathrm{ord}_Z f} f$ with $f \in I_X$, where E denotes the exceptional divisor. If Z is contained in X, X' equals the blowup of X in Z.

Weak transform: Subscheme of W' defined by the ideal $E^{-\mathrm{ord}_Z I_X} I_X$ [Hi 1, III.2].

Tschirnhaus transformation: Local coordinate change which allows to realize a hypersurface of maximal contact by $x_1 = 0$ [Ab 3, Ab 8, AM]. In essence, equivalent to the process of completing the square. For hypersurfaces of local equation $f = 0$

and order o at $a = 0$, given in Weierstrass form $f = x_1^o + \sum_{i=0}^{o-1} a_i(x_2, \dots, x_n)x_1^i$, the coordinate change is equivalent to replacing x_1 by $x_1 - \frac{1}{o}a_{o-1}(x_2, \dots, x_n)$.

Uniformization, local: Resolves a variety along a valuation, i.e., a proper birational map $X' \to X$ such that the given valuation has a smooth center on X' [Za 2, Ab 1, Co 3, Va, sec. 8, Te 1, Ku 1, Ku 2, Ur, Cu 3].

Valuation: Given an infinite sequence of blowups, a valuation at a point a of W corresponds to a sequence of points, each in one blowup, and mapping to a. Equivalently, a valuation is given by a valuation subring of the function field of X [Za 2, Za 4, Sp 4, Va, Te 1, Pt, Ku 1, Ku 2, Kr, Cs, Bo, chap. VI].

Zariski-Riemann surface (or manifold): Set of all valuations made into a compact topological space. Parametrizes all paths of points on infinite sequences of blowups [ZS, vol. II, p. 110, Za 5].

Surveys

B. Segre [Sg]: Describes various attempts to resolve surfaces, many references.

O. Zariski [Za 9]: Contains long bibliography on the period 1860–1935, extended in the second edition till 1970.

J. Lipman [Lp 4]: Discusses various resolution problems (local, global, embedded). Describes Jung's, Albanese's and Abhyankar's approach. Good bibliography.

J. Lipman, H. Hauser [Lp 8, Ha 6]: Describe and compare the constructive resolutions of Bierstone-Milman and Encinas-Villamayor. Survey historical development.

P. Berthelot [Be]: Describes in detail de Jong's proof on alterations, and discusses various applications to arithmetic geometry. Bibliography.

S. Abhyankar [Ab 4]: Describes situation in the field in the middle of the sixties.

S. Abhyankar [Ab 10]: Surveys various resolution problems and sketches briefly state of the art (p. 253). Long bibliography, listing quite completely Abhyankar's contributions in the field.

U. Orbanz [Or]: Introduction to embedded resolution of plane curves in arbitrary characteristic, and of surfaces in characteristic 0 after Abhyankar.

J. Giraud [Gi 4]: Surveys Jung's method for surfaces, Lipman's paper and Zariski's and Hironaka's approach.

V. Cossart [Co 5]: Compares the methods of Jung, Zariski, Abhyankar and Hironaka (brief) for surfaces.

K. Paranjape [Pa]: Describes the weak resolution of Bogomolov and Pantev.

E. Bierstone, P. Milman [BM 3]: Outlines canonical resolution of hypersurfaces in zero characteristic.

Miscellanea

H. Hironaka [Hi 12]: Preface to Zariski's collected papers.

A. Grothendieck [Gr 2]: Surveys very briefly Hironaka's paper [Hi 1] on the occasion of the International Congress 1970.

A. Grothendieck [Gr 1, IV.7.8]: Describes in detail notion of excellence. States resolution problem and necessity of excellence.

J. Giraud [Gi 1]: Provides several basic techniques for singularities.

V.I. Danilov [Dn]: Very brief definition of main resolution conçepts and results.

H. Hauser [Ha 5]: Brief geometric and algebraic description of blowup-algebra.

M. Brodmann [Bm 1]: Real pictures of blowup.

M. Galbiati (ed.) [Ga]: Collects various articles and surveys on the topic.

E. Brieskorn, H. Knörrer [BK]: Describes in detail resolution of plane curves in characteristic zero.

G. Regensburger [Rg]: Describes in detail resolution of plane curves in arbitrary characteristic.

E. Casas [Cs]: Treatise on the theory of plane curves and their resolution.

R. Hartshorne [Hs, II.7]: Treats notion of blowup.

M. Herrmann, S. Ikeda, U. Orbanz [HIO]: Detailed introduction to the notions of multiplicity, Hilbert-Samuel function and blowup.

M. Brandenberg [Bb]: Detailed description of blowups.

E. Kunz [Kz]: Reproduces the main steps of Hironaka's proof of resolution in characteristic zero.

H. Hauser [Ha 2]: Lists and discusses various counterexamples appearing in resolution of singularities when extending results to more general situations.

J.-M. Aroca, H. Hironaka, J.-L. Vicente [AHV 3]: Comprehensive book on resolution of singularities in the analytic category.

R. Reitberger [Re]: Collects published false proofs together with their reviews.

Selection of references before 1930

[Al] Albanese, G.: Transformazione birazionale di una superficie algebrica qualunque in un altra priva di punti multipli. Rend. Circ. Mat. Palermo **48** (1924).

[Ch] Chisini, O.: La risoluzione delle singolarità di una superficie mediante transformazioni birazionali dello spazio. Mem. Accad. Sci. Bologna VII **8** (1921).

[DW] Dedekind, R., Weber, H.: Theorie der algebraischen Functionen einer Veränderlichen. Crelle J. **92** (1882), 181–290.

[Hl] Halphen, G.: Etudes sur les points singuliers des courbes algébriques planes. Appendix to the French edition of Salmon: Higher plane curves. Dublin 1952. (1884), 535–648.

[He] Hensel, K.: *Theorie der algebraischen Zahlen.* Teubner 1908.

[Hu] Hudson, H.: *Cremona transformations.* Cambridge University Press 1927.

[Ju] Jung, H.: Darstellung der Funktionen eines algebraischen Körpers zweier unabhängiger Veränderlicher x, y in der Umgebung einer Stelle $x = a$, $y = b$. J. Reine Angew. Math. **133** (1908), 289–314.

[Lv 1] Levi, B.: Sulla risoluzione delle singolarità puntuali delle superficie algebriche dello spazio ordinario per transformazioni quadratiche. Ann. Mat. Pura Appl. II **26** (1897).

[Lv 2] Levi, B.: Risoluzione delle singolarità puntuali delle superficie algebriche. Atti Acad. Sci. Torino **33** (1897), 66–86.

[No 1] Noether, M.: Zur Theorie des eindeutigen Entsprechens algebraischer Gebilde von beliebig vielen Dimensionen. Math. Ann. **2** (1870), 293–316.

[No 2] Noether, M.: Über einen Satz aus der Theorie der algebraischen Funktionen. Math. Ann. **6** (1873), 351–359.

[NB] Noether, M., Brill, A.: Die Entwicklung der Theorie der algebraischen Funktionen in älterer und neuerer Zeit. Kap. IV. Jahresber. Dt. Math. Verein. III (1892/93), 107–566.

[Ri] Riemann, B.: Theorie der Abelschen Funktionen. J. Reine Angew. Math. **64** (1865), 115–155.

Selection of references after 1930

[Ab 1] Abhyankar, S.: Local uniformization of algebraic surfaces over ground fields of characteristic $p \neq 0$. Ann. Math. **63** (1956), 491–526.

[Ab 2] Abhyankar, S.: Resolution of singularities of arithmetical surfaces. In: Arithmetical Algebraic Geometry, Harper and Row 1965, 111–152.

[Ab 3] Abhyankar, S.: *Resolution of singularities of embedded algebraic surfaces.* Acad. Press 1966. 2nd edition, Springer 1998.

[Ab 4] Abhyankar, S.: An algorithm on polynomials in one determinate with coefficients in a two dimensional regular local domain. Annali Mat. Pura Appl., Serie 4, **71** (1966), 25–60.

[Ab 5] Abhyankar, S.: Three dimensional embedded uniformization in characteristic p. Lectures at Purdue University, Notes by M.F. Huang 1968.

[Ab 6] Abhyankar, S.: Good points of a hypersurface. Adv. Math. **68** (1988), 87–256.

[Ab 7] Abhyankar, S.: Current status of the resolution problem. Summer Institute on Algebraic Geometry 1964. Proc. Amer. Math. Soc.

[Ab 8] Abhyankar, S.: Desingularization of plane curves. In: Summer Institute on Algebraic Geometry. Arcata 1981, Proc. Symp. Pure Appl. Math. **40** Amer. Math. Soc.

[Ab 9] Abhyankar, S.: Analytic desingularization in characteristic zero. Appendix to the second edition of [Ab 3] .

[Ab 10] Abhyankar, S.: *Algebraic Geometry for Scientists and Engineers.* Math. Surveys and Monographs **35**. Amer. Math. Soc. 1990.

[Ab 11] Abhyankar, S.: Ramification theoretic methods in algebraic geometry. Princeton Univ. Press 1959.

[Ab 12] Abhyankar, S.: On the valuations centered in a local domain. Amer. J. Math. **78** (1956), 321–348.

[Ab 13] Abhyankar, S.: Resolution of Singularities of Algebraic Surfaces. In: Proceedings of the 1968 Bombay International Colloquium, Oxford University Press 1969, 1–11.

[AHV 1] Aroca, J.-M., Hironaka, H., Vicente, J.-L.: The theory of the maximal contact. Mem. Mat. Inst. Jorge Juan Madrid **29** (1975).

[AHV 2] Aroca, J.-M., Hironaka, H., Vicente, J.-L.: Desingularization theorems. Mem. Mat. Inst. Jorge Juan Madrid **29** (1975).

[AHV 3] Aroca, J.-M., Hironaka, H., Vicente, J.-L.: Bimeromorphic smoothing of complex-analytic spaces. Springer. To appear.

[AJ] Abramovich, D., de Jong, J.: Smoothness, semi-stability and toroidal geometry. J. Alg. Geometry **6** (1997), 789–801.

[AK] Abramovich, D., Karu, K.: Weak semistable reduction in characteristic 0. Preprint 1997.

[AMWK] Abramovich, D., Karu, K., Matsuki, K., Włodarczyk, J.: Torification and factorization of birational maps. Preprint 1999.

[AMR] Abramovich, D., Matsuki, K., Rashid, S.: A note on the factorization theorem of toric birational maps after Morelli and its toroidal extension. Preprint 1998.

[AM] Abhyankar, S., Moh, T.T.: Newton-Puiseux expansion and generalized Tschirnhausen transformation I, II. J. Reine Angew. Math. **260** (1973), 47–83, **261** (1973), 29–54.

[AO] Abramovich, D., Oort, F.: Alterations and resolution of singularities. This volume.

[Ar 1] Aroca, J.-M.: Reduction of singularities for differential equations. This volume.

[Ar 2] Aroca, J.-M.: Puiseux solutions of singular differential equations. This volume.

[At] Artin, M.: Lipman's proof of resolution of singularities. In: Arithmetic Geometry (eds. G. Cornell, J. H. Silverman), Springer 1986.

[AtW] Artin, M., Winters, G.: Degenerate fibres and reduction of curves. Topology **10** (1971), 373–383.

[AW] Abramovich, D., Wang, J.: Equivariant resolution of singularities in characteristic 0. Math. Res. Letters **4** (1997), 427–433.

[Bb] Brandenberg, M.: Aufblasungen affiner Varietäten. Thesis Zürich 1992.

[Be] Berthelot, P.: Altérations de variétés algébriques (d'après A.J. de Jong). Sém. Bourbaki, exp. 815 (1995/96), Astérisque **241** (1997), 273–311.

[BE] Ban, C., MacEwan, L.: Canonical resolution of a quasi-ordinary surface singularity. Preprint Ohio State University 1998.

[BK] Brieskorn, E., Knörrer, H.: *Ebene algebraische Kurven.* Birkhäuser 1981. English translation: *Plane algebraic curves.* Birkhäuser 1986.

[Bm 1] Brodmann, M.: Computerbilder von Aufblasungen. El. Math. **50** (1995), 149–163.

[Bm 2] Brodmann, M.: *Algebraische Geometrie.* Birkhäuser 1989.

[Bm 3] Brodmann, M.: Two types of birational models. Comment. Math. Helv. **58** (1983), 388–415.

[BM 1] Bierstone, E., Milman, P.: Canonical desingularization in characteristic zero by blowing up the maximum strata of a local invariant. Invent. Math. **128** (1997), 207–302.

[BM 2] Bierstone, E., Milman, P.: Uniformization of analytic spaces. J. Amer. Math. Soc. **2** (1989), 801–836.

[BM 3] Bierstone, E., Milman, P.: A simple constructive proof of canonical resolution of singularities. In: Effective Methods in Algebraic Geometry (eds. T. Mora, C. Traverso). Progress in Math. **94**, Birkhäuser 1991, 11–30.

[BM 4] Bierstone, E., Milman, P.: Semianalytic and subanalytic sets. Publ. Math. IHES **67** (1988), 5–42.

[Bn] Bennett, B.-M.: On the characteristic function of a local ring. Ann. Math. **91** (1970), 25–87.

[Bo] Bourbaki, N.: *Commutative Algebra*. Springer 1989.

[BP] Bogomolov, F., Pantev, T.: Weak Hironaka Theorem. Math. Res. Letters. **3** (1996), 299–307.

[Br 1] Brieskorn, E.: Über die Auflösung gewisser Singularitäten von holomorphen Abbildungen. Math. Ann. **166** (1966), 76–102.

[Br 2] Brieskorn, E.: Die Auflösung der rationalen Singularitäten holomorpher Abbildungen. Math. Ann. **178** (1968), 255–270.

[BS 1] Bodnár, G., Schicho, J.: A computer program for the resolution of singularities. This volume.

[BS 2] Bodnár, G., Schicho, J.: Automated resolution of singularities for hypersurfaces. Preprint 1999.

[Ca] Campillo, A.: Algebroid curves in positive characteristic. Springer Lecture Notes in Math. 813, 1980.

[CF] Cano, F.: Reduction of the singularities of non-dicritical singular foliations. Dimension three. Amer. J. Math. **115** (1993), 509–588.

[CGL] Campillo, A., Gonzalez-Sprinberg, G., Lejeune, M.: Clusters of infinitely near points. Math. Ann. **306** (1996), 169–194.

[CGO] Cossart, V., Giraud, J., Orbanz, U.: Resolution of surface singularities. Lecture Notes in Math. vol. 1101, Springer 1984.

[CJ] Cano, J.: Construction of invariant curves for singular holomorphic vector fields. Proc. Amer. Math. Soc. **125** (1997), 2649–2650.

[CC] Cano, F., Cerveau, D.: Desingularization of non-dicritical holomorphic foliations and existence of separatrices. Acta Math. **169** (1992), 1–103.

[CLS] Camacho, C., Lins Neto, A., Sad, P.: Topological invariants and equidesingularization for holomorphic vector fields. J. Diff. Geom. **20** (1984), 143–174.

[Co 1] Cossart, V.: Desingularization of embedded excellent surfaces. Tôhoku Math. J. **33** (1981), 25–33.

[Co 2] Cossart, V.: Polyèdre caractéristique d'une singularité. Thèse d'Etat, Orsay 1987.

[Co 3] Cossart, V.: Uniformisation et désingularisation des surfaces d'après Zariski. This volume.

[Co 4] Cossart, V.: Contact maximal en caractéristique positive et petite multiplicité. Duke Math. J. **63** (1991), 57–64.

[Co 5] Cossart, V.: Desingularization in dimension two. In: *Resolution of surface singularities.* Lecture Notes in Math. vol. 1101, Springer 1984.

[CP] Cano, F., Piedra, R.: Characteristic polygon of surface singularities. In: *Géométrie algébrique et applications* II (eds. J.-M. Aroca, T. Sánchez-Giralda, J.-L. Vicente), Proc. of Conference on Singularities, La Rábida 1984. Hermann 1987.

[Cr] Christensen, C.: Strong domination / weak factorization of three dimensional local rings. J. Indian Math. Soc. **45** (1984), 21–47.

[Cs] Casas, E.: *Singularities of plane curves.* Lecture Notes of London Math. Soc., Cambridge University Press. To appear.

[CS] Camacho, C., Sad, P.: Invariant varieties through singularities of holomorphic vector fields. Ann. Math. **115** (1982), 579–595.

[Cu 1] Cutkosky, D.: Local factorization of birational maps. Adv. Math. **132** (1997), 167–315.

[Cu 2] Cutkosky, D.: Local monomialization and factorization of birational maps. To appear in Astérisque.

[Cu 3] Cutkosky, D.: Simultaneous resolution of singularities. To appear in Proc. Amer. Math. Soc.

[Cx] Cox, D.: Toric varieties and toric resolutions. This volume.

[Dd] Dade, E.C.: Multiplicity and monoidal transformations. Thesis, Princeton 1960.

[DM] Deligne, P., Mumford, D.: The irreducibility of the space of curves of given genus. Publ. Math. IHES **36** (1969), 75–109.

[Dn] Danilov, V.I.: Resolution of singularities. Encyclopaedia Math. vol 4 (1995), 862–863.

[E 1] Encinas, S.: Constructive resolution of singularities of families of schemes. Thesis, Valladolid 1996.

[E 2] Encinas, S.: On constructive desingularization of non-embedded schemes. Preprint 1999.

[EV 1] Encinas, S., Villamayor, O.: Good points and constructive resolution of singularities. Acta Math. **181** (1998), 109–158.

[EV 2] Encinas, S., Villamayor, O.: A course on constructive desingularization and equivariance. This volume.

[EV 3] Encinas, S., Villamayor, O.: A new theorem of desingularization over fields of characteristic zero. Preprint 1999.

[Fa] Faltings, G.: Über Macaulayfizierung. Math. Ann. **238** (1978), 175–192.

[Ful] Fulton, W.: *Algebraic Curves.* Benjamin 1969.

[Fu 2] Fulton, W.: *Introduction to Toric Varieties.* Princeton Univ. Press 1993.

[Ga] Galbiati, M. (ed.): Proceedings of the workshop on resolution of singularities 1991, Univ. of Pisa.

[Gb] Gröbner, W.: Die birationalen Transformationen der Polynomideale. Monatshefte **58** (1954), 266–286.

[Ge] Geisser, T.: Applications of de Jong's theorem on alterations. This volume.

[Gi 1] Giraud, J.: Etude locale des singularités. Cours de 3e Cycle, Orsay 1971/72.

[Gi 2] Giraud, J.: Sur la théorie du contact maximal. Math. Z. **137** (1974), 285–310.

[Gi 3] Giraud, J.: Contact maximal en caractéristique positive. Ann. Scient. Ec. Norm. Sup. Paris **8** (1975), 201–234.

[Gi 4] Giraud, J.: Desingularization in low dimension. In: *Resolution of surface singularities*. Lecture Notes in Math. vol. 1101, Springer 1984, 51–78.

[GO] van Geemen, B., Oort, F.: A compactification of a fine moduli spaces of curves. This volume.

[Gr 1] Grothendieck, A.: Eléments de Géométrie Algébrique IV. Publ. Math. IHES **24** (1965).

[Gr 2] Grothendieck, A.: Traveaux de Heisouke Hironaka sur la résolution des singularités. Actes Congrès International Math., Nice 1970.

[GS 1] Gonzalez-Sprinberg, G.: Eventails en dimension deux et transformé de Nash. Mimeographed notes Ec. Norm. Sup. Paris (1977), 67 pages.

[GS 2] Gonzalez-Sprinberg, G.: Résolution de Nash des points doubles rationnels. Ann. Inst. Fourier **32** (1982), 111–178.

[GS 3] Gonzalez-Sprinberg, G.: Désingularisation des surfaces par des modifications de Nash normalisées. Séminaire Bourbaki, exp. 661 (1985/86). Astérisque **145–146** (1987), 187–207.

[GT] Goldin, R., Teissier, B.: Resolving singularities of plane analytic branches with one toric morphism. This volume.

[Ha 1] Hauser, H.: Resolution techniques. Preprint 1999.

[Ha 2] Hauser, H.: Seventeen obstacles for resolution of singularities. In: Singularities, The Brieskorn Anniversary volume (ed.: V.I. Arnold, G.-M. Greuel, J. Steenbrink). Birkhäuser 1998.

[Ha 3] Hauser, H.: Triangles, prismas and tetrahedra as resolution invariants. Preprint 1998.

[Ha 4] Hauser, H.: Excellent surfaces and their taut resolution. This volume.

[Ha 5] Hauser, H.: Blowup algebra. In: Encyclopaedia of Mathematics, Supplement I, 131–132, Kluwer 1997.

[Ha 6] Hauser, H.: Featured review on [Ev 1], Math. Reviews 99i:14020, september 1999.

[Hi 1] Hironaka, H.: Resolution of singularities of an algebraic variety over a field of characteristic zero. Ann. Math. **79** (1964), 109–326.

[Hi 2] Hironaka, H.: Characteristic polyhedra of singularities. J. Math. Kyoto Univ. **7** (1967), 251–293.

[Hi 3] Hironaka, H.: Certain numerical characters of singularities. J. Math. Kyoto Univ. **10** (1970), 151–187.

[Hi 4] Hironaka, H.: Desingularization of excellent surfaces. Notes by B. Bennett at the Conference on Algebraic Geometry, Bowdoin 1967. Reprinted in: Cossart, V., Giraud, J., Orbanz, U.: *Resolution of surface singularities*. Lecture Notes in Math. 1101, Springer 1984.

[Hi 5] Hironaka, H.: Additive groups associated with points of a projective space. Ann. Math. **92** (1970), 327–334.

[Hi 6] Hironaka, H.: Idealistic exponents of singularity. In: *Algebraic Geometry*, the Johns Hopkins Centennial Lectures. Johns Hopkins University Press 1977.

[Hi 7] Hironaka, H.: Schemes etc. In: 5th Nordic Summer School in Mathematics (ed. F. Oort), Oslo 1970, 291–313.

[Hi 8] Hironaka, H.: Desingularization of complex analytic varieties. Actes Congrès International Math., Nice 1970, 627–631.

[Hi 9] Hironaka, H.: On the presentations of resolution data (Notes by T.T. Moh). In: *Algebraic Analysis, Geometry and Number Theory* 1988 (ed. J.I. Igusa), The Johns Hopkins University Press 1989.

[Hi 10] Hironaka, H.: Bimeromorphic smoothing of a complex analytic space. Preprint University of Warwick 1971.

[Hi 11] Hironaka, H.: Stratification and flatness. In: Nordic Summer School, Oslo 1976 (ed. P. Holm), Sijthoff and Noordhoff 1977.

[Hi 12] Hironaka, H.: Preface to Zariski's collected papers I, 307–312. MIT Press 1979.

[Hi 13] Hironaka, H.: On the theory of birational blowing-up. Thesis, Harvard University 1960.

[HIO] Herrmann, M., Ikeda, S., Orbanz, U.: *Equimultiplicity and blowing up.* Springer 1988.

[Hs] Hartshorne, R.: *Algebraic Geometry.* Springer 1977.

[HSV] Herrmann, H., Schmidt, R., Vogel, W.: *Theorie der normalen Flachheit.* Teubner-Texte Leipzig 1977.

[Hz] Hirzebruch, F.: Über vierdimensionale Riemann'sche Flächen mehrdeutiger analytischer Funktionen von zwei komplexen Veränderlichen. Math. Ann. **126** (1953), 1–22.

[dJ] de Jong, A.J.: Smoothness, semi-stability and alterations. Publ. Math. IHES **83** (1996), 51–93.

[Ka] Kawasaki, T.: On Macaulayfication of schemes. Preprint 1999.

[KKMS] Kempf, G., Knudson, F., Mumford, D., Saint-Donat, B.: *Toroidal Embeddings* I. Lect. Notes Math. **339**, Springer 1973.

[Kr] Krull, W.: Allgemeine Bewertungstheorie. J. Reine Angew. Math. **167** (1931), 160–196.

[Kt 1] Kato, K.: Logarithmic structures of Fontaine-Illussie. In: Algebraic analysis, geometry and number theory. Proceedings of the JAMI Inaugural Conference, 191–224, Johns Hopkins Univ. Press, Baltimore, MD, 1989.

[Kt 2] Kato, K.: Toric singularities. Amer. J. Math. **116** (1994), 1073–1099.

[Ku 1] Kuhlmann, F.-V.: Valuation theoretic and model theoretic aspects of local uniformization. This volume.

[Ku 2] Kuhlmann, F.-V.: On local uniformization in arbitrary characteristic I. Preprint 1998.

[Kz] Kunz, E.: Algebraische Geometrie IV. Vorlesung Regensburg.

[La 1] Laufer, H.: Normal two-dimensional singularities. Ann. Math. Studies 71, 1971.

[La 2] Laufer, H.: Strong simultaneous resolution for surface singularities. In: Complex Analytic Singularities, North-Holland 1987, 207–214.

[Lê] Lê, D.T.: Les singularités sandwich. This volume.

[LO] Lê, D.T., Oka, M.: On resolution complexity of plane curves. Kodai Math. J. **18** (1995), 1–36.

[Lp 1] Lipman, J.: Desingularization of 2-dimensional schemes. Ann. Math. **107** (1978), 151–207.

[Lp 2] Lipman, J.: Rational singularities, with applications to algebraic surfaces and unique factorization. Publ. Math. IHES **36** (1969), 195–279.

[Lp 3] Lipman, J.: Equisingularity and equiresolution. This volume.

[Lp 4] Lipman, J.: Introduction to resolution of singularities. Proceedings Symp. Pure Appl. Math. **29** Amer. Math. Soc. 1975, 187–230.

[Lp 5] Lipman, J.: Topological invariants of quasi-ordinary singularities. Memoirs Amer. Math. Soc. **388** (1988).

[Lp 6] Lipman, J.: On complete ideals in regular local rings. In: Algebraic Geometry and Commutative Algebra, vol. 1, Kinokuniya, Tokyo 1988, 203–231.

[Lp 7] Lipman, J.: Oscar Zariski 1899–1986. This volume.

[Lp 8] Lipman, J.: Featured review of [BM 1]. Math. Reviews 98e:14010, May 1998.

[LT] Lejeune, M., Teissier, B.: Contribution à l'étude des singularités du point de vue du polygone de Newton. Thèse d'Etat, Paris 1973.

[Mo 1] Moh, T.-T.: On a stability theorem for local uniformization in characteristic p. Publ. RIMS Kyoto Univ. **23** (1987), 965–973.

[Mo 2] Moh, T.-T.: Canonical uniformization of hypersurface singularities of characteristic zero. Comm. Alg. **20** (1992), 3207–3249.

[Mo 3] Moh, T.-T.: On a Newton polygon approach to the uniformization of singularities of characteristic p. In: *Algebraic Geometry and Singularities* (eds. A. Campillo, L. Narváez). Proc. Conf. on Singularities La Rábida. Birkhäuser 1996.

[Mr] Morelli, R.: The birational geometry of toric varieties. J. Alg. Geom. **5** (1996), 751–782.

[Mü] Müller, G.: Resolution of weighted homogeneous surface singularities. This volume.

[MZ] Muhly, H.T., Zariski, O.: The resolution of singularities of an algebraic curve. Amer. J. Math. **61** (1939), 107–114.

[Na] Nagata, M.: *Local rings.* John Wiley & Sons, Interscience Publishers 1962.

[Nb] Nobile, A.: Some properties of Nash blowing-up. Pacific J. Math. **60** (1975), 297–305.

[Ne] Neto, O.: A desingularization theorem for Legendrian curves. In: Singularity theory, Trieste 1991, 527–548, World Scientific Publishing 1995.

[Od 1] Oda, T.: Infinitely very near singular points. Adv. Studies Pure Math. **8** (1986), 363–404.

[Od 2] Oda, T.: Hironaka's additive group scheme. In: *Number Theory, Algebraic Geometry and Commutative Algebra*, in honor of Y. Akizuki, Kinokuniya, Tokyo, 1973, 181–219.

[Od 3] Oda, T.: Hironaka group schemes and resolution of singularities. In: Proc. Conf. on Algebraic Geometry, Tokyo and Kyoto 1982. Lecture Notes in Math. 1016, Springer 1983, 295–312.

[Od 4] Oda, T.: *Convex Bodies and Algebraic Geometry.* Springer 1988.

[Ok] Oka, M.: Geometry of plane curves via toroidal resolution. In: *Algebraic Geometry and Singularities* (eds. A. Campillo, L. Narváez). Proc. Conf. on Singularities La Rábida. Birkhäuser 1996.

[Or] Orbanz, U.: Embedded resolution of algebraic surfaces after Abhyankar (characteristic 0). In: *Resolution of surface singularities.* Lecture Notes in Math. vol. 1101, Springer 1984.

[OW] Orlik, P., Wagreich, P.: Isolated singularities of algebraic surfaces with $\mathbf{C}^*$-action. Ann. Math. **93** (1971), 205–228.

[Pa] Paranjape, K.: Bogomolov-Pantev resolution – an expository account. In: Proceedings of Warwick Algebraic Geometry Conference 1996.

[Pf] Pfeifle, J.: Aufblasung monomialer Ideale. Masters Thesis Innsbruck 1996.

[Po] Pop, F.: Alterations and birational anabelian geometry. This volume.

[Pt] Piltant, O.: Graded algebras associated with a valuation. Preprint 1994.

[Ra] Raynaud, M.: Anneaux locaux henséliens. Lecture Notes in Math. 169, Springer 1970.

[Rb] Rebassoo, V.: Desingularization properties of the Nash blowing-up process. Thesis, Univ. of Washington 1977.

[Rg] Regensburger, G.: Auflösung von ebenen Kurvensingularitäten. Masters Thesis Innsbruck 1999.

[Re] Reitberger, H.: The turbulent fifties in resolution of singularities. This volume.

[Ro] Rosenberg, J.: Blowing up nonreduced toric subschemes of $\mathbf{A}^n$. Preprint 1998.

[RS] Rodríguez Sánchez, C.: Good points and local resolution of threefold singularities. Thesis, Univ. Léon 1998.

[RY] Reichstein, Z., Youssin, B.: Essential dimensions of algebraic groups and a resolution theorem for G-varieties. Preprint 1999.

[Sa] Sally, J.: Regular overrings of regular local rings. Trans. Amer. Math. Soc **171** ((1972), 291–300.

[Sd 1] Seidenberg, A.: Reduction of the singularities of the differential equation $Ady = Bdx$. Amer. J. Math. **90** (1968), 248–269.

[Sd 2] Seidenberg, A.: Construction of the integral closure of a finite integral domain II. Proc. Amer. Math. Soc. **52** (1975), 368–372.

[Se] Serre, J.P.: *Algèbre locale – multiplicités.* Lecture Notes in Math. 11, Springer 1965.

[Sf] Shavarevich, I.: *Basic Algebraic Geometry 2.* 2nd edition, Springer 1994.

[Sg] B. Segre: Sullo scioglimento delle singolarità delle varietà algebriche. Ann. Mat. Pura Appl. **33** (1952), 5–48.

[SG 1] Sánchez-Giralda, T.: Teoría de singularidades de superficies algebroides sumergidas. Monografias y memorias de Matemática IX, Publ. del Instituto Jorge Juan de Matemáticas. Madrid 1976.

[SG 2] Sánchez-Giralda, T.: Caractérisation des variétés permises d'une hypersurface algébroide. C.R. Acad. Sci. Paris **285** (1977), 1073–1075.

[Sh] Shannon, D.: Monoidal transforms. Amer. J. Math. **95** (1973), 294–320.

[Si 1] Singh, B.: Effect of permissible blowing up on the local Hilbert function. Invent. Math. **26** (1974), 201–212.

[Si 2] Singh, B.: Formal invariance of local characteristic functions. In: Seminar D. Eisenbud, B. Singh, W. Vogel, vol. 1, 44–59. Teubner-Texte zur Math. 29, 1980.

[Sp 1] Spivakovsky, M.: A solution to Hironaka's Polyhedra Game. In: *Arithmetic and Geometry.* Papers dedicated to I.R. Shafarevich on the occasion of his sixtieth birthday, vol II (eds.: M. Artin, J. Tate). Birkhäuser 1983, 419–432.

[Sp 2] Spivakovsky, M.: A counterexample to Hironaka's 'hard' polyhedra game. Publ. RIMS, Kyoto University **18** (1983), 1009–1012.

[Sp 3] Spivakovsky, M.: A counterexample to the theorem of Beppo Levi in three dimensions. Invent. Math. **96** (1989), 181–183.

[Sp 4] Spivakovsky, M.: Valuations in function fields of surfaces. Amer. J. Math. **112** (1990), 107–156.

[Sp 5] Spivakovsky, M.: Sandwiched singularities and desingularization of surfaces by normalized Nash transformations. Ann. Math. **131** (1990), 411–491.

[St] Stolzenberg, G.: Constructive normalization of an algebraic variety. Bull. Amer. Math. Soc. **74** (1968), 595–599.

[Su] Sussmann, H.: Real-analytic desingularization and subanalytic sets: an elementary approach. Trans. Amer. Math. Soc. **317** (1990), 417–461.

[Te 1] Teissier, B.: Valuations, deformations and toric geometry. To appear.

[Te 2] Teissier, B.: Résolution simultanée. In: *Séminaire sur les singularités des surfaces.* Lecture Notes in Math. 777, Springer 1980.

[Tr] Traverso, C.: A study on algebraic algorithms: the normalization. Rend. Sem. Mat. Univ. Politec. Torino (1986), 111–130.

[Ur] Urabe, T.: Resolution of singularities of germs in characteristic positive associated with valuation rings of iterated divisor type. Preprint 1999.

[Va] Vaquié, M.: Valuations. This volume.

[Vi 1] Villamayor, O.: Constructiveness of Hironaka's resolution. Ann. Scient. Ec. Norm. Sup. Paris **22** (1989), 1–32.

[Vi 2] Villamayor, O.: Patching local uniformizations. Ann. Scient. Ec. Norm. Sup. Paris **25** (1992), 629–677.

[Vi 3] Villamayor, O.: An introduction to the algorithm of resolution. In: *Algebraic Geometry and Singularities* (eds. A. Campillo, L. Narváez). Proc. Conf. on Singularities La Rábida. Birkhäuser 1996.

[Vi 4] Villamayor, O.: On equiresolution and a question of Zariski. Preprint 1998.

[Vi 5] Villamayor, O.: On the multiplicity of embedded hypersurfaces. Preprint 1999.

[Vs] Vasconcelos, W.: *Arithmetic of blowup algebras.* Cambridge University Press 1994.

[Wa 1] Walker, R.J.: Reduction of singularities of an algebraic surface. Ann. Math. **36** (1935), 336–365.

[Wa 2] Walker, R.J.: *Algebraic Curves.* Princeton Univ. Press 1950. Reprinted: Dover 1962.

[W 1] Włodarczyk, J.: Decomposition of birational toric maps in blow-ups and blowdowns. Trans. Amer. Math. Soc. **349** (1997), 373–411.

[W 2] Włodarczyk, J.: Combinatorial structures on toroidal varieties and a proof of the weak factorization theorem. Preprint 1999.

[Yo] Youssin, B.: Newton polyhedra without coordinates. Memoirs Amer. Math. Soc. **433** (1990), 1–74, 75–99.

[Za 1] Zariski, O.: The reduction of singularities of an algebraic surface. Ann. Math. **40** (1939), 639–689.

[Za 2] Zariski, O.: Local uniformization on algebraic varieties. Ann. Math. **41** (1940), 852–896.

[Za 3] Zariski, O.: A simplified proof for resolution of singularities of an algebraic surface. Ann. Math. **43** (1942), 583–593.

[Za 4] Zariski, O.: Reduction of singularities of algebraic three dimensional varieties. Ann. Math. **45** (1944), 472–542.

[Za 5] Zariski, O.: The compactness of the Riemann manifold of an abstract field of algebraic functions. Bull. Amer. Math. Soc. **45** (1945), 683–691.

[Za 6] Zariski, O.: Exceptional singularities of an algebraic surface and their reduction. Atti Accad. Naz. Lincei Rend. Cl. Sci. Fis. Mat. Natur., serie VIII, **43** (1967), 135–196.

[Za 7] Zariski, O.: A new proof of the total embedded resolution theorem for algebraic surfaces. Amer. J. Math. **100** (1978), 411–442.

[Za 8] Zariski, O.: Polynomial ideals defined by infinitely near base points. Amer. J. Math. **60** (1938), 151–204.

[Za 9] Zariski, O.: *Algebraic Surfaces.* Ergebnisse der Mathematik, vol 61, 2nd edition, Springer 1971.

[Za 10] Zariski, O.: The concept of a simple point of an abstract algebraic variety. Trans. Amer. Math. Soc. **62** (1947), 1–52.

[Za 11] Zariski, O.: Normal varieties and birational correspondences. Bull. Amer. Math. Soc. **48** (1942), 402–413.

[Za 12] Zariski, O.: *Collected papers*, vol. I. Foundations of algebraic geometry and resolution of singularities. MIT Press 1979.

[Za 13] Zariski, O.: *Collected papers*, vol. IV. Equisingularity. MIT Press 1979.

[ZS] Zariski, O., Samuel P.: *Commutative Algebra*, vol. I, II. Van Nostrand 1958, 1960. Reprints: Graduate Texts in Mathematics, Vol. 28, 29. Springer 1975.

Mathematisches Institut
Universität Innsbruck
A-6020 Austria
herwig.hauser@uibk.ac.at

Part 1
Classes of the Working Week

Progress in Mathematics, Vol. 181,

Alterations and Resolution of Singularities

Dan Abramovich and Frans Oort

Introduction

H. Hironaka, 1964:

In characteristic zero
any variety can be modified into a nonsingular variety.

A. J. de Jong, 1995:

Any variety can be altered into a nonsingular variety.

On July 26, 1995, at the University of California, Santa Cruz, a young Dutch mathematician by the name Aise Johan de Jong made a revolution in the study of the arithmetic, geometry and cohomology theory of varieties in positive or mixed characteristic. The talk he delivered, first in a series of three entitled "Dominating Varieties by Smooth Varieties", had a central theme: a systematic application of fibrations by nodal curves. Among the hundreds of awe-struck members of the audience, participants of the American Mathematical Society Summer Research Institute on Algebraic Geometry, many recognized the great potential of Johan de Jong's ideas even for *complex* algebraic varieties, and indeed soon more results along these lines began to form.

0.1. The alteration paradigm

A.J. de Jong's main result was, that for any variety X, there is a nonsingular variety Y and an *alteration,* namely a proper, surjective and generically finite morphism, $Y \to X$ (see Theorem 2.3 for a precise statement). This is in contrast with Hironaka's result, which uses only a *modification,* namely a proper birational morphism.

Here is the basic structure of the proof by de Jong:

- **Projection.** For a given variety X of dimension d we produce a morphism $f : X \to P$ with $\dim P = d - 1$, and all fibers of f are curves (we may first have to apply a modification to X).
- **Desingularization of fibers.** After an alteration of the base P, we arrive at a new morphism $f : X \to P$ where all fibers are curves with only ordinary nodes as singularities. The main tool here is the theory of moduli of curves.
- **Desingularization of base.** After a further alteration on the base P, we arrive at a new morphism $f : X \to P$ as above, where P is regular. Here we use

induction, i.e. supposing that the theorem is already true for varieties of dimension $d-1$. So here we "desingularize the base".

- **Desingularization of total space.** Given the last two steps, an explicit and easy method of resolution of singularities finishes the job.

0.2. The purpose of this paper

This paper is an outgrowth of our course material prepared for the Working Week on Resolution of Singularities, which was held during September 7–14, 1997 in Obergurgl, Tirol, Austria. As we did in the workshop, we intend to explain Johan de Jong's results in some detail, and give some other results following the same paradigm, as well as a few applications, both arithmetic and in characteristic zero. We hope that the reader will come to share some of the excitement we felt on that beautiful July day in Santa Cruz.

In the rest of this introduction we give an overview of the proof and the material involved. We hope that this introduction will give most readers a general feeling of what the results are about. The body of the paper is divided in two parts. We begin part I by expanding on some of the preliminary material necessary for understanding the proofs by any student of algebraic geometry. Then we go back to the proof of de Jong's main theorem, as well as some generalizations. Proofs of some variants and generalizations of de Jong's theorems are indicated in the form of exercises, with sufficient hints and references, which we hope will enable the reader to appreciate de Jong's work. Part II is an introduction to an ingredient of the proof – the theory of moduli of curves. We aim to indicate the main ideas behind the proofs of the main theorems about existence and properties of moduli spaces, again accompanied with a collection of exercises.

As a result, this account is mostly expository. The only point where some novelty appears is in Section 13, where we show the existence of tautological families of stable curves over the moduli spaces of *stable pointed curves with level structure.* This has been "well known to the experts" for years, and can be collected from the literature. However a complete account under one roof has not been published. For the definition of a "tautological curve" we refer to Section 10.4.

0.3. Historical context

There are many cases in geometry in which one wants to transform a singular variety into a non-singular one: once arrived in such a situation, various technical steps can be performed, not possible on singular varieties.

Since the beginning of the century, partial results in this direction appeared, crowned by Hironaka's theorem on resolution of singularities in characteristic zero, in 1964.

Hironaka's ingenuous proof had many applications, but it was not easy to understand the fine details of his proof. Generalizing that method to varieties in positive characteristic has failed up to now. Indeed, resolution of singularities in positive characteristic has been a topic to which many years of intensive research have been devoted, and up to now the status is not yet clear: for the general question of resolution of singularities in positive characteristic we have neither a fully verified theorem nor a counterexample. In addition, the algorithms involved in Hironaka's theory were difficult to generalize, even in characteristic 0, to some important more complicated situations.

It seemed that a lull in development of this subject had been reached, until a totally new idea came about. In 1995 Johan de Jong approached the problem above, of transforming a variety into a nonsingular one, from a different angle. The idea of the proof is surprisingly easy, and for many applications his result is sufficient. His approach is very geometric, and hence it works in a wide range of situations. The alteration paradigm automatically works in all characteristics, and a suitable version works in mixed characteristic as well. It easily gives rise to some new "semistable reduction" type results which are new even over the complex numbers. Moreover, without much effort it give birth to new, "conceptually easy" proofs of a weaker form of Hironaka's theorem.

0.4. Comparison of approaches

Let us take a moment to make a qualitative comparison of Hironaka's result and de Jong's result.

In the approach taken by Hironaka, singularities of a variety are studied closely, invariants measuring the difficulty of the singularities are defined, and a somewhat explicit algorithm is applied in order to improve the singularities, in the sense that the given invariants get "better". One needs to show that the algorithm terminates (and indeed in characteristic zero it does), resulting in the construction of a regular variety. A big advantage of this process developed by Hironaka (and by many others) is the fact that usually it is very explicit, it is canonical in a certain sense and once it works, the result is in its strongest form, see [70] and [11], as well as [19] in this volume.

In the approach by Johan de Jong, the singularities are, at first, completely ignored. The idea is to first bring the variety to a special form: a fibration by nodal curves. Here one pays a big price: in order to arrive at this special form one needs to use an operation – called alteration – which extends the function field of the variety. However, once we arrive at this form, we can use induction on the dimension for the base space of the fibration, and automatically arrive at a situation where the variety has very mild singularities. Only then, finally, attention is paid to the singularities. But these are so mild that an easy and explicit blowing up finishes the job.

0.5. A sketch of the construction of an alteration giving a regular variety

Here we give a much simplified form of the proof of A.J. de Jong's main Theorem (Theorem 2.3 in this text). We break up the proof in steps. A star attached to a step means that in that phase of the proof a finite extension of the function field might be involved, i.e. the alteration constructed might not be a modification. In steps without a star only modifications are used.

Before starting, a small technical point is necessary. In the course of the proof we use induction on the dimension of the variety X, and it turns out that for the induction to work we need the statement of the theorem to involve a closed subset $Z \subset X$ as well. Our final goal will be to find an alteration $f : Y \to X$ such that $f^{-1}Z$ is a normal crossings divisor.

We start with a field k, a variety X and a closed subset $Z \subset X$, over the field k.

STEP 0. *We can reduce to the case where k is algebraically closed, the variety X is projective and normal, and the closed subset Z is the support of an effective Cartier divisor.*
We intend to say: if we prove the theorem with this new additional data, then the theorem in the original, more general form follows. Reducing to the algebraically closed field case is standard – in the main body of the paper we avoid it, assuming k is algebraically closed. The main ingredient for projectivity is Chow's Lemma (see [Red Book], pp. 85-89, or [HAG], Exercise II.4.10): *for a variety X over k, there exists a modification $X' \to X$, such that X' is quasi-projective.* To make Z into a divisor we simply blow it up inside X.

Replacement convention. From now on, in each step, we shall replace X by a new variety X' over k which admits a modification or an alteration $X' \to X$, arriving finally at a regular variety and an alteration of the variety produced in Step 0.

STEP 1. *After modifying X, construct a morphism $f : X \to P$ of projective varieties whose generic fiber is an irreducible, complete, non-singular curve.*

Note that $\dim(P) = \dim(X) - 1$, which suggests using induction later. Actually we need a little more, but the technical details will be discussed in the main text.

This step follows a classical, geometric idea. Set $\dim(X) = d$, and assume $X \subset \mathbb{P}^N$. Using Bertini's theorem we see that we can find a linear subvariety $L \subset \mathbb{P}^N$ "in general position" with $\dim(L) = N - d$ such that the projection with center L gives a rational map $X \dashrightarrow \mathbb{P}^{d-1}$ where the generic fiber is a regular curve. After modifying X we can make this rational map into a morphism.

The strict transform. We will use an operation which de Jong called the "strict transform". (In [10] the terminology "strict alteration" is used). Consider a morphism $X \to S$, and a base change $T \to S$. Assume T to be integral, and let $\eta \in T$ be its generic point. Then define $X' \subset T \times_S X$ as the closure of the generic fiber $(T \times_S X)_\eta$ in $(T \times_S X)$. A more thorough discussion of this operation will follow in Section 3.1.

In our situation $X \to P$, we will often replace P by an alteration, and then simply replace X by its strict transform.

STEP 2*. *After applying alterations to X and to P we can arrive at a morphism $f : X \to P$ as in Step 1, and sections $\sigma_1, \ldots, \sigma_n : P \to X$, such that every geometric component C' of every geometric fiber of f meets at least three of these sections in the smooth locus of f, i.e. in $C' \cap \mathrm{Sm}(f)$.*

There is a "multi-section" in the situation of Step 1 having this property. After an alteration on Y and on X this becomes a union of sections.

Stable pointed curves. Here we follow Deligne-Mumford and Knudsen. An algebraic curve is called *nodal* if it is complete, connected and if the singularities of C are not worse than ordinary double points. Its arithmetic genus is given by $g = \dim_k \mathrm{H}^1(C, \mathcal{O}_C)$.

Suppose C is a nodal curve of genus g over a field k, and let $P_1, \ldots, P_n \in C(k)$ with $2g - 2 + n > 0$; we write $\mathcal{P} = \{P_1, \ldots, P_n\}$; this is called a *stable n-pointed curve* if:

- the points are mutually different, $i < j \Longrightarrow P_i \neq P_j$,
- none of these marked points is singular, $P_i \notin \mathrm{Sing}(C)$,
- and $\mathrm{Aut}(C, \mathcal{P})$ is a finite group; under the previous conditions (and k algebraically closed) this amounts to the condition that for every regular rational irreducible component

$$\mathbb{P}_1 \cong C' \subset C, \quad \text{then} \quad \#(C' \cap (\mathcal{P} \cup \mathrm{Sing}(C))) \geq 3.$$

A flat family of curves is called "a family of stable n-pointed curves" if all geometric fibers are stable n-pointed curves in the sense just defined, the markings given by sections.

Historically, stable curves and stable pointed curves were introduced in order to construct, in a natural way, compactifications of moduli spaces (see [17]). Certainly the following names should be mentioned: Zariski, A. Mayer, Deligne, Mumford, Grothendieck, Knudsen, and many more. It came a bit as a surprise when de Jong used these for a desingularization-type problem!

STEP 3*. *After an alteration on the base P, we can assume that $X \to P$ is a projective family of stable n-pointed curves.*

We briefly sketch the heart of the proof of this step – it will be discussed in detail later.

Extending families of curves. We need the following fundamental fact: suppose we are given a variety P, an open dense subset $U \subset P$, and a family of stable curves $C_U \to U$:

$$\begin{array}{ccc} C_U & \subset & ? \\ \downarrow & & \downarrow \\ U & \subset & P. \end{array}$$

Then there is an alteration $a : P_1 \to P$ such that the pullback family $C_{U_1} \to U_1$ over the open set $U_1 = a^{-1}U$ can be extended to a family of stable curves $C_1 \to P_1$:

$$\begin{array}{ccc} C_{U_1} & \subset & C_1 \\ \downarrow & & \downarrow \\ U_1 & \subset & P_1. \end{array}$$

The first result behind this is the existence of a moduli space of stable curves ([39], see also Section 12). Then one needs the fact that a finite cover $M \to \overline{M_{g,n}}$ of the moduli space admits a "tautological family" – namely, a family $C \to M$ such that the associated morphism $M \to \overline{M_{g,n}}$ is the given finite cover. One could consult [16] (the precise statement we need follows from that paper), or use [21], where a tautological family of nodal curves is constructed over a moduli space of stable curves with a level structure.

The sectionss of the family $X \to P$ correspond to those of the stable n-pointed curve $\mathcal{C} \to P$, under the birational transformation thus defined. We want to show this extends to a morphism $\mathcal{C} \to X$.

Flattening of the graph. We take the closure $T \subset X \times_P \mathcal{C}$ of the graph of $\beta_0 : \mathcal{C}_U \to X_U$, and apply the "Flattening Lemma", see 3.2 below. We arrive at new X, T, and $\mathcal{C}$ flat over P. All we have to show (modulo some technicalities) is that no point of a fiber of $\mathcal{C} \to P$ is blown up to a component of a fiber of $X \to P$.

The Three Point Lemma. Using the markings, and studying carefully the geometry we show that indeed β_0 extends to a morphism β. The crucial point here was that every component of every fiber of X over P has at least three nonsingular points marked by the sections σ_i (see 4.18–4.20 of [Alteration]).

STEP 4*. *After an alteration of P, we may assume that P is nonsingular.*

We simply apply induction on the dimension of the base: we suppose that the theorem we want to prove is valid for all varieties having dimension less than dim X. Thus after an alteration of the base P we can suppose P is regular and the strict transform of X has all the previous properties.
Following Z. The argument for the previous two steps should be carried through with a proper care given to the divisor Z. At the end, we can guarantee that Z is contained in the union of two types of sets:

- the images of the sections σ_i, and
- the inverse image of a normal crossings divisor $\Delta \subset P$.

Moreover, in the induction hypothesis we can guarantee that the final family of curves $X \to P$ degenerates only over the normal crossings divisor Δ.

STEP 5. *The singularities of the resulting family $X \to P$ are so mild that it is very easy to resolve them explicitly.*

Indeed, each singular point can be described in formal coordinates by the equation $xy = t_1^{k_1} \cdots t_r^{k_r}$. It is a fairly straightforward exercise to resolve these singularities.

Part I. The alteration theorem

1. Some preliminaries and generalities on varieties

1.1. Varieties

To fix notation, we use the following definition of a variety:

Definition. By a *variety* defined over k we mean a separated geometrically integral scheme of finite type over k. If $k \subset k_1$ we write X_{k_1} for $X \times_{\operatorname{Spec} k} \operatorname{Spec} k_1$.

In more down to earth terms this means: an *affine* variety defined over k is given as a closed subvariety of an affine space $\mathbb{A}_k^n$ defined by an ideal $I \subset k[T_1, \ldots, T_n] = k[T]$ such that $k_1 \cdot I \subset k_1[T]$ is a prime ideal for every (equivalently, for some) algebraically closed field k_1 containing k. In general, a variety then is defined by gluing a finite number of affine varieties in a separated way. See [Red Book], I.5, Definition 1 (p. 35) and I.6, Definition 2 (p. 52).

Remark. This definition differs slightly from that in [Alteration]. De Jong requires the algebraic scheme to be integral, and we require that the schemes stay integral after extending the field. For example for any finite field extension $k \subset K$, the scheme $\operatorname{Spec}(K)$ is called a k-variety by de Jong, but we only say it is a variety defined over k if $k = K$. For most geometric situations the differences will not be important.

1.2. Operations on varieties

Definition. A morphism of varieties $Y \to X$ is called a *modification* if it is proper and birational.

A modification is the type of "surgery operation" usually associated with resolution of singularities. Johan de Jong introduced the following important variant:

Definition (de Jong). A morphism of varieties $Y \to X$ is called an *alteration* if it is proper, surjective and generically finite. This notion of alteration will also be used for integral schemes.

See [Alteration], 2.20.

Remark. A modification is a birational alteration.

Exercise 1.1. Show that an alteration $\varphi : Y \to X$ can be factored as

$$Y \xrightarrow{\pi} Z \xrightarrow{f} X,$$

where π is a modification, and f is a finite morphism.

Exercise 1.2. Suppose moreover that a finite group G acts on Y by automorphisms such that the field of invariants $K(Y)^G$ contains the function field $K(X)$. Formulate other factorizations of φ.

Remark. Given a variety X and a nonzero coherent ideal sheaf $\mathcal{I} \subset \mathcal{O}_X$, the blowing up $Bl_{\mathcal{I}}X = \mathrm{Proj}_X(\oplus_{j\geq 0}\mathcal{I}^j) \to X$ gives naturally a modification, such that the inverse image of $\mathcal{I}$ becomes invertible. If $Z \subset X$ is a subscheme with ideal sheaf $\mathcal{I}$, the blowing up $Bl_Z(X)$ of X with *center* Z is defined to be the blowing up $Bl_{\mathcal{I}}X$. See [HAG], II.7, p. 163.

1.3. Smooth morphisms and regular varieties

The terminology "smooth" will only be used in a relative situation. Thus a morphism can be smooth. The terminology "regular", or "non-singular", will be used in the absolute sense. Thus a variety can be regular. This means that for every point P in the variety the local ring at P is a regular local ring. If a morphism $X \to \mathrm{Spec}(K)$ is smooth, then X is regular. It is not recommended to use the terminology "a smooth variety", which can be misleading and confusing.

1.4. Resolution, weak and strong

We state what we mean by a resolution of singularities. There are two variants we will use:

Definition. Let X be a variety. A *resolution of singularities in the weak sense* is a modification $Y \to X$ such that Y is nonsingular.

Definition. Let X be a variety. A *resolution of singularities in the strong sense* is a modification $Y \to X$, which is an isomorphism over the nonsingular locus X_{reg}, such that Y is nonsingular.

1.5. Normal crossings

The following type of "nice subschemes" of a variety are quite useful in desingularization problems and applications:

Definition. Let X be a variety. A subscheme $Z \subset X$ is called a *strict normal crossings divisor* if for each point $x \in Z$, there is a regular system of parameters $y_1, \ldots, y_k$ for x in X (in particular the point $x \in X$ is supposed to be a regular point on X), such that Z is given on a Zariski neighborhood of x by the equation $y_1 \cdots y_l = 0$.

Suppose furthermore we have a finite group acting on Z and X equivariantly: $G \subset \mathrm{Aut}(Z \subset X)$. We say that Z is a *G-strict normal crossings divisor* if it has normal crossings, and for any irreducible component $Z' \subset Z$, the orbit $\cup_{g\in G}\, g(Z')$ is normal.

We say that a closed subset $Z \subset X$ is a strict normal crossings divisor, if ths reduced subscheme it supports is a strict normal crossings divisor. See [Alteration], 7.1. Strict normal crossings divisors have played an important role in resolution of singularities, and are essential in the proof of de Jong's result.

1.6. Flatness

A crucial idea for studying "families of schemes" is Serre's notion of *flatness* (see [HAG], III.9).

Definition. Let A be a ring and M and A-module. Recall that M is said to be a *flat* A-module if the functor $N \mapsto M \otimes_A N$ is exact.

A morphism of schemes $X \to Y$ is *flat* if at any point $x \in X$, whose image is $y \in Y$, the local ring $\mathcal{O}_{X,x}$ is a flat $\mathcal{O}_{Y,y}$-module.

There are many important examples of flat morphisms which we will discuss later. The reader is advised to consult [HAG] or [43] for a more detailed discussion. The general picture should be that in a proper flat morphism, many essential numerical invariants (e.g. dimension, degree ...) are "constant" from fiber to fiber, so we should really think about it as a "family".

Here are some instructive examples of morphisms which are *not* flat:

Example 1.3. (See [HAG], III 9.7.1.) Let Y be a curve with a node (say, the locus $xy = 0$ in the affine plane). Let $X \to Y$ be the normalization (in the specific example, the disjoint union of two lines mapping onto the locus $xy = 0$). Then $f : X \to Y$ is not flat. The idea one should have in mind is that since over a general point in Y we have one point in X, and over the node we have two points in X, this is not really a nice family – it jumps in degree.

The same reasoning gives a more general example:

Example 1.4. Let $f : X \to Y$ be a modification. Then f is flat if and only if it is an isomorphism. In particular, a nontrivial blowup is not flat.

1.7. Stable curves

We give a formal definition of the fundamental notion introduced in the introduction:

Definition. An S scheme $C \to S$ is called a *family of nodal curves* over S if it is of finite presentation, proper and flat, and all geometric fibers are connected reduced curves with at most ordinary double points (locally $xy = 0$) as singularities.

Remark. The terminology *a nodal curve over S* can be used interchangeably with *a family of nodal curves over S*. Indeed, if $C \to S$ comes by way of an extension of a nodal curve C_η over the generic point η_S of S, it may be natural to call it *a nodal curve over S*.

Definition. The *discriminant locus* $\Delta \subset S$ is the closed subset over which $C \to S$ is not smooth.

Definition (Deligne and Mumford). A family of nodal curves

$$f : C \to S,$$

together with sections $s_i : S \to C, \quad i = 1, \dots, n$ with image schemes $S_i = s_i(S)$, is called a *family of stable n-pointed curves of genus g* if

1. The schemes S_i are mutually disjoint.
2. The schemes S_i are disjoint from the non-smooth locus $\mathrm{Sing}(f)$.
3. All the geometric fibers have arithmetic genus g.
4. The sheaf $\omega_{C/S}(\sum S_i)$ is f-ample (namely, it is ample on fall fibers of f).

In case $n = 0$ we simply call these *stable curves* (rather than stable 0-pointed curves).

The definition is made so that a stable pointed curve has a finite automorphism group (relative over S). It agrees with that made (informally) in the introduction. It is discussed in detail in [17].

Remark. In the literature one sometimes finds the terminology "n-pointed stable curve" instead of "stable n-pointed curve". We try to stick to the latter, since it effectively conveys the idea that the curve *with* the points is stable. The other terminology might give the impression we are dealing with stable curve with some points on them. This would be a different notion in general!

1.8. Minimal models, existence and uniqueness

An important stepping stone for understanding moduli of stable curves is the notion of minimal models of 1-parameter families of curves.

Let K be a field, and C a complete, geometrically irreducible algebraic curve smooth over K; suppose the genus of C is at least 2. Let v be a discrete valuation on K, and $R \subset K$ its valuation ring. Pick a projective model $\mathcal{C}_0$ of C over R. Following Abhyankar (1963) we can resolve singularities in dimension 2, therefore we may assume $\mathcal{C}_0$ is nonsingular. Following Shafarevich (1966) and Lipman (1969) we have the notion of the *minimal model* of C over $S := \mathrm{Spec}(R)$ (see Lichtenbaum, [41], Th. 4.4, and see [17], p. 87). We thus arrive at a family of curves $\mathcal{C} \to \mathrm{Spec}(R)$ which is a regular 2-dimensional scheme, and which is relatively minimal.

Remark. Here we use a special case of resolution of singularities, namely in the case of schemes of dimension 2.

2. Results

First recall Hironaka's theorem:

Theorem 2.1 (Hironaka). *Let X be a variety over a field k of characteristic 0. Then there exists a sequence of modifications*

$$X_n \to X_{n-1} \to \cdots \to X_1 \to X_0 = X,$$

where each $X_i \to X_{i-1}$ is a blowing up with nonsingular center, and the center lies over the singular locus $\mathrm{Sing}(X)$. In particular, $X_n \to X$ is a resolution of singularities in the strong sense.

See the original [29]. Hironaka's theorem and its refinements will be discussed in [19] in this volume. Our main goal is to prove the following result, due to A. J. de Jong:

Theorem 2.2. *Let X be a variety over an algebraically closed field. There is a separable alteration $Y \to X$ such that Y is quasi projective and regular.*

Corollary. *Let X/k be a variety. There is a finite extension $k \subset k_1$ and a separable alteration $Y \to X_{k_1}$ such that Y is quasi projective and regular.*

In order for the induction in the proof to work, de Jong's theorem gives more:

Theorem 2.3 (de Jong). *Let X be a variety over an algebraically closed field, $Z \subset X$ a proper closed subset. There is a separable alteration $f : Y \to X$, and an open immersion $j : Y \subset \bar{Y}$, such that $\bar{Y}$ is projective and regular, and the subset $j(f^{-1}Z) \cup (\bar{Y} \smallsetminus Y)$ is the support of a strict normal crossings divisor.*

See [Alteration], 4.1. The proof of this result will be given in Section 4. De Jong's theorem has a few important variants. First, a theorem of semistable reduction up to alteration over a one dimensional base:

Theorem 2.4 (de Jong). *Let R be a discrete valuation ring, with fraction field K and residue field k. Let $X \to \operatorname{Spec} R$ be an integral scheme of finite type such that X_K is a variety. There exists a finite extension $R \subset R_1$, where R_1 is a discrete valuation ring with residue field k_1, and an alteration $Y \to X_{R_1}$, such that Y is nonsingular, and the special fiber Y_{k_1} is a reduced, strict normal crossings divisor.*

See [Alteration], 6.5. The proof is detailed in Section 5. This theorem belongs to a class of theorems about "desingularization of morphisms". A "dual" case, which can actually serve as a building block in proving the alteration type theorems, is the case where the base is arbitrary dimensional, and the fibers are curves. A proof can be found in [31].

Theorem 2.5 (de Jong). *Let $\pi : X \to B$ be a proper surjective morphism of integral schemes, with $\dim X = \dim B + 1$. Let $Z \subset X$ be a proper closed subset. There exists an alteration $B_1 \to B$, a modification $X_1 \to \tilde{X}_{B_1}$ of the strict transform $\tilde{X}_{B_1}$ (see Section 3.1), sections $s_i : B_1 \to X_1$, and a proper closed subset $\Sigma \subset B_1$ such that*

1. *$\pi_1 : X_1 \to B_1$ is a family of pointed nodal curves,*
2. *s_i are disjoint sections, landing in the smooth locus of π_1, and*
3. *the inverse image Z_1 of Z in X_1 is contained in the union of $\pi_1^{-1}\Sigma$ (the "vertical part") and $s_i(B_1)$ (the "horizontal part").*

The reader who has solved the exercises in Section 5 will be able to complete the proof of this theorem. From this de Jong deduced the following refinement of Theorem 2.4:

Theorem 2.6 (de Jong). *Let $\pi : X \to B$ be a proper surjective morphism of integral schemes, $\dim X = \dim B + r$. Assume that B admits a proper morphism to an excellent two-dimensional scheme S. Then there are alterations $B_1 \to B$ and $X_1 \to \tilde{X}_{B_1}$, a factorization $X_1 \to X_2 \to \cdots \to X_r \to X_{r+1} = B$, and subschemes $\Sigma_i = \Sigma_i^{hor} \cup \Sigma_i^{ver}$, such that*

1. *X_i are nonsingular and Σ_i are normal crossings divisors, $i = 1, \ldots, r+1$;*
2. *$\pi_i : X_i \to X_{i+1}$ are families of nodal curves, smooth away from Σ_{i+1}, and*
3. *Σ_i^{hor} is the union of disjoint sections of π_i, lying in the smooth locus of π_i.*

See [31]. Alternative proofs of different versions of this theorem were provided in [1] and [44]. Next, we consider a finite group action:

Theorem 2.7 (de Jong). *Let X be a variety over an algebraically closed field, $Z \subset X$ a proper closed subset, $G \subset \mathrm{Aut}(Z \subset X)$. There is an alteration $f : Y \to X$, and a finite subgroup $G_1 \subset \mathrm{Aut}\, Y$, satisfying:*

1. *there is a surjection $G_1 \to G$ such that f is G_1 equivariant, and the field extension $K(X)^G \subset K(Y)^{G_1}$ is purely inseparable;*
2. *Y is quasi projective and nonsingular; and*
3. *$f^{-1}Z$ is the support of a G-strict normal crossings divisor.*

See [Alteration], 7.3. The proof is detailed in exercises in Section 5. Note that, taking $G = \{id\}$, this implies:

Corollary. *Let X be a variety over an algebraically closed field. There is a purely inseparable alteration $Y \to X$ where Y is a quotient of a nonsingular variety by the action of a finite group.*

For generalizations which combine both Theorem 2.4 and Theorem 2.7, see [31].

In characteristic 0, any purely inseparable alteration is birational, and the quotient singularities can be improved:

Theorem 2.8 (See [2] and [12]). *Let X be a variety over an algebraically closed field of characteristic 0. Then there is a projective resolution of singularities in the weak sense $Y \to X$.*

Remark. This is a rather weak version of Hironaka's theorem. The point is, that new proofs, by Abramovich and de Jong [2], and by Bogomolov and Pantev [12], were given based on de Jong's ideas. The proof by Bogomolov and Pantev is extremely simple, drawing only on toric geometry. Its proof is detailed in Section 7.

Question 2.9. *Can we improve the methods and obtain a weak resolution of singularities in all characteristics? Or, at least weak resolution up to purely inseparable alterations?*

The proof by Abramovich and de Jong, detailed in Section 8, lends itself to generalizations in the flavor of de Jong's semistable reduction theorem, such as the following two results:

Theorem 2.10 (Abramovich-Karu). *Let $X \to B$ be a dominant morphism of complex projective varieties. There exists a commutative diagram*

$$\begin{array}{ccccc} U_{X'} & \subset & X' & \to & X \\ \downarrow & & \downarrow & & \downarrow \\ U_{B'} & \subset & B' & \to & B \end{array}$$

such that

1. *$X' \to X$ and $B' \to B$ are modifications,*
2. *X' and B' are nonsingular,*
3. *$U_{X'} \subset X'$ and $U_{B'} \subset B'$ are toroidal embeddings, and the morphism $X' \to B'$ is a toroidal morphism (see definition in 6).*

Theorem 2.11 (Abramovich-Karu). *Let $X \to B$ be a dominant morphism of complex projective varieties. There exists a commutative diagram*

$$\begin{array}{ccccc} U_X & \subset & X_1 & \to & X \\ \downarrow & & \downarrow \pi_1 & & \downarrow \pi \\ U_B & \subset & B_1 & \to & B \end{array}$$

where $B_1 \to B$ is an alteration, $X_1 \to \tilde{X}_{B_1}$ is a modification *of the strict transform, $U_X \subset X_1$ and $U_B \subset B_1$ are toroidal, the morphism $\pi_1 : X_1 \to B_1$ is toroidal with $\pi_1^{-1} U_B = U_X$, the variety B_1 is nonsingular and*

1. *the morphism π_1 is equidimensional and*
2. *all fibers of π_1 are reduced.*

See [3] for details. A refinement is given in [33], and an application in [34].

3. Some tools

In this section we gather some basic tools which we are going to use. Some of these tools seem to be of vital importance in algebraic geometry, and it is instructive to see them functioning in the context of de Jong's theorem. We have included some indications of proofs for the interested reader. For the proof of the alteration theorem only the following will be necessary: Section 3.1, Lemmas 3.1, 3.2 and 3.4, and Theorem 3.6.

3.1. The strict transform

See [Alteration], 2.18. As mentioned in the introduction, we need an operation called the "strict transform". Let us recall the definition.

Definition. Consider a morphism $X \to S$, and a base change $T \to S$. Assume T to be integral, and let $\eta \in T$ be its generic point. Then define the *strict transform* $\tilde{X}_T \subset T \times_S X$ as the Zariski closure of the generic fiber $\eta \times_S X$:

$$\begin{array}{ccccccc} \tilde{X}_T & \stackrel{\text{def}}{=} & \overline{\eta \times_S X}^{Zar} & \subset & T \times_S X & \longrightarrow & X \\ & & & \searrow & \downarrow & & \downarrow \\ & & & & T & \longrightarrow & S. \end{array}$$

Note that if the image of η is not in the image of $X \to S$ (i.e. if $T \times_S X \to T$ is not dominant), then the strict transform in the sense explained here is empty.

Remark. The notion given here is different from the usual notion of the "strict transform" of a subvariety under a modification (compare with [HAG], II.7, the definition after 7.15). For example consider a blowing up $T \to S$ of a surface S in a point $P \in S$, and let $C \subset S$ be a curve in S passing through P. The "strict alteration" (or "strict transform" in the terminology above) of C under $T \to S$ is empty; the "strict transform" of C under $T \to S$ in the classical sense, as explained in [HAG], II.7, is a curve in T.

Some people have suggested the use of terminology "essential pullback of X along $T \to S$", which may have some merits. After all, $\tilde{X}_T$ contains only the "part" of $T \times_S X$ which dominates T, which is in some sense its essential part.

3.2. Chow's lemma

An algebraic curve and a regular algebraic surface are quasi-projective. However in higher dimension an "abstract variety" need not be quasi-projective. A beautiful example by Hironaka (of a variety of dimension three) is described in [HAG], Appendix B, Example (3.4.1). However in certain situations (such as the alteration method described below) we like to work with projective varieties.

Lemma 3.1. *Given a variety X, there is a modification $Y \to X$ such that Y is quasi-projective.*

See [Red Book], I.10, p. 85, or [HAG], Exc. II.4.10 p. 107. □

3.3. The flattening lemma

In some situations we want to replace a morphism by a flat morphism. One can show this is possible after a *modification* of the base. The general situation is studied in [60]. We only need this in an easier, special situation, as follows:

Lemma 3.2 (The Flattening Lemma). *Let X and Z be varieties over a perfect field K (more generally, integral schemes of finite presentation) and $X \to Z$ a projective, dominant morphism. There exists a modification $f : Y \to Z$ such that the strict transform $f' : \tilde{X}_Y \to Y$ is flat.*

The main ingredient in the proof is the existence and projectivity of the Hilbert scheme. Hilbert schemes were introduced and constructed by Grothendieck in [24], Exp. 221 (see [47] for simplified proofs, [18] for discussion). We will come back to them in Section 10. Their purpose is to parametrize all subschemes of a fixed projective space $\mathbb{P}^N$. Of course, the set of all subschemes of a projective space is rather large, so we cut it down into bounded pieces by fixing the Hilbert polynomial $P_W(T) = \chi(W, \mathcal{O}_W(T))$ for a subscheme $W \subset \mathbb{P}^N$. Grothendieck's result may be summarized as follows:

Theorem 3.3. *There is a projective scheme $\mathcal{H}_{\mathbb{P}^N,P(T)}$ over* $\operatorname{Spec}\mathbb{Z}$ *and a closed subscheme $\mathcal{X}_{\mathbb{P}^N,P(T)} \subset \mathbb{P}^N \times \mathcal{H}_{\mathbb{P}^N,P(T)}$ which is flat over $\mathcal{H}_{\mathbb{P}^N,P(T)}$, such that $\mathcal{H}_{\mathbb{P}^N,P(T)}$ parametrizes subschemes of $\mathbb{P}^N$ with Hilbert polynomial $P(T)$, and where $\mathcal{X}_{\mathbb{P}^N,P(T)} \to \mathcal{H}_{\mathbb{P}^N,P(T)}$ is a universal family, in the following sense:*

Given a scheme T, let $X \subset \mathbb{P}^N \times T$ be a closed subscheme which is flat over T and such that the Hilbert polynomial of the fibers is $P(T)$. Then there exists a unique morphism $h : T \to \mathcal{H}_{\mathbb{P}^N,P(T)}$, such that

$$X = T \underset{\mathcal{H}_{\mathbb{P}^N,P(T)}}{\times} \mathcal{X}_{\mathbb{P}^N,P(T)}$$

Back to the proof of Lemma 3.2. Since $X \to Z$ is projective, we can choose an embedding $X \subset \mathbb{P}^N \times T$ for some N. Note that the generic fiber of f is reduced. By generic flatness, there exists a dense, open subset $i : U \hookrightarrow Z$ such that

$$f_U : X_U := X|_U \longrightarrow U$$

is flat. Let P be Hilbert polynomial of the fibers of f_U (all fibers in a flat family over an irreducible base have the same Hilbert polynomial), and let $\mathcal{X} \to \mathcal{H}$ be the universal family over the Hilbert scheme associated to this polynomial. We have a cartesian commutative diagram:

$$\begin{array}{ccc} X_U & \longrightarrow & \mathcal{X} \\ \downarrow & & \downarrow \\ U & \xrightarrow{g} & \mathcal{H}. \end{array}$$

Note that $\mathcal{X} \to \mathcal{H}$ is a flat morphism. We have a morphism $i \times g : U \to Z \times \mathcal{H}$. Define

$$Z' := \overline{i \times g(U)}^{Zar} \quad \subset \quad Z \times \mathcal{H},$$

and let $X' \to Z'$ be the pull back:

$$X' = Z' \underset{\mathcal{H}}{\times} \mathcal{X}.$$

Note that the base change of a flat morphism is flat, hence $X' \to Z'$ is flat. It follows from [HAG] III.9.8 that X' is the strict transform under $Z' \to Z$ of $X \to Z$. □ 3.2

Remark. We cited [HAG] III.9.8, which is in fact an important building block in the construction of Hilbert schemes.

Remark. We can delete the word "dominant" in the flattening lemma, and still prove the conclusion, but we do not gain much: if $X \to Z$ is not dominant, the identity on Z gives a strict transform (in the sense explained above) of X such that $X' = \emptyset$, and flatness trivially follows.

Remark. In the proof above we note a general method, which will also be used in the question of extending curves below: suppose we study a certain property (e.g. flatness of a map). Suppose there is a "universal family" having this property (e.g., the Hilbert scheme). Suppose also that in a given family the property holds over a dense open subset U in the base. Then, after a modification, or an alteration of the base, depending on the situation, we can achieve that property by mapping U to the base of the universal family, taking the closure of the graph, and pulling back the universal family.

We encounter a similar situation, in the context of extending stable curves, in Section 3.6 below.

3.4. Deforming a node

An important fact underlying the role of stable curves, which is implicitly invoked in several places in this paper, is that a node $uv = 0$ can only deform in a certain way. To be precise:

Lemma 3.4. *Let R be a complete local ring with maximal ideal m and algebraically closed residue field. Let $S = \operatorname{Spec} R$ and denote the special point by s. Let $X \to S$ be the completion of a nodal curve at a closed point x on the fiber X_s over s, so $X_s = \operatorname{Spec}(R/m[\bar{u}, \bar{v}]/(\bar{u}\bar{v}))^\wedge$. Then there is an element $f \in m$, and liftings u of $\bar{u}$ and v of $\bar{v}$, such that $X \simeq \operatorname{Spec}(R[u,v]/(uv - f))^\wedge$.*

One can prove this using the deformation theory of a node: the versal deformation space (see [6]) of the completion X_s of a nodal curve has dimension $\dim \operatorname{Ext}^1(\Omega^1_{X_s}, \mathcal{O}_{X_s}) = 1$, and it is easy to see that the equation $uv = t$ is versal. An elementary proof by lifting the equation is sketched in [Alteration], Section 2.23.

3.5. Serre's lemma

A critical result in the theory of moduli of curves is, that a 1-parameter family of curves admits stable reduction after a base change (see Theorem 11.2). A crucial point in the proof is the relationship between the automorphisms of a curve and the automorphisms of its jacobian, as in the following lemma.

Lemma 3.5. *Let C be a stable curve defined over an algebraically closed field k, let $m \in \mathbb{Z}_{\geq 3}$, not divisible by the characteristic of k, and let $\varphi \in \operatorname{Aut}(C)$ such that the induced map on locally free sheaves of order m*

$$\varphi_* : \mathbf{Pic}^0_C[m] \longrightarrow \mathbf{Pic}^0_C[m]$$

is the identity map. Then $\varphi = 1_C$, the identity morphism on C.

Proof (see [65], *or* [16], *3.5.1).* Let $\tilde{C} \to C$ be the normalization of C (namely the disjoint union of the normalizations of all irreducible components). Denote $J := \mathbf{Pic}^0_C$ and $X := \mathbf{Pic}^0_{\tilde{C}}$. Consider the "Chevalley decomposition" (as in [14]):

$$0 \to T \longrightarrow J \longrightarrow X \to 0,$$

i.e. $T \subset J$ is the maximal connected linear subgroup in J, the quotient is an abelian variety, and $J/T \cong \mathbf{Pic}^0_{\tilde{C}}$. Note that $T \cong (\mathbb{G}_m)^s$ is a split torus. Define $f := \varphi_* - 1_J \in \mathrm{End}(J)$. Using

$$\mathrm{Hom}(T, X) = 0$$

we obtain a commutative diagram

$$\begin{array}{ccccccccc} 0 & \to & T & \longrightarrow & J & \longrightarrow & X & \to & 0 \\ & & g\downarrow & & f\downarrow & & h\downarrow & & \\ 0 & \to & T & \longrightarrow & J & \longrightarrow & X & \to & 0. \end{array}$$

By the original lemma of Serre we deduce that $h = 0$; let us sketch the argument. The automorphism φ is of finite order (because C is stable), hence the induced $\psi \in \mathrm{Aut}(\tilde{C})$ is of finite order, hence $\psi_* = 1 + h$ is of finite order. Note that the ring $\mathrm{End}(X)$ is torsion free, and since ψ is of finite order the subring $\mathbb{Z}[\psi_*] \subset \mathrm{End}(X)$ is cyclotomic. By assumption the element $\psi_* \otimes_{\mathbb{Z}} \mathbb{Z}/m = 1$ in $\mathrm{End}(X) \otimes_{\mathbb{Z}} \mathbb{Z}/m$. Since ψ_* is a root of unity, and $m \geq 3$, this implies $\psi_* = 1_X$, hence $h = 0$.

Moreover an analogous reasoning implies that $g = 0$: use that $\mathrm{End}(T) = \mathrm{Mat}(s, \mathbb{Z})$ is torsion-free. From $h = 0$ and $g = 0$ we deduce that $f : J \to J$ factors as

$$J \to X \xrightarrow{f'} T \to J.$$

Using

$$\mathrm{Hom}(X, T) = 0$$

we conclude $f' = 0$, hence $f = 0$. Hence $\varphi_* = 1_J$, and this implies that $\varphi = 1_C$. □ 3.5

Remark. We have used the fact that for $p \geq 3$, even modulo p, the root of unity ζ_p is not equal to 1: indeed, the ring $\mathcal{O}_{\mathbb{Q}(\zeta_p)}/p$ is artinian, with generator ζ_p.

3.6. Extending stable curves

Suppose we are given a stable curve $\mathcal{C}_U \to U$ over an open set $U \subset S$ of a base scheme S. Can it be extended to a stable curve $\mathcal{C} \to S$? In general the answer is negative. This question is discussed in [32] and [44], where we find criteria which ensure that in certain cases this is possible. The general situation has the following answer: an extension to a stable curves is possible after an *alteration* on the base. Note the difference from the Flattening Lemma, which has to do with extending families of flat subschemes of a fixed scheme.

Theorem 3.6 (Stable Extension Theorem). *Let S be a locally noetherian integral scheme, let $U \subset S$ be a dense open subset, and let $\mathcal{C} \to U$ with sections $s_i^U : U \to \mathcal{C}$ be a stable pointed curve. There exists an alteration $\varphi : T \to S$, and a stable pointed* *curve $\mathcal{D} \to T$ with sections $\tau_i : T \to \mathcal{D}$, such that, if we write $\varphi^{-1}(U) =: U' \subset T$, we have an isomorphism*

$$\mathcal{D}|_{U'} \xrightarrow{\phi} U' \times_U \mathcal{C},$$

such that $\phi^* s_i^U = \tau_i$.

Remark. A proof for unpointed curves can be found in [16], Lemma 1.6. We present here a somewhat different proof. For simplicity of notation the proof is stated in the case of unpointed curves.

The first step is to extend *isomorphisms* of stable curves. The first lemma is the following:

Lemma 3.7. *Suppose T is the spectrum of a discrete valuation ring, and $\mathcal{D} \to T$ and $\mathcal{D}' \to T$ are stable (pointed) curves, such that the generic fibers are isomorphic: $\mathcal{D}_\eta \cong \mathcal{D}'_\eta$. Then this extends a unique isomorphism: $\mathcal{D} \cong_T \mathcal{D}'$.*

For the proof see [17], Lemma 1.12. The main point is that the *minimal models* of $\mathcal{D}$ and $\mathcal{D}'$ coincide, and $\mathcal{D}$ or $\mathcal{D}'$ are obtained from the minimal model in a unique way by blowing down (-2)-curves. This lemma implies the following (see [17], 1.11):

Lemma 3.8. *Suppose T is a scheme, and $\mathcal{D} \to T$ and $\mathcal{D}' \to T$ are stable (pointed) curves. Then* $\mathrm{Isom}_T(\mathcal{D}, \mathcal{D}') \to T$ *is finite and unramified.*

Indeed, the previous lemma implies that $\mathrm{Isom}_T(\mathcal{D}, \mathcal{D}') \to T$ is proper. Since stable curves have a finite automorphism groups, the morphism is finite. And since stable curves have no nonzero vector fields, the morphism is unramified.

As a consequence we get the following general result about extending isomorphisms:

Lemma 3.9. *Suppose T is an integral* normal *scheme, and $\mathcal{D} \to T$ and $\mathcal{D}' \to T$ are stable (pointed) curves, such that the generic fibers are isomorphic: $\mathcal{D}_\eta \cong \mathcal{D}'_\eta$. Then this induces an isomorphism: $\mathcal{D} \cong_T \mathcal{D}'$.*

Proof. The given isomorphism over the generic point η gives a lifting $\eta \to \mathrm{Isom}_T(\mathcal{D}, \mathcal{D}')$. The closure of its image in $\mathrm{Isom}_T(\mathcal{D}, \mathcal{D}')$ maps finitely and birationally to T. By Zariski's Main theorem it is isomorphic to T, and therefore gives a section of $\mathrm{Isom}_T(\mathcal{D}, \mathcal{D}') \to T$. □

Exercise 3.10. We show that the condition "normal" in the previous lemma is needed. To this end, choose a regular curve T_0, and a smooth curve $\mathcal{D}_0 \to T_0$. Choose it in such a way that the geometric generic fiber has only the identity as automorphism, and such that there exist closed points $x, y \in T_0$ and two *different* isomorphisms

$$\alpha, \beta : (\mathcal{D}_0)_x \xrightarrow{\sim} (\mathcal{D}_0)_y.$$

Let $T_0 \to T$ be the nodal curve obtained by identifying x and y as a nodal point $P \in T$ (and the curves isomorphic outside these points). Construct $\mathcal{D}_\alpha \to T$ by "identifying $(\mathcal{D}_0)_x$ and $(\mathcal{D}_0)_y$ via α". Analogously $\mathcal{D}_\beta \to T$. Show that

$$\mathcal{D}_\alpha \not\cong_T \mathcal{D}_\beta, \quad \text{and} \quad (\mathcal{D}_\alpha)_{\eta_T} = (\mathcal{D}_0)_{\eta_{T'}} = (\mathcal{D}_\beta)_{\eta_T}.$$

It is instructive to describe $\mathrm{Isom}_T(\mathcal{D}_\alpha, \mathcal{D}_\beta)$.

Remark. The phenomenon described in the exercise is characteristic of situations where one has a *coarse* moduli space rather than a *fine* one. See Section 10.4 for details.

The following is an analogous lemma about isomorphisms of the geometric generic fibers:

Lemma 3.11. *Suppose T is an integral scheme, $\mathcal{D} \to T$ and $\mathcal{D}' \to T$ stable (pointed) curves, such that the* geometric *generic fibers are isomorphic:*

$$\mathcal{D}_{\overline{\eta}} \cong \mathcal{D}'_{\overline{\eta}}.$$

Then there exists a finite surjective morphism $T' \to T$ and an isomorphism

$$\mathcal{D} \underset{T}{\times} T' \underset{T'}{\cong} \mathcal{D}' \underset{T}{\times} T'.$$

Remark. It is easy to give examples where the isomorphism requested does not exist even over the generic point of T.

Proof. As in [17], 1.10 we consider $\mathrm{Isom}_T(\mathcal{D}, \mathcal{D}')$. The condition in the lemma assures that this is not empty, it is finite and dominant over T, and the lemma follows. □

Proof of Theorem 3.6: Here we use the fact that there exists a "tautological family" of curves over the compactified moduli space of curves with a level structure. For stable curves without points, this is given in [21]. For the case of stable pointed curves, use Theorem 13.2. Another proof, more in the line of [17], is sketched in Section 13.4

Let us suppose that there exists $m \in \mathbb{Z}_{\geq 3}$ such that $S \to \mathrm{Spec}(\mathbb{Z}[1/m])$. Hence the family $\mathcal{C} \to U$ defines a moduli morphism

$$f : U \to \overline{M_g}[1/m] := \overline{M_g} \underset{\mathrm{Spec}\,\mathbb{Z}}{\times} \mathrm{Spec}\,\mathbb{Z}[1/m].$$

We write $M := \overline{M_g^{(m)}}$ (after having fixed g and m) for the moduli scheme of stable curves of genus g with level-m structure (see [21] and Section 13.2). We have a curve $\mathcal{Z} \to M$ such that the associated moduli morphism to $\overline{M_g}[1/m]$ is the natural morphism $\pi : M \to \overline{M_g}[1/m]$ (we say that $\mathcal{Z} \to M$ is a *tautological family*; see Section 10.4). Let $U'' := U \times_{\overline{M_g}} M$, let $U' \subset U''$ be a reduced, irreducible component of U'' dominant over U, and let $\mathcal{C}'$ be the pull back $\mathcal{C}' = \mathcal{C} \times_U U'$. Let $\mathcal{Z}' \to U'$ be the pull back of the tautological family, $\mathcal{Z}' = \mathcal{Z} \times_M U'$. The very fact that $\overline{M_g}$ os a coarse moduli scheme (see Section 10.4, condition 1) guarantees that over the *geometric* generic point we have $\mathcal{C}'_{\overline{\eta}} \simeq \mathcal{Z}'_{\overline{\eta}}$. By the previous lemma we can replace U' by a finite cover (call it again U') for which there is a U' isomorphism $\mathcal{C}' \simeq \mathcal{Z}'$. Let S' be the normalization of S in the function field of U'. We define $V \subset S' \times M$ to be the image of U' by the two morphisms into S' and M, and let $T = \overline{V}^{Zar} \subset S' \times M$.

By construction there is a stable curve over T, obtained by pulling back $\mathcal{Z} \to M$, which moreover by construction extends the pull back of $\mathcal{C}' \to V$. This proves the theorem in case $S \to S_m := \mathrm{Spec}(\mathbb{Z}[1/m])$.

In case $S \to \mathrm{Spec}(\mathbb{Z})$ is surjective, one does the construction for two different values of m, and then one pastes the result using Lemma 3.11. □ 3.6

3.7. Contraction and stabilization

In [39], II, Section 3, pp. 173–179, we find a description of the following two constructions.

1. Consider a stable $(n+1)$-pointed curve $(\mathcal{X}, \mathcal{P}) \to S$ with $2g - 2 + n > 0$. Deleting one section gives a nodal n-pointed curve (with an extra section), which need not be a stable n-pointed curve. However, if necessary one can contract "non-stable components" of fibers (regular rational curves containing not enough singularities and marked points). After this blowing down one obtains a stable n-pointed curve $(X', \mathcal{Q}) \to S$, and an S-morphism $\mathcal{X} \to \mathcal{X}'$ mapping the first n sections of $\mathcal{P}$ to $\mathcal{Q}$. This process, which arrives at a unique solution to this problem, is called *"contraction"*.
2. Consider a stable n-pointed curve $(\mathcal{Y}, \mathcal{Q}) \to S$ plus an extra section $\sigma : S \to \mathcal{Y}$ not in $\mathcal{Q}$. This extra section may meet sections in $\mathcal{Q}$, or meet the nodes of $\mathcal{Y} \to S$. One can blow up $\mathcal{Y}$ in such a way that the strict transforms (in the old sense) of elements of $\mathcal{Q}$ and of the extra section give a stable $(n+1)$-pointed curve $(\mathcal{X}, \mathcal{P}) \to S$, and an S-morphism $\mathcal{X} \to \mathcal{Y}$ mapping the first n sections of $\mathcal{P}$ to $\mathcal{Q}$. This process, which arrives at a unique solution to this problem, is called *"stabilization"*.

4. Proof of de Jong's main theorem

One striking feature of the proofs of de Jong's theorem and its derivatives is, that all the ingredients, with the exception of one subtle, but still natural, result (the Three Point Lemma), were known and understood nearly two decades before. The way they are put together is quite ingenious.

4.1. Preparatory steps and observations

The proof of de Jong's theorem 2.3 starts with a series of simple reduction steps. We are given a variety X defined over an algebraically closed field k, and a Zariski-closed subset $Z \subset X$. We perform some elementary reductions:

Replacing X by an alteration. In order to prove the theorem for a variety X and a closed subset Z, it is enough to prove it for an alteration X' of X while replacing

Z by its inverse image Z' in X'. Thus in several stages of the proof, once we find an alteration $X' \to X$ which we like better than X, we simply replace the pair (X, Z) by (X', Z').

Making Z into a divisor. By blowing up Z in X, and using the observation above, we may assume that Z is the support of an effective Cartier divisor. We will slightly abuse terminology, and say that "Z is a divisor" when we mean that Z is a closed subset supporting an effective Cartier divisor.

Enlarging Z. Suppose $Z_i \subset X$ are divisors and $Z_1 \subset Z_2$, then to prove the theorem for (X, Z_1) it suffices to prove it for (X, Z_2). Indeed, if $f : Y \to X$ is an alteration such that Y is nonsingular and $f^{-1}(Z_2)$ is a strict normal crossings divisor, then $f^{-1}(Z_1)$ is a Cartier divisor contained in $f^{-1}(Z_2)$, and it is clear from the definition that it is a strict normal crossings divisor as well. Thus we may always enlarge the divisor Z.

Making X quasi-projective. Using Chow's Lemma 3.1, we may assume X is quasi-projective. Indeed, by Chow's lemma there is a modification $X' \to X$ such that X' is quasi-projective. We may replace X by X'.

Enlarging X. Suppose $X \subset X_1$ is an open embedding of varieties, $Z_1 \subset X_1$ a divisor which containing $X_1 \smallsetminus X$, and $Z = X \cap Z_1$. Then evidently to prove the theorem for (X, Z) it suffices to prove it for (X_1, Z_1).

Making X projective. Since X is quasi-projective, there is an open embedding $X \subset \overline{X}$ where $\overline{X}$ is projective. Denote $Z_1 = \overline{Z} \cup (\overline{X} \smallsetminus X)$. We may replace $\overline{X}$ by the blowup of Z_1, thus we may assume that Z_1 is the support of a Cartier divisor. By the previous observation it is enough to prove the result for $(\overline{X}, Z_1)$.

We may assume X is normal. Indeed, we can simply replace X by its normalization.

To summarize, one may assume that the variety X is projective and normal, and the subset Z is the support of an effective Cartier divisor. Moreover, one may always replace Z by a larger divisor.

4.2. Producing a projection

The next step is to produce a projection with some nice properties. We first start with some general facts about projections in projective spaces. Let $Y \subset \mathbb{P}^N$ be a projective variety over an algebraically closed field (in fact, separably closed would suffice). For any closed point $p \in \mathbb{P}^N \smallsetminus Y$ we have a projection $pr_p : Y \to \mathbb{P}^{N-1}$.

Lemma 4.1. *Suppose* $\dim Y < N - 1$. *Then there is a nonempty open set* $U \subset \mathbb{P}^N$, *such that if* $p \in U$ *then* pr_p *sends* Y *birationally to its image.*

Proof. Let $q \in Y$ be a regular point. Define the cone $C_{Y,q}$ over Y with vertex q to be the Zariski closure of the union of all secant lines lines containing q and another q', for all $q' \in Y$. It is easy to see that $C_{Y,q}$ has dimension $\leq \dim Y + 1 < N$. Note that $C_{Y,q}$ contains (as "limit points") the projective tangent space $\mathbb{T}_{Y,q}$ at q.

Therefore if $p \in \mathbb{P}^N \smallsetminus C_{Y,q}$ then the line through p and q meets Y transversally, at q only. This property holds as well for the line through p and q', for any $q' \in Y$ in a neighborhood of q. Hence the lemma. □

Lemma 4.2. *Suppose* $\dim Y = N - 1$. *Then there is a nonempty open set* $U \subset \mathbb{P}^N$, *such that if* $p \in U$ *then* pr_p *maps* Y *generically étale to* $\mathbb{P}^{N-1}$.

Proof. Same as before, using $\mathbb{T}_{Y,q}$ instead of $C_{Y,q}$. □

We go back to our X and Z.

Lemma 4.3. *There exists a modification* $\phi : X' \to X$ *and a morphism* $f : X' \to \mathbb{P}^{d-1}$ *such that*

1. *There exists a finite set of nonsingular closed points* $S \subset X_{ns}$ *disjoint from* Z, *such that* X' *is the blowup of* X *at the points of* S.
2. f *is equidimensional of relative dimension 1.*
3. *The smooth locus of* f *is dense in all fibers.*
4. *Let* $Z' = \phi^{-1}Z$. *Then* $f|_{Z'}$ *is finite and generically étale.*
5. *At least one fiber of* f *is smooth.*

Proof. First project $\pi : X \to \mathbb{P}^d$ using the previous lemmas $N - d - 1$ times. Let $B \subset \mathbb{P}^d$ be the locus over which π is not étale. If we choose a general $p \in \mathbb{P}^d$, then $pr_p : \pi(Z) \to \mathbb{P}^{d-1}$ is generically étale – simply use the lemma above for all irreducible components of $\pi(Z)$. We choose such a p away from B. By the local description of blowing up, we can identify the variety

$$X' = \{(x, \ell) \in X \times \mathbb{P}^{d-1} | \pi(x) \in \ell\}$$

with the blowing up of X at the points in $\pi^{-1}(p)$. We define $f : X' \to \mathbb{P}^{d-1}$ to be the second projection. We can identify the fibers: the fiber over a point ℓ is the scheme theoretic inverse image $\pi^{-1}(L)$ where L is the line corresponding to ℓ. It follows immediately that f is equidimensional: all fibers have dimension at most 1, and are defined by $d-1$ equations (the equations of L). Since no line through p is contained in B, every fiber has a dense smooth locus. The last assertion follows by Bertini's theorem, since the fibers are obtained by intersecting X with linear subspaces. □

Lemma 4.4. *The morphism* f *has connected fibers.*

Proof. Since the smooth locus is dense in every fiber, the Stein factorization is étale. Since projective space has no nontrivial finite étale covers, the Stein factorization is trivial. □

Remark. (1) The last assertion is not really necessary: if f did not have connected fibers, we could replace $f : X' \to \mathbb{P}^{d-1}$ by its Stein factorization.
(2) The projection above is the only point where it is crucial that X should be normal, to guarantee that the generic fiber is smooth. From here on we will allow ourselves to make reductions after which X might not be normal.

To summarize, one may assume that we have a morphism of varieties $X \to P$, for some variety P, which makes X into a generically smooth family of curves, satisfying some nice properties, in particular $Z \to P$ is finite and generically étale.

4.3. Enlarging the divisor Z

In order to "rigidify" the situation, it will be useful to enlarge Z so it meets every fiber "sufficiently". This is done as follows:

Lemma 4.5. *Let $X \to P$ be as above. There exists a divisor $H \subset X$ such that*

1. *$f|_H : H \to P$ is finite and generically étale, and*
2. *for any irreducible component C of a geometric fiber of f, we have*

$$\#sm(X/P) \cap C \cap H \geq 3.$$

Here we count the points without *multiplicities.*

Proof. Let $n \geq 1$ be an integer. Given a very ample line bundle $\mathcal{L}$ on X, consider the embedding

$$i : X \hookrightarrow \mathbb{P} = \mathbb{P}(\Gamma(X, \mathcal{L}^{\otimes n}))$$

associated to $\mathcal{L}^{\otimes n}$.

Claim. Given any irreducible curve $C \subset X$, the image $i(C) \subset \mathbb{P}$ is not contained in any linear subspace of dimension $n-1$.

Proof of claim. Since $\mathcal{L}$ is very ample, the image of $\Gamma(X, \mathcal{L}) \to \Gamma(C, \mathcal{L}_{|C})$ contains a rank-2 subspace $V \subset \Gamma(C, \mathcal{L}_{|C})$ such that the corresponding linear series (of dimension 1) has no base points. The map $\mathrm{Sym}^m V \to \Gamma(C, \mathcal{L}_{|C}^{\otimes n})$ has rank $\geq n+1$, therefore $\Gamma(X, \mathcal{L}^{\otimes n}) \to \Gamma(C, \mathcal{L}_{|C}^{\otimes n})$ has rank $\geq n+1$, which is what we claimed. □ (Claim)

The divisors of sections of $\mathcal{L}^{\otimes n}$ are parametrized by the dual projective space $\mathbb{P}^\vee$. We consider the collection of "bad" divisors and show that there are "good" ones left. So consider

$$T = \{(H, y) \in \mathbb{P}^\vee \times P | \dim f^{-1}y \cap H = 1\} \subset \mathbb{P}^\vee \times P.$$

It is clear that T is a Zariski closed subset. We can describe the fibers of $pr_2 : T \to P$ using irreducible components of the fibers:

$$pr_2^{-1}(y) = \bigcup_{C \subset f^{-1}y} \{H | i(C) \subset H\}.$$

But by the fact that $i(C)$ is not contained in any linear subspace of dimension $n-1$, we have

$$\mathrm{codim}(pr_2^{-1}(y), \mathbb{P}^\vee) \geq n.$$

Therefore $\dim T \leq \dim P + \dim \mathbb{P}^\vee - n$. Thus if n is large enough, $pr_1(T) \subset \mathbb{P}^\vee$ is of large codimension (at least $n - \dim P$). In particular $pr_1(T) \neq \mathbb{P}^\vee$. We fix such large n. Thus there are plenty of H which map finitely to P.

For a fixed closed point $y \in P(k)$ consider the set

$$U(y) = \left\{ H \in \mathbb{P}^\vee(k) \;\middle|\; \begin{array}{l} H \notin pr_1(T) \\ H \cap f^{-1}y \subset sm(X/P) \\ H \cap f^{-1}y \text{ is reduced} \end{array} \right\}.$$

This is clearly a nonempty open set of $\mathbb{P}^\vee$. Moreover, if $H \in U(y)$ then $H \in U(y')$ for all y' in a neighborhood of y. If moreover $n \geq 3$, then we have that $\#H \cap f^{-1}y \geq 3$. so we are done for all points in a neighborhood V of y. We deal with points in $P \smallsetminus V$ in the same way. Using Noetherian induction we are done. □ (Lemma)

Summarizing, one may assume that Z meets every irreducible component of every geometric fiber in at least three smooth points.

4.4. The idea of simplifying the fibers

De Jong's idea is to simplify the fibers of the morphism $X \to P$. Then by induction on dimension one can simplify the base P, and finally put these simplifications together.

The method of simplifying the fibers uses the deepest ingredient in the program: the theory of moduli of curves (see Section 10 for discussion).

Here is the general plan. First, as we will see below, it is easy to make an alteration of P, and replace X and Z by their pullbacks, such that Z becomes the union of sections of $X \to P$.

We can think of the generic fiber of $X \to P$ as a smooth curve with a number of points marked on it. Say the genus of this curve is g, and the number of points is n. By the Stable Extension Theorem 3.6, the generic fiber can be extended, *after an alteration* $P_1 \to P$, to a family of stable curves $X_1 \to P_1$:

$$\begin{array}{ccc} X_1 & \dashrightarrow & X \\ \downarrow & & \downarrow \\ P_1 & \to & P \end{array}$$

The new morphism $f_1 : X_1 \to P_1$ is much nicer than f, since at least the fibers are as nice as one can expect: they are nodal curves. Moreover, Z was made much nicer: it is replaced by n sections which are mutually disjoint, and pass through the smooth locus of f_1.

If we can resolve P_1 (say using induction on dimension), then it is easy to resolve X_1 as well.

There is a problem though: if we want to repeat this inductively, we cannot allow a rational map $X_1 \dashrightarrow X$ which is not a morphism – since we cannot pull back nicely along rational maps. So we want to find a way to make sure that $X_1 \dashrightarrow X$ is actually a regular map.

Remark. If one is satisfied with proving a weaker result, namely that every variety admits a "rational alteration" by a nonsingular variety, then there is an alternative way to avoid the issue. This is carried out by S. Mochizuki in [44].

Remark. Another way to circumvent the issue of extending β to a morphism, is to ensure that it extends automatically, by using a moduli space into which a morphism is built in: the space of stable maps. This was carried out, in characteristic 0, in [1], Lemma 4.2. Unfortunately the details of constructing moduli spaces of stable maps have not yet been written out in positive or mixed characteristics, although this would not be difficult: the results of [9] imply that the moduli of stable maps forms a proper Artin stack, and the results of [37] imply that this stack admits a proper algebraic space as a coarse moduli space. One should even be able to modify the argument of [40], Proposition 4.5 and show that this space is projective, but this is not essential for the argument.

Let us go into details.

4.5. Straightening out Z

Lemma 4.6. *There exists a normal variety P_1 and a separable finite morphism $P_1 \to P$ satisfying the following property:*

Let $X_1 = \tilde{X}_{P_1}$ be the strict transform (see Section 3.1), and let Z_1 be the inverse image of Z in X_1. Then there is an integer $n \geq 3$, and n distinct sections $s_i : P_1 \to X_1$, $i = 1, \dots, n$ such that

$$Z_1 = \bigcup_{i=1}^{n} s_i(P_1).$$

Proof. This can be proven by induction on the degree n of $Z \to P$ as follows: Let Z_1 be an irreducible component of Z and let $P' := Z_1^\nu$ be its normalization. We have a generically étale morphism $P' \to P$. Denote $X' = \tilde{X}_{P'}$ and let Z' be the inverse image of Z. The morphism $P' \to Z$ gives rise to a section $s_{k+1} : P' \to Z'$, and therefore we can write $Z' = s_{k+1}(P') \cup Z''$. We have $\deg(Z'' \to P') = \deg(Z \to P) - 1$, and therefore the inductive assumption holds for Z''. $\square$

Thus one can assume Z is the union of sections of $X \to P$.

4.6. Producing a family of stable pointed curves

Let $X \to P$, $s_i : P \to X$ be the new family. Let $U \subset P$ be an open set satisfying the following assumptions:

1. $X_U \to U$ is smooth;
2. the sections $s_i|_U : U \to X_U$ are disjoint.

Such an open set clearly exists.

Since $n \geq 3$ this gives the morphism $X_U \to U$ the structure of a family of stable n-pointed curves.

And here comes the point where moduli theory is used: by Theorem 3.6, there exists an alteration $P_1 \to P$, a family of stable pointed curves $\mathcal{C} \to P_1$, with sections $\tau_i : P_1 \to \mathcal{C}$, such that over the open set $U_1 = P_1 \times_P U \subset P_1$ we have an isomorphism $\beta : \mathcal{C}_{U_1} \to U_1 \times_P X$, satisfying $\beta^* s_i = \tau_i$.

4.7. The three point lemma

As usual, we replace P by P_1 and X by its strict transform. Thus we may assume that we have a diagram as follows:

$$\begin{array}{ccc} C & \overset{\beta}{\dashrightarrow} & X \\ & \searrow & \downarrow \\ & & P \end{array}$$

The crucial point, for which we needed to "enlarge Z" in a previous step, is the following:

Lemma 4.7 (Three Point Lemma). *Suppose Z meets the smooth locus of every irreducible component of every fiber in at least three points. Then, at least after a modification of P, the rational map $\beta : C \dashrightarrow X$ extends to a morphism.*

The proof of this lemma, which is detailed in the next few paragraphs, is probably the most subtle point in this chapter.

4.8. Flattening the graph

Let $T \subset X \times_P C$ be the closure of the graph of the rational map β. We have two projection maps $pr_1 : T \to C$ and $pr_2 : T \to X$.

Claim. *There exists a modification P' of P such that the strict transform of X, and the closure of the graph of $C \dashrightarrow X$ are both flat. Thus we might as well assume $X \to P$ and $T \to P$ are flat.*

Proof. By the Flattening Lemma 3.2 there exists a modification $P' \to P$ such that $\tilde{X}_{P'}$ and $\tilde{T}_{P'}$ are both flat. Evidently the closure of the graph of the rational map $C \times_P P' \to \tilde{X}_{P'}$ is contained in $\tilde{T}_{P'}$, and since $\tilde{T}_{P'}$ is flat they coincide by [HAG] III.9.8. □ (Claim)

Let p be a geometric point on P, and denote by X_p, T_p, C_p the fibers over p. There exists a finite set $W \subset X_p$ such that $T_p \to X_p$ is finite away from W. Indeed, the flatness implies that $\dim T_p = \dim X_p = 1$. Thus, for any $x \in X_p \smallsetminus W$, there is an open neighborhood $x \in V \subset X$ such that $pr_2^{-1}V \to V$ is finite and birational. In case $x \in \mathrm{Sm}(X_p) \smallsetminus W$, we may choose $V \subset \mathrm{Sm}(X \to P)$. Using the assumption that P is normal, it follows that V is normal. In this case, by Zariski's main theorem, $pr_2^{-1}V \to V$ is an isomorphism. Note that the assumption that $x \in \mathrm{Sm}(X_p)$ excludes only finitely many points, since our projection $X \to P$ is smooth at the generic point of each component of the geometric fiber X_p. Therefore we conclude that the following lemma holds:

Lemma 4.8. *If $X' \subset X_p$ is an irreducible component, then there is a* unique *irreducible component T' of T_p mapping finitely onto X' via $pr_2 : T \to X$. Moreover, $T' \to X'$ is birational.*

Repeating the argument for $pr_1 : T \to C$, we also have:

Lemma 4.9. *If $C'' \subset C_p$ is an irreducible component, then there is a unique irreducible component T'' of T_p mapping finitely onto C'' via $pr_1 : T \to C$. Moreover, $T'' \to C''$ is birational.*

Let $x \in \mathrm{Sm}(X_p)$ be a closed point. Considering the Stein factorization $T \to \tilde{X} \to X$, we have that the fiber $pr_2^{-1}(x)$ is connected. Indeed, since X is normal at x, we have that $\tilde{X} \to X$ is an isomorphism at x.

4.9. Using the three point assumption

Let $X' \subset X_p$ be an irreducible component, and $T' \subset T_p$ the unique component mapping finitely (and birationally) onto it, as in Lemma 4.8 above. We will prove that $pr_1 : T' \to C$ is non-constant. Assume by contradiction that $pr_1(T') = \{c\}$ is a point.

We will use the three point assumption. Let $s_i : P \to X, i = 1, \dots, 3$ be three of the given sections such that $s_i(p) = x_i$ are three distinct points on $\mathrm{Sm}(X')$. Let $T_i = pr_2^{-1}x_i$. Let $\tau_i(p) = c_i \in C_p$. Note that the point $t_i = (c_i, x_i) \in C \times X$ is in T. Assume $c \notin \{c_i, i = 1, \dots, 3\}$. Then each of $T_i, i = 1, \dots, 3$ contains an irreducible component T_i' whose image in C is again a curve passing through c. These image components are *distinct.* Indeed, T_i are disjoint subschemes of T_p, whose images in C connect c with c_i, and therefore each has an irreducible component whose image contains c. These components are distinct, and by Lemma 4.9 their images are distinct. This contradicts the fact that C_p is nodal. Thus c is among the c_i.

Assume, without loss of generality, $c = c_1$. Repeating the argument of the previous paragraph we conclude that there are two distinct components of C_p passing through c. This contradicts the fact that C_p has a marked point at $c = c_1$. Thus we conclude that $pr_1 : T \to C$ is finite and birational. By Serre's criterion C is normal: it is clearly regular in codimension 1, and condition S_2 follows since $C \to P$ has reduced one-dimensional fibers and P is normal. We conclude that $T \to C$ is an isomorphism, hence β extends as a morphism! □ 4.7

4.10. Induction

We arrived at the following situation:

$$\begin{array}{ccc} C & \xrightarrow{\beta} & X \\ & \searrow & \downarrow \\ & & P \end{array}$$

We may replace X by C, and Z by its inverse image in C. Note that Z is no longer finite over P: it has a"finite part", the union of the sections $\tau_i : P \to C = X$, but there is a "vertical" part Z_{vert}, which is the union of irreducible components of *singular* fibers of $X \to P$.

Let $\Sigma \subset P$ be the closed subset over which $f : X \to P$ is not smooth. By the induction assumption there is a projective alteration $P_1 \to P$ such that P_1 is nonsingular and the inverse image of Σ is a strict normal crossings divisor. We may replace X by its pullback to P_1, and replace P by P_1. It is convenient to replace Z by its union with $f^{-1}(\Sigma)$.

We arrived at a situation where both P, and the morphism $f : X \to P$, are simplified. The resulting variety has very simple singularities, and its desingularization results from the following exercises.

4.11. Exercises on blowing up of nodal families

The exercises below, which aim at completing the proof, are adapted from De Jong's complete exposition in [Alteration]. We have not reproduced his proofs here. The reader may consult [Alteration], pages 63-64 (Section 3.4) and 75-76 (Sections 4.23-4.28). We find it hard to improve upon that text, but we hope the reader will enjoy unraveling the details by following the exercises below.

Let $f : X \to S$ be a flat morphism of varieties over an algebraically closed field k, with $n = \dim X = \dim S + 1$. Let $D \subset S$ be a reduced divisor. We make the following assumptions.

N1 The base S is nonsingular.
N2 The divisor D has strict normal crossings.
N3 The morphism f is smooth over $S \setminus D$.
N4 The morphism $f : X \to S$ is a nodal curve.

Let $x \in X$ be a closed point and let $s = f(x) \in S$. By assumption we may choose a regular system of parameters $t_1, \ldots, t_{n-1}$ at s such that D coincides on a neighborhood with the zero locus of $t_1 \cdots t_r$ for some $r \leq n-1$. It can be seen that if x is a singular point of X, then the completed local ring of X at x can be described as

$$k[[u,v]]/(uv - t_1^{n_1} \cdot \cdots \cdot t_r^{n_r})\,. \qquad (*)$$

Step 1: Assume $\operatorname{codim}_X \operatorname{Sing}(X) = 2$.

1. Show that there is an irreducible component $D_1 \subset D$ and $\Sigma_1 \subset \operatorname{Sing}(X)$ such that $f(\Sigma_1) = D_1$.
2. Fix a point $x \in \Sigma_1$, and use formal coordinates as in $(*)$, such that $D_1 = V(t_1)$. Show that the power n_1 of t_1 is > 1.
3. Show that the ideal of Σ_1 in the formal completion is (u, v, t_1).
4. Conclude that $\Sigma_1 \to D_1$ is étale, in particular Σ_1 is nonsingular.
5. Let $X_1 = Bl_{\Sigma_1} X$. Show that $X_1 \to S$ satisfies conditions N1-N4, there is at most one component of $\operatorname{Sing}(X_1)$ over Σ_1, with the exponent n_1 replaced by $n_1 - 2$
6. Conclude by induction that there is a blowup $X' \to X$ centered above $\operatorname{Sing}(X)$, such that X' satisfies N1-N4, and $\operatorname{codim}_X \operatorname{Sing}(X) > 2$.
7. Show that each component of $\operatorname{Sing}(X')$ is defined by $u = v = t_i = t_j$ in equation $(*)$, in particular it is nonsingular.

Step 2: Assume $\operatorname{codim}_X \operatorname{Sing}(X) > 2$. Define $Z = f^{-1}D$. Unfortunately here we need to abandon the structure $X \to B$ of a family of nodal curves. Instead we look at X itself. The situation is as follows:

T1 whenever x is a nonsingular point of X, Z has normal crossings at x.
T2 whenever $x \in \mathrm{Sing}(X)$, we have formal description

$$(**) \quad k[[u,v]]/(uv - t_1 \cdots t_s), \quad 2 \leq s \leq r \leq n-1$$

and $Z = V(t_1 \cdots t_r)$.
T3 All components of $\mathrm{Sing}(X)$ are nonsingular.

1. Let $E \subset \mathrm{Sing}(X)$ be an irreducible component. Show that the blowup $Bl_E X$ satisfies T1-T3, and its singular locus has one fewer irreducible component.
2. Conclude by induction that there is a resolution of singularities $X' \to X$.

This concludes the proof of Theorem 2.3. □

5. Modifications of the proof for Theorems 2.4 and 2.7

5.1. Exercises on removing the conditions on the projection

An important step in the proof of de Jong's theorem was, that given the projection $X \to P$, one could construct an alteration $P_1 \to P$ and a diagram

$$\begin{array}{ccc} C & \stackrel{\beta}{\to} & X \\ \downarrow & & \downarrow \\ P_1 & \to & P \end{array}$$

where $C \to P_1$ was a family of nodal curves. In order for the proof to go through, we made several assumptions on the projection $X \to P$. Here we will show that even if these conditions fail, we can still reduce to the case where they do hold.

Exercise 5.1. Using an alteration, show that the condition that $Z \to P$ be finite in Lemma 4.3 (4) is unnecessary for the rest of the proof.

Exercise 5.2. Show that, if one is willing to accept inseparable alterations in the theorem, the condition that $Z \to P$ be generically étale in Lemma 4.3 (4) is unnecessary for the rest of the proof.

Exercise 5.3. * By reviewing the arguments, show that the condition that every component of every fiber of $X \to P$ be generically smooth is unnecessary.
Here a modification of the three point lemma is be necessary! In [31], de Jong uses a trick of "raising the genus of the curves" with finite covers. Another way goes as follows: in the proof of the Three Point Lemma 4.7, after flattening X and T, one works with fibers of *the normalizations* X^ν and T^ν. This way one avoids the need for $\mathrm{Sm}(X_p)$ to be dense. One notes that the sections s_i lift to X^ν, and at least three meet every component of every fiber, since Z is the support of a Cartier divisor! The details are left to the reader.

Exercise 5.4. Using the flattening lemma and the previous exercise, show that the condition that $X \to P$ be equidimensional is unnecessary.

Exercise 5.5. Show that, if one is willing to accept inseparable alterations in the theorem, the condition that the generic fiber of $X \to P$ be smooth is unnecessary.

5.2. Exercises on Theorem 2.4

Let us address Theorem 2.4 on semistable reduction up to alteration. Suppose $S = \operatorname{Spec} R$ where R is a discrete valuation ring, $X \to S$ is a morphism as in the theorem, and Z a proper closed subset.

Exercise 5.6. Show that we may assume X projective over S, that Z is the support of a Cartier divisor.

Exercise 5.7. Show that we may assume that the generic fiber is a normal variety, and that X is a normal scheme. (You may need an inseparable base change!)

Exercise 5.8. Let d be the dimension of X_η. Produce a projection $X \to \mathbb{P}_S^{d-1}$ with connected fibers.

Exercise 5.9. Use the semistable reduction argument, with the Three Point Lemma, and the results of Section 5.1 to replace X by a nice family of curves $X \to P \to S$.

Exercise 5.10. Use induction on the dimension to conclude the proof of the theorem.

Exercise 5.11. Can you think of other situations where a similar theorem can be proven, where S is not necessarily the spectrum of a discrete valuation ring? (This is interesting even in characteristic 0!)

5.3. Exercises on Theorem 2.7

We address the equivariant version of the theorem. Suppose X is a variety, Z a proper closed subset, and a finite group G acts on X stabilizing Z. We wish to prove Theorem 2.7.

Exercise 5.12. Produce an equivariant version of Chow's lemma, so that we may assume X is projective.

Exercise 5.13. Show that, to prove the theorem, it suffices to consider the case where Z is a divisor.

Exercise 5.14. Show that we may replace Z by a bigger *equivariant* divisor; in particular we may assume Z contains the fixed point loci of elements in G.

Exercise 5.15. Using a projection of X/G, show that we may assume we have an equivariant projection $X \to P$ making X into a nice family of curves.

Exercise 5.16. Consider the case $X = \mathbb{A}_k^2$ where $\operatorname{char} k = p$, and $G = \mathbb{Z}/p\mathbb{Z}$ acting via $(x, y) \mapsto (x, x+y)$. Show that the fixed point set maps inseparably to the image. In particular, the map $Z \to P$ in the previous exercise might be inseparable!

Exercise 5.17. Making an alteration "Galois": Given a variety W, a finite group action $H \subset \operatorname{Aut} W$ and an alteration $V_0 \to W$, show that there exists an alteration $V \to V_0$, and a finite group H' with a surjection $H' \to H$, and a lifting of the H action $H' \subset \operatorname{Aut}(V \to W)$ such that the extension of fixed fields $K(W)^H \subset K(V)^{H'}$ is purely inseparable.

Exercise 5.18. $*$ Use the uniqueness in the stable reduction theorem to show that there is an alteration $P' \to P$, a family of stable pointed curves $C \to P'$ and a finite group G' with a quotient $G' \to G$ and a diagram

$$\begin{array}{ccc} C & \to & X \\ \downarrow & & \downarrow \\ P' & \to & P \end{array}$$

on which G' acts equivariantly, such that C is birational to $\widetilde{X_{P'}}$, and the extensions $K(X)^G \subset K(C)^{G'}$ and $K(P)^G \subset K(P')^{G'}$ are purely inseparable.

Exercise 5.19. Use induction on the dimension and a suitable modification of the elementary blowups argument to conclude the theorem.

6. Toroidal geometry

Toroidal geometry is a generalization of the more well known geometry of toric varieties. In this section we will show that various aspects of toric varieties generalize with few difficulties to the toroidal case. The reader is assumed to be familiar with the basic facts about toric varieties, as given in [15] in this volume.

6.1. Basic definitions

For simplicity we work over an algebraically closed field. We recall the notion of a toric variety (a more thorough discussion is available in [15]):

Definition. A variety X together with an open dense embedding $T \subset X$ is called a *toric variety* if X is normal, T is a torus (geometrically isomorphic to $\mathbb{G}_m^k$), and the action of T on itself by translations extends to an action on X.

To get an intuitive idea about the singularities of a toric variety, it is worth noting that a normal, affine variety, defined by equations between monomials (such as $z^2 = xy$) is toric, and every toric variety is locally of this type.

For many purposes toric varieties are too restrictive. A more general notion was introduced by Mumford in [38]:

Definition. A variety X together with an open embedding $U \subset X$ is called a *toroidal embedding* if any point $x \in X$ has an étale neighborhood X' such that X' is isomorphic to an étale neighborhood of a point on a toric variety, and the isomorphism carries the open subset $U' = X' \times_X U \subset X'$ to the torus of the toric variety.

Thus a toroidal embedding looks locally like a toric variety, and the big open set U is a device which ties together these "local pictures". In a sense, this notion is suitable for studying varieties whose singularities are like those of toric varieties.

In this section we recall facts about toric varieties and briefly indicate how one can obtain analogous facts about toroidal embeddings. The details are available in [38].

Remark. A more sheaf-theoretic approach was introduced by K. Kato, see [35], [36].

Definition. A toroidal embedding is said to be *strict* (or a *toroidal embedding without self intersections*) if every irreducible component of $X \smallsetminus U$ is normal.

For instance, if X is a nonsingular variety, $D \subset X$ is a strict normal crossings divisor, and $U = X \smallsetminus D$, then $U \subset X$ is a strict toroidal embedding.

We will only work with strict toroidal embeddings.

Definition. If $G \subset \mathrm{Aut}(U \subset X)$ is a finite group, we say that *G acts toroidally* if for any point $x \in X$, the stabilizer G_x of x can be identified with a subgroup of the torus in an appropriate étale neighborhood of x.

Definition. A morphism between toric varieties iș called a *toric morphism*, if it is surjective and torus-equivariant. A morphism of toroidal embeddings $(U_X \subset X) \to (U_Y \subset Y)$ is called a *toroidal morphism* if locally on X it looks like a toric morphism.

6.2. The cone

First recall some notation (see [15]):

$M = \mathrm{Hom}(T, \mathbb{G}_m)$ – this is the group of algebraic characters of T;
$M_{\mathbb{R}} = M \otimes \mathbb{R}$;
$N = \mathrm{Hom}(M, \mathbb{Z}) = \mathrm{Hom}(\mathbb{G}_m, T)$ – the group of 1-parameter subgroups on T;
$N_{\mathbb{R}} = N \otimes \mathbb{R}$.

It is common to call the functions defined by elements of M the *monomials*. One uses the notation x^m for the monomial associated with the element $m \in M$. Recall the basic correspondence between

$$\{\text{affine toric varieties } T \subset X\}$$
and
$$\{\text{strictly convex rational polyhedral cones } \sigma \subset N_{\mathbb{R}}\}$$

which can be defined in one direction via

$$X = V_\sigma := \mathrm{Spec}\, k[\sigma^\vee \cap M],$$

and in the other direction by

$$\sigma = \begin{array}{l}\text{the cone spanned by the 1-parameter subgroups } \phi : \mathbb{G}_m \to T \\ \text{such that "the limit } \lim_{z\to 0} \phi(z) \text{ exists in } X\text{", that is, } \phi \text{ extends} \\ \text{to a morphism } \mathbb{A}^1 \to X.\end{array}$$

There is another, less well-known characterization of σ, which is less dependent on the torus action, and is therefore useful for toroidal embeddings: Any monomial $m \in M$ defines a Cartier divisor $\mathrm{Div}(x^m)$ supported on $X \smallsetminus T$. If $\sigma^\vee$ contains a line through the origin, then for any m on this line the divisor is easily seen to be trivial (both m and $-m$ give regular functions).

We use the following notation:

$\sigma^\perp = \{m \in M_\mathbb{R} | \langle m, \sigma \rangle = 0\}$

$M^\sigma =$ Cartier divisors supported on $X \smallsetminus T$.

One can easily see that $M^\sigma = M/\sigma^\perp \cap M$.

$N_\sigma = span(\sigma)$. Clearly $N_\sigma = \mathrm{Hom}(M^\sigma, \mathbb{Z})$.

Let $M^\sigma_+ \subset M^\sigma$: the *effective* Cartier divisors.

We have that $M^\sigma_+ = \sigma^\vee \cap M/\sigma^\perp \cap M$.

It is not hard to see that $\sigma = (M^\sigma_+)^\vee_\mathbb{R}$, the dual cone of the cone spanned by M^σ_+. In short: σ is the dual cone to the cone of effective Cartier divisors supported on $X \smallsetminus T$.

6.3. The toroidal picture

We wish to mimic the construction of the cone in the toroidal case. We follow [38], Chapter II.

Let $U \subset X$ be a strict toroidal embedding, $X \smallsetminus U = \cup D_i$, where D_i normal. We decompose $\cap_{i \in I} D_i = \cup X_\alpha$; the locally closed subsets X_α are called *strata.* Each stratum has its star: $\mathrm{Star}(X_\alpha) = \cup_{X_\alpha \subset \overline{X_\beta}} X_\beta$. Note that X_α is the unique closed stratum in $\mathrm{Star}(X_\alpha)$. In a sense it is analogous to the unique closed orbit in an affine toric variety. Define:

$M^\alpha =$ group of Cartier divisors supported on $\mathrm{Star}(X_\alpha) \smallsetminus U$;

$M^\alpha_+ =$ subset of effective Cartier divisors;

$N_\alpha = \mathrm{Hom}(M_\alpha, \mathbb{Z})$;

$\sigma_\alpha = (M^\alpha_+)^\vee_\mathbb{R}$.

Thus, to each stratum we associated a strictly convex rational polyhedral cone.

Remark. The cone σ_α has a description analogous to the toric one using 1-parameter subgroups, in terms of valuations. Let $RS(X)$ be the discrete valuations on X. Let v be a valuation centered in $\mathrm{Star}(X_\alpha)$. Let f_j be rational function defining generators of M_α on a small affine open. Then $v(f_j)$ is a vector in σ_α, and in fact σ_α can be described as a set of equivalence classes of discrete valuations centered in $\mathrm{Star}(X_\alpha)$, the equivalence being defined by equality of the valuations of these functions f_j.

6.4. Birational affine morphisms

Recall: if $\tau \subset \sigma$ are two strictly convex rational polyhedral cones, then $\tau^\vee \supset \sigma^\vee$ gives rise to a morphism $V_\tau \to V_\sigma$, which is birational and affine.

Note that $V_\tau \to V_\sigma$ can be described in the following invariant manner:

$$V_\tau = \mathrm{Spec}_{V_\sigma} \sum_{E \in M_+^\tau} \mathcal{O}_{V_\sigma}(-E),$$

where the sum is taken inside the field of rational functions of V_σ. This clearly works over $\mathrm{Star}(X_\alpha)$ in the toroidal case as well.

6.5. Principal affine opens

If $m \in \sigma^\vee$ then $\tau = \{n \in \sigma| < n, m >= 0\}$ is a face of σ. We have $\tau^\vee = \sigma^\vee + \mathbb{R} \cdot m$, and therefore V_τ is the principal open set on V_σ obtained by inverting the monomial x^m. Again, this can be described divisorially in terms of $\mathrm{Div}(m)$. Thus the same is true for $\mathrm{Star}(X_\alpha)$: given a face τ of σ_α, we get an open set

$$\mathrm{Star}(X_\beta) \subset \mathrm{Star}(X_\alpha)$$

such that $\tau = \sigma_\beta$. The most important face of a cone is the vertex. It corresponds to the open set $T \subset V$. In the toroidal case you get U.

6.6. Fans and polyhedral complexes

Recall: if σ_1 and σ_2 intersect along a common face τ, then V_{σ_1} and V_{σ_2} can be glued together along the common open set V_τ, forming a new toric variety. In general, whenever you have a *fan* Σ in N, namely a collection of cones σ_i intersecting along faces, you can glue together the V_{σ_i} and get a toric variety V_Σ. It is not hard to see that *every* toric variety is obtained in this way in a unique manner. The point is that every toric variety is covered by affine open toric varieties. In the toroidal case, X is covered by the open sets $\{\mathrm{Star}(X_\alpha)\}_\alpha$. In general $\mathrm{Star}(X_\alpha) \cap \mathrm{Star}(X_\beta) = \cup \mathrm{Star}(X_{\gamma_i})$, so σ_{γ_i} are possibly several faces of both σ_α and σ_β. Still these can be glued together, as a rational conical polyhedral complex. The main difference from the toric case, is that it is abstractly defined, and in general it is not linearly contained in some vector space $N_\mathbb{R}$.

6.7. Modifications and subdivisions

Let Σ be a fan, and $\Sigma' \to \Sigma$ a (complete) subdivision. This corresponds to a toric modification $V_{\Sigma'} \to V_\Sigma$.

Since the construction is local (the Spec construction, as in Section 6.4, and gluing) it works word for word in the toroidal case. There is a small issue in checking that the resulting modification is still a strict toroidal embedding; this is discussed in detail in [38].

In [38] (see also [15]) it is shown that a modification is *projective* if and only if the subdivision is induced by a *support function* – one associates to a support function an ideal, whose blowup gives the modification. This works in the toroidal case as well.

6.8. Nonsingularity

Recall: an affine toric variety V_σ is nonsingular if and only if σ is simplicial, generated by a basis of N_σ (namely, part of a basis of N). Such a cone is called nonsingular. In general, a toric variety V_Σ is nonsingular if and only if every cone $\sigma \in \Sigma$ is nonsingular. This is a local fact, so it is true in the toroidal case as well.

6.9. Desingularization

Recall that it is easy to resolve singularities of a toric variety: one finds a simplicial subdivision such that every cone is nonsingular. Obviously, the same works in toroidal case! We obtained:

Theorem 6.1. *For any toroidal embedding $U \subset X$ there is a projective toroidal modification $U \subset X' \to X$ such that X' is nonsingular.*

See [38], Theorem 11*, page 94.

6.10. Exercises on toric varieties and toroidal embeddings

1. Show that $\mathbb{G}_m^n \subset \mathbb{A}^n$ is a toric variety. Describe its cone.
2. Show that $\mathbb{G}_m^n \subset \mathbb{P}^n$ is a toric variety. Describe its fan.
3. Let $X \subset \mathbb{A}^n$ be a normal variety defined by monic monomial equations of type
$$\prod x_j^{n_j} = \prod x_j^{m_j}.$$
Show that X is toric. (Identify the torus!)
4. Do the same if the monomial equations are not necessarily with coefficients $= 1$.
5. Describe the cone associated to the affine toric variety defined by
$$xy = t_1^{k_1} \cdots t_r^{k_r}.$$
6. Look at the affine 3-fold $xy = zw$. Let $X' \to X$ be the blowup of X at the ideal (x, z). Describe this blowup, show that it is toric, and describe the cone subdivision associated to it.
7. Let $X = \mathbb{A}^2$, $D = \{xy(x + y - 1) = 0\}$, $U = X \smallsetminus D$. Show that $U \subset X$ is a toroidal embedding. Describe its conical polyhedral complex. (Compare with the fan of $\mathbb{P}^2$!)
8. Do the same for $D = \{y(x^2 + y^2 - 1) = 0\}$. Show that the resulting complex can not be linearly embedded in a vector space.
9. Consider the surface $X = \{z^2 = xy\}$, $U = \{z \neq 0\}$. Show that $U \subset X$ is toric and describe its cone.
10. Consider the surface X above, let $D_1 = \{x = 0\}$, $D_2 = \{y = x(x-1)^2 \text{ and } z = x(x-1)\}$. Let $U = X \smallsetminus (D_1 \cup D_2)$. Show that $U \subset X$ is toroidal. Describe its conical polyhedral complex. Make sure to describe the integral structure!

6.11. Abhyankar's lemma in toroidal terms

Abhyankar's lemma about fundamental groups (see [25], [26]) describes the local tame fundamental group of a variety around a normal crossings divisor. Let $X = \operatorname{Spec} k[[t_1, \ldots, t_n]]$ and let $D = V(t_1 \cdots t_n)$. Let $Y \to X$ be a finite alteration which is tamely branched along D, and étale away from D. For m prime to char k, denote $X_m = \operatorname{Spec} k[[t_1^{1/m}, \ldots, t_n^{1/m}]]$. Abhyankar's lemma says that the normalization of $Y \times_X X_m$ is étale over X_m.

In the following exercises we interpret this in toroidal terms.

Exercise 6.2. Let $U \subset X$ be a *nonsingular* strict toroidal embedding. Let $f : Y \to X$ be a finite cover, which is tame, and étale over U. Then $f^{-1}U \subset Y$ is a strict toroidal embedding.

Exercise 6.3. Suppose further that $Y \to X$ is Galois, with Galois group G. Show that G acts toroidally on Y.

7. Weak resolution of singularities I

Given the existence of toroidal resolution, the proof of weak resolution of singularities in characteristic 0 by Bogomolov and Pantev is arguably the simplest available. It does not even require surface resolution.

We will go through this proof. The steps of proof here include some simplifications on the arguments in [12], which came up in discussions with T. Pantev. These and additional simplifications were discovered independently by K. Paranjape [53], and we have used his exposition in some of the following exercises. The version given in [53] has the advantage that it does not even require moduli spaces.

7.1. Projection

Let X be a variety over an algebraically closed field of characteristic 0, and $Z \subset X$ a proper closed subset. Let $n = \dim X$, and again assume we know the weak resolution theorem for varieties of dimension $n - 1$.

First a few reduction steps:

1. Show, as in 4.1 that we may assume X projective and normal, and Z the support of a Cartier divisor.
2. Show that there is a finite projection $X \to \mathbb{P}^n$.
3. Let $P \to \mathbb{P}^n$ be the blowup at a closed point. Show that
$$P \simeq \mathbb{P}_{\mathbb{P}^{n-1}}(\mathcal{O}_{\mathbb{P}^{n-1}} \oplus \mathcal{O}_{\mathbb{P}^{n-1}}(1)).$$
Denote by E the exceptional divisor of $P \to \mathbb{P}^n$.
4. By blowing up a general point on $\mathbb{P}^n$, and blowing up X at the points above, show that we may assume we have a finite morphism $f : X \to P$, which is étale along E, such that the image of Z is disjoint from E, and maps finitely to $\mathbb{P}^{n-1}$.

5. By the Nagata-Zariski purity theorem, note that the branch locus of $X \to P$ is a divisor B, disjoint from E, mapping finitely to $\mathbb{P}^{n-1}$.

   We replace Z by $Z \cup f^{-1}B$.

7.2. Vector bundles

The next steps are aimed at replacing P by another $\mathbb{P}^1$-bundle $Q \to \mathbb{P}^{n-1}$, such that the branch locus in Q of $X \to Q$ becomes simpler. Let Y be any variety, F a rank-2 vector bundle on Y, $P = \mathbb{P}_Y(E)$. Let $E \subset P$ be a divisor which is a section of $\pi : P \to Y$ and let $D \subset P$ be another effective divisor disjoint from E. Denote by $\mathcal{O}_P(1)$ the tautological bundle, and by d the relative degree of D over Y.

1. Consider the exact sequence
$$0 \to \mathcal{I}_D(d) \to \mathcal{O}_P(d) \to \mathcal{O}_P(d)|_D \to 0.$$
   Use this to show that there is an invertible sheaf $\mathcal{L}_D$ on Y such that $\mathcal{I}_D(d) \simeq \pi^*\mathcal{L}_D$.
2. If $D_1, D_2 \subset P$ are any two *disjoint* divisors finite of degree d over Y, show that there is an embedding of vector bundles $\mathcal{L}_{D_1} \oplus \mathcal{L}_{D_2} \subset sym^d F$ inducing a *surjection* $\pi^*(\mathcal{L}_{D_1} \oplus \mathcal{L}_{D_2}) \to \mathcal{O}_P(d)$.
3. Assume the characteristic is 0. Consider the case $D_1 = dE, D_2 = D$. show that the resulting morphism $P \to \mathbb{P}_Y(\mathcal{L}_{D_1} \oplus \mathcal{L}_{D_2}) = P'$ maps E to a section E' and D to a disjoint section; and its branch locus is of the form $(d-1)E+D'$ where D' has degree $d-1$ over Y and is disjoint from E'.
4. Continue by induction to show that there is a $\mathbb{P}^1$ bundle $Q \to Y$ and a morphism $g : P \to Q$ over Y, such that the image of D and the branch locus of g form a union of sections of $\pi_Q : Q \to Y$.

7.3. Conclusion of the proof

Back to our theorem, where $Y = \mathbb{P}^{n-1}$. Composing with the morphism $f : X \to P$, we obtain that the image $g(f(Z)) \subset Q$ is the image D_1 of a section $s_1 : \mathbb{P}^{n-1} \to Q$ of $Q \to \mathbb{P}^{n-1}$ and the branch locus of $g \circ f$ is the union of images D_i sections $s_i : \mathbb{P}^{n-1} \to Q$ as well. Denote $\Delta = \pi_Q(\cup_{i \neq j} D_i \cap D_j)$.

The following steps use moduli theory; however it has been shown (in the preprint version of [12], and in Paranjape's exposition [53]) that the use of moduli theory can be circumvented within a few pages of work.

1. $*$ Use the stable reduction argument to show that there is a *modification* $Y' \to Y$, and a modification $Q' \to Q \times_Y Y'$ such that $Q' \to Y$ is a family of nodal curves of genus 0, and the sections lift to *disjoint* sections $s_i' : Y' \to Q'$. We replace Y by Y', Δ by its inverse image, etc.

Hints. The point is that the generic fiber of $Q \to Y$ is a projective line with a number (say k) of points marked by the sections we obtained. This gives a rational map $Y \dashrightarrow \overline{M_{0,k}}$, which can be replaced by a morphism after a modification $Y' \to Y$.

Since the moduli schemes in genus 0 are *fine* moduli schemes, there is a family of pointed rational curves $Q' \to Y'$. one would like to use the Three Point Lemma to get a morphism $Q' \to Q$. However, the argument above only guarantees that every fiber of Q has two marked points, and not necessarily three. This is easy to correct by adding sections on the $\mathbb{P}^1$-bundle $Q \to Y$ before applying the moduli argument.

Another approach is to use Knudsen's stabilization method directly. The details of this can be found in [12].

2. Use induction on the dimension to replace Y by a nonsingular variety such that Δ becomes a strict divisor of normal crossings.
3. Use either toroidal geometry, or Section 4.11, to replace Q by a nonsingular variety, such that the inverse image of D is a strict normal crossings divisor. (Note that at this point $Q \to Y$ is a family of nodal pointed curves, degenerating over the divisor of normal crossings D.)
4. Let $\tilde{X}$ be the normalization of Q in the function field of X. Use Abhyankar's lemma (Section 6.11) to show that $\tilde{X}$ has a toroidal structure, such that the inverse image $\tilde{Z}$ of Z is a toroidal divisor.
5. Conclude that there is a weak resolution of singularities $r : X' \to X$ such that $r^{-1}Z$ is a strict divisor of normal crossings.

8. Weak resolution of singularities II

The weak resolution argument according to Abramovich-de Jong starts very much like de Jong's theorem: a projection $X \to P$ is produced, and a Galois alteration $P_1 \to P$ over which one has stable reduction $X_1 \to P_1$, equivariant under the Galois group G, is produced. Induction on the dimension for P allows one to assume that X_1 and P_1 are toroidal, and the Galois action on P_1 is toroidal as well. The only point left is to make the group action on X_1 toroidal, so that the quotient should be toroidal, and therefore admit toroidal resolution.

Let us go through the steps. Let X be a variety over an algebraically closed field of characteristic 0, and let $Z \subset X$ be a Zariski-closed subset. We want to find a nonsingular, quasi-projective variety X' and a modification $f : X' \to X$ such that $f^{-1}Z$ is a divisor with simple normal crossings.

8.1. Reduction steps

Exercise 8.1. Show that it is enough to prove the result when X is projective and normal, and Z a Cartier divisor.

Exercise 8.2. Reduce to the case where there is a projection $X \to P$, such that the generic fiber is a smooth curve.

Exercise 8.3. *Using the trick of enlarging Z and stable reduction, show that there is a diagram as follows:

$$\begin{array}{ccc} X_1 & \rightarrow & X \\ \downarrow & & \downarrow \\ P_1 & \rightarrow & P \end{array}$$

such that $P_1 \rightarrow P$ is an alteration, $X_1 \rightarrow \tilde{X}_{P_1}$ is birational, and $X_1 \rightarrow P_1$ has section $s_i : P_1 \rightarrow X_1$ making it a family of stable pointed curves, and the image of these sections in X contains Z.

Exercise 8.4. Show that you can make $P_1 \rightarrow P$ a Galois alteration. Call the Galois group G. Show, using the uniqueness of stable reduction3.7, that the action of G on P_1 lifts to an action on X_1, which permutes the sections s_i.

You can replace X by X_1/G and P by P_1/G

Exercise 8.5. Use induction on the dimension to reduce to the case where:

1. P is nonsingular, with a normal crossings divisor Δ;
2. The branch locus of $P_1 \rightarrow P$ is contained in δ;
3. The locus where $X_1 \rightarrow P$ is not smooth is contained in δ.

Exercise 8.6. Show that in this case $P_1 \rightarrow P$ is a toroidal morphism, G acts toroidally on X, and $X_1 \rightarrow P_1$ is a toroidal morphism as well.

The only point left is to make the action of G on X_1 toroidal – if it were, then X would be toroidal and we could easily resolve its singularities.

Looking locally, the question boils down to the following situation:

Let $T_0 \subset X_0$ be an affine torus embedding, $X_0 = \text{Spec } R$. Let $G \subset T_0$ be a finite subgroup of T_0, let $p_0 \in X_0$ be a fixed point of the action of G, and let ψ_u be a character of G. Consider the torus embedding of $T = T_0 \times \text{Spec } k[u, u^{-1}]$ into $X = X_0 \times \text{Spec } k[u]$, and let G act on u via the character ψ_u. Write $p = (p_0, 0) \in X$ and write $D = (X_0 \smallsetminus T_0) \times \text{Spec } k[u]$. We wish to find a canonical blowup $X_1 \rightarrow X$, such that if $U \subset X_1$ is the inverse image of T_0, then it is a toroidal embedding, and the group G acts toroidally.

8.2. The ideal

Let $M \subset R[u]$ be the set of monomials. For each $t \in M$ let χ_t be the associated character of T, and let let $\psi_t : G \rightarrow k^*$ be the restriction of χ to G. Define $M_u = \{t | \psi_t = \psi_u\}$, the set of monomials on which G acts as it acts on u. Define $I_G = \langle M_u \rangle$, the ideal generated by M_u.

Exercise 8.7 (canonicity). * Show that if X'_0, T'_0, G', p'_0 and ψ'_u is a second set of such data, and if we have an isomorphism of formal completions

$$\varphi : \hat{X}_p \xrightarrow{\sim} \hat{X'}_{p'},$$

which induces isomorphisms $G \cong G'$ and $\hat{D}_p \cong \hat{D'}_{p'}$, then φ pulls back I_G to the ideal $I_{G'}$.

Exercise 8.8 (gluing property). $*$ If q_0 is any point of X_0 and if $G_q \subset G$ is the stabilizer of $q = (q_0, 0)$ in G, then the stalk of I_G at q is the same as the stalk of I_{G_q} at q.

Exercise 8.9. Show that I_G is generated by u and a finite number of monomials $t_1, \ldots, t_m$ in $M_u \cap R$.

Exercise 8.10. Let $X' = B_{I_G}(X)$ be the blowup. Let X'_u be the chart with coordinates $u, t_j/u$. Show that the action of G on X'_u is toroidal.

Exercise 8.11. Let X'_i be the chart on X' with coordinates $t_i, v = u/t_i, s_j = t_j/t_i$. Show that G acts trivially on v, and that $X'_i = \text{Spec } R'_i[v]$ where R'_i is generated over R by s_j.

Exercise 8.12. Let X_1 be the normalization of X'. Show that if $U \subset X_1$ is the inverse image of T_0, then it is a toroidal embedding, and the group G acts toroidally.

9. Intersection multiplicities

Intersection theory has a long history, and certainly we are not going to say much about it here. One aspect is, that it is not so easy to have a good definition for intersection multiplicities.

Remark, exercise: let $C \subset \mathbb{P}^2_k$ be a plane algebraic curve, $P \in C$ a closed point at which C is regular, and $D = \mathcal{Z}(F) \subset \mathbb{P}^2$ a plane curve given by a homogeneous polynomial F; suppose F is not identical zero on a neighborhood of P in C (i.e. no component of D contains the component of C containing P). Show that the following two definitions of *the intersection multiplicity* $i(C, D; P)$ *of* C *and* D *at* P are equivalent:

- the dimension of the k-vector space

$$\mathcal{O}_{C,P} \otimes_{\mathcal{O}_{\mathbb{P}^2,P}} \mathcal{O}_{D,P},$$

- the value of the valuation $v = v_{C,P}$ defined by the discrete valuation ring $\mathcal{O}_{C,P}$ computed on the function on C given by F,

see [HAG], Exercise (5.4) on page 36, and Remark (7.8.1) on page 54.

Consider two varieties $V, W \subset \mathbb{P}^n$ which have an isolated point of intersection at $P \in V \cap W$. One could try to define the intersection of V and W at P as the length of

$$\mathcal{O}_{V,P} \otimes_{\mathcal{O}_{\mathbb{P}^n,P}} \mathcal{O}_{W,P}.$$

Analogous situations of intersections of arbitrary schemes in some regular ambient scheme can be considered.

Exercise 9.1. (See Gröbner [23], 144.10/11, also see [66], [62], [10], see [HAG], I.7):
a) Let $C \subset \mathbb{P}^3$ be the space curve with parameterization

$$(x_1 : x_2 : x_3 : x_4) = (t^4 : t^3{\cdot}s : t{\cdot}s^3 : s^4)$$

(we work over some field K). Show that the prime ideal given by this curve equals
$j := (T_1^2T_3 - T_2^3\ ,\ T_1T_4 - T_2T_3\ ,\ T_1T_3^2 - T_2^2T_4\ ,\ T_2T_4^2 - T_3^3) \subset K[T_1, T_2, T_3, T_4]$.
b) Consider C as a curve embedded in $\mathbb{P}^4$: choose the hyper plane $\mathbb{P}^3 \cong \mathcal{Z}(T_0) = H$, and we get $C \subset H \subset \mathbb{P}^4$. Let $P := (x_0 = 1 : 0 : 0 : 0 : 0) \in \mathbb{P}^4$. Define $V \subset \mathbb{P}^4$ as the cone with vertex P over the curve $C \subset \mathbb{P}^4$, i.e. V is defined by the ideal

$$J := K[T_0, T_1, T_2, T_3, T_4] \cdot j, \quad V = \mathcal{Z}(J).$$

Note that the dimension of V equals two, that the degree of $V \subset \mathbb{P}^4$ equals four.
c) Let W be the 2-plane given by

$$I := (T_1, T_4) \subset K[T_0, T_1, T_2, T_3, T_4], \quad W := \mathcal{Z}(I).$$

Note that $P \in W$. Remark that (set-theoretically):

$$W \cap V = \{P\}$$

(use the geometric situation, or give an algebraic computation).
We like to have a Bézout type of theorem for the situation $W \cap V \subset \mathbb{P}^4$, however:
d) Define

$$M := \mathcal{O}_{W,P}, \quad A := \mathcal{O}_{\mathbb{P}^4,P}, \quad N := \mathcal{O}_{V,P},$$

and compute

$$\dim_K (M \otimes_A N)$$

(surprise: this is not equal to four).
e) Compute

$$\dim_K \left(\mathrm{Tor}_i^A(M, N)\right), \quad \forall i$$

(either using, or reproving $\chi_A(M, N) = 4$, for notation see below).

Hence we see that just the length of the appropriate tensor product does not define necessarily the correct concept. Serre proposed in 1957/58 to define the intersection multiplicity as the alternating sum of the lengths of the Tor_i (note that $\mathrm{Tor}_0 = \otimes$), i.e. by

$$\chi_A(M, N) := \sum_{i \geq 0} (-1)^i \ \mathrm{length}_A \mathrm{Tor}_i^A(M, N)$$

(we follow notation of [66], also see [10], 6.1, see [62]), here A is a regular local ring, and M and N are A-modules such that $M \otimes_A N$ has finite length. In equal characteristic this is the right geometric concept (i.e. satisfies Bézout's theorem, coincides with previously defined intersection multiplicities etc.).

The following theorem was conjectured by Serre, proved by Gabber (using de Jong's alteration result), and written up by Berthelot (in [10], 6.1):

Theorem 9.2. *Let the characteristic of A be equal to zero. Suppose $p \in \mathfrak{m}^2$, hence its residue field $A/\mathfrak{m}$ has characteristic $p > 0$. Then:*

$$\chi_A(M, N) \geq 0.$$

Part II. Moduli of curves

10. Introduction to moduli of curves

It is an important feature of algebraic geometry, that the set of all objects (e.g. smooth projective curves) of the same a fixed geometric nature (e.g. genus) often has the structure of an algebraic variety itself. Such a space is a "moduli space", which gives a good algebraic meaning to the problem of "classification". It is fair to say that this "self referential" nature of algebraic geometry is one of the main reasons for the depth of the subject – it is impossible to overestimate its importance.

The first instances of this phenomenon to be discovered were those of *embedded* variety: the projective space as a parameter space for lines in a vector space; Grassmannians parametrizing vector subspaces of arbitrary dimension; the projective space (of dimension $(d^2+3d)/2$) parametrizing all plane curves of degree d, and so on. The case of abstract varieties, such as smooth curves, had to await for some technical advances, although already Riemann knew that algebraic curves of genus g "vary in $3g-3$ parameters"; see [61], page 124:

> "Die $3p-3$ übrigen Verzweigungswerthe in jenen Systemen gleichverzweigter μ-werthiger Functionen können daher beliebige Werthe annehmen; und es hängt also eine Klasse von Systemen gleichverzweigter $(2p+1)$-fach zusammenhängender Functionen und die zu ihr gehörende Klasse algebraischer Gleichungen von $3p-3$ stetig veränderlichen Grössen ab, welche die Moduln dieser Klasse genannt werden sollen."

Historically moduli spaces of curves, or of curves with points on them, were constructed with more or less ad hoc methods. Moduli spaces for 3 or 4 points on rational curves have been known for ages, using the so called "cross ratio" (see exercise 10.9 below). For genus 1, the modular function j was used (see exercise 10.12). The case of genus 2 was already quite difficult to achieve by algebraic methods [30]. For years, moduli spaces for higher genus were only known to exist using Teichmüller theory.

One problem which took years to solve was, that no good understanding of "what moduli spaces really are" was available. Then Grothendieck introduced the notion of "representable functor", describing the best possible meaning for moduli spaces. This had a great success with the development of Hilbert schemes. For a while one hoped that nature would be as ideal as expected (see Grothendieck hopeful Conjecture 8.1 in [24] 212–18, and its retraction in the Additif of [24] 221–28). But it was soon seen that in general moduli functors are not representable, or as we say now, some moduli functors do not give rise to "fine moduli spaces" due to existence of automorphisms. Finally, Mumford pinned down the compromise notion of a "coarse moduli scheme", which enables us to have a good insight in various aspects of moduli theory. This is what we shall try to describe here. It should be said that since then, other good approaches were developed, by way of "enlarging the category of schemes" to include some "moduli objects", called stacks. For details see [17], Section 4. We will not pursue this direction here.

In this section we gather some basic definitions on functors of moduli for curves. In the next sections we discuss existence theorems for moduli spaces of curves, and for complete moduli spaces with extra structure carrying a "tautological family". Section 14 is devoted to some further questions, examples and facts, not needed for the methods of alterations, but in order to give a more complete picture of this topic.

10.1. The functor of points and representability

To any scheme M one naturally associates a contravariant functor

$$\mathcal{F}_M : \{\text{Schemes}\} \to \{\text{Sets}\} \qquad \text{via} \qquad X \mapsto \operatorname{Mor}(X, M).$$

This is known as *the functor of points of* M, see [Red Book], II §6.

We say that a contravariant functor $\mathcal{F} : \{\text{Schemes}\} \to \{\text{Sets}\}$ is *representable* by a scheme M, if it is isomorphic to $\mathcal{F}_M$, i.e. there is a functorial isomorphism

$$\xi : \mathcal{F}(-) \xrightarrow{\sim} \operatorname{Mor}(-, M).$$

Remark. Strictly speaking, it is the *pair* (M, ξ), consisting of the object M *and* the isomorphism ξ, which represent $\mathcal{F}$. But it has become customary to say "M represents $\mathcal{F}$", suppressing ξ.

Already in the early Bourbaki literature one finds this notion in the disguise of a "universal property". The question of representability of functors can also be posed in categories other than the category of schemes.

Exercise 10.1.

1. Fix an integer N, and let V be a vector space of dimension $N+1$ over $\mathbb{C}$. Consider the functor $\mathcal{F}_{\mathbb{C}}$ that associates to a scheme T over $\mathbb{C}$, the set $\{\mathcal{L} \subset T \times V\}$ of all line sub-bundles of the trivial vector bundle $T \times V$. Show that $\mathcal{F}_{\mathbb{C}}$ is represented by $\mathbb{P}^N_{\mathbb{C}}$.
2. Let $\mathcal{F}$ be the functor that associates to any scheme T (over $\mathbb{Z}$) the set $\{\mathcal{L} \subset \mathcal{O}_T^{N+1}\}$ of all locally free subsheaves of rank 1 of the trivial sheaf $\mathcal{O}_T^{N+1}$ having locally free quotient. Show that $\mathcal{F}$ is represented by the projective scheme $\mathbb{P}^N_{\mathbb{Z}}$.
3. In general, show that the Grassmannian scheme $\mathbf{Grass}(n, r)$ represents the functor of locally free subsheaves of rank r of the trivial free sheaf $\mathcal{O}_T^n$ of rank n having locally free quotients.

Exercise 10.2. Fix integers N and d, and let $\mathcal{G}$ be the functor that associates to a scheme T the set $\{\mathcal{X} \subset \mathbb{P}^N_T\}$ of all flat families of hypersurfaces of degree d in projective N-space over T. Show that $\mathcal{G}$ is represented (over $\mathbb{Z}$) by a projective space $\mathbb{P}^{M-1}$, where $M = \binom{N+d}{d}$ is the dimension of the space of homogeneous polynomials of degree d in $N+1$ variables.

Exercise 10.3. Show that the Hilbert scheme $\mathcal{H}_{\mathbb{P}^N,P(T)}$ represents the "Hilbert" functor, that associates to a scheme T, the set of all subschemes $X \subset \mathbb{P}^N_T$, which are flat over T and such that the geometric fibers have Hilbert polynomial $P(T)$.

10.2. Moduli functors and fine moduli schemes

Suppose a contravariant functor $\mathcal{F}$ has the nature of a *moduli functor*, namely, it assigns to a scheme S the set $\{C \to S\}/\cong$ of isomorphism classes of certain families of objects over S. As a guiding example, let us fix an integer g, with $g \in \mathbb{Z}_{\geq 0}$, and define the moduli functor for smooth curves:

$$\mathcal{M}_g(S) = \{\text{isom. classes of families of curves of genus } g \text{ over } S\}.$$

A morphism $T \to S$ defines (by pulling back families) a map of sets in the opposite direction: $\mathcal{M}_g(T) \leftarrow \mathcal{M}_g(S)$, and we have obtained a contravariant functor.

Assume the functor $\mathcal{F}$ were represented by a scheme M. Then we would call M a *fine moduli scheme* for this functor $\mathcal{F}$, and the object

$$C \to M \quad \text{corresponding to the identity} \quad id \in= \mathrm{Mor}(M, M)$$

would be called a *universal family.*

Remark. Note that in the exercise above on the Hilbert scheme, we can view it as a fine moduli scheme, if we agree that "families up to isomorphism" means "up to isomorphisms as subfamilies of the fixed $\mathbb{P}^N_T$, namely up to equality.

It is a fact of life that *for every $g \geq 0$ the functor $\mathcal{M}_g$ is not representable.* We will explain later why this is true in general, but for the moment let us consider the easiest case:

Exercise 10.4. Let us say that C is a "curve of genus 0", if it is an algebraic curve defined over a field K, and over some extension of $K \subset L$ it is isomorphic with $\mathbb{P}^1_L$. In other words: C is geometrically irreducible, reduced, it is complete and of genus equal to zero.

1. Let K be a field. Show there exist an extension $K \subset K'$, and two curves of genus 0 over K' which are not isomorphic.
2. For every algebraically closed field k, the set $\mathcal{M}_0(k)$ consists of one element, $\mathcal{M}_0(k) = \{\mathbb{P}^1_k\}$.
3. Show that the moduli functor $\mathcal{M}_0$ is not representable.

10.3. Historical interlude

The first case of a highly nontrivial algebraic construction of a moduli space of curves in all charactersitics, appeared in Igusa's work [30]. This is a construction of a "moduli scheme for non-singular curves of genus two in all characteristics", which would now be denoted by $M_2 \to \mathrm{Spec}(\mathbb{Z})$. This happened almost concurrently with Grothendieck's study of representability of functors. But notice that, when Samuel discussed these beautiful results by Igusa in Séminaire Bourbaki (see [63]), his very first comment was:

"Signalons aussitôt que le travail d' IGUSA *ne résoud pas pour les courbes de genre 2, le "problème des modules" tel qu'il a été posé par* GROTHENDIECK *à diverses reprises dans ce Séminaire."*

It really seemed that Nature was working against algebraic geometers, refusing to provide us with these fine moduli schemes ...

The truth is, Nature does provide us with a replacement. Indeed, not much later, Mumford (see [GIT], 5.2) discovered how to follow nature's dictations and come to a good working definition, requiring that the scheme gives geometrically what you want, and does it in the best possible way.

10.4. Coarse moduli schemes

Here is the definition:

Definition. A scheme M and a morphism of functors

$$\varphi : F \to \mathrm{Mor}_S(-, M)$$

is called a *coarse moduli scheme* for F if:

1. for every algebraically closed field k the map
$$\varphi(k) : F(\mathrm{Spec}(k)) \to \mathrm{Mor}_S(\mathrm{Spec}(k), M) = M(k)$$
is bijective, and
2. for any scheme N and any morphism $\psi : F(-) \to \mathrm{Mor}_S(-, N)$ there is a unique $\chi : M \to N$ factoring ψ.

By definition, a coarse moduli scheme does not carry a universal family, *unless* it is a fine moduli scheme. A replacement, called a tautological family, is defined as follows:

Definition. Let $\mathcal{F}$ be a moduli functor. Suppose T is a scheme, and let $f : T \to M$ be a morphism. A family $\mathcal{C} \to T$ giving an element of $\mathcal{F}(T)$, is called a *tautological family* if it defines f, namely $\psi(\mathcal{C} \to T) = f$. In particular this implies that for every geometric point $t \in T$ the fiber $\mathcal{C}_t$ is an object whose isomorphism class defines the image under f, i.e.: $[\mathcal{C}_t] = f(t)$.

Remark. There exist cases (and we shall give examples), where a moduli functor is not representable, where there is no (unique) universal family, but where a tautological family does exist. In such cases the use of the word "tautological", and the distinction between "universal" and "tautological" is necessary, and it pins down the differences.

The terminology "tautological" will also be used in cases such as pointed curves, curves with a level structure, and so on.

Here is the first triumphant success of the notion of coarse moduli scheme:

Theorem 10.5 (Mumford). *Suppose $g \geq 2$. The functor $\mathcal{M}_g$ of smooth curves of genus g admits a quasi-projective coarse moduli scheme.*

See [GIT], Th. 5.11 and Section 7.4, or [17], Coroll. 7.14. We will denote the coarse moduli scheme of $\mathcal{M}_g$ by $M_g \to \operatorname{Spec} \mathbb{Z}$.
We note some properties of M_g:

- As we mentioned before, for every $g \geq 2$ the functor $\mathcal{M}_g$ is not representable: there does not exist a universal family of curves over M_g which can give an isomorphism between $\mathcal{M}_g$ and M_g.
- For every $g \geq 2$ and for any field K, the variety $(M_g)_K = M_g \times_{\operatorname{Spec} \mathbb{Z}} \operatorname{Spec} K$ is not complete. A fortiori, the morphism $M_g \to \operatorname{Spec}(\mathbb{Z})$ is not proper.
- At least for the sake of de Jong's theorem, we need a moduli space of curves with points on them.

The first problem is solved by introducing a finite covering $M \to M_g$ admitting a *tautological family*, namely a family realizing the morphism $M \to M_g$ as its moduli morphism. The nicest way of doing this is by introducing a new moduli functor, of smooth curves "enriched" with a finite amount of "extra structure", which does admit a fine moduli scheme. See Section 13.

In order to "compactify" these spaces, the notion of *stable curves* was invented. Historically, the influential paper [17] by Deligne and Mumford seems to be one of the first printed versions in which the concept of stability, especially in the case of algebraic curves is explained and used. In [46] we see that already in 1964 Mumford was trying to find the appropriate notions assuring good compactifications. In [GIT], page 228, Mumford attributes the notion of a stable curve to unpublished joint work with Alan Mayer.

As it turns out, the third problem was solved almost concurrently with the second.

First, the moduli space of smooth pointed curves:

Theorem 10.6. *Let $g \in \mathbb{Z}_{\geq 0}$, and $n \in \mathbb{Z}_{\geq 0}$ such that $2g - 2 + n > 0$. Consider the functor $\mathcal{M}_{g,n}$ of isomorphism classes of families of stable smooth n-pointed curves of genus g. This functor admits a quasi-projective coarse moduli scheme.*

We will denote this moduli space by $M_{g,n} \to \operatorname{Spec}(\mathbb{Z})$.

Remark.

1. Note that this includes the previous theorem.
2. For $g = 0$ and $n = 3$ this space is proper over $\operatorname{Spec}(\mathbb{Z})$. However in all other cases in the theorem $M_{g,n}$ is not proper. In many cases it will not represent the functor (see Section 14 for a further discussion), in other words, in general this is not a fine moduli scheme.
3. It is important to note that these spaces exist over $\operatorname{Spec}(\mathbb{Z})$, which is useful for arithmetical applications.
4. The litterature poses difficulties in choosing notations. In [GIT] the subscript n denotes a level structure, but in [39] it indicates the number of marked points. We have chosen to indicate the markings as lower index, using n, and the level structure as upper index, using (m).

Finally the moduli space of stable pointed curves:

Theorem 10.7 (Knudsen and Mumford). *Let $g \in \mathbb{Z}_{\geq 0}$, and $n \in \mathbb{Z}_{\geq 0}$ such that $2g-2+n>0$. Consider the functor $\overline{\mathcal{M}_{g,n}}$ of isomorphism classes of families of stable n-pointed curves of genus g. This functor admits a* projective *coarse moduli scheme.*

See [39], part II, Theorem 2.7 and part III, Theorem 6.1, or [22], Theorem 2.0.2. We will denote this moduli scheme by $\overline{M_{g,n}} \to \mathrm{Spec}(\mathbb{Z})$.
The following exercise should give you an idea why the moduli space $\overline{\mathcal{M}_{0,n}}$ is complete. This is discussed in further detail in the next section.

Exercise 10.8. Let K be a field, and let $R \subset K$ be a discrete valuation ring having K as field of fractions. Consider the projective line $\mathbb{P}^1$ over K and suppose $n \geq 3$, let $P_1, \dots, P_n \in \mathbb{P}^1(K)$ are distinct points. Write $P = \{P_1, \dots, P_n\}$. Construct a stable n-pointed curve $(\mathcal{C}, \mathcal{P}) \to \mathrm{Spec}(R)$ extending $(\mathbb{P}^1, P)$. (You will need to blow up closed points over the special fiber where the Zariski closures of P_i meet. Then you may need to blow down some components! See [39].)

Exercise 10.9.

1. Let K be a field. Given three distinct finite points P_1, P_2 and P_3 on $\mathbb{P}^1_k$ consider the cross ratio
$$\lambda(P_1, P_2, P_3, z) = \frac{(z-P_1)(P_2-P_3)}{(z-P_3)(P_2-P_1)}.$$
Show that, as a function of z, the cross ratio is an automorphism of $\mathbb{P}^1$ carrying P_1, P_2, P_3 to $0, 1$, and ∞, respectively. Show that this automorphism is the unique one with this property. Check that this definition can be extended to the case where one of the points is ∞.
2. Using the cross ratio λ defined above, describe $M_{0,3}$.
3. Show that $M_{0,3}$ is a fine moduli scheme by exhibiting a universal family over it!
4. Show that $M_{0,3} = \overline{M_{0,3}}$.
5. Use the cross ratio to give an explicit description of $M_{0,4}$. Show that it is a fine moduli scheme by explicitly constructing a universal family.
6. Use the above (possibly together with the previous exercise) to describe $M_{0,4} \subset \overline{M_{0,4}}$.
7. Show that $\overline{M_{0,4}}$ is a fine moduli scheme, and give explicit descriptions of the universal family.
8. Show that the universal family over $\overline{M_{0,4}}$ is canonically isomorphic to $\overline{M_{0,5}}$.

Exercise 10.10. Give an alternative description of $\overline{M_{0,4}}$ as follows: consider the projective space $\mathbb{P}$ of dimension 5 parametrizing conics in $\mathbb{P}^2$. Choose four points in general position in $\mathbb{P}^2$ (for instance $(1:0:0), (0:1:0), (0:0:1), (1:1:1)$ will do). Let $M \subset \mathbb{P}$ be the subscheme parametrizing conics passing through these four points. Show that $M = \overline{M_{0,4}}$ and the universal family of conics is a universal family for $\overline{M_{0,4}}$.

Exercise 10.11.

1. Show that $M_{0,n}$ exists and is a fine moduli scheme (you may exhibit it as an open subscheme of $(\mathbb{P}^1)^{n-3}$).
2. Show that, assuming $\overline{M_{0,n}}$ is a fine moduli scheme, then there is a canonical morphism $\overline{M_{0,n+1}} \to \overline{M_{0,n}}$ which exhibits $\overline{M_{0,n+1}}$ as the universal family over $\overline{M_{0,n}}$.
3. * Show that for every $n \geq 3$, the scheme $\overline{M_{0,n}}$ is a fine moduli scheme. (You may want to use Knudsen's stabilization technique.)

Remark. For every $n \geq 3$, let (C, P) be a stable n-pointed rational curve. Then $\mathrm{Aut}(C, P) = \{id\}$. You do not need to know this in the previous exercise, but it "explains" why the result should be true.

Exercise 10.12. Let k be a field of characteristic $\neq 2$ and let (E, O) be an elliptic curve, namely a projective, smooth and connected curve E of genus 1 with a k-rational point O on it.

1. Considering the linear series of $\mathcal{O}_E(2O)$, show that E can be exhibited as a branch covering of $\mathbb{P}^1$ of degree 2.
2. Show that the branch divisor B on $\mathbb{P}^1$ is reduced and has degree 4.
3. If k is algebraically closed, show that E is determined up to isomorphism by the divisor B.
4. Conclude that $M_{1,1}$ is isomorphic to the quotient of $M_{0,4}$ by the action of the symmetric group S_4, permuting the four points.
5. Assume further that $\mathrm{char}\, k \neq 3$, so that every elliptic curve can be written in affine coordinates as $y^2 = x^3 + ax + b$. Show that
$$\jmath(E) = 1728\frac{4a^3}{4a^3 + 27b^2}$$
is an invariant characterizing the $\bar{k}$-isomorphism class of E, exhibiting $M_{1,1} = \mathbb{A}^1$.

11. Stable reduction and completeness of moduli spaces

11.1. General theory

In order to understand the reason why $\overline{M_{g,n}}$ is projective, let us recall the following:

Theorem 11.1 (The valuative criterion for properness). *A morphism*
$$f : X \to Y$$
of finite type is proper, if and only if the following holds:

Let R be a discrete valuation ring, and let $S := \mathrm{Spec}(R)$ be the corresponding "germ of a non-singular curve", with generic point η. Let $\varphi : S \to Y$ and let $\psi_\eta : \eta \to X$ be a lifting:

$$\begin{array}{ccc} \eta & \stackrel{\psi_\eta}{\to} & X \\ \downarrow & & \downarrow \\ S & \stackrel{\varphi}{\to} & Y. \end{array}$$

Then there is an extension $\psi : S \to X$, lifting φ:

$$\begin{array}{ccc} \eta & \stackrel{\psi_\eta}{\to} & X \\ & \psi & \\ \downarrow & \nearrow & \downarrow \\ S & \stackrel{\varphi}{\to} & Y. \end{array}$$

See [HAG], II, Theorem (4.7) for a precise formulation.
Let us translate this to our moduli scheme. Keeping in mind the relationship between the functor $\overline{\mathcal{M}_{g,n}}$ and the space $\overline{M_{g,n}}$, one might hope that every family of stable pointed curves over η as in the theorem above might extend to R. This is not the case, as we shall see later. However, a weaker result, sometimes called "the *weak* valuative criterion for properness", does hold for the functor $\overline{\mathcal{M}_{g,n}}$, and it does imply the valuative criterion for $\overline{M_{g,n}}$. The first case to consider is when the generic fiber is *smooth* and $n = 0$. This is the content of the following result, the Stable Reduction Theorem for a one parameter family of curves:

Theorem 11.2. *Let $S = \mathrm{Spec}(R)$ be the spectrum of a discrete valuation ring R, $\eta \in S$ the generic point, corresponding with the field of fractions K of R. Let $C_\eta \to \eta$ be a smooth stable curve of genus $g > 1$. There exists a finite extension of discrete valuation rings $R \hookrightarrow R_1$, with $S_1 = \mathrm{Spec}\, R_1$ and generic point η_1, and an extension*

$$\begin{array}{ccc} C_{\eta_1} & \hookrightarrow & C_1 \\ \downarrow & & \downarrow \\ \{\eta_1\} & \hookrightarrow & S_1, \end{array}$$

such that $C_1 \to S_1$ is a family of stable curves.

Proofs of this theorem, using different methods, may be found in various references. One proof which works in pure characteristic 0 is relatively simple. As the reader will notice, none of the general proofs is easy or elementary.

Most proofs of this theorem use resolution of singularities of 2-dimensional schemes (Abhyankar).

Exercise 11.3. Suppose R is of pure characteristic 0. Let $s \in \mathrm{Spec}\, R$ be the closed point.

1. Show that there exists an extension $\pi : C \to S$ such that π is proper and flat, C is nonsingular, and $C_s \subset C$ is a normal crossings divisor.
2. Let $x \in C_s$ be a singular point. After passing to the algebraic closure of the field of constant, let $\bar{x} \in \hat{C}_{\bar{k}}$ be the completion. Show that one can find local

parameters u, v at $\bar{x}$ and t at $\bar{s} \in \hat{S}_{\bar{k}}$, and positive integers k_x, l_x, such that $t = u^{k_x} v^{l_x}$.

3. Let $S_1 \to S$ be a finite cover obtained by extracting the n-th root of a uniformizer, where n is divisible by all the non-zero k_x, l_x given above. Let C_1' be the normalization of $C \times_S S_1$. Show that the special fiber is reduced and nodal.
4. Show that the minimal model C_1 of $C_1' \to S_1$ is stable.

We list some approaches for positive and mixed characteristic:

Artin-Winters. This proof can be found in [7]. A precise and nice description and analysis of the proof is given by Raynaud, see [59].

In this proof one attaches an numerical invariant to a given genus, and one proves that by choosing a prime number q larger than this invariant, and not equal to the residue characteristic, and by extending the field of definition of a curve of that genus such that all q-torsion point on the jacobian are rational over the extension, then one acquires stable reduction. The proof consists of a careful numerical analysis of the possible intersection matrices of components of degenerating curves. The proof does not rely on a lot of theory, but is quite subtle.

Grothendieck, Deligne-Mumford. This proof can be found in [17], Theorem (2.4) and Corollary (2.7).

In this proof one shows that a curve has stable reduction if and only if its jacobian has stable reduction. Then one shows following Grothendieck that eigenvalues of algebraic ℓ-adic monodromy are roots of unity (see [67], Appendix). Moreover, again following Grothendieck one shows that these eigenvalues are all equal to one iff the abelian variety in question has stable reduction. The advantage of this proof is that it has a more conceptual basis. The big disadvantage is that it relies on the theory of Néron models, whose foundations are quite difficult.

Hilbert schemes and GIT-Gieseker. See [22], Chapter 2, Proposition(0.0.2). He says on the first page of the introduction: " ... we use results of Chapter 1 to give an indirect proof that the n-canonical embedding of a stable curve is stable if $n \geq 10$, and to construct the projective moduli space for stable curves. As corollaries, we obtain proofs of the stable reduction theorem for curves, and of the irreducibility for smooth curves." The proof uses Geometric Invariant Theory to prove directly that $\overline{M_g}$ exists and is projective, and then one can easily derive the theorem. This proof does not use resolution of singularities for surfaces in any explicit manner.

Remark. This theorem is an instance of the *semistable reduction problem.* In [10], 1.3, the definition of semistable reduction, over a one-dimensional base, and *arbitrary* fiber dimension, is recalled. As we have seen above, it is true that if the relative dimension is one, stable reduction, hence semistable reduction, exists over a one-dimensional base. For higher relative dimension an analogous result holds in pure characteristic zero – see [38]. The general case is an important open problem, which seems difficult.

Once Theorem 11.2 is known, it is easy to generalize it. The pointed case can be easily proven using Knudsen's stabilization technique:

Exercise 11.4. Let R be a discrete valuation ring, with field of fractions K, suppose $(C, p_1, \ldots, p_n)$ is a *smooth, stable n-pointed curve* of genus $g > 1$ defined over K. There exists a finite extension $R \subset R_1$ of valuation rings, with K_1 the field of fractions of R_1, such that $C_1 = C \otimes K_1$ extends to a stable n-pointed curve $\mathcal{C}_1 \to \operatorname{Spec}(R_1)$.

The case of genus zero follows from Exercise 10.9. We will discuss the case of genus 1 in Section 11.2 below.

We can also consider the case when the generic fiber is not necessarily smooth:

Exercise 11.5. Let R be a discrete valuation ring, with field of fractions K, suppose C is a stable curve defined over K. There exists a finite extension $R \subset R_1$ of valuation rings, with K_1 the field of fractions of R_1, such that $C_1 = C \otimes K_1$ extends to a stable curve $\mathcal{C}_1 \to \operatorname{Spec}(R_1)$. [Below we formulate a generalization to stable pointed curves of this.]

We give a full generalization of (11.2):

Exercise 11.6. Let S be the spectrum of a discrete valuation ring, $\eta \in S$ the generic point. Let $(C, P) \to \{\eta\}$ be a stable n-pointed curve of genus g, i.e., C is a complete, nodal curve defined over a field K, and $P := \{P_1, \ldots, P_n\}$ are distinct closed points $P_j \in C(K)$, with such that (C, P) is stable n-pointed over K.

Then there exists a finite extension of discrete valuation rings $S \hookrightarrow S_1$, with generic point η_1, and an extension

$$\begin{array}{ccc} (C_{\eta_1}, P) & \hookrightarrow & (\mathcal{C}, \mathcal{P}) \\ \downarrow & & \downarrow \\ \operatorname{Spec}(K_1) & \hookrightarrow & S_1 \end{array}$$

such that $(\mathcal{C}, \mathcal{P}) \to S_1$ is a family of stable n-pointed curves.

This is the "weak valuative criterion for properness" of the functor $\overline{\mathcal{M}_{g,n}}$.

Remark. Here is a hint about a technical detail which can be used in solving the previous exercises, *"The normalization of a stable n-pointed curve"*: Suppose given a stable n-pointed curve (C, P) over a field K, with $P = \{P_1, \ldots, P_n\}$. There exists a finite extension $K \subset L$, a finite disjoint union (D, Q) of stable pointed curves, and a morphism ("the normalization") $\varphi : (D, Q) \to (C, P)_L = (C, P) \otimes_K L$ such that: $D = \coprod D^{(t)}$, let the singular points of C_L be: $R_j \in C(L)$, with $1 \leq j \leq d$, moreover $Q = \{Q_1, \ldots, Q_n\} \cup \{S_j, T_j \mid 1 \leq j \leq d\}$, for every irreducible component of C_L there is a unique component of D mapping birationally onto it, the morphism φ is an isomorphism outside $\operatorname{Sing}(C_L)$, the markings Q_i corresponds with the markings P_i of the pointed curve (C, P), and the markings $\{S_j, T_j\}$ are precisely the points mapping to R_j.

You need to show this choice can be made, and show it is unique in case $K = k$ is an algebraically closed field.

Corollary. *Let $g \in \mathbb{Z}_{\geq 0}$, and $n \in \mathbb{Z}_{\geq 0}$ such that $2g-2+n>0$. The coarse moduli scheme $\pi : \overline{M_{g,n}} \to \mathrm{Spec}(\mathbb{Z})$ is proper over $\mathrm{Spec}(\mathbb{Z})$.*

Proof. We use the valuative criterion for properness setting $X = \overline{M_{g,n}}$ and $Y = \mathrm{Spec}(\mathbb{Z})$. Suppose R is a discrete valuation ring, with field of fractions K, and suppose given

$$\begin{array}{ccc} \mathrm{Spec}\, K & \stackrel{\psi_K}{\to} & X \\ \downarrow & & \downarrow \\ \mathrm{Spec}\, R & \stackrel{\varphi}{\to} & Y. \end{array}$$

By the definition of a coarse moduli scheme, there is a finite extension $K \subset K'$ such that the point $\psi_K(\mathrm{Spec}(K)) \in X$ corresponds to a stable pointed curve (C, P) over K'. By the stable reduction theorem there is a finite extension $K' \subset K_1$ such that $(C, P) \times_{\mathrm{Spec}\, K'} \mathrm{Spec}\, K_1$ extends to a stable pointed curve; this defines a morphism $\tau : \mathrm{Spec}(R_1) \to X$, "extending" φ and ψ_K. It factors over $\mathrm{Spec}(R)$, because $R = K \cap R_1$. This shows that the condition for the valuative criterion holds in our situation, hence that $\pi : \overline{M_{g,n}} \to \mathrm{Spec}(\mathbb{Z})$ is proper. □

11.2. Stable reduction for elliptic curves

In these exercises we illustrate the concept of stable reduction by studying the case of elliptic curves. The concepts, ideas and examples below can be found in Silverman's book [69]. In this case examples are easy to give because in many cases we can choose plane models (Weierstrass equations). These exercises can be used at motivation for more abstract methods which apply for higher genus. You can do the exercises by explicit methods and calculations.

For details on elliptic curves, Weierstrass equations, the j-function, and related issues, see [69], Chapters III and VII.

A non-singular one-pointed curve of genus one is called an elliptic curve. In other words: an elliptic curve is an algebraic curve E defined over a field, absolutely irreducible, non-singular, of genus one, with a marked point $P \in E(K)$. Morphisms are supposed to respect the marked point.

The following exercise is an easy exercise using the theorem of Riemann-Roch.

Exercise 11.7. Show the following are equivalent:

1. (E, P) is an elliptic curve over K.
2. $E \subset \mathbb{P}^2_K$ is a plane, nonsingular cubic curve, with a marked point $P \in E(K)$.
3. (E, P) is an abelian variety of dimension one over K.

Definition. Let $R = R_v$ be a discrete valuation ring, with $K = \mathrm{fract}(R)$ its field of fractions, and $k = R_v/m_v$ the residue class field.

1. An elliptic curve E defined over K is said to have *good reduction* (at the given valuation) if there exists a smooth proper morphism $\mathcal{E} \to \mathrm{Spec}(R)$ with generic fiber isomorphic to $E \to \mathrm{Spec}(K)$. If E does not have good reduction, we say that it has *bad* reduction.

2. We say E has *stable reduction* at v if either it has good reduction, or there exists a *nodal* $\mathcal{E} \to \operatorname{Spec}(R)$ with generic fiber isomorphic to $E \to \operatorname{Spec}(K)$.

Definition. We say that E has *potentially good reduction*, if there exists a finite extension $K \subset L$, where B is the integral closure of R in L, and w a valuation over v, such that $E \otimes L$ has good reduction at w.

We define *potentially stable reduction* analogously.

Here are some exercises to warm up:

Exercise 11.8. Suppose $R = k[T]$, with $\operatorname{char}(k) \neq 2, \neq 3$, and let E over $K = k(T)$ be given by the equation $Y^2 = X^3 + T^6$. Show that E has good reduction at the valuation given by $v(T) = 1$.

Exercise 11.9. Suppose $R = k[S]$, with $\operatorname{char}(k) \neq 2, \neq 3$, and let E over $K = k(S)$ be given by the equation $Y^2 = X^3 + S$. Show that E has bad reduction at the valuation v given by $v(S) = 1$.

Exercise 11.10. Suppose $R = k[S]$, with $\operatorname{char}(k) \neq 2, \neq 3$, and let E over $K = k(S)$ be given by the equation $Y^2 = X^3 + S$. Show that E has potentially good reduction at the valuation v given by $v(S) = 1$.

Suppose that E is given over K by a Weierstrass equation with coefficients in R (see [69], III). Such an equation defines an *affine* plane curve $\mathcal{E} \subset \mathbb{A}^2_B$ over $\operatorname{Spec}(R) = B$, and it is easy to see that the curve $E_0 := \mathcal{E} \otimes_R k$ is irreducible and has at most one singular point. The curve E is obtained by adding the point at infinity to E_0. Suppose the Weierstrass equation is *minimal* at v. If this singular point is a cusp, we say that this reduction is of *additive type*, if it is a node we say that this reduction is of *multiplicative type*, or we say in this case the reduction is stable.

Exercise 11.11. Show that the notion of "good reduction" as defined earlier is equivalent by saying there is Weierstrass equation defining good reduction. Show that a reduction of multiplicative type is a stable reduction.

A reduction given by a minimal Weierstrass equation of additive type is bad reduction which is non-stable; non-stable bad reduction is sometimes called cuspidal reduction.

Exercise 11.12. Suppose $R = k[T]$, with $\operatorname{char} k \neq 2$, and let E over $K = k(T)$ be given by the equation $Y^2 = X \cdot (X-1) \cdot (X-T)$. Show that any model of this curve given by a Weierstrass equation has stable reduction at the valuation given by $v(T) = 1$. Show that this curve does not have potentially good reduction.

Exercise 11.13. Let R be a discrete valuation ring, with residue characteristic $\neq 2$, and fraction field K. Let E be an elliptic curve over K.

1. Show that after a suitable extension of R, the curve E admits a minimal Weierstrass equation of the form
$$y^2 = x(x-1)(x-\lambda)$$
for some $\lambda \in R$.
2. Conclude that this curve has potentially stable reduction.

Exercise 11.14. $*$ Let R_v be a DVR, with residue characteristic $\neq 3$. Suppose E is an elliptic curve over K given by a Weierstrass equation $E = V(F) \subset \mathbb{P}^2_K$ such that all flex points of E have coordinates in K.

1. Show that this curve admits a plane equation (not a Weierstrass equation!) over K of the form
$$\lambda(X^3 + Y^3 + Z^3) = 3\mu XYZ,$$
for some $(\lambda : \mu) \in \mathbb{P}^1_{R_v}$.
2. Show that $E \otimes_K L$ has stable reduction at v.
3. Show that E has potentially stable reduction.

Conclusion. Every elliptic curve over a field K with a discrete valuation has potentially stable reduction at that valuation.

This is a special case of 11.6, the stable reduction theorem for stable curves of arbitrary genus.

Exercise 11.15. $*$

1. Let R_v be a DVR, and E an elliptic curve of the fraction field K. Show that E has potentially good reduction at R_v if and only if $j(E) \in R_v$.
2. Can you formulate (and prove?) the same result for curves of arbitrary genus?
3. If E is an elliptic curve over a field K, and $\mathrm{End}_K(E) \neq \mathbb{Z}$, then E has potentially good reduction at every place of K.

11.3. Remarks about monodromy

Let C be a non-singular curve over a field K, and let v be a discrete valuation of K. Consider properties of good reduction, bad reduction at v, and so on. We have quoted that C has stable reduction at v iff $J := \mathrm{Jac}(C)$ has stable reduction, see [17], Proposition (2.3).

However note that it may happen that J has good reduction, and C has bad reduction; this is the case if the special fiber C_0 of the minimal model of C at v has a generalized jacobian $J_0 = \mathrm{Jac}(C_0)$ which is an abelian variety. Such a curve C_0 is called a curve of "compact type", or a "nice curve" (and sometimes called a "good curve", but we do not like that terminology, because a curve reducing to a "good curve" may not have good reduction ...). In this case the special fiber C_0 is a tree of non-singular curves, i.e. every irreducible component is non-singular, and in the dual graph of C_0 there are no cycles. The easiest example is: a join of two non-singular curves, each of genus at least one, meeting transversally at one singular point. For example a curve of genus two degenerating to a transversal crossing of two curves of genus one is the easiest example. Here is another example: take $\mathbb{P}^1$ with three marked points, and attach three elliptic tails via normal crossings at the markings, arriving at a nice curve of genus three

Monodromy (action of the local fundamental group of the base on cohomology) decides about the reduction of an abelian variety being bad or good, see [67], Theorem 1 on page 493. In the analytic context one can take the local fundamental group of a punctured disc acting on cohomology; in all cases one considers the inertia-Galois group of v acting on ℓ-adic cohomology, where ℓ is a prime number not equal to the residue characteristic of v.

Note that algebraic monodromy has *eigenvalues which are roots of unity.* This was proved by Landman, Steenbrink, Brieskorn in various settings, and we find a proof by Grothendieck in the appendix of [67]. For a sketch of that proof, see [49], for further references, see [52].

Algebraic monodromy is trivial iff $X = J$ has good reduction, iff C has compact type reduction (which may be either good reduction or bad but "nice" as explained above).

The algebraic monodromy is unipotent (all eigenvalues are equal to one) if and only if X has stable reduction, if and only if C has stable reduction.

But, how can we distinguish for curves the difference between good reduction and bad compact type reduction? As we have seen, this is not possible via algebraic monodromy on cohomology. But, in a beautiful paper, [8] we find a method which for curves unravels these subtle differences for curves: the local fundamental group of the base acts via outer automorphisms on the fundamental group of the generic fiber (again, here one can work in the analytic-topological context, or in the ℓ-adic algebraic context). *This action is trivial iff C has good reduction.*

12. Construction of moduli spaces

Early constructions of the moduli spaces of smooth curves M_g included a complex-analytic constructions via Teichmüller theory and via the construction of moduli of abelian variety using locally symmetric spaces. These constructions are not algebraic in nature and therefore cannot be generalized to positive or mixed characteristics.

A first algebraic approach, which is still commonly used today, was given by Mumford using his Geometric Invariant Theory [GIT]. We will sketch one version of this approach, due to Gieseker, which automatically gives also the moduli spaces of stable curves $\overline{M_g}$. There is another approach, due to Artin and Kollár [40], which circumvents the use of Geometric Invariant Theory. Nowadays both approaches work over $\mathbb{Z}$.

How does one start? It is evident that if we want to parametrize *all* stable curves of a certain genus, we had better have *some* family of curves in which all these curves appear. We know of two general approaches for that. One method uses parameter spaces for curves embedded in projective space, such as Hilbert schemes (or Chow varieties). We will follow this approach. The other approach, due to Artin [6], uses versal deformation spaces. It works in greater generality but involves a number of technicality which we would rather avoid here.

It is easy to see that for any stable curve C of genus $g > 1$, and any $\nu \geq 3$, the ν-canonical series $H^0(C, \omega_C^\nu)$ gives an embedding of C as a curve of degree $d := \nu(2g-2)$ in a projective space of dimension $N := \nu(2g-2) - g$. Thus the Hilbert scheme $\mathcal{H}_{\mathbb{P}^N, P(T)}$ (over $\mathbb{Z}$!) parametrizing subschemes of $\mathbb{P}^N$ with Hilbert polynomial $P(T) := dT + 1 - g$ carries a universal family $\mathcal{C}_{\mathbb{P}^N, P(T)} \to \mathcal{H}_{\mathbb{P}^N, P(T)}$ in which each stable curve of genus g appears at least once.

There are two problems with this family:

1. Each curve appears more than once in the family. Indeed, the embedding of the curve C in $\mathbb{P}^N$ involves two choices: a choice of a line bundle of degree d, and a choice of a basis for the linear series. And of course the curves could also be embedded in a projective subspace using a subseries.
2. There are many curves in $\mathbb{P}^N$ with Hilbert polynomial $P(T)$ which are far from stable.

Since a nodal curve can only deform into nodal curves, it is easy to see that there is an open subset $\mathcal{H}_{st} \subset \mathcal{H}_{\mathbb{P}^N, P(T)}$ which parametrizes *stable* curves, embedded by a *complete* linear system in $\mathbb{P}^N$. Denote the restriction of the universal family to $\mathcal{H}_{st}$ by $\pi : \mathcal{C}_{st} \to \mathcal{H}_{st}$. Considering the locus in $\mathcal{H}_{st}$ where $R^1\pi_*(\mathcal{O}(1) \otimes \omega^\nu_{\mathcal{C}_{st} \to \mathcal{H}_{st}})$ jumps in dimension, we immediately see that there is a *closed* subscheme $\mathcal{H}_g \subset \mathcal{H}_{st}$ parametrizing stable curves embedded by a complete ν-canonical series. The restriction of the universal family will be denoted $\mathcal{C}_g \to \mathcal{H}_g$.

There is a natural action of the projective linear group $PGL(N+1)$ on $\mathcal{H}_g$ via changing coordinates on $\mathbb{P}^N$. It is easy to see that the "ambiguity" for choosing the embedding of a curve C in the latter universal family is fully accounted for by the action of this group. In other words, stable curves correspond in a one-to one manner with $PGL(N+1)$ orbits in $\mathcal{H}_g$. Thus, at least set theoretically, $\overline{M}_g = \mathcal{H}_g / PGL(N+1)$.

12.1. Geometric Invariant Theory and Gieseker's approach

We arrived at the following questions:

1. Does the quotient $\mathcal{H}_g / PGL(N+1)$ exist as a scheme?
2. Can we show that it is projective?
3. Does it satisfy the requirements of a coarse moduli scheme?

Geometric Invariant Theory is a method which allows one to approach the first two questions simultaneously. The third question then becomes an easy gluing exercise.

The general situation is as follows: Let $X \subset \mathbb{P}^n$ be a quasi-projective scheme and suppose G is an algebraic group acting on $\mathbb{P}^n$ and stabilizing X. One wants to know whether or not a quotient X/G exists as a scheme and whether or not it is projective.

A natural approach is to look for a space of invariant sections of some line bundle. Thus assume that the action of G on $\mathbb{P}^n$ lifts to $\mathcal{O}_{\mathbb{P}^n}(l)$. Then it also lifts to any power $\mathcal{O}_{\mathbb{P}^n}(l \cdot m)$, and we can look at the ring of invariants $R := \oplus(\mathcal{O}_{\mathbb{P}^n}(l \cdot m)^G$. We have a natural rational map $q : \mathbb{P}^n \dashrightarrow \operatorname{Proj} R$. We would like to know whether or not this map is well defined along X, and what the image is like.

First, an easy observation. For any point $x \in X$, the map q is well defined at x if and only if there exists a nonconstant invariant $f \in R$ such that $f(x) \neq 0$.

We want to check whether q is a quotient map in a neighborhood of x. To go any further, we need to assume that the group G is *reductive.* Assuming that G is reductive, then the question whether map q is a quotient map at a neighborhood of x can be translated to a question about the closure $\overline{Gx}$ of the orbit of x: one needs to check that for any point $y \in \overline{Gx} \smallsetminus Gx$ there is an invariant $f \in R$ which vanishes at y but not at x. A point x is called *GIT-stable* if it satisfies this condition.

Mumford's numerical criterion for stability (see [GIT]) gives a way to check GIT-stability in some situations.

Let us consider our situation. The scheme $\mathcal{H}_g$ is quasi projective - from its construction one sees that it naturally sits inside a Grassmannian, which has a Plücker embedding in some $\mathbb{P}^n$. It is easy to see that the action of $PGL(N+1)$ extends to $\mathbb{P}^n$, and lifts to some line bundle $\mathcal{O}_{\mathbb{P}^n}(k)$. Applying this criterion systematically, Gieseker verified in [22] that

1. If a point $x \in \mathcal{H}_{\mathbb{P}^N, P(T)}$ corresponds to a scheme which is not a stable curve, or to a curve which is not embedded by a complete linear series, then *every* nonconstant invariant vanishes at x.
2. If a point $x \in \mathcal{H}_{\mathbb{P}^N, P(T)}$ corresponds to a stable curve embedded by the complete ν-canonical linear series, then x is GIT-stable.

Using the two statements, and the fact that G is reductive, it is not difficult to realize that

1. the map $\mathcal{H}_g \to \operatorname{Proj} R$ is a quotient map, and
2. the image of $\mathcal{H}_g$ is projective.

This proves the existence and projectivity of $\overline{M_g}$.

12.2. Existence of $\overline{M_{g,n}}$

There is no known analogue of Gieseker's result for stable pointed curves. It is not difficult to construct a Hilbert-type scheme for stable pointed curves, with a reductive group action, and such that the quotient is set-theoretically $\overline{M_{g,n}}$. But in order to tell that the quotient is isomorphic to $\overline{M_{g,n}}$ as a scheme, we first need to construct $\overline{M_{g,n}}$ in some other way.

But there is a very useful trick, which reduces the construction of $\overline{M_{g,n}}$ to the existence of $\overline{M_g}$ for some larger value of g. We give the reduction over a field, but it works similarly over $\mathbb{Z}$.

Fix n irreducible stable curves C_i of genus $g_i > g$, all nonisomorphic to each other, and fix a rational point $x_i \in C_i$. For any stable n-pointed curve $(C, p_1, \ldots, p_n)$ of genus g, we can construct a stable curve C' of genus $g' = g + \sum g_i$ as follows: $C' = (\cup C_i) \cup C$, where we glue together C and C_i by identifying p_i with x_i.

Clearly this construction gives a set theoretic embedding $\overline{M_{g,n}} \to \overline{M_{g'}}$. The image set is easily seen to be a scheme, and by working the construction in a family it is easy to see that it is a coarse moduli scheme.

13. Existence of tautological families

For almost any application of moduli spaces of curves including the alteration theorem, it is necessary to know that there exists a family $C \to M$ over a scheme M such that the associated morphism to the moduli space is finite and surjective. Such a family is called a *tautological family*; see Section 10.4. Various authors have devised general methods of showing this, but for the moduli spaces of curves there is a "very nice" way to find such a cover, using level structures. The case of the moduli space $\overline{M_g}$ of stable (unpointed) curves is discussed in detail in [21]. In this section we describe how this can be generalized for sable *pointed* curves as well. We rely throughout on the treatment in [21]. In Section 13.4 we outline another way to construct tautological families, which works in greater generality.

13.1. Hilbert schemes and level structures

Fix:

- an integer $g \in \mathbb{Z}_{\geq 0}$ (the genus),
- an integer $n \in \mathbb{Z}_{\geq 0}$ (the number of marked points),
- such that $2g - 2 + n > 0$,
- and an integer $m \in \mathbb{Z}_{\geq 1}$ (the level).
- Fix an integer $\nu \in \mathbb{Z}_{\geq 5}$, which will be used to study ν-canonical embeddings of curves into a projective space.

Remark. If $n = 0$ or $m = 1$ these data will be omitted from the notation, e.g. $M_{g,0} = M_g$. If $g = 0$, the level structure is irrelevant, $M_{0,n}^{(m)} = M_{0,n}$.

Let C be a curve whose jacobian is an abelian variety. By a level-m structure on C we mean a symplectic level structure as explained in [21]. If a level-m structure is considered we assume that all schemes, varieties are over a base on which m is invertible, i.e. are schemes over $\mathrm{Spec}(\mathbb{Z}[1/m])$.

Recall that there is a Hilbert scheme H_P parametrizing curves $C \subset \mathbb{P}^N$, where $N = \nu\cdot(2g-2+n)-g$, with Hilbert polynomial $P(t) = \nu\cdot(2g-2+n)\cdot t - g+1$. We want to find a scheme parametrizing *pointed* curves – this is done in a standard way as follows. Observe that there is a closed subscheme $H_{P,n} \subset H_P \times (\mathbb{P}^N)^n$ parametrising pairs $(C, (p_1, \dots, p_n))$ where $p_i \in C$. There is an open subscheme $H_{st} \subset H_{P,n}$ where the curves are nodal, the points are distinct and regular points on the curves, and the pairs $(C, (p_1, \dots, p_n))$ are stable. Last, there is a closed subscheme $H_{g,n} \subset H_{st}$ where the embedding line bundle of $C \subset \mathbb{P}^N$ is isomorphic to $(\omega_C(p_1 + \dots + p_n))^\nu$.

Over $H_{g,n}$ there is a universal family $C_{g,n} \to H_{g,n}$ with sections $s_i : H_{g,n} \to C_{g,n}$ of stable pointed curves, embedded in $\mathbb{P}^N$ by the chosen line bundle. The linear group PGL $= PGL(N)$ acts on $C_{g,n} \to H_{g,n}$ equivariantly, and $\overline{M_{g,n}} = H_{g,n}/\mathrm{PGL}$ is the quotient.

Note that there is an open subset $H^0_{st} \subset H_{g,n}$ parametrizing *smooth* stable pointed curves.

13.2. Moduli with level structure

Theorem 13.1. *For $m \geq 3$, and $2g-2+n > 0$, there exists a* fine *moduli scheme $M^{(m)}_{g,n}$ for smooth stable n-pointed curves with level-m structure. In particular there exists a universal curve with level structure over $M^{(m)}_{g,n}$. This moduli scheme is smooth over* $\mathrm{Spec}(\mathbb{Z}[1/m])$.

Note in particular that $M^{(m)}_{g,n}$ is a normal scheme, and that $M^{(m)}_{g,n} \to M_{g,n}$ is a Galois cover with Galois group $\mathrm{Sp}(2g, \mathbb{Z}/m)$.

We use the notation $S_m := \mathrm{Spec}(\mathbb{Z}[1/m])$.

Definition. Let $g \in \mathbb{Z}_{\geq 1}$. Fix $n \in \mathbb{Z}_{\geq 0}$, with $2g-2+n > 0$. For any $m \in \mathbb{Z}_{\geq 3}$, the scheme

$$\overline{M^{(m)}_{g,n}} \longrightarrow S_m$$

is defined as the normalization of $\overline{M_{g,n}}[1/m] = \overline{M_{g,n}} \times_{\mathbb{Z}} S_m$ in $M^{(m)}_{g,n}$.

For simplicity of notation in this section, we write $M = \overline{M^{(m)}_{g,n}}$ and $M^0 = M^{(m)}_{g,n} \subset M$.

Theorem 13.2. *Fix g, n, and m as above. Suppose $m \geq 3$. There exist a stable n-pointed curve $(\mathcal{C}, \mathcal{P}) \to M$, and a level-$m$-structure α on $\mathcal{C}^0 := \mathcal{C}|_{M^0}$ such that*

$$(\mathcal{C}, \mathcal{P}) \to M \quad \text{is tautological for} \quad M \to \overline{M_{g,n}},$$

and such that

$$(\mathcal{C}^0, \mathcal{P}^0, \alpha) \to M^0$$

represents the functor $\mathcal{M}^{(m)}_{g,n}$.

We give an argument for 13.1 and 13.2 following the line of [21]. This is a kind of "boot-strap" argument, which uses the idea that *once one quotient space exists, many others follow.* We also sketch another argument which reduces the problem to the case of [21].

There is a relative jacobian scheme $J(C^0_{g,n}) \to H^0_{g,n}$. This is an abelian scheme, so we can look at its group-subscheme of m-torsion points. Taking a symplectic rigidification of this group scheme we arrive at $H^{(m),0}_{g,n}$ – the Hilbert scheme of smooth stable n-pointed curves with symplectic level-m structure – embedded in projective space as above.

The action of PGL on $H^0_{g,n}$ clearly lifts to $H^{(m),0}_{g,n}$. This immediately implies that

$$M^{(m)}_{g,n} = H_{g,n} \;/\; \mathrm{PGL}$$

exists, since it is finite over $M_{g,n}$. By Serre's lemma this action has no fixed points, and it also lifts to $C^{(m),0}_{g,n} = C_{g,n} \times_{H_{g,n}} H^{(m),0}_{g,n}$. This means that the quotient $\mathrm{PGL}\backslash C^{(m),0}_{g,n} \to M^{(m)}_{g,n}$ is a universal family of smooth stable pointed curves with level structure.

This proves Theorem 13.1.

The normalization of $H_{g,n}$ in $H^{(m),0}_{g,n}$ will be denoted by $H^{(m)}_{g,n}$. The argument of [21], (2.6) works word for word, and shows that PGL still acts without fixed points on $H^{(m)}_{g,n}$. This gives the existence of the quotient

$$\overline{M^{(m)}_{g,n}} = H^{(m)}_{g,n} \;/\; \mathrm{PGL}.$$

Again the universal family over $H^{(m)}_{g,n}$ descends to a family over $\overline{M^{(m)}_{g,n}}$, this extends the universal family over $M^{(m)}_{g,n}$, and clearly it is tautological. This proves Theorem 13.2.

□

13.3. Proof by reduction to the unpointed case

Starting from $\overline{M^{(m)}_g}$ and its tautological family we can construct $\overline{M^{(m)}_{g,n}}$ and its tautological family by induction on the number of points n in the manner described below.

Denote by $D \to \overline{M^{(m)}_{g,n}}$ the tautological family. It is easy to see that in fact $D = \overline{M^{(m)}_{g,n+1}}$. So $D \times_{\overline{M^{(m)}_{g,n}}} D \to \overline{M^{(m)}_{g,n+1}}$ is a family of stable n-pointed curves with level structure, but with an additional section given by the diagonal. Using the stabilization process as described in [39] (see Section 3.7 above) one blows this scheme up, to obtain the tautological family over $\overline{M^{(m)}_{g,n+1}}$ as desired.

Remark. The moduli space $M^{(m)}_{g,n}$ is smooth over S_m for $m \geq 3$; this follows from Serre's lemma and deformation theory. However, the moduli space $\overline{M^{(m)}_{g,n}}$ is singular if $g > 2$; Serre's lemma holds also in this situation, but the space is not the coarse (or fine) moduli space of a moduli functor whose deformation spaces coincide with the deformations of stable curves. For more explanation, see [45] or [21].

The argument above works for $g > 1$ when $\overline{M_g^{(m)}}$ exists. For rational curves these theorems are relatively easy, and known, since the moduli spaces are fine moduli spaces in genus 0. For elliptic curves these theorems are known by the theory of modular curves.

13.4. Artin's approach via slicing

A general approach for constructing tautological families over finite covers of coarse moduli space was developed by Artin (see description in [40]). Here we present a version of this approach adapted to stable pointed curves.

Step 1: Slicing Consider the locally closed subset of the Hilbert scheme $H_{g,n}$ discussed above. It carries a universal family of stable pointed curves $C_{g,n} \to H_{g,n}$ suitably embedded in a projective space. This family induces a natural morphism $H_{g,n} \to \overline{M_{g,n}}$. The fibers coincide with the G-orbits associated to the embedded curves, where $G = \mathrm{PGL}$.

Fix a point $x \in H_{g,n}$. By repeatedly taking hyperplane sections, we can find a locally closed subscheme $V_x \subset H_{g,n}$ such that

1. $Gx \cap V_x \neq \emptyset$;
2. If $x' \in H_{g,n}$ and $Gx' \cap V_x \neq \emptyset$, then there exists a neighborhood $x' \in U$ such that for any $y \in U$ we have that $Gy \cap V \neq \emptyset$; and
3. for any $y \in H_{g,n}$ we have that $V_x \cap Gy$ consists of finitely many closed points.

These V_x are "multi-sections" of the map $H_{g,n} \to \overline{M_{g,n}}$ in a neighborhood of Gx. The essential point is that all orbits in $H_{g,n}$ are of the same dimension.

Using the Noetherian property, we can choose finitely many of these, say $V_1, \ldots, V_l$, such that every orbit meets at least one of them.

Step 2: Normalization. Let K be the join of the function fields of V_i over $\overline{M_{g,n}}$. Let V be the normalization of $\overline{M_{g,n}}$ in the Galois closure of K. The scheme V admits many rational maps to the V_i. It is not hard to see that for every point $v \in V$ at least one of these maps is well defined at v! Pulling back the families on V_i, we see that V is covered by open sets, each of which carries a family of stable pointed curves, compatible with the given morphism $V \to \overline{M_{g,n}}$.

Step 3: Gluing. Now we can use Lemma 3.11 inductively. We obtain a finite surjective $M \to V$ over which the families glue together to a family $C \to M$ such that the associated moduli morphism is the composition $M \to V \to \overline{M_{g,n}}$. Since V is finite over $\overline{M_{g,n}}$, this forms a tautological family.

Remark. It is not hard to construct a tautological as above *without using the existence of $\overline{M_{g,n}}$!* One can use this to construct the moduli space "from scratch" as a proper algebraic space, which is roughly speaking a quotient of a scheme by a finite equivalence relation. Kollár in [40] has shown how to use this to prove, without GIT, that $\overline{M_{g,n}}$ is projective.

14. Moduli, automorphisms, and families

This section will not be needed in the proofs above. The central theme here is the relationship between automorphisms, coarseness of moduli, and the existence of families. The main principle which will emerge is:

$$\text{a moduli space } M \text{ is a fine moduli space}$$
$$\Updownarrow$$
$$\text{objects parametrized by } M \text{ have no nontrivial automorphisms}$$
$$\Uparrow$$
$$M \text{ carries a unique tautological family.}$$

We also touch on the issue of singularities of moduli spaces.

For rational curves, and $n \geq 3$, the moduli schemes $M_{0,n}$ and $\overline{M_{0,n}}$ exist, these are smooth over $\mathrm{Spec}(\mathbb{Z})$, these are fine moduli schemes, i.e. they carry a universal family.

However, the moduli space $M_{1,1}$ and the moduli spaces M_g for $g > 1$ are not fine for the related moduli functor.

Exercise 14.1 (Deuring). Let K be a field, let $x \in K$. Then there exists an elliptic curve E defined over k with $j(E) = x$. [Suppose $\mathrm{char}(K) \neq 2, \neq 3$, suppose E is given over K by the equation $Y^2 = X^3 + AX + B$, with $4A^3 + 27B^2 \neq 0$. Then define

$$j(E) := 1728{\cdot}4{\cdot}A^3/(4A^3 + 27B^2).$$

For the definition of the j-invariant, see [69].]

This can partly be made more precise as follows:

Exercise 14.2. Consider $M_{0,1} \cong \mathbb{A}^1_{\mathbb{Z}}$, and remove the sections $j = 0$ and $j = 1728$:

$$U := \mathbb{A}^1_{\mathbb{Z}} \smallsetminus \{0, 1728\}_{\mathbb{Z}}.$$

There exists a tautological curve

$$\mathcal{E} \to U.$$

1. This cannot be extended over any of the deleted points.
2. This family is not at all unique.

Exercise 14.3. Consider $U := \mathbb{C} \smallsetminus \{0, 1728\}$. Show: up to isomorphisms there exist exactly 4 tautological curves (stable, one pointed smooth curves of genus 1 with j invariant different from 0 and 1728) over this moduli space. Show that for the ground field $K = \mathbb{Q}$ there are *infinitely many* tautological curves over the moduli space $\mathbb{A}^1_{\mathbb{Q}} \smallsetminus \{0, 1728\}$. Characterize them all.

We have seen the difference between a universal curve and a tautological curve: the moduli problem for elliptic curves with geometrically no non-trivial automorphisms admits a coarse moduli scheme; over that scheme there is a tautological curve, but the scheme is not a fine moduli scheme (not every family is a pull-back from one chosen tautological curve). Here is another example:

Definition. A curve $\mathcal{C} \to S$ is called a *hyperelliptic curve* if it is smooth, of relative genus g with $g \geq 2$, and if there exists an involution $\iota \in \mathrm{Aut}(\mathcal{C}/S)$ such that the quotient $\mathcal{C}/ < \iota > \to S$ is a smooth family of rational curves.

Remark. Elliptic curves and rational curves are not called "hyperelliptic", but sometimes the terminology "quasi-hyperelliptic" is used for curves having an involution with rational quotient.

Theorem 14.4. *Consider the moduli space* Hip_g *of hyperelliptic curves of genus* $g \geq 2$ *(even over* $\mathbb{C}$*). If* g *is even there* does not exist *a curve defined over the function field* $\mathbb{C}(\mathrm{Hip}_g)$ *having as moduli point the generic point of* Hip_g.

(See Shimura [68], Theorem 3.)

In different terminology: For no open dense subset $U \subset H_g$ does there exist a tautological curve when g is *even.*

There does exist a open dense subset $U \subset H_g$ and a tautological curve $\mathcal{C}_U$ when g is *odd.*

Corollary. *No dense open subset in* M_2 *or in* $M_2 \otimes K$ *carries a tautological curve.*

Exercise 14.5. Choose $g \in \mathbb{Z}_{>2}$, and consider nonsingular curves of genus g.

1. Show that there exists such a curve which has no nontrivial automorphisms.
2. (variant:) Show that a general curve of genus > 2 has no nontrivial automorphisms.

Remark. There is a morphism $M_{g,n+1} \to M_{g,n}$ ("forgetting the last marking"). Sometimes this is called the "universal curve over $M_{g,n}$", but we think in general this terminology is not justified in all cases possible.

Theorem 14.6. *Let* $U \subset M_g$ *with* $g \geq 3$ *fixed, be the set of points corresponding with curves which have geometrically no non-trivial automorphisms. This set is dense and open. Let* $\mathcal{M}_U$ *be the corresponding moduli functor. This functor is representable.*

In other terminology: there does exist a (unique) universal curve $\mathcal{C}_U \to U$ for the moduli problem of curves of genus $g \geq 3$ with geometrically no non-trivial automorphisms.

In particular: Let K be a field, $g \in \mathbb{Z}_{\geq 3}$, and η be the generic point of $M_g \otimes K$. There exists an algebraic curve defined over $K(\eta)$ having η as moduli point. However the universal family as indicated above over $U \subset M_g$ does not extend to any smooth family of curves over M_g.

Exercise 14.7. Formulate and prove a generalization of previous theorems to the case of stable pointed curves.

Exercise 14.8. Let $n > 2g+2$ and let $(C, P_1, \ldots, P_n)$ be any stable n-pointed curve of genus g. Suppose that C is *regular* (and hence irreducible). Show that

$$\mathrm{Aut}((C, P_1, \ldots, P_n)) = \{1\}$$

(if you want, assume that $\mathrm{char}(k) = 0$).

Exercise 14.9. Let $g \in \mathbb{Z}_{\geq 1}$ and $2 - 2g < n \leq 2g + 2$ and $0 \leq n$. Show that $M_{g,n}$ is a coarse, but not a fine moduli space.

Exercise 14.10. Choose $g \in \mathbb{Z}_{\geq 0}$, and let $n > 2g + 2$. Show that $M_{g,n}$ is a fine moduli space. Show that the universal curve over $M_{g,n}$ is smooth if $n \geq 2$.

Exercise 14.11. Consider all stable n-pointed curves of genus g. Suppose that

$$2g - 2 + n \geq 3.$$

1. Show that there exists such a curve which has no nontrivial automorphisms.
2. (variant:) Show that a general curve as above has no nontrivial automorphisms.

Exercise 14.12. Choose some g (e.g. $g = 3$), choose a very large integer n (e.g. $n = 1997$), and construct a stable n-pointed curve of genus g which has a non-trivial group of automorphisms.

Variant: Let $2g - 2 + n \geq 2$; show that there exist stable n-pointed curves of genus $g \geq 3$ in codimension two in the moduli space with non-trivial groups of automorphisms.

Exercise 14.13. Let $g \in \mathbb{Z}_{\geq 1}$, and $n > 2 - 2g$ and $n \geq 0$. Show that $\overline{M_{g,n}}$ is not a fine moduli space.

Choose $2g - 2 + n > 0$, choose $m \geq 1$ and let M be one of the following spaces: $M_{g,n}^{(m)}$, or $\overline{M_{g,n}}$ (all these spaces are defined by a moduli functor). Let $x \in M(k)$ be a geometric point, and let $X_0 := (C, P, \alpha)$ be the corresponding object over k (if C is non-smooth there is no level structure, the genus of C is g, we have $P = \emptyset$ if $n = 0$, we have $\alpha = id$ if $m = 1$). Let $D = \mathrm{Def}(X_0)$ be the universal deformation space; i.e. consider $\Lambda = k$ if $\mathrm{char}(k) = 0$, and $\Lambda = W_\infty(k)$ in case of positive characteristic, consider all local artin Λ-algebras, and consider the object prorepresenting all deformations of X_0 over such algebras (see [64]). This universal deformation object exists, and it is formally smooth over Λ on $3g-3+n$ variables; in case $n = 0$ this can be found in [17], page 81, the case of pointed curves follows along the same lines; in case $m > 1$, we have required that m is invertible in k, finite, flat group schemes of m-power order on such bases are étale, and deformations of level structures are unique by EGA IV[4], 18.1. Let $G := \mathrm{Aut}(X_0)$. Note that G *is a finite group* (because we work with stable curves). Note that G acts in a natural way on $D = \mathrm{Def}(X_0)$ by "transport of structure".

Theorem 14.14. *In the cases described, the formal completion of M at x is canonically isomorphic with the quotient*

$$\mathrm{Def}(X_0)\ /\ G \xrightarrow{\sim} M_x^{\wedge}.$$

This is well known, e.g. see [27], §1.

Exercise 14.15. (Rauch, Popp): Let $g \in \mathbb{Z}_{\geq 4}$, and let $A \subset M_g$ be an irreducible component of the set of all points corresponding with curves with non-trivial automorphisms. Show that the codimension of $A \subset M_g$ is ≥ 2. (In positive characteristic this is also correct, but you might need some extra insight to prove also those cases.)

Remark. Stable *rational* pointed curves have no non-trivial automorphisms. For *elliptic* curves there are curves with more than 2 automorphisms in codimension one. For curves *of genus two* we find a description of all curves with "many automorphisms" in [30]. Note that hyperelliptic curves of genus three are in codimension one.

Exercise 14.16. Show that non-hyperelliptic curves of genus three with non-trivial automorphisms are in codimension at least two.

Exercise 14.17. (Rauch [58], Popp [57]): Let $g \in \mathbb{Z}_{\geq 4}$, and let $[C] = x \in M_g$ be a geometric point. Show that x is a singular point on M_g iff $\mathrm{Aut} \neq \{id\}$. [You might like to use: [5], Coroll. 3.6 on page 95: A quasi-finite local homomorphism of regular local rings having the same dimension is flat. Also you might like to use purity of branch locus: a ramified *flat* covering is ramified in codimension one.]

Remark. For singularities of M_2 see [30]. Show that for genus three non-hyperelliptic points are singular iff there are non-trivial automorphisms, e.g. see [50]. For singularities of moduli schemes of abelian varieties, see [51].

Remark. As we have seen in [21], the moduli schemes $\overline{M_g^{(m)}}$ have singularities for all $g \geq 3$ and $m \geq 3$ (these spaces cannot be handled with the methods just discussed, these spaces are not given by "an obvious" moduli functor !). As Looijenga, see [42], in characteristic zero, and Pikaart and De Jong, see [54] showed, there exist a finite map $M \to M_g$ with M regular (using non-abelian level structures) (it is even true that M is smooth over $\mathbb{Q}$, or smooth over $\mathbb{Z}[1/r]$ for some natural number $r > 1$).

Summary about

$$M_{g,n}^{(m)} \hookrightarrow \overline{M_{g,n}^{(m)}} \longrightarrow \mathrm{Spec}(\mathbb{Z}[1/m]) =: S_m$$

for

$$g \in \mathbb{Z}_{\geq 0},\quad n \in \mathbb{Z}_{\geq 0},\quad m \in \mathbb{Z}_{\geq 1},\quad \text{with}\quad 2g-2+n > 0,$$

$M_{g,n}^{(m)}$ and $\overline{M_{g,n}}$ exist as coarse moduli schemes, we have constructed $\overline{M_{g,n}^{(m)}}$. We have seen:

- For $g \geq 2$ the coarse moduli scheme $M_g \to S = \mathrm{Spec}(\mathbb{Z})$ exists. These are not fine moduli spaces. They do not carry a tautological family. For every g this is singular.
- For $g \geq 2$ the coarse moduli scheme $\overline{M_g} \to S = \mathrm{Spec}(\mathbb{Z})$ exists. These are not fine moduli spaces. They do not carry a tautological family. They are singular.
- A dense open set in $M_{1,1}$ carries a tautological family, and it is not universal.
- No dense open set in M_2 carries a tautological family.
- For $g \geq 3$ a dense open set in $M_{g,n}$ carries a universal family.
- For $n \geq 3$ the moduli spaces $M_{0,n} \subset \overline{M_{0,n}}$ exist, they are fine moduli spaces, they are smooth over $S = \mathrm{Spec}(\mathbb{Z})$.
- For $2g-2+n > 0$, and $m \geq 0$ the moduli spaces $M_{g,n} \to S$, and $\overline{M_{g,n}} \to S$ and $M_g^{(m)} \to S_m$ exist, they coarsely represent a moduli functor. For $n > 2g+2$ the moduli space $M_{g,n}$ is fine, and smooth over $\mathrm{Spec}(\mathbb{Z})$ (but the universal family is not smooth for $n > 1$). For $m \geq 3$ the space $M_g^{(m)}$ is fine and smooth over S_m.
- For $2g-2+n > 0$, and $m \geq 0$ there is a moduli space, and a tautological family, with properties as in 13.2. For $g \geq 3$ the morphism $\overline{M_{g,n}^{(m)}} \to S_m$ is not smooth.

References

[Alteration] A. J. de Jong, *Smoothness, semistability, and alterations*, Publications Mathématiques I.H.E.S. **83**, 1996, pp. 51–93.

[GIT] D. Mumford, J. Fogarty and F. Kirwan *Geometric invariant theory.* Springer, Berlin, 1994.

[HAG] R. Hartshorne, *Algebraic geometry.* Springer, New York, 1977.

[Red Book] D. Mumford, *The red book of varieties and schemes.* Lecture Notes in Math., 1358, Springer, Berlin, 1988.

[SGA] A. Grothendieck (with M. Raynaud and D. S. Rim), *Groupes de monodromie en géométrie algébrique I.* (Séminaire de géométrie algébrique du Bois-Marie.) Lecture Notes in Math., 288, Springer, Berlin, 1972.

[1] D. Abramovich, *A high fibered power of a family of varieties of general type dominates a variety of general type,* Invent. Math. **128** (1997), no. 3, 481–494.

[2] D. Abramovich and A. J. de Jong, *Smoothness, semistability and toroidal geometry,* J. Algebraic Geom. **6** (1997), no. 4, 789–801.

[3] D. Abramovich and K. Karu, *Weak semistable reduction in characteristic 0,* preprint. `alg-geom/9707012`.

[4] D. Abramovich and J. Wang, *Equivariant resolution of singularities in characteristic 0,* Math. Res. Lett. **4** (1997), no. 2–3, 427–433.

[5] A. Altman and S. Kleiman, *Introduction to Grothendieck duality theory.* Lecture Notes in Math., 146, Springer, Berlin, 1970.

[6] M. Artin, *Versal deformations and algebraic stacks,* Invent. Math. **27** (1974), 165–189.

[7] M. Artin and G. Winters, *Degenerate fibres and reduction of curves.* Topology **10** (1971), 373–383.

[8] M. Asada, M. Matsumoto and T. Oda, *Local monodromy on the fundamental groups of algebraic curves along a degenerated stable curve.* Journ. Pure Appl. Algebra **103** (1995), no. 3, 235–283.

[9] K. Behrend and Yu. Manin, *Stacks of stable maps and Gromov-Witten invariants,* Duke Math. J. **85** (1996), no. 1, 1–60.

[10] P. Berthelot, *Altérations de variétés algébriques (d'après A. J. de Jong),* Astérisque No. 241 (1997), Exp. No. 815, 5, 273–311.

[11] E. Bierstone and P. D. Milman, *Canonical desingularization in characteristic zero by blowing up the maximum strata of a local invariant,* Invent. Math. **128** (1997), no. 2, 207–302.

[12] F. Bogomolov and T. Pantev, *Weak Hironaka theorem*, Math. Res. Lett. **3** (1996), no. 3, 299–307.

[13] J-L. Brylinski, *Propriétés de ramification à l'infini du groupe modulaire de Teichmüller.* Ann. Sci. Ecole Norm. Sup. (4) **12** (1979), 295–333.

[14] C. Chevalley, *Une démonstration d'un théorème sur les groups algébriques,* Journ. Math. Pures Appl. (9) **39** (1960), 307–317.

[15] D. Cox, *Toric varieties and toric resolutions,* this volume.

[16] P. Deligne, *Le lemme de Gabber.* In: Sém. sur les pinceaux arithmétiques: la conjecture de Mordell (Ed. L.. Szpiro). Astérisque **127** (1985), 131–150.

[17] P. Deligne and D. Mumford, *The irreducibility of the space of curves of given genus,* Inst. Hautes Études Sci. Publ. Math. No. **36** (1969), 75–109.

[18] D. Eisenbud and J. Harris, *Schemes: the language of modern algebraic geometry.* Wadsworth & Brooks/Cole Adv. Books Software, Pacific Grove, CA, 1992. Forthcoming edition as: *Why Schemes?*, Springer.

[19] S. Encinas and O. Villamayor, *A course on constructive desingularization and equivariance*, this volume.

[20] W. Fulton, *Introduction to toric varieties*, Annals of Math. Studies 131, Princeton Univ. Press, Princeton, NJ, 1993.

[21] B. van Geemen and F. Oort, *A compactification of a fine moduli spaces of curves,* this volume.

[22] D. Gieseker, *Lectures on moduli for curves.* Published for the Tata Institute of Fundamental Research, Bombay; Springer-Verlag, Berlin-New York, 1982.

[23] W. Gröbner, *Moderne algebraische Geometrie. Die idealtheoretischen Grundlagen,* Springer, Wien und Innsbruck, 1949.

[24] A. Grothendieck, *Fondements de la géométrie algébrique.* Extraits du Sém. Bourbaki, 1957-1962. Secrétariat Math., Paris, 1962.

[25] A. Grothendieck, *Revêtements étales et groupe fondamental*, Lecture Notes in Math., 224, Springer, Berlin, 1971.

[26] A. Grothendieck and J. P. Murre, *The tame fundamental group of a formal neighbourhood of a divisor with normal crossings on a scheme*, Lecture Notes in Math., 208, Springer, Berlin, 1971.

[27] J. Harris and D. Mumford, *On the Kodaira dimension of the moduli space of curves.* Invent. Math. **67** (1982), 43–70.

[28] J. Harris, *Algebraic Geometry – a first course.* Grad. texts in Math. **133**, Springer-Verlag, 1992.

[29] H. Hironaka, *Resolution of singularities of an algebraic variety over a field of characteristic zero: I, II,* Ann. of Math. (2) **79** (1964), 109–326.

[30] J.-I. Igusa, *Arithmetic variety of moduli for genus two.* Ann. Math. **72** (1960), 612–649.

[31] A. J. de Jong, *Families of curves and alterations*, Ann. Inst. Fourier (Grenoble) **47** (1997), no. 2, 599–621.

[32] A. J. de Jong and F. Oort, *On extending families of curves*, Journ. Algebr. Geom. **6** (1997), 545-562.

[33] K. Karu, *Semistable reduction in characteristic 0 for families of surfaces and threefolds,* preprint. alg-geom/9711020.

[34] K. Karu, *Minimal models and boundedness of stable varieties,* preprint. math.AG/9804049.

[35] K. Kato, *Logarithmic structures of Fontaine-Illusie*, in *Algebraic analysis, geometry, and number theory (Baltimore, MD, 1988)*, 191–224, Johns Hopkins Univ. Press, Baltimore, MD.

[36] K. Kato, *Toric singularities.* Amer. J. Math. **116** (1994), no. 5, 1073–1099.

[37] S. Keel and S. Mori, *Quotients by groupoids,* Ann. of Math. (2) **145** (1997), no. 1, 193–213.

[38] G. Kempf, F. Knudsen, D. Mumford and B. Saint-Donat, *Toroidal Embeddings I,* Lecture Notes in Math., 339, Springer, Berlin, 1973.

[39] F. F. Knudsen, *The projectivity of the moduli space of stable curves, II: the stacks $M_{g,n}$. III: The line bundles on $M_{g,n}$ and a proof of the projectivity of $\overline{M_{g,n}}$ in characteristic zero.* Math. Scand. **52** (1983). 161–199, 200–221.

[40] J. Kollár, *Projectivity of complete moduli,* J. Differential Geom. **32** (1990), no. 1, 235–268.

[41] S. Lichtenbaum, *Curves over discrete valuation rings.* Amer. Journ. Math. **90** (1968), 380–405.

[42] E. Looijenga, *Smooth Deligne-Mumford compactifications by means of Prym level structures.* Jour. Algebr. Geom. **3** (1994), 283–293.

[43] H. Matsumura, *Commutative ring theory,* Translated from the Japanese by M. Reid, Second edition, Cambridge Univ. Press, Cambridge, 1989; MR 90i:13001.

[44] S. Mochizuki, *Extending Families of Curves I, II.* RIMS preprints 1189 and 1188, 1998.

[45] M. Mostafa, *Die Singularitäten der Modulmannigfaltigkeiten $\overline{M}_g(n)$ der stabilen Kurven vom Geschlecht $g \geq 2$ mit n-Teilungspunktstruktur.* Journ. Reine Angew. Math., **343** (1983), 81–98.

[46] D. Mumford, *The boundary of moduli problems.* In: Lect. Notes Summ. Inst. Algebraic Geometry, Woods Hole 1964; 8 pp., *unpublished.*

[47] D. Mumford, *Lectures on curves on an algebraic surface*, Princeton Univ. Press, Princeton, N.J., 1966.

[48] D. Mumford, *Curves and their Jacobians*, The University of Michigan Press, Ann Arbor, Mich., 1975.

[49] F. Oort, *Good and stable reduction of abelian varieties.* Manuscr. Math. **11** (1974), 171–197.

[50] F. Oort, *Singularities of the moduli scheme for curves of genus three.*, Indag. Math. **37** (1975), 170–174.

[51] F. Oort, *Singularities of coarse moduli schemes.* In *Séminaire d'Algèbre Paul Dubreil, 29ème année (Paris,1975–1976)*, 61–76, Lecture Notes in Math., 586, Springer, Berlin, 1977.

[52] F. Oort, *The algebraic fundamental group.* In *Geometric Galois actions, 1*, 67–83, Cambridge Univ. Press, Cambridge 1997.

[53] K. Paranjape, *Bogomolov-Pantev resolution – An expository account.* To appear in: proceedings of the Warwick Algebraic Geometry Conference July/August 1996.

[54] M. Pikaart and A. J. de Jong, *Moduli of curves with non-abelian level structure.* In: The moduli space of curves (Ed. R. Dijkgraaf, C. Faber and G. van der Geer), Proc. 1994 Conference Texel, PM **129**, Birkhäuser, 1995, pp. 483–510.

[55] H. Popp, *Moduli theory and classification theory of algebraic varieties.* Lecture Notes in Math., 620, Springer, Berlin, 1977.

[56] H. Popp, *On the moduli of algebraic varietes III, Fine moduli spaces.* Compos. Math. **31** (1975), 237–258.

[57] H. Popp, *The singularities of moduli schemes of curves.* Journ. Number theory **1** (199), 90–107.

[58] H. E. Rauch, *The singularities of the modulus space.* Bull. Amer. Math. Soc. **68** (1962), 390–394.

[59] M. Raynaud, *Compactification du module des courbes.* In *Séminaire Bourbaki (23ème année, 1970/1971), Exp. No. 385*, 47–61. Lecture Notes in Math., 244, Springer, Berlin, 1971.

[60] M. Raynaud and L. Gruson, *Critères de platitude et de projectivité, Technique de "platification" d'un module,* Invent. Math. **13** (1971), 1–89.

[61] B. Riemann, *Theorie der Abel'schen Funktionen.* Journ. reine angew. Math. **54** (1857), pp. 115–155.

[62] P. C. Roberts, *Intersection theory and the homological conjectures in commutative algebra.* Proceed. ICM, Kyoto 1990. Math. Soc. Japan & Springer-Verlag, 1991; Vol. I, pp. 361–368.

[63] P. Samuel, *Invariants arithmétiques des courbes de genre 2* (d'après Jun Ichi Igusa), in *Séminaire Bourbaki, Vol. 7*, (1961/62) Exp. 228, 81–93, Soc. Math. France, Paris.

[64] M. Schlessinger, *Functors of Artin rings,* Trans. Amer. Math. Soc. **130** (1968), 208–222.

[65] J.-P. Serre, *Rigidité du foncteur de Jacobi d'echelon* $n \geq 3$. Appendix of Exp. 17 of Séminaire Cartan 1960/61.

[66] J.-P. Serre, *Algèbre locale – multiplicités.* Cours au Collège de France, 1957–1958, rédigé par Pierre Gabriel. Seconde édition, 1965. Lecture Notes in Math., 11, Springer, Berlin, 1965.

[67] J.-P. Serre and J. Tate, *Good reduction of abelian varieties.* Ann. of Math. (2) **88** (1968), 492–517. [Serre ŒII, 79.]

[68] G. Shimura, *On the field of rationality for an abelian variety.* Nagoya Math. Journ. **45**, (1971), 167–178.

[69] J. Silverman, *The arithmetic of elliptic curves.* Grad. Texts in Math. **106**, Springer-Verlag, 1986.

[70] O. Villamayor, *Constructiveness of Hironaka's resolution,* Ann. Sci. École Norm. Sup. (4) **22** (1989), no. 1, 1–32.

Department of Mathematics
Boston University
111 Cummington Street
Boston, MA 02215 USA
abrmovic@math.bu.edu
http://math.bu.edu/people/abrmovic

Mathematisch Instituut
University of Utrecht
Budapestlaan 6
NL-3508 TA Utrecht, The Netherlands
oort@math.uu.nl
http://www.math.uu.nl/staff/oort.html

Progress in Mathematics, Vol. 181, © 2000 Birkhäuser Verlag Basel/Switzerland

Reduction of Singularities for Differential Equations

José Manuel Aroca

1. Introduction

The reduction of the singularities of an algebraic or analytic variety has, at least in its local formulation, close relation with another problem, that of parametrizing the neighborhood of a point on the variety, i.e. solving, in some sense, the system of equations defining the variety.

Given a system (S) of algebraic or analytic equations, let V be the germ of the zero set of (S) at some point P (assume we are dealing with well chosen systems of equations, that is with a basis of the ideal of all functions which vanish on V). Then we have two possibilities, either the implicit function theorem is applicable to S or not. In the first situation the implicit function theorem gives us a solution of the system on a neighborhood of P, that means that there are local formal parametrizations of a representative of V around P.

In the second situation we are placed in a singular point of V. In this case the resolution of the singularities of V in a neighborhood of P gives us parametrizations of some special subsets of V (called wedges in [33]), that cover a neighborhood of P.

This equivalence is the basic idea of some classical proofs of the resolution of surface singularities (Jung [25], Walker [33]) inspired by the construction of local parametrizations by Black [3] and Hensel [20].

If we have a system of ordinary differential equations in the complex plane

$$f_i(x, y, y', \dots, y^{(r)}) = 0, \ 1 \leq i \leq n,$$

solving the system is equivalent to answering the problem of finding integral curves for the family of differential forms

$$\omega_j = z_j dx - dz_{j-1}, \ 1 \leq j \leq r, z_0 = y,$$

on the variety with equations

$$f_i(x, z_0, \dots, z_r) = 0, \ 1 \leq i \leq n,$$

and projecting these curves onto the (x, z_0)-plane.

In order to solve differential equations we need to study differential forms on complex analytic manifolds. It is very natural to consider the possibility that the formal apparatus of resolution of singularities will enable us to solve differential equations. The fact that there are more than formal analogies between the problems of solving differential equations and the reduction of singularities can be easily seen in the case of differential equations of first order and first degree.

Consider the equation

$$(*) \qquad a(x,y) + b(x,y)y' = 0,$$

with a, b convergent power series on some disk D centered at 0. From a geometric point of view, this equation is given by the differential form

$$\omega = a(x,y)d(x) + b(x,y)d(y),$$

which determines an analytic field of directions on the disk D, constructed by taking at each point (x_0, y_0) the straight line through (x_0, y_0) with the direction of $(-b(x_0, y_0), a(x_0, y_0))$. To solve the equation $(*)$ with the initial conditions (x_0, y_0) means to find an analytic curve $t \mapsto (x(t), y(t))$ such that $(x(0), y(0)) = (x_0, y_0)$ and

$$a(x(t), y(t)) \, \frac{d(x(t))}{dt} + b(x(t), y(t)) \, \frac{d(y(t))}{dt} = 0, \text{ for all } t.$$

That signifies that the curve is tangent at each point to the line determined at this point by ω. Picard's theorem says that for any point (α, β) of D, with $(a(\alpha,\beta), b(\alpha,\beta)) \neq (0,0)$, there is one and only one solution. This is the regular situation. The singular situation occurs when $a(\alpha,\beta) = b(\alpha,\beta) = 0$. In this case (α,β) will be called a singular point of the differential equation.

The map which assigns to each point (α,β) the vector $(-b(\alpha,\beta), a(\alpha,\beta))$ is continuous. Then, from a naive point of view, we may think that the annihilation of $(-b(\alpha,\beta), a(\alpha,\beta))$ at a singular point $P = (\alpha,\beta)$ is a consequence of the existence of several sequences of points, $\mathbf{s_i} = \{(\alpha_{i,j}, \beta)_{i,j})\}$, converging to P from different directions, and such that the limits (in j) of the directions defined by the vectors $(-b(\alpha_{i,j}, \beta)_{i,j}), a(\alpha_{i,j}, \beta)_{i,j})$ are different, because in this situation the vector $(-b(\alpha,\beta), a(\alpha,\beta))$ would have several directions and thus must be zero. Blowing up these singular points we separate the limit directions, simplifying in this way the local structure of the differential equation. This idea was introduced by Seidenberg [30] in order to prove the existence of a sort of reduction of singularities for differential forms in dimension 2. Seidenberg's theorem has been used by C. Camacho and P. Sad [7] to prove the existence of an analytic curve solution at each singular point.

For people working on singularities, there is another reason of interest in the problem of reduction of singularities for differential forms. In characteristic $p > 0$ the hypersurface singularity defined by the equation

$$z^{p^l} = f(x_1, \dots, x_n)$$

is especially interesting. The singularities of this hypersurface are the same as the singularities of the differential form

$$\omega = df = \sum_{i=1}^{n} \frac{\partial f}{\partial x_i} dx_i.$$

The behavior under blowing up of both objects is very similar. This is the basis of a program of J. Giraud [18], [19]: a form ω on K^2 has strong similarity with a curve, and helps us to solve the singularities of a surface, closely connected to a form in K^3, etc. The singularities of forms and vector fields in any characteristic in dimension two and three have been studied by F. Cano [11], [10], [8], [9] and V. Cossart [17].

F. Cano proves a reduction theorem for forms in $\mathbb{C}^3$. With this theorem, Cano-Cerveau [12] and Cano-Mattei [14] proved in the non-dicritical case the conjecture of Thom:

There is an integral hypersurface for any differential form in $\mathbb{C}^n$.

In the dicritical case, there is a counterexample due to Jouanolou [24]. The problem of solving singularities of differential forms or differential ideals, in dimension bigger than three, remains open. Our objective in this paper is the reduction of singularities of differential forms in the complex plane, and the application of the reduction of singularities to prove the existence of integral curves at the singularities of forms in $\mathbb{C}^2$. The references [13], [15] are used freely throughout this work.

2. Singular foliations (codimension one)

Let M be a complex analytic manifold of pure dimension n, let $T(M)$ (resp. $T^*(M)$) be the tangent (resp. cotangent) bundle of M. Let Θ_M (resp. Ω^1_M) be the sheaves of these bundles. For any section $X \in \Theta_M(U)$, or $\omega \in \Omega^1_M(U)$, and for $\mathbf{p} \in U$, we will write $X(\mathbf{p})$, $\omega(\mathbf{p})$ for the values at $\mathbf{p}$ of these sections. We will write $X_{\mathbf{p}}$, $\omega_{\mathbf{p}}$ for the germs at $\mathbf{p}$ of these sections.

Definition 2.1. A holomorphic foliation of codimension one on M is an analytic atlas $\mathcal{F} = \{U_i, \phi_i\}_{i\in I}$ compatible with the structure of M, such that:
(1) $\phi_i(U_i) = P_i \times D_i$ with P_i a polydisc in $\mathbb{C}^{n-1}$ and D_i a disc in $\mathbb{C}$, both centered at the origin.
(2) If $\mathbf{x_i} = (x_{i1}, \dots, x_{in-1})$ and y_i are respectively the coordinates on P_i and D_i, then for all i, j there are functions h_{ij}, g_{ij} such that

$$\phi_j\phi_i^{-1}(\mathbf{x_i}, y_i) = (h_{ij}(\mathbf{x_i}, y_i), g_{ij}(y_i)) \text{ for all } (x_i, y_i) \in \phi_i(U_i \cap U_j).$$

Remark 2.2. The sets $T_{i,\mathbf{p}} = \phi_i^{-1}(P_i \times \{\mathbf{p}\})$, $\mathbf{p} \in D_i$, are called plaques of the foliation. One can define an equivalence relation on M by

$$\mathbf{p} \sim \mathbf{q} \;\Leftrightarrow\; \begin{array}{l}\text{there are plaques } T_1, \dots, T_r \text{ such that}\\ T_i \cap T_{i+1} \neq \emptyset, \text{ if } 0 < i < r,\ \mathbf{p} \in T_1,\ \mathbf{q} \in T_r.\end{array}$$

Each equivalence class in this relation is called a leaf of the foliation. We will denote by $L_{\mathbf{p}}$ the leave that contains the point $\mathbf{p}$. The leaves are connected analytic manifolds of dimension $n-1$. For each $\mathbf{p} \in M$ there is a local chart (U, ϕ) of M, containing $\mathbf{p}$, contained in a chart of the foliation and a countable subset A of D such that $\phi(U) = P \times D$ and

$$\phi(U \cap L_{\mathbf{p}}) = \{(\mathbf{x}, y) \in U \mid y \in A\}.$$

Example 2.3. (1) If $f : M \to N$ is a holomorphic submersion such that $\dim M = \dim N + 1$, then the connected components of the fibers of f are the leaves of a holomorphic foliation on M.
(2) Let ω be a holomorphic 1-form on M not identically zero. Let us define singular locus of ω as $\mathrm{Sing}(\omega) = \{\mathbf{p} \in M \mid \omega(\mathbf{p}) = 0\}$. Then by the canonical pairing between $T_{\mathbf{p}}(M)$ and $T^*_{\mathbf{p}}(M)$, ω defines at each point $\mathbf{p} \in M \setminus \mathrm{Sing}(\omega)$ an $(n-1)$-dimensional linear subspace of $T_{\mathbf{p}}(M)$

$$\omega(\mathbf{p})^{\perp} = \{v \in T_{\mathbf{p}}(M) \mid \omega(\mathbf{p})(v) = 0\}.$$

We obtain in this way a distribution of $(n-1)$-planes on $M \setminus \mathrm{Sing}(\omega)$. By the classical Frobenius theorem, there is a codimension one foliation on $M \setminus \mathrm{Sing}(\omega)$ such that the leaf through each point is tangent to the distribution if and only if the distribution is involutive (closed for the Lie product). It is easy to prove that the involutivity condition is equivalent to the condition $\omega \wedge d(\omega) = 0$. The 1-forms that verify this condition will bc called integrable.

Remark 2.4. By the definition of foliations, a (regular) codimension one foliation $\mathcal{F}$ is locally defined as in the above example. In fact, with the notation of definition 2.1 and setting $z_i = y_i \circ \phi_i$, $\mathcal{F}|_{U_i}$ is the foliation defined on U_i by the integrable form dz_i. The chart change gives the relation between two local descriptions of $\mathcal{F}$. There is a holomorphic function h_{ij} on $U_i \cap U_j$ such that

$$dz_j = h'_{ij}(z_i)dz_i.$$

The derivative of h_{ij} does not vanish at any point of $U_i \cap U_j$. Thus, we may also define a codimension one holomorphic foliation as a family $\{(U_i, \omega_i)\}_{i \in I}$, such that:

1. The family $\{U_i\}_{i \in I}$ is an open covering of M.
2. Each ω_i is a holomorphic integrable 1-form on U_i.
3. For all $i, j \in I$ with $U_i \cap U_j \neq \emptyset$ there exists $h_{ij} \in \mathcal{O}^*_M(U_i \cap U_j)$ such that
$$\omega_i|_{U_i \cap U_j} = h_{ij}\omega_j|_{U_i \cap U_j}.$$
4. $\mathrm{Sing}(\omega_i) = \emptyset$ for all $i \in I$.

If we omit the last condition, in order to admit singularities, this definition can be formalized in the following way:

Definition 2.5. A singular holomorphic codimension one foliation is a locally free rank one submodule $\mathcal{F}$ of the cotangent sheaf Ω^1_M on a complex analytic manifold M, such that $\mathcal{F} \wedge d\mathcal{F} = 0$.

Remark 2.6. The definition means that $\mathcal{F}$ is locally generated by the pull-back, via local charts, of a differential 1-form

$$\omega = a_1(\mathbf{x})dx_1 + \cdots + a_n(\mathbf{x})dx_n, \quad a_i(\mathbf{x}) \in \mathbb{C}\{\mathbf{x}\},\ 1 \leq i \leq n,$$

verifying the integrability condition $\omega \wedge d(\omega) = 0$. Usually we will call these forms *local generators of* $\mathcal{F}$.

The singular locus of $\mathcal{F}$ is the closed analytic subset of M locally defined by:

$$\mathrm{Sing}(\mathcal{F}) = \{\mathbf{p} \in M \mid \omega(\mathbf{p}) = 0, \text{ for any generator } \omega \text{ of } \mathcal{F}\}.$$

It is obvious that a singular foliation gives rise to a (regular) foliation outside its singular locus.

For any foliation there is another equivalent (in some sense) foliation with *small* singular locus. To construct this foliation let us observe that the duality between tangent and cotangent bundles induces a pairing

$$\Theta_M \times \Omega^1_M \longrightarrow \mathcal{O}_M.$$

If we define $S^{\perp}$ as the orthogonal submodule of a submodule S with respect to this pairing, then $\mathcal{F}^{\perp\,\perp}$ is another foliation of M such that:

1. $\mathrm{Sing}(\mathcal{F}^{\perp\,\perp}) \subset \mathrm{Sing}(\mathcal{F})$.
2. $\mathcal{F}^{\perp\,\perp}|_{M\setminus\mathrm{Sing}(\mathcal{F})} = \mathcal{F}|_{M\setminus\mathrm{Sing}(\mathcal{F})}$.
3. The irreducible components of $\mathrm{Sing}(\mathcal{F}^{\perp\perp})$ have codimension ≥ 2.
4. The $\mathcal{O}_M$-module $\Omega^1_M/\mathcal{F}^{\perp\,\perp}$ is torsion free. That means that for any local generator ω of $\mathcal{F}^{\perp\,\perp}$, there are $a_1, \ldots, a_n \in \mathbb{C}\{\mathbf{x}\}$ such that

$$\omega = a_1(\mathbf{x})dx_1 + \cdots + a_n(\mathbf{x})dx_n \text{ and } g.c.d.\{a_1(\mathbf{x}), \ldots, a_n(\mathbf{x})\} = 1.$$

From now on, we will reserve the name foliation for singular holomorphic codimension one foliations $\mathcal{F}$ such that $\Omega^1_M/\mathcal{F}$ is a torsion-free $\mathcal{O}_M$-module. We will call prefoliations (following Cano-Cerveau [13]) the foliations without this last property. We call regular foliations the foliations $\mathcal{F}$ such that $\mathrm{Sing}(\mathcal{F}) = \emptyset$. A leaf of a foliation $\mathcal{F}$ will be a leaf of the regular foliation induced by $\mathcal{F}$ on $M \setminus \mathrm{Sing}(\mathcal{F})$. If we have a prefoliation, in order to get a foliation we can take the local generators and divide each one by the g.c.d. of its coefficients. By the division lemma of K. Saito [29] two differential forms ω and ω' induce the same foliation if and only if $\omega \wedge \omega' = 0$.

Sometimes it is convenient to extend the base ring to the sheaf of meromorphic functions on M. After this extension, a foliation is a one dimensional integrable vector subspace of the meromorphic cotangent sheaf, and can be locally generated by any of its nonzero sections in a local chart.

Remark 2.7. Foliations are the natural geometric realization of ordinary differential equations of first order and first degree. In fact, in dimension two a singular foliation is locally a germ of a 1-form

$$\omega = a(x,y)dx + b(x,y)dy, \quad a(x,y), b(x,y) \in \mathbb{C}\{x,y\}$$

which is equivalent to each of the following ordinary differential equations of first order and first degree

$$f(x,y,y') = a(x,y) + b(x,y)y', \quad g(x,y,x') = a(x,y)x' + b(x,y).$$

By duality a foliation is equivalent to a complex vector field, or more precisely, to a complex one dimensional distribution on M, generated by

$$X = -a(x,y)\frac{\partial}{\partial y} + b(x,y)\frac{\partial}{\partial x}.$$

In this dimension the integrability condition is automatically verified, and the absence of common factors of the coefficients is a condition to avoid solutions outside the differential situation.

Let $\mathcal{F}$ be a regular foliation on M. A non-singular analytic hypersurface H of M is called *invariant* by $\mathcal{F}$ if its connected components are leaves of the foliation.

Proposition 2.8. *If $f \in \mathcal{O}_M(M)$ verifies that $H = \{f = 0\}$ is a regular hypersurface, and if $\mathcal{F}$ is a regular foliation globally defined by $\omega \in \Omega^1_M(M)$, H is invariant by $\mathcal{F}$ if and only if there exists $\mu \in \Omega^2_M(M)$ such that $\omega \wedge df = f\mu$.*

Proof. Let $\mathbf{p} \in H$. By taking a chart of the foliation centered at $\mathbf{p}$, we may suppose that $\mathbf{p}$ lies in a plaque of local equation $y = 0$. Since $f = 0$ on the plaque, locally at $\mathbf{p}$ $f = h(\mathbf{x}, y)y^k$, with $h(\mathbf{0}, 0) \neq 0$ and $\omega = g(\mathbf{x}, y)dy$. Then

$$\omega \wedge df = gy^k \sum \frac{\partial h}{\partial x_i} dy \wedge dx_i = f\mu, \quad \mu = \frac{g}{h} \sum \frac{\partial h}{\partial x_i} dy \wedge dx_i.$$

The compatibility condition for μ follows from the fact that f is globally defined.

Now, if $\omega \wedge df = f\mu$ and L is an irreducible component of H, there are two possibilities. Either $df|_L \equiv 0$, and then by Saito's division lemma L is a leaf of the foliation, or $df|_L \not\equiv 0$. Since L is smooth at $\mathbf{p}$, for all $\mathbf{p} \in H$, its local equation is $y = 0$, thus $f(\mathbf{x}, y) = y^k h(\mathbf{x}, y)$, $h(\mathbf{0}, 0) \neq 0$ at $\mathbf{p}$. And locally at $\mathbf{p}$

$$\omega = \sum_{i=1}^{n-1} a_i dx_i + bdy, \quad \omega \wedge \frac{df}{f} = \omega \wedge \frac{dh}{h} + k \sum_{i=1}^{n-1} a_i dx_i \wedge \frac{dy}{y}.$$

As $\frac{df}{f}$ and $\frac{dh}{h}$ are holomorphic, y must divide a_i for all i, hence $y = 0$ is locally at $\mathbf{p}$ invariant by ω. □

Definition 2.9. Let $\mathcal{F}$ be a foliation; a function $f \in \mathcal{O}_M(M)$ is a first integral of $\mathcal{F}$ if $\omega \wedge df = 0$ for all local sections ω of $\mathcal{F}$ (i. e. the leaves of the foliation are the level curves of f given by f = constant).

An analytic hypersurface H of M is an integral hypersurface if for any point $\mathbf{p} \in M$, h being a local (reduced) equation of H, and ω a generator of $\mathcal{F}$ at $\mathbf{p}$, one has

$$\omega \wedge dh = h\mu, \quad \mu \in \Omega^2_{M,\mathbf{p}}.$$

Remark 2.10. A germ ω of a 1-form on M at $\mathbf{p}$ is also an element of the $\mathcal{O}_{M,\mathbf{p}}$-module $\Omega_{M,\mathbf{p}}$ of Kähler differentials of $\mathcal{O}_{M,\mathbf{p}}$, that is a submodule of the module $\widehat{\Omega}_{M,\mathbf{p}}$ of Kähler differentials of the completion $\widehat{\mathcal{O}_{M,\mathbf{p}}} \simeq \mathbb{C}[[x_1,\ldots,x_n]]$ of $\mathcal{O}_{M,\mathbf{p}}$.

Then a local (resp. formal) solution of ω is an element $f \in \mathcal{O}_{M,\mathbf{p}}$ (resp. $f \in \widehat{\mathcal{O}_{M,\mathbf{p}}}$) such that

$$\omega \wedge df = f\eta,\ \eta \in \Omega^2_{M,\mathbf{p}}\ (\text{resp. } \eta \in \widehat{\Omega}^2_{M,\mathbf{p}}).$$

We will call the hypersurface germ defined by a reduced local solution also *integral hypersurface*. In the same way, a formal reduced solution will be called a "formal integral hypersurface". A foliation may have formal non convergent integral hypersurfaces. For instance, the differential form $x^2dy+(-y+x)dx$ (Euler equation) has $x=0$ as an integral hypersurface and has also the formal integral hypersurface

$$y=\sum_{i=0}^{\infty} n!x^{n+1}.$$

The problem we want to solve is the existence of integral curves in the two-dimensional case. Our method will be first the reduction of the singularities to some which are as simple as possible by blowing up points, and then to show that for these singularities there are solutions. In dimension 3 there is a form constructed by Darboux:

$$\omega=(x^nz-y^{n+1})dx+(y^nx-z^{n+1})dy+(z^ny-x^{n+1})dz, n\geq 3,$$

that has no formal solutions (Jouanoulou [24]). But there is also a reduction method (F. Cano [10], [11]) which gives sufficient conditions for the existence of solutions in the general case [12], [14].

3. Reduction of singularities in dimension two

Our objective is to reduce by point blowups the singularities of any foliation in dimension two to simpler ones which are stable by blowup.

Remark 3.1. Let $f: N \longrightarrow M$ be a holomorphic map between analytic varieties of the same dimension and let $\mathcal{F}$ be a foliation of M. If $f^\star(\mathcal{F})$ is a locally free codimension one submodule of Ω^1_N, it is integrable. Thus it defines a prefoliation. We call the foliation $f^\star(\mathcal{F})^{\perp\perp}$ the strict transform of $\mathcal{F}$ by f. For simplicity, we will denote by $f^\star(\mathcal{F})$ the foliation $f^\star(\mathcal{F})^{\perp\perp}$.

If $\mathcal{F}$ is defined by the generators $\{(U_i,\omega_i)\}_{i\in I}$, $f^\star(\mathcal{F})$ is defined over each open set $f^{-1}(U_i)$ by the division of $f^\star\omega_i$ by the g.c.d. of its coefficients.

Remark 3.2. Let

$$Z=\{(\mathbf{x},[\mathbf{y}])\in\mathbb{C}^n\times\mathbb{P}^{n-1}_{\mathbb{C}} \mid \mathbf{x} \text{ and } \mathbf{y} \text{ linearly dependent}\},$$

and let $\pi: Z \longrightarrow \mathbb{C}^n$ be the first projection. The set Z is a complex analytic submanifold of $\mathbb{C}^n\times\mathbb{P}^{n-1}_{\mathbb{C}}$ of dimension n. The charts $\alpha_i:\mathbb{C}^n \longrightarrow Z$, defined by

$$\alpha_i(u_1,\ldots,u_n)=((u_1u_i,\ldots,u_i,\ldots,u_nu_i),[u_1,\ldots,1,\ldots,u_n]),$$

form an atlas of Z. Set $U_i = \mathrm{Im}\ \alpha_i$, $\pi_i = \pi|_{U_i}$. Remark that

$$\pi\alpha_i(u_1, \ldots, u_n) = (u_1 u_i, \ldots, u_i, \ldots, u_n u_i).$$

Then $\pi^{-1}(0) \simeq \mathbb{P}^{n-1}_{\mathbb{C}}$ is called the exceptional divisor of π, and $\pi : Z \setminus \{\pi^{-1}(0)\} \longrightarrow \mathbb{C}^n \setminus \{0\}$ is a biholomorphic morphism. The map π is called the blowup of $\mathbb{C}^n$ with center 0. If M is some complex analytic manifold and $\mathbf{p} \in M$ one can construct the blowup $\pi : \tilde{M} \longrightarrow M$ with center $\mathbf{p}$ modifying a local chart at $\mathbf{p}$.

If $\mathcal{F}$ is a foliation on M and $\pi : \tilde{M} \longrightarrow M$ is the blowup of M with center $\mathbf{p}$, one can construct $\pi^*(\mathcal{F})$ easily. Outside $\pi^{-1}(\mathbf{p})$, π is biholomorphic, thus $\mathcal{F}$ does not change. Let

$$\omega = \sum_{i=1}^{n} a_i(\mathbf{x}) dx_i$$

be the germ of a holomorphic 1-form that generates $\mathcal{F}$. If $\mathbf{q} \in \pi^{-1}(\mathbf{p})$ and $\mathbf{q}$ is, for instance, the origin in the local chart U_1, $\pi_1^*(\omega)$ is given by

$$\begin{aligned} \pi_1^*(\omega) &= [a_1(u_1, u_1u_2, \ldots, u_1u_n) + \sum_{i=2}^{n} u_i a_i(u_1, u_1u_2, \ldots, u_1u_n)]du_1 + \\ & \quad + u_1 \sum_{i=2}^{n} a_i(u_1, u_1u_2, \ldots, u_1u_n)du_i. \end{aligned}$$

We have similar equations for any $i \in \{1, \ldots, n\}$. Then if the multiplicity of a_i is ν_i and the initial form of a_i is $\mathrm{In}(a_i)$ we have

$$a_i(u_1, u_1u_2, \ldots, u_1u_n) = u_1^{\nu_i}\mathrm{In}(a_i)(1, u_2, \ldots, u_n) + u_1^{\nu_i+1} a_i'(u_1, \ldots, u_n).$$

Let us call $\nu = \min\{\nu_1, \ldots, \nu_n\}$ and let $\mathrm{In}_\nu(a_i)$ be the homogeneous component of a_i of degree ν, then $\mathrm{In}(a_i) = 0$ if $\nu_i \neq \nu$ and $\mathrm{In}_{\nu_i}(a_i) = \mathrm{In}(a_i)$ if $\nu = \nu_i$. With this notation, it is possible to write

$$\pi_1^*(\omega) = u_1^\nu[P(1, u_2, \ldots, u_n)du_1 + u_1 \sum_{i=2}^{n} \mathrm{In}_\nu(a_i)(1, u_2, \ldots, u_n)du_i] + u_1^{\nu+1}\omega'$$

with $P(x_1, \ldots, x_n) = \sum x_i \mathrm{In}_\nu(a_i)$. There are two possibilities:

(1) $P \not\equiv 0$. In this case the strict transform $\pi_1^*(\mathcal{F})$ of $\mathcal{F}$ by π is defined by $\pi_1^*(\omega){u_1}^{-\nu}$ on a neighborhood of $\mathbf{q}$. This is the so called non-dicritical case. In this case, the exceptional divisor, given in this chart by $u_1 = 0$, is a solution. The new singularities, placed all over the exceptional divisor, are given by the homogeneous equation $P(u_1, u_2, \ldots, u_n) = 0$. Let us set $\nu_{\mathbf{p}}(\mathcal{F}) = \nu_{\mathbf{p}}(\omega) = \min\{\nu(a_i)\}$, where $\omega = \sum a_i(\mathbf{x})dx_i$ is the expression of ω in a local chart centered at $\mathbf{p}$. In the non-dicritical case it can happen that $\nu_{\mathbf{q}}(\pi^*(\mathcal{F})) > \nu_{\mathbf{p}}(\mathcal{F})$. For instance, if $\omega = dy$ in $\mathbb{C}^2$ then, in the chart U_1, $\pi_1^*(\omega) = d(uv) = udv + vdu$ and $\tilde{\omega} = \pi_1^*(\omega)$ has a singularity at 0 with $\nu_{(0)}(\tilde{\omega}) = 1 > 0 = \nu_{(0)}(\omega)$.

(2) $P \equiv 0$. This is the so-called dicritical case. In this case the strict transform of $\mathcal{F}$ is defined by $\pi_1^*(\omega)u_1^{-\nu-1}$; the exceptional divisor is not an integral hypersurface

of $\pi^\star(\mathcal{F})$. In the non singular points of $\pi^\star(\mathcal{F})$ it is transversal to the leaves of the strict transform of $\mathcal{F}$. The singularities of $\pi^\star(\mathcal{F})$ are given by

$$\mathrm{In}_\nu(a_i) = 0,\ i \geq 2, \quad \mathrm{In}_{\nu+1}(\sum x_i a_i) = 0.$$

In this case, the multiplicity does not increase, and there appears an infinite set of solutions. For instance, $\omega = xdy - ydx$ is dicritical and in the chart U_1, $\pi_1^\star(\omega) = u^2 dv$. Then $\tilde{\omega} = \pi_1^\star(\omega)u^{-2} = du$, with solutions $u =$ constant along all the points of the exceptional divisor.

Remark 3.3. In dimension two and locally at a point $\mathbf{p}$, a foliation gives a germ of a vector field, that is, a derivation X of $\mathcal{O}_{M,\mathbf{p}}$. If $\mathbf{p}$ is a singular point, $X(m_\mathbf{p}) \subset m_\mathbf{p}$, and X induces a $\mathbb{C}$-linear map

$$L_{X,\mathbf{p}} : m_\mathbf{p}/m_\mathbf{p}^2 \longrightarrow m_\mathbf{p}/m_\mathbf{p}^2.$$

In a local chart centered at $\mathbf{p}$, a local generator of $\mathcal{F}$ is a differential form

$$\omega = a(x,y)dx + b(x,y)dy, \text{ with } a(0) = b(0) = 0.$$

Hence

$$X = -b(x,y)\frac{\partial}{\partial x} + a(x,y)\frac{\partial}{\partial y}$$

and $L_{X,\mathbf{p}}$, in the basis $\{x + m_\mathbf{p}^2, y + m_\mathbf{p}^2\}$, has the matrix

$$J_{\mathcal{F},(x,y)} = \begin{pmatrix} -\partial_x b(0) & \partial_x a(0) \\ \partial_y b(0) & \partial_y a(0) \end{pmatrix}.$$

A change of coordinates, the linear map being intrinsic, gives rise to a similar matrix, but a change of the generator ω of $\mathcal{F}$ may cause the multiplication of the original matrix by some element of $\mathbb{C}^*$. Then the eigenvalues of $J_{\mathcal{F},(x,y)}$ are not invariants of $\mathcal{F}$. Their quotients $\{\lambda/\mu, \mu/\lambda\}$, when defined, are invariants of $\mathcal{F}$.

Definition 3.4. A singular point $\mathbf{p}$ for $\mathcal{F}$ is called presimple, if $L_{X,\mathbf{p}}$ is not nilpotent. A singular point $\mathbf{p}$ for $\mathcal{F}$ is called simple, if $L_{X,\mathbf{p}}$ has eigenvalues λ, μ such that one of them, for instance μ, is $\neq 0$ and $\lambda/\mu \notin \mathbb{Q}_+$.

Remark 3.5. If $\mathbf{p}$ is a simple singularity, it is possible to perform a linear change of coordinates in such a way that $J_{\mathcal{F},(x,y)}$ becomes diagonal. Then, in the new coordinate system

$$\omega = \lambda y dx - \mu x dy + a(x,y)dx + b(x,y)dy, \text{ with } \nu(a) \geq 2, \nu(b) \geq 2.$$

Now, by blowing up $\mathbf{p}$, the singularity being non-dicritical, the exceptional divisor is an integral curve, and the strict transform of $\mathcal{F}$ is generated, in the first chart, by

$$\pi_1^\star(\omega) = (\lambda - \mu)vdu - \mu udv + \delta udu + \text{higher degree terms.}$$

The only singular point is the origin, and the matrix of $L_{X,0}$ is

$$\begin{pmatrix} \mu & \delta \\ 0 & \lambda - \mu \end{pmatrix},$$

with quotient of eigenvalues

$$\{\frac{\mu}{\lambda-\mu}, \frac{\lambda-\mu}{\mu}\} = \{\frac{1}{a-1}, a-1\},$$

where $a = \lambda/\mu$.

In the chart U_2, the situation is the same, with a singularity at the origin and with the quotients of the eigenvalues being $\{1/(b-1), b-1\}$, where $b = \mu/\lambda$, or $\{a/(1-a), (1-a)/a\}$.

Assume that $\mu \neq 0$ and $\mu \neq \lambda$ ($a \notin \mathbb{Q}_+$), $a - 1 \notin \mathbb{Q}_+$, $a/(a-1) \notin \mathbb{Q}_+$, then by blowup, each simple singularity gives rise to exactly two simple singularities. Therefore the simple singularities are stable by blowup.

Remark 3.6. A divisor with normal crossings on an analytic manifold M is an analytic subset E of M such that, for any $\mathbf{p}$ in M, the ideal of E at $\mathbf{p}$ is generated by $\prod_{i\in A} x_i$, where $(x_1, \dots, x_n)$ is a suitable coordinate system at $\mathbf{p}$ and $A \subset \{1, \dots, n\}$.

If $\mathcal{F}$ is a foliation on M and E a divisor with normal crossing, we will say that an irreducible component H of E is dicritical if H is not an integral hypersurface of $\mathcal{F}$. Assume that $\mathbf{p} \notin \text{Sing}(\mathcal{F})$. Let L be the leaf of $\mathcal{F}$ at $\mathbf{p}$. We will say that $\mathcal{F}$ has a normal crossing with E at $\mathbf{p}$ if and only if $L \cup E$ has normal crossings at $\mathbf{p}$. We define the adapted singular locus of $\mathcal{F}$ w.r.t. E as

$$\text{Sing}(\mathcal{F}, E) = \text{Sing}(\mathcal{F}) \cup \{\mathbf{p} \in M \mid \mathcal{F} \text{ and } E \text{ have no normal crossing at } \mathbf{p}\}.$$

The set $\text{Sing}(\mathcal{F}, E)$ is an analytic subset of M of codimension ≥ 2. A point $\mathbf{p} \in \text{Sing}(\mathcal{F}, E)$ will be called a simple singularity of $\mathcal{F}$ adapted to E if $\mathbf{p}$ is a simple singularity of $\mathcal{F}$; $\mathbf{p} \in E$ and the irreducible components of E through $\mathbf{p}$ are non-dicritical.

Now we are going to prove the theorem of reduction of singularities of foliations in dimension two. This theorem was first proved by Seidenberg [30], and the proof that we will give here is the one of Cano-Cerveau [13].

Theorem 3.7. Reduction theorem. *Let $\mathcal{F}$ be a germ of a foliation on $(\mathbb{C}^2, 0)$. There exist a neighborhood U of 0 in $\mathbb{C}^2$ and a holomorphic map $\pi : Z \longrightarrow U$, such that:*

1. *π is composition of blowups $\pi = \pi_n \cdots \pi_1$, where π_1 is the blowup of $0 = \mathbf{p}_0$, and π_i the blowup with center $\mathbf{p}_i$ for some $\mathbf{p}_i \in \pi_i^{-1}(\mathbf{p}_{i-1})$.*
2. *If we denote $\tilde{\mathcal{F}} = \pi^\star\mathcal{F}$, $\tilde{E} = \pi^{-1}(0)$, then any point $\mathbf{p} \in Sing(\tilde{\mathcal{F}}, \tilde{E})$ is a simple singularity of $\tilde{\mathcal{F}}$ adapted to $\tilde{E}$.*

Remark 3.8. The proof will be divided into three steps. Each step is based on the construction of a system of invariants adapted to that step. First we will show how to reduce all the singularities of $\mathcal{F}$ to presimple ones. In the second step we will reduce the singularities to simple ones. The third step allows us the adaptation of the singularities to the exceptional divisor by some extra blowups, possibly with non singular centers.

3.1. Local invariants and blowup

Let $\mathcal{F}$ be a foliation on M (dim M) = 2), $\mathbf{p} \in M$, and set

$$\nu_{\mathbf{p}}(\mathcal{F}) = \max\{r \mid \text{ there exists a generator } \omega \in m_{\mathbf{p}}^r \cdot \Omega^1_{M,\mathbf{p}} \text{ of } \mathcal{F} \text{ at } \mathbf{p}\}.$$

If, in a local chart, $\omega = a(x,y)dx + b(x,y)dy$, then

$$\nu_{\mathbf{p}}(\mathcal{F}) = \min\{\nu(a(x,y)), \nu(b(x,y))\}.$$

The integer $\nu_{\mathbf{p}}(\mathcal{F})$ is called the order, or the multiplicity, of $\mathcal{F}$ at $\mathbf{p}$. Set $\nu_{\mathbf{p}}(\mathcal{F}) = \nu$. The subset $C(\mathcal{F}, \mathbf{p})$ of the exceptional divisor $\pi^{-1}(\mathbf{p})$ defined by the polynomial P of Remark 3.2 is intrinsically defined. This subset of $\mathbb{C}^2$ is defined by the equation

$$xa_\nu(x,y) + yb_\nu(x,y) = 0.$$

It will be called the tangent cone to $\mathcal{F}$ at $\mathbf{p}$. It can be either a finite set of straight lines or all $\mathbb{C}^2$. As we saw before, the projectivization of this cone is equal to $\pi^{-1}(\mathbf{p}) \cap \operatorname{Sing}(\pi^\star \mathcal{F})$.

Definition 3.9. If $\omega = a(x,y)dx + b(x,y)dy$ is a local generator of $\mathcal{F}$ at $\mathbf{p}$,

$$\mu_{\mathbf{p}}(\mathcal{F}) = \dim_{\mathbb{C}} \mathbb{C}\{x,y\}/a\mathbb{C}\{x,y\} + b\mathbb{C}\{x,y\}$$

is called the Milnor number of $\mathcal{F}$ at $\mathbf{p}$.

Remark 3.10. The conditions, $\nu_{\mathbf{p}}(\mathcal{F}) = 0$, $\mu_{\mathbf{p}}(\mathcal{F}) = 0$ and $\mathcal{F}$ non-singular at $\mathbf{p}$, are equivalent. Moreover,

$$\mu_{\mathbf{p}}(\mathcal{F}) = 1 \Rightarrow \mathbf{p} \text{ is presimple} \Rightarrow \nu_{\mathbf{p}}(\mathcal{F}) = 1.$$

There are singularities with order 1 that are not presimple (e.g. $x^2dx + ydy$ at 0). There are presimple singularities whose Milnor number is greater than 1, (e.g. $xdy + y^2dx$ at 0). If a and b have no common factors, $\mu_{\mathbf{p}}(\mathcal{F}) < \infty$.

Let us observe that $\mu_{\mathbf{p}}(\mathcal{F})$, which is independent of the local chart used in its construction, is the intersection number of $\{a(x,y) = 0\}$ and $\{b(x,y) = 0\}$. Then its properties are well known; we recall Noether's formula, which gives the relation between the intersection number of two curves and the intersection number of their strict transforms by blowup:

$$\sharp(a,b)_{\mathbf{p}} = \nu_{\mathbf{p}}(a)\nu_{\mathbf{p}}(b) + \sum_{\mathbf{q} \in \pi^{-1}(\mathbf{p})} \sharp(\tilde{a}_{\mathbf{q}}, \tilde{b_{\mathbf{q}}})_{\mathbf{q}},$$

where $\tilde{a}_{\mathbf{q}}$, $\tilde{b_{\mathbf{q}}}$ are the strict transforms at $\mathbf{q}$ of a and b and $\sharp(a,b)_{\mathbf{p}}$ is the intersection number of a and b.

Proposition 3.11. *Let* $\pi : \tilde{M} \longrightarrow M$ *be a blowup with center* $\mathbf{p}$*; let* $\tilde{\mathcal{F}} = \pi^\star \mathcal{F}$, $E = \pi^{-1}(0)$, $\nu = \nu_{\mathbf{p}}(\mathcal{F})$. *Then*
(1) *If* π *is not dicritical*

$$\mu_{\mathbf{p}}(\mathcal{F}) = \nu^2 - (\nu + 1) + \sum_{\mathbf{q} \in E} \mu_{\mathbf{q}}(\tilde{\mathcal{F}}).$$

(2) *If π is dicritical*

$$\mu_{\mathbf{p}}(\mathcal{F}) = (\nu+1)^2 - (\nu+2) + \sum_{\mathbf{q}\in E} \mu_{\mathbf{q}}(\tilde{\mathcal{F}}).$$

Proof. We may choose the coordinates in such a way that, for a local generator $\omega = a(x,y)dx + b(x,y)dy$

1. $\nu = \nu_{\mathbf{p}}(a) = \nu_{\mathbf{p}}(b)$.
2. 0 in U_2 is non singular for $\tilde{\mathcal{F}}$.
3. x^2 does not divide $b_\nu(x,y)$ in the dicritical case

The proposition follows from Noether's formula. □

Remark 3.12. Let us suppose that all the irreducible components of E through $\mathbf{p}$ are non-dicritical, and let F be one of them. We will define the residual order of $\mathcal{F}$ adapted to E as

$$\rho(\mathcal{F}, E, F; \mathbf{p}) = \nu(f(y)).$$

Here $f(y)$ is constructed as follows. We choose a local chart (x,y) at $\mathbf{p}$, in such a way that F is defined by the equation $x=0$, and E by the equation $xy^\varepsilon = 0$ with $\varepsilon \in \{0,1\}$. Then

$$\frac{\omega}{xy^\varepsilon} = f(y)\frac{dx}{x} + g(x)\frac{dy}{y} + \mu, \quad \text{for some } \mu \in \Omega^1_{\mathbb{C}^2}.$$

If $\mathbf{p}$ is regular, $\rho(\mathcal{F}, E, F; \mathbf{p}) = 0$. If $\mathbf{p}$ is singular and $\rho(\mathcal{F}, E, F; \mathbf{p}) = 0$, then $\mathbf{p}$ is presimple. One can prove easily that

$$\tilde{\mathcal{F}} = \pi^\star \mathcal{F}, \quad \tilde{E} = \pi^{-1}(E), \quad E' = \pi^{-1}(\mathbf{p}), \quad \nu = \nu_{\mathbf{p}}(\mathcal{F}).$$

If F' is the strict transform of F and $\{\mathbf{q}\} = F' \cap E'$, then

$$\begin{aligned} \nu + 1 &= \sum_{\mathbf{p}'\in E'} \rho(\mathcal{F}', E', E'; \mathbf{p}'), \\ \rho(\mathcal{F}, E, F; \mathbf{p}) &= \rho(\tilde{\mathcal{F}}, \tilde{E}, F'; \mathbf{q}) + \nu. \end{aligned}$$

Remark 3.13. Let Γ be a nonsingular curve through $\mathbf{p}$. Let us choose a local chart at $\mathbf{p}$ such that Γ is defined by $y=0$; if a local generator of $\mathcal{F}$ in this chart is $\omega = a(x,y)dx + b(x,y)dy$ the number

$$t(\mathcal{F}, \Gamma; \mathbf{p}) = \nu(a(x,0))$$

is independent of the choice of the chart (always with $y=0$ being the equation of Γ) and is called the restricted order of $\mathcal{F}$ at $\mathbf{p}$. This number is infinity if and only if Γ is an integral curve of $\mathcal{F}$.

If $\tilde{\Gamma}$ is the strict transform of Γ by a blowup π, and $\mathbf{p}' = \pi^{-1}(\mathbf{p}) \cap \tilde{\Gamma}$, it is easy to prove that

$$\begin{aligned} t(\tilde{\mathcal{F}}, \tilde{\Gamma}; \mathbf{p}') &= t - \nu, \text{ if } \pi \text{ is non dicritical}, \\ t(\tilde{\mathcal{F}}, \tilde{\Gamma}; \mathbf{p}') &= t - \nu - 1, \text{ if } \pi \text{ is dicritical}. \end{aligned}$$

3.2. First step: reduction to presimple singularities

Lemma 3.14. *Let $\mathcal{F} = \mathcal{F}_0$ be a foliation germ on $(\mathbb{C}^2, 0)$, let $\Gamma = \Gamma_0$ be a regular algebroid curve (i.e. a regular formal curve) at 0, and let*

$$U_0 \overset{\pi_1}{\leftarrow} U_1 \overset{\pi_2}{\leftarrow} U_2 \cdots$$

be an infinite sequence of blowups defined in the following way:

1. *U_0 is a representative of $(\mathbb{C}^2, 0)$ contained in the domain of $\mathcal{F}_0$.*
2. *π_1 is the blowup with center $\mathbf{p}_0 = 0$.*
3. *π_{i+1} is the blowup with center $\mathbf{p}_i = \Gamma_i \cap \pi_i^{-1}(\mathbf{p}_{i-1})$ of U_i.*
4. *Γ_i is the strict transform of Γ_{i-1} by π_i.*

Then, if $\mathcal{F}_i$ is the strict transform of $\mathcal{F}_{i-1}$ by π_i, and $\mathbf{p}_i \in Sing(\mathcal{F}_i)$, Γ is an integral curve for $\mathcal{F}$. Moreover if all the π_i's are non dicritical, there is an r such that $\mathbf{p}_r$ is presimple.

Proof. If Γ is not an integral curve of $\mathcal{F}$, $t(\mathcal{F}, \Gamma; 0) < \infty$. Since

$$t(\mathcal{F}_i, \Gamma_i; \mathbf{p}_i) < t(\mathcal{F}_{i-1}, \Gamma_{i-1}; \mathbf{p}_{i-1}), \text{ for all } i \geq 1,$$

we get a contradiction. For the second statement, π_1 being non-dicritical, Γ_1 and $\pi_1^{-1}(0)$ are transversal integral curves of Γ. Then in some local chart (possibly formal), $\mathcal{F}_1$ has a generator

$$\omega_1 = x_1 y_1 (a_1(x_1, y_1)\frac{dx_1}{x_1} + b_1(x_1, y_1)\frac{dy_1}{y_1}),$$

where $\{x_1 = 0\} = \pi_1^{-1}(0)$, $\{y_1 = 0\} = \Gamma_1$ and y_1 does not divide b_1 because π_1 is non-dicritical. If $\nu = \nu(\mathcal{F}_1) = 1$, $\mathbf{p}_1$ is presimple, and, if $\nu > 1$, $\mathcal{F}_2$ is generated by

$$\omega_2 = x_2 y_2 (a_2(x_2, y_2)\frac{dx_2}{x_2} + b_2(x_2, y_2)\frac{dy_2}{y_2}),$$

with

$$b_2 = \frac{b_1(x_2, x_2 y_2)}{x_2^{\nu-1}}.$$

Now, we begin again with $\mathcal{F}_2$. As y_1 does not divide b_1 we can iterate this only a finite number of times. We conclude that in some step $\nu(\mathcal{F}_r) = 1$ and $\mathbf{p}_r$ is presimple. □

Remark 3.15. Let us consider a collection (S) of data, constructed as follows

(1) $U_0 \overset{\pi_1}{\leftarrow} U_1 \overset{\pi_2}{\leftarrow} U_2 \cdots$ is a sequence of blowups with center $\mathbf{p}_i \in U_i$; U_0 is an open neighborhood of 0 in $\mathbb{C}^2$, $\mathbf{p}_0 = 0$ and $\pi_i(\mathbf{p}_i) = \mathbf{p}_{i-1}$.

(2) $\mathcal{F} = \mathcal{F}_0$ is a foliation in U_0 with $\mathrm{Sing}(\mathcal{F}_0) = \{0\}$ and $\mathcal{F}_i$ is the strict transform of $\mathcal{F}_{i-1}$ by π_i, assuming $\mathbf{p}_i \in \mathrm{Sing}(\mathcal{F}_i)$ for all i.

(3) $E_i = (\pi_1 \cdots \pi_i)^{-1}(0)$, E_i' is the union of all the non-dicritical components of E_i.

Let $e(E_i', \mathbf{p}_i)$ denote the number of irreducible components of E_i' that contain $\mathbf{p}_i$. If $e(E_i', \mathbf{p}_i) = 2$, there is a system of local coordinates (x, y) on a neighborhood of

$\mathbf{p}_i$ such that $x(\mathbf{p}_i) = y(\mathbf{p}_i) = 0$, E_i' equals $\{xy = 0\}$ and $\nu_{\mathbf{p}_i}(\mathcal{F}_i) = 1$. Therefore $\mathcal{F}_i$ has a generator

$$(*) \qquad \omega = xy(\frac{dx}{x} + b(x,y)\frac{dy}{y}).$$

Hence $\mathbf{p}_i$ is a presimple singularity of $\mathcal{F}_i$. If π_{i+1} is dicritical, there are two integral non singular curves Γ_1 and Γ_2 which are transversal to $\pi_{i+1}^{-1}(\mathbf{p}_i)$ in two different points. We may choose a coordinate system at $\mathbf{p}_i$ such that $\pi_{i+1}(\Gamma_1 \cup \Gamma_2)$ has the equation $xy = 0$. In this system, there is a local generator like $(*)$ and if $\mu_{\mathbf{p}_i}(\mathcal{F}_i) = 1$, $\mathbf{p}_i$ is presimple.

Remark 3.16. Let us set $I_i = \mu_{\mathbf{p}-i}(\mathcal{F}_i) - e(E_i', \mathbf{p}_i)$. If $\mathbf{p}_i$ is not presimple and $\nu_{\mathbf{p}_i}(\mathcal{F}_i) \geq 2$, the formulas which give the behavior of $\mu_{\mathbf{p}}(\mathcal{F})$ imply $I_{i+1} < I_i$. If $\nu_{\mathbf{p}_i}(\mathcal{F}_i) = 1$, π_{i+1} is dicritical and $\mathbf{p}_i$ is not presimple, then π_{i+1} must be non-dicritical. Using the formula for $\mu_{\mathbf{p}}(\mathcal{F})$ and looking at the four cases:

$$e_i = 0, e_{i+1} = 1; e_i = 1, e_{i+1} = 1; e_i = 1, e_{i+1} = 2; e_i = 2, e_{i+1} = 2$$

one obtains also $I_{i+1} < I_i$.

Theorem 3.17. *Given (S) as in 3.15, there is an index r such that $\mathbf{p}_r$ is presimple with respect to $\mathcal{F}_r$.*

Proof. (1) If there are infinitely many dicritical π_i's, we get a contradiction, because $\{I_i\}$ cannot decrease infinitely. Then, by supressing the dicritical blowup, we may suppose that all of them are non-dicritical.
(2) Since $I_0 < \infty$, after some steps, $I_i = I_r$, for all $i \geq r$; we may forget the first r blowups and suppose $I_i = I_0$ for all i. Then $e(E_i', \mathbf{p}_i) = 1$ for all i, and the π_i's being non-dicritical, the chain of blowups gives a formal non singular branch Γ, that must be an integral curve. Hence there is an i such that $\mathbf{p}_i$ is presimple. □

3.3. Adapted reduction

Now, using theorem 3.17, it suffices to consider the case of a presimple singularity of a foliation on $(\mathbb{C}^2, 0)$.

Proposition 3.18. *If (S) is a collection of data as in Remark 3.15, and 0 is presimple, there is an index i such that Sing$(\mathcal{F}_i)$ consists only of simple singularities.*

Proof. Let $\mathbf{p}$ be a presimple singularity of $\mathcal{F}$. Consider the invariant of resonance given by

$$\mathrm{Res}(\mathcal{F}, \mathbf{p}) = \begin{cases} 0 & \text{if } \alpha \notin \mathbb{Q}_+ \\ p+q & \text{if } \alpha = \frac{p}{q} \in \mathbb{Q}_+, \text{ g.c.d.}(p,q) = 1, \end{cases}$$

where $\alpha = \lambda/\mu$ with λ, μ ($\mu \neq 0$) the eigenvalues of $L_{X,\mathbf{p}}$. By the formula about the behavior of α by blowup, we know that $\mathrm{Res}(\mathcal{F}, \mathbf{p})$ decreases after blowup. After a finite number of steps, all the singular points of $\mathcal{F}_i$ are simple. □

Remark 3.19. In order to adapt the situation to a convenient divisor, we need only to separate at some simple singularity $\mathbf{p}$ the components of E' through $\mathbf{p}$ and the formal integral curves of $\mathcal{F}$ through $\mathbf{p}$, in order to have normal crossing. The possibility to do this is a well known result of the theory of curves.

This completes the proof of the theorem of reduction. □

4. Existence of integral curves

This section is dedicated to J. Cano's [15] proof of Camacho-Sad's theorem [7]: *Any foliation of* $(\mathbb{C}^2, 0)$ *has a convergent integral curve.*

Remark 4.1. Cano-Cerveau's proof of Briot-Bouquet's theorem. Let $\mathcal{F}$ be a foliation on $(\mathbb{C}^2, 0)$ with a non singular convergent integral curve E through 0, and let 0 be a simple singularity of $\mathcal{F}$. Then, as we saw before, there is a local chart such that the generator of $\mathcal{F}$ is

$$\omega = (\lambda y + a(x,y))dx - (\mu x + xb(x,y))dy,$$

with $\nu(a(x,y)) \geq 2$, $\nu(b(x,y)) \geq 1$. We want to know if the formal regular branch, Γ parametrized by

$$\gamma(z) = (z, \sum_{n\geq 2} t_n z^n)$$

is an integral curve. That means that there are polynomials $P_n(X_2, \ldots, X_{n-1}), n \geq 2$, depending only on ω such that

$$(*) \qquad 0 = \gamma^*\omega \Leftrightarrow (\lambda - \mu\nu)t_n = P_n(t_2, \ldots, t_{n-1}), n \geq 2.$$

Now, since 0 is simple, $\lambda - \mu\nu \neq 0$, and there is only one solution of $(*)$. After a (formal) change of coordinates, we may suppose that Γ is given by $y = 0$, then

$$(**) \qquad \omega = xy[(\lambda + \alpha(x,y))\frac{dx}{x} - (\mu + \beta(x,y))\frac{dy}{y}] \text{ with } \alpha(0) = \beta(0) = 0.$$

The same system as before proves that there is no Puiseux solution of $\mathcal{F}$. Then:

There is one and only one formal integral curve Γ *of* $\mathcal{F}$ *at* 0 *different from* E, *and this curve is non-singular, transversal to* E, *and with tangent direction generated by an eigenvector of* $L_{X,\mathbf{p}}$.

It only remains to prove that if the eigenvalue corresponding to the tangent direction to Γ is non zero, then Γ is a convergent curve. To do that we may write for $\mu \neq 0$, since the coefficient of xdy in $(**)$ is irreducible,

$$\omega = (\alpha y - \sum_{i+j\geq 2} c_{ij}x^i y^j)dx - dy, c = \inf\{\mid \alpha - \mu \mid\}.$$

If the solution of

$$cv = \sum_{i+j\geq 2} \mid c_{ij} \mid u^i v^j$$

is $v = \sum_{n\geq 2} v_n u^n$, then $\mid c_n \mid\leq v_n$, for all $n \geq 2$. And $\sum v_n u^n$ being convergent, Γ is also convergent.

Then, if $\mathcal{F}$ is a foliation germ on $(\mathbb{C}^2, 0)$ with 0 as a simple singularity, $\mathcal{F}$ has exactly two formal integral curves, both are non singular and transversal, and their tangent directions are given by the eigenvectors of $L_{X,\mathbf{p}}$. If we take eigenvalues of $L_{X,\mathbf{p}}$ different from zero, the corresponding curve is convergent.

Remark 4.2. Let $\tilde{\mathcal{F}}$ be a reduction of singularities of $\mathcal{F}$, with exceptional divisor $\tilde{E}$. Putting together the theorem of reduction of singularities and Briot-Bouquet's theorem, we conclude that the following statements are equivalent:

1. $\mathcal{F}$ is non-dicritical ($\tilde{E}$ has no dicritical components).
2. There is only a finite number of formal integral curves for $\mathcal{F}$.
3. There is only a finite number of convergent integral curves for $\mathcal{F}$.

Remark 4.3. Camacho-Sad index. Let $\mathcal{F}$ be a foliation on a surface M. Let $\mathbf{p} \in M$. Let Γ be a regular integral curve of $\mathcal{F}$. In a suitable chosen local chart centered at $\mathbf{p}$, $\mathcal{F}$ admits a generator

$$\omega = y(a(x,y)dx + b(x,y)\frac{dy}{y}),$$

with $a, b \in \mathbb{C}\{x, y\}$. We will define an index of $\mathcal{F}$ at $\mathbf{p}$ with respect to Γ as

$$i_{\mathbf{p}}(\mathcal{F}, \Gamma) = \mathrm{Res}_0([\frac{\partial}{\partial y}(\frac{-ya}{b})]_{(x,0)}) = \mathrm{Res}_0(\frac{-a(x,0)}{b(x,0)}).$$

This index is independent of the generator and of the local chart.

Let $\pi : \tilde{M} \longrightarrow M$ be a non-dicritical blowup centered at $\mathbf{p}$. Let $\tilde{\mathcal{F}}$ and $\tilde{\Gamma}$ be the strict transforms of $\mathcal{F}$ and Γ respectively. Let E be the exceptional divisor of π, and $\mathbf{q} = \tilde{\Gamma} \cap E$. As π is non-dicritical, $\tilde{\Gamma}$ is an integral curve of $\mathcal{F}$ at $\mathbf{q}$. It is easy to prove that:

1. $i_{\mathbf{q}}(\tilde{\mathcal{F}}, \tilde{\Gamma}) = i_{\mathbf{p}}(\mathcal{F}, \Gamma) - 1$,
2. $\sum_{\mathbf{r} \in E} i_{\mathbf{r}}(\tilde{\mathcal{F}}, \tilde{\Gamma}) = -1$.

Then, if $\mathbf{p}$ is a simple singularity and Γ_1, Γ_2 are the integral curves at $\mathbf{p}$, and if the eigenvalues of $L_{X,\mathbf{p}}$ are both different from zero (and thus Γ_1 and Γ_2 are convergent)

$$i_{\mathbf{p}}(\mathcal{F}, \Gamma_1).i_{\mathbf{p}}(\mathcal{F}, \Gamma_2) = 1.$$

And if a eigenvalue is zero, the index for the corresponding curve is zero.

Definition 4.4. Let $\mathcal{F}$ be a foliation on a two dimensional complex analytic variety M. Let D be a normal crossing divisor over M, and let $\mathbf{q} \in D$. We say that $(\mathcal{F}, D, \mathbf{q})$ has the property $(\star)$ if one of the following conditions holds:

1. -$(\star)$ The point $\mathbf{q}$ is exactly in one irreducible component S of D; this component is an integral curve of $\mathcal{F}$ and $i_{\mathbf{q}}(\mathcal{F}, S) \notin \mathbb{Q}_{\geq 0}$.

2. -($\star$) The point $\mathbf{q}$ is in two irreducible components S_+ and S_- of D (let us call this point a corner), both are integral curves and there is a real number $a > 0$ such that

$$\begin{array}{rcl} i_{\mathbf{q}}(\mathcal{F}, S_+) & \in & \mathbb{Q}_{\leq -a}, \\ i_{\mathbf{q}}(\mathcal{F}, S_-) & \notin & \mathbb{Q}_{\geq -1/a}. \end{array}$$

3. -($\star$) The point $\mathbf{q}$ is exactly in one irreducible component S of D, it is a non singular point of $\mathcal{F}$ and S is transversal to $\mathcal{F}$ at $\mathbf{q}$.

Remark 4.5.

1. If we have the property 2-($\star$), then $\mathbf{q}$ is not a simple singularity.
2. If we have either 3-($\star$) or 1-($\star$), and $\mathbf{q}$ is a simple singularity, then there is an analytic integral curve Γ through $\mathbf{q}$, non-singular and transversal to D.

Theorem 4.6. *Assume that* $(\mathcal{F}, D, \mathbf{q})$ *satisfies either* $1-(\star)$ *or* $2-(\star)$. *Let* $\pi : \tilde{M} \longrightarrow M$ *be the blowup of* M *at the point* $\mathbf{q}$. *Let* $\tilde{\mathcal{F}}$ *be the strict transform of* $\mathcal{F}$ *by* π. *Let* $E = \pi^{-1}(\mathbf{q})$ *and* $\tilde{D} = \pi^{-1}(D)$. *Then there is a point* $\mathbf{q}' \in E$ *such that* $(\tilde{\mathcal{F}}', \tilde{D}, \mathbf{q}')$ *satisfies the property* $(\star)$.

Proof. If π is a dicritical blowup, there is a point $\mathbf{q}' \in E$ such that $(\tilde{\mathcal{F}}, \tilde{E}, \mathbf{q}')$ satisfies the property $3-(\star)$.

If π is non-dicritical then E is an invariant curve for $\tilde{\mathcal{F}}$. If for all $\mathbf{q}' \in E$, $(\tilde{\mathcal{F}}, \tilde{E}, \mathbf{q}')$ does not verify the property $(\star)$, follows a contradiction.

(1) Suppose that $(\mathcal{F}, D, \mathbf{q})$ satisfies $1-(\star)$. Let $\tilde{S}$ be the strict transform of S by π and let be $\{\mathbf{p}'\} = E \cap \tilde{S}$. For any $\mathbf{q}' \in E, \mathbf{q}' \neq \mathbf{p}'$, we have that $i_{\mathbf{q}'}(\tilde{\mathcal{F}}, E) \in \mathbb{Q}_{\geq 0}$, and hence

$$i_{\mathbf{p}'}(\tilde{\mathcal{F}}, E) = -1 - \sum_{\mathbf{q}' \neq \mathbf{p}'} i_{\mathbf{q}'}(\tilde{\mathcal{F}}, E) \in \mathbb{Q}_{\leq -1}.$$

Since $(\tilde{\mathcal{F}}, \tilde{D}, \mathbf{p}')$ does not satisfy 2-($\star$), we obtain that for $a = 1$

$$i_{\mathbf{p}'}(\tilde{\mathcal{F}}, \tilde{S}) \in \mathbb{Q}_{\geq -1}.$$

Thus $i_{\mathbf{p}}(\mathcal{F}, S) = i_{\mathbf{p}'}(\tilde{\mathcal{F}}, \tilde{S}) + 1 \in \mathbb{Q}_{>0}$. This is a contradiction.

(2) Suppose that $(\mathcal{F}, D, \mathbf{q})$ satisfies $2-(\star)$. Let $\tilde{S_+}$ (resp. $\tilde{S_-}$) be the strict transform of S_+ (resp. S_-) by π. Let $\mathbf{p}_+ = E \cap \tilde{S_+}$, $\mathbf{p}_- = E \cap \tilde{S_-}$. Then

$$i_{\mathbf{p}_+}(\tilde{\mathcal{F}}, E) + i_{\mathbf{p}_-}(\tilde{\mathcal{F}}, E) = -1 - \sum_{\mathbf{q}' \notin \{\mathbf{p}_+, \mathbf{p}_-\}} i_{\mathbf{q}'}(\tilde{\mathcal{F}}, E) \in \mathbb{Q}_{\leq -1}.$$

By hypothesis $(\mathcal{F}, D, \mathbf{q})$ satisfies $2-(\star)$. Therefore

$$i_{\mathbf{q}}(\mathcal{F}, S_+) \in \mathbb{Q}_{\leq -a} \Rightarrow i_{\mathbf{p}_+}(\tilde{\mathcal{F}}, \tilde{S_+}) = i_{\mathbf{q}}(\mathcal{F}, S_+) - 1 \in \mathbb{Q}_{\leq -a-1}.$$

Since $\mathbf{p}_+ \in D$, $2-(\star)$ does not happen at $\mathbf{p}_+$. Therefore

$$i_{\mathbf{p}_+}(\tilde{\mathcal{F}}, \tilde{S_+}) \in \mathbb{Q}_{\geq -1/(a+1)}.$$

Hence

$$i_{\mathbf{p}_-}(\tilde{\mathcal{F}}, \tilde{S_+}) \in \mathbb{Q}_{\leq -a/(a+1)}.$$

Since $2-(\star)$ does not happen at $\mathbf{p}_-$,

$$i_{\mathbf{p}_-}(\tilde{\mathcal{F}}, \tilde{S_+}) \in \mathbb{Q}_{\geq -(a+1)/a} \Rightarrow i_{\mathbf{q}}(\mathcal{F}, S_-) = i_{\mathbf{p}_-}(\tilde{\mathcal{F}}, \tilde{S_-}) + 1 \notin \mathbb{Q}_{-1/a}.$$

We get a contradiction with $2-(\star)$ at $\mathbf{q}$. □

Remark 4.7. Algorithm. We start at a singularity $\mathbf{p_0} \in M = M_0$, let us set $\mathcal{F} = \mathcal{F}_0$, and $E_0 = \emptyset$. Let $\pi_1 : M_1 \to M_0$ be the blowup with center $\mathbf{p_0}$, and let $E_1 = \pi_1^{-1}(\mathbf{p_0})$ and let $\mathcal{F}_1$ be the strict transform of $\mathcal{F}$ by π_1. If π_1 is dicritical, we take a point $\mathbf{p_1} \in E_1$ with the property 3-($\star$) and we have a convergent integral curve at $\mathbf{p_1}$ which gives via π_1 a convergent integral curve at $\mathbf{p_1}$ and we finish.

If π_1 is non-dicritical, since the sum of indices is -1, we have a point $\mathbf{p_1} \in E_1$ such that 1-($\star$) holds. If $\mathbf{p_1}$ is simple, we do the same thing as above and we finish.

If $\mathbf{p_1}$ is not simple, let $\pi_2 : M_2 \to M_1$ be the blowup with center $\mathbf{p_1}$, and let $E_2 = \pi_2^{-1}(E_1)$ and let $\mathcal{F}_2$ be the strict transform of $\mathcal{F}_1$ by π_2. We take a point $\mathbf{p_2} \in E_2$ with the ($\star$)-property, if it is 3-($\star$) or if $\mathbf{p_2}$ is simple we may finish as before, otherwise we repeat.

Then by the theorem of reduction of singularities, the algorithm ends after a finite number of steps.

References

[1] Aroca, J.M., Hironaka, H., Vicente, J.L.: *The theory of the maximal contact.* Mem. Mat. Inst. Jorge Juan 29, Madrid, 1975.

[2] Aroca, J.M., Hironaka, H., Vicente, J.L.: *Desingularization theorems.* Mem. Mat. Inst. Jorge Juan 30, Madrid, 1975.

[3] Black, C.W.M.: *The parametric representation of the neighborhood of a singular point of an analytic surface.* Proc. Amer. Acad. Arts and Sci. vol. 37 (1901).

[4] Briot, C.A., Bouquet, J.C.: *Propriétés des fonctions définies par les équations différentielles.* Journal de l'Ecole Polytechnique 36 (1856), 133–198.

[5] Camacho, C.: *Quadratic forms and holomorphic foliations on singular surfaces.* Math. Ann. 282 (1988), 177–184.

[6] Camacho, C., Lins Neto, A.: *Geometric theory of foliations.* Birkhäuser, Berlin 1985.

[7] Camacho, C., Sad, P.: *Invariant Varieties Through Singularities of Holomorphic Vector Fields.* Ann. of Math. 115 (1982), 579–595.

[8] Cano, F.: *Dicriticalness of a singular foliation.* Proceedings México 1986, (X. Gómez-Mont, J. Seade, A. Verjovski (Eds.)), Lect. Notes in Math. 1345 (1988), 73–95.

[9] Cano, F.: *Foliaciones singulares dicríticas.* Mem. Real Acad. Ciencias Exactas, Físicas y Naturales, Serie Ciencias Exactas, t. XXIV, 1989.

[10] Cano, F.: *Reduction of the Singularities of non-dicritical Singular Foliations. Dimension three.* Amer. J. of Math. 115, 3 (1993), 509–588.

[11] Cano, F.: *Reduction of Singularities of Codimension One Foliations in Dimension Three.* To be published in the collection of papers in honor S. Lojasievicz. Cracow.

[12] Cano, F., Cerveau, D.: *Desingularization of non-dicritical Holomorphic Foliations and Existence of Separatrices.* Acta Mathematica 169 (1992), 1–103.

[13] Cano, F., Cerveau, D.: *Feuilletages holomorphes singuliers.* Book in preparation.

[14] Cano, F., Mattei, J.F.: *Hypersurfaces intégrales des feuilletages holomorphes.* Colloque de Géométrie à la mémoire de Claude Godbillon et Jean Martinet. Ann. Inst. Fourier, t. 42, 1–2, (1992), 49–72.

[15] Cano, J.: *Construction of Invariant Curves for Holomorphic Vector Fields.* Proc. Amer. Math. Soc. vol. 125, num. 9 (1997), 2649–2650.

[16] Cerveau, D., Mattei, J.-F.: *Formes intégrables holomorphes singulières.* Astérisque 97, 1982.

[17] Cossart, V.: *Forme normale pour une fonction sur une variété de dimension trois en caractéristique positive.* Thèse d'Etat, Orsay, 1988.

[18] Giraud, J.: *Forme normale d'une fonction sur une surface de caractéristique positive.* Bull. Soc. Math. France 111 (1983), 109–124.

[19] Giraud, J.: *Condition de Jung pour les revêtements radiciels de hauteur un.* Proc. Algr. Geom. Tokyo/Kyoto 1982, Lect. Notes in Math. 1016, Springer-Verlag 1983, 313–333.

[20] Hensel: *Über eine neue Theorie der algebraischen Funktionen zweier Variablen.* Acta Math. vol. 23 (1900), 339–416.

[21] Hironaka, H.: *Introduction to the Theory of Infinitely Near Singular Points.* Mem. Mat. Inst. Jorge Juan 28, Madrid, 1975.

[22] Hironaka, H.: *La voûte étoilée.* Singularités à Cargèse. Astérisque 7–8 (1973).

[23] Hironaka, H.: *Desingularization of Excellent Surfaces.* Adv. Sci. Seminar (1967). Bowdoin College. Appeared in Lect. Notes in Mathematics 1101, Springer-Verlag (1984).

[24] Jouanolou, J.P.: *Equations de Pfaff algébriques.* Lect. Notes in Math. 708, Springer-Verlag, 1979.

[25] Jung, H.: *Darstellung der Funktionen eines algebraischen Körpers zweier unabhängigen Veränderlichen x, y in der Umgebung einer Stelle $x = a$, $y = b$.* Jour. für Math. 136 (1908), 289–306.

[26] Lacombe, E., Sienra, G.: *Blow up techniques in the Kepler problem.* Preprint.

[27] Lins Neto, A., Scardua, B.: *Folheacoes algebricas complexas.* IMPA, Rio de Janeiro, 1997.

[28] Malgrange, B.: *Frobenius avec singularités 1. Codimension un.* Publ. Math. I. H. E.S. 46 (1976), 162–173.

[29] Saito, K.: *On a generalization of de Rham Lemma.* Ann. Inst. Fourier 26, 2 (1976), 165–170.

[30] Seidenberg, A.: *Reduction of the singularities of the differential equation $Ady = Bdx$.* Amer. J. of Math. 90 (1968), 248–269.

[31] Siu, Y.-T.: *Techniques of Extension of Analytic Objects.* Lecture Notes in Pure and Appplied Math., vol. 8, Marcel Dekker Inc., New York, 1974.

[32] Trifonov, S.I.: *Resolution of singularities of one parameter analytic families of differential equations.* Amer. Math. Soc. Transl. (2), vol. 151 (1992).

[33] Walker: *Reduction of the singularities of an algebraic surface.* Ann. of Math. 36 (1935), 336–365.

Instituto de Estudios
de Iberoamerica y Portugal
Casas del Tratado
Tordesillas
Valladolid, Spain
aroca@cpd.uva.es

Progress in Mathematics, Vol. 181,

Puiseux Solutions of Singular Differential Equations

José Manuel Aroca

1. Introduction

The purpose of this paper is to show how Newton-Puiseux techniques can be used to solve some ordinary differential equations (ODE in the sequel) in the complex domain. We will show how by using these techniques we will be able to find solutions of complex ordinary differential equations having convergent or formal power series as coefficients. That is, we consider equations as

$$f(x,y,y') = \sum_{i=0}^{n} a_i(x,y)(y')^i = 0,$$

where $a_i(x,y) \in \mathbb{C}\{x,y\}$ or $a_i(x,y) \in \mathbb{C}[[x,y]]$ for all i. In the simplest case $n=1$, we will write the equation as

$$a_0(x,y)dx + a_1(x,y)dy = 0,$$

a_0 and a_1 being convergent in a disk U centered at 0. The equation gives us at any point $\mathbf{p} = (x_0, y_0) \in U$ the line $r_{\mathbf{p}}$ passing through $\mathbf{p}$ with direction $\mathbf{v_p} = (-a_1(x_0,y_0), a_0(x_0,y_0))$ provided that this vector is different from zero. The points of U with this property are called regular points of the equation, and the non regular points are called singular.

From a geometric point of view, the solutions of our equations are regular analytic curves γ, contained in the open set of regular points and such that the tangent line to γ at any point $\mathbf{p}$ is $r_{\mathbf{p}}$. As the singular points are limit points of regular ones it seems to be natural to try to find analytic, or at least formal, branches as solutions in the singular points. Then a solution of our equation will be a formal or convergent Puiseux series $y(x) = \sum_{j>0} b_j x^{j/n}$ such that $f(x,y(x),y'(x)) \equiv 0$.

In the general case, if U is a disk where all the coefficients converge, the singular points are the zeroes of the discriminant with respect to y' of f and the points $a_0(x,y) = a_n(x,y) = 0$. Outside these points, the equation gives n different directions, and we have a web of solutions. But there is the possibility, as we will see below, that there are no solutions in the singular points.

The resolution of algebraic equations has many points in common with the resolution of differential equations. For instance, it is well known that we may find a solution $y = y(x)$ for the equation

$$g(x,y) = 0, g(x,y) \in \mathbb{C}\{x,y\}, (\partial_y g)(0) \neq 0,$$

just by computing its Taylor expansion at 0

$$\begin{aligned} g(x,y(x)) &\equiv 0 \Rightarrow \frac{dg}{dx} = \frac{\partial g}{\partial x} + \frac{\partial g}{\partial y}\frac{dy}{dx} = 0 \\ \frac{dy}{dx} &= -\frac{\frac{\partial g}{\partial x}}{\frac{\partial g}{\partial y}}. \end{aligned}$$

As $\frac{\partial g}{\partial y}$ is a unit we turn to the simplest case of ODE, that is the case of a single equation of first order and first degree

$$y' = f(x,y) \text{ with } y' = \frac{dy}{dx}, \tag{1}$$

where $f(x,y)$ is a germ at $0 \in \mathbb{C}^2$ of a complex analytic function, i.e. a convergent complex power series in the variables x, y. In this situation, the so called "method of limits" gives us, as follows, the formal solution corresponding to the initial conditions $x = 0, y = 0$.

The relation (1) gives by repeated differentiation

$$\begin{aligned} \frac{d^2y}{dx^2} &= \frac{\partial f}{\partial x} + \frac{\partial f}{\partial y}\frac{dy}{dx} \\ \frac{d^3y}{dx^3} &= \frac{\partial^2 f}{\partial x^2} + 2\frac{\partial^2 f}{\partial x \partial y} f + \frac{\partial^2 f}{\partial y^2} f^2 + \frac{\partial f}{\partial y}\frac{d^2y}{dx^2} \\ &\text{etc.} \end{aligned}$$

Using these equalities we can build a formal series

$$y(x) = \sum_{i \geq 1} (\frac{d^i y}{dx})(0)x^i.$$

It is clear that this series formally verifies the differential equation. The point is to prove the convergence, but if

$$f(x,y) = \sum f_{ij} x^i y^j$$

is convergent, there are M, a, b in $\mathbb{R}_+$ such that

$$| f_{ij} | < \frac{M}{a^i b^j} \quad \text{for all } i, j.$$

Since

$$F(x,y) = \sum \frac{M}{a^i b^j} x^i y^j = \frac{M}{(1 - x/a)(1 - y/b)} \quad \text{(with real coefficients)},$$

we have that

$$(\frac{\partial^{i+j} F}{\partial x^i \partial y^j})(0) \geq |(\frac{\partial^{i+j} f}{\partial x^i \partial y^j})(0)|. \tag{2}$$

If we now consider the equation

$$(3) \qquad y' = F(x,y)$$

and $y = \phi(x)$ is the solution, constructed by the same method that we used before for the equation (1), an easy computation shows (by (2)) that

$$(\frac{d^r\phi}{dx^r})(0) \geq |(\frac{d^r y}{dx^r})(0)| \quad \text{for all } r.$$

Then to prove that $y(x)$ converges, it is sufficient to do it for $\phi(x)$. As $\phi(x)$ is a solution of the equation (3) we have

$$\phi'(x) = F(x,\phi) = \frac{M}{(1-x/a)(1-\phi/b)} \Leftrightarrow$$

$$\Leftrightarrow (1-\phi/b)\cdot\phi' = \frac{M}{(1-x/a)}.$$

And the solution of this equation, with initial conditions $x = 0$, $y = 0$, is

$$\phi(x) = b - b\sqrt{1 + 2Ma/b \cdot log(1-x/a)}.$$

The radius of convergence of the power series expansion of $\phi(x)$ is

$$\rho = a(1 - e^{-b/(2Ma)}) > 0.$$

The situation is not so easy in the singular case, that means for functions

$$f(x,y) = 0, f \in \mathbb{C}\{x,y\}, \partial_x f(0) = \partial_y f(0) = 0,$$

and for first order and first degree ODE's

$$(4) \qquad \begin{array}{l} f(x,y)y' + g(x,y) = 0 \Leftrightarrow y' = -\frac{g(x,y)}{f(x,y)} \\ f(0) = g(0) = 0, \quad f,g \in \mathbb{C}\{x,y\}. \end{array}$$

(In the case of degree > 1, the singularity conditions are given by the annihilation of the discriminant with respect to y' of f, or by the annihilation of the coefficients of the terms of degree zero or n).

Now the situation is less clear: It can be better than in the case of regular equations (in the sense of having more solutions), for instance, the equation

$$xy' = y$$

has an infinite set of solutions $y = \lambda x$, for all $\lambda \in \mathbb{C}$, with initial conditions $x = 0$, $y = 0$. Or can be worse, for example, the equation

$$2yy' = 3x^2$$

has no solution in $\mathbb{C}[[x]]$ (but observe that $y = x^{3/2}$ is a Puiseux solution). For a general algebraic differential equation

$$(5) \qquad f(x,y,\frac{dy}{dx},\dots,\frac{d^n y}{dx^n}) = 0$$

with

$$f(x,y,y_1,\dots,y_n) \in \mathbb{C}\{x,y\}[y_1,\dots,y_n]$$

the construction of a solution

$$y = \sum_{i \geq n} b_i x^i$$

with initial conditions

$$x = 0,\ y(0) = \frac{dy}{dx}(0) = \ldots = \frac{d^{n-1}y}{dx^{n-1}}(0) = 0$$

is easy when

$$(\frac{\partial f}{\partial y_n})(0) \neq 0. \tag{6}$$

Under this hypothesis, the implicit function theorem, or a similar computation to the one which we made before, gives us a formal series solution, majorized by a convergent one, thus convergent.

The case of equation (1) or equation (5), verifying (6), is called regular, and equation (4), which is a particular case of (5), without (6), is called singular. The problem is to solve differential equations in the singular case. We will do this in some cases and by two different methods:

- By direct computation using a sort of Newton polygon to systematize the construction.
- By analyzing the geometric meaning of a differential equation, it is possible to attach to any equation of first order and first degree some geometric singular object. Then by solving the singularities of this object, we get a solution of the equation (see the second paper of the author in this volume). This method can be generalized to more variables.

When we try to solve ODE's by the first method, our idea is to construct a Puiseux series solution of the equation at the singular points. The reason for using such a series is the existence of a flow of analytic solutions of the equation in the regular set. As we will see, the results of the application of Newton's method to the construction of solutions of differential equations present serious differences compared with the case of algebraic or analytic equations. The most interesting are:

- The possibility of non-existence of solutions.
- The existence of divergent formal solutions of analytic equations.
- The existence, sometimes, of an infinite set of solutions, even for first order and first degree equations.

A natural issue is to ask for the possibility of reducing a first order and degree n equation $(n > 1)$, $f(x, y, y') = 0$, to n first order and first degree equations just by solving the algebraic equation in the differential variable. This can be made in some cases, for instance in the case of quasi-ordinary equations.

Let us recall that an equation $f(x,y,y') = 0$ is called quasi-ordinary with respect to the projection

$$(x,y,y') \longmapsto (x,y),$$

if the projection is transversal to the equation, that is

$$f(x,y,y') = (y')^n + \sum_{r=0}^{n-1} f_r(x,y)(y')^r \text{ with } \operatorname{ord}(f_r) \geq n-r$$

and the discriminant has normal crossings. Under these hypotheses the equation admits a local parametrisation

$$\begin{aligned} x &= u^s \\ y &= v^s \\ y' &= g(u,v). \end{aligned}$$

For each local parametrisation (because in principle, the solutions depend on the parametrisation) our equation gives rise to a first order and first degree equation

$$g(u,v) = y' = \frac{dy}{dx} = \frac{sv^{s-1}dv}{su^{s-1}du} \Rightarrow v' = \frac{dv}{du} = \frac{u^{s-1}g(u,v)}{v^{s-1}},$$

and each solution of the second equation gives a solution of the first one. For more general equations there are several possibilities: either they can be transformed in quasi-ordinary ones by using bimeromorphic transformations [1] or can be solved in more general rings [22].

Mc Donald's solution in [22] is nothing but a special case of Zariski's local uniformisation theorem, and the parametrisations obtained by his Newton polyhedra can be considered as parametrisations of wedges on the surface $f(x,y,z) = 0$, attached in the obvious way to the differential equation [29].

2. The Newton polygon of a differential equation

Using the Newton polygon to build Puiseux solutions of an ordinary differential equation over the complex domain goes back to H. B. Fine [13, 14]; he proves that any (sufficiently general in some undetermined sense) ordinary differential equation has a convergent Puiseux solution. The first complete proof of the existence of a solution for a singular first order and first degree differential equation, by using the Newton polygon, is due to J. Cano [7, 8]. Given a polynomial

$$\begin{aligned} f(x,\mathbf{y}) &= f(x,y_0,\ldots,y_n) = \\ &= \sum f_{ij_0\ldots j_n} x^i y_0^{j_0} \ldots y_n^{j_n} \in \mathbb{C}[[x^{1/s},y_0]][y_1,\ldots,y_n],\ s \in \mathbb{N}, \end{aligned}$$

we define an operator $\hat{f} : \mathcal{U} \longrightarrow \mathcal{U}'$ where

$$\begin{aligned} \mathcal{U} &= \{\sum_{i=1}^{\infty} a_i x^{\mu_i} \mid a_i \in \mathbb{C},\ \mu_i \in \mathbb{Q}_{>0}, \quad \mu_i < \mu_{i+1}, \text{ for all } i\} \\ \mathcal{U}' &= \{\sum_{i=1}^{\infty} a_i x^{\mu_i} \mid a_i \in \mathbb{C},\ \mu_i \in \mathbb{Q}, \quad \mu_i < \mu_{i+1}, \text{ for all } i\} \end{aligned}$$

by the obvious substitution in f

$$y_0 = \sum_{i=1}^{\infty} a_i x^{\mu_i}, \quad y_j = \frac{d^j}{dx^j}(\sum_{i=1}^{\infty} a_i x^{\mu_i}) =$$

$$= \sum_{i=j}^{\infty} (\mu_i)_j a_i x^{\mu_i - j}, \quad (\mu)_j = \mu(\mu-1)\ldots(\mu-j+1),$$

and we will call any $s(x) \in \mathcal{U}$, such that $\hat{f}(s(x)) = 0$, a solution of the differential equation $f(x, \mathbf{y}) = 0$. Clearly we may use convergent equations and convergent solutions in the same way.

Let us set:

- for all $\mathbf{j} = (j_0, \ldots, j_n)$, $\mid \mathbf{j} \mid = j_0 + \cdots + j_n$, $<\mathbf{j}> = j_1 + 2j_2 + \cdots + nj_n$
- $\alpha : \frac{1}{s}\mathbb{Z} \times \mathbb{N}^{n+1} \longrightarrow \frac{1}{s}\mathbb{Z} \times \mathbb{N}$, $\alpha(i, \mathbf{j}) = (i- <\mathbf{j}>, \mid \mathbf{j} \mid)$
- $M_{f,(a,b)}(x, y) = \sum_{\alpha(i,\mathbf{j})=(a,b)} f_{i\mathbf{j}} x^i \mathbf{y}^{\mathbf{j}}$
- $\Delta(f) = \{(a, b) \mid M_{f,(a,b)}(x, y) \neq 0\} \subset \mathbb{R} \times \mathbb{R}_{\geq 0}$.

We call $\Delta(f)$ the Newton diagram of $f(x, \mathbf{y})$. The border $N(f)$ of the convex hull of

$$\bigcup_{P \in \Delta(f)} (P + \mathbb{R}^2_{\geq 0})$$

is called the Newton polygon of $f(x, \mathbf{y})$. The sides and vertices of $N(f)$ are defined as the obvious sides and vertices of its border, with the only exception for the vertical side, and also for the horizontal one (if it is placed on the x-axis). For brevity we will say *straight line of coslope* μ instead of *straight line of slope* $-1/\mu$.

For any $\mu \in \mathbb{Q}_+$ the unique straight line of coslope μ which intersects $N(f)$ either on a side, or on a vertex, will be denoted by $L(f, \mu)$. Also for any f we will use the polynomials

$$\begin{aligned} \Psi_{f,(a,b)}(u) &= \sum_{\alpha(i,\mathbf{j})=(a,b)} f_{i\mathbf{j}} (u)_1^{j_1} \cdots (u)_n^{j_n}, \text{ for all } (a, b) \in \frac{1}{s}\mathbb{Z} \times \mathbb{N}, \\ \Phi_{f,\mu}(c) &= \sum_{(a,b) \in L(f,\mu)} \Psi_{f,(a,b)}(\mu) \cdot c^b. \end{aligned}$$

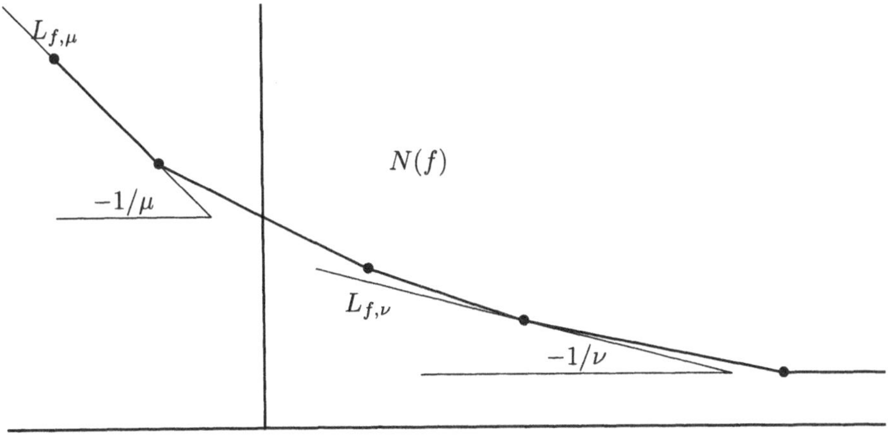

FIGURE 1.

Proposition 2.1. *Let $z \in \mathcal{U}$, $z = c_1 x^{\mu_1} +$ higher degree terms, $c_1 \neq 0$, be a solution of f. Then*

(1) *$L(f, \mu_1)$ intersects the border of $N(f)$ either along a side, or in a vertex (a,b) such that there is more than one element $(i, \mathbf{j})$ with $\alpha(i, \mathbf{j}) = (a, b)$ and $f_{i,\mathbf{j}} \neq 0$. Also the sides of $N(f)$ are contained in a single half-plane with respect to $L(f, \mu_1)$.*
(2) *If $L(f, \mu_1)$ intersects $N(f)$ only in a vertex (a, b), then $\Psi_{f,(a,b)}(\mu_1) = 0$.*
(3) *If $L(f, \mu_1)$ intersects $N(f)$ in a side (of coslope μ_1), then $\Phi_{f,\mu_1}(c_1) = 0$.*

Proof. This is a straightforward calculation of the terms of minimum degree in $\hat{f}(z)$. □

In order to have a solution in $\mathcal{U}$ of f, we need either a vertex or a side of $N(f)$ which verifies (2) or (3) in Proposition 2.1 for some μ_1 and c_1, and then $c_1 x^{\mu_1}$ may be the first term of a solution.

Example 2.2. We will give some examples to show the limits of this construction:
(1) The equation $f(x, y, y_1, y_2) = xyy_2 - xy_1^2 + yy_1$ has the Newton Polygon of Fig. 2. For any line through $(-1, 2)$, all the Newton polygon is contained in the same half-plane, and for the vertex $(-1, 2)$

$$\Psi_{f,(-1,2)}(u) \equiv 0.$$

In fact, $y = cx^\mu$ is a solution of f for all $\mu \in \mathbb{Q}_+$ and even for all $\mu \in \mathbb{C}$.
(2) The equation $f(x, y, y_1) = y_1^3 + x^2 y_1^2 - 7xyy_1 + 12y^2 + x^8$ has the Newton polygon of Fig. 3. This polygon has two sides with slopes $-1/3$, $-1/4$, three

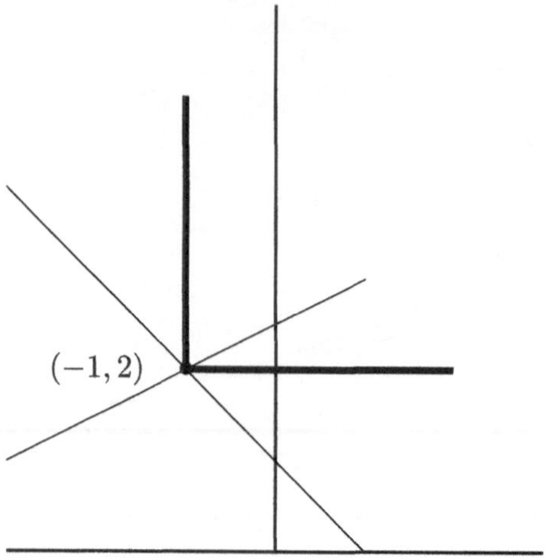

FIGURE 2.

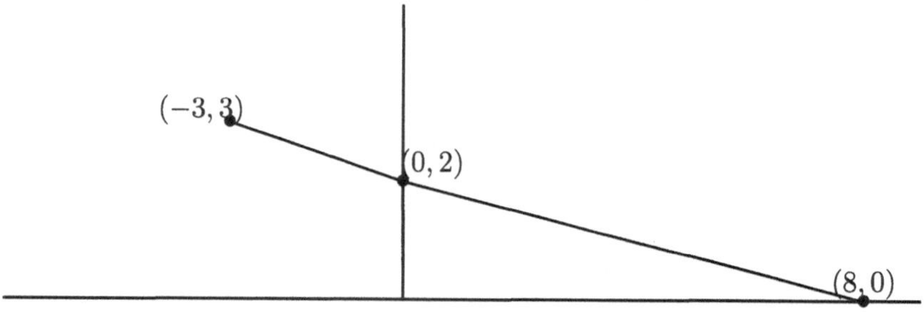

FIGURE 3.

vertices, $(-3,3), (0,2), (8,0)$, and

$$\begin{aligned}
\Psi_{f,(-3,3)}(u) &= u^3 \\
\Psi_{f,(0,2)}(u) &= u^2 - 7u + 12 = (u-3)(u-4) \\
\Psi_{f,(8,0)}(u) &= 1 \\
\Phi_{f,3}(c) &= 27c^3 \\
\Phi_{f,4}(c) &= 1.
\end{aligned}$$

Then there are neither solutions $y = cx^3 + h.d.t.$ nor $y = cx^4 + h.d.t.$, corresponding to the sides. The only vertex with possibilities according (1) of Proposition 2.1 is $(0,2)$, and for $(0,2)$ the only possible values for u are $u = 3$ and $u = 4$, which are not useful. Therefore f has no solution in $\mathcal{U}$.

Remark 2.3. Once we have (c_1, μ_1) satisfying either (2) or (3) of 2.1, we may apply a change of variable $y = c_1 x^{\mu_1} + z$. This gives the new equation

$$f[c_1 x^{\mu_1} + z] = f(x, c_1 x^{\mu_1} + z_0, (\mu_1)_1 c_1 x^{\mu_1 - 1} + z_1, \dots, (\mu_1)_n c x^{\mu_1 - n} + z_n).$$

We now want to repeat the process by computing for the new equation c_2 and μ_2 with $\mu_2 > \mu_1$. Observe that:

(1) $\sum_{i=1}^{\infty} c_i x^{\mu_i}$ is a solution of f if and only if $\sum_{i=2}^{\infty} c_i x^{\mu_i}$ is a solution of $f[c_1 x^{\mu_1} + z]$.

(2) If $M_{i\mathbf{j}}(x, \mathbf{y}) = x^i \mathbf{y}^{\mathbf{j}}$ then for all $c \in \mathbb{C}$ and all $\mu \in \mathbb{Q}_+$ we have (see Fig. 4)

$$\alpha(i, \mathbf{j}) \in \Delta(M_{i,\mathbf{j}}[cx^{\mu} + y]) \subset \{(a, b) \in L(M_{i\mathbf{j},\mu}) | b \leq | \mathbf{j} | \}.$$

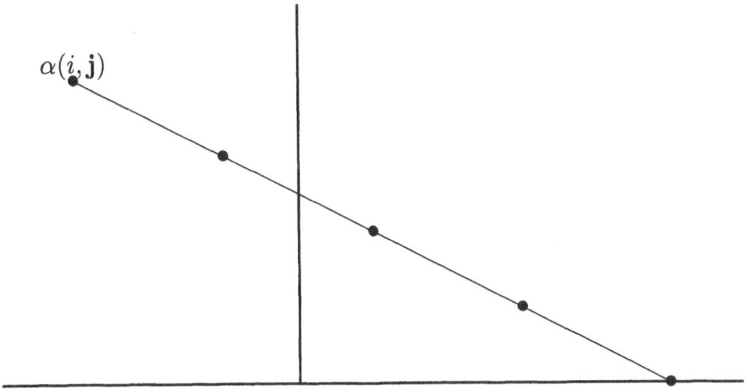

FIGURE 4.

That means that, after the substitution, each point of the Newton polygon of f gives rise to a collection of points, placed on the line of slope $-1/\mu$, passing through the original point, and including it, and points placed below it.

As a consequence, the border of the Newton polygon of $f[cx^{\mu} + y]$ coincides with the original one on the region placed to the left of the higher point $Q(f, \mu)$ in $L(f, \mu) \cap N(f)$ (see Fig. 5).

(3) Under the hypothesis that $L(f, \mu)$ corresponds to a side

$$L(f, \mu) \cap \{y = 0\} = (t, 0) \Rightarrow M_{f[cx^{\mu}+y],(t,0)} = \Phi_{f,\mu}(c) x^t.$$

Then, in $N(f[cx^{\mu} + y])$, the side of slope $-1/\mu$ does not reach the x-axis. Consequently, in this new polygon, there is at least a side with coslope bigger than μ.

Example 2.4. Let $f(x, y, y_1) = (y^6 + xy^2 - 3x^2 y + 2x^3) y_1 + xy^2 - x^2 y + x^5$ Then $N(f)$ has a side of coslope 1 and $\Phi_{f,1}(c) = c(c-1)^2$. For $c = 1$, $\mu = 1$ we get

$$f[x + y] = (x + y)^6 (1 + y_1) + xy^2 y_1 - x^2 y y_1 + 2xy^2 + x^5.$$

And $N(f[x + y])$ has only a side with coslope $2 > 1$, but

$$\Phi_{f[x+y],2}(c) = 1$$

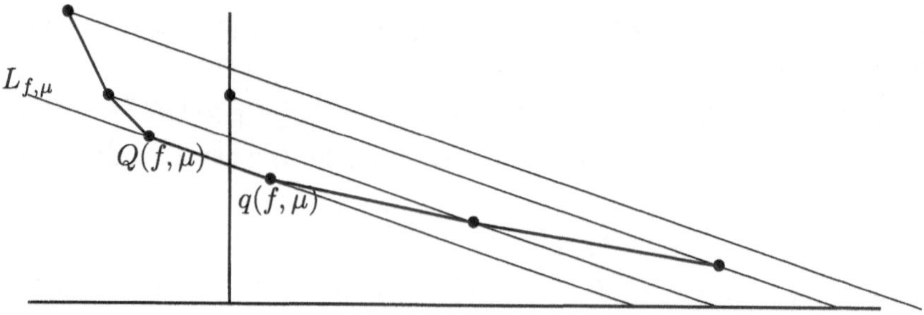

FIGURE 5.

and f has no solutions $y = x + h.d.t.$. Thus, there is no guaranty for the possibility of continuing the process.

Example 2.5. Let us consider Euler's equation

$$f(x, y, y_1) = x^2 y_1 - y + x.$$

Its Newton polygon is given in Fig. 6. This equation has the convergent solution

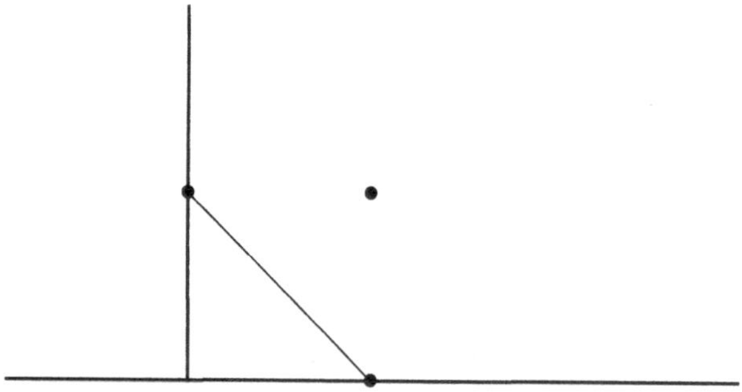

FIGURE 6.

$x = 0$ (in the following sense: the equation corresponds to the differential form $x^2 dy + (-y + x)dx$ and may be also written as $x^2 + (-y + x)x' = 0$. Now it is clear that $x = 0$ is a solution) and also, corresponding to the side of slope -1, the non-convergent solution

$$y = \sum_{i=0}^{\infty} n! x^{n+1}.$$

This example proves that a solution constructed by using Newton Polygons may be a non-convergent series.

Remark 2.6. (1) If $z = \sum_{i=1}^{\infty} c_i x^{\mu_i} \in \mathcal{U}$ is a solution of f, we have a sequence of Newton polygons, $N(f)$, $N(f_1), \ldots$, with

$$f_1 = f[c_1 x^{\mu_1} + y],\ f_2 = f_1[c_2 x^{\mu_2} + y],\ \ldots.$$

After each step in the process the y-coordinate of the higher point $Q_i(f_i, \mu_i)$ in $L(f_i, \mu_i) \cap N(f_i)$ either remains the same or decreases. Then, since the y-coordinate of $Q_1(f_1, \mu_1)$ is finite, after a finite number of steps, the point $Q_i(f_i, \mu_i)$ remains fixed. This fixed point will be called pivot point of the solution.

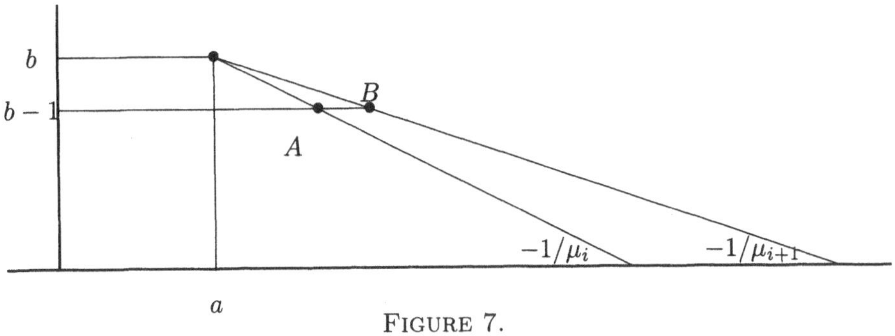

FIGURE 7.

(2) With the same notations as above, if $Q = (a, b)$ is the pivot point, there is an index i_0 such that $(a + \mu_i, b - 1) \in \Delta(f_i)$ for all $i \geq i_0$. For, otherwise, there would be an infinite set of indices $\mu_i \in \mathbb{Q}_+$ such that

$$M_{f_i[c_{i+1}x^{\mu_{i+1}}+y],(a,b)}(x, \mathbf{y}) = M_{f_i,(a,b)}(x, \mathbf{y}) + T,$$

where all the monomials in T have total exponents on $\mathbf{y}$ not exceeding $b - 2$. It is easy to see, by elementary linear algebra, that this is not possible.

Then, if $i \geq i_0$ and $\Delta(f_i) \subset \frac{1}{s}\mathbb{Z} \times \mathbb{N}$, we have $A = (a + \mu_i, b - 1) \in \frac{1}{s}\mathbb{Z} \times \mathbb{N}$ and $\mu_i \in \frac{1}{s}\mathbb{Z}$ (see Fig. 7). Therefore, after the substitution, $B = (a + \mu_{i+1}, b - 1) \in \frac{1}{s}\mathbb{Z} \times \mathbb{N}$ and $\mu_{i+1} \in \frac{1}{s}\mathbb{Z}$. Then if $z = \sum_{i=1}^{\infty} a_i x^{\mu_i} \in \mathcal{U}$ is a solution of f, z is a Puiseux series.

3. Solutions for first order, first degree equations

Our problem now is to seek Puiseux solutions of

$$f(x, y, y') = A(x, y) + B(x, y)y' = \sum_{(i,j) \in 1/s\mathbb{Z} \times \mathbb{N}} (A_{ij} x^i y^j + B_{ij} x^{i+j} y^{j-1} y').$$

Let us observe that we have written f as a sum of the monomials $M_{f,(i,j)}$. As we did before, for any $L(f, \mu)$ corresponding to a side of $N(f)$, we will define $Q(f, \mu)$ and $q(f, \mu)$ respectively as the higher and lower points of $L(f, \mu) \cap N(f)$. We will also denote the side contained in this line by $L(f, \mu)$.

Definition 3.1. $L(f, \mu)$ is a pregood side of f or μ is a pregood coslope of f, if $L(f, \mu)$ contains a side of $N(f)$ and $\Psi_{f;Q(f,\mu)}(u)$ has no roots in $\mathbb{Q}_{\geq \mu}$. $L(f, \mu)$ is a good side (or μ is a good coslope) for f if it is pregood and $\Psi_{f;q(f,\mu)}(u)$ either has no roots or has at least a root in $\mathbb{Q}_{> \mu}$.

Proposition 3.2. *Let $L(f,\mu)$ be a pregood side which is not good, and suppose that $y=0$ is not a solution of f. Then there is $\mu' > \mu$, $\mu' \in \mathbb{Q}_{>0}$ such that $L(f,\mu')$ is a good side.*

Proof. In fact, $\Psi_{f,(i,j)}(u) = A_{ij} + B_{ij}u$, then $\Psi_{f,(i,j)}(u)$ has no roots if $B_{ij} = 0$, and, if $B_{ij} \neq 0$, its only root is $-A_{ij}/B_{ij}$. The hypothesis means that:

(1) The point $q(f,\mu)$ is not placed on the x-axis.
(2) The next side to the one of coslope μ (which is not parallel to the x-axis because $y=0$ is not a solution) is pregood.

Then we can follow the polygon until reaching the side with a point on the axis. This side, being pregood, is automatically good. □

Remark 3.3. Let us call

$$\begin{aligned}\mathrm{In}(f,\mu) &= \sum_{(i,j)\in L(f,\mu)\cap N(f)} M_{f,(i,j)(x,y)} = \\ &= A_0x^\alpha + \sum_{i=1}^{t}(A_ix^{\alpha-\mu i}y^i + B_ix^{\alpha-\mu i+1}y^{i-1}y_1)\end{aligned}$$

(in this expression we omit the second subindex in the coefficients). Then for any $c \in \mathbb{C}$

$$\begin{aligned}\mathrm{In}(f[cx^\mu + y],\mu) &= \sum_{(i,j)\in L(f,\mu)\cap N(f)} M_{f,(i,j)}(x,\mathbf{y}) = \\ &= A_0(c)x^\alpha + \sum_{i=1}^{t}(A_i(c)x^{\alpha-\mu i}y^i + B_i(c)x^{\alpha-\mu i+1}y^{i-1}y_1).\end{aligned}$$

And we have that:

1. $A_0(c) = \Phi_{f,\mu}(c)$
2. $A_i(c) = \frac{1}{i!}\Phi^i_{f,\mu}(c) - \frac{\mu}{(i-1)!}\Phi^{i-1}_{0,1}(c)$
3. $B_i(c) = \frac{1}{(i-1)!}\Phi^{i-1}_{0,1}(c)$,

where $\Phi_{0,1} = \Phi_{\frac{\partial f}{\partial y_1},\mu}$ and the superindex j means $\frac{d^j}{dc^j}$. These results are trivial for first order, first degree equations and can be extended to more general ones (see Farto [12]).

Remark 3.4. Let us write

$$\begin{aligned}\Phi_{f,\mu}(c) &= a_t(f,\mu)c^t + \cdots + a_s(f,\mu)c^s \\ \Phi_{0,1} &= b_t(f,\mu)c^{t-1} + \cdots + b_2(f,\mu)c + b_1(f,\mu)\end{aligned}$$

and set $b_0(f,\mu) = 0$. Then $L(f,\mu)$ is a good side if and only if:

1. $a_t(f,\mu) \neq 0, a_s(f,\mu) \neq 0$ and $t > s$
2. $-\frac{b_t(f,\mu)}{a_t(f,\mu)} \notin \mathbb{Q}_{>0}$
3. $b_j(f,\mu) = 0$ when $j < s$
4. $-\frac{b_s(f,\mu)}{a_s(f,\mu)} \in \mathbb{Q}_{\geq 0}$.

Definition 3.5. Let γ be a root of $\Phi_{f,\mu}(c) = 0$ and set $k = \min\{\nu_\gamma(\Phi_{f,\mu}), \nu_\gamma(\Phi_{0,1})+1\}$, where $\nu_\gamma(\Phi)$ means the multiplicity of γ as a root of Φ. We will say that γ is a good root for f, μ if $\gamma \neq 0$ and either $\Phi_{0,1}^{k-1}(\gamma) = 0$ or $\Phi_{0,1}^{k-1}(\gamma) \neq 0$ and

$$-\frac{\Phi_{f,\mu}^{k}(\gamma)}{\Phi_{0,1}^{k-1}(\gamma)} \notin \mathbb{Q}_{>0}.$$

Proposition 3.6. *Let $L(f,\mu)$ be a side of $N(f)$ and let c_0 be a good root, then either $y = 0$ is a solution of $f[c_0x^\mu + y]$ or the side of $N(f[c_0x^\mu + y])$ which follows the one of coslope μ is pregood.*

Proof. If $y = 0$ is not a solution of $f[c_0x^\mu + y]$ this equation has a side of coslope $\mu' > \mu$ contiguous to the one of coslope μ. Now, c_0 being a good root, if

$$k = \min\{\nu_{c_0}(\Phi_{f,\mu}), \nu_{c_0}(\Phi_{0,1}) + 1\}$$

then

$$\begin{aligned}\Psi_{f,q(f,\mu)}(u) &= \Psi_{f,Q(f,\mu')}(u) = B_k(c_0)u + A_k(c_0)\\ &= \frac{1}{(k-1)!}\Phi_{0,1}^{k-1}(c_0)u + \frac{1}{k!}\Phi_{f,\mu}^{k}(c_0) - \frac{\mu}{(k-1)!}\Phi_{0,1}^{k-1}(c_0)\end{aligned}$$

has roots in $\mathbb{Q}_{>0}$ and *a fortiori* in $\mathbb{Q}_{\geq 0}$. □

Proposition 3.7. *If $L(f,\mu)$ is a good side for f, then $\Phi_{f,\mu}(c)$ has at least one good root.*

Proof. Since $L(f,\mu)$ is a good side, with the notations of Remark 3.4, we have $t > s$ and there are non zero roots of $\Phi_{f,\mu}(c) = 0$. Let these roots be $\gamma_1, \ldots, \gamma_r$. Let us suppose that there are no good roots. Then setting

$$k_i = \min\{\nu_{\gamma_i}(\Phi_{f,\mu}), \nu_{\gamma_i}(\Phi_{0,1}) + 1\},\ 1 \leq i \leq r,$$

we have

$$\Phi_{0,1}^{k_i-1}(\gamma_i) \neq 0, \text{ and } -\frac{\Phi_{f,\mu}^{k_i}(\gamma_i)}{\Phi_{0,1}^{k_i-1}(\gamma_i)} \in \mathbb{Q}_{>0},\ 1 \leq i \leq r,$$

and then

$$\nu_{\gamma_i}(\Phi_{f,\mu}) = k_i, \text{ and } \nu_{\gamma_i}(\Phi_{0,1}) = k_i - 1.$$

First we may suppose, dividing by a convenient factor, that all the $k_i - s$ are 1, and then $\Phi_{f,\mu}(c) = c^s\prod_{i=1}^r(c-\gamma_i)$, and in consequence, $a_s(f,\mu) = a_t(f,\mu)\prod_{i=1}^r(-\gamma_i)$. Using now the equalities of Remark 3.3, we have

$$\sum_{i=1}^r \frac{\Phi_{0,1}(\gamma_i)}{\Phi'_{f,\mu}(\gamma_i)} = \frac{b_t(f,\mu)}{a_t(f,\mu)} - \frac{b_s(f,\mu)}{a_s(f,\mu)}.$$

Then

$$\sum_{i=1}^{r} \frac{-\Phi_{0,1}(\gamma_i)}{\Phi'_{f,\mu}(\gamma_i)} + \frac{-b_s(f,\mu)}{a_s(f,\mu)} = \frac{-b_t(f,\mu)}{a_t(f,\mu)}$$

and the L.H.S. is in $\mathbb{Q}_{>0}$ and the R.H.S. is not, in this way we get a contradiction. □

Theorem 3.8. *Any first order and first degree equation f, formal or convergent, has a formal Puiseux solution.*

Proof. In $N(f)$ the first side is pregood, then there is a good side $L(f,\mu_1)$. For this side there is at least one good root c_1. Then, by applying the change of variable $z = c_1 x^{\mu_1} + y$, we get a new equation f_1. In $N(f_1)$ the side that follows the one of coslope μ_1 is pregood, and then we may continue the process. The algorithm has been implemented in Maple by M. Farto [12]. □

Remark 3.9. The solutions obtained using this algorithm are in general not convergent. There exists another algorithm (as we will see below) which gives us a very precise convergent solution, but neither exists a characterization of the solutions obtained using this algorithm over all the Newton polygon of a differentiable equation, nor is it known if there are convergent solutions which do not come from a Newton polygon.

Theorem 3.10. *Any convergent differential equation of first order and first degree has a convergent Puiseux solution.*

Proof. We need some more restrictive criteria, for the selection of the good side and the good root in order to get the convergence of the solution.

We will define a c-pregood side as a side $L(f,\mu)$ such that $\Psi_{f,Q(f,\mu)}(u) = 0$ has a root which is not in $\mathbb{Q}_{\geq\mu}$; as above, a c-good side is a c-pregood one with the same additional conditions which make the difference between pregood and good. The only difference in the characterization (Remark 3.4) is that $\mathbb{Q}_{>0}$ is replaced by $\mathbb{Q}_{\geq 0}$ in condition 2. In the definition of good root the only condition to add is

$$\nu_\gamma(\Phi_{f,\mu}) \geq \nu_\gamma(\Phi_{0,1}) + 1.$$

Then k equals the right-hand side of the inequality.

By the same arguments as above there are c-good sides and c-good roots, and any pair of them gives rise, by the same process, to one solution of the equation. The solutions obtained now can also be divergent, but if we choose at each step the maximum of the $\mu - s$ which give c-good sides (whose corresponding sides are called principal ones) the solution is convergent.

To prove the convergence we consider the solution constructed above and let $Q = (a,b)$ be its pivot point, reached at the step r_0. Since obviously $b > 0$, there are two possibilities

(1) $b \geq 2$. In this case we consider the curve

$$g(x,y) = \frac{\partial^{b-1} f}{\partial y^{b-2} \partial y'} = 0$$

and, as we do with f, define

$$g_0 = g, \; g_i = g_{i-1}[c_i x^{\mu_i} + y].$$

Then

$$g_i(x,y) = \frac{\partial^{b-1} f_i}{\partial y^{b-2} \partial y'}, \text{ for all } i \geq 0$$

and consequently

$$(i,j) \in \Delta(g_r) \Rightarrow (i-1, j+b-1) \in \Delta(f_r), \text{ and } j \geq 0.$$

Since $L(f, \mu_r)$ is principal , the coefficient of $x^{a+1}y^{b-1}y'$ in f_r is different from zero for $r \geq r_0$. Then, $(a+1,1) \in \Delta(g_r)$, and lies consequently above $L(g_r, \mu_r)$, for $r \geq r_0$. Then the Puiseux series solution of of f is also a branch of $g = 0$ and is convergent.
(2) $b = 1$. In this case, after a ramification we may suppose that the solution z is a formal series in x, and that the pivot point is $(a, 1)$. Now the coefficient of $x^{a+1}y'$ in f_r, $r \geq r_0$ is fixed and different from zero. Then

$$\nu_x(\frac{\partial f}{\partial y'}[z]) = a+1, \; \nu_x(\frac{\partial f}{\partial y}[z]) \geq a,$$

and the differential operator

$$(\frac{\partial f}{\partial y'}[z])\partial + \frac{\partial f}{\partial y}[z]$$

has a regular singularity at the origin. Thus z converges (Malgrange [25]). □

Remark 3.11. We define a Gevrey series of index s for $s \in [0, \infty]$ as a formal power series $\sum a_i x^i$ for which the series

$$\sum \frac{a_i}{(i!)^s} x^i$$

converges. Gevrey series of zero (resp. infinite) index are the convergent series (resp. all the formal ones). If the differential equation

$$f(x,y,y') = \sum_{i=0}^{n} a_i(x,y)(y')^i = 0,$$

where $a_i(x,y) \in \mathbb{C}\{x,y\}$ has some Puiseux series as solutions, we may assume, after a ramification, that all of them are entire series. By a theorem of Maillet [24] proved also by Mahler [23], there is a number $s, 0 \leq s < \infty$, such that all of these series are Gevrey of index s.

If we take a differential equation of first order such that all their coefficients are series with Gevrey index r, can be proved by using a modification of the last proof, that all their series solutions are also Gevrey series (in general with Gevrey index greater than r).

References

[1] Aroca, F.: *Resolución de ecuaciones diferenciales ordinarias de primer orden.* Ph.D. Thesis Univ. Valladolid 1998.

[2] Briot, C.A., Bouquet, J.C.: *Propriétés des fonctions définies par les équations différentielles.* Journal de l'Ecole Polytechnique 36 (1856), 133–198.

[3] Camacho, C., Sad, P.: *Invariant Varieties Through Singularities of Holomorphic Vector Fields.* Ann. of Math. 115 (1982), 579–595.

[4] Cano, F., Cerveau, D.: *Desingularization of non-dicritical Holomorphic Foliations and Existence of Separatrices.* Acta Mathematica 169 (1992), 1–103.

[5] Cano, F., Cerveau, D.: *Feuilletages holomorphes singulières.* Book in preparation.

[6] Cano, F., Mattei, J.F.: *Hypersurfaces intégrales des feuilletages holomorphes.* Colloque de Géométrie à la mémoire de Claude Godbillon et Jean Martinet. Ann. Inst. Fourier 42, 1–2 (1992), 49–72.

[7] Cano, J.: *An Extension of the Newton-Puiseux Polygon Construction to Give Solutions of Pfaffian Forms.* Ann. Inst. Fourier 43 (1993), 125–142.

[8] Cano, J.: *On the series defined by differential equations with an extension of the Puiseux polygon construction to these equations.* Analysis 3 (1993), 103–117.

[9] Cano, J.: *Construction of Invariant Curves for Holomorphic Vector Fields.* Proc. Amer. Math. Soc. vol.125, num. 9 (1997), 2649–2650.

[10] Cerveau, D., Mattei, J.F.: *Formes intégrables holomorphes singulières.* Astérisque 97, 1982.

[11] Cossart, V.: *Forme normale pour une fonction sur une variété de dimension trois en caractéristique positive.* Thèse d'Etat, Orsay, 1988.

[12] Farto, M.: *Resolución efectiva de ecuaciones diferenciales.* Tesis doctoral, Univ. de Valladolid 1995.

[13] Fine, J.: *On the functions defined by differential equations with an extension of the Puiseux polygon construction to these equations.* Am. J. of Math. XI (1889), 317–328.

[14] Fine, J.: *Singular solutions of ordinary differential equations.* Amer. J. of Math. XII (1890), 293–322.

[15] Giraud, J.: *Forme normale d'une fonction sur une surface de caractéristique positive.* Bull. Soc. Math. France 111 (1983), 109–124.

[16] Giraud, J.: *Condition de Jung pour les revêtements radiciels de hauteur un.* Proc. Alg. Geom. Tokyo/Kyoto 1982, Lecture Notes in Math. 1016, Springer-Verlag, 1983, 313–333.

[17] Grigoriev, D. Yu, Singer, M.: *Solving ordinary differential equations in terms of series with real exponents.* Trans. Amer. Math. Soc. 327 (1991), 339–351.

[18] Hironaka, H.: *Introduction to the Theory of Infinitely Near Singular Points.* Mem. Mat. Inst. Jorge Juan 28, Madrid, 1975.

[19] Hironaka, H.: *Characteristic Polyhedra of Singularities.* J. Math. Kyoto Univ. 7-3 (1967), 251–293.

[20] Hironaka, H.: *Desingularization of Excellent Surfaces.* Adv. Sci. Seminar (1967), Bowdoin College. Appeared in Lect. Notes in Mathematics 1101, Springer-Verlag (1984).

[21] Ince, P.: *Ordinary differential equations.* Dover, New York, 1956.

[22] McDonald, J.: *Fiber polytopes and fractional power series.* Journal of Pure and Appl. Algebra 104 (1955), 213–233.

[23] Mahler, K.: *On formal power series defined by algebraic differential equations.* Rend. Linc., Vol. L (1971), 76–89.

[24] Maillet, E.: *Sur les séries divergentes et les équations différentielles.* L'Ens. Math. 20 (1903), 487–518.

[25] Malgrange, B.: *Sur le théorème de Maillet.* Assymp. Anal. 2 (1989), 1–4.

[26] Seidenberg, A.: *Reduction of the singularities of the differential equation $Ady = Bdx$.* Amer. J. Math. 90 (1968), 248–269.

[27] Siu, Y.-T.: *Techniques of Extension of Analytic Objects.* Lecture Notes in Pure and Appplied Math., vol. 8, Marcel Dekker Inc., New York, 1974.

[28] Trifonov, S.I.: *Resolution of singularities of one parameter analytic families of differential equations.* Amer. Math. Soc. Transl. (2), vol. 151, (1992).

[29] Walker: *Reduction of the singularities of an algebraic surface.* Ann. of Math. 36, No. 2 (1935), 336–365.

Instituto de Estudios de
Iberoamerica y Portugal
Casas del Tratado
Tordesillas
Valladolid, Spain
aroca@cpd.uva.es

Progress in Mathematics, Vol. 181, © 2000 Birkhäuser Verlag Basel/Switzerland

A Course on Constructive Desingularization and Equivariance

Santiago Encinas and Orlando Villamayor

Abstract. We study a constructive proof of desingularization, as the outcome of a process obtained by successively blowing up the maximum stratum of a function f_X. We focus on canonical properties of this desingularization such as compatibility with change of base field and that of equivariance, namely the lifting of any group action on X to an action on the desingularization defined by this procedure.

Introduction

Let W denote a regular variety or a separated irreducible scheme of finite type and smooth over a field $\mathbf{k}$ of characteristic zero. Set $d = \dim W$, W is equipped with a locally free sheaf of relative differentials Ω^1_W. If $\Pi : W_1 \longrightarrow W$ is a monoidal transformation (on a closed smooth subscheme Y), then W_1 is also smooth over $\mathbf{k}$, Π introduces an exceptional hypersurface $H = \Pi^{-1}(Y)$, also regular, and both W and W_1 carry locally free sheaves: Ω^1_W and $\Omega^1_{W_1}$.

We consider here $W \longleftarrow W_1$ to be a monoidal transformation or a composition of monoidal transformations, in the latter case the exceptional locus is a union of hypersurfaces $H_i \in E = \{H_1, \dots, H_s\}$, and we shall require that these hypersurfaces have only normal crossings.

If $X \subset W$ is a reduced closed subscheme then $(X \subset W) \longleftarrow (X_1 \subset W_1)$ is said to be an embedded desingularization of X if setting $W \longleftarrow W_1$ as before, and X_1 the strict transform of X, then X_1 is regular, the induced morphism $X \longleftarrow X_1$ is an isomorphism over $\mathrm{Reg}(X)$ and, in addition, X_1 has normal crossings with E.

In various and very natural ways one can classify singularities of a singular scheme X. Roughly speaking you want to assign a value at each point in order to weigh and compare singularities of X, and this criterion is expected to be algebraic in the sense that it defines a partition of X into finitely many locally closed subsets or strata, each subset or stratum corresponding to points with the same assigned value. An example is $\mathrm{mult}_X : X \longrightarrow \mathbb{Z}$ where X is a hypersurface and $\mathrm{mult}_X(\xi)$ is the multiplicity of X at ξ. The induced partition corresponds to the equimultiple points.

A function $f : X \longrightarrow (I, \leq)$ is said to be upper-semi-continuous (u. s. c.) if $(I, \leq)$ is a totally ordered set, the image $f(X)$ is a finite subset of I and for any value $\alpha \in I$ the set $\{\xi \in X \mid f(\xi) \geq \alpha\}$ is closed in X. A criterion of valuation of singularities should be formulated by a suitable $(I, \leq)$ and f, and the partition of X is defined by subsets of the form $\{f = \alpha\}$. Note that $\underline{\mathrm{Max}}\, f = \{\xi \in X \mid f(\xi) = \max f\}$ (where $\max f$ is the maximum value achieved by f), called the set of f-worst points, is closed in X.

We shall consider $I = \mathbb{Z}^{\mathbb{Z}}$ with the lexicographic ordering and define $\mathrm{HS}_X : X \longrightarrow (I, \leq)$ as an u. s. c. function where whenever ξ is a closed point, $\mathrm{HS}_X(\xi)$ is the full set of values of the Hilbert-Samuel function at ξ, the induced partition being the so called Hilbert-Samuel stratification.

The constructive proof of Hironaka's desingularization relies on a suitable definition of $(I, \leq)$ and f, so that:

1. $\underline{\mathrm{Max}}\, f$ is regular (and any f-stratum is regular).
2. If $(X \subset W) \longleftarrow (X_1 \subset W_1)$ is defined by blowing up $\underline{\mathrm{Max}}\, f$, then $\max f > \max f_1$ (f_1 defined in terms of X_1, strict transform of X).
3. Embedded desingularization is achieved by blowing up successively at $\underline{\mathrm{Max}}\, f_i$.

This is what these notes are about, so we shall construct the set $(I, \leq)$, and define the u. s. c. function f mentioned above. The stratification defined by f is a refinement of the Hilbert-Samuel stratification, but with the nice property that the stratum of worst points is regular, and that by blowing up at the worst points the singularities of the strict transform have improved.

Note that the values of the functions HS or the function mult_X at a closed point ξ are defined in terms of the local ring of X at ξ, so both functions are intrinsic to X and not related at all to the relative structure of X over the field $\mathbf{k}$. We shall draw attention on this fact. If you take $X = V(Z_1^2 - Z_2^2 Z_3) \subset \mathbb{A}^3_{\mathbb{Q}}$, we say that X is a hypersurface in the *ambient space* $\mathbb{A}^3_{\mathbb{Q}}$. We may define a *change of the ambient space* in various ways, for instance extending $\mathbb{Q}$ by $\mathbb{C}$. There is a natural morphism $\Pi : \mathbb{A}^3_{\mathbb{C}} \longrightarrow \mathbb{A}^3_{\mathbb{Q}}$ inducing a surjective morphism $\Pi : X_1 \longrightarrow X$, which is not a morphism of finite type, and however the multiplicity is not affected by such change, in fact $\mathrm{mult}_X(\Pi(\xi)) = \mathrm{mult}_{X_1}(\xi)$ for any $\xi \in X_1$. So to some extent, in computing the multiplicity at a given point, we are allowed to make suitable changes of the ambient space. In our case we may replace $X \subset \mathbb{A}^3_{\mathbb{Q}}$ by $X_1 \subset \mathbb{A}^3_{\mathbb{C}}$. The multiplicity is also not affected by changing W by $W_1 = W \times \mathbb{A}^1_{\mathbf{k}}$, here Π is the projection on the first coordinate, and X_1 is the pullback of X. The same holds if W is changed by W_1 where $\Pi : W_1 \longrightarrow W$ is smooth or étale.

This possibility of changing the ambient space is very convenient. If, for instance, we want to compute the multiplicity or the Hilbert-Samuel function of X at a closed point, we may always reduce to the case when the point is rational. Note also that mult_{X_1} is compatible with $\Theta : X_1 \longrightarrow X_1$ ($\mathrm{mult}_X(\xi) = \mathrm{mult}_X(\Theta(\xi))$) if Θ is any isomorphism, even if such Θ is not compatible with the relative structure.

The natural questions are:

1. What role does the relative structure over **k** and the existence of the locally free sheaf Ω^1_W play, if the very notion of singularity is not a relative notion?
2. To what extent is our resolution function f intrinsic to the scheme, and what kind of changes of the ambient space are for free?

In relation to the first question, let us say that the relative structure is not intrinsic to singularities but it helps. The singular locus of a hypersurface X and the very notion of singularity is of course intrinsic to X, but it is the relative structure that allows us to provide equations defining the singular locus in terms of the equation defining X, namely

$$\left(h, \frac{\partial h}{\partial x_1}, \ldots, \frac{\partial h}{\partial x_n}\right).$$

Note here that $\frac{\partial}{\partial x_i} \in \left(\Omega^1_W\right)^*$ (dual of Ω^1_W). The desingularization defined by the function f_X is such that the local equations of $\underline{\mathrm{Max}}\, f_X$, and of any f_X-stratum, can be obtained in terms of the local equations defining $X \subset W$.

On the one hand we want the stratification defined by f_X not to be affected by suitable changes of the ambient space. On the other hand we want to profit from this extra relative structure in order to achieve ultimately and by means of an algorithm, equations defining the center $\underline{\mathrm{Max}}\, f_X$ in terms of the equations defining X. The first aspect will lead us to the canonical properties of our desingularization, such as lifting of group actions (3.1) or showing that formally isomorphic points undergo the "same" desingularization [V2], whereas the second to the constructive nature of this procedure.

A list of exercises is included to motivate the development, to clarify the difficulties that one must overcome in order to argue by induction on the dimension of the ambient space. The exercises should also help the reader get a better grasp of the spirit of our proof, namely how the subtle equivariant functions w-ord (4.11, 7.6) and n (4.14) are suitably defined so as to make this induction possible, thereby leading to equivariant desingularization.

The notes are organized around our main theorem 3.1 of constructive desingularization proved by induction on the dimension of the ambient space, say $n = \dim W$. However, the way we go about this *induction* requires some clarification. We first introduce a class of objects of the form $B = (W, (J, b), E)$ where W is a smooth scheme, J is a coherent sheaf of ideals, b is an integer, and E is a set of smooth hypersurfaces in W having only normal crossings. We say that B is an n-dimensional basic object if $n = \dim W$. To each such B we assign a closed subset $\mathrm{Sing}(J, b) \subset W$, called the singular locus of B, setting $\mathrm{Sing}(J, b)$ to be the points of W where the ideal J has order at least b. This is a closed set in the Zariski topology of W. We will define a notion of *transformation* of basic objects and also one of *resolution* of basic objects. So in these notes we never say *resolution* for *desingularization*.

The notion of *resolution* of basic objects plays a priviledged role, in fact, the algorithm of *desingularization* of embedded schemes in theorem 3.1 will follow from an algorithm of *resolution* of basic objects.

The functions w-ord(ξ) and $n(\xi)$ are the key to show how an algorithm of *resolution* of n-dimensional basic objects follows by induction on n. Incidentally, it is in this sense that embedded *desingularization* follows by induction on the dimension of the ambient space W.

The functions w-ord and n are defined so as to mimic a very simple but fundamental property of the multiplicity of an embedded hypersurface. Let $X \subset W$ be a hypersurface, set $J = I(X)$, the invertible sheaf of ideals defining the hypersurface, let b the highest possible multiplicity at points of X. In this case $(W, (J, b), E)$ will be an n-dimensional basic object ($n = \dim W$) and Sing(J, b) consists exactly of the b-fold points of the hypersurface. The striking feature of the multiplicity of a hypersurface is expressed by the following property: if Sing(J, b) is of codimension 1 in W, then such closed subset is already regular. And in case Sing(J, b) has codimension at least 2 in W, then it is locally the singular locus of an $(n-1)$-dimensional basic object. This is the main motivation for induction in algorithmic or constructive *desingularization* and the starting point of our notes.

Sections 10 to 15 are devoted to exercises. A first look at exercise 13.1 is strongly recommended to motivate the notion of transformation and *resolutions* of basic objects. Examples 4.16 and 4.18 are the first contact with the behaviour of the two inductive invariants w-ord(ξ) and $n(\xi)$, examplified in the *resolution* of 2-fold points of the hypersurface $X = V(Z^2 + (X^2 - Y^3)^2) \subset \mathbb{A}^3_{\mathbf{k}}$. More examples will arise both in the notes and the list of exercises.

We refer to [V1, V2, V4] to other presentations of the algorithm treated in these notes, and for other explicit examples on how the algorithm works. We also refer to [BM2] and [EV1] for other algorithms, always within Hironaka's line of proof and to [AJ], [AW] and [BP] for algorithms in a different line.

1. First definitions

In what follows $\mathbf{k}$ will denote a field of characteristic zero.

Definition 1.1. If $(R, \mathfrak{m})$ is a local regular ring, and $J \neq 0$ is an ideal in R, we say that the order of J at R is b ($\nu_R(J) = b$) if $J \subset \mathfrak{m}^b$ and $J \not\subset \mathfrak{m}^{b+1}$. The order of an element $f \in R$ is defined as the order of $\langle f \rangle = fR$.

1.2. The ring of formal power series $\mathbf{k}[[X_1, \dots, X_n]]$ is a local regular ring of dimension n. If $0 \neq J \subset \mathbf{k}[[X_1, \dots, X_n]]$ is an ideal, then $J = \langle f_1, \dots, f_r \rangle$ (is finitely generated) and we define $\hat{\Delta}(J) \supseteq J$ (an extension of J)

$$\hat{\Delta}(J) = \left\langle f_1, \dots, f_r, \frac{\partial f_i}{\partial x_j} \begin{array}{l} i = 1, \dots, r \\ j = 1, \dots, n \end{array} \right\rangle \tag{1.2.1}$$

And we set $\hat{\Delta}^d(J) = \hat{\Delta}(\hat{\Delta}^{d-1}(J))$ for $d \in \mathbb{Z} > 0$.

Take $f = X^2 + Y^3 \in \mathbf{k}[[X,Y]]$ and $J = \langle f \rangle$, then $J \subset \hat{\Delta}(J) = \langle X, Y^2 \rangle \subset \hat{\Delta}(\hat{\Delta}(J)) = \mathbf{k}[[X,Y]]$.

Note that, whenever an ideal $J \subset \mathbf{k}[[X_1, \dots, X_n]]$ is proper (i.e. whenever $\nu(J) > 0$) then $\nu(\hat{\Delta}(J)) = \nu(J) - 1$ but this formula fails in positive characteristic (e.g. check the example $f = X^2 + Y^3$ in case of characteristic two). It is for this reason that we exclude fields of positive characteristic.

1.3. We state now some properties on $\hat{\Delta}$:

1. $\hat{\Delta}(J)$ is independent of the choice of the generators $\{f_1, \dots, f_r\}$ and of the regular system of coordinates in 1.2.1 (see 10.2).
2. $\nu(J) = b > 0$ iff $\nu(\hat{\Delta}(J)) = b - 1$.
3. $\nu(J) = b > 0 \iff \nu(\hat{\Delta}^{b-1}(J)) = 1$
   $\iff \nu(\hat{\Delta}^{b-1}(J)) = 1$ and $\nu(\hat{\Delta}^{b}(J)) = 0$
4. $\nu(J) \geq b$ iff $\hat{\Delta}^{b-1}(J)$ is a proper ideal.

1.4. In what follows W is a regular variety over $\mathbf{k}$ (irreducible scheme separated of finite type and smooth over $\mathbf{k}$), of dimension n. So W is irreducible and at any point $\xi \in W$ (here all points are closed, see exercise 10.6) there is a local regular ring $\mathcal{O}_{W,\xi}$, and

$$\hat{\mathcal{O}}_{W,\xi} = \mathbf{k}'[[X_1, \dots, X_n]]$$

where $\{x_1, \dots, x_n\}$ can be chosen as a regular system of parameters of $\mathcal{O}_{W,\xi}$, and $\mathbf{k}'$ is a finite extension of $\mathbf{k}$ (e.g. $W = \mathbb{A}^n_{\mathbb{Q}}$, $W = \mathbb{P}^n_{\mathbb{R}}$). We do not assume that $\mathbf{k}$ is algebraically closed.

1.5. **Notation:**

1. If $J \subset \mathcal{O}_W$ is an ideal (which means a coherent sheaf of ideals), then $V(J) = \{\xi \in W \mid J_\xi \subsetneq \mathcal{O}_{W,\xi}\} \subset W$ is the closed set defined by J.
2. If $F \subset W$ is closed, $I(F) \subset \mathcal{O}_W$ is the ideal of functions vanishing on F.

In our setting there is a locally free sheaf of modules of differentials $\Omega^1_{W/\mathbf{k}}$ which will allow us to define, for an ideal $J \subset \mathcal{O}_W$, an extension $\Delta(J) \subset \mathcal{O}_W$ (see 10.2) so that for any $\xi \in W$

$$\Delta(J)\hat{\mathcal{O}}_{W,\xi} = \hat{\Delta}(J\hat{\mathcal{O}}_{W,\xi}) \tag{1.5.1}$$

Given $J \subset \mathcal{O}_W$ and $\xi \in W$, set $\nu_\xi(J)$ to be the order of J_ξ in the local regular ring $\mathcal{O}_{W,\xi}$.

1.6. **Properties.** Fix $0 \neq J \subset \mathcal{O}_W$, then:

P1 $J \subset \Delta(J) \subset \cdots \subset \Delta^b(J) = \mathcal{O}_W$ for some b.

P2 $\nu_\xi(J) = b > 0$ iff $\nu_\xi(\Delta(J)) = b - 1$.

P3 $\nu_\xi(J) = b > 0$ iff $\nu_\xi(\Delta^{b-1}(J)) = 1$.

P4 $\nu_\xi(J) = b > 0$ iff $\xi \in V(\Delta^{b-1}(J)) \setminus V(\Delta^b(J))$.

P5 $\nu_\xi(J) \geq b > 0$ iff $\xi \in V(\Delta^{b-1}(J))$.

Definition 1.7. Let X be a noetherian topological space [Har, p. 5] and $(I, \leq)$ a totally ordered set. A function

$$f : X \longrightarrow (I, \leq)$$

is said to be upper-semi-continuous (u. s. c.) if

i. $\operatorname{Im} f = \{\alpha_1 < \cdots < \alpha_r\}$ is a finite subset of I.

ii. For any $i = 1, \ldots, r$, $\{\xi \in X \mid f(\xi) \geq \alpha_i\}$ is closed.

In such case we denote by

(a) $\max f (= \alpha_r)$ the biggest value achieved, and

(b) $\underline{\operatorname{Max}} f = \{\xi \in X \mid f(\xi) = \max f\}$, which is closed in X.

So an u. s. c. function f, defines a partition of X

$$X = X_{\alpha_1} \sqcup \cdots \sqcup X_{\alpha_r} \qquad \text{(disjoint union)}$$

setting $X_{\alpha_i} = \{\xi \in X \mid f(\xi) = \alpha_i\}$ which is locally closed, and $X_{\alpha_r} = \underline{\operatorname{Max}} f$ is closed.

1.8. *Examples.*

1. Let X be a proper closed set in the n-dimensional variety W. Define $f : X \longrightarrow (\mathbb{Z}, \leq)$ by setting for $\xi \in X$, $f(\xi) =$ local dimension of X at ξ. Then $f(\xi) < \dim W = n$ since X is proper, and we can check that f is an u. s. c. function.

   We shall denote by $R(1)(X)(\subset X)$ the set of points where the local dimension of X is $n-1$ (i.e. where the local codimension of X in W is 1). Note that $R(1)(X)$ is a (possibly empty) closed subset since f is u. s. c.

2. We attach to an ideal $0 \neq J \subset \mathcal{O}_W$ a function $\nu_J : W \longrightarrow (\mathbb{Z}, \leq)$ by setting $\nu_J(\xi) = \nu_\xi(J)$ (1.5). It follows from P1 and P5 in 1.6 that ν_J is u. s. c. If $\max \nu_J = b$ then

   $$\underline{\operatorname{Max}}\, \nu_J = V(\Delta^{b-1}(J))$$

   and 1.6(P3) asserts that $\nu_{\Delta^{b-1}(J)}(\xi) = 1$ for any $\xi \in \underline{\operatorname{Max}}\, \nu_J$.

3. Let $X \subset W$ be a hypersurface, and set

   $$\operatorname{mult}_X : X \longrightarrow (\mathbb{Z}, \leq)$$

   $\operatorname{mult}_X(\xi) =$ multiplicity of X at $\xi = \nu_\xi(I(X))$.

   Note that mult_X is the restriction of ν_J to $V(J)$ for $J = I(X)$. In particular mult_X is also an u. s. c. function.

   If $\max \operatorname{mult}_X = b$ then $\underline{\operatorname{Max}} \operatorname{mult}_X$ is the closed set of b-fold points. Note that

   $$\underline{\operatorname{Max}} \operatorname{mult}_X = V(\Delta^{b-1}(I(X)))$$

   and $\nu_{\Delta^{b-1}(I(X))}(\xi) = 1$ for any $\xi \in \underline{\operatorname{Max}} \operatorname{mult}_X$. In particular $\underline{\operatorname{Max}} \operatorname{mult}_X$ is locally included in a regular hypersurface.

1.9. **Important remark.** Fix an ideal $0 \neq \mathfrak{a} \subset \mathcal{O}_W$, assume that $\nu_{\mathfrak{a}}(\xi) = 1$ for any $\xi \in V(\mathfrak{a})$, then

1. $R(1)(V(\mathfrak{a}))$ is open and closed in $V(\mathfrak{a})$.
2. $R(1)(V(\mathfrak{a}))$ is regular.
3. $V(\mathfrak{a}) \setminus R(1)$ is locally included in regular varieties of dimension $n-1$.

Proof. Recall that $R(1)(V(\mathfrak{a}))$ is closed and possibly empty (1.8(1)). If $V(\mathfrak{a})$ has codimension 1 at $\xi \in V(\mathfrak{a})$ then $\mathfrak{a}_\xi \subset \mathcal{O}_{W,\xi}$ must be locally principal. This last condition is open, which proves (1). Note that if $\mathfrak{a}_\xi$ is principal, it is an ideal defining (locally) a regular hypersurface, this proves (2). Also (3) follows from the fact that $\mathfrak{a}_\xi$ has order 1 for any $\xi \in V(\mathfrak{a})$. ☐

Example 1.10. $I(X) = \langle X^2 - Y^3Z^3\rangle \subset \mathbf{k}[X,Y,Z]$, $\max \mathrm{mult}_X = 2$,

$$\underline{\mathrm{Max}}\,\mathrm{mult}_X = V(\Delta(I(X)))$$

$\Delta(I(X)) = \langle X, Y^2Z^3, Y^3Z^2\rangle$, so $\underline{\mathrm{Max}}\,\mathrm{mult}_X$ is the union of two axes in $\mathbb{A}^3_{\mathbf{k}}$, and $R(1)(\underline{\mathrm{Max}}\,\mathrm{mult}_X) = \emptyset$. Condition (3) in 1.9 holds (globally in this case) choosing $\langle X\rangle \subset \Delta(I(X))$ as the ideal defining a regular variety of dimension $n-1=2$.

2. Pairs

Definition 2.1.

(a) Let $W \longleftarrow W_1$ be a morphism of varieties. An ideal $J \subset \mathcal{O}_W$ induces naturally the sheaf $J\mathcal{O}_{W_1} \subset \mathcal{O}_{W_1}$, called the *total transform* of J.

(b) If W is a regular variety, and $Y \subset W$ is a closed and regular subscheme,

$$\begin{array}{ccc} W & \xleftarrow{\Pi} & W_1 \\ Y & & \end{array}$$

will denote the blow up of W along Y. $H = \Pi^{-1}(Y)$ is a hypersurface (i.e. $I(H)$ is invertible) and Π induces an isomorphism $W \setminus Y \xleftarrow{\sim} W_1 \setminus H$ (see exercises 11.1 to 11.5).

(c) If, in the setting of (b), $Y \subset X \subset W$ where X is a subscheme, then

$$\begin{array}{ccc} W & \xleftarrow{\Pi} & W_1 \\ Y \subset X & & X_1 \end{array}$$

will denote the blow-up, as in (b), and X_1 denotes the strict transform of X. The induced morphism $X \longleftarrow X_1$ is the blow-up along $Y \subset X$.

Theorem 2.2. *With the setting of 2.1(b) W_1 is regular and $H \subset W_1$ is a regular hypersurface defined by $I(H) = I(Y)\mathcal{O}_{W_1}$.*

The proof is developed in the exercises 11.1 and 11.2.

Theorem 2.3. *Let $Y \subset X \subset W$, X a hypersurface and Y a closed regular subscheme. Assume that $b = \max \operatorname{mult}_X$ and $Y \subset \underline{\operatorname{Max}} \operatorname{mult}_X$ and set*

$$\begin{array}{ccc} W & \xleftarrow{\Pi} & W_1 \\ X & & X_1 \\ Y & & \end{array}$$

X_1 the strict transform of X. Then

A *$I(X)\mathcal{O}_{W_1} = I(H)^b I(X_1)$ (left-hand side is the total transform).*

B *For any $x_1 \in X_1$, $\operatorname{mult}_{X_1}(x_1) \leq \operatorname{mult}_X(\Pi(x_1))$.*

Theorem 2.4. *Set $J \subset \mathcal{O}_W$, $b = \max \nu_J$. If $Y \subset \underline{\operatorname{Max}} \nu_J$ then*

A *There is an ideal $J_1 \subset \mathcal{O}_{W_1}$ so that $J\mathcal{O}_{W_1} = I(H)^b J_1$.*

B *For any $x_1 \in W_1$, $\nu_{J_1}(x_1) \leq \nu_J(\Pi(x_1))$.*

Remark 2.5. For B) to hold in theorem 2.4 it suffices that the orders of ideals be locally constant along Y, where locally constant means that the order is constant along each irreducible component of Y.

Example 2.6. (where the conditions of theorem 2.3 hold).

$$I(X) = X^2 + Y^3 Z^3 \qquad I(Y) = \mathfrak{p} = \langle X, Y \rangle$$

Example 2.7. (where the conditions don't hold and 2.3(B) fails).

$$I(X) = X^6 Y^2 + Z^3 \qquad I(Y) = \mathfrak{p} = \langle Y, Z \rangle$$

Proof of theorems 2.3 and 2.4. We refer to exercises 11.2 up to 11.6 for a characteristic free proof of these theorems. We shall also indicate that, over fields of characteristic zero, both theorems also follow from Giraud's lemma 6.6. □

2.8. **Main remark**

1. In the setting of theorem 2.3, $\max \operatorname{mult}_X \geq \max \operatorname{mult}_{X_1}$.
2. In the setting of remark 2.5 (in particular of theorem 2.4), $\max \nu_J \geq \max \nu_{J_1}$.

2.9. We say that a set $E = \{H_1, \dots, H_s\}$ of regular hypersurfaces has normal crossings if at any point $\xi \in \cup_{i=1}^s H_i$ there is a regular system of parameters $\{x_1, \dots, x_n\}$ at $\mathcal{O}_{W,\xi}$ so that

$$\bigcup_{i=1}^{s} H_i = V(x_1 \cdots x_r)$$

in some suitable neighborhood of ξ and for some $r \leq n$.

2.10. Set a pair $(W, E = \{H_1, \dots, H_r\})$, where W is a regular variety (irreducible scheme separated of finite type and smooth over a field $\mathbf{k}$), each H_i a regular hypersurface in W and we assume that $\cup_{i=1}^r H_i$ has normal crossings. A regular closed subscheme Y has normal crossings with E if at any $x \in Y$ there is a regular system of parameters $\{x_1, \dots, x_n\}$ such that $I(Y)_\xi = \langle x_1, \dots, x_s \rangle$ and for any H_i containing ξ, $I(H_i) = \langle x_{i_j} \rangle$. In such case we say that Y is permissible and we define a transformation of the pair

$$(W, E) \longleftarrow (W_1, E_1)$$

by blowing up Y and setting $E_1 = \{H_1', \dots, H_r', H_{r+1}\}$, H_i' the strict transform of H_i, $H_{r+1} = \Pi^{-1}(Y)$ the exceptional hypersurface in W_1 (see exercise 12.1). We will also consider a sequence of transformations of pairs:

$$(W_0, E_0) \longleftarrow (W_1, E_1) \longleftarrow \cdots \longleftarrow (W_k, E_k)$$

2.11. We shall say that an isomorphism $\Theta : W \longrightarrow W$ acts on the pair (W, E), $E = \{H_1, \dots, H_r\}$ if $\Theta(H_i) = H_i$, $i = 1, \dots, r$. If Y is permissible and $\Theta(Y) = Y$ there is a natural lifting of Θ to an action on the transform

$$(W, E) \xleftarrow{\Pi_1} (W_1, E_1)$$

and we say that Π_1 is Θ-equivariant. A sequence

$$(W_0, E_0) \xleftarrow{\Pi_1} \cdots \xleftarrow{\Pi_k} (W_k, E_k)$$

is Θ-equivariant if each Π_i is Θ-equivariant.

We shall say that Θ acts on $(X \subset W, E)$ if it acts on (W, E) and $\Theta(X) = X$. If now $Y \subset X$ and $\Theta(Y) = Y$, then Θ acts on $(X_1 \subset W_1, E_1)$, X_1 the strict transform of X.

Remark 2.12.

1. If X is a hypersurface in W and Θ acts on $(X \subset W, E)$ then $\operatorname{mult}_X(x) = \operatorname{mult}_X(\Theta(x))$. In fact isomorphic points have the same multiplicity. Note in particular that
$$\Theta(\underline{\operatorname{Max}} \operatorname{mult}_X) = \underline{\operatorname{Max}} \operatorname{mult}_X$$
2. We do not consider $\Theta : W \longrightarrow W$ as an isomorphism of $\mathbf{k}$-varieties, but simply an isomorphism of schemes, for instance, complex conjugation defines an isomorphism on $\mathbb{P}^2_{\mathbb{C}}$ acting on $X = V_{\text{proj}}(XY^5 - Z^6) \subset \mathbb{P}^2_{\mathbb{C}}$.
3. We shall ultimately see that constructive desingularization extends naturally to non-embedded schemes X and that any automorphism $\Theta : X \longrightarrow X$ (non-embedded) will lift to such desingularization.

3. Constructive desingularization

We now state the constructive theorem of desingularization.

Theorem 3.1. *For a fixed integer $n \geq 0$ we define constructively the following:*

1. *A totally ordered set $(I_n, \leq)$.*
2. *Given a triple $(X_0 \subset W_0, E_0)$, where (W_0, E_0) is a pair (2.10) with* $\dim W_0 =$ *n and X_0 is a reduced closed subscheme of W_0; we define functions $f_0, f_1, f_2, \ldots$, depending on $(X_0 \subset W_0, E_0)$, satisfying the following properties:*

   2.a *The function $f_0 : X_0 \longrightarrow I_n$ is u. s. c. and $\underline{\operatorname{Max}} f_0$ is permissible for (W_0, E_0) (2.10).*

   2.b *If $(X_0 \subset W_0, E_0) \overset{\Pi_1}{\longleftarrow} (X_1 \subset W_1, E_1)$ is the transformation with center $\underline{\operatorname{Max}} f_0$, then $f_1 : X_1 \longrightarrow I_n$ is u. s. c. and $\underline{\operatorname{Max}} f_1$ is permissible for (W_1, E_1).*

   2.c *Repeating (2.b) we obtain a sequence*

   $$(X_0 \subset W_0, E_0) \overset{\Pi_1}{\longleftarrow} (X_1 \subset W_1, E_1) \overset{\Pi_2}{\longleftarrow} \cdots \overset{\Pi_r}{\longleftarrow} (X_r \subset W_r, E_r) \tag{3.1.1}$$

   then $f_r : X_r \longrightarrow I_n$ is u. s. c. and $\underline{\operatorname{Max}} f_r$ is permissible for (W_r, E_r).

   2.d *There exists an integer r (depending on $(X_0 \subset W_0, E_0)$) so that the following conditions hold for the sequence 3.1.1:*

   (i) *X_r is regular and the induced morphism $X \overset{\Pi}{\longleftarrow} X_r$ defines an isomorphism over* $\operatorname{Reg}(X) \subset X$ *(i.e.* $\operatorname{Reg}(X) \cong \Pi^{-1}(\operatorname{Reg}(X)) \subset$ *X_r via Π).*

   (ii) *X_r has normal crossings with E_r.*

   (iii) *Any Θ acting on $(X_0 \subset W_0, E_0)$ will also act on $(X_r \subset W_r, E_r)$.*

Proof: See 7.16 for the hypersurface case and 8.18 for the general case.

Remark 3.2.

1. Each function f_i will be defined also in terms of the previous steps of the sequence 3.1.1, but this does not affect the uniqueness of the chain.
2. Each $f_i : X_i \longrightarrow I_n$ will have the property that $f_i(x) = f_i(\Theta(x))$ for any Θ acting on $(X_i \subset W_i, E_i)$. In particular $\Theta(\underline{\operatorname{Max}} f_i) = \underline{\operatorname{Max}} f_i$ for any such Θ. This proves (iii).
3. *Notation:* (i) is the condition of desingularization, (i)+(ii) is embedded desingularization and (iii) is the condition for equivariance.

Example 3.3. Take $X = V(XY - Z^2) \subset \mathbb{C}^3$. By blowing up the origin all three conditions (i), (ii) and (iii) are fulfilled.

Example 3.4. Let$X = V(Z^2 - X^3Y^3) \subset \mathbb{C}^3$. Here the singular locus is the union of two axes in $\mathbb{C}^3$. There is an action of $\mathbb{C}^*$ by setting, for $\lambda \in \mathbb{C}^*$, $\Theta_\lambda(a, b, c) = (\lambda a, \lambda^{-1} b, c)$. There is also an action of $\mathbb{Z}/2\mathbb{Z}$ setting $\Theta(a, b, c) = (b, a, c)$. Note that the *only* possible equivariant center is the origin.

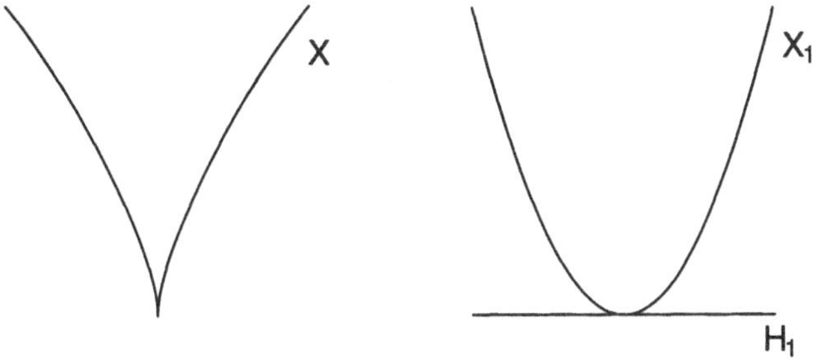

Example 3.5. Set $X = V(X^2 - Y^3) \subset \mathbb{C}^2$. By blowing up the origin we get a desingularization (X_1 is regular) but not an embedded desingularization.

4. Basic objects

4.1. Fix a pair (W_0, E_0), $E_0 = \{H_1, \dots, H_r\}$ (2.10), an ideal $0 \neq J \subset \mathcal{O}_{W_0}$ and consider a sequence of transformations of pairs

$$\begin{array}{ccccccc} (W_0, E_0) & \stackrel{\Pi_1}{\longleftarrow} & \cdots & \stackrel{\Pi_{s-1}}{\longleftarrow} & (W_{s-1}, E_{s-1}) & \stackrel{\Pi_s}{\longleftarrow} & (W_s, E_s) \\ Y_0 & & & & Y_{s-1} & & \end{array} \tag{4.1.1}$$

setting $\Pi_i^{-1}(Y_{i-1}) = H_{r+i} \in E_i = \{H_1, \dots, H_r, H_{r+1}, \dots, H_{r+i}\}$. We shall now assign for each index $1 \leq i \leq s$ an expression

$$J\mathcal{O}_{W_i} = I(H_{r+1})^{c_{r+1}} \cdots I(H_{r+i})^{c_{r+i}} \bar{J}_i \tag{4.1.2}$$

where the left-hand side is the total transform of J (2.1(a)) and all terms in 4.1.2 are ideals in $\mathcal{O}_{W_i}$. This will be done by a suitable definition of locally constant functions c_i. In the particular case in which all regular centers Y_i are irreducible, then such expression can (and will) be defined by setting inductively $c_{r+i} \in \mathbb{Z} \geq 0$ to be the order of $J\mathcal{O}_{W_{i-1}}$ at the local regular ring $\mathcal{O}_{W_{i-1}, y_{i-1}}$, where y_{i-1} denotes the generic point of $Y_{i-1} \subset W_{i-1}$ [Har, Ex 2.9, p. 80].

Note that if Y_{i-1} is irreducible, also H_{r+i} is irreducible. In general Y_{i-1} is regular and therefore a disjoint union of its irreducible components. There will also be a natural one to one correspondence between irreducible components of H_{r+i} and irreducible components of Y_{i-1}.

We finally agree to consider

$$c_{r+i} : H_{r+i} \longrightarrow \mathbb{Z} \geq 0$$

as a locally constant function (i.e. constant on each irreducible component of H_{r+i}) so that if H'_{r+i} is an irreducible component of H_{r+i} in correspondence with an irreducible component Y'_{i-1} of Y_{i-1}, then

$$c_{r+i}|_{H'_{r+i}} \in \mathbb{Z} \geq 0$$

is the order of $J\mathcal{O}_{W_{i-1}}$ at the generic point of Y'_{i-1}.

Example 4.2. Assume $\mathbf{k}$ algebraically closed and take $\xi \neq \zeta$ two points in $W_0 = \mathbb{A}^2_{\mathbf{k}}$, set $E_0 = \{H_1, \dots, H_r\}$, $Y_0 = \{\xi, \zeta\}$ and

$$\begin{array}{ccc}(W_0, E_0) & \longleftarrow & (W_1, E_1) \\ Y_0 & & \end{array}$$

Here H_{r+1} has two irreducible components: $H^{(\xi)}_{r+1} \cong \mathbb{P}^1_{\mathbf{k}}$ mapping to ξ and $H^{(\zeta)}_{r+1} \cong \mathbb{P}^1_{\mathbf{k}}$ mapping to ζ. Set $J = \langle f(x,y) \rangle$, $f(x,y) = \ell_1\ell_2\ell_3\ell_4$, $\ell_i = a_i + b_i x + c_i y$, $a_i, b_i, c_i \in \mathbf{k}$. Assume that $\xi \in V(\ell_i)$ iff $i = 1,2$ and that $\zeta \in V(\ell_j)$ iff $j = 2,3,4$.

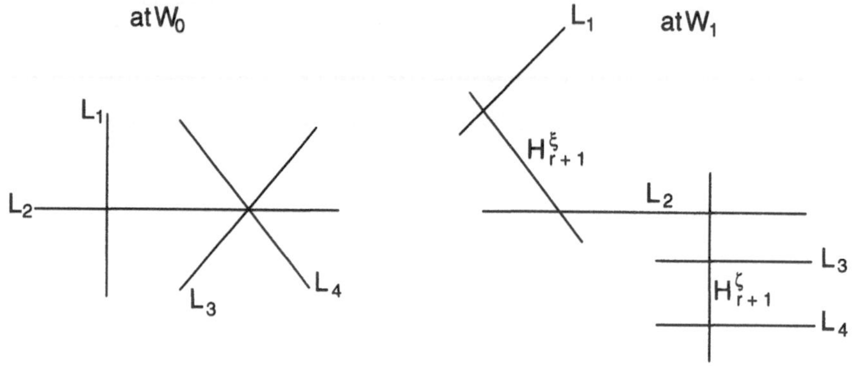

In this case $J\mathcal{O}_{W_1} = I(H_{r+1})^{c_{r+1}} \bar{J}_1$, where $c_{r+1}|_{H^\xi_{r+1}} = 2$ and $c_{r+1}|_{H^\zeta_{r+1}} = 3$.

Definition 4.3. Fix $J_0 \subset \mathcal{O}_{W_0}$ as in 4.1. For any sequence (4.1.1) and any index $i \in \{1, \dots, s\}$ we define the function

$$\nu_{\bar{J}_i} : W_i \longrightarrow (\mathbb{Z}, \leq)$$

as in 1.8(2) for $\bar{J}_i$ as in 4.1.2.

Proposition 4.4. *Suppose that the sequence 4.1.1 has been constructed inductively with $Y_i \subset \underline{\mathrm{Max}}\, \nu_{\bar{J}_i}$, then*

$$\max \nu_J \geq \max \nu_{\bar{J}_1} \geq \cdots \geq \max \nu_{\bar{J}_s} \tag{4.4.1}$$

in $\mathbb{Z} \geq 0$.

Proof. This is a corollary of 2.4(B). □

Remark 4.5. Note that $\max \nu_{\bar{J}_s} = 0$ iff $J\mathcal{O}_{W_s} = I(H_{r+1})^{c_{r+1}} \cdots I(H_{r+s})^{c_{r+s}}$.

Definition 4.6. If $\max \nu_{\bar{J}_s} = 0$, $J\mathcal{O}_{W_s}$ is an invertible sheaf of ideals and 4.1.1 is said to define a principalization if in addition $Y_i \subset V(J\mathcal{O}_{W_i})$ for all $i = 0, \dots, s-1$.

Definition 4.7. **Basic objects.** A basic object is $(W_0, (J_0, b), E_0)$, where (W_0, E_0) is a pair, $J_0 \subset \mathcal{O}_{W_0}$ is an ideal such that $(J_0)_\xi \neq 0$ for any $\xi \in W_0$, and b is a positive integer. We define the singular locus

$$F_0 = \mathrm{Sing}(J_0, b) = \{\xi \in W_0 \mid \nu_\xi(J_0) \geq b\} = V(\Delta^{b-1}(J_0)) \subset W_0$$

which is closed.

We shall say that Y_0 is *permissible* for the basic object if Y_0 is permissible for (W_0, E_0) (2.10) and $Y_0 \subset F_0$. In such case let

$$\begin{array}{ccc} W_0 & \longleftarrow & W_1 \\ Y_0 & & H \end{array}$$

be the blow-up with center Y_0. Then there is an ideal $\bar{J}_1 \subset \mathcal{O}_{W_1}$ such that $J_0\mathcal{O}_{W_1} = I(H)^{c_1}\bar{J}_1$ where $c_1 \geq b$ is locally constant on H (4.1). We define $J_1 = I(H)^{c_1-b}\bar{J}_1$ and set

$$(W_0, (J_0, b), E_0) \longleftarrow (W_1, (J_1, b), E_1)$$

as a *transformation* of the basic object.

Remark 4.8.

1. If Y_0 is irreducible $c_1 = \nu_{Y_0}(J_0)$ (order at generic point) and $c_1 \geq b$. The integer b also plays a role on the notion of transformation of basic objects. In fact, note that J_1 is not the total transform of J_0, say $J_0\mathcal{O}_{W_1} = I(H)^{c_1}\bar{J}_1$ (4.1.2), but $J_1 = I(H)^{c_1-b}\bar{J}_1$. In general, if Y_0 is not irreducible then c_1 is locally constant and so is the function $a_1 = c_1 - b$.
2. A transformation as above with center $Y \subset \mathrm{Sing}(J_0, b)$ induces an isomorphism $W_0 \setminus Y_0 \cong W_1 \setminus \Pi^{-1}(Y_0)$, and the corresponding restrictions of J_0 and J_1 coincide via this isomorphism. Now $\Pi(\mathrm{Sing}(J_1, b)) \subset \mathrm{Sing}(J_0, b)$ follows easily from the fact that $Y \subset \mathrm{Sing}(J_0, b)$.

We consider now a sequence of transformations of basic objects:

$$(W_0, (J_0, b), E_0) \longleftarrow \cdots \longleftarrow (W_k, (J_k, b), E_k) \tag{4.8.1}$$

always with the additional constraint that $Y_i \subset \mathrm{Sing}(J_i, b)$. For each index i there is an expression for the total transform of J_0, say

$$J\mathcal{O}_{W_i} = I(H_{r+1})^{c_{r+1}} \cdots I(H_{r+i})^{c_{r+i}}\bar{J}_i$$

in terms of locally constant functions c_j (4.1.2). We define now, for each J_i an expression

$$J_i = I(H_{r+1})^{a_{r+1}} \cdots I(H_{r+i})^{a_{r+i}}\bar{J}_i \tag{4.8.2}$$

in terms of locally constant functions a_{r+i}, so that $a_{r+i} \leq c_{r+i}$. Assume first that such expression is defined for $i \geq 0$ and, for simplicity, that $Y_i \subset \mathrm{Sing}(J_i, b)$ is irreducible. Set $\alpha_{r+i+1} = \nu_{y_i}(J_i)$ where y_i denotes the generic point of Y_i. Note from 4.8.2 that

$$\alpha_{r+i+1} = \sum_{j=1}^{i} a_{r+j}(y_i) + \nu_{y_i}(\bar{J}_i) \leq c_{r+i}$$

Now check that 4.8.2 is naturally defined at level $i+1$ by setting

$$a_{r+i+1} = \alpha_{r+i+1} - b.$$

Definition 4.9. The sequence of transformations 4.8.1 is a *resolution* of the basic object $(W_0, (J_0, b), E_0)$ if $\mathrm{Sing}(J_k, b) = \emptyset$.

Remark 4.10. We shall reduce our theorem of constructive desingularization to a sufficiently strong theorem of resolution of basic objects. Recall that if $b = \max \mathrm{mult}_{X_0}$ for a hypersurface $X_0 \subset W_0$, then $\mathrm{Sing}(I(X_0), b) = \underline{\mathrm{Max}}\,\mathrm{mult}_{X_0}$ and a resolution

$$(W_0, (I(X_0), b), E_0) \longleftarrow \cdots \longleftarrow (W_s, (I(X_s), b), E_s)$$

induces

$$(X_0 \subset W_0, E_0) \longleftarrow \cdots \longleftarrow (X_s \subset W_s, E_s)$$

(see theorem 2.3) and $\max \mathrm{mult}_{X_0} = \cdots = \max \mathrm{mult}_{X_{s-1}} > \max \mathrm{mult}_{X_s}$.

Definition 4.11. Recall that for any sequence of transformations, say 4.8.1, and any index i, we have fixed an expression of J_i in 4.8.2.

We shall study the functions $\nu_{\bar{J}_i}$ but now restricted to $\mathrm{Sing}(J_i, b)$, say $\nu_{\bar{J}_i} : \mathrm{Sing}(J_i, b) \longrightarrow \mathbb{Z}$, and slightly modified, namely set:

$$\text{w-ord}_i : \mathrm{Sing}(J_i, b) \longrightarrow \frac{1}{b}\mathbb{Z} \subset \mathbb{Q} \qquad \mathrm{ord}_i : \mathrm{Sing}(J_i, b) \longrightarrow \frac{1}{b}\mathbb{Z} \subset \mathbb{Q}$$

$$\text{w-ord}_i(\xi) = \frac{\nu_\xi(\bar{J}_i)}{b} = \frac{\nu_{\bar{J}_i}(\xi)}{b} \qquad \mathrm{ord}_i(\xi) = \frac{\nu_\xi(J_i)}{b} = \frac{\nu_{J_i}(\xi)}{b}$$

Since $\nu_{\bar{J}_i}$ is u. s. c. (1.8(2)), also w-ord$_i$ is u. s. c.

For $i = 0$, w-ord$_0$ is the same as Hironaka's function ord$_0$ [H2, p. 56]. Note that w-ord$_0$ = ord$_0$.

Remark 4.12. Although there is a constraint on the centers Y_i in the definition of 4.8.1, if

$$Y_i \subset \underline{\mathrm{Max}}\,\text{w-ord}_i \subset \mathrm{Sing}(J_i, b)$$

then remark 2.8 asserts that

$$\text{w-ord}_{i-1}(\Pi_i(\xi_i)) \geq \text{w-ord}_i(\xi_i) \tag{4.12.1}$$

for any $\xi_i \in \mathrm{Sing}(J_i, b)$ (see remark 2.5). So if $Y_i \subset \underline{\mathrm{Max}}\,\text{w-ord}_i$ then

$$\max \text{w-ord}_0 \geq \cdots \geq \max \text{w-ord}_k \tag{4.12.2}$$

Remark 4.13. Note that $\max \text{w-ord}_k \in \frac{1}{b}\mathbb{Z}$. We shall show in section 5 that if $\max \text{w-ord}_k = 0$ in 4.12.2 then it is simple to "extend" 4.8.1 to a resolution of $(W_0, (J_0, b), E_0)$ (4.9).

Definition 4.14. If $\max \text{w-ord}_k > 0$ in 4.12.2, set k_0 the smallest index so that $\max \text{w-ord}_{k_0-1} > \max \text{w-ord}_{k_0} = \max \text{w-ord}_k$ ($k_0 = 0$ if $\max \text{w-ord}_0 = \cdots = \max \text{w-ord}_k$), set $E_k = E_k^+ \sqcup E_k^-$, E_k^- hypersurfaces of E_k which are strict transforms of hypersurfaces of E_{k_0}, and define

$$t_k : \mathrm{Sing}(J_k, b) \longrightarrow (\mathbb{Q} \times \mathbb{Z}, \leq) \text{ (lexicographic order)}$$

$$t_k(\xi) = (\text{w-ord}_k(\xi), n_k(\xi))$$

$$n_k(\xi) = \begin{cases} \#\{H \in E_k \mid \xi \in H\} & \text{if } \text{w-ord}_k(\xi) < \max \text{w-ord}_k \\ \#\{H \in E_k^- \mid \xi \in H\} & \text{if } \text{w-ord}_k(\xi) = \max \text{w-ord}_k \end{cases}$$

In the same way we define functions $t_{k-1}, t_{k-2}, \ldots, t_0$.

4.15. **Properties of the function t.**

1. Each $t_i : \mathrm{Sing}(J_i, b) \longrightarrow \mathbb{Q} \times \mathbb{Z}$ is u. s. c.

   Proof. Fix $(\alpha, \beta) \in \mathbb{Q} \times \mathbb{Z}$, we must show that

   $$F_{(\alpha,\beta)} = \{\xi \in \mathrm{Sing}(J_i, b) \mid t_i(\xi) \geq (\alpha, \beta)\}$$

   is closed. Since w-ord$_i$ is u. s. c. and its image is a finite subset in $\mathbb{Q}$, then

   $$G_\alpha = \{\xi \in \mathrm{Sing}(J_i, b) \mid \text{w-ord}_i(\xi) \geq \alpha\}$$

   $$G_\alpha^+ = \{\xi \in \mathrm{Sing}(J_i, b) \mid \text{w-ord}_i(\xi) > \alpha\}$$

   are also closed sets. Now check that

   $$F_{(\alpha,\beta)} = G_\alpha^+ \cup \bigcup_{i_1,\dots,i_\beta} (G_\alpha \cap H_{i_1} \cap \cdots H_{i_\beta})$$

   the union for $i_1 < \cdots i_\beta$ and $H_{i_j} \in E_i^-$. □

2. If 4.8.1 is constructed with the additional constraint $Y_i \subset \underline{\mathrm{Max}}\, t_i \subset \underline{\mathrm{Max}}\, \text{w-ord}_i$, then

   $$t_{i-1}(\Pi_i(\xi_i)) \geq t_i(\xi_i) \qquad \forall \xi_i \in \mathrm{Sing}(J_i, b)$$

   in particular

   $$\max t_0 \geq \cdots \geq \max t_k$$

   Proof. First check that if $\Pi_i(\xi_i) \notin Y_{i-1}$ then $t_{i-1}(\Pi_i(\xi_i)) = t_i(\xi_i)$. If $\Pi_i(\xi_i) \in Y_{i-1}(\subset \underline{\mathrm{Max}}\, t_{i-1} \subset \underline{\mathrm{Max}}\, \text{w-ord}_{i-1})$ it suffices to consider the case $\text{w-ord}_{i-1}(\Pi_i(\xi_{i-1})) = \text{w-ord}_i(\xi_i)$ (see 4.12.1). In this case $\xi_i \in \underline{\mathrm{Max}}\, \text{w-ord}_i$ and $\max \text{w-ord}_{i-1} = \max \text{w-ord}_i$ (see 4.12.2), and therefore E_i^- consist of the strict transforms of the hypersurfaces in E_{i-1}^-. The inequality follows now by looking at the second coordinates. □

3. We say that $\max t$ drops at i_0 if $\max t_{i_0-1} > \max t_{i_0}$. If $\max \text{w-ord}_0 = \frac{b'}{b}$ and $\dim W_0 = d$, note that $\max t_i = \left(\frac{e}{b}, m\right)$, $0 \leq e \leq b'$, $0 \leq m \leq d$. So it is clear that $\max t$ can drop at most $b'd$ times.

4. Via some form of induction (to be defined in section 6) we shall construct a unique enlargement of 4.8.1, say

   $$\begin{gathered}(W_0, (J_0, b), E_0) \longleftarrow \cdots \longleftarrow (W_k, (J_k, b), E_k) \longleftarrow \\ \longleftarrow (W_{k+1}, (J_{k+1}, b), E_{k+1}) \longleftarrow \cdots \longleftarrow (W_N, (J_N, b), E_N)\end{gathered} \tag{4.15.1}$$

   so that $\max t_k = \max t_{k+1} = \cdots = \max t_{N-1}$ and either

(a) $\operatorname{Sing}(J_N, b) = \emptyset$.

(b) $\operatorname{Sing}(J_N, b) \neq \emptyset$ and $\max \text{w-ord}_N = 0$.

(c) $\operatorname{Sing}(J_N, b) \neq \emptyset$, $\max \text{w-ord}_N > 0$ and $\max t_{N-1} > \max t_N$.

Now 3 says that for some index N either (a) or (b) will hold. If (b) holds the extension to a resolution is simple. This is the strategy for constructive resolution of basic objects.

Example 4.16. Set $W_0 = \operatorname{Spec} \mathbb{C}[X, Y]$, $J_0 = \langle X^3 - Y^{11} \rangle$, $b = 2$, $E_0 = \{H_1, H_2\}$, $H_1 = V(X)$ and $H_2 = V(Y)$. Consider the basic object $(W_0, (J_0, b), E_0)$, the singular locus $\operatorname{Sing}(J_0, 2)$ is the origin of W_0, say $(0,0)$, and $\text{w-ord}_0(0,0) = \frac{3}{2}$. As in 4.14 we set $E_0^- = E_0$, here

$$t_0(0,0) = \left(\frac{3}{2}, 2\right) = \max t_0$$

In this case $\underline{\operatorname{Max}} t_0$ is the origin $(0,0)$.

Consider the transformation with center $(0,0)$:

$$(W_0, (J_0, 2), E_0) \longleftarrow (W_1, (J_1, 2), E_1)$$

Set $L = V(X^3 - Y^8) \subset W_0$ and still denote by $L \subset W_1$ the strict transform of $L \subset W_0$. Let also $H_1, H_2 \subset W_1$ denote the strict transforms of $H_1, H_2 \subset W_0$. If $H_3 \subset W_1$ is the exceptional divisor of the transformation, then $E_1 = \{H_1, H_2, H_3\}$ and

$$J_1 = I(H_3) I(L)$$

The singular locus is a point, $\operatorname{Sing}(J_1, 2) = H_1 \cap H_3 = \{\xi_1\}$ and $\text{w-ord}_1(\xi_1) = \frac{3}{2} = \max \text{w-ord}_0$. Now $E_1^- = \{H_1, H_2\}$ and the function t_1 at the point ξ_1 is

$$t_1(\xi_1) = \left(\frac{3}{2}, 1\right) = \max t_1 < \max t_0$$

Consider now the transformation with center ξ_1

$$(W_1, (J_1, 2), E_1) \longleftarrow (W_2, (J_2, 2), E_2)$$

where $E_2 = \{H_1, H_2, H_3, H_4\}$ and $J_2 = I(H_3) I(H_4)^2 I(L)$. The singular locus is $\operatorname{Sing}(J_2, 2) = H_4$. At the point $\xi_2 = H_1 \cap H_4$, $\text{w-ord}_2(\xi_2) = \frac{3}{2} > 0$, and at $\eta \in H_4 \setminus H_1$, $\text{w-ord}_2(\eta) = 0$. We consider here, the function t_2 defined only for ξ_2. Now $E_2^- = \{H_1, H_2\}$, since $\max \text{w-ord}_2 = \max \text{w-ord}_0$, and

$$t_2(\xi_2) = \left(\frac{3}{2}, 1\right) = \max t_2 = \max t_1 < \max t_0$$

We note that the values $\max t_i$ cannot grow by transformations with center included in $\underline{\operatorname{Max}} t_i$.

After the transformation with center ξ_2, one can check that $\max \text{w-ord}_3 < \max \text{w-ord}_2$ and $E_3^- = E_3$.

Remark 4.17. **On the inductive nature of the function t.**

We fix a basic object $(W_0, (J_0, b), E_0)$ where $\dim W_0 = n$. We wish to define a constructive resolution of $(W_0, (J_0, b), E_0)$. The outcome should be a sequence 4.8.1 with $\mathrm{Sing}(J_k, b) = \emptyset$. The function w-ord$_k$ is defined so that $\underline{\mathrm{Max}}\,\text{w-ord}_k = 0$ iff $\left(\bar{J}_k\right)_\xi = \mathcal{O}_{W_k,\xi}$ (4.8.2) at any $\xi \in \mathrm{Sing}(J_k, b)$. In such case we say that $(W_k, (J_k, b), E_k)$ is a monomial basic object and a constructive extension of 4.8.1 to a resolution will be defined in the next section.

So assume now that $\underline{\mathrm{Max}}\,\text{w-ord}_k > 0$ and that the sequence 4.8.1 was defined with the additional constraint $Y_i \subset \underline{\mathrm{Max}}\, t_i$, so that $\max t_0 \geq \max t_1 \geq \cdots \geq \max t_k$ (4.15(2)). Our strategy is to define an enlargement of the sequence 4.8.1, say

$$(W_0, (J_0, b), E_0) \longleftarrow \cdots \longleftarrow (W_k, (J_k, b), E_k) \longleftarrow \cdots \longleftarrow (W_N, (J_N, b), E_N)$$

so that

$$\max t_k = \max t_{k+1} = \cdots = \max t_{N-1} > \max t_N$$

If so, 4.15(3) says that by repeating such procedure we come to the monomial case and ultimately to a resolution.

It is in the construction of this enlargement where the suitably defined function t plays its role. In fact if $\underline{\mathrm{Max}}\, t_k$ is of codimension one in W_k (i.e. if $\underline{\mathrm{Max}}\, t_k = R(1)(\underline{\mathrm{Max}}\, t_k)$), then one can easily check that $\underline{\mathrm{Max}}\, t_k$ is a smooth permissible center, and setting the transformation

$$(W_k, (J_k, b), E_k) \longleftarrow (W_{k+1}, (J_{k+1}, b), E_{k+1})$$

at such center, then $\underline{\mathrm{Max}}\, t_k > \underline{\mathrm{Max}}\, t_{k+1}$ (see 6.24(P2)).

On the other hand, if $R(1)(\underline{\mathrm{Max}}\, t_k) = \emptyset$ then, locally at any point $\xi \in \underline{\mathrm{Max}}\, t_k$ we will attach to $\underline{\mathrm{Max}}\, t_k$ a $n-1$-dimensional basic object (6.24(P3)), so that a resolution of the locally defined basic objects defines an enlargement with the required property.

The following example illustrates this general fact:

Example 4.18. Part I

Set $W_0 = \mathbb{A}^3_{\mathbf{k}}$, $J_0 = \langle Z^2 + (X^2 - Y^3)^2 \rangle$, $b = 2$ and $E_0 = \emptyset$. Show that for any sequence 4.8.1:

1. All $a_i = 0$ in the expression 4.8.2 (i.e. $J_i = \bar{J}_i$) for $i = 0, \ldots, k$.
2. $\max \text{w-ord}^3_0 = \cdots = \max \text{w-ord}^3_k = 1$
   $\max t^3_0 = \cdots = \max t^3_k = (1, 0)$

   where we set w-ord $=$ w-ord^3 and $t = t^3$ since we are dealing with a 3-dimensional basic object.
3. $\underline{\mathrm{Max}}\, t^3_i = \mathrm{Sing}(J_i, 2)$.

Apply now remark after exercise 13.1 together with (3) above to show:

a Any sequence of transformations of $(\mathbb{A}^2_{\mathbf{k}}, (C(f), 2), \emptyset)$ induces a sequence of transformations of $(W_0, (J_0, 2), E_0)$ (see exercise 13.1(4), here $C(f) = \langle (X^2 - Y^3)^2 \rangle$).

b $\underline{\mathrm{Max}}\, t_i^3 = \mathrm{Sing}(C(f), 2)$.

c If the sequence 13.1.4 is a resolution then the induced sequence is such that $\max t_0^3 = \cdots = \max t_{r-1}^3 > \max t_r^3$. Actually $\mathrm{Sing}(J_r, 2) = \emptyset$ in this case.

Part II. On the resolution of $(\mathbb{A}^2_{\mathbf{k}}, (C(f), 2), \emptyset)$.
Set the two dimensional basic object $(W_0^2, (\mathfrak{a}_0, 2), E_0^2) = (\mathbb{A}^2_{\mathbf{k}}, (C(f), 2), \emptyset)$ and set t_0^2 the function t on this basic object. Now we will define a resolution with the constraint $Y_i \subset \underline{\mathrm{Max}}\, t_i^2$ as in 4.15(2). This last condition defines uniquely the first three steps

$$(W_0^2, (\mathfrak{a}_0, 2), E_0^2) \longleftarrow (W_1^2, (\mathfrak{a}_1, 2), E_1^2) \longleftarrow$$
$$\longleftarrow (W_2^2, (\mathfrak{a}_2, 2), E_2^2) \longleftarrow (W_3^2, (\mathfrak{a}_3, 2), E_3^2)$$

To check this note that $\underline{\mathrm{Max}}\, t_0^2$, $\underline{\mathrm{Max}}\, t_1^2$ and $\underline{\mathrm{Max}}\, t_2^2$ are closed points. Here $\max t_0^2 = (2,0)$, $\max t_1^2 = (1,1)$ and $\max t_2^2 = (1,1)$. In fact the expressions in 4.8.2 are as follows:

$$\mathfrak{a}_0 = \bar{\mathfrak{a}}_0 \qquad \mathfrak{a}_2 = I(H_1)^2 I(H_2)^2 \bar{\mathfrak{a}}_2$$
$$\mathfrak{a}_1 = I(H_1)^2 \bar{\mathfrak{a}}_1 \qquad \mathfrak{a}_3 = I(H_1)^2 I(H_2)^2 I(H_3)^4 \bar{\mathfrak{a}}_3$$

Note that $\underline{\mathrm{Max}}\,\text{w-ord}_1^2 = R(1)(\underline{\mathrm{Max}}\,\text{w-ord}_1^2)$ is smooth and however it is not a permissible center.

Now one can check that

$$\underline{\mathrm{Max}}\, t_3^2 = R(1)(\underline{\mathrm{Max}}\, t_3^2)$$

is a permissible center for $(W_3^2, (\mathfrak{a}_3, 2), E_3^2)$ and if

$$(W_3^2, (\mathfrak{a}_3, 2), E_3^2) \longleftarrow (W_4^2, (\mathfrak{a}_4, 2), E_4^2)$$

is the transformation with center $R(1)(\underline{\mathrm{Max}}\, t_3^2)$ then

$$\mathfrak{a}_4 = I(H_1)^2 I(H_2)^2 I(H_3)^4$$

so the basic object is monomial, namely $\bar{\mathfrak{a}}_4 = \mathcal{O}_{W_4^2}$.

Part III. Epilogue.
Let us bring to the surface the role of the functions w-ord(ξ) and $n(\xi)$ (i.e. of the function $t(\xi)$) in the canonical construction of a resolution of $(\mathbb{A}^3_{\mathbf{k}}, (\langle Z^3 + (X^2 - Y^3)^2 \rangle, 2), E_0 = \emptyset)$, via induction on the ambient space. So this is a 3-dimensional basic object, we define the function t_0^3, we consider the value $\max t_0^3$ and the closed set $\underline{\mathrm{Max}}\, t_0^3$. Part I says that there is a closed embedding $\mathbb{A}^2_{\mathbf{k}} \to \mathbb{A}^3_{\mathbf{k}}$ and a 2-dimensional basic object $(\mathbb{A}^2_{\mathbf{k}}, (C(f), 2), \emptyset)$ so that $\mathrm{Sing}(C(f), 2)(\mathbb{A}^2_{\mathbf{k}} \subset \mathbb{A}^3_{\mathbf{k}})$ is $\underline{\mathrm{Max}}\, t_0^3$.

So, in principle, we have attached a two dimensional basic object to the value $\underline{\mathrm{Max}}\, t_0^3$.

Let us stress on the use of the word *any* in Part I(a). It says that *any* permissible center for the two dimensional basic object is permissible for the 3-dimensional basic object. In particular, that any such center has normal crossings with the hypersurfaces in E_0. Of course in this example $E_0 = \emptyset$, but as soon as you choose *any* such center you will have a transformation of 3-dimensional basic objects.

$$(W_0, (J_0, b), E_0 = \emptyset) \longleftarrow (W_1, (J_1, b), E_1)$$

a transformation of 2-dimensional basic objects

$$(\mathbb{A}^2_{\mathbf{k}}, (C(J_0), b), E^2_0) \longleftarrow (W^2_1, (C(J_0)_1, b!), E^2_1)$$

where $(W^2_1, (C(J_0)_1, b!), E^2_1)$ relates to $(W_1, (J_1, b), E_1)$ as before, and now $E_1 \neq \emptyset$. The point is that *any* permissible center for $(W^2_1, (C(J_0)_1, b!), E^2_1)$ has normal crossings with E_1 and is also permissible for $(W_1, (J_1, b), E_1)$.

The same holds after any sequence of transformations of the 2-dimensional basic object. This property is guaranteed by the combination of first and second coordinate of $t(\xi)$ (i.e. by w-ord(ξ) and $n(\xi)$). This is induction on the dimension of the ambient space.

Of course we ultimately are not interested on *any* sequence of transformations. Part II is there to enlighten how to produce a unique sequence via induction.

Exercise 4.19. Let (W_0, E_0) be a pair, X_0 a regular hypersurface. Assume that no component of X_0 is included in a component of $H \in E$. Set $J_0 = I(X_0)$, $b = 1$ in 4.8.1 and assume that $Y_i \subset \underline{\mathrm{Max}}\, t_i$.

1. Show that $J_i = I(X_i)$ where X_i is the strict transform of X_{i-1}.
2. Show that max w-ord$_i$ $= 1$ for $i = 0, \dots, r$ and that if $\max t_k = (1, 0)$ then $\underline{\mathrm{Max}}\, t_k = X_k$ and X_k (is regular and) has normal crossings with E_k. This last condition will be important for embedded desingularization.

Exercise 4.20. Apply 4.19 to X_1 and $E_1 = \{H_1\}$ obtained in example 3.5.

5. Monomial case

Let $B = (W, (J, b), E)$ be a basic object, $E = \{H_1, \dots, H_r\}$, and assume that

$$J = I(H_1)^{\alpha_1} \cdots I(H_r)^{\alpha_r}$$

where $\alpha_i : H_i \longrightarrow \mathbb{Z} \geq 0$ are locally constant functions (see also 4.8.2). In this case we call B a **monomial** basic object.

We attach to such basic object B the function:

$$\Gamma(B) : \mathrm{Sing}(J, b) \longrightarrow I_M = \mathbb{Z} \times \mathbb{Q} \times \mathbb{Z}^{\mathbb{N}}$$

$$\Gamma(B)(\xi) = (-\Gamma_1(\xi), \Gamma_2(\xi), \Gamma_3(\xi))$$

where I_M is considered as a total ordered set with the lexicographic order.

Define at a point $\xi \in \operatorname{Sing}(J,b)$:

$$\Gamma_1(\xi) = \min\{p \mid \exists i_1,\dots,i_p,\ \alpha_{i_1}(\xi)+\cdots+\alpha_{i_p}(\xi) \geq b, \xi \in H_{i_1}\cap\cdots\cap H_{i_p}\}$$

$$\Gamma_2(\xi) = \begin{array}{l} \max\{\dfrac{\alpha_{i_1}(\xi)+\cdots+\alpha_{i_p}(\xi)}{b} \mid \\ p=\Gamma_1(\xi), \alpha_{i_1}(\xi)+\cdots+\alpha_{i_p}(\xi) \geq b, \xi \in H_{i_1}\cap\cdots\cap H_{i_p}\} \end{array}$$

$$\Gamma_3(\xi) = \begin{array}{l} \max\{(i_1,\dots,i_p,0,\dots) \mid \\ \Gamma_2(\xi) = \dfrac{\alpha_{i_1}(\xi)+\cdots+\alpha_{i_p}(\xi)}{b}, \xi\ in H_{i_1}\cap\cdots\cap H_{i_p}\} \end{array}$$

Where for Γ_3, max means the maximum for the lexicographic order in $\mathbb{Z}^{\mathbb{N}}$.

These functions $\Gamma(B)$ will naturally lead us to a resolution of the basic object B. In fact, $\underline{\operatorname{Max}}\,\Gamma(B)$ is a permissible center and setting $B \longleftarrow B'$ the transformation with center $\underline{\operatorname{Max}}\,\Gamma(B)$, then B' is also monomial. The following examples 5.1 and 5.2 illustrate that

$$\max\Gamma(B) > \max\Gamma(B') \tag{5.0.1}$$

Note also that the third coordinate takes values at $\mathbb{N}^n$, where $n = \dim W$, since at most n hypersurfaces with normal crossings can contain a point. It finally follows from 5.0.1 that a resolution of B is achieved by successive blowing-ups at $\underline{\operatorname{Max}}\,\Gamma$.

Example 5.1. Let $B = (W,(J,2),E)$ the monomial basic object where $W = \mathbb{C}^2$ with coordinates X and Y, $E = \{H_1,H_2\}$, $H_1 = V(X)$, $H_2 = V(Y)$ and $J = I(H_1)^3 I(H_2)^3$. The singular locus of B is $\operatorname{Sing}(B) = H_1 \cup H_2$ and the values of $\Gamma(B)$ are:

$$\Gamma(B)(\xi) = \begin{cases} (-1,\frac{3}{2},(1,0)) & \text{if} \quad \xi \in H_1 \setminus H_2 \\ \\ (-1,\frac{3}{2},(2,0)) & \text{if} \quad \xi \in H_2 \end{cases}$$

Then $\max\Gamma(B) = (-1,\frac{3}{2},(2,0))$ and $\underline{\operatorname{Max}}\,\Gamma(B) = H_2$. Set

$$B = (W,(J,2),E) \xleftarrow{\Pi_1} B_1 = (W_1,(J_1,2),E_1)$$

the transformation with center H_2, so $\Pi_1 : W_1 \longrightarrow W$ is the identity map. Let H_1, H_2 denote the strict transforms of H_1 and H_2, then $H_1 = H_1$ in W_1 and $H_2 = \emptyset$ in W_1. Note that $E_1 = \{H_1, H_2, H_3\}$, where $H_3 \subset W_1$ coincides with $H_2 \subset W$, and $J_1 = I(H_1)^3 I(H_3)$.

The singular locus is now $\mathrm{Sing}(B_1) = H_1$ and the function $\Gamma(B_1)$ is constant along H_1 and equal to $(-1, \frac{3}{2}, (1,0))$, so $\underline{\mathrm{Max}}\,\Gamma(B_1) = H_1$. Set

$$B_1 = (W_1, (J_1, 2), E_1) \xleftarrow{\Pi_2} B_2 = (W_2, (J_2, 2), E_2)$$

the transformation with center H_1. Again Π_2 is an isomorphism and using the same notation, $E_2 = \{H_1, H_2, H_3, H_4\}$, where $H_1 = H_2 = \emptyset$ in W_2, $H_4 \subset W_2$ coincides with $H_1 \subset W_1$, and $J_2 = I(H_3)I(H_4)$.

Now $\mathrm{Sing}(B_2) = H_3 \cap H_4$ is a unique point and the function $\Gamma(B_2)$ take the value $(-2, 1, (4,3))$, so $\underline{\mathrm{Max}}\,\Gamma(B_2) = H_3 \cap H_4$. After the transformation with center $H_3 \cap H_4$:

$$B_2 = (W_2, (J_2, 2), E_2) \xleftarrow{\Pi_3} B_3 = (W_3, (J_3, 2), E_3)$$

One can check that $\mathrm{Sing}(B_3) = \emptyset$ and we get a resolution. Note that in the process:

$$\max\Gamma(B) = (-1, \frac{3}{2}, (2,0)) >$$
$$\max\Gamma(B_1) = (-1, \frac{3}{2}, (1,0)) > \max\Gamma(B_2) = (-2, 1, (4,3))$$

Example 5.2. Let $B = (W, (J, 3), E)$ the monomial basic object given by $W = \mathbb{C}^3$, $E = \{H_1, H_2, H_3\}$, $H_1 = V(X)$, $H_2 = V(Y)$, $H_3 = V(Z)$ and $J = I(H_1)^2 I(H_2)^2 I(H_3)^2$.

$$\mathrm{Sing}(B) = (H_1 \cap H_2) \cup (H_1 \cap H_3) \cup (H_2 \cap H_3)$$

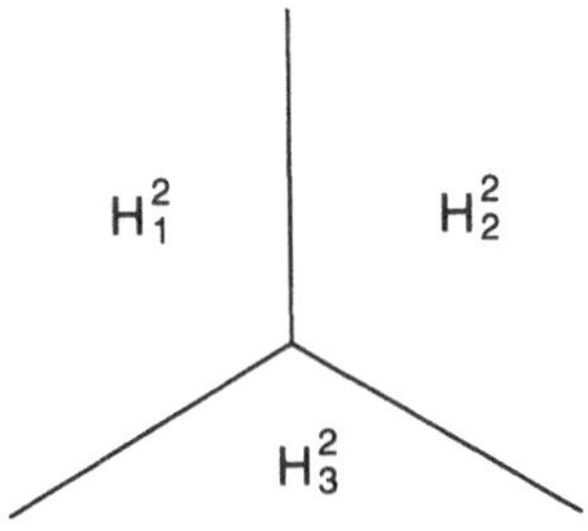

Note that considering only the first two coordinates of $\Gamma(B)$, that $\max(-\Gamma_1(B), \Gamma_2(B)) = \left(-2, \frac{4}{2}\right)$ and the set

$$\underline{\mathrm{Max}}(-\Gamma_1(B), \Gamma_2(B)) = (H_1 \cap H_2) \cup (H_1 \cap H_3) \cup (H_2 \cap H_3)$$

is a union of three irreducible components. The third coordinate of Γ provides an ordering of these components. Here

$$\max \Gamma(B) = (-2, \frac{4}{3}, (3, 2, 0))$$

and

$$\underline{\mathrm{Max}}\, \Gamma(B) = H_2 \cap H_3$$

which is regular.

After the transformation with center $H_2 \cap H_3$, we obtain the monomial basic object $B_1 = (W_1, (J_1, 3), E_1)$, $E_1 = \{H_1, H_2, H_3, H_4\}$,
$J_1 = I(H_1)^2 I(H_2)^2 I(H_3)^2 I(H_4)$.

$$\mathrm{Sing}(B_1) = (H_1 \cap H_2) \cup (H_1 \cap H_3) \cup (H_1 \cap H_4) \cup (H_2 \cap H_4) \cup (H_3 \cap H_4)$$

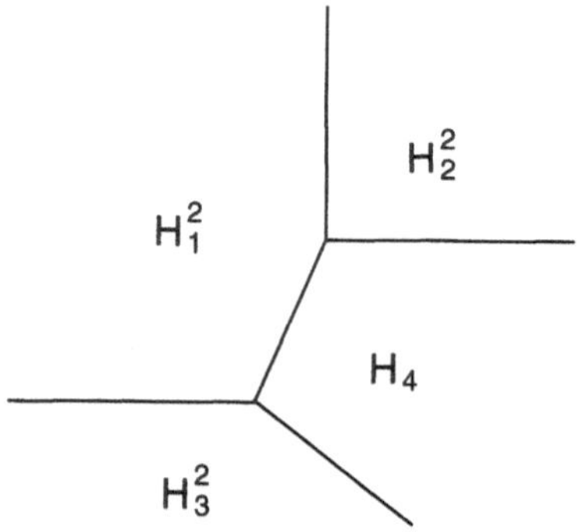

Note that the first two coordinates of $\max \Gamma$ have not dropped. Now

$$\underline{\mathrm{Max}}(-\Gamma_1(B_1), \Gamma_2(B_1)) = (H_1 \cap H_2) \cup (H_1 \cap H_3)$$

is the strict transform of $\underline{\mathrm{Max}}(-\Gamma_1(B), \Gamma_2(B))$ (and this is a general fact). Note that the set $\underline{\mathrm{Max}}(-\Gamma_1(B_1), \Gamma_2(B_1))$ has less irreducible components than the original one,

$$\max \Gamma(B_1) = (-2, \frac{4}{3}, (3, 1, 0))$$

$$\underline{\mathrm{Max}}\, \Gamma(B_1) = H_1 \cap H_3$$

Set $B_1 \longleftarrow B_2$ the transformation with center $\underline{\mathrm{Max}}\, \Gamma(B_1)$.

$$J_2 = I(H_1)^2 I(H_2)^2 I(H_3)^2 I(H_4) I(H_5)$$

$$\mathrm{Sing}(B_2) = \\ (H_1 \cap H_2) \cup (H_1 \cap H_4) \cup (H_1 \cap H_5) \cup (H_2 \cap H_4) \cup (H_3 \cap H_4) \cup (H_3 \cap H_5)$$

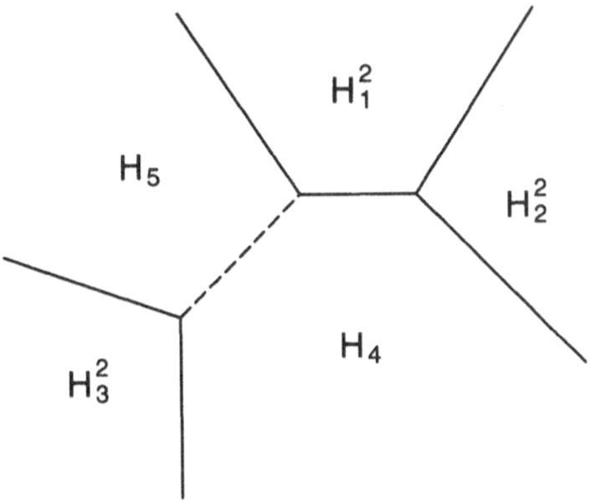

Again the first two coordinates of $\max\Gamma$ have not dropped, and $\underline{\mathrm{Max}}(-\Gamma_1(B_2),\Gamma_2(B_2))$ is the strict transform of $\underline{\mathrm{Max}}(-\Gamma_1(B_1),\Gamma_2(B_1))$.

$$\max\Gamma(B_2)=(-2,\frac{4}{3},(2,1,0)) \qquad \underline{\mathrm{Max}}\,\Gamma(B_2)=H_1\cap H_2$$

Set $B_2 \longleftarrow B_3$ the transformation with center $\underline{\mathrm{Max}}\,\Gamma(B_2)$.

$$J_3=I(H_1)^2I(H_2)^2I(H_3)^2I(H_4)I(H_5)I(H_6)$$

$$\mathrm{Sing}(B_3)=(H_1\cap H_4)\cup(H_1\cap H_5)\cup(H_1\cap H_6)\cup(H_2\cap H_4)\cup (H_2\cap H_6)\cup(H_3\cap H_4)\cup(H_3\cap H_5)$$

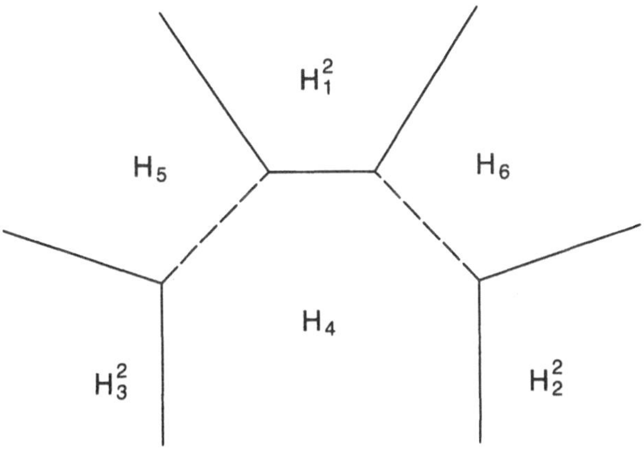

Now the second coordinate of $\max\Gamma$ has dropped:

$$\max\Gamma(B_3)=(-2,1,(6,2)) \qquad \underline{\mathrm{Max}}\,\Gamma(B_3)=H_2\cap H_6$$

Seven steps more are still required, each of them defined by blowing-up at $\underline{\mathrm{Max}}\,\Gamma(B_i)$, in order to achieve a resolution of B.

6. Compatibility of the function t with induction

Desingularization is an application of analytic methods in algebraic geometry, so extensions of the results to the completion at a point or the analytic context are quite natural [V4].

Analytic methods will lead to resolution of basic objects by induction on the dimension of the ambient space. However this form of induction is of local nature. The strength of lemma 6.6 is that the very formulation of induction is possible within the class of schemes of finite type over a field (of characteristic zero), avoiding Weierstrass theorem and Tschirnhausen transformation.

In what follows $\{W^\lambda\}_{\lambda\in\Lambda}$ will denote an open covering of the regular variety W. Since W has a noetherian topology, we may and we will assume that Λ is a finite set.

6.1. Let (W,E), $E=\{H_1,\dots,H_s\}$ be a pair, $\{W^\lambda\}_{\lambda\in\Lambda}$ an open covering of W. Define (W^λ,E^λ) by setting $E^\lambda=\{H_i\cap W^\lambda \mid H_i\in E\}$, we shall say that $\{(W^\lambda,E^\lambda)\}$ is a covering of (W,E).

If $(W,E)\xleftarrow{\Pi}(W_1,E_1)$ is a transformation of pairs (2.10), one can easily check that:

1. $\{\Pi^{-1}(W^\lambda)=W_1^\lambda\}$ is a covering of W_1.
2. Π induces, for each λ a transformation (2.10): $\Pi^\lambda:(W^\lambda,E^\lambda)\longleftarrow(W_1^\lambda,E_1^\lambda)$.
3. $\{(W_1^\lambda,E_1^\lambda)\}$ is a covering of (W_1,E_1). And for any sequence

   $$(W_0,E_0)\longleftarrow\cdots\longleftarrow(W_r,E_r)$$

   there is an induced sequence

   $$(W_0^\lambda,E_0^\lambda)\longleftarrow\cdots\longleftarrow(W_r^\lambda,E_r^\lambda) \tag{6.1.1}$$

   so that $\{(W_r^\lambda,E_r^\lambda)\}$ is a covering of (W_r,E_r).

6.2. If W^λ is open in W we have defined a natural notion of restriction of a pair (W,E) as (W^λ,E^λ) in 6.1. A restriction of a basic object $B=(W,(J,b),E)$ will be defined as $B^\lambda=(W^\lambda,(J^\lambda,b),E^\lambda)$, $J^\lambda=J|_{W^\lambda}$. If $\{W^\lambda\}$ is an open covering of W we will associate to B the basic objects B^λ. Note that

1. For each λ, $\mathrm{Sing}(J^\lambda,b)=\mathrm{Sing}(J,b)\cap W^\lambda$.
2. A transformation $B=(W,(J,b),E)\longleftarrow B_1=(W_1,(J_1,b),E_1)$ induces, for each λ, a transformation

   $$B^\lambda=(W^\lambda,(J^\lambda,b),E^\lambda)\longleftarrow B_1^\lambda=(W_1^\lambda,(J_1^\lambda,b),E_1^\lambda)$$

   The open sets $\{W_1^\lambda\}$ define a covering of W_1 and each B_1^λ is the restriction of B_1.
3. A sequence of transformations is a resolution of B iff it induces a resolution for each B^λ.

Definition 6.3. We shall say that $(W,(J,b),E)$ is a simple basic object if one of the following equivalent conditions hold (see P2 in 1.6):

(i) $\nu_\xi(J)=b$ for any $\xi\in \mathrm{Sing}(J,b)=V(\Delta^{b-1}(J))$.
(ii) $\nu_\xi(\Delta^{b-1}(J))=1$ for any $\xi\in V(\Delta^{b-1}(J))$.

6.4. If now $B_0=(W_0,(J_0,b),E_0)$ is any simple basic object, it follows from 6.3 that there is an open covering $\{W_0^\lambda\}_{\lambda\in\Lambda}$ defining $B_0^\lambda=(W_0^\lambda,(J_0^\lambda,b),E_0^\lambda)$ (6.2), and for each λ a regular hypersurface $(W_0)_h^\lambda\subset W_0^\lambda$ so that $I\left((W_0)_h^\lambda\right)\subset\Delta^{b-1}\left(J_0^\lambda\right)$.

If $X_0\subset W_0$ is a hypersurface and $b=\max\mathrm{mult}_X$ (see 4.10) then $\mathrm{Sing}(I(X_0),b)=V(\Delta^{b-1}(I(X_0)))$ and $\Delta^{b-1}(I(X_0))$ has order 1 at each point of $\mathrm{Sing}(I(X_0),b)$. So locally at $x\in\mathrm{Sing}(I(X_0),b)$ we may choose a regular hypersurface W_h so that $I(W_h)\subset\Delta^{b-1}(I(X_0))$, hence $\mathrm{Sing}(I(X_0),b)\subset W_h$ at a suitable neighborhood.

In case $X_0=V(Z^2-X^3Y^3)\subset\mathbb{C}^3$ we may choose $W_h=V(Z)$ since $Z\in\Delta(I(X_0))$. Note that $\mathrm{Sing}(I(X_0),2)\subset W_h$ can be identified with $\mathrm{Sing}(X^3Y^3,2)$ for the basic object $(W_h,(X^3Y^3,2),\emptyset)$.

Note that the closed set of 2-fold points of the hypersurface X_0 is included in the regular hypersurface W_h. Lemma 6.6 will show that this inclusion is preserved by blowing up a regular center included in the set of 2-fold points of X_0 (replacing X_0 and W_h by their strict transforms). Here the hypersurface W_h is a particular case of "hypersurface of maximal contact" in Hironaka's notation and 6.6 a particular setting of his "stability of maximal contact".

Remark 6.5. Let $B=(W,(J,b),E)$ be a basic object and $W_h\subset W$ a regular hypersurface. Suppose that $I(W_h)\subset\Delta^{b-1}(J)$, then one can check that:

(i) $\mathrm{Sing}(J,b)(=V(\Delta^{b-1}(J)))\subset W_h$
(ii) $\nu_\xi(\Delta^{b-1}(J))=1$ for any $\xi\in V(\Delta^{b-1}(J))$.

Lemma 6.6. *(Giraud) Given B, $W_h\subset W$ so that $I(W_h)\subset\Delta^{b-1}(J)$ as before, then any transformation*

$$(W,(J,b),E)\longleftarrow(W_1,(J_1,b),E_1)$$

induces

$$(W,(I(W_h),1),E)\longleftarrow(W_1,(I((W_h)_1),1),E_1)$$

where $(W_h)_1$ denotes the strict transform of W_h and

$$I((W_h)_1)\subset\Delta^{b-1}(J_1)$$

Proof. Is a corollary of lemma 9.1 to be proved in 9.2. □

Corollary 6.7. *Any sequence of transformations*

$$(W,(J,b),E)\longleftarrow\cdots\longleftarrow(W_s,(J_s,b),E_s) \tag{6.7.1}$$

induces a sequence

$$(W,(I(W_h),1),E) \longleftarrow \cdots \longleftarrow (W_s,(I((W_h)_s),1),E_s)$$

($(W_h)_i$ strict transform of $(W_h)_{i-1}$) and

$$I((W_h)_s) \subset \Delta^{b-1}(J_s)$$

Definition 6.8. Let $W \overset{\Pi}{\longleftarrow} W_1 = W \times \mathbb{A}^1_{\mathbf{k}}$ be the projection pr_1. Let $(W, E = \{H_1, \ldots, H_r\})$ be a pair and $B = (W,(J,b),E)$ a basic object. There is a natural definition of *restriction* of pairs and of basic objects:

$$(W,E) \longleftarrow (W_1,E_1)$$
$$(W,(J,b),E) \overset{\Pi}{\longleftarrow} (W_1,(J_1,b),E_1)$$

setting $J_1 = J\mathcal{O}_{W_1}$ and $E_1 = \{H'_1, \ldots, H'_r\}$, $H'_i = \Pi^{-1}(H_i)$.

Use 1.5.1 and 1.2.1 to check that $\Delta^{b-1}(J_1) = \Delta^{b-1}(J)\mathcal{O}_{W_1}$. Note in particular that the statement of Giraud's lemma 6.6 and corollary 6.7 also hold if some of the transformations in 6.7.1 are replaced by restrictions defined by smooth morphisms. Note also that $\mathrm{Sing}(J_1,b) = \Pi^{-1}(\mathrm{Sing}(J,b))$.

Proposition 6.9. *Any transform of a simple basic object is a simple basic object.*

Proof. Follows easily from 6.3(ii) and 6.6. □

Remark 6.10. We leave it as an exercise to show that both 2.3(B) and 2.4(B) follow from 6.9.

Definition 6.11. **Immersions.** Assume now that $\widetilde{W} \hookrightarrow W$ is a closed immersion of regular varieties, that $\widetilde{W}$ has normal crossings with $E = \{H_1, \ldots, H_s\}$, that $\widetilde{W} \cap H_i = \widetilde{H}_i$ is a hypersurface in $\widetilde{W}$ and set $\widetilde{E} = \{H_i \cap \widetilde{W} \mid H_i \in E\}$. Then $(\widetilde{W}, \widetilde{E})$ is a pair and we say that

$$(\widetilde{W},\widetilde{E}) \hookrightarrow (W,E)$$

is an immersion of pairs.

One can check that any sequence of transformation over $(\widetilde{W},\widetilde{E})$ induces a diagram

$$\begin{array}{ccccc} (\widetilde{W},\widetilde{E}) & \longleftarrow & \cdots & \longleftarrow & (\widetilde{W}_r,\widetilde{E}_r) \\ \downarrow & & & & \downarrow \\ (W,E) & \longleftarrow & \cdots & \longleftarrow & (W_r,E_r) \end{array} \tag{6.11.1}$$

of transformations and closed immersions.

In particular, if $(\widetilde{W},(\mathfrak{a},d),\widetilde{E})$ is a basic object, any sequence of transformations

$$(\widetilde{W},(\mathfrak{a},d),\widetilde{E}) \longleftarrow \cdots \longleftarrow ((\widetilde{W})_r,(\mathfrak{a}_r,d),(\widetilde{E})_r)$$

induces a diagram as in 6.11.1 together with closed subsets $F_i \subset W_i$ setting $F_i = \mathrm{Sing}(\mathfrak{a}_i,d) \subset (\widetilde{W})_i \subset W_i$.

If $(\widetilde{W}, \widetilde{E}) \longrightarrow (W, E)$ is an immersion and $(W, E) \longleftarrow (W_1, E_1)$ is a restriction (6.8) then there is a naturally defined immersion in (W_1, E_1) defined by taking pull-backs. So if some of the Π_i are restrictions as in 6.8 then we also get a diagram as in 6.11.1 with transformations and restrictions.

Lemma 6.12. Inductive lemma. (Induction on the embedded dimension). *Let $B = (W, (J, b), E)$ be a basic object, $\{W^\lambda\}_{\lambda \in \Lambda}$ be a covering of W. We assume now that:*

C1 *B is a simple basic object.*

C2 *For each $\lambda \in \Lambda$ there is a regular irreducible hypersurface $W_h^\lambda \subset W^\lambda$ such that $I(W_h^\lambda) \subset \Delta^{b-1}(J^\lambda)$.*

C3 *For each $\lambda \in \Lambda$ there is an immersion of pairs (6.11):*

$$(W_h^\lambda, E_h^\lambda) \hookrightarrow (W^\lambda, E^\lambda)$$

In these conditions we claim that

A *If $R(1)(\mathrm{Sing}(J, b)) \neq \emptyset$ (see 1.8(1) for notation), then $R(1)$ is regular, open and closed in $\mathrm{Sing}(J, b)$ and has normal crossings with E. Furthermore setting*

$$B = (W, (J, b), E) \longleftarrow B_1 = (W_1, (J_1, b), E_1)$$

with center $Y = R(1)$, then $R(1)(\mathrm{Sing}(J_1, b)) = \emptyset$.

B *If $R(1)(\mathrm{Sing}(J, b)) = \emptyset$, for each $\lambda \in \Lambda$ we define a basic object*

$$B_h^\lambda = (W_h^\lambda, (C(J^\lambda), b!), E_h^\lambda)$$

that satisfies the following properties:

B1 $\mathrm{Sing}(J^\lambda, b) = \mathrm{Sing}(C(J^\lambda), b!)$.

B2 *For any sequence of transformations and restrictions, say*

$$B = B_0 = (W_0, (J_0, b), E_0) \longleftarrow \cdots \longleftarrow B_s = (W_s, (J_s, b), E_s)$$

and setting for each λ the induced sequence

$$B_0^\lambda = (W_0^\lambda, (J_0^\lambda, b), E_0^\lambda) \longleftarrow \cdots \longleftarrow B_s^\lambda = (W_s^\lambda, (J_s^\lambda, b), E_s^\lambda)$$

there is a sequence of transformations and restrictions

$$\begin{aligned} B_h^\lambda = (B_h)_0^\lambda = ((W_h)_0^\lambda, (C(J_0^\lambda), b!), (E_h)_0^\lambda) \longleftarrow \cdots \\ \longleftarrow (B_h)_s^\lambda = ((W_h)_s^\lambda, (C(J_0^\lambda)_s, b!), (E_h)_s^\lambda) \end{aligned}$$

and a diagram of closed immersions

$$\begin{array}{ccccc} (W_0^\lambda, E_0^\lambda) & \longleftarrow & \cdots & \longleftarrow & (W_s^\lambda, E_s^\lambda) \\ \uparrow & & & & \uparrow \\ ((W_h)_0^\lambda, (E_h)_0^\lambda) & \longleftarrow & \cdots & \longleftarrow & ((W_h)_s^\lambda, (E_h)_s^\lambda) \end{array} \tag{6.12.1}$$

such that

$$\mathrm{Sing}(C(J_0^\lambda)_i, b!) = \mathrm{Sing}(J_i^\lambda, b) \tag{6.12.2}$$

Proof. For the proof of (A) see exercise 13.14. The general proof of the lemma will be developed in 9.4. For the time being, a look at exercise 13.1 is strongly recommended. □

6.13. We are aiming to define a canonical resolution of basic objects (4.9). Our development will show that such canonical procedure reduces to:

i. A canonical resolution of monomial basic objects (section 5).

ii. A canonical resolution of simple basic objects (6.3).

These reductions are due to our suitably defined functions w-ord and n (4.14)

Lemma 6.12, which applies for simple basic objects, is a lemma of induction on the dimension of the ambient space. We must prove that we can profit of this *local* form of induction. This will lead to a generalization of the notion of basic object.

Fix (W_0, E_0), a covering $\{W_0^\lambda\}$ and for each λ data $\mathcal{D}_\lambda$ consisting of

$\mathcal{D}_\lambda$: (i) an immersion $j_0^\lambda : (\widetilde{W}_0^\lambda, \tilde{E}_0^\lambda) \hookrightarrow (W_0^\lambda, E_0^\lambda)$.
(ii) a basic object $(\widetilde{W}_0^\lambda, (\mathfrak{a}_0^\lambda, b^\lambda), \tilde{E}_0^\lambda)$.

Note that for each index λ we have defined a closed set F_0^λ in W_0^λ:

$$F_0^\lambda = \mathrm{Sing}(\mathfrak{a}_0^\lambda, b^\lambda) \subset \widetilde{W}_0^\lambda \subset W_0^\lambda$$

and we now say that the singular loci $\{F_0^\lambda\}$ patch if there is a closed set $F_0 \subset W_0$ so that $F_0 \cap W_0^\lambda = F_0^\lambda$.

If this condition holds, then for any center $Y \subset F_0$ permissible for (W_0, E_0), the transformation

$$(W_0, E_0) \underset{Y}{\longleftarrow} (W_1, E_1)$$

induces for each λ a diagram

$$\begin{array}{ccc} (\widetilde{W}_0^\lambda, \tilde{E}_0^\lambda) & \longleftarrow & (\widetilde{W}_1^\lambda, \tilde{E}_1^\lambda) \\ \downarrow & & \downarrow \\ (W_0^\lambda, E_0^\lambda) & \longleftarrow & (W_1^\lambda, E_1^\lambda) \end{array}$$

and $(\widetilde{W}_0^\lambda, (\mathfrak{a}_0^\lambda, b^\lambda), \tilde{E}_0^\lambda) \longleftarrow (\widetilde{W}_1^\lambda, (\mathfrak{a}_1^\lambda, b^\lambda), \tilde{E}_1^\lambda)$.

So $\mathcal{D}_\lambda$ can be lifted changing the index 0 to 1. In particular we obtain, for each λ

$$F_1^\lambda = \mathrm{Sing}(\mathfrak{a}_1^\lambda, b^\lambda) \subset \widetilde{W}_1^\lambda \subset W_1^\lambda$$

If the singular loci F_1^λ patch (i.e. if there is a closed set F_1 such that $F_1 \cap W_1^\lambda = F_1^\lambda$) we may repeat this argument once more for any center $Y_1 \subset F_1$ permissible for (W_1, E_1).

Exercise 6.14. If $(W_0, E_0) \overset{\Pi}{\longleftarrow} (W_1, E_1)$ is a restriction, as defined in 6.8, then prove that we obtain a diagram as before and the singular loci F_1^λ patch in $F_1 = \Pi^{-1}(F_0)$.

We are interested in the case in which the singular loci of the data $\mathcal{D}_\lambda$ patch after *any sequence of transformations and restrictions.* This motivates the following definition:

Definition 6.15. A *general basic object* (g. b. o.) over (W_0, E_0), say $(\mathcal{F}_0, (W_0, E_0))$ consists of defining several closed sets F_i:

A A closed set $F_0 \subset W_0$.

B For a morphism $(W_0, E_0) \xleftarrow{\Pi} (W_1, E_1)$ as in 6.8, set $F_1 = \Pi^{-1}(F_0) \subset W_1$ and we denote the restriction by:

$$\begin{array}{ccc} (\mathcal{F}_0, (W_0, E_0)) & \longleftarrow & (\mathcal{F}_1, (W_1, E_1)) \\ F_0 & & F_1 \end{array}$$

C If $(W_0, E_0) \xleftarrow{\Pi} (W_1, E_1)$ is defined as in 2.10 by blowing-up Y_0, and if $Y_0 \subset F_0$, then $F_1 \subset W_1$ is a closed set such that $F_0 \setminus Y_0 \cong F_1 \setminus \Pi^{-1}(Y_0)$ via the isomorphism $W_0 \setminus Y_0 \cong W_1 \setminus \Pi^{-1}(Y_0)$. We denote the transformation by

$$\begin{array}{ccc} (\mathcal{F}_0, (W_0, E_0)) & \longleftarrow & (\mathcal{F}_1, (W_1, E_1)) \\ F_0 & & F_1 \end{array}$$

By composition of restrictions and transformations we get

$$\begin{array}{ccccccc} (\mathcal{F}_0, (W_0, E_0)) & \longleftarrow & \cdots & \longleftarrow & (\mathcal{F}_{r-1}, (W_{r-1}, E_{r-1})) & \longleftarrow & (\mathcal{F}_r, (W_r, E_r)) \\ F_0 & & & & F_{r-1} & & F_r \end{array} \tag{6.15.1}$$

called a permissible sequence for the g. b. o. or a $\mathcal{F}_0$-permissible sequence.

We should think of a g. b. o. as an assignment of closed sets $F_i \subset W_i$ for any sequence 6.15.1.

Example 6.16. (of a g. b. o.) Let $X_0 \subset W_0$ be a hypersurface, set $b = \max \operatorname{mult}_{X_0}$ and E_0 any finite set of hypersurfaces having normal crossings. We now define a g. b. o. say $(\mathcal{F}_0, (W_0, E_0))$ in terms of the set of b-fold points. Let us formulate conditions A, B and C in 6.15:

A Set $F_0 \subset W_0$ to be the closed set of b-fold points.

B Set X_1 as the pull-back of X_0. Note that F_1 is the set of b-fold points of X_1.

C In this case set X_1 the strict transform of X_0 and set F_1 to be the set of b-fold points of X_1 (eventually the empty set).

Finally, for any sequence 6.15.1 we define F_r as the set of b-fold points of an inductively defined hypersurface $X_r \subset W_r$.

Note that in this case, the closed sets F_i do not depend on the set of hypersurfaces E_0. But the allowed transformations 6.15(C) do depend on E_0.

Definition 6.17. With the settings of 6.13, if after any sequence of restrictions and transformations the singular loci F_i^λ patch and define a closed set F_i then the data $\mathcal{D}_\lambda$ define a g. b. o. $(\mathcal{F}_0, (W_0, E_0))$ as in 6.15.

We say that a g. b. o. $(\mathcal{F}_0, (W_0, E_0))$, has a d-dimensional structure if there is a covering and data $\mathcal{D}_\lambda$ as in 6.13 where $\dim \widetilde{W_0^\lambda} = d$ for each λ.

We will sometimes abuse notation and say simply that the g. b. o. has dimension d.

Remark 6.18. Note that if $(W_0, (J_0, b), E_0)$ is a basic object and $d = \dim W_0$, it defines in a trivial way a g. b. o. with a d-dimensional structure, by setting $\{W^\lambda\} = \{W_0\}$ and $F_i = \operatorname{Sing}(J_i, b)$ in 6.15.1.

Example 6.19. Let X_0 be a hypersurface in a smooth scheme W_0 of dimension d. Set $b = \max \operatorname{mult}_{X_0}$ and $B_0 = (W_0, (I(X_0), b), E_0)$. We check now that the d-dimensional g. b. o. defined by B_0 (6.18) is the g. b. o. of b-fold points introduced in 6.16. We follow A,B and C as in 6.15:

- **A** $F_0 = \operatorname{Sing}(I(X_0), b)$ is the set of b-fold points of X_0.
- **B** Let X_1 be the pull back of X_0.
  Now $(W_0, (I(X_0), b), E_0) \longleftarrow (W_1, (I(X_1), b), E_1)$ is as in 6.8 and $F_1 = \operatorname{Sing}(I(X_1), b)$ is the set of b-fold points of X_1.
- **C** Set X_1 as the strict transform of X_0.
  Now $(W_0, (I(X_0), b), E_0) \longleftarrow (W_1, (I(X_1), b), E_1)$ is a transformation as in 4.7 and $F_1 = \operatorname{Sing}(I(X_1), b)$ is the set of b-fold points of X_1.

After any sequence of transformations and restrictions

$$(W_0, (I(X_0), b), E_0) \longleftarrow \cdots \longleftarrow (W_r, (I(X_r), b), E_r)$$

then $F_r = \operatorname{Sing}(I(X_r), b)$ is the set of b-fold points of X_r.

Definition 6.20.

1. We say that 6.15.1 is a resolution of the g. b. o. if $F_r = \emptyset$.
2. We say that an isomorphism $\Theta : W_0 \longrightarrow W_0$ acts or defines an isomorphism on the g. b. o. $(\mathcal{F}_0, (W_0, E_0))$, if:
   (a) Θ acts on (W_0, E_0) and $\Theta(F_0) = F_0$.
   (b) Whenever 6.15.1 is defined by a Θ-equivariant sequence 2.11, then $\Theta(F_i) = F_i$, $i = 0, 1, \dots, r$.

   Note here that some Π_i might be as in 6.8 (see 6.15(B)). In such case Θ is lifted to $\Theta \times \mathrm{Id}_{\mathbb{A}^1_k}$.
3. And finally, we say that 6.15.1 is an equivariant sequence if it is Θ-equivariant for any Θ acting on the g. b. o.

Remark 6.21. If Θ acts on $(X \subset W, E)$ (2.11), then Θ acts on the basic object $(W, (I(X), b), E)$, for $b = \max \operatorname{mult}_X$ as in 4.10 (see also 6.19). Note here that the only point to be checked is that for any $\xi \in X$, $\operatorname{mult}_X = \operatorname{mult}_X(\Theta(\xi))$. Note also that if W is smooth over k we do not assume that Θ is a k-morphism.

Remark 6.22. (Not used elsewhere on the text) **On non-embedded isomorphisms.** In 2.11 we introduced a notion of action or isomorphism Θ on $(X \subset W, E)$, where $E = \{H_1, \ldots, H_r\}$. There $\Theta : W \longrightarrow W$ was an isomorphism, inducing an isomorphism on X and on each $X \cap H_i$. There is a natural lifting of Θ to an isomorphism on $(X_1 \subset W_1, E_1)$ if $(X \subset W, E) \overset{\Pi}{\longleftarrow} (X_1 \subset W_1, E_1)$ is defined by a transformation (2.10) at a Θ-invariant center. If now Π is defined by a restriction (6.8), there is also an action of $\Theta \times \mathrm{Id}_{\mathbb{A}^1_k}$ on $(X_1 \subset W_1, E_1)$.

Definition 6.23. (Not used elsewhere on the text)

1. We say that Θ is a non-embedded isomorphism on $(X \subset W, E)$ if $\Theta : X \longrightarrow X$ is an isomorphism of schemes and, for each $H_i \in E$, it defines an isomorphism of schemes $\Theta : X \cap H_i \longrightarrow X \cap H_i$. Note that there is a natural lifting of Θ to a non-embedded isomorphism on $(X_1 \subset W_1, E_1)$, both in case $(X \subset W, E) \longleftarrow (X_1 \subset W_1, E_1)$ is defined by a transformation (2.10) at a Θ-invariant center, or in case it is defined by a restriction (6.8).
2. Within this frame we can also define a notion of non-embedded isomorphism on a g. b. o. $(\mathcal{F}_0, (W_0, E_0))$ reformulating 6.20(2):
   An isomorphism $\Theta : F_0 \longrightarrow F_0$ is a non-embedded isomorphism of the g. b. o. $(\mathcal{F}_0, (W_0, E_0))$ if
   (a) Θ defines isomorphisms $\Theta : F_0 \cap H_i \longrightarrow F_0 \cap H_i$ for each $H_i \in E$.
   (b) For any restriction or any transformation with Θ-invariant center (i.e. Θ-equivariant)
   $$(\mathcal{F}_0, (W_0, E_0)) \longleftarrow (\mathcal{F}_1, (W_1, E_1))$$
   there is a lifting of Θ to an isomorphism $\Theta : F_1 \longrightarrow F_1$ such that it defines isomorphisms $\Theta : F_1 \cap H \longrightarrow F_1 \cap H$ for each $H \in E_1$.
   (c) Whenever a sequence 6.15.1 is defined by a non-embedded Θ-equivariant sequence, then we have isomorphisms $\Theta : F_i \longrightarrow F_i$ for $i = 0, 1, \ldots, r$ which define isomorphisms
   $$\Theta : F_i \cap H \longrightarrow F_i \cap H \qquad \forall H \in E_i$$

Note that remark 6.21 still holds for a non-embedded isomorphism Θ on $(X \subset W, E)$, so that Θ induces a non-embedded isomorphism on the g. b. o. defined by $(W, (I(X), b), E)$.

Proposition 6.24. The function t revisited and general basic objects. *Fix a basic object $(W_0, (J_0, b), E_0)$, $d = \dim W_0$ and we now assume that the following conditions (C1), (C2) and (C3) hold for the sequence*

$$(W_0, (J_0, b), E_0) \overset{\Pi_1}{\longleftarrow} \cdots \overset{\Pi_r}{\longleftarrow} (W_r, (J_r, b), E_r) \qquad (6.24.1)$$

C1 Π_{i+1} *is defined with center* $Y_i \subset \underline{\mathrm{Max}}\, t_i \subset \underline{\mathrm{Max}}\, \text{w-ord}_i \subset \mathrm{Sing}(J_i, b)$, *for* $i = 0, \dots, r-1$.

C2 *The sequence 6.24.1 is equivariant.*

C3 $\max \text{w-ord}_r > 0$.

Condition (C1) asserts that $\text{w-ord}_i(x_i) \le \text{w-ord}_{i-1}(\Pi_i(x_i))$, $t(x_i) \le t(\Pi_i(x_i))$ *for* $x_i \in \mathrm{Sing}(J_i, b)$. *In particular*

$$\begin{array}{c} \max \text{w-ord}_0 \ge \max \text{w-ord}_1 \ge \cdots \ge \max \text{w-ord}_r \\ \max t_0 \ge \max t_1 \ge \cdots \ge \max t_r \end{array} \tag{6.24.2}$$

If you are skeptical about (C2) take $r = 0$.

Under conditions (C1), (C2) and (C3) we claim that the following properties hold:

P1 *For any* Θ *acting on* $(W_0, (J_0, b), E_0)$

$$\begin{array}{c} \text{w-ord}_i(x) = \text{w-ord}_i(\Theta(x)) \\ t_i(x) = t_i(\Theta(x)) \end{array} \qquad \forall x \in \mathrm{Sing}(J_i, b) \quad i = 0, \dots, r$$

Remark. *Note, in particular, that for any* Θ *as before* $\Theta(\underline{\mathrm{Max}}\, t_i) = \underline{\mathrm{Max}}\, t_i$.

P2 **Canonical choice of centers.** *Set* $R(1)(\underline{\mathrm{Max}}\, t_r)$ *the points where* $\underline{\mathrm{Max}}\, t_r$ *has dimension* $d-1$, *then* $R(1)(\underline{\mathrm{Max}}\, t_r)$ *is open and closed in* $\underline{\mathrm{Max}}\, t_r$ *and regular in* W_r, $R(1)(\underline{\mathrm{Max}}\, t_r)$ *has normal crossings with* E_r, *and defining a transformation with center* $R(1)(\underline{\mathrm{Max}}\, t_r)$, *say*

$$(W_r, (J_r, b), E_r) \stackrel{\Pi_{r+1}}{\longleftarrow} (W_{r+1}, (J_{r+1}, b), E_{r+1})$$

and we may assume that $R(1)(\underline{\mathrm{Max}}\, t_{r+1}) = \emptyset$.

Remark. $\Theta(\underline{\mathrm{Max}}\, t_r) = \underline{\mathrm{Max}}\, t_r$ *implies that for any* Θ *as before* $\Theta(R(1)(\underline{\mathrm{Max}}\, t_r)) = R(1)(\underline{\mathrm{Max}}\, t_r)$, *in particular* Π_{r+1} *is equivariant.*

P3 **Inductive step.** *If* $R(1)(\underline{\mathrm{Max}}\, t_r) = \emptyset$ *we will define a g. b. o. of dimension* $d-1$ *on the pair* (W_r, E_r) *which is linked to the sequence 6.24.1 with the following property:*

If

$$\begin{array}{ccccccc} (W_r, E_r) & \longleftarrow & \cdots & \longleftarrow & (W_{N-1}, E_{N-1}) & \longleftarrow & (W_N, E_N) \\ F_r & & & & F_{N-1} & & F_N \end{array} \tag{6.24.3}$$

is a resolution of the defined g. b. o. (i.e. $F_N = \emptyset$*), then 6.24.3 induces a sequence of transformations over the basic object* $(W_r, (J_r, b), E_r)$ *in 6.24.1, say*

$$(W_r, (J_r, b), E_r) \longleftarrow \cdots \longleftarrow (W_N, (J_N, b), E_N) \tag{6.24.4}$$

so that:

P3(a) $\max t_r = \max t_{r+1} \cdots = \max t_{N-1}$.

P3(b) $\underline{\mathrm{Max}}\, t_i = F_i$ *for* $i = r, \dots, N-1$,

P3(c) *And finally at level* N, *either:*

1. $\mathrm{Sing}(J_N, b) = \emptyset$, *or*
2. $\mathrm{Sing}(J_N, b) \neq \emptyset$ *and* $\max \text{w-ord}_N = 0$, *or*
3. $\mathrm{Sing}(J_N, b) \neq \emptyset$, $\max \text{w-ord}_N > 0$ *and* $\max t_{N-1} > \max t_N$.

Comments: The strength of the function t relies on the fact that properties P2 and P3 hold. The point is that we may describe $\underline{\mathrm{Max}}\, t$ locally as the singular locus of a simple basic object which fulfils all three conditions of 6.12. A property which is not fulfilled by $\underline{\mathrm{Max}}\,\text{w-ord}$. This is the core of the proof of algorithmic resolution and the power of our invariant $t = (\text{w-ord}, n)$.

Although 6.24.1 is a sequence of transformations of basic objects, there is a form of induction in P3. Namely 6.24.3 is defined in terms of a g. b. o. of dimension $d-1$. This illustrates that a correct inductive proof should start by replacing 6.24.1 by a g. b. o. of dimension d. This we shall do in the next section (see 7.10) where we shall prove P1, P2 and P3 after defining the functions n and w-ord on a g. b. o.

7. Equivariant desingularization

7.1. **On Hironaka's trick.** Set $\mathbb{A}^2_k = \mathrm{Spec}(k[X,Y])$ and $B_0 = (W_0, (J_0, b), E_0)$ by taking $W_0 = \mathbb{A}^2_k$, $J_0 = \left\langle X^{b'} - Y^{b''} \right\rangle$, $0 < b \leq b' < b''$ and $E_0 = \emptyset$.

Recall that a basic object defines a g. b. o. $(\mathcal{F}_0, (W_0, E_0))$ with a $\dim W_0$-dimensional structure (6.18), by setting for a sequence of transformations and restrictions (4.7 and 6.8), say:

$$(W_0, (J_0, b), E_0) \longleftarrow \cdots \longleftarrow (W_r, (J_r, b), E_r)$$

the sequence

$$(\mathcal{F}_0, (W_0, E_0)) \longleftarrow \cdots \longleftarrow (\mathcal{F}_r, (W_r, E_r))$$

where $F_i = \mathrm{Sing}(J_i, b)$ (6.15).

Let $B'_0 = (W_0, (\mathfrak{a}_0, d), E_0)$ (same W_0 and same E_0 as for B_0) and assume that B_0 and B'_0 define the same g. b. o. This amounts to saying that such equality is preserved by any sequence of transformations and restrictions. Under these conditions Hironaka shows that, at any closed point $x_0 \in \mathrm{Sing}(J_0, b) = \mathrm{Sing}(\mathfrak{a}_0, d)$,

$$\frac{\nu_{x_0}(J_0)}{b} = \frac{\nu_{x_0}(\mathfrak{a}_0)}{d}$$

Question: How does Hironaka go about proving this?

Answer: By showing that such rational number is expressible in terms of the g. b. o.

To be precise, he shows that such rational number can be characterized in terms of the existence (or not) of a certain sequence of permissible transformations for the g. b. o.

In our example, let x_0 be the origin at $\mathbb{A}_k^2$, thus $x_0 \in \mathrm{Sing}(J_0, b)$ and

$$\frac{\nu_{x_0}(J_0)}{b} = \frac{b'}{b}$$

For the purpose of our discussion we will only use that $\nu_{x_0}(J_0) = b'$.

Define

$$\mathrm{Spec}(k[X,Y]) = W_0 = \mathbb{A}_k^2 \xleftarrow{\Pi_1} W_1 = W_0 \times \mathbb{A}_k^1 = \mathrm{Spec}(k[X,Y,Z])$$

via the inclusion $k[X,Y] \subset k[X,Y,Z]$. Here $W_1 = \mathbb{A}_k^3$ and the fiber over x_0 is a line, say $L_1 = \{x_0\} \times \mathbb{A}_k^1$. Set $x_1 = (x_0, 0) \in L_1$, so x_1 is the origin at $\mathbb{A}_k^3$.

The morphism Π_1 defines a restriction (6.8)

$$(W_0, (J_0, b), E_0) \longleftarrow (W_1, (J_1, b), E_1)$$

where $J_1 = \left\langle X^{b'} - Y^{b''} \right\rangle \subset k[X,Y,Z]$ and $E_1 = \emptyset$. Note that $L_1 = V(X,Y) \subset \mathrm{Sing}(J_1, b)$.

Since x_1 is a closed point on the line L_1, one can define for any integer N, a canonical sequence of transformations of pairs

$$(W_1, E_1) \xleftarrow{\Pi_2} (W_2, E_2) \xleftarrow{\Pi_3} \cdots \xleftarrow{\Pi_N} (W_N, E_N) \tag{7.1.1}$$

defined as follows: Let Π_2 be the blow-up at x_1. For any index $i > 1$, set H_i the exceptional locus of Π_i, L_i the strict transform of L_{i-1} and $x_i = L_i \cap H_i$, finally define Π_{i+1} as the blow-up at the closed point x_i.

In our example $x_1 \in L_1 \subset \mathrm{Sing}(J_1, b)$. One can check by induction that for any index i, $x_i \in L_i \subset \mathrm{Sing}(J_i, b)$. So the sequence 7.1.1 induces a sequence of transformations of basic objects:

$$(W_1, (J_1, b), E_1) \xleftarrow{\Pi_2} \cdots \xleftarrow{\Pi_N} (W_N, (J_N, b), E_N) \tag{7.1.2}$$

In the example $W_1 = \mathbb{A}_k^3 = \mathrm{Spec}(k[X,Y,Z])$ and there is an affine chart $\mathrm{Spec}(k[X_N, Y_N, Z]) \subset W_N$, where $X_N = \dfrac{X}{Z^{N-1}}$, $Y_N = \dfrac{Y}{Z^{N-1}}$, so that $L_N = V(X_N, Y_N)$, $H_N = V(Z)$ and the closed point x_N is the origin of $\mathrm{Spec}(k[X_N, Y_N, Z])$. Now check that locally at x_N:

$$J_N = I(H_N)^{(N-1)(b'-b)} \bar{J}_N \tag{7.1.3}$$

where, in our case, $\bar{J}_N = \left\langle X_N^{b'} - Y_N^{b''} Z^{(N-1)(b''-b')} \right\rangle$.

In terms of the g. b. o. $(\mathcal{F}_0, (W_0, E_0))$, defined by the basic object B_0, we obtain a sequence induced by 7.1.2:

$$(\mathcal{F}_0, (W_0, E_0)) \xleftarrow{\Pi_1} (\mathcal{F}_1, (W_1, E_1)) \xleftarrow{\Pi_2} \cdots \xleftarrow{\Pi_N} (\mathcal{F}_N, (W_N, E_N)) \tag{7.1.4}$$

where $F_i = \mathrm{Sing}(J_i, b)$, Π_1 is a restriction and Π_i for $i > 1$ is the blow-up at a closed point. Since 7.1.1 is a sequence of quadratic transformations over $W_1 = \mathbb{A}_k^3$, it is clear that $\dim W_N = 3$, $\dim H_N = 2$ and $\dim F_N \leq 2$. Formula 7.1.3 tells us that

$$(N-1)(b'-b) \geq b \iff \dim(F_N \cap H_N) = 2 \iff H_N \subset F_N \quad (7.1.5)$$

Note that $b' - b = 0$ (i.e. $\frac{b'}{b} = 1$) if and only if for *any* N, $\dim(F_N \cap H_N) < 2$ (a formula that involves only the g. b. o. $(\mathcal{F}_0, (W_0, E_0))$ defined by B_0).

On the other hand, if $b' - b > 0$, for any integer N big enough $H_N \subset F_N$ and we may define a transformation with center H_N:

$$(W_N, (J_N, b), E_N) \overset{\Pi_{N+1}}{\longleftarrow} (W_{N+1}, (J_{N+1}, b), E_{N+1})$$

Note that $H_N \subset W_N$ is a hypersurface and the blow-up at a hypersurface is an isomorphism, so W_N may be identified with W_{N+1}, H_N with the exceptional locus of Π_{N+1}, say H_{N+1}, and x_N with a unique point, say $x_{N+1} \in W_{N+1}$. Note that locally at x_{N+1}

$$J_{N+1} = I(H_{N+1})^{(N-1)(b'-b)-b} \bar{J}_{N+1}$$

Set $F_{N+1} = \mathrm{Sing}(J_{N+1}, b)$ and note that

$$(N-1)(b'-b) - b \geq b \iff \dim(F_{N+1} \cap H_{N+1}) = 2 \iff H_{N+1} \subset F_{N+1}$$

If these equivalent conditions hold we may blow-up again along the hypersurface H_{N+1}. So whenever possible set

$$(W_N, (J_N, b), E_N) \longleftarrow \cdots \longleftarrow (W_{N+S}, (J_{N+S}, b), E_{N+S}) \quad (7.1.6)$$

by blowing-up the same hypersurface. One can check that locally at x_{N+S} (mapping to x_N via the identity map):

$$J_{N+S} = I(H_{N+S})^{(N-1)(b'-b)-bS} \bar{J}_{N+S}$$

Now sequences 7.1.6 and 7.1.2 induce a sequence over the g. b. o. $(\mathcal{F}_0, (W_0, E_0))$:

$$\begin{aligned} &(\mathcal{F}_0, (W_0, E_0)) \longleftarrow \\ &\qquad (\mathcal{F}_1, (W_1, E_1)) \longleftarrow \cdots \longleftarrow (\mathcal{F}_N, (W_N, E_N)) \\ &\quad (\mathcal{F}_{N+1}, (W_{N+1}, E_{N+1})) \longleftarrow \cdots \longleftarrow (\mathcal{F}_{N+S}, (W_{N+S}, E_{N+S})) \quad (7.1.7) \end{aligned}$$

where $F_i = \mathrm{Sing}(J_i, b)$ for $i = 0, 1, \ldots, N+S$.

Finally the sequence 7.1.7 can be defined if and only if:

$$\begin{aligned} &(N-1)(b'-b) - (S-1)b \geq b \iff \\ &\qquad (N-1)(b'-b) \geq Sb \iff \\ &\qquad\qquad S \leq \left[\frac{(N-1)(b'-b)}{b}\right] \end{aligned}$$

where [ ] denotes the integer part.

This means that, for a fixed integer N, $\left[\frac{(N-1)(b'-b)}{b}\right]$ is the biggest integer S so that 7.1.7 is defined. Thus for any integer N the integer $\left[\frac{(N-1)(b'-b)}{b}\right]$ is determined by the g. b. o. defined by B_0. Finally note that

$$\frac{\nu_{x_0}(J_0)}{b} - 1 = \frac{b'}{b} - 1 = \lim_{N\to\infty} \frac{1}{N-1}\left[\frac{(N-1)(b'-b)}{b}\right]$$

In particular the rational number $\frac{\nu_{x_0}(J_0)}{b}$ is defined in terms of the g. b. o. $(\mathcal{F}_0, (W_0, E_0))$ defined by B_0.

7.2. **Induction on the dimension. What difficulty should we overcome.**

Let X_0 be a hypersurface in a smooth scheme W_0 of dimension d. Set $b = \max \operatorname{mult}_{X_0}$ and say $E_0 = \emptyset$.

In 6.16 we attached to these data a g. b. o. $(\mathcal{F}_0, (W_0, E_0))$. We went one step further to show that this g. b. o. was defined by a simple basic object $B_0 = (W_0, (I(X_0), b), E_0)$ in 6.19, the g. b. o. has a d-dimensional structure and let us denote it $(\mathcal{F}_0^d, (W_0, E_0))$.

If we assume that $b > 1$ and that X_0 is reduced, it is easy to check that $R(1)(\operatorname{Sing}(I(X_0), b)) = \emptyset$. Now our inductive lemma 6.12 asserts that $(\mathcal{F}^d, (W_0, E_0))$ has a $(d-1)$-dimensional structure. In fact the smooth subschemes $W_h^\lambda \subset W^\lambda$ are of dimension $d-1$ and 6.12.1 and 6.12.2 say that

$$B_0^\lambda = ((W_h)_0^\lambda, (C(J_0^\lambda), b!), E_0^\lambda)$$

play the role of $(\widetilde{W}_0^\lambda, (\mathfrak{a}_0^\lambda, b^\lambda), \widetilde{E}_0^\lambda)$ in 6.13(ii).

Now we want to define a resolution of the g. b. o. $(\mathcal{F}_0, (W_0, E_0))$ (6.20(1)) in terms of these local data. This will be achieved defining resolutions of the $(d-1)$-dimensional basic objects B_0^λ so that all these resolutions patch to define a resolution of $(\mathcal{F}_0, (W_0, E_0))$ (i.e. of $(W_0, (I(X_0), b), E_0)$). The difficulty arises since the choice of the covering and the basic objects B_0^λ are not unique.

Now set $F_0 = \operatorname{Sing}(I(X_0), b) \subset W_0$. Fix a closed point $x_0 \in F_0$ and assume that $x_0 \in F_0 \cap W_0^\lambda \cap W_0^\beta$. In such case $x_0 \in \operatorname{Sing}(C(J_0^\lambda), b!)$ and $x_0 \in \operatorname{Sing}(C(J_0^\beta), b!)$. We will prove in 7.3 that

$$\frac{\nu_{x_0}(C(J_0^\lambda))}{b!} = \frac{\nu_{x_0}(C(J_0^\beta))}{b!} \tag{7.2.1}$$

Equality that will be achieved by showing that both rational numbers can be defined in terms of the g. b. o. $(\mathcal{F}_0, (W_0, E_0))$ which admits a $(d-1)$-dimensional structure.

Equality 7.2.1 says that the functions

$$\operatorname{ord}_\lambda^{d-1} : \operatorname{Sing}(C(J_0^\lambda), b!) \longrightarrow \mathbb{Q}$$

$$\operatorname{ord}_\lambda^{d-1}(x_0) = \frac{\nu_{x_0}(C(J_0^\lambda))}{b!}$$

patch to define a function

$$\text{ord}^{d-1} : F_0 \longrightarrow \mathbb{Q}$$

So once 7.2.1 is proved, our task is to define resolutions of basic objects in terms of the functions ord.

In section 4 we fixed a basic object $B_0 = (W_0, (J_0, b), E_0)$ and defined functions ord; w-ord and t. We now define ord, w-ord and t for any general basic object 6.13 with a d-dimensional structure (6.17).

Proposition 7.3. *Let $(\mathcal{F}_0, (W_0, E_0))$ be a g. b. o. which admits a d-dimensional structure (6.17) and fix notation as in 6.13.*

There is a function $\text{ord}^d : F_0 \longrightarrow \mathbb{Q}$ so that, for each λ, the composition

$$F_0^\lambda \xrightarrow{j_0^\lambda} F_0 \xrightarrow{\text{ord}^d} \mathbb{Q}$$

is the function $\text{ord}^\lambda : \text{Sing}(\mathfrak{a}_0^\lambda, b^\lambda) = F_0^\lambda \longrightarrow \mathbb{Q}$ as defined in 4.11.

Proposition 7.4. *With the setting as above, if Θ is an isomorphism on W_0 acting on $(\mathcal{F}_0, (W_0, E_0))$ (6.20(2)) then*

$$\text{ord}^d(x) = \text{ord}^d(\Theta(x)) \ \forall x \in F_0$$

(recall that $\Theta(F_0) = F_0$).

Proof of proposition 7.3. Fix the setting and notation as in 6.13 and a point $x_0 \in F_0$. Suppose that $x_0 = j_0^\lambda(x_0^\lambda)$, $x_0^\lambda \in \text{Sing}(\mathfrak{a}_0^\lambda, b^\lambda) \subset \widetilde{W}_0^\lambda \subset W_0^\lambda$, and $x_0 = j_0^\beta(x_0^\beta)$, $x_0^\beta \in \text{Sing}(\mathfrak{a}_0^\beta, b^\beta) \subset \widetilde{W}_0^\beta \subset W_0^\beta$.

It suffices to show that $\dfrac{\nu_{x_0^\lambda}(\mathfrak{a}_0^\lambda)}{b^\lambda} = \dfrac{\nu_{x_0^\beta}(\mathfrak{a}_0^\beta)}{b^\beta}$. We shall prove this by expressing both rational numbers in an intrinsic way, namely in terms of the g. b. o.

Set $(\mathcal{F}_0, (W_0, E_0)) \xleftarrow{\Pi_1} (\mathcal{F}_1, (W_1, E_1))$ where $W_1 = W_0 \times \mathbb{A}_k^1$, and set also $x_1 = (x_0, 0)$. Hence x_1 maps to x_0 and

$$x_1 \in \{x_0\} \times \mathbb{A}_k^1 \subset F_1 = \Pi_1^{-1}(F_0) \qquad \text{(see 6.15(B))} \tag{7.4.1}$$

This map induces for any λ:

$$(\widetilde{W}_0^\lambda, (\mathfrak{a}_0^\lambda, b^\lambda), \widetilde{E}_0^\lambda) \longleftarrow (\widetilde{W}_1^\lambda(\mathfrak{a}_1^\lambda, b^\lambda), \widetilde{E}_1^\lambda)$$

where $\widetilde{W}_1^\lambda = \widetilde{W}_0^\lambda \times \mathbb{A}_k^1$ and $(x_0^\lambda, 0)(\in\ x_0^\lambda \times \mathbb{A}_k^1)$ maps to x_1 via the immersion j_1^λ. Note also that $\dim \widetilde{W}_1^\lambda = \dim \widetilde{W}_0^\lambda + 1 = d + 1$.

We first define, for each integer $N > 1$, a permissible sequence:

$$T_N : (\mathcal{F}_0(W_0, E_0)) \xleftarrow{\Pi_1} (\mathcal{F}_1, (W_1, E_1)) \xleftarrow{\Pi_2} \cdots \xleftarrow{\Pi_N} (\mathcal{F}_N, (W_N, E_N)) \tag{7.4.2}$$

For $i > 1$ each Π_i will be a quadratic transformation at a point $x_{i-1} \in F_{i-1}$, and $x_1 = (x_0, 0)$ is as above. The sequence T_N is defined as follows. Set $L_1 = x_0 \times \mathbb{A}_k^1$ (a line in W_1), and for any index $1 < j \leq N$, set at W_j:

- H_j the exceptional hypersurface of Π_j.
- L_j the strict transform of L_{j-1}.
- and finally set $x_j = L_j \cap H_j$.

Since $L_1 \subset F_1$, it follows inductively that Π_j induces an isomorphism from $L_j \setminus H_j$ to $L_{j-1} \setminus \{x_{j-1}\}$, so that $x_j \in F_j$ and T_N is indeed well defined (i.e. permissible (6.15.1)).

Now for each index λ, T_N induces sequences of basic objects:

$$(\widetilde{W}_0^\lambda, (\mathfrak{a}_0^\lambda, b^\lambda), \widetilde{E}_0^\lambda) \longleftarrow (\widetilde{W}_1^\lambda, (\mathfrak{a}_1^\lambda, b^\lambda), \widetilde{E}_1^\lambda) \longleftarrow \cdots \longleftarrow (\widetilde{W}_N^\lambda, (\mathfrak{a}_N^\lambda, b^\lambda), \widetilde{E}_N^\lambda) \tag{7.4.3}$$

As said before $\widetilde{W}_1^\lambda = \widetilde{W}_0^\lambda \times \mathbb{A}_k^1$ and $x_1^\lambda = (x_0^\lambda, 0)$ maps to $x_1 \in W_1$. Now check that 7.4.3 is, as 7.4.2 and for $j > 0$, a sequence of quadratic transformations with centers $x_j^\lambda = L_j \cap \widetilde{H}_j^\lambda$, same L_j as before and here $\widetilde{H}_j^\lambda = H_j \cap \widetilde{W}_j^\lambda$.

Note that $\dim \widetilde{W}_j^\lambda = d+1$ for $j = 1, \dots, N$ and $\dim \widetilde{H}_j^\lambda = d$. Set $b' = \nu_{x_0^\lambda}(\mathfrak{a}_0^\lambda)$. Induction on N shows that $\mathfrak{a}_N^\lambda$ can be expressed as a product (see also 7.1.3):

$$\mathfrak{a}_N^\lambda = I(\widetilde{H}_N^\lambda)^{(N-1)(b'-b^\lambda)} K_N^\lambda \subset \mathcal{O}_{\widetilde{W}_N^\lambda} \tag{7.4.4}$$

and $(N-1)(b' - b^\lambda)$ is the highest possible exponent in such expression.

We now go back to the very definition of d-dimensional structure of a g. b. o. in 6.13. There is a closed immersion $\widetilde{W}_0^\lambda \subset W_0^\lambda$, here W_0^λ is an open neighborhood of x_0. The basic object $(\widetilde{W}_0^\lambda, (\mathfrak{a}_0^\lambda, b^\lambda), \widetilde{E}_0^\lambda)$ has been defined so that $F_0 \cap W_0^\lambda \subset \widetilde{W}_0^\lambda$ and moreover $F_0 \cap W_0^\lambda = \mathrm{Sing}(\mathfrak{a}_0^\lambda, b^\lambda)$.

The sequence 7.4.2 induces naturally a sequence over the open subset $W_0^\lambda \subset W_0$, say

$$\begin{array}{ccccccc} T_N^\lambda: & (W_0^\lambda, E_0^\lambda) & \longleftarrow & (W_1^\lambda, E_1^\lambda) & \longleftarrow \cdots \longleftarrow & (W_N^\lambda, E_N^\lambda) \\ & F_0^\lambda & & F_1^\lambda & & F_N^\lambda \end{array} \tag{7.4.5}$$

where each W_i^λ is an open subset of W_i and $F_i^\lambda = W_i^\lambda \cap F_i$. Moreover there is a closed immersion from terms in 7.4.3 in those in 7.4.5 so that: $F_i^\lambda \subset \widetilde{W}_i^\lambda$ and moreover $F_i^\lambda = \mathrm{Sing}(\mathfrak{a}_i^\lambda, b^\lambda)$.

In order to profit from 7.4.4 note first that $H_N \subset W_N$ is the exceptional hypersurface and via the closed immersion $\widetilde{W}_N^\lambda \subset W_N^\lambda$

$$\widetilde{H}_N^\lambda = H_N \cap \widetilde{W}_N^\lambda \tag{7.4.6}$$

where $\widetilde{H}_N^\lambda \subset \widetilde{W}_N^\lambda$ is a hypersurface in $\widetilde{W}_N^\lambda$ and has dimension d. Since $F_N^\lambda \subset \widetilde{W}_N^\lambda$ it follows that:

$$\dim(F_N^\lambda \cap H_N) \leq \dim \widetilde{H}_N^\lambda = d$$

and moreover

$$\dim(F_N^\lambda \cap H_N) = d \Longleftrightarrow \widetilde{H}_N^\lambda \subset F_N^\lambda \Longleftrightarrow (N-1)(b' - b^\lambda) \geq b^\lambda$$

Note now that $b' - b^\lambda = 0$ (i.e. $\frac{b'}{b^\lambda} = 1$) iff for *any* N $\dim(F_N \cap H_N) < d$. This shows that the condition $\frac{\nu_{x_0(\mathfrak{a}_0^\lambda)}}{b^\lambda} = \frac{b'}{b^\lambda} = 1$ is independent of λ.

On the other hand, if $b' - b^\lambda > 0$ it is clear that choosing N big enough such that $(N-1)(b' - b^\lambda) \geq b^\lambda$, then $F_N \cap H_N$ is a permissible center (normal crossing with E_N) of dimension d. In fact $F_N \cap H_N = \widetilde{H}_N^\lambda$.

We now define, for some integers S, an enlargement of T_N, say

$$T_{N,S} : (\mathcal{F}_0, (W_0, E_0)) \xleftarrow{\Pi_1} \cdots \xleftarrow{\Pi_N} (\mathcal{F}_N, (W_N, E_N)) \xleftarrow{\Pi_{N+1}}$$

$$\xleftarrow{\Pi_{N+1}} (\mathcal{F}_{N+1}, (W_{N+1}, E_{N+1})) \xleftarrow{\Pi_{N+2}} \cdots \xleftarrow{\Pi_{N+S}} (\mathcal{F}_{N+S}, (W_{N+S}, E_{N+S})) \quad (7.4.7)$$

setting H_{N+i} the exceptional hypersurface of Π_{N+i} and defining Π_{N+i+1} as the monoidal transformation with center

$$Y_{N+i} = F_{N+i} \cap H_{N+i} \quad (7.4.8)$$

where we assume $\dim Y_{N+i} = d$, for $i = 0, 1, \ldots, S-1$. Note that

$$\dim Y_{N+i} = d \quad \text{iff} \quad \widetilde{H}_{N+i}^\lambda = H_{N+i} \cap \widetilde{W}_{N+i}^\lambda \subset F_{N+i} \quad (7.4.9)$$

In fact the sequence 7.4.7 induces an enlargement of the sequence of basic objects 7.4.3:

$$(\widetilde{W}_N^\lambda, (\mathfrak{a}_N^\lambda, b^\lambda), \widetilde{E}_N^\lambda) \longleftarrow \cdots \longleftarrow (\widetilde{W}_{N+S}, (\mathfrak{a}_{N+S}^\lambda, b^\lambda), \widetilde{E}_{N+S}^\lambda) \quad (7.4.10)$$

One can check that 7.4.10 is a sequence of monoidal transformations at centers $\widetilde{H}_j^\lambda$, $j = N, \ldots, N+S-1$. Now $\dim \widetilde{W}_j^\lambda = d+1$ and $\dim \widetilde{H}_j^\lambda = d$ for $j = N, \ldots, N+S-1$, so that each morphism in the sequence 7.4.10 is in fact the identity map.

An inductive argument shows that

$$\mathfrak{a}_{N+i}^\lambda = I(\widetilde{H}_{N+i}^\lambda)^{(N-1)(b'-b^\lambda)-ib^\lambda} K_N^\lambda$$

so that $T_{N,S}$ is defined as a permissible sequence if $(N-1)(b' - b^\lambda) \geq Sb^\lambda$, in other words, with $\xi_{N+i} \in F_{N+i}$ as in 7.4.9

$$\dim(F_{N+i} \cap H_{N+i}) = d \quad \text{for } i = 0, 1, \ldots, S-1$$

$$\text{iff} \quad S \leq \ell_N = \left[\frac{(N-1)(b' - b^\lambda)}{b^\lambda}\right] \quad (7.4.11)$$

where brackets denote integer part.

This shows that the integer ℓ_N is determined both by N and the general basic object $(\mathcal{F}_0, (W_0, E_0))$. Finally note that

$$\lim_{N \to \infty} \frac{\ell_N}{N-1} = \frac{b'}{b^\lambda} - 1 \quad (7.4.12)$$

is defined in terms of the g. b. o., hence independent of λ. □

Proof of proposition 7.4. Now $\theta(F_0) = F_0$ and we want to prove that $\mathrm{ord}^d(x_0) = \mathrm{ord}^d(\theta(x_0))$ ($x_0 \in F_0$ as before). Set $y_0 = \theta(x_0)$, $y_1 = (y_0, 0) \in y_0 \times \mathbb{A}^1_k$, and L_1 now as the union of $x_0 \times \mathbb{A}^1_k$ and $y_0 \times \mathbb{A}^1_k$. Now $\theta \times \mathrm{Id}$ acts on (W_1, E_1) interchanging both components of L_1 and interchanging $y_1 = (y_0, 0)$ with $x_1 = (x_0, 0)$.

Now modify the sequence 7.4.7 setting $\{y_1, x_1\}$ as center of Π_2. For any index j, L_j will have two components interchanged by θ and so will H_j, thus $L_j \cap H_j$ will consist of two points interchanged by θ.

The centers Y_{N+i} in 7.4.8 are expressed now as an intersection of θ invariant sets; Y_{N+i} has now by construction two components and θ interchanges them, in particular both have the same dimension. The assertion follows now from the proof of Proposition 7.3. □

Remark 7.5. Note that the proof of proposition 7.4 can also be adapted for a non-embedded isomorphism θ on the g. b. o. as defined in 6.23(2).

Theorem 7.6. *Let $(\mathcal{F}_0, (W_0, E_0))$ be a g. b. o. with a d-dimensional structure as in 7.3. Consider any permissible sequence*

$$(\mathcal{F}_0, (W_0, E_0)) \xleftarrow{\Pi_1} \cdots \xleftarrow{\Pi_r} (\mathcal{F}_r, (W_r, E_r))$$

1. *There are functions* $\text{w-ord}^d_i : F_i \longrightarrow \mathbb{Q}$ *such that the composition*

$$F_i^\lambda \xrightarrow{j_i^\lambda} F_i \xrightarrow{\text{w-ord}^d_i} \mathbb{Q}$$

is the function $\text{w-ord}^\lambda_i : F_i^\lambda = \mathrm{Sing}(\mathfrak{a}_i^\lambda, b^\lambda) \longrightarrow \mathbb{Q}$ *defined in 4.11.*

2. *Now if in addition the centers Y_i of the transformations Π_{i+1} are such that* $Y_i \subset \underline{\mathrm{Max}}\,\text{w-ord}^d_i$ *then there are functions* $t^d_i : F_i \longrightarrow \mathbb{Q} \times \mathbb{Z}$ *such that the composition*

$$F_i^\lambda \xrightarrow{j_i^\lambda} F_i \xrightarrow{t_i^d} \mathbb{Q} \times \mathbb{Z}$$

is the function $t_i^\lambda : F_i^\lambda = \mathrm{Sing}(\mathfrak{a}_i^\lambda, b^\lambda) \longrightarrow \mathbb{Q} \times \mathbb{Z}$ *defined in 4.14.*

Proof. The function t was defined in 4.14 entirely in terms of the function w-ord, with the additional constraint $Y_i \subset \underline{\mathrm{Max}}\,\text{w-ord}_i$ for any i, so it suffices to prove that a function w-ord^d can be defined for general basic objects with d-dimensional structure.

Set $\text{w-ord}^d_0 : F_0 \longrightarrow \mathbb{Q}$ where $\text{w-ord}^d_0 = \mathrm{ord}^d$; now 7.3 says that it is obtained by patching the functions $\text{w-ord}^\lambda_0 = \mathrm{ord}^\lambda : \mathrm{Sing}(\mathfrak{a}_0^\lambda, b^\lambda) \longrightarrow \mathbb{Q}$ for all λ.

Recall from 4.8.2 that for each λ we have expressions:

$$\mathfrak{a}_i^\lambda = I(\widetilde{H}_1^\lambda)^{a_1} \cdots I(\widetilde{H}_i^\lambda)^{a_i} \bar{\mathfrak{a}}_i^\lambda \tag{7.6.1}$$

as a product of ideals in $\mathcal{O}_{\widetilde{W}_i^\lambda}$. Here $\widetilde{H}_j^\lambda$ denotes the exceptional locus of the transformation $\Pi_j^\lambda : \widetilde{W}_j^\lambda \longrightarrow \widetilde{W}_{j-1}^\lambda$ with center $Y_{j-1} \cap W_{j-1}^\lambda \subset \widetilde{W}_{j-1}^\lambda$.

One can check that at a point $\xi_i \in \mathrm{Sing}(\mathfrak{a}_i^\lambda, b^\lambda)$ one can recover Hironaka's function $\mathrm{ord}^d(\xi_i) = \frac{\nu_{\xi_i}(\mathfrak{a}_i^\lambda)}{b^\lambda}$, in fact:

$$\mathrm{ord}_i^d(\xi_i) = \text{w-ord}_i^\lambda(\xi_i) + \frac{1}{b^\lambda}\sum_{j=1}^{i} a_j(\xi_i)$$

The expression of $\mathfrak{a}_i^\lambda$ in 7.6.1 has been achieved by induction. If $\xi_i \in H_i$, set $\xi_{i-1} = \Pi_i(\xi_i)$ its image in Y_{i-1} and η_{i-1} the generic point of the irreducible component $Y'_{i-1} \subset Y_{i-1}$ which contains ξ_{i-1}, then $a_i(\xi_i) = \nu_{\eta_{i-1}}(\mathfrak{a}_{i-1}) - b^\lambda$ so that

$$\frac{a_i(\xi_i)}{b^\lambda} = \left(\text{w-ord}_{i-1}^\lambda(\eta_{i-1}) + \sum_{j=1}^{i-1} \frac{a_j(\eta_{i-1})}{b^\lambda}\nu_{\eta_{i-1}}\left(I\left(H_j\right)\right)\right) - 1$$

and

$$\text{w-ord}_i^\lambda(\xi_i) = \mathrm{ord}_i^d(\xi_i) - \sum_{j=1}^{i} \frac{a_j(\xi_i)}{b^\lambda}$$

Now attach to any $\xi_i \in \mathrm{Sing}(\mathfrak{a}_i^\lambda, b^\lambda)$ the sets of points:

$$\begin{array}{cc} \{\xi_0, \xi_1, \dots, \xi_i\} & \{\xi'_0, \xi'_1, \dots, \xi'_i\} \\ \{\eta_0, \eta_1, \dots, \eta_{i-1}\} & \{\eta'_0, \eta'_1, \dots, \eta'_{i-1}\} \end{array} \tag{7.6.2}$$

where ξ_{s-1} is the image of ξ_s at W_{s-1}^λ, $\xi'_s = j_s^\lambda(\xi_s) \in F_s$, for $s = 0, 1, \dots, i$, and η_s is the generic point of the irreducible component $Y'_s \subset Y_s$ which contains ξ_s, $\eta'_s = j_s^\lambda(\eta_s) \in F_s$, for $s = 0, 1, \dots, i-1$.

An inductive argument shows that the values $\frac{a_j(\xi_i)}{b^\lambda}$, $j = 1, \dots, i$ and $\text{w-ord}_i^\lambda(\xi_i)$ are specified by:

1. The rational numbers

$$\mathrm{ord}_0^d(\xi'_0), \mathrm{ord}_1^d(\xi'_1), \dots, \mathrm{ord}_i^d(\xi'_i) \quad \text{and} \quad \mathrm{ord}_0^d(\eta'_0), \mathrm{ord}_1^d(\eta'_1), \dots, \mathrm{ord}_i^d(\eta'_i) \tag{7.6.3}$$

2. The boolean functions

$$YH(\xi'_s, j) = \begin{cases} 1 & \text{if} \quad \xi'_s \in Y_s \text{ and } Y_s \subset H_j \text{ locally at } \xi_s \\ 0 & \text{if} \quad \xi'_s \notin Y_s \text{ or } Y_s \not\subset H_j \text{ locally at } \xi_s \end{cases} \tag{7.6.4}$$

Our theorem follows now from 7.3. In fact we want to show that if two points, say

$$\xi_i^\lambda \in \mathrm{Sing}(\mathfrak{a}_i^\lambda, b^\lambda) \qquad \xi_i^\mu \in \mathrm{Sing}(\mathfrak{a}_i^\mu, b^\mu)$$

map to the same point $\xi'_i \in F_i$ (i.e. $j_i^\lambda(\xi_i^\lambda) = j_i^\mu(\xi_i^\mu) = \xi'_i$), then

$$\text{w-ord}_i^\lambda(\xi_i^\lambda) = \text{w-ord}_i^\mu(\xi_i^\mu)$$

This follows now from 7.6.3 (see 7.3) and 7.6.4 since ξ_i^λ and ξ_i^μ define the same sets $\{\xi'_0, \dots, \xi'_i\}$ and $\{\eta'_0, \dots, \eta'_i\}$. □

Theorem 7.7. *If Θ acts on $(\mathcal{F}_0, (W_0, E_0))$ (6.20(2)) and*

$$\begin{array}{ccccc} (W_0, E_0) & \longleftarrow & \cdots & \longleftarrow & (W_r, E_r) \\ F_0 & & & & F_r \end{array}$$

is Θ-equivariant, then for any $\xi \in F_r$

$$\text{w-ord}_r(\xi) = \text{w-ord}_r(\Theta(\xi)) \qquad t_r(\xi) = t_r(\Theta(\xi))$$

In particular $\Theta(\underline{\text{Max}}\, t_r) = \underline{\text{Max}}\, t_r$ and $\Theta(\underline{\text{Max}}\, \text{w-ord}_r) = \underline{\text{Max}}\, \text{w-ord}_r$.

Proof. The same argument applied in 7.6 will say now that it suffices to prove that $\text{w-ord}_r(\xi) = \text{w-ord}_r(\Theta(\xi))$. This follows from 7.4 applied to 7.6.3 and from the fact that $YH(\xi'_s, j) = YH(\Theta(\xi'_s), j)$ for any j and s in 7.6.4, since by assumption $\Theta(Y_s) = Y_s$ and $\Theta(H_j) = H_j$. □

Remark 7.8. Apply now remark 7.5 to show that theorem 7.7 also holds for non-embedded isomorphisms Θ on $(\mathcal{F}_0, (W_0, E_0))$ (see remark 6.23(2)).

Proposition 7.9. *Let $(\mathcal{F}_0, (W_0, E_0))$ be a g. b. o. with a d-dimensional structure as in 7.3 and consider a sequence of transformations*

$$(\mathcal{F}_0, (W_0, E_0)) \longleftarrow \cdots \longleftarrow (\mathcal{F}_r, (W_r, E_r))$$

such that $\underline{\text{Max}}\, \text{w-ord}_r = 0$. Then there is a function $\Gamma_r : F_r \longrightarrow I_M$, where I_M is the set defined in section 5, such that the composition

$$F_r^\lambda \xrightarrow{j_r^\lambda} F_r \xrightarrow{\Gamma_r} \mathbb{Q}$$

is the function $\Gamma : F_r^\lambda = \text{Sing}(\mathfrak{a}_r^\lambda, b^\lambda) \longrightarrow \mathbb{Q}$ defined in section 5 in terms of the basic object $(\widetilde{W}_r^\lambda, (\mathfrak{a}_r^\lambda, b^\lambda), \widetilde{E}_r^\lambda)$.

Proof. Note first that the function Γ of section 5 depends only on the values $\dfrac{a_r(\xi_r)}{b^\lambda}$ (notation as in the proof of 7.6) and the result follows now from the proof of 7.6. □

Theorem 7.10. *Fix a general basic object with a d-dimensional structure, say $(\mathcal{F}_0^d, (W_0, E_0^d))$, and we now assume that the following conditions (C1), (C2) and (C3) hold for a permissible sequence*

$$\begin{array}{ccccc} (\mathcal{F}_0^d, (W_0, E_0^d)) & \xleftarrow{\Pi_1} & \cdots & \xleftarrow{\Pi_r} & (\mathcal{F}_r^d(W_r, E_r^d)) \\ F_0^d & & & & F_r^d \end{array} \qquad (7.10.1)$$

C1 *Π_{i+1} is defined with center $Y_i \subset \underline{\text{Max}}\, t_i \subset \underline{\text{Max}}\, \text{w-ord}_i \subset F_i^d$, for $i = 0, 1, \ldots, r-1$.*

C2 *The sequence 7.10.1 is equivariant.*

C3 $\max \text{w-ord}_r > 0$.

Condition (C1) asserts that $\text{w-ord}_i(\xi_i) \leq \text{w-ord}_{i-1}(\Pi_i(\xi_i))$, $t_i(\xi_i) \leq t_{i-1}(\Pi_i(\xi_i))$ *for* $\xi_i \in F_i$. *In particular*

$$\begin{gathered} \max \text{w-ord}_0 \geq \max \text{w-ord}_1 \geq \cdots \geq \max \text{w-ord}_r \\ \max t_0 \geq \max t_1 \geq \cdots \geq \max t_r \end{gathered} \tag{7.10.2}$$

Under conditions (C1), (C2) and (C3) we claim that the following properties hold:

P1 *For any* Θ *acting on* $(\mathcal{F}_0^d, (W_0, E_0^d))$

$$\begin{aligned} \text{w-ord}_i(\xi_i) &= \text{w-ord}_i(\Theta(\xi_i)) \\ t_i(\xi_i) &= t_i(\Theta(\xi_i)) \end{aligned} \qquad \forall \xi_i \in F_i^d \quad i = 0, \ldots, r$$

Note, in particular, that for any Θ *as before* $\Theta(\underline{\text{Max}}\, t_i) = \underline{\text{Max}}\, t_i$.

P2 Canonical choice of centers. *Set* $R(1)(\underline{\text{Max}}\, t_r)$ *the points where* $\underline{\text{Max}}\, t_r$ *has dimension* $d-1$, *then* $R(1)$ *is open and closed in* $\underline{\text{Max}}\, t_r$ *and regular in* W_r, $R(1)$ *is a permissible center for*

$$(\mathcal{F}_r^d, (W_r, E_r^d)) \stackrel{\Pi_{r+1}}{\longleftarrow} (\mathcal{F}_{r+1}^d, (W_{r+1}, E_{r+1}^d))$$

and either $R(1)(\underline{\text{Max}}\, t_{r+1}) = \emptyset$ *or* $\max t_r > \max t_{r+1}$.
Note that $\Theta(\underline{\text{Max}}\, t_r) = \underline{\text{Max}}\, t_r$ *implies that for any* Θ *as before* $\Theta(R(1)) = R(1)$, *in particular* Π_{r+1} *is equivariant.*

P3 Inductive step. *If* $R(1)(\underline{\text{Max}}\, t_r) = \emptyset$, *there is a g. b. o. of dimension* $d-1$, *say* $(\mathcal{F}_r^{d-1}, (W_r, E_r^{d-1}))$, *so that if*

$$\begin{array}{ccccc} (\mathcal{F}_r^{d-1}, (W_r, E_r^{d-1})) & \longleftarrow & (\mathcal{F}_{r+1}^{d-1}, (W_{r+1}, E_{r+1}^{d-1})) & \longleftarrow & \cdots \\ F_r^{d-1} & & F_{r+1}^{d-1} & & \\ \cdots \longleftarrow & (\mathcal{F}_{N-1}^{d-1}, (W_{N-1}, E_{N-1}^{d-1})) & \longleftarrow & (\mathcal{F}_N^{d-1}, (W_N, E_N^{d-1})) & \\ & F_{N-1}^{d-1} & & F_N^{d-1} & \end{array} \tag{7.10.3}$$

is a resolution (i.e. $F_N^{d-1} = \emptyset$*), then 7.10.3 induces a sequence*

$$(\mathcal{F}_r^d, (W_r, E_r^d)) \longleftarrow \cdots \longleftarrow (\mathcal{F}_{N-1}^d, (W_{N-1}, E_{N-1}^d)) \longleftarrow (\mathcal{F}_N^d, (W_N, E_N^d)) \tag{7.10.4}$$

so that $\max t_r = \cdots = \max t_{N-1}$, $\underline{\text{Max}}\, t_i = F_i$ *for* $i = r, \ldots, N-1$, *and either*

1. $F_N^d = \emptyset$.
2. $F_N^d \neq \emptyset$ *and* $\max \text{w-ord}_N = 0$.
3. $F_N^d \neq \emptyset$, $\max \text{w-ord}_N > 0$ *and* $\max t_{N-1} > \max t_N$.

Proof: see 9.5.

7.11. **Description of the invariant.** Given a g. b. o. $(\mathcal{F}^d, (W^d, E^d))$ of dimension d, we want to describe an invariant f^d for any point $\xi \in \text{Sing}(\mathcal{F}^d)$. The value of $f^d(\xi)$ will have d coordinates and it can be classified within one of the following types:

A $f^d(\xi) = (t^d(\xi), t^{d-1}(\xi), \dots, t^{d-r}(\xi), \infty, \infty, \dots, \infty)$.
B $f^d(\xi) = (t^d(\xi), t^{d-1}(\xi), \dots, t^{d-r}(\xi), \Gamma(\xi), \infty, \dots, \infty)$.
C $f^d(\xi) = (t^d(\xi), t^{d-1}(\xi), \dots, t^1(\xi))$.

Each coordinate is a function defined for a g. b. o. of the corresponding dimension. The first coordinates $t^d, \dots, t^{d-r}$ are function as in 7.6 where $t^i = (\text{w-ord}^i, n^i)$, $i = d, \dots, d-r$ and always $\text{w-ord}^i(\xi) > 0$.

In case of type (A), the function t^{d-r} will be such that $\xi \in R(1)(\underline{\text{Max}}\, t^{d-r})$, here the codimension is taken in a $(d-r)$-dimensional g. b. o. If $f^d(\xi)$ is within type (B), then $\text{w-ord}^{d-(r+1)}(\xi) = 0$ and Γ is the function defined in section 5 for the monomial case. Note finally that type (C) is within case (A) where $r = d-1$.

This invariant f^d will be an u. s. c. function so that $\underline{\text{Max}}\, f^d$ is a closed set and it will be the center of the next transformation. Of course value of $\max f^d$ is of one the types (A), (B) or (C). We note now that the dimension of $\underline{\text{Max}}\, f^d$ can be described in terms of the value $\max f^d$:

Case A $\dim(\underline{\text{Max}}\, f^d) = (d-r) - 1$.

Case B $\dim(\underline{\text{Max}}\, f^d) = d - r - \Gamma_1$, where Γ_1 is the first coordinate of Γ (see section 5).

Case C $\dim(\underline{\text{Max}}\, f^d) = 0$ (a closed point).

Example 7.12. Set $W = \operatorname{Spec} \mathbb{C}[X, Y, Z]$, $f = X^2 + Y^2 Z$ and $X \subset W$ the hypersurface defined by f (the Whitney umbrella).

As $\max \text{mult}_X = 2$, first we consider the basic object $(W, (I(X), 2), \emptyset)$. In fact we are considering the 3-dimensional general basic object, say $\mathcal{F}_0^3$, defined by this basic object. The singular locus F_0^3 is $\text{Sing}(I(X), 2) = V(X, Y)$ (the Z-axe).

The function $t_0^3 : F_0^3 \longrightarrow \mathbb{Q} \times \mathbb{Z}$ is such that $\max t_0^3 = (1, 0)$ and $\underline{\text{Max}}\, t_0^3 = F_0^3$. We are in the case $R(1)(\underline{\text{Max}}\, t_0^3) = \emptyset$, so we must construct a new g. b. o. $\mathcal{F}_0^2$ of dimension 2.

As $X \in \Delta(f)$, we can take $\widetilde{W} = V(X)$ a hypersurface in W and the basic object $(\widetilde{W}, (\mathfrak{a}_0, 2), \widetilde{E}_0) = (\widetilde{W}, (Y^2 Z, 2), \emptyset)$ which defines a g. b. o. $\mathcal{F}_0^2$ of dimension 2. Note that the singular locus F_0^2 is $\text{Sing}(Y^2 Z, 2) = V(Y) \subset \widetilde{W}$.

Now the function $t_0^2 : F_0^2 \longrightarrow \mathbb{Q} \times \mathbb{Z}$ has maximum value $\max t_0^2 = \left(\frac{3}{2}, 0\right)$ and $\underline{\text{Max}}\, t_0^2$ is the origin of $\widetilde{W}$. Denote by $L_1 \subset \widetilde{W}$ the variety defined by $Y = 0$ and $L_2 \subset \widetilde{W}$ the variety defined by $Z = 0$. So $\mathfrak{a}_0 = I(L_1)^2 I(L_2)$.

Since $R(1)(\underline{\text{Max}}\, t_0^2) = \emptyset$, we should construct a g. b. o. of dimension 1, $\mathcal{F}_0^1$. Here the invariant of 7.11 is

$$(t_0^3, t_0^2, t_0^1) \qquad \text{(Case C)}$$

It turns out that the transformation with center $L_1 \cap L_2$ is a resolution of $\mathcal{F}_0^1$, the permissible sequence in dimension 2 is

$$(\mathcal{F}_0^2, (W_0, E_0^2)) \longleftarrow (\mathcal{F}_1^2, (W_1, E_1^2))$$

and we obtain the basic object $(\widetilde{W}_1, (\mathfrak{a}_1, 2), \widetilde{E}_1)$, $\mathfrak{a}_1 = I(H_1)I(L_1)^2I(L_2)$ (where L_i still denotes the strict transform of L_i). As for dimension 1 we have achieved a resolution of $\mathcal{F}_0^1$, so that one of the possibilities of 7.10(P3) must hold, in this case:

$$\max t_0^2 = \left(\frac{3}{2}, 0\right) > \max t_1^2 = (1,1)$$

and $\underline{\mathrm{Max}}\, t_1^2 = L_1 \cap \widetilde{H}_1$. Following the notation of 4.14, $\widetilde{E}_1^- = \widetilde{E}_1 = \{H_1\}$.

Again $R(1)(\underline{\mathrm{Max}}\, t_1^2) = \emptyset$ and the second step is the transformation with center $L_1 \cap \widetilde{H}_1$:

$$(\mathcal{F}_1^2, (W_1, E_1^2)) \longleftarrow (\mathcal{F}_2^2, (W_2, E_2^2))$$

Now $\max t_2^2 = (1,0)$ and $\underline{\mathrm{Max}}\, t_2^2 = L_1$, so that $R(1)(\underline{\mathrm{Max}}\, t_2^2) = L_1$ is not empty and this is the next center in the resolution process.

Theorem 7.13. *Fix an integer $d \geq 0$. There exists an algorithm of resolution of general basic objects with d-dimensional structure, which means that one can define:*

- **A** *A totally ordered set $(I_d, \leq)$.*
- **B** *For each g. b. o. $(\mathcal{F}_0^d, (W_0, E_0^d))$ with a d-dimensional structure:*
  - **i.** *An equivariant function*

$$f_0^d : F_0^d \longrightarrow I_d$$

such that $\underline{\mathrm{Max}}\, f_0^d$ is permissible for (W_0, E_0^d).
If an equivariant sequence with centers Y_i:

$$\begin{array}{ccccc} (\mathcal{F}_0^d, (W_0, E_0^d)) & \longleftarrow & \cdots & & \\ Y_0 \subset F_0 & & & & \\ & \longleftarrow & (\mathcal{F}_{r-1}^d, (W_{r-1}, E_{r-1}^d)) & \longleftarrow & (\mathcal{F}_r^d, (W_r, E_r^d)) \\ & & Y_{r-1} \subset F_{r-1} & & F_r \neq \emptyset \end{array} \qquad (7.13.1)$$

and equivariant functions $f_i^d : F_i^d \longrightarrow I_d$, $i = 0, \ldots, r-1$, are defined so that $Y_i = \underline{\mathrm{Max}}\, f_i^d$, then there is:

  - **ii.** *An equivariant function*

$$f_r^d : F_r^d \longrightarrow I_d$$

such that $\underline{\mathrm{Max}}\, f_r^d$ is permissible for (W_r, E_r^d).

Note that (B(i)) asserts that the setting of 7.13.1 holds for $r = 1$, whereas (B(ii)) says that whenever $F_r \neq \emptyset$ there is an equivariant enlargement of 7.13.1 with center $Y_r = \underline{\mathrm{Max}}\, f_r^d$

C *For each g. b. o. $(\mathcal{F}_0^d, (W_0, E_0^d))$ with a d-dimensional structure, there is an index N so that the equivariant sequence constructed by (B) is a resolution:*

$$\begin{array}{ccccc} (\mathcal{F}_0^d, (W_0, E_0^d)) & \longleftarrow & \cdots & & \\ Y_0 \subset F_0 & & & & \\ \longleftarrow & (\mathcal{F}_{N-1}^d, (W_{N-1}, E_{N-1}^d)) & \longleftarrow & (\mathcal{F}_N^d, (W_N, E_N^d)) & \\ & Y_{N-1} \subset F_{N-1} & & F_N = \emptyset & \end{array} \quad (7.13.2)$$

Proof: See 9.6.

Example 7.14. Let (W, E) be a pair (2.10) with $d = \dim W$ and $E = \emptyset$. Let X be a regular subvariety of pure codimension r in W. If we consider the g. b. o. of dimension d defined by $(W, (I(X), 1), E)$, we claim that the function f_0^d given by theorem 7.13 is constant and equal to

$$f_0^d = \left((1,0), \overset{(r)}{\ldots}, (1,0), \infty, \ldots, \infty\right)$$

So that $\underline{\mathrm{Max}}\, f_0^d = X$, the codimension of $\underline{\mathrm{Max}}\, f_0^d$ is r and this codimension can be described in terms of the value $\max f_0^d$.

If $r = 1$, the function $t^d : X \longrightarrow \mathbb{Q} \times \mathbb{Z}$ is constant and equal to $(1,0)$, so that $R(1)(\underline{\mathrm{Max}}\, t^d) = X$ and we are in case 7.10(P2). Assume that the claim holds for $r-1$, where $r > 1$. Finally choose (locally) a hypersurface W_h such that $I(W_h) \subset I(X)$ and we set $X_h = W_h \cap X$. Now apply induction for $(W_h, (I(X_h), 1), \emptyset)$.

Note that after the transformation with center $\underline{\mathrm{Max}}\, f_0^d = X$ one achieves a resolution of the b. o. $(W, (I(X), 1), \emptyset)$.

7.15. **Properties on theorem 7.13.**

P1 Fix a g. b. o. and its resolution 7.13.2. If $\xi \in \mathrm{Sing}(\mathcal{F}_i^d)$, $i = 0, \ldots, N-1$, and if $\xi \notin Y_i$ then one can identify the point ξ with a point ξ' of $\mathrm{Sing}(\mathcal{F}_{i+1}^d)$ and

$$f_i^d(\xi) = f_{i+1}^d(\xi')$$

P2 The resolution is obtained by transformations with centers $\underline{\mathrm{Max}}\, f_i^d$, for $i = 0, \ldots, N-1$, and

$$\max f_0^d > \max f_1^d > \cdots > \max f_{N-1}^d$$

P3 If the g. b. o. $(\mathcal{F}_0^d, (W_0, E_0^d))$ is defined by the b. o. $(W_0, (J_0, b), E_0^d)$ as in example 7.14 (where J is the ideal of a regular subvariety X of pure dimension), then the function f_0^d is constant.

P4 For any $i = 0, \ldots, N-1$, the closed set $\underline{\mathrm{Max}}\, f_i^d$ is equidimensional and its dimension is determined by the value $\max f_i^d$ (see 7.11).

7.16. **Proof of theorem 3.1 within the hypersurface case.** Given a hypersurface X embedded in W, where W is smooth over a field k of characteristic zero, we have defined an u. s. c. function $\mathrm{mult}_X : X \longrightarrow \mathbb{Z}$. Set now $(X_0 \subset W_0, E_0) = (X \subset W, E)$, where E is empty and $b = \max \mathrm{mult}_X$.

The sequence 3.1.1 will be constructed by blowing up along regular subvarieties of b-fold points so that for some index N, $\max \mathrm{mult}_{X_N} \leq b-1$. If $\max \mathrm{mult}_{X_N} = b' < b$ and $b' < 1$ we continue, as before replacing b by b' and so on.

Following example 6.19 we attach to b the g. b. o. of dimension $d = \dim W_0$ defined by $B_0 = (W_0, (I(X_0), b), E_0)$, where $E_0 = \emptyset$, and apply now the resolution is provided by theorem 7.13. So we obtain for some index N, $(X_N \subset W_N, E_N)$ with $\max \mathrm{mult}_{X_N} = b' < b$. Now attach to $(X_N \subset W_N, E_N)$ the g. b. o. defined by $(W_N, (I(X_N), b'), E_N)$, note that here E_N is not empty, and apply again 7.13.

Note also that if $b = 1$ then $\mathrm{Sing}(I(X), b) = X$. In this last case we do not want a resolution of $(W, (I(X), b), E)$, but we want to stop at the first index k so that $\max t_k = (1, 0)$ (see 4.19).

Now the constructive theorem 3.1 is a direct consequence of the theorem 7.13.

Apply exercise 13.15 setting here $I_1 = \mathbb{Z}$, $I_2 = I_n$ (as defined in 7.13), $n = \dim W$,

$$g_i : X_i \longrightarrow I_1 \qquad g_i = \mathrm{mult}_{X_i}$$

and

$$g_i' : \underline{\mathrm{Max}}\, \mathrm{mult}_{X_i} \longrightarrow I_2$$

where $g_i' = f_i^n$ as defined in 7.13 in terms of the g. b. o. attached to $\underline{\mathrm{Max}}\, \mathrm{mult}_{X_i}$ (see 4.10).

8. Change of base field and generalization to the non-hypersurface case

Recall that $X \subset W$ is a hypersurface in a regular variety of pure dimension n over a field $\mathbf{k}$. We showed that desingularization reduces to a sequence of resolutions of basic objects, namely those of the form $(W, (I(X), b), E)$ where $b = \max \mathrm{mult}_X$. Since this is an n-dimensional basic object (4.7), and $\mathbf{k}$ a field of characteristic zero, such resolutions are given by theorem 7.13.

8.1. There are two important and related questions on which to focus in this section. The answer will illustrate the strength of our algorithm.

Question 1.
Suppose that X is a singular hypersurface $X \subset W = \mathbb{P}^n_{\mathbb{Q}}$. Now $X \subset W$ induces a hypersurface $X^\lambda \subset W^\lambda = \mathbb{P}^n_{\mathbb{R}}$ (resp. $W^\lambda = \mathbb{P}^n_{\mathbb{C}}$) via $\mathbb{Q} \subset \mathbb{R}$ (resp. $\mathbb{Q} \subset \mathbb{C}$). How does the constructive embedded desingularization of $X \subset W$ relate to that of $X^\lambda \subset W^\lambda$? To be precise, fix $\mathbf{k} \longrightarrow \mathbf{k}^\lambda$ a field extension and $(X \subset W, E)$. Define $(X^\lambda \subset W^\lambda, E^\lambda)$ naturally by change of base field. Set 3.1.1 the embedded desingularization of $(X \subset W, E)$ as defined by the algorithm.

We first note that 3.1.1 induces an embedded desingularization of $X^\lambda \subset W^\lambda$ by change of base field. In fact, since such change induces flat maps, a step by step argument reduces the claim to showing that regularity is preserved by change of the base field. And this holds in our context, where $\mathbf{k}$ is a field of characteristic zero (in general for perfect fields).

The claim, which we address in this section, is that the embedded desingularization of $X^\lambda \subset W^\lambda$ induced by 3.1.1 (as before), is exactly the embedded desingularization of $X^\lambda \subset W^\lambda$ defined by the algorithm (by the constructive resolution).

Question 2.
Suppose now that W (that W^λ) is a regular scheme of finite type over a field $\mathbf{k}$ (over a field $\mathbf{k}^\lambda$): What should we require, on a morphism $W^\lambda \longrightarrow W$ (of schemes) so that setting $(X^\lambda \subset W^\lambda, E^\lambda)$ the pull-back of $(X \subset W, E)$, then:

i. X^λ is a hypersurface.

ii. The embedded desingularizations defined by the algorithm, both for $(X \subset W, E)$ and $(X^\lambda \subset W^\lambda, E^\lambda)$ be related as before?

Set $b = \max \operatorname{mult}_X$ in question 1 and let us go back to 7.2 for the definition of the g. b. o. $(\mathcal{F}_0, (W_0, E_0))$ defined by $(W, (I(X), b), E)$ and its $(n-1)$-dimensional structure via a covering $\{W^\lambda\}$ and basic objects

$$B_0^\lambda = ((W_h)_0^\lambda, (C(J_0^\lambda), b!), (E_h)_0^\lambda)$$

The resolution of $(\mathcal{F}_0, (W_0, E_0))$ in 7.13 is defined by resolutions of the basic objects B_0^λ and the resolution of B_0^λ is defined in terms of invariants that grow from the functions

$$\operatorname{ord}^\lambda : F_0 \cap W^\lambda = \operatorname{Sing}(C(J_0^\lambda), b!) \longrightarrow \mathbb{Q} \qquad \text{(see 7.6.3 and 7.6.4)}$$

Set now $W' = \mathbb{P}^n_{\mathbb{R}}$ and $B_0' = (W', (I(X'), b), E')$ by change of base field. B_0' is a simple basic object, it also defines a g. b. o. of b-fold points $(\mathcal{F}', (W', E'))$ and again this g. b. o. has a $(n-1)$-dimensional structure defined by, say:

$$(B_0')^\lambda = ((W_h')_0^\lambda, (C((J_0')^\lambda), b!), (E_h')_0^\lambda)$$

We want to show that the resolution of $(\mathcal{F}', (W', E'))$ is the pull-back of that of $(\mathcal{F}, (W, E))$.

The resolutions are defined by the functions ord, we claim that if $\xi' \in \operatorname{Sing}(I(X'), b)$ maps to $\xi \in \operatorname{Sing}(I(X), b)$ then $\operatorname{ord}^{n-1}(\xi') = \operatorname{ord}^{n-1}(\xi)$.

If $\xi \in \operatorname{Sing}(C(J_0^\lambda), b!)$ and $\xi' \in \operatorname{Sing}(C((J_0')^\lambda), b!)$, the value $\operatorname{ord}^{n-1}(\xi)$ was defined in proposition 7.3 in terms of certain sequences of transformations over $(\mathcal{F}_0, (W_0, E_0))$ (see 7.4.7).

We claim that such sequence induces a sequence, as 7.4.7, over $(\mathcal{F}_0', (W_0', E_0'))$. Recall that the values both of $\operatorname{ord}^{n-1}(\xi)$ and $\operatorname{ord}^{n-1}(\xi')$ are defined in terms of certain dimensions. In order to prove that $\operatorname{ord}^{n-1}(\xi) = \operatorname{ord}^{n-1}(\xi')$ it suffices to show that setting $F \subset W$ and $F' \subset W'$ as the pull-back, then F regular implies F' regular (to lift a sequence of transformations) and that $\dim F = \dim F'$. We address both conditions at once by comparing the Hilbert-Samuel functions at corresponding points in F and F'.

8.2. We consider here a closed immersion of schemes $F \subset W$, where W is regular of pure dimension n, and has structure of scheme of finite type over a field.

At a closed point $\xi \in F$ we set

$$h(s) = \ell\left(\mathcal{O}_{F,\xi}/\mathfrak{m}^{s+1}\right) \qquad s \in \mathbb{N}$$

where $\mathfrak{m}$ is the maximal ideal of $\mathcal{O}_{F,\xi}$. So that $h : \mathbb{N} \longrightarrow \mathbb{N}$ is an element of $\mathbb{N}^{\mathbb{N}}$ and we consider $\mathbb{N}^{\mathbb{N}}$ as a totally ordered set by taking lexicographic ordering.

We finally set

$$\mathrm{HS}_F : F \longrightarrow \mathbb{N}^{\mathbb{N}} \tag{8.2.1}$$

so that at any closed point $\xi \in F$, $\mathrm{HS}_F(\xi) = h$ as before, and that conditions of exercise 10.6 hold. It turns out that HS_F is an u. s. c. function [Gi3].

Note that if Θ is an isomorphism acting on a pair (W, E) (2.11) and inducing an isomorphism on the subscheme F, then

$$\mathrm{HS}_F(\xi) = \mathrm{HS}_F(\Theta(\xi)) \qquad \forall \xi \in F \tag{8.2.2}$$

Let us point out that if W is of finite type over a field $\mathbf{k}$, we do not need to require that Θ be compatible with the $\mathbf{k}$-structure for 8.2.2 to hold.

8.3. Let W^λ, W be two schemes and $j^\lambda : W^\lambda \longrightarrow W$ a morphism, subject to the following conditions:

C1 W (resp. W^λ) is smooth of pure dimension n over a field $\mathbf{k}$ (resp. over a field $\mathbf{k}^\lambda$).

C2 $j^\lambda : W^\lambda \longrightarrow W$ is flat.

C3 For any closed point ξ of the scheme W, the fiber $j^{-1}(\xi)$ is reduced of dimension zero.

Example 8.4. 1. $j^\lambda : W^\lambda \longrightarrow W$ defined by a change of base field $\mathbf{k} \longrightarrow \mathbf{k}^\lambda$ (e.g. $\mathbb{P}^n_{\mathbb{C}} \longrightarrow \mathbb{P}^n_{\mathbb{Q}}$ which is not of finite type!).

2. $j^\lambda : W^\lambda \longrightarrow W$ a smooth morphism of relative dimension zero [Har, p. 268].

Lemma 8.5. *Set $F \subset W$ as in 8.2, $j^\lambda : W^\lambda \longrightarrow W$ as in 8.3, $F^\lambda = (j^\lambda)^{-1}(F) \subset W^\lambda$ (the scheme theoretical pull-back), and denote also $j^\lambda : F^\lambda \longrightarrow F$ the morphism induced by j^λ. Then*

$$j^\lambda(\xi^\lambda) = \xi \quad \Longrightarrow \quad \mathrm{HS}_F(\xi) = \mathrm{HS}_{F^\lambda}(\xi^\lambda) \tag{8.5.1}$$

Proof. Conditions C2 and C3 in 8.3 ensure that if ξ^λ maps to ξ, then local rings are of the same dimension ([Ma1, 13 B] or [Ma2, Th. 15.1]). We may assume that both are closed points (10.6). If $\{y_1, \dots, y_n\}$ is a regular system of parameters at $\mathcal{O}_{W,\xi}$ and $j^* : \mathcal{O}_{W,\xi} \longrightarrow \mathcal{O}_{W^\lambda,\xi^\lambda}$ is the corresponding homomorphism of local rings, then $\{j^*(y_1), \dots, j^*(y_n)\}$ is a regular system of parameters at $\mathcal{O}_{W^\lambda,\xi^\lambda}$. So the maximal ideal generates the maximal ideal and j^* induces $\mathbf{k}(\xi) \longrightarrow \mathbf{k}(\xi^\lambda)$, extension of residue fields. Therefore the homomorphism

$$\operatorname{gr} j^* : \operatorname{gr}(\mathcal{O}_{W,\xi}) \longrightarrow \operatorname{gr}(\mathcal{O}_{W^\lambda,\xi^\lambda})$$

say

$$\operatorname{gr} j^* : \mathbf{k}(\xi)[Z_1,\dots,Z_n] \longrightarrow \mathbf{k}(\xi^\lambda)[Z'_1,\dots,Z'_n]$$

where $Z_i = \operatorname{In}(y_i)$, $Z'_i = \operatorname{In}(j^*(y_i))$ and $\operatorname{gr} j^*(Z_i) = Z'_i$, is simply defined as the change of base field in the polynomial rings.

If the subscheme F is defined by $I(F) \subset \mathcal{O}_W$, set $j^{-1}(F) = F^\lambda \subset W^\lambda$ the pullback, and fix closed points $\xi \in F$, $\xi^\lambda \in F^\lambda$ so that $j(\xi^\lambda) = \xi$. Recall that $\operatorname{gr}(\mathcal{O}_{F,\xi}) = \operatorname{gr}(\mathcal{O}_{W,\xi})/\operatorname{In}(I(F)_\xi)$ and $\operatorname{gr}(\mathcal{O}_{F^\lambda,\xi^\lambda}) = \operatorname{gr}(\mathcal{O}_{W^\lambda,\xi^\lambda})/\operatorname{In}(I(F^\lambda)_{\xi^\lambda})$. The flatness of the homomorphism $\mathcal{O}_{F,\xi} \longrightarrow \mathcal{O}_{F^\lambda,\xi^\lambda}$ ensures that

$$\mathfrak{m}^k/\mathfrak{m}^{k+1} \otimes_{\mathcal{O}_{F,\xi}} \mathcal{O}_{F^\lambda,\xi^\lambda} = \mathfrak{m}_\lambda^k/\mathfrak{m}_\lambda^{k+1}$$

($\mathfrak{m}$, $\mathfrak{m}_\lambda$ corresponding maximal ideals), so that also $\operatorname{gr}(\mathcal{O}_{F,\xi}) \longrightarrow \operatorname{gr}(\mathcal{O}_{F^\lambda,\xi^\lambda})$ is obtained by the change of base field $\mathbf{k}(\xi) \longrightarrow \mathbf{k}(\xi^\lambda)$.

Set now $\operatorname{HS}_{F^\lambda} : F^\lambda \longrightarrow \mathbb{N}^{\mathbb{N}}$ and $\operatorname{HS}_F : F \longrightarrow \mathbb{N}^{\mathbb{N}}$ as in 8.2. Set $h = \operatorname{HS}_F(\xi)$, $h' = \operatorname{HS}_{F^\lambda}(\xi^\lambda)$ and note that for any integer $k \geq 0$:

$$\begin{aligned} h(k) &= \ell(\mathcal{O}_{F,\xi}/\mathfrak{m}^{k+1}) = \sum_{r=0}^{k+1} \dim_{\mathbf{k}(\xi)} \left[\operatorname{gr}(\mathcal{O}_{F,\xi})\right]_r \\ &= \sum_{r=0}^{k+1} \dim_{\mathbf{k}(\xi^\lambda)} \left[\operatorname{gr}(\mathcal{O}_{F^\lambda\xi^\lambda})\right]_r = \ell(\mathcal{O}_{F^\lambda,\xi^\lambda}/\mathfrak{m}^{k+1}) \\ &= h'(k) \end{aligned}$$

which proves the lemma. □

Corollary 8.6. $\dim_{\xi^\lambda}(F^\lambda) = \dim_\xi(F)$ *and* $\xi^\lambda \in F^\lambda$ *is a regular point iff* $\xi \in F$ *is regular.*

Definition 8.7. Let the setting be as in 8.3, so that W is a regular scheme of finite type over a field. Fix a finite set Λ, we say that the data

$$\{j^\lambda : W^\lambda \longrightarrow W \mid \lambda \in \Lambda\}$$

define a **covering** of W if all three conditions C1, C2, C3 of 8.3 hold, and also

C4 Each $j^\lambda : W^\lambda \longrightarrow W$ is of finite type.

C5 $\bigcup_{\lambda\in\Lambda} \operatorname{Im} j^\lambda = W$.

8.8.

A Assume that conditions C1, C2 and C3 of 8.3 hold for $j^\lambda : W^\lambda \longrightarrow W$, and set $j^\lambda : F^\lambda \longrightarrow F$ as in 8.5. We now fix a totally ordered set $(I, \leq)$ and we say that an u. s. c. function

$$g^\lambda : F^\lambda \longrightarrow I$$

is a restriction of an u. s. c. function $g : F \longrightarrow I$ if $g^\lambda = g \circ j^\lambda$.

Note that lemma 8.5 now says that the Hilbert-Samuel function, as defined in 8.2, is compatible with pull-backs by morphism as in 8.3. In other words, if C1, C2 and C3 hold, then HS_{F^λ} is the restriction of HS_F and also the local dimension $\dim^\lambda : F^\lambda \longrightarrow \mathbb{Z}$, $\dim^\lambda(\xi^\lambda) = \dim_{\xi^\lambda}(F^\lambda)$ is the restriction of the local dimension $\dim : F \longrightarrow \mathbb{Z}$ (8.6).

B Fix now:

- A covering $\{j^\lambda : W^\lambda \longrightarrow W \mid \lambda \in \Lambda\}$ as in 8.7.
- A closed reduced subscheme $F \subset W$.
- A totally ordered set $(I, \leq)$ and for each $\lambda \in \Lambda$, an u. s. c. function

$$g^\lambda : F^\lambda \longrightarrow I$$

We will say that the functions **patch** if $g^\lambda(\xi^\lambda) = g^\mu(\xi^\mu)$ whenever $\xi^\lambda \in F^\lambda$ and $\xi^\mu \in F^\mu$ map to the same point $\xi \in F$. In such case we define

$$g : F \longrightarrow I$$

by setting $g(\xi) = g^\lambda(\xi^\lambda)$.

Lemma 8.9. *With the setting of 8.8(B):*

1. *Each F^λ is a reduced scheme.*
2. *g is an u. s. c. function.*
3. *If $(j^\lambda)^{-1}(\underline{\mathrm{Max}}\, g) \neq \emptyset$ then $(j^\lambda)^{-1}(\underline{\mathrm{Max}}\, g) = \underline{\mathrm{Max}}\, g^\lambda$.*

Proof. Condition C4 of 8.7 asserts that $j^\lambda : F^\lambda \longrightarrow F$ is of finite type, and flat morphisms of finite type are open morphisms (the image of an open set is also open) ([Ma1, 6.I, p. 48]). Now $\mathrm{Reg}(F)$ (regular points in F) is dense in F, so the same will hold at each F^λ (see 8.6). This proves 1. As for 2 note that for a fixed element $\alpha \in I$

$$\{\xi \in F \mid g(\xi) < \alpha\} = \bigcup_{\lambda \in \Lambda} j^\lambda\left(\{\xi^\lambda \in F^\lambda \mid g^\lambda(\xi^\lambda) < \alpha\}\right)$$

and therefore open in F. □

Remark 8.10. The fact that $F^\lambda \subset W^\lambda$ is reduced if $F \subset W$ is reduced (8.9(1)) also holds for any $j^\lambda : W^\lambda \longrightarrow W$ as in 8.3, and follows only from C2 and C3 (see [Mal, 21.E. p. 156]).

An extension of the definition of pairs (W^λ, E^λ) in terms of (W, E) (see 6.1) is clear. We now want to define a g. b. o. over (W^λ, E^λ), in terms of one on (W, E) for any morphism in the setting of 8.3 (definition 6.15).

Lemma 8.11. *Fix a diagram:*

$$\begin{array}{ccccc} & W^\lambda & \stackrel{\Pi_1^\lambda}{\longleftarrow} & W_1^\lambda & \\ j^\lambda & \downarrow & & \downarrow & j_1^\lambda \\ & W & \stackrel{\Pi_1}{\longleftarrow} & W_1 & \end{array}$$

which is the fiber product of j^λ and Π_1.

1. *If the conditions of 8.3 hold for j^λ, and Π_1 is the blow-up of a closed and regular center F, then:*
   - *j_1^λ fulfills the three conditions of 8.3.*
   - *Π_1^λ is the blow-up at the regular center F^λ.*
   - *If $\{j^\lambda : W^\lambda \longrightarrow W \mid \lambda \in \Lambda\}$ is a covering then $\{j_1^\lambda : W_1^\lambda \longrightarrow W_1 \mid \lambda \in \Lambda\}$ is a covering.*
2. *Ditto, but now replacing Π_1 by a restriction as defined in 6.8.*

8.12. Note that if $j^\lambda : W^\lambda \longrightarrow W$ is as in 8.3 and if $(\mathcal{F}, (W, E))$ is a g. b. o. (6.15), then 8.10 and 8.11 say that any sequence of transformations and restrictions

$$\begin{array}{ccccc} (W_0, E_0) & \longleftarrow & \cdots & \longleftarrow & (W_r, E_r) \\ F_0 & & & & F_r \end{array}$$

induces by successive pull-backs, a sequence of transformations and restrictions

$$\begin{array}{ccccc} (W_0^\lambda, E_0^\lambda) & \longleftarrow & \cdots & \longleftarrow & (W_r^\lambda, E_r^\lambda) \\ F_0^\lambda & & & & F_r^\lambda \end{array}$$

Definition 8.13. Let $(\mathcal{F}_0, (W_0, E_0))$ be a g. b. o. and $W_0^\lambda \longrightarrow W_0$ as in 8.3. Set $(W_0^\lambda, E_0^\lambda)$ as in 8.10. We will say that a g. b. o. $(\mathcal{F}_0^\lambda, (W_0^\lambda, E_0^\lambda))$ is a **restriction** of $(\mathcal{F}_0, W_0, E_0))$, if any $\mathcal{F}_0$-permissible sequence, induces an $\mathcal{F}_0^\lambda$-permissible sequence as in 8.12.

Theorem 8.14.

A *The theorem of constructive and equivariant resolution of d-dimensional general basic objects (7.13) also applies if we replace the notion of open covering in 6.13 by that of covering in 8.7.*

B *Fix $W^\lambda \longrightarrow W$ as in 8.3. If $(\mathcal{F}, (W, E))$ and $(\mathcal{F}^\lambda, (W^\lambda, E^\lambda))$ are two g. b. o. with a d-dimensional structure, and if $(\mathcal{F}^\lambda, (W^\lambda, E^\lambda))$ is a restriction of $(\mathcal{F}, (W, E))$ (8.13), then the resolution of $(\mathcal{F}^\lambda, (W^\lambda, E^\lambda))$ defined by theorem 7.13 is the pull-back of that of $(\mathcal{F}, (W, E))$.*

Proof. As for (A) note that the resolution functions $(f_i^d)^\lambda$ are certainly defined for each W_i^λ. We want to prove that the different $(f_i^d)^\lambda$ patch and define $f_i^d : F_i^d \longrightarrow I_d$, in which case (A) would follow from 8.9.

Recall that, up to induction, the functions f_i^d were defined in terms of t_i, and these functions were defined in terms of the function ord.

A similar argument applies for (B). So that the proof of both (A) and (B) reduces to showing that if $(\mathcal{F}^\lambda, (W^\lambda, E^\lambda))$ and $(\mathcal{F}, (W, E))$ are as in 8.13 then the function ord is defined and restricts in the sense of 8.8(A). Recall finally that the value of the function ord was defined in terms of the *dimension* of the singular locus for certain sequences of transformations (see 7.4.11 and 7.4.12). Apply this argument over $(\mathcal{F}, (W, E))$ and $(\mathcal{F}^\lambda, (W^\lambda, E^\lambda))$, and the result follows now from 8.13. □

Three main theorems and embedded desingularization

Fix now W smooth of dimension n, over a perfect field (not characteristic zero in general) and set X a closed subscheme of W.

We will attach to $X \subset W$ a general basic object, as done for a hypersurface, say $\mathcal{F}_0, (W, E)$.

The first main theorem, due to Bennett, is the same as 2.3 but replacing the function mult_X by the function HS_X

Theorem 8.15. *Let $Y \subset X \subset W$, X a subvariety and Y a regular irreducible variety. Assume that $h = \max \text{HS}_X$ and $Y \subset \underline{\text{Max}}\, \text{HS}_X$ and set*

$$\begin{array}{ccc} W & \xleftarrow{\Pi} & W_1 \\ X & & X_1 \\ Y & & \end{array}$$

X_1 the strict transform of X. Then for any $x_1 \in X_1$,

$$\text{HS}_{X_1}(x_1) \leq \text{HS}_X(\Pi(x_1))$$

Theorem 8.16. *Now we mention the second main theorem, due to Aroca (see also* [H2, p. 100], [Gi3, Th. 3.12] *and* [O], *and* [Gi4, p. 233] *for the case of characteristic $p > 0$), which we state by saying that there is a general basic object with a n-dimensional structure, $(\mathcal{F}_h, (W, E))$, as modified in 8.14(A), for $n = \dim W$, such that if*

$$(\mathcal{F}_h, (W, E)) \longleftarrow \cdots \longleftarrow (\mathcal{F}_{h,r}, (W_r, E_r))$$

is any permissible sequence:

- *If $F_r \neq \emptyset$ then $\max \text{HS}_{X_r} = \max \text{HS}_X = h$ and $F_r = \underline{\text{Max}}\, \text{HS}_{X_r}$.*
- *If $F_r = \emptyset$ then $\max \text{HS}_{X_r} < \max \text{HS}_X = h$.*

8.17. If Θ acts on (W, E) and $\Theta(X) = X$, then 8.2.2 ensures that Θ acts on $(\mathcal{F}_h, (W, E))$.

So in case of characteristic zero, the second main theorem 8.16, together with theorem 8.14 says that there is a sequence of monoidal transformations, so that on the one hand max HS will drop, and on the other hand any Θ as before, will naturally lift via these monoidal transformations. Repeating this procedure again and again, we can always force max HS to drop.

We state now the third main theorem, due to Hironaka, saying that if X is reduced, after finitely many procedures as above, each forcing max HS to drop, the final strict transform of X will be regular.

Finally, and in order to achieve embedded desingularization as defined in 3.1, we apply theorem 8.14 to the basic object $(W,(I(X),1),E)$ so as to force $\max t = (1,0)$ (see 4.19).

Corollary 8.18. Theorem 3.1 in the non-hypersurface case. *There is a totally ordered set I_n so that theorem 3.1 holds (recall: for any $(X \subset W, E)$, $\dim W = n$). Furthermore, whenever the three conditions in 8.3 hold for $j^\lambda : W^\lambda \longrightarrow W$, the constructive desingularization defined by theorem 3.1 on $(X^\lambda \subset W^\lambda, E^\lambda)$ is the pull-back of that of $(X \subset W, E)$.*

Proof. Argue as in 7.16 but replacing $I_1 = \mathbb{Z}$ by $I_1 = \mathbb{N}^{\mathbb{N}}$ and mult_X by HS_X. □

9. Proofs

Lemma 9.1. *Consider a transformation of basic objects:*

$$B = (W,(J,b),E) \xleftarrow{\Pi} B_1 = (W_1,(J_1,b),E_1)$$

and denote by H_{r+1} the exceptional divisor (2.10).

Then for all $i \in \{1,\dots,b\}$, $\Delta^{b-i}(J)\mathcal{O}_{W_1} \subset I(H_{r+1})^i$ and

$$\frac{1}{I(H_{r+1})^i}\Delta^{b-i}(J) \subset \Delta^{b-i}(J_1)$$

Proof. If $i = b$, $\Delta^0(J) = J$, $\Delta^0(J_1) = J_1$ and the claim is trivial. We argue by decreasing induction on i, so assume that the inclusion holds for some $i > 1$. Let $\xi_1 \in H_{r+1}$ be any closed point, $\xi = \Pi(\xi_1)$ and choose $y \in \mathcal{O}_{W_1,\xi_1}$ such that $I(H_{r+1})_{\xi_1} = (y)$.

By 1.5.1, we can consider the situation at the corresponding complete local rings $\hat{\mathcal{O}}_{W,\xi}$ and $\hat{\mathcal{O}}_{W_1,\xi_1}$. So we identify at this moment Δ with $\hat{\Delta}$ (1.2.1). It suffices to show that for generators f of $\Delta^{b-(i-1)}(J_0)$, $\frac{f}{y^{i-1}} \in \Delta^{b-(i-1)}(J_1)$. If $f \in \Delta^{b-i}(J)\left(\subset \Delta^{b-(i-1)}(J)\right)$ then the assertion follows by induction. Therefore, by 1.2.1, it suffices to treat the case $f = D(g)$, for $g \in \Delta^{b-i}(J)_\xi$ and D a derivation. By induction we have:

$$\frac{g}{y^i} \in \frac{1}{I(H_{r+1})^i}\Delta^{b-i}(J)_\xi \subset \Delta^{b-i}(J_1)_{\xi_1} \left(\subset \Delta^{b-(i-1)}(J_1)_{\xi_1}\right)$$

Set $D' = yD$, we claim that D' is a derivation on $\xi_1 \in W_1$. To check this note first that y can be chosen so that $\{y = x_1, x_2, \dots, x_n\}$ is a regular system of parameters at $\mathcal{O}_{W,\xi}$ and $\left\{y, \frac{x_2}{y}, \dots, \frac{x_r}{y}, x_{r+1}, \dots, x_n\right\}$ is a regular system of parameters at $\mathcal{O}_{W_1,\xi_1}$. It suffices now to check that $D'\left(\frac{x_i}{y}\right) \in \mathcal{O}_{W_1,\xi_1}$, which is a straightforward computation.

Since $\frac{g}{y^i} \in \Delta^{b-i}(J_1)_{\xi_1}$, it follows that $D'\left(\frac{g}{y^i}\right) \in \Delta^{b-(i-1)}(J_1)_{\xi_1}$. Finally

$$D'\left(\frac{g}{y^i}\right) = \frac{D(g)}{y^{i-1}} - iD(y)\frac{g}{y^i}$$

hence $\frac{f}{y^{i-1}} = \frac{D(g)}{y^{i-1}} = \left(D'\left(\frac{g}{y^i}\right) + iD(y)\frac{g}{y^i}\right)$ belongs to $\Delta^{b-(i-1)}(J_1)_{\xi_1}$. □

9.2. *Proof of lemma 6.6.*

$I(W_h) \subset \Delta^{b-1}(J)$ implies that

$$\mathrm{Sing}(J, b) \subset \mathrm{Sing}(I(W_h), 1)$$

So any transformation $(W, (J, b), E) \longleftarrow (W_1, (J_1, b), E_1)$ induces naturally a transformation $(W, (I(W_h), 1), E) \longleftarrow (W_1, (I((W_h)_1), 1), E_1)$, and $I((W_h)_1) \subset \Delta^{b-1}(J_1)$ follows from 9.1 applied for $i = 1$.

Definition 9.3. (See also exercise 13.1) Let $B = (W, (J, b), E)$ be a basic object and $\widetilde{W}$ a regular closed subvariety of W such that $I(\widetilde{W}) \subset \Delta^{b-1}(J)$. We define the coefficient ideal of B on $\widetilde{W}$ as:

$$C(J) = \sum_{i=0}^{b-1} \Delta^i(J)^{\frac{b!}{b-i}} \mathcal{O}_{\widetilde{W}}$$

9.4. *Proof of the inductive lemma (6.12).*

We may assume that the covering $\{W^\lambda\}$ reduce to W. If $R(1)(\mathrm{Sing}(J, b)) \neq \emptyset$ then $R(1)(\mathrm{Sing}(J, b))$ is a union of some connected components of W_h and the assertion is clear.

Let $H_k \subset W_k$ be the exceptional hypersurface corresponding to the transformation $(W_{k-1}, (J_{k-1}, b), E_{k-1}) \longleftarrow (W_k, (J_k, b), E_k)$.

We set for $k > 0$:

$$\left[\Delta^{b-i}(J_0)\right]_k = \frac{1}{I(H_k)^i} \left[\Delta^{b-i}(J_0)\right]_{k-1} \mathcal{O}_{W_k}$$

and

$$C(J_0)_k = \sum_{i=1}^{b} \left[\Delta^{b-i}(J_0)\right]_k^{\frac{b!}{i}} \mathcal{O}_{(W_h)_k}$$

We begin by formulating a claim, say:

Claim(s): For any index $k = 0, \dots, s$

1. $\left[\Delta^{b-i}(J_0)\right]_k \subset \Delta^{b-i}(J_k)$.
2. At any closed point $\xi_k \in \mathrm{Sing}(C(J_0)_k, b!)$ there is a regular system of parameters $z_k, x_{k,1}, \dots, x_{k,n-1}$ such that
   (a) $I((W_h)_k)_{\xi_k} = \langle z_k \rangle$.
   (b) Setting $R_k = \hat{\mathcal{O}}_{W_k,\xi_k}$, $\bar{R}_k = \hat{\mathcal{O}}_{(W_h)_k,\xi_k}$, there is a set of generators $\{f_k^{(\lambda)}\}$ of $J_k R_k$

$$f_k^{(\lambda)} = \sum_\alpha a_{k,\alpha}^{(\lambda)} Z_k^\alpha \qquad a_{k,\alpha}^{(\lambda)} \in k'[[X]] = \bar{R}_k$$

so that

$$\left(a_{k,\alpha}^{(\lambda)}\right)^{\frac{b!}{b-\alpha}} \in C(J_0)_k \bar{R}_k \tag{9.4.1}$$

for all α with $\alpha < b$.

Before we proceed with the proof of our claim, let us point out that if (1) holds then $C(J_0)_k \subset C(J_k)$ and in particular

$$(\mathrm{Sing}(J_k, b) =)\, \mathrm{Sing}(C(J_k), b!) \subset \mathrm{Sing}(C(J_0)_k, b!)$$

On the other hand, if (2) holds at any $\xi_k \in \mathrm{Sing}(C(J_0)_k, b!)$, it follows from 9.4.1 that $\xi_k \in \mathrm{Sing}(J_k, b)$, so

$$\mathrm{Sing}(C(J_0)_k, b!) \subset \mathrm{Sing}(J_k, b)$$

As for claim(0), (1) is trivial and (2) follows from the fact that $a_{k,\alpha}^{(\lambda)} \in \Delta^{b-\alpha}(J_0)\bar{R}_0$ if $|\alpha| < b$.

We now assume claim(s) and consider a sequence of transformations of length $s+1$. Since $\left[\Delta^{b-i}(J_0)\right]_s \subset \Delta^{b-i}(J_s)$, then

$$\left[\Delta^{b-i}(J_0)\right]_{s+1} = \frac{1}{I(H_{s+1})^i}\left[\Delta^{b-i}(J_0)\right]_s \subset \frac{1}{I(H_{s+1})^i}\Delta^{b-i}(J_s) \subset \Delta^{b-i}(J_{s+1})$$

see 9.1 for the last inclusion.

Let $\xi_{s+1} \in \mathrm{Sing}(C(J_0)_{s+1}, b!)$ be a closed point, $\xi_s \in \mathrm{Sing}(C(J_0)_s, b!)$ the image in W_s. After a finite extension of the base field and a linear change involving only the variables $x_{s,j}$ in R_s, we may assume at $R_{s+1} = \hat{\mathcal{O}}_{W_{s+1},\xi_{s+1}}$ a regular system of parameters $z_{s+1}, x_{s+1,1}, \dots, x_{s+1,n-1}$ with $I(H_{s+1})_{\xi_{s+1}} = \langle x_{s+1,1} \rangle$, $x_{s+1,1} = x_{s,1}$, $I((W_h)_{s+1})_{\xi_{s+1}} = \langle z_{s+1} \rangle$, $z_{s+1} = \dfrac{z_s}{x_{s,1}}$, and define

$$f_{s+1}^{(\lambda)} = \frac{f_s^{(\lambda)}}{x_{s,1}^b} = \sum_\alpha a_{s+1,\alpha}^{(\lambda)} Z_{s+1}^\alpha$$

so that $a^{(\lambda)}_{s+1,\alpha} = \dfrac{a^{(\lambda)}_{s,\alpha}}{x^{b-\alpha}_{s,1}}$. In particular

$$\left(a^{(\lambda)}_{s+1,\alpha}\right)^{\frac{b!}{b-\alpha}} \in C(J_0)_{s+1}\bar{R}_{s+1}$$

This proves claim(s+1).

9.5. *Proof of theorem 7.10.*

Property P1 was treated in 7.7, so we address the proofs of properties P2 and P3. We first attach to the sequence 7.10.1 and to the value $\max t_r$ the integer r_0 (see 7.10.2)

$$r_0 = \text{smallest index so that } \max \text{w-ord}_{r_0} = \max \text{w-ord}_r \tag{9.5.1}$$

and we will construct a g. b. o. with a d-dimensional structure over the pair $(W_r, (E^d_r)^+)$:

$$(\mathcal{F}''_r, (W_r, (E^d_r)^+)), \qquad F''_r = \underline{\text{Max}}\, t_r \tag{9.5.2}$$

where F''_r is the closed set in W_r defined by the basic object (see 6.15). Here W_r is as in 7.10.1, but $(E^d_r)^+ \subset E^d_r$ is the subset $E^d_r \setminus (E^d_r)^-$ where $(E^d_r)^-$ denotes the strict transform of hypersurfaces in $E^d_{r_0}$. The construction of 9.5.2 will be done by defining the basic objects $(\widetilde{W}^\lambda_r, (\mathfrak{a}^\lambda_r, b^\lambda), \widetilde{E}^\lambda_r)$ (see 6.13), note that the condition $\underline{\text{Max}}\, t_r \cap W^\lambda_r = \text{Sing}(\mathfrak{a}^\lambda_r, b^\lambda)$ will assure that these basic objects define a g. b. o. (6.15). Then we will prove that:

- **A** The three conditions in lemma 6.12 hold for each $(\widetilde{W}^\lambda_r, (\mathfrak{a}^\lambda_r, b^\lambda), \widetilde{E}^\lambda_r)$.
- **B** Any center Y permissible for $(\mathcal{F}''_r, (W_r, (E^d_r)^+))$ is also permissible for $(\mathcal{F}^d_r, (W_r, E^d_r))$ and setting

$$\begin{array}{ccc} (\mathcal{F}''_r, (W_r, (E^d_r)^+)) & \longleftarrow & (\mathcal{F}''_{r+1}, (W_{r+1}, (E^d_{r+1})^+)) \\ (\mathcal{F}^d_r, (W_r, E^d_r)) & \longleftarrow & (\mathcal{F}^d_{r+1}, (W_{r+1}, E^d_{r+1})) \end{array}$$

  either $\max t_r > \max t_{r+1}$, in which case $F''_{r+1} = \emptyset$; or $\max t_r = \max t_{r+1}$ and $F''_{r+1} = \underline{\text{Max}}\, t_{r+1}$.

So that P2 and P3 of 7.10 will follow from properties (A), (B) and an inductive argument.

Let the first term of 7.10.1, namely $(\mathcal{F}^d_0, (W_0, E^d_0))$ be defined in terms of $\mathcal{D}_\lambda$ in 6.13. In order to simplify notation we denote $(\widetilde{W}^\lambda_0, (\mathfrak{a}^\lambda_0, b^\lambda), \widetilde{E}^\lambda_0)$ in 6.13 simply by $(W_0, (J_0, b), E_0)$ and the sequence of transformations induced by 7.10.1 will be identified with the sequence 6.24.1. In doing so we certainly abuse notation, but the point is that theorem 7.6 allow us to reduce our proofs of properties P2 and P3 within the context of 6.24.

Consider a sequence of transformations of basic objects:

$$B_0 = (W_0, (J_0, b), E_0) \longleftarrow \cdots \longleftarrow B_r = (W_r, (J_r, b), E_r) \tag{9.5.3}$$

and assume that all centers are such that $Y_i \subset \underline{\text{Max}}\, t_i \subset \underline{\text{Max}}\,\text{w-ord}_i$. Since the terms in 9.5.3 are charts in 7.10.1; $\max t_i(\text{in } 9.5.3) \leq \max t_i(\text{in } 7.10.1)$ for $i =$

$0, \ldots, r$. We will assume that equality holds at r, in which case inequalities 7.10.2 and 6.24.2 show that the integer r_0 in 9.5.1 plays the same role in 9.5.3.

Recall from 4.11 that there is an expression

$$J_r = I(H_1)^{a_1} \cdots I(H_r)^{a_r} \bar{J}_r$$

Set b_r such that $\max \text{w-ord}_r = \frac{b_r}{b}$, we are assuming $b_r > 0$ by condition C3 of 7.10.

Define a basic object $B'_r = (W_r, (J'_r, b'), E_r)$ as follows:

$$J'_r = \begin{cases} \bar{J}_r & \text{if } \ b_r \geq b \\ \bar{J}_r^{b-b_r} + (I(H_1)^{a_1} \cdots I(H_r)^{a_r})^{b_r} & \text{if } \ b_r < b \end{cases}$$
$$b' = \begin{cases} b_r & \text{if } \ b_r \geq b \\ b_r(b-b_r) & \text{if } \ b_r < b \end{cases} \tag{9.5.4}$$

One can check that

a. $\operatorname{Sing}(J'_r, b') = \underline{\operatorname{Max}}\, \text{w-ord}_r$.

b. A transformation $B'_r = (W_r, (J'_r, b'), E_r) \longleftarrow B'_{r+1} = (W_{r+1}, (J'_{r+1}, b'), E_{r+1})$ induces naturally a transformation

$$B_r = (W_r, (J_r, b), E_r) \longleftarrow B_{r+1} = (W_{r+1}, (J_{r+1}, b), E_{r+1})$$

c. $\operatorname{Sing}(J'_{r+1}, b') = \emptyset$ iff $\max \text{w-ord}_r > \max \text{w-ord}_{r+1}$; and if $\max \text{w-ord}_r = \max \text{w-ord}_{r+1}$ then

$$\operatorname{Sing}(J'_{r+1}, b') = \underline{\operatorname{Max}}\, \text{w-ord}_{r+1} = \{\xi \in \operatorname{Sing}(J_{r+1}, b) \mid \text{w-ord}_{r+1}(\xi) = \frac{b_r}{b}\}$$

Remark: One can check that (J'_{r+1}, b') satisfies 9.5.4.

d. $B'_r = (W_r, (J'_r, b'), E_r)$ is a simple basic object, so 6.4 holds.

Set now $\max t_r = \left(\frac{b_r}{b}, N\right)$. Locally at a point $\xi \in \underline{\operatorname{Max}}\, t_r$ (so $t_r(\xi) = (\max \text{w-ord}_r, N)$) there exists N hypersurfaces $H_{i_1}, \ldots, H_{i_N} \in E_r^-$ (4.14) so that ξ belongs to

$$H_{i_1} \cap \cdots \cap H_{i_N} \cap \underline{\operatorname{Max}}\, \text{w-ord}_r \neq \emptyset \tag{9.5.5}$$

hence 9.5.5 is not empty, and the maximality of N asserts that $N+1$ hypersurfaces of E_r^- have empty intersection with $\underline{\operatorname{Max}}\, \text{w-ord}_r$. The fact that N is maximal also implies that at a suitable open neighborhood, say W_r^λ of ξ

$$\underline{\operatorname{Max}}\, t_r \cap W_r^\lambda = H_{i_1}^\lambda \cap \cdots \cap H_{i_N}^\lambda \cap \underline{\operatorname{Max}}\, \text{w-ord}_r^\lambda \tag{9.5.6}$$

with right-hand terms restricted to W_r^λ. Set now

$$\begin{aligned}(B_r'')^\lambda &= (W_r^\lambda, ((J_r'')^\lambda, b''), (E_r'')^\lambda) \\ (J_r'')^\lambda &= (J_r')^\lambda + I(H_{i_1}^\lambda)^{b''} + \cdots + I(H_{i_N}^\lambda)^{b''} \\ b'' &= b' \\ E_r'' &= E_r^+\end{aligned} \tag{9.5.7}$$

where the superscript λ means restriction to W_r^λ.

Check that

a′ $\mathrm{Sing}((J_r'')^\lambda, b'') = \underline{\mathrm{Max}}\, t_r \cap W_r^\lambda$ (see 9.5.6).

b′ A transformation $(B_r'')^\lambda \longleftarrow (B_{r+1}'')^\lambda$ induces a transformation

$$B_r^\lambda = (W_r^\lambda, (J_r^\lambda, b), E_r^\lambda) \longleftarrow B_{r+1}^\lambda = (W_{r+1}^\lambda, (J_{r+1}^\lambda, b), E_{r+1}^\lambda)$$

c′ $\mathrm{Sing}((J_{r+1}'')^\lambda, b'') = \emptyset$ iff ($\max t_r > \max t_{r+1}$) or ($\max t_r = \max t_{r+1}$ and $\underline{\mathrm{Max}}\, t_{r+1} \cap W_{r+1}^\lambda = \emptyset$). And if $\max t_r = \max t_{r+1}$ and $\underline{\mathrm{Max}}\, t_{r+1} \cap W_{r+1}^\lambda \neq \emptyset$ then

$$\mathrm{Sing}((J_{r+1}'')^\lambda, b'') = \underline{\mathrm{Max}}\, t_{r+1} \cap W_{r+1}^\lambda = \left\{\xi \in \mathrm{Sing}(J_{r+1}, b) \cap W_r^\lambda \mid t_{r+1}(\xi) = \left(\frac{b_r}{b}, N\right)\right\}$$

Remark: One can check also that (J_r'', b'') satisfies 9.5.7.

d′ Each $(B_r'')^\lambda$ is a simple basic object, and so 6.4.

Note that properties (b), (c), (b′) and (c′) are formulated in terms of transformations, there is a natural extension of these properties in terms of restrictions (see 13.12). Once this is checked, it follows that the $(B_r'')^\lambda$ define the general basic object 9.5.2.

If Y is a permissible center for 9.5.2 then Y has normal crossings with E_r^+. However the local description in 9.5.6 leads to (b′), namely Y has normal crossings with E_r (see 13.7). This proves (B).

We prove now property (A). Define B_{r_0}' exactly as B_r' was. now (c) says that B_r' is the transform of the simple basic object B_{r_0}' via all intermediate transformations. On the other hand E_r^+ are the exceptional hypersurfaces arising in these steps.

As in 6.4 we define an open covering $\{W_{r_0}^\lambda\}$ of W_{r_0} and hypersurfaces $(W_h)_{r_0}^\lambda$ so that $I((W_h)_{r_0}^\lambda) \subset \Delta^{b'-1}((J_{r_0}')^\lambda)$.

This covering induces a covering of W_r (6.1) and furthermore, setting $(W_h)_r^\lambda$ as the strict transform of $(W_h)_{r_0}^\lambda$, then 6.7 asserts that

i. $I((W_h)_r^\lambda) \subset \Delta^{b'-1}((J_r')^\lambda)$.

ii. Each $(W_h)_r^\lambda$ has normal crossings with $(E_r^+)^\lambda$.

After a suitable refinement we may assume that both (i) and (ii) also hold at each W_r^λ in 9.5.6, and since $(J_r')^\lambda \subset (J_r'')^\lambda$ it follows that

iii. $I((W_h)_r^\lambda) \subset \Delta^{b'-1}((J_r')^\lambda) \subset \Delta^{b''-1}((J_r'')^\lambda)$.

So that (A) follows from (i), (ii) and (iii).

Property (P2) of 7.10 follows from (A), 9.5.2 and 6.12(A).

Now if $R(1)(\underline{\mathrm{Max}}\, t_r)$ is empty, then by (A) and 6.12(B) the g. b. o. $(\mathcal{F}_r'', (W_r, (E_r^d)^+))$ (which we have defined with a d-dimensional structure (9.5.7)) has also a $(d-1)$-dimensional structure, and we set:

$$(\mathcal{F}_r^{d-1}, (W_r, E_r^{d-1})) = (\mathcal{F}_r'', (W_r, (E_r^d)^+))$$

Given a resolution 7.10.3, it follows from remark in (c') and an inductive argument that (B) holds for $(\mathcal{F}_{r+i}'', (W_{r+i}, (E_{r+i}^d)^+))$, $i = 0, 1, \dots, N$. So that the resolution 7.10.3 induces the required sequence 7.10.4, and this proves the theorem 7.10.

9.6. *Proof of theorem 7.13.*

The proof is by induction on d. Assume $d = 1$ and set $I_1 = \mathbb{Q} \times \mathbb{Z} \sqcup \infty$ ordered lexicographically, where ∞ is the maximum of I_1.

Given any g. b. o. $(\mathfrak{F}_0^1, (W_0, E_0^1))$ of dimension one, define $f_0^1 = t_0^1$ (7.6). The closed set F_0 is locally properly included in a one dimensional smooth variety (6.13), so $\dim F_0 = 0$. In this case always $R(1)(\underline{\mathrm{Max}}\, t_0^1) \neq \emptyset$, so we are in case P2 of 7.10 and after the permissible transformation with center $\underline{\mathrm{Max}}\, t_0^1$:

$$(\mathfrak{F}_0^1, (W_0, E_0^1)) \longleftarrow (\mathfrak{F}_1^1, (W_1, E_1^1))$$

we have that $\max t_0^1 > \max t_1^1$. Now we can define a permissible sequence:

$$(\mathfrak{F}_0^1, (W_0, E_0^1)) \longleftarrow \cdots \longleftarrow (\mathfrak{F}_r^1, (W_r, E_r^1))$$

were each transformation has center $\underline{\mathrm{Max}}\, f_i^1$, setting $f_0^1 = t_0^1$ and for $i > 0$

$$f_i^1(\xi) = \begin{cases} t_i^1(\xi) & \text{if } \text{w-ord}_i^1(\xi) > 0 \\ \Gamma_i(\xi) & \text{if } \text{w-ord}_i^1(\xi) = 0 \end{cases}$$

the function Γ_i being the function defined in section 5. And we have that

$$\max f_0^1 > \cdots > \max f_r^1$$

It follows from 4.15(3) that there is an index r such that the above sequence is a resolution. The equivariance follows from 6.20(3).

Set now $d > 1$ and assume that 7.13 holds for $d-1$, so there is a totally ordered set $(I_{d-1}, \leq)$ as in 7.13.

Consider the set $\mathbb{Q} \times \mathbb{Z}$ ordered lexicographically. Set $I_d' = (\mathbb{Q} \times \mathbb{Z}) \sqcup I_M \sqcup \{\infty\}$ (disjoint union), where I_M is the totally ordered set defined in section 5 and I_d' is totally ordered by means of:

$$\alpha \in \mathbb{Q} \times \mathbb{Z},\ \beta \in I_M \implies \beta < \alpha$$

and the element ∞ is the maximum of I_d'.

Define now $I_d = I'_d \times I_{d-1}$ with the lexicographic ordering. We shall apply exercise 13.15 to define the functions f_i^d.

Let $(\mathfrak{F}_0^d, (W_0, E_0^d))$ be a g. b. o. of dimension d. Note that the conditions C1, C2 and C3 of 7.10 hold (in fact only C3).

Set the function $g_0 : F_0^d \longrightarrow I'_d$ as

$$g_0(\xi) = t_0^d(\xi) \qquad \forall \xi \in F_0^d$$

and the function $g'_0 : \underline{\mathrm{Max}}\, g_0 \longrightarrow I_{d-1}$ as

$$g'_0(\xi) = \begin{cases} \infty & \text{if} \quad \xi \in R(1)(\underline{\mathrm{Max}}\, t_0^d) \\ f_0^{d-1}(\xi) & \text{if} \quad \xi \notin R(1)(\underline{\mathrm{Max}}\, t_0^d) \end{cases}$$

where f_0^{d-1} is the function defined by the g. b. o. of dimension $d-1$, $(\mathcal{F}_0^{d-1}, (W_0, E_0^{d-1}))$, given by P3 of 7.10.

Assume now inductively that we have defined a permissible sequence as in 7.10.1, such that the conditions (C1), (C2) and (C3) hold and such that we have defined functions

$$g_i : F_i^d \longrightarrow I'_d \quad \text{and} \quad g'_i : \underline{\mathrm{Max}}\, g_i \longrightarrow I_{d-1}$$

which satisfy the conditions (a), (b), (c), (d) and (e) of 13.15 (setting $I_1 = I'_d$ and $I_2 = I_{d-1}$) for $i = 0, \dots, r-1$.

Now if $F_r^d \neq \emptyset$, define the function $g_r : F_r^d \longrightarrow I'_d$:

$$g_r(\xi) = \begin{cases} \Gamma(B_r)(\xi) & \text{if} \quad \text{w-ord}_r^d(\xi) = 0 \\ t_r^d(\xi) & \text{if} \quad \text{w-ord}_r^d(\xi) > 0 \end{cases}$$

where $\Gamma(B_r)$ is the function defined in 7.9 for the g. b. o. $B_r = (\mathcal{F}_r^d, (W_r, E_r^d))$.

And define the function $g'_r : \underline{\mathrm{Max}}\, g_r \longrightarrow I_{d-1}$:

$$g'_r(\xi) = \begin{cases} \infty & \text{if} \quad \text{w-ord}_r^d(\xi) = 0 \\ \infty & \text{if} \quad \xi \in R(1)(\underline{\mathrm{Max}}\, t_r^d) \quad \text{and} \quad \text{w-ord}_r(\xi) > 0 \\ f_r^{d-1}(\xi) & \text{if} \quad \xi \notin R(1)(\underline{\mathrm{Max}}\, t_r^d) \quad \text{and} \quad \text{w-ord}_r(\xi) > 0 \end{cases}$$

where f_r^{d-1} is the function defined by the g. b. o. of dimension $d-1$, $(\mathcal{F}_r^{d-1}, (W_r, E_r^{d-1}))$, given by P3 of 7.10.

Now 7.10 and 4.15(3 and 4) show that the sequence 7.10.1 extends uniquely to a resolution within this setting and together with the functions g_i and g'_i. We refer now to exercise 13.15 for the construction and properties of the functions

$$f_i^d : F_i^d \longrightarrow I_d$$

10. Exercises: Order of ideals and upper-semi-continuity

Here W denotes a separated scheme of finite type over a field $\mathbf{k}$ and smooth over $\mathbf{k}$, so there is a locally free coherent sheaf $\Omega^1_{W/\mathbf{k}}$ and a derivation $d : \mathcal{O}_W \longrightarrow \Omega^1_{W/\mathbf{k}}$.

Exercise 10.1. Let $\mathfrak{p}$ be a regular prime ideal in the local regular ring $(R, \mathfrak{m})$, $R = \mathcal{O}_{W,x}$. Set, for an element $f \in R$, $\nu_{\mathfrak{p}}(f) = b$ if $f \in \mathfrak{p}^b$ and $f \notin \mathfrak{p}^{b+1}$ (the same replacing f by an ideal $J \subset R$). We compare now $\nu_{\mathfrak{p}}$ with $\nu_{\mathfrak{p}R_{\mathfrak{p}}}$ (usual order at the local regular ring $R_{\mathfrak{p}}$) and with $\nu_{\mathfrak{m}}(f)$ (order at $(R, \mathfrak{m})$).

1. Show that $\operatorname{gr}_{\mathfrak{p}}(R) \otimes_R R_{\mathfrak{p}} = \operatorname{gr}_{\mathfrak{p}R_{\mathfrak{p}}}(R_{\mathfrak{p}})$ and that $\operatorname{gr}_{\mathfrak{p}}(R) \longrightarrow \operatorname{gr}_{\mathfrak{p}}(R) \otimes_R R_{\mathfrak{p}}$ is injective. Conclude that $\nu_{\mathfrak{p}}(f) = \nu_{\mathfrak{p}R_{\mathfrak{p}}}(f)$.
2. Prove that $\nu_{\mathfrak{p}}(fg) = \nu_{\mathfrak{p}}(f) + \nu_{\mathfrak{p}}(g)$.
3. Show that $\mathfrak{p}\hat{R}$ is a regular prime ideal.
4. Show that $\nu_{\mathfrak{p}R_{\mathfrak{p}}}(f) \leq \nu_{\mathfrak{m}}(f)$. (suggestion: $\nu_{\mathfrak{p}R_{\mathfrak{p}}}(f) = \nu_{\mathfrak{p}}(f)$, now apply (3) and formulate the problem in a ring of formal power series).

Assume now that the characteristic of $\mathbf{k}$ is zero. By means of Fitting ideals one can define an operator on $\mathcal{O}_W$-ideals: for $J \subset \mathcal{O}_W$, an ideal $\Delta(J) \subset \mathcal{O}_W$ is defined so that if $\{x_1, \dots, x_n\}$ is a regular system of parameters at a closed point $x \in W$, and if $J_x = J\mathcal{O}_{W,x}$ is generated by elements $f_1, \dots, f_r \in J_x$ then

$$\Delta(J)_x = \left\{ f_i, \frac{\partial f_i}{\partial x_j} \mid \begin{array}{l} i = 1, \dots, r \\ j = 1, \dots, n \end{array} \right\}$$

Note that $J \subset \Delta(J) \subset \Delta^2(J) = \Delta(\Delta(J)) \subset \cdots$

Exercise 10.2. With notation as above:

1. Prove that
$$\Delta(J)\hat{\mathcal{O}}_{W,x} = \left\{ f_i, \frac{\partial f_i}{\partial x_j} \mid \begin{array}{l} i = 1, \dots, r \\ j = 1, \dots, n \end{array} \right\}$$
at a ring of formal power series.
2. Conclude that the order of J at the regular ring $\mathcal{O}_{W,x}$ is b, ($\nu_x(J) = b$), iff $x \in V(\Delta^{b-1}(J)) \setminus V(\Delta^b(J))$.
3. Shows that $\nu_x(J) = b$ iff $\nu_x(\Delta^{b-1}(J)) = 1$.
4. For $J \subset \mathcal{O}_W$, $J \neq 0$, set $\nu_J : W \longrightarrow \mathbb{Z}$ the function that assigns to each closed point x, $\nu_J(x) = \nu_x(J) \in \mathbb{Z}$. Show that ν_J is an upper-semi-continuous function along the closed points.

Exercise 10.3. Let $Y \subset W$ be an irreducible subvariety of W, $x \in Y$ a (closed) point, maybe a singular point of Y, and set $R = \mathcal{O}_{W,x}$, $\mathfrak{p} = I(Y)_x$. Show that $\nu_{\mathfrak{p}R_{\mathfrak{p}}}(J) \leq \nu_{\mathfrak{m}}(J)$ for any ideal $J \subset \mathcal{O}_W$.
Suggestion: $\operatorname{Reg}(Y)$ is dense in Y and ν_J is upper-semi-continuous on closed points.

Exercise 10.4. Set $(R, \mathfrak{m})$, $R = \mathcal{O}_{W,x}$, $\mathfrak{p} \subset R$ a regular prime ideal. Show that $\nu_{\mathfrak{p}R_{\mathfrak{p}}}(J) = b$ iff $\mathfrak{p} \supset (\Delta^{b-1}(J))_x$ and $\mathfrak{p} \not\supset (\Delta^b(J))_x$.

Suggestion: $\nu_{\mathfrak{p}R_{\mathfrak{p}}}(f) = \nu_{\mathfrak{p}}(f)$ (as defined in 10.1) and now formulate the problem in a formal power series ring.

Exercise 10.5. Same as exercise 10.4 but with $\mathfrak{p} \subset R$ not regular. Conclude that the function $\nu_J : \mathrm{Spec}(R) \longrightarrow \mathbb{N}$, $\nu_J(\mathfrak{p}) = \nu_{\mathfrak{p}R_{\mathfrak{p}}}(J)$, is u. s. c.

Exercise 10.6. Let A be a $\mathbf{k}$-algebra of finite type and $\mathrm{SpecMax}(A)$ the maximal spectrum of A, included in $\mathrm{Spec}(A)$. Let $(I, \leq)$ be a totally ordered set and $f : \mathrm{SpecMax}(A) \longrightarrow I$ an u. s. c. function. For any point $\xi \in \mathrm{Spec}(A)$ set $\bar{\xi} \subset \mathrm{SpecMax}(A)$ the irreducible closed set defined by ξ.

1. Show that there is a unique extension of f to $\tilde{f} : \mathrm{Spec}(A) \longrightarrow I$ subject to the following condition:

   - For any $\xi \in \mathrm{Spec}(A)$, there is a non empty open set $U \subset \mathrm{SpecMax}(A)$ such that $U \cap \bar{\xi} \neq \emptyset$.

$$f(\zeta) = \tilde{f}(\xi) \qquad \forall \zeta \in U \cap \bar{\xi}$$

2. Show that $\tilde{f}$ is an u. s. c. function.
3. Set $\mathrm{Sch}(W)$ the scheme attached to a regular variety W, show that the extension of the function ν_J of 10.2(4) is defined as in 10.5.

Exercise 10.7. The total degree of a monomial $X_1^{\alpha_1} \cdots X_n^{\alpha_n} \in \mathbf{k}[X_1, \dots, X_n]$ is $\alpha_1 + \cdots + \alpha_n \in \mathbb{N}$ and the total degree of a polynomial $f = \sum a_\alpha X^\alpha \in \mathbf{k}[X_1, \dots, X_n]$ ($\alpha = (\alpha_1, \dots, \alpha_n)$, $X^\alpha = X_1^{\alpha_1} \cdots X_n^{\alpha_n}$) is the highest degree of X^α with $a_\alpha \neq 0$.

1. Let X^β, $\beta = (\beta_1, \dots, \beta_n)$, be a monomial of highest degree in f, show that

$$\frac{\partial^{\beta_1}}{\partial {X_1}^{\beta_1}} \cdots \frac{\partial^{\beta_n}}{\partial {X_n}^{\beta_n}}(f) \in \mathbf{k} \setminus \{0\}$$

   and conclude that the hypersurface defined by f in $\mathbb{A}^n_{\mathbf{k}}$ has no point of multiplicity $\geq \deg(f) + 1$.
2. Set $F_N \in \mathbf{k}[X_1, \dots, X_n]$ homogeneous of degree N. Show that the hypersurface $X \subset \mathbb{P}^n_{\mathbf{k}}$ defined by F_N has no points of multiplicity $> N$.
3. Suppose that the characteristic of $\mathbf{k}$ is not zero. Prove that a polynomial $f \in \mathbf{k}[X_1, \dots, X_n]$ cannot be included in the $\deg(f)+1$-th power of a maximal ideal and conclude the same as in (1).

   Suggestion: Assume $\mathbf{k}$ algebraically closed, so all points are rational, and then generalize.
4. Prove (2) with no condition on the characteristic of $\mathbf{k}$.

11. Exercises: Blow-ups

For this part of our exercises, we impose no condition on the characteristic of $\mathbf{k}$.

Let A be a noetherian ring, $K \subset A$ an ideal. The homomorphism

$$i : A \longrightarrow A[Kt] = A \oplus K \oplus K^2 \oplus \cdots \oplus K^n \oplus \cdots$$

induces a morphism

$$W_1 = \text{Proj}(A[Kt]) \xrightarrow{\Pi} \text{Spec}(A) = W$$

and for any ideal $I \subset A$, the total transform $I\mathcal{O}_{W_1}$ defined via Π, is the ideal induced by the graded ideal $i(I)A[Kt]$ in $Z = \text{Proj}(A[Kt])$.

For the case $I = K$, $KA[Kt]$ and $A[Kt](1)$ induce the same ideal in W_1 known as Serre's $\mathcal{O}(1)$, locally free of rank one. So $K\mathcal{O}_{W_1}$ is locally principal and (locally) generated by a regular element (a non zero divisor), defining a hypersurface

$$H = \text{Proj}\,(A[Kt]/KA[Kt]) = \text{Proj}\,(\text{gr}_K(A)) \subset W_1$$

We shall denote $K\mathcal{O}_{W_1}$ by $I(H)$.

Set now $\bar{A} = A/J$, $J \subset A$ an ideal, $\bar{K} = K\bar{A}$. There are natural surjections of graded rings

$$A[Kt] \longrightarrow \bar{A}[\bar{K}t] \longrightarrow 0 \qquad \text{gr}_K(A) \longrightarrow \text{gr}_{\bar{K}}(\bar{A}) \longrightarrow 0$$

and the first induces a square diagram

$$\begin{array}{ccccc} X_1 = & \text{Proj}(\bar{A}[\bar{K}t]) & \longrightarrow & \text{Proj}(A[Kt]) & = W_1 \\ & \downarrow & & \downarrow & \\ X = & \text{Spec}(\bar{A}) & \longrightarrow & \text{Spec}(A) & = W \end{array}$$

both horizontal morphisms being closed immersions. X_1 is called the strict transform of $X = \text{Spec}(\bar{A})$.

Exercise 11.1. Let $(A, \mathfrak{m})$ be a local ring, $\{x_1, \dots, x_s\} \subset \mathfrak{m}$ and suppose that $\dim\,(A/\langle x_1, \dots, x_s\rangle) = \dim A - s$. Show that if $A/\langle x_1, \dots, x_s\rangle$ is regular then A is regular and $\{x_1, \dots, x_s\}$ extends to a regular system of parameters. The converse also holds.

We now blow up a regular variety W along a regular subvariety $Y \subset W$:

$$\begin{array}{ccc} W & \xleftarrow{\Pi} & W_1 \\ Y & & H \end{array}$$

so that H is the closed immersion defined by the locally principal ideal $I(Y)\mathcal{O}_{W_1}$. We want to show that H and W_1 are regular and so is $\Pi^{-1}(\xi)$ for any $\xi \in Y$.

Exercise 11.2. Set $(R, \mathfrak{m})$ to be the local regular ring $\mathcal{O}_{W,x}$, $\{x_1, \dots, x_n\}$ a regular system of parameters such that $\mathfrak{p} = \langle x_1, \dots, x_s\rangle = I(Y)_\xi$.

1. Show that there are natural surjections

$$R[\mathfrak{p}t] \longrightarrow \mathrm{gr}_{\mathfrak{p}}(R) \longrightarrow 0 \qquad \mathrm{gr}_{\mathfrak{p}}(R) \longrightarrow R/\mathfrak{m} \otimes_R \mathrm{gr}_{\mathfrak{p}}(R) \longrightarrow 0$$

   inducing closed immersions $H \subset W$ and $\Pi^{-1}(\xi) \subset H$.
2. Show that $\mathrm{Proj}(\mathrm{gr}_{\mathfrak{p}}(R)) = H$ is regular.
   (suggestion: $\mathrm{gr}_{\mathfrak{p}}(R) = R/\mathfrak{p}[X_1, \dots, X_s]$).
3. Fix $\xi_1 \in W_1$ mapping to ξ, apply (2) and exercise 11.1 to the natural homomorphism

$$\mathcal{O}_{W_1,\xi_1} \longrightarrow \mathcal{O}_{H,\xi_1} \longrightarrow 0$$

   to show that W_1 is regular.
4. Show that $\Pi^{-1}(\xi) = \mathrm{Proj}\,(R/\mathfrak{m}[X_1, \dots, X_s]) = \mathbb{P}^{s-1}_{\mathbf{k}}$.

We now study how the multiplicity of a hypersurface $X \subset W$ relates to that of the strict transform $X_1 \subset W_1$ (see theorem 2.3 B).

Exercise 11.3. With the setting as in 11.2.

1. Show that the natural inclusion $R[\mathfrak{p}t] \longrightarrow R[\mathfrak{m}t]$ is a graded homomorphism inducing naturally a diagram

$$\begin{array}{cccc} & R[\mathfrak{p}t] & \longrightarrow & R[\mathfrak{m}t] \\ \alpha & \downarrow & & \downarrow \\ & \mathrm{gr}_{\mathfrak{p}}(R) & \xrightarrow{\beta} & \mathrm{gr}_{\mathfrak{m}}(R) \end{array}$$

2. Fix $f \in R$, set $b = \nu_{\mathfrak{p}}(f)$ and $f_b = f$ as a homogeneous element of degree b in $R[\mathfrak{p}t]$. Show that:
   (a) $\alpha(f_b) = \mathrm{In}_{\mathfrak{p}}(f)$.
   (b) If $\nu_{\mathfrak{p}}(f) < \nu_{\mathfrak{m}}(f)$ then $\beta(\mathrm{In}_{\mathfrak{p}}(f)) = 0$.
   (c) If $\nu_{\mathfrak{p}}(f) = \nu_{\mathfrak{m}}(f)$ then $\beta(\mathrm{In}_{\mathfrak{p}}(f)) = \mathrm{In}_{\mathfrak{m}}(f)$.

3. Show that there is a natural graded diagram

$$\begin{array}{ccc} \mathrm{gr}_{\mathfrak{p}}(R) & \xrightarrow{\beta} & \mathrm{gr}_{\mathfrak{m}}(R) \\ \downarrow & \nearrow & \\ R/\mathfrak{m} \otimes_R \mathrm{gr}_{\mathfrak{p}}(R) & & \end{array}$$

   and if $\nu_{\mathfrak{p}}(f) = \nu_{\mathfrak{m}}(f)$, the class of $\mathrm{In}_{\mathfrak{p}}(f)$ in $R/\mathfrak{m} \otimes_R \mathrm{gr}_{\mathfrak{p}}(R)$ is (homogeneous of degree b) non zero.

We show now that the strict transform of X (assuming $I(X) = f$) is the hypersurface defined by $f_b A[\mathfrak{p}t]$, $b = \nu_{\mathfrak{p}}(f)$.

Exercise 11.4. With the setting as in 11.3, set $\bar{R} = R/f$, $X = \mathrm{Spec}(\bar{R}) \longrightarrow \mathrm{Spec}(R)$ and let $\Pi : X_1 \longrightarrow X$ denote the blow-up with center $\mathfrak{p}$. We study now the graded homomorphisms:

$$a : R[\mathfrak{p}t] \longrightarrow \bar{R}[\bar{\mathfrak{p}}t] \longrightarrow 0$$
$$b : \mathrm{gr}_{\mathfrak{p}}(R) \longrightarrow \mathrm{gr}_{\bar{\mathfrak{p}}}(\bar{R}) \longrightarrow 0$$

1. Show that $f_b R[\mathfrak{p}t] \subset \mathrm{Ker}\, a$.
2. $\mathrm{Ker}\, a$ and $f_b R[\mathfrak{p}t]$ define the same sheaf of ideals at $\mathrm{Proj}(R[\mathfrak{p}t])$.
3. $\mathrm{Ker}\, b = \langle \mathrm{In}_{\mathfrak{p}}(f) \rangle$.

   (suggestion for (2) and (3), 10.1(2)).
4. Conclude from (2) that $f_b R[\mathfrak{p}t]$ induces the sheaf of ideals defining X_1 (the strict transform of X) if X is defined by f.

Exercise 11.5. (see 2.3) With the setting as before:

1. Prove that $I(X)\mathcal{O}_{W_1} = I(H)^b I(X_1)$ (use 11.4).
2. Fix $\xi_1 \in X_1$ mapping to $\xi \in X$ and assume that the center Y is such that $Y \subset \underline{\mathrm{Max}}\,\mathrm{mult}_X$. Apply 10.7 and 11.3 to show that

$$\mathrm{mult}_X(\xi) \geq \mathrm{mult}_{X_1}(\xi_1)$$

Exercise 11.6. Show now that the proof of 2.4 and remark 2.5 follow from a local version of theorem 2.3.

12. Exercises: Desingularization

See 3 in remark 3.2 for notation.

Exercise 12.1. With the setting and notation of 2.10, show that E_1 has normal crossings in W_1.

Exercise 12.2. Show that a desingularization of $X = V(X^2+Y^3) \subset \mathbb{C}^2$ is obtained by blowing up the origin, but it is not an embedded desingularization.

Exercise 12.3. Show that $X = V(XY - ZW) \subset \mathbb{A}^4_{\mathbf{k}}$ has an isolated singularity at the origin and that an embedded desingularization is obtained by blowing up the origin.

Exercise 12.4. Set $Y = V(X,Z) \subset X = V(XY - ZW) \subset \mathbb{A}^4_{\mathbf{k}}$, $\mathbb{A}^4_{\mathbf{k}} \longleftarrow W_1$ the blow up at Y, X_1 the strict transform of X.

1. Show that $Y \cap \mathrm{Reg}(X) \neq \emptyset$
2. Show that X_1 is regular and $X_1 \longrightarrow X$ induces an isomorphism over $\mathrm{Reg}(X)$ (i.e. it is a desingularization).

Exercise 12.5. Define an action of $\mathbb{C}^*$ on $W = \mathbb{C}^3$ setting, for $\lambda \in \mathbb{C}^*$, $\Theta_\lambda(a,b,c) = (a,b,\lambda c)$. Define an action of $\mathbb{Z}/2\mathbb{Z}$ by $\Theta(a,b,c) = (a,c,b)$. Show that:

1. Both $\mathbb{C}^*$ and $\mathbb{Z}/2\mathbb{Z}$ define an action on the pair (W, E) (2.11) where $E = \{H_1\}$, $H_1 = V(X)$.
2. Describe the actions defined in the transform (W_1, E_1) of (W, E) after the blow-up of $Y = V(Y, Z)$ (note that Y is equivariant by both actions).

Exercise 12.6. Define an action of $\mathbb{C}^*$ on $\mathbb{C}^3$ setting, for $\lambda \in \mathbb{C}^*$, $\Theta_\lambda(a, b, c) = (a, \lambda b, \lambda^{-1} c)$, and consider the action of $\mathbb{Z}/2\mathbb{Z}$ as above, given by $\Theta(a, b, c) = (a, c, b)$. Show:

1. that both are also actions on $X = V(X^2 - Y^3 Z^3) \subset \mathbb{C}^3$.
2. that $\max \operatorname{mult}_X = 2$ and $\underline{\operatorname{Max}} \operatorname{mult}_X$ is the union of two lines through the origin.
3. that the origin is the only center invariant for any isomorphism $\Theta : X \longrightarrow X$.
4. Set $X_1 \xrightarrow{\Pi} X$ the blow up of the origin 0. Show that $\Pi^{-1}(0) = \mathbb{P}^1 \subset X_1$ is equivariant for any isomorphism $\Theta : X \longrightarrow X$.

Exercise 12.7. Set $X = V(X^6 Y^2 + Z^3) \supset Y = V(Y, Z)$ in $\mathbb{C}^3$ and $X_1 \longrightarrow X$ the blow up at Y. Show that X has multiplicity 3 at the origin but there is a point of X_1 of multiplicity 4 mapping to the origin.

Exercise 12.8. Set $W \xleftarrow{\Pi} W_1$ by blowing up $Y \subset W$ regular of pure codimension 2.

1. For any closed point $x \in Y \subset W$ show that $\Pi^{-1}(x) = \mathbb{P}^1$.
2. If $Y \subset X \subset W$, where X is a hypersurface and the multiplicity of X is constant along Y, show that $X_1 \longrightarrow X$ is a finite morphism.

13. Exercises: On basic objects

Exercise 13.1. Set $W = \mathbb{A}^n_{\mathbf{k}}$ the n-dimensional affine space.

1. Fix the ring of coordinates $\mathbf{k}[Z_1, \dots, Z_n]$. Let $Y \subset W$ be defined by $I(Y) = \langle Z_1, \dots, Z_r \rangle$ for some $r \leq n$, and set $W \longleftarrow W_1$ the monoidal transformation with center Y. Show that W_1 can be covered by r affine charts, each of them isomorphic to $\mathbb{A}^n_{\mathbf{k}}$.

   Definition: Let us say that a polynomial $f \in \mathbf{k}[Z_1, \dots, Z_n]$ *fulfills condition (T)* on the variable Z_1 if:

$$(\mathbf{T}) \left\{ \begin{array}{l} f = Z_1^b + a_1 Z_1^{b-1} + \dots + a_b \\ a_i \in \mathbf{k}[Z_2, \dots, Z_n] \text{ and } a_1 = 0 \end{array} \right.$$

2. Let $X \subset \mathbb{A}^n_{\mathbf{k}}$ be the hypersurface defined by f, note that X has no points of multiplicity $> b$, and that $\operatorname{Sing}(\langle f \rangle, b)$ is the set of b-fold points of the hypersurface X. Show that if (T) holds then $Z_1 \in \Delta^{b-1}(f)$ (1.5), in particular $\operatorname{Sing}(\langle f \rangle, b) \subset W_h = V(Z_1)$ (4.7).

3. If (T) holds, show that $R(1)(\operatorname{Sing}(\langle f\rangle, b)) = \emptyset$ (1.8) iff some $a_i \neq 0$.

Definition: If (T) holds and some $a_i \neq 0$, define

$$C(f) = \left\langle a_2^{\frac{b!}{2}}, a_3^{\frac{b!}{3}}, \dots, a_b^{\frac{b!}{b}} \right\rangle \subset \mathcal{O}_{W_h}$$

Note that $C(f) \neq 0$.

4. Show that $\operatorname{Sing}(\langle f\rangle, b) = \operatorname{Sing}(C(f), b!)$ via the immersion $W_h \subset W$.
5. We now set $W = \mathbb{A}^n_{\mathbf{k}}$, $(W, (J, b), E = \emptyset) \longleftarrow (W_1, (J_1, b), E_1)$ a transformation with center Y (4.7) and assume that Y is as in (1). Here $J = \langle f\rangle$ and f and b are as in condition (T).
   (a) Show that $\operatorname{Sing}(J_1, b) = \emptyset$ or $\operatorname{Sing}(J_1, b)$ can be covered by at most $r-1$ affine charts, each one isomorphic to $\mathbb{A}^n_{\mathbf{k}}$.
   (b) At each affine chart as above, $J_1 = \langle f_1\rangle$ where f_1 is a strict transform of f. Show that f_1 fulfills condition (T) on the strict transform of Z_1. Note that in particular $C(f_1)$ is well defined.
   (c) Show that there is a naturally defined diagram

$$\begin{array}{ccc} (W, E = \emptyset) & \longleftarrow & (W_1, E_1) \\ \uparrow & & \uparrow \\ (W_h, E_h = \emptyset) & \longleftarrow & ((W_h)_1, (E_h)_1) \end{array}$$

   as in 6.11.1, and that $(W_h)_1$ can be covered by $r-1$ affine charts isomorphic to $\mathbb{A}^{n-1}_{\mathbf{k}}$, each of them embedded in a chart ($\cong \mathbb{A}^n_{\mathbf{k}}$) of W_1.
   (d) Since $Y \subset \operatorname{Sing}(J, b) = \operatorname{Sing}(C(J), b!)$ and $E = \emptyset$ and $E_h = \emptyset$, note that the same center Y induces both

$$(W, (J, b), E) \longleftarrow (W_1, (J_1, b), E_1) \tag{13.1.1}$$

$$(W_h, (C(J), b!), E_h) \longleftarrow ((W_h)_1, ([C(J)]_1, b!), (E_h)_1) \tag{13.1.2}$$

   Show that the restriction of $[C(J)]_1$ to the mentioned affine charts of $(W_h)_1$ is exactly the ideal $C(f_1)$.
   (e) Show that a resolution of $(W, (J, b), E)$ with the constraint (1), induces and is equivalent to a resolution of $(W_h, (C(J), b!), E_h)$.
   (f) Set $X \subset W = \mathbb{A}^n_{\mathbf{k}}$ the hypersurface defined by f and $X_1 \subset W_1$ the strict transform of X via the transformation $W \longrightarrow W_1$ in 13.1.1. Note that J_1 is the sheaf of ideals defining X_1 and $\operatorname{Sing}(J_1, b)$ is the set of b-fold points of X_1. In particular 13.1.1 is a resolution of basic objects iff X_1 has no b-fold points.

Remark: Lemma 6.12 will show that property (5e) above holds with independence of the construction (1) on the choice of permissible centers. In particular, setting $(W, (J, b), E = \emptyset) = (\mathbb{A}^n_{\mathbf{k}}, (\langle f\rangle, b), \emptyset)$ a resolution of the basic object $(W, (J, b), E)$

(4.9) is equivalent to a resolution of $(W_h, (C(J), b!), E_h)$, and setting such resolutions

$$(W, (J, b), E) \longleftarrow \cdots \longleftarrow (W_r, (J_r, b), E_r) \qquad (13.1.3)$$

$$(W_h, (C(J), b!), E_h) \longleftarrow \cdots \longleftarrow ((W_h)_r, (C(J)_r, b!), (E_h)_r) \qquad (13.1.4)$$

then:

1. For each index k there is an immersion $(W_h)_i \subset W_i$ (smooth hypersurface).
2. $\mathrm{Sing}(C(J)_i, b!) = \mathrm{Sing}(J_i, b)$, via $(W_h)_i \subset W_i$.
3. $\mathrm{Sing}(J_r, b) = \mathrm{Sing}(C(J)_r, b!) = \emptyset$.
4. Setting $X = V(J)$ and $X_i = V(J_i)$ then each $X_i \subset W_i$ is the strict transform of $X \subset W$, and $X_r \subset W_r$ has no b-fold points.

This is how the notion of *resolution of basic objects* relate with another notion, that of *desingularization*. But for the time being we can say, for example, that *resolution of 4-fold points* of the non-reduced hypersurface in $\mathbb{A}^3_{\mathbf{k}}$ defined by $f = (Z^2 + X^3 + Y^4)^2$ is equivalent to the resolution of the basic object $(\mathbb{A}^3_{\mathbf{k}}, (\langle f \rangle, 4), E = \emptyset)$ (to the resolution of $(\mathbb{A}^2_{\mathbf{k}}, (C(f), 4!), E = \emptyset)$, $C(f) = \left\langle (X^3 + Y^4)^{12} \right\rangle$).

Exercise 13.2. Apply 13.1 to define a resolution of $(\mathbb{A}^2_{\mathbf{k}}, (\langle X^2 - Y^5 \rangle, 2), E = \emptyset)$ in terms of a resolution of $(\mathbb{A}^1_{\mathbf{k}}, (\langle Y^5 \rangle, 2), E_h = \emptyset)$, via the immersion $\mathbb{A}^1_{\mathbf{k}} \cong V(X) \subset \mathbb{A}^2_{\mathbf{k}}$.

Exercise 13.3. The group $\mathbb{Z}/2\mathbb{Z}$ acts on $X = V(X^2 - Y^3 Z^3) \subset \mathbb{A}^3_{\mathbf{k}}$ interchanging Y with Z. Apply 13.1 to define a resolution of $(\mathbb{A}^3_{\mathbf{k}}, (\langle X^2 - Y^3 Z^3 \rangle, 2), E = \emptyset)$ in terms of a resolution of $(\mathbb{A}^2_{\mathbf{k}}, (\langle Y^3 Z^3 \rangle, 2), E_h = \emptyset)$ which lifts the action of $\mathbb{Z}/2\mathbb{Z}$ to the final transform of X.

Exercise 13.4. Set $0 \neq J \subset \mathcal{O}_W$, assume that $\nu_x(J) = 1$ for any $x \in V(J)$, set $R(1)(V(J)) \subset V(J)$ the subset of points of codimension 1 in W. Show that $R(1)(V(J))$ is open and closed in $V(J)$ and it is a regular hypersurface in W.

Exercise 13.5. Let (W, E) be a pair, $X \subset W$ a regular subvariety, and set $J = I(X) \subset \mathcal{O}_W$, show that:

1. $\mathrm{Sing}(J, 1) = X$.
2. If $(W, (J, 1), E) \longleftarrow (W_1, (J_1, 1), E_1)$ is a transformation 4.7, then $J_1 = I(X_1)$, $X_1 \subset W_1$ the strict transform of X.

Remark: For the case that X is a regular hypersurface 1 and 2 follow from 4.10 with $b = 1$.

Exercise 13.6. With the setting as in 4.10 where $J_i = I(X_i)$, X_i the strict transform of X_{i-1} and $b = \max \mathrm{mult}_X$; show that the expressions 4.8.2 reduce to $J_i = \bar{J}_i$ (all $a_i = 0$).

Exercise 13.7. Let (W, E) be a pair, assume that $E = E' \sqcup E''$ is a disjoint union. Let $Y \subset W$ be a closed and regular set such that for any hypersurface $H \in E'$, either $Y \subset H$ or $Y \cap H = \emptyset$.

Show that Y has normal crossings with E if and only if it has normal crossings with E''.

Exercise 13.8. Let (W, E) be a pair, and $E = E' \sqcup E''$ be a disjoint union, where $E = \{H_1, \dots, H_s\}$. Define a function $n : W \longrightarrow \mathbf{Z} \geq 0$:

$$n(\xi) = \#\{H \in E' \mid \xi \in H\}$$

1. Show that n is an u. s. c. function.
2. Show that $\max n \leq \dim W$.
3. If $m = \max n$, given indices $i_1 < \dots < i_m$ and $j_1 < \dots < j_m$, such that
$$H_{i_1}, \dots, H_{i_m}, H_{j_1}, \dots, H_{j_m} \in E'$$
then the sets $H_{i_1} \cap \dots \cap H_{i_m}$ and $H_{j_1} \cap \dots \cap H_{j_m}$ are both non empty and either equal or disjoint.
4. If $m = \max n$ then
$$\underline{\mathrm{Max}}\, n = \bigcup_{i_1 < \dots < i_m} (H_{i_1} \cap \dots \cap H_{i_m})$$
the union over $i_1 < \dots < i_m$ and $H_{i_1}, \dots, H_{i_m} \in E'$.
5. If $Y \subset \underline{\mathrm{Max}}\, n$ is closed and regular, then Y has normal crossings with E iff it has normal crossings with E''.
6. Set $(W, E) \stackrel{\Pi}{\longleftarrow} (W_1, E_1)$ a transformation with center $Y \subset \underline{\mathrm{Max}}\, n$, set $E_1 = E_1' \sqcup E_1''$ where E_1' consists of the strict transforms of hypersurfaces in E'. Define $n_1 : W_1 \longrightarrow \mathbb{Z} \geq 0$
$$n_1(\xi) = \#\{H \in E_1' \mid \xi \in H\}$$
Show that $n_1(\xi) \leq n(\Pi(\xi))$ for any $\xi \in W_1$, in particular $\max n_1 \leq \max n$.

Exercise 13.9. Let $(\widetilde{W}, \widetilde{E}) \longrightarrow (W, E)$ be a closed immersion (6.11). Show that if $\widetilde{Y} \subset \widetilde{W}$ is permissible for $(\widetilde{W}, \widetilde{E})$, then it is also permissible for (W, E).

Exercise 13.10. Let $W_1 \longrightarrow W$ be smooth and $\xi_1 \in W_1$ map to $\xi \in W$. Show that:

1. A regular system of parameters at $\mathcal{O}_{W,\xi}$ extends to a regular system of parameters at $\mathcal{O}_{W_1,\xi_1}$.
2. For an ideal $J \subset \mathcal{O}_W$, then
$$\Delta(J)\mathcal{O}_{W_1} = \Delta_1(J_1)$$
Where Δ_1 is defined for W_1 (see 1.2.1), and $J_1 = J\mathcal{O}_{W_1}$ (suggestion, see 1.5.1).

Exercise 13.11.

1. Extend 6.7 to the case that the sequence 6.7.1 consists now of transformations and restrictions (6.8).
2. With the setting as in 6.7, show that if $E = \emptyset$ then $(W_h)_s$ has normal crossings with E_s and the conditions of 6.11hold for $(W_h)_s \subset W_s$.

Exercise 13.12. The functions w-ord$_i$ and t_i were defined (4.11 and 4.14) in terms of the expressions of the J_i in 4.8.2. We now want to extend the definition of w-ord$_i$ and t_i to the case that 4.8.1 consists now of transformations and restrictions (6.8). So if for some index i, $W_i \stackrel{\Pi_{i+1}}{\longleftarrow} W_{i+1}$ in 4.8.1 is defined by a smooth map, set

$$(W_i, (J_i, b), E_i) \stackrel{\Pi_{i+1}}{\longleftarrow} (W_{i+1}, (J_{i+1}, b), E_{i+1})$$

as in 6.8, and we agree to define the expression 4.8.2 for J_{i+1} by setting

$$J_{i+1} = J_i \mathcal{O}_{W_{i+1}} = I(H_{r+1})^{a_1} \cdots I(H_{r+i})^{a_i} \bar{J}_i \mathcal{O}_{W_{i+1}}$$

1. Show that if Π_{i+1} is smooth as above, then

   $$\text{w-ord}_i(\Pi_{i+1}(\xi)) = \text{w-ord}_{i+1}(\xi)$$

   for any $\xi \in \mathrm{Sing}(J_{i+1}, b)$.
2. Show that inequalities 4.12.1 and 4.12.2 still hold.
3. Extend now the definition of the function t_i to this case.

Exercise 13.13.

1. Show that the basic object $(W_0, (I(X_0), b), E_0)$ defined in 4.10 is a simple basic object (6.3).
2. If $(W_0, (J_0, b), E_0)$ is a simple basic object, and

   $$(W_0, (J_0, b), E_0) \longleftarrow (W_1, (J_1, b), E_1)$$

   is either a transformation (4.7) or a restriction (6.8), then $(W_1, (J_1, b), E_1)$ is a simple basic object.
3. If $(W_0, (J_0, b), E_0)$ is a simple basic object, and

   $$(W_0, (J_0, b), E_0) \longleftarrow \cdots \longleftarrow (W_r, (J_r, b), E_r)$$

   is a sequence of transformations and restrictions, then for any index $i = 0, \dots, r$: $\max \text{w-ord}_i = 1$ and $\underline{\mathrm{Max}}\,\text{w-ord}_i = \mathrm{Sing}(J_i, b)$

Exercise 13.14. (On 6.12(A)) Here $B = (W, (J, b), E)$ is a simple basic object.

1. Show that C1 implies C2 (in 6.12) for a suitable covering of W.
2. Show that $\xi \in R(1)(\mathrm{Sing}(J, b))$ iff the ideal $\Delta^{b-1}(J)$ is invertible, locally at ξ.
3. Show that $R(1)(\mathrm{Sing}(J, b))$ is open and closed in $\mathrm{Sing}(J, b)$ (see 1.8(1)).
4. Set $H = R(1)(\mathrm{Sing}(J, b))$. If $H \neq \emptyset$ then H is a regular hypersurface and $\Delta^{b-1}(J) = I(H)$ in an open neighborhood of H.
5. Prove now that $J = I(H)^b$ in an open neighborhood of H.

Suggestion: Fix $\xi \in H$ and a regular system of coordinates $\{x_1, \dots, x_n\}$, so that $\Delta^{b-1}(J) = \langle x_1 \rangle$. Express any $f \in J\hat{\mathcal{O}}_{W,\xi}$ as $f = \sum a_s X_1^s$, $a_s \in \mathbf{k}'[[X_2, \dots, X_n]]$, and show that $a_{b-1} = a_{b-2} = \dots = a_0 = 0$.

Exercise 13.15. Let the sequence

$$\begin{array}{ccccccc} (\mathcal{F}_0, (W_0, E_0)) & \xleftarrow{\Pi_1} & (\mathcal{F}_1, (W_1, E_1)) & \xleftarrow{\Pi_2} & \cdots & \xleftarrow{\Pi_N} & (\mathcal{F}_N, (W_N, E_N)) \\ Y_0 \subset F_0 & & Y_1 \subset F_1 & & & & F_N = \emptyset \end{array} \tag{13.15.1}$$

be a resolution of the general basic object $(\mathcal{F}_0, (W_0, E_0))$, and let $(I_1, \leq)$ and $(I_2, \leq)$ be totally ordered sets.

Assume that

a. For each index $i = 0, \dots, N-1$ there is an u. s. c. function $g_i : F_i \longrightarrow I_1$.

b. For each index $i = 0, \dots, N-1$ there is an u. s. c. function $g'_i : \underline{\mathrm{Max}}\, g_i \longrightarrow I_2$.

c. $Y_i = \underline{\mathrm{Max}}\, g'_i$ in 13.15.1, $i = 0, \dots, N-1$.

d. For any $\xi \in F_i$

$$\begin{array}{lll} g_i(\Pi_{i+1}(\xi)) \geq g_{i+1}(\xi) & \text{if} & \Pi_{i+1}(\xi) \in Y_i \\ g_i(\Pi_{i+1}(\xi)) = g_{i+1}(\xi) & \text{if} & \Pi_{i+1}(\xi) \notin Y_i \end{array}$$

Note that (d) implies that $\max g_0 \geq \max g_1 \geq \dots \geq \max g_{N-1}$ and, in case $\max g_i = \max g_{i+1}$, that $\Pi_{i+1}(\underline{\mathrm{Max}}\, g_{i+1}) \subset \underline{\mathrm{Max}}\, g_i$.

e. If $\max g_i = \max g_{i+1}$ and $\xi \in \underline{\mathrm{Max}}\, g_{i+1}$ then

$$\begin{array}{lll} g'_i(\Pi_{i+1}(\xi)) > g'_{i+1}(\xi) & \text{if} & \Pi_{i+1}(\xi) \in Y_i \\ g'_i(\Pi_{i+1}(\xi)) = g'_{i+1}(\xi) & \text{if} & \Pi_{i+1}(\xi) \notin Y_i \end{array}$$

Prove that

1. The function $f_{N-1} : F_{N-1} \longrightarrow I_1 \times I_2$, $f_{N-1} = (g_{N-1}, g'_{N-1})$ is constant. (Note first that $\underline{\mathrm{Max}}\, g_{N-1} = \underline{\mathrm{Max}}\, g'_{N-1} = Y_{N-1} = F_{N-1}$ since $F_N = \emptyset$).
2. For each $\xi_i \in F_i$ there is an index $j \geq i$ so that ξ_i can be identified via a local isomorphism with $\xi_j \in F_j$, and $\xi_j \in \underline{\mathrm{Max}}\, g_j$.
3. There are well-defined functions

$$f_i : F_i \longrightarrow I_1 \times I_2$$

$f_i(\xi_i) = (\max g_i, \max g_i')$ if $\xi_i \in Y_i$, $f_i(\xi_i) = f_{i+1}(\xi_{i+1})$ if $\xi_i \notin Y_i$ (here $\Pi_{i+1}(\xi_{i+1}) = \xi_i$ is a natural identification).
(Suggestion: See (1) and prove by decreasing induction).

4. The first coordinate of $f_i(\xi)$ is $g_i(\xi)$ for any $\xi \in F_i$. And $f_i(\xi) = (g_i(\xi), g_i'(\xi))$ for any $\xi \in \underline{\text{Max}}\, g_i$ (apply decreasing induction).
5. $f_i(\Pi_{i+1}(\xi)) > f_i(\xi) \quad$ if $\quad \Pi_{i+1}(\xi) \in Y_i$
   $f_i(\Pi_{i+1}(\xi)) = f_i(\xi) \quad$ if $\quad \Pi_{i+1}(\xi) \notin Y_i$
   $\max f_i = (\max g_i, \max g_i') \qquad \underline{\text{Max}}\, f_i = \underline{\text{Max}}\, g_i'$
   $\max f_i > \max f_{i+1}$
6. The functions $f_i : F_i \longrightarrow I_1 \times I_2$ are u. s. c. (apply decreasing induction).

14. Exercises: A do-it-yourself help guide of theorem 7.10

Exercise 14.1. Fix a pair (W_0, E_0) (2.10), $\dim W_0 = d$. For any non zero sheaf of ideals $0 \neq J \subset \mathcal{O}_{W_0}$ and any integer $b \geq 1$, we defined a basic object (4.7). Now take all basic objects over the fixed pair (W_0, E_0) and define an equivalence relation by setting:
Definition: $(W_0, (J_0, b), E_0) \sim (W_0, (J_0', b'), E_0)$ iff the following conditions hold:

- $\text{Sing}(J_0, b) = \text{Sing}(J_0', b')(\subset W_0)$.
- Any sequence of transformations and restrictions (4.7 and 6.8) over one of the basic objects induces a sequence over the other one, say:

$$\begin{array}{ccccc} (W_0, (J_0, b), E_0) & \longleftarrow & \cdots & \longleftarrow & (W_s, (J_s, b), E_s) \\ (W_0, (J_0', b'), E_0) & \longleftarrow & \cdots & \longleftarrow & (W_s, (J_s', b'), E_s) \end{array} \tag{14.1.1}$$

1. Show that the relation is indeed an equivalence relation on the set of basic objects over the same pair (W_0, E_0).
2. Show that if $W \longleftarrow W_1$ is smooth (e.g. an open immersion) and $(W, (J, b), E) \sim (W, (J', b'), E)$ then
   $$(W_1, (J_1, b), E_1) \sim (W_1, (J_1', b'), E_1)$$
   (notation as in 6.8).
3. Show that a basic object $(W_0, (J_0, b), E_0)$ defines a general basic object of dimension $d = \dim W_0$, and prove that $(W_0, (J_0, b), E_0) \sim (W_0, (J_0', b'), E_0)$ if and only if both define the same general basic object (6.15).
4. Show that if $B = (W_0, (J_0, b), E_0)$ is a simple basic object and $E_0 = \emptyset$, then the general basic object defined by B has structure of dimension $d - 1$ (see 6.12, 13.1(1) and 13.11).
5. Fix an integer $r \geq 1$, define $J_0' = J_0^r$ and show that
   (a) $(W_0, (J_0, b), E_0) \sim (W_0, (J_0', rb), E_0)$.
   (b) For any sequence 14.1.1 and any index $0 \leq i \leq s$: $J_i' = J_i^r$.

6. Fix a pair (W, E) and an open covering $\{W^\lambda\}_{\lambda\in\Lambda}$ of W. Suppose that for each index λ there is a basic object
$$B_\lambda = (W^\lambda, (J^\lambda, b^\lambda), E^\lambda)$$
and that for any two indices λ and ν, the restrictions of B_λ and B_ν to $W_\lambda \cap W_\nu$ (6.2) are equivalent. Show that the B_λ define a general basic object of dimension $d = \dim W$, each j_0^λ in 6.13 being the identity map.

Remark: An important property of our notion of equivalence relies on the behavior of Hironaka's function ord (see 4.11). In fact, if $(W,(J,b),E) \sim (W,(J',b'),E)$, the proof of 7.3 will show (in particular) that for any $\xi \in \operatorname{Sing}(J,b) = \operatorname{Sing}(J',b')$
$$\frac{\nu_\xi(J)}{b} = \frac{\nu_\xi(J')}{b'}$$
And the proof of this equality is what forces us to add restrictions (and not only transformations).

Exercise 14.2. Show that in 6.15.1:

1. Each $F_i \subset W_i$ maps to $F_{i-1} \subset W_{i-1}$.
2. If $W_{i-1} \xleftarrow{\Pi_i} W_i$ is a smooth morphism, then $F_i = \Pi_i^{-1}(F_{i-1})$.

Exercise 14.3. *Inclusion of general basic objects.* Fix notation as in 6.15. We will say that $(\mathcal{F}_0,(W_0,E_0)) \subset (\mathcal{F}_0',(W_0,E_0))$ (over the same pair (W_0,E_0)) if:

a. $F_0 \subset F_0'$.

b. Any sequence over the first, say
$$(\mathcal{F}_0,(W_0,E_0)) \longleftarrow \cdots \longleftarrow (\mathcal{F}_r,(W_r,E_r)) \tag{14.3.1}$$
induces a sequence over the second, say
$$(\mathcal{F}_0',(W_0,E_0)) \longleftarrow \cdots \longleftarrow (\mathcal{F}_r',(W_r,E_r)) \tag{14.3.2}$$
and $F_i \subset F_i'$, $i = 1, \dots, r$.

Show that:

1. If $(\mathcal{F}_0,(W_0,E_0)) \subset (\mathcal{F}_0',(W_0,E_0))$ and $(\mathcal{F}_0',(W_0,E_0)) \subset (\mathcal{F}_0'',(W_0,E_0))$ then $(\mathcal{F}_0,(W_0,E_0)) \subset (\mathcal{F}_0'',(W_0,E_0))$.
2. $(\mathcal{F}_0,(W_0,E_0)) \subset (\mathcal{F}_0',(W_0,E_0))$ and $(\mathcal{F}_0',(W_0,E_0)) \subset (\mathcal{F}_0,(W_0,E_0))$ if and only if $(\mathcal{F}_0,(W_0,E_0)) = (\mathcal{F}_0',(W_0,E_0))$.
3. In the setting of 6.6, corollary 6.7 asserts that the general basic object defined by $(W,(J,b),E)$ is included on that defined by $(W,(I(W_h),1),E)$.
4. If $0 \neq J \subset J' \subset \mathcal{O}_W$ are sheaves of ideals, then the g. b. o. defined by $(W,(J',b),E)$ is included in the g. b. o. defined by $(W,(J,b),E)$.
5. If $W_h \subset W$ is a regular hypersurface and $I(W_h) \subset \Delta^{b-1}(J)$, then $I(W_h) \subset \Delta^{b-1}(J')$ ($J \subset J'$ as above).

Exercise 14.4. *Intersection of basic objects.* Fix two basic objects over the same pair (W_0, E_0), say

$$B_1 = (W_0, (\mathfrak{a}_0, b_1), E_0) \qquad B_2 = (W_0, (\mathfrak{b}_0, b_2), E_0)$$

and set

$$B = (W_0, (J_0, b_1 b_2), E_0), \qquad J_0 = \mathfrak{a}_0^{b_2} + \mathfrak{b}_0^{b_1}$$

1. Show that $\mathrm{Sing}(J_0, b_1 b_2) = \mathrm{Sing}(\mathfrak{a}_0, b_1) \cap \mathrm{Sing}(\mathfrak{b}, b_2)$.
2. If $(\mathcal{F}_0^i, (W_0, E_0))$ is the g. b. o. defined by B_i, $i = 1, 2$, and $(\mathcal{F}_0, (W_0, E_0))$ the g. b. o. defined by B, show that:
$$(\mathcal{F}_0, (W_0, E_0)) \subset (\mathcal{F}_0^i, (W_0, E_0)) \qquad i = 1, 2$$
3. For any sequence over $(\mathcal{F}_0, (W_0, E_0))$ say 14.3.1, inducing now two sequences as in 14.3.2 over $(\mathcal{F}_0^i, (W_0, E_0))$, $i = 1, 2$, then show that for each index $j = 0, \ldots, r$:
$$F_j = F_j^1 \cap F_j^2$$
4. B is called the intersection of the basic objects B_1 and B_2. Show that if B_1 is a simple basic object 6.3, then B is a simple basic object.
5. Set $d = \dim W_0$, define now a notion of intersection of two d-dimensional g. b. o. (6.17).

15. Exercises: Constructive desingularization of locally embedded schemes. Compatibility with group actions and formal isomorphisms

Here X will be a separated scheme of finite type over a field $\mathbf{k}$ of characteristic zero, and $X_1, \ldots, X_r$ an open affine covering of X.

Exercise 15.1. If $\Theta \in \mathrm{Aut}(X)$ (automorphisms of X) show that there is an open covering of any X_i, say $X_{i,1}, \ldots, X_{i,r_i}$, so that $\Theta(X_{i,j}) \subset X_j$.

Exercise 15.2. Each X_i is defined by a $\mathbf{k}$-algebra of finite type. Each $\mathbf{k}$-algebra of finite type is a quotient of a polynomial ring. Show that for d big enough there are closed immersions $X_i \subset \mathbb{A}_{\mathbf{k}}^d$, for $i = 1, \ldots, r$.

So far we have defined closed embeddings $X_i \subset W_i = \mathbb{A}_{\mathbf{k}}^d$. For each index $i = 1, \ldots, r$ there is a g. b. o. with a d-dimensional structure, defined in terms of the Hilbert-Samuel function HS_{X_i} (see theorem 8.16). Furthermore the three theorems at the of section 8 (see Three Main Theorems) say that the function HS_X plays the same role as mult_X did in the hypersurface case: "There is a g. b. o. with a d-dimensional structure attached to the function and embedded desingularization is achieved by successive resolutions of those g. b. o."

So for each index i, there is an embedded desingularization of $X_i \subset W_i$. Such embedded desingularization defines, of course, a (non-embedded) desingularization of each X_i. We want to show that these desingularizations of X_i patch to define a desingularization of X. We also want this desingularization to lift the action of $\mathrm{Aut}(X)$ on X.

Definition 15.3. Let us set (W, E) as in 2.10 and $(X \subset W, E)$, with $X \subset W$ closed subscheme. Set $(X \subset W, E) \stackrel{\Pi}{\longleftarrow} (X_1 \subset W_1, E_1)$ where either:

- **a.** $(W, E) \stackrel{\Pi}{\longleftarrow} (W_1, E_1)$ is as in 2.10 and X_1 is the strict transform of X.
- **b.** or $(W, E) \stackrel{\Pi}{\longleftarrow} (W_1, E_1)$ is as in 6.8 and $X_1 = \Pi^{-1}(X)$.

Exercise 15.4. Given $(X \subset W, E)$, set $h = \max \mathrm{HS}_X$ and let $(\mathcal{F}_h, (W, E))$ be the g. b. o. defined in terms of h (8.16), so $F_0 = F = \underline{\mathrm{Max}}\, \mathrm{HS}_X \subset W$.

Set $(X \subset W, E) \stackrel{\Pi}{\longleftarrow} (X_1 \subset W_1, E_1)$ as in 15.3(b). Here $F_1 = \Pi^{-1}(F_0)$ (6.15(B)). Prove that $F_1 = \underline{\mathrm{Max}}\, \mathrm{HS}_{X_1}$, but however $h \neq \max \mathrm{HS}_{X_1}$ (note, for instance, that $\dim X < \dim X_1$).

Remark: The sequence of transformations in theorem 8.16 involves only transformations as those in 6.15(C). This was done just to simplify the formulation, but suitably formulated, the theorem holds both for transformations of type 6.15 (B) and (C).

Exercise 15.5. We consider here $(X \subset W, E)$, $(X' \subset W', E')$ and an isomorphism $\Theta : X \longrightarrow X'$ and we will assume that $\dim W = \dim W'$ that $E = \{H_1, \ldots, H_r\}$, that $E' = \{H'_1, \ldots, H'_r\}$ and furthermore that $\Theta(H_i \cap X) = H'_i \cap X'$. In other words we reformulate 6.23 with the only change that X and X' might embedded in different ambient spaces W and W', both of the same dimension.

1. Lift Θ to an isomorphism $\Theta_1 : X_1 \longrightarrow X'_1$ (in the sense of 6.23) if both $(X_1 \subset W_1, E_1)$ and $(X'_1 \subset W'_1, E'_1)$ arise from transformations as in 15.3(b).
2. Ditto for transformations as in 15.3(a) if $\Theta(Y) = Y'$ in 2.10.
3. Note that $\mathrm{HS}_X(\xi) = \mathrm{HS}_{X'}(\Theta(\xi))$ for any $\xi \in X$, in particular $h = \max \mathrm{HS}_X$ $= \max \mathrm{HS}_{X'}$ and $\Theta(\underline{\mathrm{Max}}\, \mathrm{HS}_X) = \underline{\mathrm{Max}}\, \mathrm{HS}_{X'}$.
4. With the setting of 15.5 and 15.5, if H (resp. H') denote the exceptional hypersurface that arises by blowing up Y (resp. Y'), then

$$\Theta\left((\underline{\mathrm{Max}}\, \mathrm{HS}_{X_1}) \cap H\right) = (\underline{\mathrm{Max}}\, \mathrm{HS}_{X'_1}) \cap H'$$

Exercise 15.6. Fix an isomorphism $\Theta : X \longrightarrow X'$ for $(X \subset W, E)$ and $(X' \subset W', E')$ as in exercise 15.5. Set $h = \max \mathrm{HS}_X = \max \mathrm{HS}_{X'}$ and $(\mathcal{F}_h, (W, E))$, $(\mathcal{F}'_h, (W', E'))$ as in theorem 8.16. Prove that $\Theta : X \longrightarrow X'$ defines an isomorphism of $(\mathcal{F}_h, (W, E))$ with $(\mathcal{F}'_h, (W', E'))$ in the sense of 6.23(2).

Exercise 15.7. Assume that Θ is an isomorphism of g. b. o. from $(\mathcal{F}_0, (W_0, E_0))$ to $(\mathcal{F}'_0, (W'_0, E'_0))$ (6.23(2)), and assume also that both g. b. o. have a d-dimensional structure (same d). Prove that

$$\mathrm{ord}^d(\xi) = \mathrm{ord}^d(\Theta(\xi))$$

for any $\xi \in F_0$ (Hint: see proof of proposition 7.4).

Exercise 15.8. With the setting as in exercise 15.7, show that the resolutions of $(\mathcal{F}_0, (W_0, E_0))$ and $(\mathcal{F}'_0, (W'_0, E'_0))$ defined by theorem 7.13 are step by step compatible with Θ (i.e. $\Theta(Y_i) = Y'_i$ for each index i in 7.13.1). (Hint: see theorem 7.7).

Exercise 15.9. Let $\Theta : X \longrightarrow X'$ be an isomorphism for $(X \subset W, E)$ and $(X' \subset W', E')$ as in 15.5. Recall that $d = \dim W = \dim W'$. Apply now exercise 15.8 to the g. b. o. with d-dimensional structure $(\mathcal{F}_h, (W, E))$ and $(\mathcal{F}'_h, (W', E'))$ given by theorem 8.16 and show that the constructive desingularization of $X \subset W$ and of $X' \subset W'$ (8.18) are compatible with Θ.

Exercise 15.10. (desingularization of locally embedded schemes)
Set $X_i \subset W_i = \mathbb{A}^d_{\mathbf{k}}$ as in exercise 15.2, so $X_{i,j} = X_i \cap X_j$ is open in X_i and in X_j.

1. Show that there are open sets $W'_i \subset W_i$ and $W'_j \subset W_j$ so that $X_{i,j} \subset W'_i$ and $X_{i,j} \subset W_j$ are closed immersions.
2. Apply theorem 8.14(B) (for open immersions) to show that:
   (a) The constructive embedded desingularization of $(X_{i,j} \subset W'_i, E'_i)$ (8.18) is the restriction of that of $(X_i \subset W_i, E_i)$.
   (b) The constructive embedded desingularization of $(X_{i,j} \subset W'_j, E'_j)$ is the restriction of that of $(X_j \subset W_j, E_j)$.
3. Apply exercise 15.9 to id : $X_{i,j} \longrightarrow X_{i,j}$ for $(X_{i,j} \subset W'_i, E'_i)$ and $(X_{i,j} \subset W'_j, E'_j)$ to show that the desingularizations of the X_i patch and define a non-embedded desingularization of X.

Exercise 15.11. Use exercises 15.8 and 15.9 to show that any $\Theta \in \mathrm{Aut}(X)$ will lift to an automorphism on the non-embedded desingularization given by exercise 15.10.

Exercise 15.12. Compatibility with formal isomorphisms
Let X_1 and X_2 be schemes of finite type over a field $\mathbf{k}$ and let $\xi_i \in X_i$, $i = 1, 2$, points. A corollary of Artin's approximation theorem says that if the completions of the local rings at $\xi_i \in X_i$ are isomorphic, then there is a common étale neighborhood (X', ξ') of $\xi_i \in X_i$, i.e. étale maps $X' \longrightarrow X_i$ sending ξ' to ξ_i, for $i = 1, 2$ [Ar, 2.6 p. 28]. This will allow us to give a precise meaning to the fact that "the constructive resolutions of $\xi_i \in X_i$ are locally the same".

Étale maps are smooth maps of relative dimension zero, these are flat morphisms of finite type, in particular they are open maps [Mal, p. 48]. So to make this concept precise it suffices to show that after suitable open restrictions, the constructive resolution of X' is the restriction (the pull-back) of that of X_i, for $i = 1, 2$.

Since matter are local, assume that $X_i = \mathrm{Spec}(A_i)$, $i = 1, 2$, $X' = \mathrm{Spec}(A')$ and that A_1, A_2 and A' are $\mathbf{k}$-algebras of finite type. After suitable open restrictions we may assume that

$$A' = (A_i[Z]/p_i(Z))_{g_i(Z)} \qquad i = 1, 2$$

with $g_i(Z) \in A_i[Z]$ chosen so that $\frac{\partial p_i}{\partial Z}$ is invertible at such ring (see [R, Th1 Chap. V p. 51]).

1. Show that after a change of base field we may define surjective morphisms
$$\varphi_i : S_n = \mathbf{k}[Z_1, \ldots, Z_n] \longrightarrow A_i \qquad i = 1, 2$$
(same n), so that the points ξ_i map to the origin via $X_i \subset \mathbb{A}^n_{\mathbf{k}} = \mathrm{Spec}(S_n)$.
2. Set $\tilde{p}_i(Z), \tilde{g}_i(Z) \in S_n[Z]$, for $i = 1, 2$ such that $\tilde{p}_i$ maps to $p_i \in A_i[Z]$ and $\tilde{g}_i$ maps to $g_i \in A_i[Z]$. Set
$$T_i = (S_n[Z]/\tilde{p}_i(Z))_{\tilde{g}_i(Z)}, \qquad i = 1, 2$$
and show that the natural induced morphism $\mathrm{Spec}(T_i) \longrightarrow \mathrm{Spec}(S_n)$ is étale for $i = 1, 2$.
3. Show that $T_i \otimes_{S_n} A_i = A'$ for $i = 1, 2$.
4. Set $W_i = \mathrm{Spec}(T_i)$. Show that
   (a) $\dim W_1 = \dim W_2 = n$.
   (b) There are closed immersions $X' \subset W_1$ and $X' \subset W_2$.
   (c) The setting for both
$$(X' \subset W_1) \longrightarrow (X_1 \subset \mathbb{A}_{\mathbf{k}^n}) \qquad (X' \subset W_2) \longrightarrow (X_2 \subset \mathbb{A}_{\mathbf{k}^n})$$
   are such that the embedded desingularization of $X' \subset W_i$ is the pull-back of that of $X_i \subset \mathbb{A}^n_{\mathbf{k}}$ (see theorem 8.16 and theorem 8.14).
5. Apply exercise 15.9 to id : $X' \longrightarrow X'$, $X' \subset W_1$ and $X' \subset W_2$ to show that both embedded desingularizations induce the same (non-embedded) desingularization on X'.

Exercise 15.13. Compatibility of constructive desingularization with group actions Fix $Y \subset X$ and assume that $\Theta(Y) = Y$ for any $\Theta \in \mathrm{Aut}(X)$.

1. Show that $\mathrm{Aut}(X)$ acts naturally on $X \times \mathbb{A}_{\mathbf{k}^1}$ (on the $\mathcal{O}_X$-sheaf of algebras $\mathcal{O}_X[Z]$).
2. Set $\mathcal{O}_X[I(Y)Z] \subset \mathcal{O}_X[Z]$ as a subsheaf of graded algebras. Show that $\mathrm{Aut}(X)$ acts on this subsheaf and the action preserves the grading.
3. Set $X_1 = \mathrm{Proj}\,(\mathcal{O}_X[I(Y)Z]) \longrightarrow X$ and show that there is a natural lifting of $\mathrm{Aut}(X)$ to an action on X_1.
4. Show that 15.13 holds for any group G acting on X.
5. Show that the constructive desingularization of X lifts any group action.

Added in proof:

1. In the meanwhile we have proved in [EV2] that the algorithm of resolution of general basic objects (theorem 7.13) is strong enough to provide a direct and simple proof of the main Theorem 3.1 with minor modifications, avoiding all the theory of normal flatness, the use of the Hilbert-Samuel function, and

hence: theorems 8.15, 8.16 and 8.17 in chapter 8. As in our chapter 15, the new proof in [EV99] also extends to locally embedded schemes. But leaving these important matters of globalization and equivariance aside, let us sketch here how our new proof works for the very particular case of an *irreducible* variety X *globally* included in a smooth variety W. Recall that Theorem 7.13 states an algorithm with properties (P1), (P2), (P3) and (P4) described in 7.15. Fix $(X_0 \subset W_0, E_0)$ and set $J_0 = I(X_0)\,(\subset \mathcal{O}_{W_0})$. We may assume that E_0 is empty.
Set b the maximum order of the ideal J_0. Theorem 7.13 provides a resolution of the basic object $(W_0, (J_0, b), E_0)$, say

$$(W_0, (J_0, b), E_0 \longleftarrow \cdots \longleftarrow (W_N, (J_N, b), E_N)$$

which means that the maximum order of J_N is smaller than b. Repeating this procedure one may achieve a chain of transformations as above such that, for some index M, $J_M = \mathcal{O}_{W_M}$ (i.e. the maximum order of J_M is zero). Property (P3) says that the function $f^d_{=}$ is constant along the dense open set $\mathrm{Reg}(X_0)$. From (P1) and (P2) we see that there must be a unique index $r' \leq M-1$ such that $\max f^d_{r'}$ is that particular constant value. It follows from (P3) and (P1) that there is an open dense set in W_0, containing $\mathrm{Reg}(X_0)$, which is isomorphic to an open dense set in $W_{r'}$. So the strict transform $X_{r'}$, of X_0 in $W_{r'}$ and the closed set $\underline{\mathrm{Max}}\, f^d_{r'}$ coincide in this dense open set of $W_{r'}$, therefore $X_{r'}$ must be a component of the regular scheme $\underline{\mathrm{Max}}\, f^d_{r'}$. This shows that the first r' steps define an embedded desingularization, and an isomorphism over the open set $\mathrm{Reg}(X_0)$.

2. One of the participants to the Working Week, J. Schicho, jointly with G. Bodnár, has recently implemented in Maple this algorithm for resolution of basic objects [BS]. In particular, their implementation provides desingularization in any dimension.

References

[Ab1] S.S. Abhyankar, *Good points of a hypersurface*, Advances in Math. 68, (1988), pp. 87–256.

[Ab2] S.S. Abhyankar, *Resolution of Singularities of Embedded Algebraic Surfaces*, Second enlarged edition, Springer-Verlag, 1998.

[AHV1] J.M. Aroca, H. Hironaka and J.L. Vicente, *The theory of maximal contact*, Memorias de Matemática del Instituto "Jorge Juan" (Madrid), 29, (1975).

[AHV2] J.M. Aroca, H. Hironaka and J.L. Vicente, *Desingularization theorems*, Memorias de Matemática del Instituto "Jorge Juan" (Madrid), 30, (1977).

[AM] S.S. Abhyankar and T.T. Moh, *Newton-Puiseux expansion and generalized Tschirnhausen transformation I-II*. J. Reine Angew. Math. 260, (1973), pp. 47–83, ibid. 261, (1973), pp. 29–54.

[AJ] D. Abramovich and A.J. de Jong, *Smoothness, semistability and toroidal geometry*. Journal of Algebraic Geometry, 6 (1997), pp. 789–801.

[AW] D. Abramovich and J. Wang, *Equivariant resolution of singularities in characterisitic 0.* Mathematical Research Letters 4 (1997), pp. 427–433.

[Ar] M. Artin, *Algebraic approximation of structures over complete local rings*, Pub. Math. I.H.E.S., 36, (1969), pp. 23–58.

[Ben] B.M. Bennett, *On the characteristic function of a local ring*, Ann. of Math., 91, (1970), pp. 25–87.

[Ber] P. Berthelot, *Altérations des variétés algébriques*, Sem. Bourbaki 48, 815, (1995–96).

[BM1] E. Bierstone and P. Milman, *A simple constructive proof of canonical resolution of singularities*, Effective Methods in Algebraic Geometry. Prog. in Math, 94, Birkhäuser, Boston, (1991), pp. 11–30.

[BM2] E. Bierstone and P. Milman, *Canonical desingularization in characteristic zero by blowing-up the maximal strata of a local invariant*, Inv. Math. 128(2), (1997), pp. 207–302.

[BS] G. Bodnár, J. Schicho, *Automated resolution of singularities for hypersurfaces.* Preprint 1999, available at http:/www.risc.uni-linz.ac.at/projects/basic/adjoints/blowup/

[BP] F. Bogomolov and T. Pantev, *Weak Hironaka Theorem* Mathematical Research Letters 3 (1996) pp. 299–307.

[CGO] V. Cossart, J. Giraud and U. Orbanz, *Resolution of Surface Singularities.* Lectures Notes in Mathematics, Springer Verlag 1101, (1984).

[E1] S. Encinas, *Resolución Constructiva de Singularidades de Familias de Esquemas*, PhD thesis, Universidad de Valladolid, (1996).

[E2] S. Encinas, *On constructive desingularization of non-embedded schemes.*, Preprint, (1998).

[EV1] S. Encinas and O. Villamayor, *Good points and constructive resolution of singularities*, Acta Math. Vol. 181:1 (1998).

[EV2] S. Encinas and O. Villamayor, *A new theorem of desingularization over fields of characteristic zero.* Preprint 1999.

[Gi1] J. Giraud, *Analysis Situs*, Sem. Bourbaki, 256, (1962–63).

[Gi2] J. Giraud, *Etude locale des singularités*, Pub. Math. Orsay, France 1972.

[Gi3] J. Giraud, *Sur la théorie du contact maximal*, Math. Zeit., 137, (1974), pp. 285–310.

[Gi4] J. Giraud, *Contact maximal en caractéristique positive*, Ann. Scien. de l'Ec. Norm. Sup., 4 série, 8 (2), (1975), pp. 201–234.

[Gi5] J. Giraud, *Remarks on desingularization problems*, Nova Acta Leopoldina, 52, (1981), pp. 103–107.

[Har] R. Hartshorne, *Algebraic Geometry*, Graduate Texts in Mathematics 52, Springer Verlag, New York, 1983.

[H1] H. Hironaka, *Resolution of singularities of an algebraic variety over a field of characteristic zero I–II*, Ann. Math., 79, (1964), pp. 109–326.

[H2] H. Hironaka, *Idealistic exponent of a singularity*, Algebraic Geometry. The John Hopkins centennial lectures, Baltimore, Johns Hopkins University Press (1977), pp. 52–125.

[J] A.J. de Jong, *Smoothness, semi-stability and alterations*, Pub. Math. I.H.E.S. (1996) no. 83, pp. 51–93.

[L] J. Lipman, *Introduction to resolution of singularities*, Proc. Symp. in Pure Math., 29, (1975), pp. 187–230.

[LT] M. Lejeune and B. Teissier, *Quelques calculs utiles pour la résolution des singularités*, Centre de Mathématique de l'Ecole Polytechnique, (1972).

[Ma1] H. Matsumura. *Commutative Algebra*. W. A. Benjamin Co., New York, 1970.

[Ma2] H. Matsumura. *Commutative Ring Theory*. Cambridge University Press, Londres, 1986.

[Mo] T. T. Moh. *Quasi-canonical uniformization of hypersurface singularities of characteristic zero*. Communications in Algebra, 20(11), (1992), pp. 3207–3251.

[R] M. Raynaud. *Anneaux locaux henséliens*, volume 169 of *Lecture Notes in Mathematics*. Springer-Verlag, 1970.

[O] T. Oda, *Infinitely Very Near-Singular Points*, Complex Analytic Singularities, Advanced Studies in Pure Mathematics, 8, (1986), pp. 363–404.

[V1] O.E. Villamayor, *Constructiveness of Hironaka's resolution*, Ann. Scient. Ec. Norm. Sup., 4[e] serie, 22, (1989) pp. 1–32.

[V2] O.E. Villamayor, *Patching local uniformizations*, Ann. Scient. Ec. Norm. Sup., 25, (1992), pp. 629–677.

[V3] O.E. Villamayor, *On good points and a new canonical algorithm of resolution of singularities (Announcement)*, Real Analytic and Algebraic Geometry, Walter de Gruyter Berlin, New York, 1995.

[V4] O.E. Villamayor. *Introduction to the algorithm of resolution*. Proceedings of the 1991 La Rábida Meeting. Progress in Mathematics, Vol 134, (1996), Birkhäuser Verlag Basel/Switzerland.

[Z] O. Zariski. *Local uniformization of algebraic varieties*. Ann. of Math., 41(4), (1940), pp. 852–860.

Dpto. Matemática Aplicada Fundamental
E.T.S. Arquitectura, Universidad de Valladolid
47014 Valladolid Spain
sencinas@cpd.uva.es

Dpto. Matemáticas
Universidad Autónoma de Madrid
28049 Madrid Spain
villamayor@uam.es

Part 2
Contributions

Progress in Mathematics, Vol. 181,

A Computer Program for the Resolution of Singularities

Gábor Bodnár and Josef Schicho

1. Introduction

Since recently we have constructive proofs for resolution of singularities in characteristic zero (see [V1, V2, BM1, BM2, EV]). These proofs do contain, at least implicitly, algorithms for computing the desingularization automatically. The authors have implemented an algorithm based on [EV] in Maple V R5.

The program takes a polynomial in n variables as input. The output is a tree of charts, where

- the root node is k^n, the ambient space of the given hypersurface (k is a field of algebraic numbers),
- the children of each node are obtained from the parent node by blowing up or by some other operation described below (cover and exchange),
- the proper transform of the given hypersurface in the leaf nodes is nonsingular, and the total transform is a normal crossing divisor.

The output tree may be viewed by "clicking through the resolution" in an html hierarchy (Figure 1).

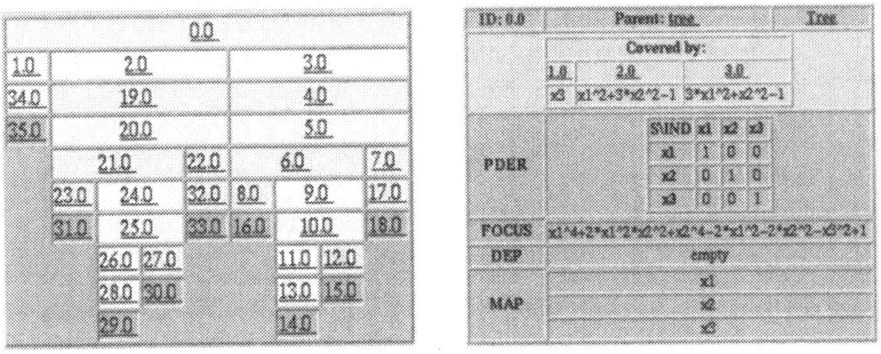

FIGURE 1. Resolution tree with hyperlinks to the charts

Actually, there was a long way to go from the algorithm in [EV] to a working implementation, but space restrictions do not permit to go into technical details. Instead, we explain two typical computational problems and their solutions in the implementation.

First, it is technically difficult to represent blowups of arbitrary regular varieties in affine space. Using the general construction for blowups of arbitrary ideals, the blown up variety gets embedded in some high dimensional projective space (see [H], p. 163). This is not convenient for performing the differentiating operations that are necessary in the algorithm. This problem is computationally mastered by working simultaneously with affine coordinates on the one side and global regular parameters on the other side. The concept of global regular parameters also appears in [BM2], p. 232, in another context.

Second, the constructive proofs are formulated in a local language, i.e. using power series. Theoretically, it is possible to treat the most difficult point in the next blowup in a similar local fashion. However, a computer program has to resolve the variety as a whole (covering it by a finite number of charts), without making explicit reference to points. There were also other problems which had to be solved (redundancy in the charts, growth of exponents), but these will not be treated here.

The full program including illustrative examples is available via the web site http://www.risc.uni-linz.ac.at/projects/basic/adjoints/blowup.
A detailed description may be found in [BS].

This work was supported by the Austrian Science Fund FWF. We are indebted to S. Encinas and O. Villamayor for their help in the clarification of details, and to H. Hauser for helpful remarks.

2. How to blow up a circle

Suppose you want to resolve the surface S in three-space with equation

$$z^2 + (x^2 + y^2 - 1)^2 y = 0.$$

The circle $x^2 + y^2 - 1 = 0$, $z = 0$ is a twofold curve of S which has to be blown up. Actually, the algorithm [EV] tells us to first blowup the two points $(\pm 1, 0, 0)$. But for our purposes it suffices to consider what happens in the open chart $y \neq 0$. In this chart, the algorithm [EV] will blow up the circle.

Since the ideal of the circle has two generators, the blown up variety can be covered by two affine charts. One is isomorphic to the zero set of $u(x^2+y^2-1)-z$ in k^4. The variable z can be eliminated, so that this chart is actually isomorphic to k^3, with coordinates x, y, u. The blowing down map is given by $(x, y, u) \mapsto (x, y, u(x^2 + y^2 - 1))$.

The second chart W is isomorphic to the zero set of $vz-(x^2+y^2-1)$ in k^4. We do not know if this variety is isomorphic to k^3. In any case, such an isomorphism would be hard to find. Hence we have to continue computation with 4 variables, the ambient space W being itself a hypersurface. The next step, computing the singularities of the proper transform of S in W, is already much more difficult than in the case of charts isomorphic to k^3. One can imagine that these difficulties accumulate in cases where more blowups are necessary.

The problem is that we need to represent varieties in such a way that the blowup can be covered by charts represented in a systematic way. On the other hand, this way of representing varieties should allow to perform various tasks (like computing multiplicities or coefficient ideals) easily and fast.

The solution is to use two different "coordinate systems", so that a chart is described by the following data.

1. A list of variables (VAR): For each chart W, we have the affine coordinates of the affine ambient space, say $x_1, \ldots, x_m$.
2. A list of polynomials (DEP): The chart W is an algebraic subset of k^m. Its defining ideal is generated by polynomials $D_1, \ldots, D_r \in k[x_1, \ldots, x_m]$, which we also call the *dependencies*, since they describe the algebraic relations between the coordinate functions restricted to W. In general, the number of affine ambient coordinates is larger than $\dim W$.
3. A list of polynomials (IND): We require for each chart the existence of a global system of regular parameters, i.e. an n-tuple $(P_1, \ldots, P_n)$ of functions in $k[W]$, where $n = \dim W$, forming a regular system of parameters of each local ring of points in W. It is easy to see that every nonsingular variety can be covered by a finite number of charts which have such a global system of parameters (see also [BM2], p. 232).
4. An $m \times n$ matrix (PDER): For each chart, we need to know the partial derivations with respect to the parameter system. Note that these are functions from $k[W]$ to $k[W]$. They are uniquely determined by their value at the generators $x_1, \ldots, x_m$. This allows to store the information of all partial derivatives in a finite matrix. Computing partial derivatives is easy (using this matrix), and we can now perform the various tasks mentioned above easily and fast.

During the resolution process, there are three operations of forming new charts from a given one. Sometimes, we need to *cover* a chart by smaller charts. Sometimes, we need to *exchange* some function in the list IND with another one. And sometimes, most notably, we need to *blow up* a chart. In the blowup operation, we require that the center is a nonsingular algebraic set defined by a subset of IND. Note that this implies that the center is nonsingular and has normal crossing with the hypersurfaces of W defined by the elements of IND.

To proceed with our example from above, we first do a cover operation, producing a chart C_1 isomorphic to the open set $y \neq 0$:

1. VAR $= [x, y, z, v_1]$,
2. DEP $= [yv_1 - 1]$,
3. IND $= [x, y, z]$,

4. PDER $= \begin{pmatrix} 1 & 0 & 0 \\ 0 & 1 & 0 \\ 0 & 0 & 1 \\ 0 & -v_1^2 & 0 \end{pmatrix}.$

The entry of the last row of the matrix PDER is introduced by the cover operation and comes from the formula $\partial_y v_1 = \partial_y \frac{1}{y} = -\frac{1}{y^2} = -v_1^2$. The cover operation also produces another chart C_2, isomorphic to the open sets $y^2 - 1 \neq 0$, but we shall only treat C_1.

Next, we exchange y by $x^2 + y^2 - 1$ in IND. This is possible because $\partial_y(x^2 + y^2 - 1)$ is invertible in C_1. We obtain the chart E:

1. VAR $= [x, y, z, v_1]$,
2. DEP $= [v_1 y - 1]$,
3. IND $= [x, x^2 + y^2 - 1, z]$,
4. PDER $= \begin{pmatrix} 1 & 0 & 0 \\ -v_1 x & \frac{1}{2}v_1 & 0 \\ 0 & 0 & 1 \\ v_1^3 x & -\frac{1}{2}v_1^3 & 0 \end{pmatrix}.$

Now, we are ready for blowing up. This will provide two charts. The first, say B_1, is given by

1. VAR $= [x, y, v_1, v_2]$,
2. DEP $= [v_1 y - 1]$,
3. IND $= [x, x^2 + y^2 - 1, v_2]$,
4. PDER $= \begin{pmatrix} 1 & 0 & 0 \\ -v_1 x & \frac{1}{2}v_1 & 0 \\ v_1^3 x & -\frac{1}{2}v_1^3 & 0 \\ 0 & 0 & 1 \end{pmatrix}.$

The variable z has been eliminated. The proper transform of S has equation $v_2^2 + y$, hence it is nonsingular. The exceptional divisor is $x^2 + y^2 - 1$, and one can show easily that the total transform is a normal crossing divisor.

The second chart B_2 is given by

1. VAR $= [x, y, z, v_1, v_2]$,
2. DEP $= [v_1 y - 1, v_2 z - x^2 - y^2 + 1]$,
3. IND $= [x, v_2, z]$,
4. PDER $= \begin{pmatrix} 1 & 0 & 0 \\ -v_1 x & -\frac{1}{2}v_1 z & -\frac{1}{2}v_2 v_1 \\ 0 & 0 & 1 \\ v_1^3 x & -\frac{1}{2}v_1^3 z & -\frac{1}{2}v_2 v_1^3 \\ 0 & 1 & 0 \end{pmatrix}.$

Again the proper transform v_2^2y+1 is nonsingular and has normal crossing with the exceptional divisor z.

Here are Maple programs for the operations blowup, cover, and exchange (actually these are not the subprograms of the downloadable package mentioned above, but simplified variants written for demonstration purposes).

```
#Input:  VAR, DEP, IND, PDER: corresponding data of a chart,
#        C:   center--list of indices to IND.
#Output: OUT: a list of new charts generated by the blowup
blowup := proc(VAR, DEP, IND, PDER, C)
local i, j, k, VAR1, DEP1, IND1, PDER1, OUT;
OUT := [];
for i from 1 to nops(C) do
  VAR1:= VAR; DEP1:= DEP; IND1:= IND; PDER1:= PDER;
  for j from 1 to nops(C) do
    if j<>i then
      VAR1:= [op(VAR1), v.(nops(VAR1)+1)];
      DEP1:= [op(DEP1), VAR1[nops(VAR1)]*IND[C[i]]-IND[C[j]]];
      IND1[C[j]]:= VAR1[nops(VAR1)];
      for k from 1 to nops(VAR) do
      PDER1[k][C[i]]
            := PDER1[k][C[i]]+VAR1[nops(VAR1)]*PDER1[k][C[j]];
      PDER1[k][C[j]]:= PDER1[k][C[j]]*IND[C[i]];
      od;
      PDER1:= [op(PDER1), [seq(0, k=1..nops(IND))]];
      PDER1[nops(PDER1)][C[j]]:= 1;
    fi;
  od;
  OUT:= [op(OUT), [VAR1, DEP1, IND1, PDER1]];
od;
RETURN(OUT);
end:
#Input:  VAR, DEP, IND, PDER: corresponding data of a chart,
#        F:   a list of polynomials
#Output: OUT: a list of new charts
cover := proc(VAR, DEP, IND, PDER, F)
local i, j, VAR1, DEP1, PDER1, OUT;
OUT := [];
for i from 1 to nops(F) do
  VAR1:= [op(VAR), v.(nops(VAR)+1)];
  DEP1:= [op(DEP), VAR1[nops(VAR1)]*F[i]-1];
  PDER1:= [op(PDER), [seq(0, j=1..nops(IND))]];
  for j from 1 to nops(IND) do
  PDER1[nops(VAR1)][j]
```

```
        := -VAR1[nops(VAR1)]^2*pdiff(F[i], j, VAR, PDER);
  od;
  OUT:= [op(OUT), [VAR1, DEP1, IND, PDER1]];
od;
RETURN(OUT);
end:

#Input:  VAR, DEP, IND, PDER: corresponding data of a chart,
#        F:   a polynomial,    p:   an index to IND
#        a:   the inverse of pdiff(F, p, VAR, PDER)
#Output: the chart with IND[p] and F exchanged
exchange := proc(VAR, DEP, IND, PDER, F, p, a)
local i, j, IND1, PDER1;
IND1:= IND;  IND1[p]:= F;  PDER1:= PDER;
for i from 1 to nops(VAR) do
  PDER1[i][p]:= a*PDER[i][p];
  for j from 1 to nops(IND) do
   if j<>p then
    PDER1[i][j]:= PDER[i][j]-a*PDER[i][p]*pdiff(F, j, VAR, PDER);
   fi;
  od;
od;
RETURN([VAR, DEP, IND1, PDER1]);
end:

#Input:  VAR, PDER: corresponding data of a chart,
#        F:   a polynomial,    p:   an index to IND
#Output: dF/dIND[p]
pdiff := proc(F, p, VAR, PDER)
local i;
RETURN(add(diff(F,VAR[i])*PDER[i][p], i=1..nops(VAR)));
end:
```

3. How to compute the centers of the blowups

The most natural idea for computing the centers is to use the upper semicontinuous function that defines the stratification, see [EV]. One would have to express the constraints of lying in the maximum stratum by algebraic equations, and, computing the radical of the ideal generated by these equations, one would obtain the ideal of the center. Of course, in order to apply the efficient blowup procedure described above, one would then need some cover and exchange operations as to make this ideal generated by parameters. All these tasks can be done algorithmically, but the computations are rather costly.

In our implementation, we ignore the upper semicontinuous function of [EV]. Instead we use the inductive construction directly. Adjusting the charts by cover and exchange operations as described below, we achieve that the center arises as the zero set of a subset of IND almost automatically.

In the algorithm [EV], the centers of blowup arise in two possible ways: as a hypersurface component of the singular locus or as an intersection of exceptional divisors (in the "monomial case", see [EV]). Normally, a blowup operation happens after applying some induction steps that reduce the dimension. To get the center of the blowup which effects the given hypersurface, one has to intersect with all the intermediate hypersurfaces in the inductive steps. Therefore, when we want to compute centers, we have to perform three tasks:

1. find a hypersurface to reduce the dimension (i.e. a hypersurface of "maximal contact", see [EV]);
2. find the hypersurface components of the singular locus;
3. intersect exceptional divisors or/and previously found hypersurfaces of maximal contact.

In task 1, we also want to perform an exchange operation in order to have a local equation of the hypersurface of maximal contact in IND. Doing this, we observe that task 3 requires no real computation at all (i.e. it is purely combinatorial), because the center is already defined by some subset of IND.

Suppose we have given a chart W and an ideal in $J \subset k[W]$ of order $b > 0$, generated by a finite set B of elements in $k[W]$. Let B' be the set of partial derivatives of elements in B up to order $b-1$. (The set of common zeroes of B' is the singular locus of (J, b).) Then the partial derivatives of B' do not have any common zeroes. Hence we can cover W by the subsets $F \neq 0$ $\partial_P F \neq 0$, $F \in B'$, $P \in$ IND. For any of these subsets, one can show that either the singular locus is empty or (IND $\setminus \{P\}) \cup \{F\}$ is a global parameter system: we can replace P by F. Moreover, F defines a hypersurface of maximal contact. Hence task 1 can be solved by cover and exchange operations. The theory also tells us that the cover operation is necessary because there might not exist a global element of order 1 in the ideal generated by B' (see [EV]).

Task 2 is the most difficult one. Let us assume that we have to resolve an ideal $J \subset k[U]$ of order b. First, we have to perform cover and exchange operations to achieve that $P \in$ IND is of maximal contact, i.e. that P is in the ideal J' generated by the partial derivatives of J up to order $b-1$. Assume that this is done. Because the singular locus of (J, b) is contained in the zero set V of P, any hypersurface component of the singular locus is a component of V. Note that V might be reducible, even though it is a "parameter hyperplane": we only know that it is locally irreducible, but it may have several connected components. In this situation, task 2 requires a cover that separates the components of V contained in the singular locus from the remaining components.

In order to do a cover operation, we need a finite set of elements that do not have a common zero. Here is the idea for finding such a set.

Let $I \subset k[V]$ be the coefficient ideal of (J, b) in V. (For the precise definition of the coefficient ideal, we refer to [EV]. It should be mentioned that the singular locus of the coefficient ideal coincides with the singular locus of (J, b).) In any component of V in the singular locus, the stalk of I is zero, and the stalk of the annihilator $\mathrm{Ann}(I)$ contains 1. On the other hand, there is a number b', such that the derivatives of I of order up to b' do not have common zeroes in all other components. (The smallest such number is the order of the coefficient ideal.) Hence these derivatives together with a generating set for $\mathrm{Ann}(I)$ do not have common zeroes.

It should be noted that the algorithm requires only a very small toolbox of algebro-geometric algorithms: deciding if a finite set of elements has a common zero, computing a generating set for the annihilator, and computing the quotient of two elements. This toolbox is provided by standard techniques using Gröbner bases (see [B, BW]).

References

[BM1] E. Bierstone, P. Milman. A simple constructive proof of canonical resolution of singularities. In: Effective methods in algebraic geometry (T. Mora, C. Traverso eds.), 11–30. Birkhäuser, 1991.

[BM2] E. Bierstone, P. Milman. Canonical desingularization in characteristic zero by blowing up the maximum strata of a local invariant. Invent. Math. 128 (1997), 207–302.

[BS] G. Bodnár, J. Schicho. Automated Resolution of Singularities for Hypersurfaces. Techn. Rep. RISC 99–06, 1999.

[B] B. Buchberger. Gröbner Bases: An Algorithmic Method in Polynomial Ideal Theory. In: Recent Trends in Multidimensional Systems Theory (N. K. Bose, ed.), chapter 6. D. Riedel Publ. Comp. 1985.

[BW] T. Becker, V. Weispfenning. Gröbner bases – a computational approach to commutative algebra. Graduate Texts in Mathematics. Springer Verlag 1993.

[EV] S. Encinas, O. Villamayor. A course on constructive desingularization and equivariance. This volume.

[H] R. Hartshorne. Algebraic Geometry. Springer Verlag 1977.

[V1] O. Villamayor. Constructiveness of Hironaka's resolution. Ann. Scient. Ecole Norm. Sup. Paris 4, 22 (1989), 1–32.

[V2] O. Villamayor. Introduction to the algorithm of resolution. In: Algebraic geometry and singularities (T. Mora, C. Traverso eds.), 123–154. Birkhäuser, 1991.

RISC-Linz
Universität Linz
A-4040 Linz Austria
gbodnar@risc.uni-linz.ac.at
jschicho@risc.uni-linz.ac.at

Progress in Mathematics, Vol. 181, © 2000 Birkhäuser Verlag Basel/Switzerland

Uniformisation et désingularisation des surfaces d'après Zariski

Vincent Cossart

Introduction

Le problème de la désingularisation tel que se l'était posé Zariski peut être formulé ainsi:

Étant donnée une variété projective V sur un corps algébriquement clos k, existe-t-il une variété projective $\widehat{V}$ régulière en tous ses points et birationnellement équivalente à V?

Cet article est une application directe du papier "Valuations" de M. Vaquié que l'on trouvera dans ce volume [V]. On a l'ambition d'exposer la désingularisation des surfaces projectives en suivant "le programme de Zariski". Les ingrédients de cette démonstration se trouvent dans [Z2], et les notions fondamentales dans [V]. Le lecteur aura donc intérêt à garder ces deux articles "sous le coude" pendant la lecture de notre travail.

Ce "programme de Zariski" devait aboutir au théorème général de désingularisation des variétés projectives, malheureusement pour lui, Zariski n'a pas pu le mener à bout, ce programme est actuellement repris par Spivakovsky qui semble proche du but. Nous n'avons pas suivi l'ordre chronologique des résultats de Zariski. En effet, Zariski dans [Z2] n'avait pas encore lancé son programme et, après avoir établi l'*uniformisation*, il ne fait pas le *recollement* (cf. début de 1.2 et [V §8]) et il conclut en donnant le très beau théorème suivant: si on sait résoudre le problème d'*uniformisation* pour une surface projective, on désingularise cette surface projective par une suite finie de normalisations suivies de l'éclatement du lieu singulier. Ce n'est que plus tard, dans [Z4] que Zariski donne une démonstration du *recollement* pour les surfaces.

On trouvera une autre démonstration de la désingularisation des surfaces dans [Z5], papier dévolu à la désingularisation des variétés de dimension 3, en effet, pour résoudre le problème du *recollement*, en dimension 3, Zariski a du établir un résultat plus fort en dimension 2, à savoir, la désingularisation *plongée* des surfaces.

Cette conférence se compose de deux parties principales, dans 1 (Uniformisation et recollement), on a du d'abord reprendre quelques théorèmes de géométrie birationnelle qui sont cités *ex abrupto* par Zariski sans aucune référence (ils étaient des grands classiques à l'époque). Ensuite, on traite le problème du *recollement* en exposant la démonstration de [Z4]. Pour la démonstration de *l'uniformisation*, nous nous contentons de faire les cas simples ou ceux pour lesquels nous pensons avoir amélioré les preuves de Zariski et de citer [Z2] pour les autres avec quelques mots d'explication.

Dans 2 (désingularisation par éclatements normalisés), on donne une preuve nouvelle du théorème final de [Z2], à savoir l'implication: uniformisation $\Rightarrow$ désingularisation par éclatements normalisés.

Cette preuve est dûe à l'auteur qui utilise (comme Zariski dans [Z2]) la théorie des idéaux complets [SZ2, appendice] mais avec des arguments nouveaux pour éviter de supposer que le corps de base est algébriquement clos, hypothèse résolument utilisée dans [Z2]. Bien sûr, on aurait pu se contenter de citer [L3, Prop. 3.1], où l'implication ci-dessus est prouvée dans le cadre le plus général possible, mais avec des méthodes cohomologiques.

Ce sont les professeurs Lê et Teissier qui m'ont demandé d'exposer et de rédiger ces démonstrations, je les en remercie car en expliquant ce travail de Zariski, il m'a semblé faire à la fois œuvre de mathématicien et d'historien, cela m'a permis d'assouvir deux de mes passions: celle pour la géométrie d'une part et celle pour l'histoire d'autre part.

Je dois également remercier Olivier Piltant qui a accepté de relire mes notes et dont les remarques judicieuses m'ont permis de simplifier plusieurs démonstrations. Et un dernier merci pour les deux referees qui m'ont envoyé, pour ma plus grande confusion, une liste fort longue de "coquilles" (misprints in English).

1. Uniformisation et recollement

Géométrie birationnelle. Dans cette section, nous nous proposons de reprendre les notions de points correspondants, points fondamentaux, etc., avec les définitions de Zariski et de montrer l'équivalence avec les définitions utilisées actuellement. Notre référence pour les définitions "actuelles" est le livre de Hartshorne [H].

Dans toute cette section, V et $\widehat{V}$ sont deux variétés projectives sur un corps de base k et birationnellement équivalentes, on note $\pi\ : V \cdots\to \widehat{V}$ l'application entre ces deux variétés [H], Z est leur variété de Riemann (que Zariski appelle surface de Riemann abstraite) [V §7]. Si π est défini sur V tout entier, on dira que *V domine $\widehat{V}$.*

Définition. Un point $P \in V$ et un point $\widehat{P} \in \widehat{V}$ sont en correspondance si il existe une valuation $\nu \in Z$ dont le centre sur V est P et sur $\widehat{V}$ est $\widehat{P}$.

Proposition 1.1. *Soit F un fermé de V, l'ensemble des points $\widehat{P}$ de $\widehat{V}$ en correspondance avec les points de F est un fermé de $\widehat{V}$.*

Preuve. Soit W une variété qui domine V et $\widehat{V}$. Soit $\widehat{F}$ le fermé de $\widehat{V}$ qui est la projection sur $\widehat{V}$ de l'image réciproque de F dans W, il est clair que $\widehat{F}$ est la projection sur $\widehat{V}$ de toutes les valuations de Z dont les centres sur V sont des points de F et que c'est donc l'ensemble des points de $\widehat{V}$ en correspondance avec les points de F. □

Proposition 1.2. *Reprenons les hypothèses de la proposition 1.1 et supposons de plus que V est normale, si P est fermé, alors, l'ensemble des points de $\widehat{V}$ en correspondance avec P est un point fermé ou bien contient une courbe.*

Preuve. C'est un corollaire du "théorème principal" de Zariski [H, 11.4, p. 280] qui nous assure que le fermé des points en correspondance avec P dans W est connexe et comme l'image d'un connexe est connexe, il en est de même pour les points de $\widehat{V}$ en correspondance avec P.

Malheureusement pour lui, en 1939, Zariski ne faisait que conjecturer le "théorème principal" et il a du faire une preuve *ad hoc* de la proposition 1.2, preuve qui nécessite l'hypothèse k infini.

Voici l'argument de Zariski. Supposons la proposition fausse. Alors, les points fermés $\widehat{P}_1, \dots, \widehat{P}_m$ en correspondance avec P sont en nombre fini $m \geq 2$. Comme k est infini, on peut trouver un ouvert affine U de $\widehat{V}$ contenant tous les $\widehat{P}_i$. Posons $\Delta = O_{\widehat{V},\widehat{P}_1} \cap \dots \cap O_{\widehat{V},\widehat{P}_m}$. Comme V est normale, $O_{V,P}$ est égal à l'intersection des anneaux de valuation le dominant, on a donc les inclusions

$$O_{\widehat{V}}(U) \subset \Delta \subset O_{V,P}.$$

Notons $\mathfrak{P}_i$, $1 \leq i \leq m$ la trace sur Δ de l'idéal maximal $\mathfrak{M}_i$ de $O_{\widehat{V},\widehat{P}_i}$, ces idéaux $\mathfrak{P}_i$ sont distincts deux à deux car, U contenant les $\widehat{P}_i$, $1 \leq i \leq m$, les traces sur $O_{\widehat{V}}(U)$ des $\mathfrak{M}_i$ sont les idéaux de définition des $\widehat{P}_i$ et sont distincts deux à deux. Soit $x \in \mathfrak{P}_1$ avec $x \notin \mathfrak{P}_2$, soit ν une valuation dont le centre sur $\widehat{V}$ est $\widehat{P}_1$, donc $x \in (M_\nu \cap \Delta) \subset M$ où M_ν et M sont les idéaux maximaux de R_ν et de $O_{V,P}$. Donc x est de valuation strictement positive pour toute valuation μ dominant P, on peut prendre μ dominant $\widehat{P}_2$, et donc $x \in \mathfrak{P}_2$, contradiction. □

Proposition 1.3. *Reprenons les hypothèses et notations de la proposition 1.1 (on ne suppose plus que V est normale). Si π est définie en P, alors, $\widehat{P} = \pi(P)$ est le seul point de $\widehat{V}$ en correspondance avec P et on a $O_{\widehat{V},\widehat{P}} \subset O_{V,P}$, où l'inclusion est donnée par l'isomorphisme naturel entre les corps de fonctions de V et $\widehat{V}$ qui sont aussi les corps de fractions des anneaux locaux $O_{\widehat{V},\widehat{P}}$ et $O_{V,P}$.*

Preuve. Reprenons les notations de la preuve de la proposition 1.1, désignons par $e: W \longrightarrow V$ et par $\widehat{e}: W \longrightarrow \widehat{V}$, dans un voisinage de $e^{-1}(P)$, $\pi \circ e$ est défini et coïncide avec $\widehat{e}$, on en déduit que $\pi(P) = \widehat{e}(e^{-1}(P))$, d'après la démonstration précédente, $\widehat{e}(e^{-1}(P))$ est l'ensemble des points de $\widehat{V}$ en correspondance avec P, d'où la première assertion.

Maintenant, puisque $\widehat{P} = \pi(P)$, on a un morphisme $O_{\widehat{V},\widehat{P}} \longrightarrow O_{V,P}$ qui se prolonge à leurs corps de fractions, c'est une injection. □

Proposition 1.4. *Soient P un point de V et $\widehat{P}$ un point de $\widehat{V}$, (P et $\widehat{P}$ pas forcément fermés), si on a $O_{\widehat{V},\widehat{P}} \subset O_{V,P}$, où l'inclusion est donnée par l'isomorphisme naturel entre les corps de fonctions de V et $\widehat{V}$, alors π est défini dans un voisinage de P et $\pi(P) = \widehat{P}$.*

Preuve. Soit $\omega = \mathrm{Spec}(k[\widehat{x}_1, \ldots, \widehat{x}_n])$ un ouvert affine de $\widehat{V}$ contenant $\widehat{P}$, notons $\widehat{\mathfrak{P}}$ l'idéal de définition de P. On désigne par π^* l'isomorphisme naturel entre les corps de fonctions $K(\widehat{V}) = k(\widehat{x}_1, \ldots, \widehat{x}_n)$ et $K(V)$. Par restriction, π^* nous donne un morphisme $k[\widehat{x}_1, \ldots, \widehat{x}_n] \longrightarrow k[\pi^*(\widehat{x}_1), \ldots, \pi^*(\widehat{x}_n)]$, dont un prolongement est l'inclusion $O_{\widehat{V},\widehat{P}} \subset O_{V,P}$, donc $O_{V,P}$ est un localisé de $k[\pi^*(\widehat{x}_1), \ldots, \pi^*(\widehat{x}_n)]$. Soit U un ouvert de V où $\pi^*(\widehat{x}_1), \ldots, \pi^*(\widehat{x}_n)$ sont définis, $P \in U$, on a les flèches:

$$k[\widehat{x}_1, \ldots, \widehat{x}_n] \longrightarrow k[\pi^*(\widehat{x}_1), \ldots, \pi^*(\widehat{x}_n)] \longrightarrow O_V(U) \longrightarrow O_{V,P},$$

$\pi(P) = \widehat{P}$ et π est défini sur U. □

Ces propositions 1.3 et 1.4 impliquent que la définition de *point fondamental* utilisée par Zariski est équivalente à celle utilisée de nos jours [H, p. 410]. Constatons le en lisant les deux définitions qui suivent.

Définition. Un point P de V est fondamental pour π si, pour tout point $\widehat{P} \in \widehat{V}$, π^* ne se restreint pas à une inclusion entre les anneaux locaux $O_{\widehat{V},\widehat{P}} \not\subset O_{V,P}$ (Zariski).

Le lieu fondamental de π est le complémentaire du plus grand ouvert de V où π est défini [H].

On appelle *joint* de $\widehat{V}$ et de V l'adhérence du graphe de π dans $V \times_k \widehat{V}$.

Proposition 1.5. *Désignons par p_1 et p_2 les projections du joint sur V d'une part et $\widehat{V}$ d'autre part. Soit U l'ouvert de définition de π. Alors, $p_1^{-1}(U)$ est isomorphe à U.*

Preuve. On a deux morphismes inverses l'un de l'autre:

$$p_1^{-1}(U) \longrightarrow U, \ (x, \pi(x)) \longmapsto x,$$

$$U \longrightarrow p_1^{-1}(U), \ x \longmapsto (x, \pi(x)).$$

Ces deux ouverts sont donc isomorphes en tant qu'espaces topologiques. Il suffit de se reporter à la proposition 3 ou plutôt à sa preuve pour constater que ces isomorphismes topologiques induisent des isomorphismes sur les anneaux locaux de x et $(x, \pi(x))$. □

Le problème du recollement en dimension 2. Pour désingulariser les variétés projectives, Zariski a proposé la stratégie suivante [V §8]: on résoud d'abord le *problème d'uniformisation.* Soit V une variété projective sur un corps algébriquement clos k, pour toute valuation ν de la surface de Riemann Z, trouver un modèle birationnel V_ν de V (i.e. une variété projective V_ν birationnellement équivalente à V) "résolvant ν" (i.e. le centre de ν sur V_ν est régulier). On dira aussi que V_ν uniformise ν.

Une fois ce problème résolu, les images réciproques des ouverts de régularité des différents V_ν recouvrent Z, par quasi-compacité, on peut extraire de ce recouvrement un recouvrement fini $V_{\nu(1)}, \dots, V_{\nu(m)}$ tel que toute valuation de Z a un centre régulier dans au moins un des $V_{\nu(i)}, 1 \leq i \leq m$. Il ne reste alors plus qu'à résoudre le *problème du recollement.* Construire à l'aide de $V_{\nu(1)}, \dots, V_{\nu(m)}$ une variété $\widehat{V}$ birationnellement équivalente à V dont tous les points sont réguliers.

Avant de nous atteler au recollement, prouvons un résultat fondamental [Z2, th. 10 p. 681] que nous allons prouver ici *sans supposer k algébriquement clos.* Dorénavant, une surface est une variété projective irréductible de dimension 2 sur un corps k. La preuve que nous rédigeons est quasiment celle de Zariski [Z2].

Théorème 1.6. (Zariski) *Soit V une surface projective sur k (pas forcément normale), on définit par récurrence $V_i \longleftarrow V_{i+1}$, $0 \leq i$, $V_0 = V$ la flèche est l'éclatement suivi d'une normalisation d'un point fermé $P_i \in V_i$, P_i se projetant sur P_{i-1}, $1 \leq i$.*

Alors,

$$\Omega := \varinjlim(O_{V_i,P_i})$$

est un anneau de valuation et sa valuation est un point fermé de Z.

On remarque qu'une valuation est un point fermé de Z (pour sa topologie de Zariski [V, §7]) si tous ses centres dans toutes les surfaces projectives birationnellement équivalentes à V sont des points fermés, c'est une conséquence de [V, Th. 7.2].

Preuve. Rappelons qu'une valuation est *divisorielle* si son degré de transcendance est 1 et si l'inégalité d'Abhyankar est pour elle une égalité (et donc son groupe est $\mathbb{Z}$) [V, 9.3].

Lemme 1.7. *Soit R_ν, $\nu \in Z$, un anneau de valuation* divisorielle *contenant Ω, alors, pour i assez grand, le centre de ν dans V_i est de dimension* 1.

Preuve. En fait, il suffit de prouver que, pour un i, le centre de ν sur V_i est de dimension 1, car la dimension des centres est croissante avec i. Supposons le contraire, alors, le centre de ν dans V_i est un point fermé, puisque $O_{V_i,P_i} \subset R_\nu$, ce centre est P_i. Donc, pour tout i, $\frac{O_{V_i,P_i}}{\mathfrak{M}_i} = k_i$ est une extension algébrique de k. Mais, l'extension $k \subset \frac{R_\nu}{\mathfrak{M}_\nu}$ est transcendante, il existe donc un élément $x \in R_\nu$ de valuation nulle et dont l'image dans $\frac{R_\nu}{\mathfrak{M}_\nu}$ est transcendante sur k. Pour tout $i, 0 \leq i$, on a une égalité $x = \frac{p(i)}{q(i)}$ avec $p(i) \in O_{V_i,P_i}$, $q(i) \in O_{V_i,P_i}$. Comme l'image de

x dans le corps résiduel de R_ν est transcendante sur k_i, ces éléments $p(i)$ et $q(i)$ ne sont pas inversibles dans O_{V_i,P_i}, ils sont donc de (même) valuation strictement positive, prenons les de valuation minimale, soit $\omega_i \in \mathbb{N}$ cette valuation. Je dis que

$$(1)\ \omega_{i+1} < \omega_i,\ 0 \leq i.$$

Ce qui est évidemment impossible pour une suite d'entiers positifs et ce qui nous donne la contradiction cherchée.

Prouvons (1). Comme $p(i) \in \mathfrak{M}_i$, dans $O_{V_{i+1},P_{i+1}}$, on a des factorisation $p(i) = tp'(i)$ et $q(i) = tq'(i)$ où t est un générateur du diviseur exceptionnel. Comme P_{i+1} est sur le diviseur exceptionnel, t n'est pas inversible dans $O_{V_{i+1},P_{i+1}}$, bref

$$\omega_{i+1} = \nu(p(i+1)) \leq \nu(p(i)) - \nu(t) < \nu(p(i)) = \omega_i.$$

Ce qui prouve (1) et le lemme. □

Fin de la preuve du théorème. L'anneau Ω est intègralement clos, donc, on a $\Omega = \bigcap R_\nu$, $\nu \in Z$, $\Omega \subset R_\nu$. Je dis que

$$(2)\ \Omega = \bigcap R_\nu,\ \nu \in Z,\ \nu \text{ fermée},\ \Omega \subset R_\nu.$$

En effet, les valuations non fermées de Z sont les valuations de degré de transcendance 1 ou 2. En degré 2, on a la valuation impropre d'anneau $K(V)$ dont les centres sont les points génériques des modèles. En degré 1, le groupe de la valuation n'étant pas trivial, il est de rang au moins 1, l'inégalité d'Abhyankar est alors une égalité et donc, le groupe de la valuation est $\mathbb{Z}$, la valuation est divisorielle, c'est à dire que son groupe est $\mathbb{Z}$ et que l'inégalité d'Abhyankar est en fait une égalité. D'après le lemme, si ν est divisorielle, pour i assez grand, le centre Y_i de ν dans V_i est de dimension 1, mais, comme les V_i sont normalisées, O_{V_i,Y_i} est un anneau de valuation discrète dominé par R_ν , c'est donc R_ν. Il est clair que Y_{i+1} est le transformé strict de Y_i et donc, pour i assez grand, Y_i est une courbe régulière et tous les Y_i se projettent isomorphiquement les unes sur les autres, les P_i se correspondant dans ces isomorphismes. D'autre part, les anneaux O_{Y_i,P_i} sont tous isomorphes pour i assez grand, nous les confondons tous et notons $\mathbf{R} = O_{Y_i,P_i}$, avec i assez grand.

On définit alors une valuation μ de $K(V)$ ainsi:

$$R_\mu = \{a \in R_\nu; \overline{a} \in \mathbf{R}\},$$

où $\overline{a}$ désigne la classe de a dans $\frac{R_\nu}{\mathfrak{M}_\nu} = K(Y_i) = Frac(\mathbf{R})$.

Il est clair que R_μ est un anneau, montrons que c'est un anneau de valuation de $K(V)$. Soit $x \in K(V)$ avec $x^{-1} \notin R_\mu$; si $x^{-1} \in R_\nu$, alors, $x^{-1} \notin \mathfrak{M}_\nu$ parce que $\mathfrak{M}_\nu \subset R_\mu$, donc $x \in R_\nu$ et donc, $\overline{x} \in \mathbf{R}$ ou $\overline{x}^{-1} \in \mathbf{R}$, x ou x^{-1} est dans R_μ; si $x^{-1} \notin R_\nu$, alors, $x \in \mathfrak{M}_\nu$, et donc, $x \in R_\mu$. Maintenant, il est clair que μ est composée de ν et de la valuation d'anneau $\mathbf{R}$ et donc, le centre de μ sur V_i est P_i, pour tout i. Donc

$$O_{V_i,P_i} \subset R_\mu \subset R_\nu,\ 0 \leq i,$$

ce qui prouve (2).

Supposons que deux valuations fermées distinctes dominent Ω, soient ν_1 et ν_2, on peut trouver un modèle W où les centres de ν_1 et ν_2 sont deux points (fermés) distincts Q_1 et Q_2. Alors, d'après la proposition 1.2 de la section géométrie birationnelle, le fermé F_i des points de W en correspondance avec P_i contient une courbe, d'autre part, $F_i \supset F_{i+1}$, $0 \leq i$. Il existe donc une courbe irréductible C de W qui est dans l'intersection de tous les F_i, le point générique ξ de cette courbe est en correspondance avec tous les P_i, on peut prendre W lisse en ξ, auquel cas, la valuation divisorielle ν attachée au point ξ a pour centre P_i sur V_i, $0 \leq i$, ce qui contredit le lemme 1.7. Ainsi, un seul anneau R_ν avec ν fermée domine Ω, on a $\Omega = R_\nu$. □

Remarque 1.8. L'argument de la preuve de (2) nous permet de prouver le résultat suivant: si $O_{V,P}$ est normal, alors, $O_{V,P}$ est intersection des anneaux de valuations fermées qui le dominent.

Rappelons [V, §1] qu'un anneau de valuation R_ν *domine* $O_{V,P}$ si $O_{V,P} \subset R_\nu$ et si de plus $\mathfrak{M} = \mathfrak{M}_\nu \cap O_{V,P}$, où $\mathfrak{M}$ est l'idéal maximal de $O_{V,P}$ et $\mathfrak{M}_\nu$ celui de R_ν. Bien sûr, $O_{V,P}$ étant normal, $O_{V,P}$ est intersection des anneaux de valuation qui le contiennent, et on a prouvé que, pour toute valuation ν dont l'anneau R_ν contient $O_{V,P}$, il existe une valuation fermée μ telle que $O_{V,P} \subset R_\mu \subset R_\nu$, d'où le résultat.

Exercice. (1) Généraliser le résultat de la remarque au cas où V est une variété projective de dimension quelconque (utiliser [V 7.2, 7.11]).
(2) Montrer par un exemple que le théorème 1.6 est faux si V est une variété projective de dimension ≥ 3.

Solution du problème du recollement en dimension 2. Il suffit de résoudre le problème suivant.

(Pb.) *Étant donnés deux modèles projectifs normaux F et F' de la surface V, construire un modèle $\widehat{F}$ telle que toute valuation ayant un centre régulier sur F ou F' a son centre régulier sur $\widehat{F}$* [V, §8].

Lemme 1.9. *Si F est normale, il n'y a dans F qu'un nombre fini de points fondamentaux pour $\pi : F \cdots \longrightarrow F'$.*

De plus, on pose $F_0 = F$, pour $0 \leq i$, soit $F_i \longleftarrow F_{i+1}$ le morphisme obtenu en éclatant les points fondamentaux de F_i pour sa correspondance avec F' et en normalisant l'éclaté. Pour i assez grand, F_i domine F'.

Preuve. Soit P un point fondamental de F. Éclatons P, normalisons l'éclaté, prenons dans cet éclaté-normalisé que nous notons abusivement F_1 un point fermé P_1 au-dessus de P, recommençons. D'après le théorème 1.6, $\Omega := \varinjlim(O_{F_i,P_i})$ est un anneau de valuation, la valuation associée est fermée, soit P' le centre de cette valuation sur F', on a $O_{F',P'} \subset \Omega$, tous ces anneaux étant des localisés de k-algèbres de type fini, pour i assez grand, $O_{F',P'} \subset O_{F_i,P_i}$, donc P_i n'est pas fondamental pour F_i. Si la première assertion du lemme est vraie, cela en termine la preuve. Mais, cela prouve aussi la première assertion du lemme, en effet, si le lieu fondamental de F (qui est un fermé, cf. définition 2 bis ci-dessus) contenait une infinité de points, il serait de dimension 1, on pourrait prendre P sur une composante C de dimension 1 de ce lieu fondamental de F, on prendrait les P_i sur le transformé strict de C, on obtiendrait une contradiction car le transformé strict de C est évidemment dans le lieu fondamental de F_i (c'est ici qu'intervient l'hypothèse que F est normale, F et F_i sont isomorphes au-dessus du point générique de C). □

On se ramène donc à résoudre (Pb.) dans le cas où F domine F'. Nous ne sommes pas au bout de nos peines car il se peut qu'il y ait des points réguliers de F' au-dessus desquels se trouvent des points singuliers de F. Ces points sont fondamentaux pour la correspondance de $F' \cdots \to F$, sinon, localement, les correspondances $F \longrightarrow F'$ et $F' \cdots \to F$ seraient des isomorphismes. Notons les $P_1, \ldots, P_m$.

On recommence notre processus d'éclatements suivis de normalisation en prenant pour centres des éclatements les points fermés de F' et des transformés de F' fondamentaux pour la correspondance avec F et *se projetant sur les* P_i. J'obtiens une nouvelle surface $\overline{F}'$ qui domine F' et dont le lieu fondamental pour sa correspondance avec F ne rencontre pas les fibres des P_i. Soit $\widehat{F}$ le joint de F et de $\overline{F}'$. En dehors de l'image réciproque des P_i, $\widehat{F}$ est isomorphe à F, dans un voisinage de ces points, $\widehat{F}$ est régulière et est isomorphe à $\overline{F}'$. Il est clair que $\widehat{F}$ résoud (Pb.).

Solution du problème d'uniformisation. Nous allons répertorier les valuations $\nu \in Z$ où Z est la surface de Riemann d'une surface singulière S par le triplet (rang rat.(ν), rang(ν), deg.tr.$_k(\nu)$), de plus, nous séparerons les deux cas où rang rat.$(\nu) = 1 =$ rang(ν) suivant que le groupe de ν est isomorphe à $\mathbb{Z}$ ou pas. On rappelle les définitions et notations de [V]: deg.tr.$_k(\nu)$ le degré de transcendance de l'extension $k \subset R_\nu/\mathcal{M}_\nu$ et rang rat.(ν) désigne le *rang rationnel* de ν, c'est à dire la dimension de $\Gamma_\nu \otimes \mathbb{Q}$, où Γ_ν est le groupe des valeurs (ou des ordres) de ν [V, §3] et rang(ν) est la dimension de l'anneau R_ν [V, 3.4]. Bien sûr, R_ν est l'anneau de valuation associée à ν. Nous nous contentons de rédiger les trois premiers cas, et de donner quelques indications pour résoudre les quatrième et cinquième cas, et de citer le dernier cas. En effet, l'étude de ces cas dans [Z2] est limpide. Il y a maintenant des preuves de l'uniformisation en toute caractéristique, même en cas d'inégale caractéristique, mais elles sont fort compliquées [A2], [CGO, appendice] où l'on prouve un théorème de désingularisation plus fort: à savoir, pour toute surface définie sur un corps algébriquement clos, il existe une suite d'éclatements à centres réguliers qui permet de désingulariser cette surface, on trouvera dans [C2] une preuve de ce résultat dans le cas d'une surface excellente. On peut aussi consulter [Ha] où se trouve une nouvelle preuve de la désingularisation des surfaces définies sur un corps algébriquement clos.

Rappelons l'inégalité d'Abhyankar [V, 9.3]:

$$\text{rang rat.}(\nu) \; + \; \text{deg.tr.}_k(\nu)) \; \leq \dim A,$$

où A est un anneau local noethérien de corps des fractions $K(= K(S))$, de corps résiduel k et ν une valuation de K dont l'anneau domine A. Enfin, rang$(\nu) \leq$ rang rat.(ν) [V, 3.8].

À titre d'exercice, on laisse au lecteur le soin de vérifier que tous les cas que nous allons citer font une liste exhaustive.

Premier cas, deg.tr.$_k(\nu) = 2$. Par l'inégalité d'Abhyankar, on a rang rat.$(\nu) = 0$, ν est la valuation stupide dont l'anneau est $K(S)$, son centre sur S est le point générique de S qui est régulier.

Deuxième cas, deg.tr.$_k(\nu) = 1$. Dans ce cas, l'anneau R_ν n'est pas $K(S)$. Le groupe de ν n'est pas trivial, donc, rang rat.$(\nu) \geq 1$, par l'inégalité d'Abhyankar, on a rang rat.$(\nu) = 1$, la valuation ν est divisorielle. D'après [V, 6.6], il existe une surface S' où le centre de ν est de dimension 1, sur la normalisée $\widehat{S}'$ de S', le centre de ν est une courbe irréductible dont le point générique n'est pas singulier dans $\widehat{S}'$, donc $\widehat{S}'$ uniformise ν.

Passons aux cas difficiles où deg.tr.$_k(\nu) = 0$. Pour les cas 3, 4 et 5 qui suivent, Zariski se ramène au cas d'une singularité d'hypersurface plongée dans un espace lisse de dimension 3, en fait, plongée localement dans $\mathbb{A}_3$.

Troisième cas, rang rat.$(\nu) = 2 =$ rang(ν), deg.tr.$_k(\nu) = 0$ le groupe de ν est alors $\mathbb{Z} \oplus \mathbb{Z}$ muni de l'ordre lexicographique. (Zariski appelle une telle valuation *valuation discrète*, ce qui n'est plus la terminologie actuelle.) Bien sûr [V Th.7.2], il s'agit d'une valuation fermée qui est adhérente à la valuation divisorielle μ ainsi définie:

$$\text{pour tout } x \in K(S),\ \mu(x) = \text{première coordonnée de } \nu(x).$$

Zariski commence par construire un modèle projectif F de S avec:

1. F est dans un voisinage du centre de ν une hypersurface de $\mathbb{P}^3(k)$ d'un espace régulier.
2. Le centre C de μ sur F est une courbe irréductible qui n'est pas dans le lieu singulier de F.

Pour cela, on peut supposer qu'on a 2 pour S: on prend un modèle où le centre C de μ est de dimension 1, puis on normalise, soit F' ce modèle. Si F' est plongée dans $\mathbb{P}^3(k)$, il n'y a rien à faire. Sinon, on choisit un point fermé $M \in C \subset F' \subset \mathbb{P}^n(k)$, $4 \leq n$, et M régulier sur C et sur F'. (Donc M n'est en général pas le centre de ν.) L'ensemble des droites de passant par M est une variété de dimension n-1 dans la grassmanienne des droites de $\mathbb{P}^n(k)$, tandis que l'ensemble des tangentes et des multi-sécantes de F' contenant M est une variété de dimension 2, (l'ensemble des droites passant par M est un $\mathbb{P}^{n-1}(k)$ et l'ensemble des tangentes et des multi-sécantes à F' est isomorphe à la projection de F' depuis M sur $\mathbb{P}^{n-1}(k)$). On peut donc trouver une droite passant par M qui n'est ni tangente, ni multi-sécante à F'. On prend un point P sur cette droite et on projette depuis P sur un $\mathbb{P}^{n-1}(k)$, la projection restreinte à F' est un isomorphisme dans un voisinage de M, bref, par projections successives, une récurrence décroissante sur n finit par donner 1 sans perdre 2, F est la projection de F' dans $\mathbb{P}^3(k)$. (Cette construction est dûe à l'auteur, dans [Z2], Zariski met en référence [Z1, §16, théorème 11] où il construit F par des méthodes algébriques.)

Soit O le centre de ν sur F, si O est régulier, il n'y a rien à faire. Sinon, dans un voisinage de O, F est donné par une équation $f(x,y,z)=0$ et O est le point de coordonnées $x=y=z=0$.

La condition 2 implique que les trois dérivées partielles f'_x, f'_y, f'_z ne sont pas toutes dans l'idéal du centre de μ, on en déduit que

$$\min(\nu(f'_x),\nu(f'_y),\nu(f'_z)) = (0,n),$$

avec $n\in\mathbb{N}$. Maintenant, nous pouvons supposer que

$$0 < \nu(x) \leq \nu(y) \leq \nu(z).$$

Effectuons l'éclatement $F \longleftarrow F_1$ centré en O, le centre O_1 de ν sur F_1 est le point de paramètres

$$x_1 = x,\ y_1 = \frac{y-cx}{x},\ z_1 = \frac{z-dx}{x}$$

où c et d sont des éléments de k convenables (On utilise ici l'hypothèse k algébriquement clos). Localement en O_1, F_1 est donnée par une équation

$$f_1(x_1,y_1,z_1)=0,$$

où f_1 est donnée par

$$f(x,y,z) = x_1^r f_1(x_1,y_1,z_1),$$

r étant la multiplicité de f en O. Un calcul donne

$$f'_x = x_1^{r-1}[x_1 f'_{1,x_1} - (y_1+c)f'_{1,y_1} - (z_1+d)f'_{1,z_1}],$$
$$f'_y = x_1^{r-1} f'_{1,y_1},$$
$$f'_z = x_1^{r-1} f'_{1,z_1}.$$

D'où

$$\min(\nu(f'_x),\nu(f'_y),\nu(f'_z)) \geq (r-1)\nu(x_1) + \min(\nu(f'_{1,x_1}),\nu(f'_{1,y_1}),\nu(f'_{1,z_1})).$$

Bien sûr, O étant singulier, on a $2\leq r$, $0<\nu(x_1)$, puisque x_1 est un paramètre de O_1 et donc,

$$\min(\nu(f'_{1,x_1}),\nu(f'_{1,y_1}),\nu(f'_{1,z_1})) = (0,n_1) < (0,n),$$
$$n_1 < n.$$

Si O_1 est régulier, il n'y a rien à faire, sinon, nous avons un modèle F_1 qui est défini localement par une équation et sur lequel le centre C_1 de μ est le transformé strict de C le centre de μ sur F et donc est une courbe qui n'est pas dans le lieu singulier de F_1. On a donc les conditions 1 et 2 pour F_1, mais on a remplacé n par n_1 qui est strictement plus petit, on a le résultat par récurrence sur n.

Les démonstrations des cas 4 et 5 qui suivent se font en contrôlant les monômes du nuage de points de cette équation locale et de ses transformées strictes. Nous allons en donner des résumés très courts (les calculs sont longs, mais absolument limpides dans [Z2]) et attirer l'attention du lecteur sur un point délicat qui annonce le polyèdre caractéristique d'Hironaka (voir [CGO, appendice] ou [Ha, Figure J]). En effet, par des transformations de Cremona, Zariski essaie de faire baisser la multiplicité d'un monôme d'une équation locale de la singularité pour un modèle ad hoc. Il se peut que ses transformations ne fassent pas baisser cette multiplicité il est alors obligé de faire des translations sur une des variables pour en maximiliser la valeur pour ν, comme Hironaka le fait pour minimaliser un polygone (voir la remarque 1.10 précédant le sixième cas).

Quatrième cas, [Z2 p. 652], rang rat.$(\nu) = 2, \text{rang}(\nu) = 1, \text{deg.tr.}_k(\nu) = 0$, le groupe de ν est alors le groupe ordonné $\mathbb{Z} + \mathbb{Z}\tau \subset \mathbb{R}$ où τ est un nombre irrationnel.

Cinquième cas, [Z2 p. 656], rang rat.$(\nu) = 1 = \text{rang}(\nu), \text{deg.tr.}_k(\nu) = 0$, et le groupe de ν est un sous-groupe ordonné non discret de $\mathbb{Q}$.

Dans ces deux cas, on construit une surface affine F de k^3 d'équation

$$f(X,Y,Z) = 0,$$

avec K pour corps de fonctions. On obtient les conditions suivantes:

1. Cette surface passe par l'origine qui est le centre de ν sur la complétion projective de cette surface dans $\mathbb{P}^3$.
2. Quitte à changer de variables (en remplaçant X par $x+\lambda z^n$ et y par $y+\mu z^m$ avec m et n grands), on peut supposer que $f(0,0,Z)$ n'est pas identiquement nul.
3. Notons x, y, z les images de X, Y, Z dans $k[X,Y,Z]/(F)$, Zariski arrive en plus à obtenir que,
   3.1 dans le quatrième cas, $\nu(x) = 1, \ \nu(y) = \tau$,
   3.2 dans le cinquième cas, x et y sont des éléments algébriquement indépendants de l'anneau local de la singularité.

Ainsi, Zariski construit un modèle projectif qui est une hypersurface plongée dont une équation dans un voisinage du centre de la valuation est un polynôme

$$f(X,Y,Z) = \sum a_{m,n,p} X^m Y^n Z^p \ \in \ k[X,Y,Z],$$

avec un $a_{0,0,p} \neq 0, p \in \mathbb{N}$, et le point $X = Y = Z = 0$ est le centre de la valuation ν. Comme le devinent les "aficionados" de la désingularisation, on fait une récurrence décroissante sur

$$r = \inf\{p; a_{0,0,p} \neq 0\} \ \in \mathbb{N} \setminus \{0\}$$

qui est l'ordre de multiplicité de z comme racine du polynôme $f(0,0,Z)$. (Voir [CGO], [Ha]).

Dans le développement de $f(X,Y,Z)$, substituons X par x, Y par y et Z par z, ce qui donne:

$$0 = a_\alpha(x,y)z^\alpha + a_\beta(x,y)z^\beta + \cdots + a_\delta(x,y)z^\delta + \sum{}' a_p(x,y)z^p \in k[x,y][z],$$

$\alpha < \beta < \cdots < \delta$, où $\nu(a_\alpha(x,y)z^\alpha) = \nu(a_\beta(x,y)z^\beta) = \cdots = \nu(a_\delta(x,y)z^{p_\delta})$ est la valuation minimale des termes non nuls du développement de $f(x,y,z)$ et où la somme $\sum'$ est faite sur les termes de valuation strictement plus grande.

Exercice. Montrer que, dans le quatrième cas, les termes $a_\alpha(x,y),\ldots,a_\delta(x,y)$, sont en fait des monômes: $a_\alpha(x,y) = a_\alpha x^{m_\alpha}y^{n_\alpha}$, $a_\alpha \in k$, etc.

Dans ces quatrième et cinquième cas, on fait une transformation de Cremona de k^3 telle que, la transformée stricte de notre surface soit birationnellement équivalente à F et, on a les conditions 1, 2, 3 ci-dessus et de plus, la somme des termes de valeurs minimale pour ν de la nouvelle équation soit: $a_{\alpha'}(x_1,y_1)z_1^{\alpha'} + a_{\beta'}(x_1,y_1)z_1^{\beta'} + \cdots + a_{\delta'}z_1^{\delta'}$, $a_{\delta'} \in k, \delta' \leq \delta$. Si $\delta' < r$, on a gagné. Sinon, Zariski remarque qu'alors

$$\nu(z) = \nu(a_{r-1}(x,y)). \tag{1}$$

On recommence tout, depuis le début, (Cremona pour (x,y,z), etc.), en remplaçant z par $z_{(1)} = z - c_1a_{r-1}(x,y)$, avec $c_1 \in k$ tel que $\nu(z) < \nu(z_{(1)})$ (c_1 existe car k est algébriquement clos), si r ne baisse toujours pas, on construit par récurrence une suite $z_{(i)}$ et des équations $f_i(x,y,z_{(i)}) = 0$ de notre surface, obtenues par le simple changement de variable, avec

$$\begin{aligned} \mathrm{ord}_{z_{(i)}}(f_i(0,0,z_{(i)}) &= r, \\ \nu(z_{(i)}) &= \nu(b_i(x,y)),\ b_i(x,y) \in k[x,y], \\ \nu(z) &< \nu(z_{(1)}) < \cdots < \nu(z_{(i)}) < \cdots, \\ z_{(i+1)} &= z_{(i)} - c_ib_i(x,y). \end{aligned}$$

On a, avec des notations évidentes: $(\frac{\partial f}{\partial Z}) = (\frac{\partial f}{\partial Z_{(i)}}), 0 \leq i$, donc, on a, si la caractéristique est 0:

$$f(X,Y,Z) \not| \frac{\partial f}{\partial Z}(X,Y,Z),\ \infty > \nu(\frac{\partial f}{\partial Z}(x,y,z)) \geq \nu(b_i),$$

ce qui implique que la suite des $\nu(b_i)$ est bornée, mais c'est impossible, en effet, une suite strictement croissante de valuations de polynômes doit tendre vers $+\infty$.

Donnons la preuve de ce résultat. Supposons que la suite des $\nu(b_i)$ est bornée par un réel ρ, soient R l'anneau local de F au centre de ν et $\mathfrak{M}$ l'idéal maximal de R, on peut trouver $n \in \mathbb{N}$ tel que $n\nu(\mathfrak{M}) \geq \rho$, on pose $P_i = \{x \in R | \nu(x) \geq \nu(b_i)\}$, il est clair que, pour tout $i \in N$, on a $P_i \supset P_{i+1} \supset \mathfrak{M}^n$ et donc

$$lg(\frac{R}{P_i}) < lg(\frac{R}{P_{i+1}}) \leq lg(\frac{R}{\mathfrak{M}^n}),\ 0 \leq i,$$

où lg désigne la longueur. Ce qui fait une contradiction.

Remarque 1.10. Cette technique annonce déjà la minimalisation du polygone de Newton de f utilisée par la suite: ici, Zariski fait des translations sur z pour obtenir un z_i de valeur la plus grande possible, dans [CGO], Hironaka fait des translations sur z pour obtenir un polyèdre minimal, voir aussi l'invariant défini par Hauser dans [Ha]. Enfin, l'argument de "tuer" a_{r-1} se retrouve constamment dans les travaux d'Abhyankar.

Sixième cas, rang rat.$(\nu) = 1, \text{deg.tr.}_k(\nu) = 0$ et le groupe de ν est $\mathbb{Z}$. On laisse le lecteur se convaincre que l'argument est le même que dans le troisième cas, en plus simple, car ici le groupe est $\mathbb{Z}$ et donc, en se reportant à la fin du troisième cas, on a

$$\min(\nu(f'_{1,x_1}), \nu(f'_{1,y_1}), \nu(f'_{1,z_1})) \;=\; n_1 \;<\; n,$$

où n est donné par:

$$\min(\nu(f'_x), \nu(f'_y), \nu(f'_z)) = n,$$

on a le résultat par récurrence décroissante sur n. On peut lire [Z2, p. 647] pour avoir une autre preuve.

2. Désingularisation par éclatements normalisés

Nous allons donner ici une preuve nouvelle du résultat bien connu qu'on peut résoudre les singularités d'une surface projective sur un corps quelconque par une suite d'éclatements de points fermés suivis d'une normalisation. Rappelons que la démonstration du théorème 1.6 (cf. 1.2) n'utilise aucune hypothèse sur le corps de base k. Plus précisément, nous allons montrer le résultat suivant:

Théorème de Zariski-Lipman. *Soit V une surface normale projective sur un corps k quelconque, si on sait résoudre le problème d'uniformisation pour toute valuation fermée de V, alors, l'algorithme suivant est fini.*

Posons $V_0 = V$, on définit V_i par récurrence: si V_i est régulière, on s'arrête, sinon, on éclate le lieu singulier de V_i et on normalise cet éclaté. Alors V_{i+1} est la surface obtenue.

Avant de faire une preuve de ce théorème, effectuons quelques simplifications.

Remarque 2.1. Il suffit de montrer que, pour toute valuation ν fermée dans la surface de Riemann de V, il existe un i tel que le centre de ν dans V_i est régulier.

En effet, supposons le théorème de Zariski-Lipman faux. Alors, l'algorithme des éclatements normalisés du lieu singulier serait infini, on pourrait donc trouver une suite infinie de points fermés non réguliers P_i se projetant les uns sur les autres tels que les O_{V_i,P_i} vérifieraient les hypothèses du théorème 1.6. Donc, la limite inductive des O_{V_i,P_i} est l'anneau d'une valuation fermée ν (cf. th. 1.6), cela contredit le résultat annoncé dans la remarque.

Théorème 2.2. *On prend les hypothèses du théorème de Zariski-Lipman, on désigne par P_i le centre de ν dans V_i, où ν est une valuation fermée dans la surface de Riemann de V. Il existe un i tel que O_{V_i,P_i} domine un anneau local régulier R et est dominé par un anneau local régulier S.*

$$R \subset O_{V_i,P_i} \subset S, \; Frac(R) = Frac(O_{V_i,P_i}) = Frac(S).$$

Preuve. En effet, soit il existe i_0 avec $O_{V_{i_0},P_{i_0}}$ régulier et on prend $R = O_{V_{i_0},P_{i_0}} = S$.

Sinon, par le théorème 1.6, $R_\nu = \varinjlim O_{V_i,P_i}$. Par l'uniformisation, il existe un modèle V' de V où le centre P' de ν est régulier. L'anneau $R = O_{V',P'}$ est essentiellement de type fini sur k et inclus dans la limite inductive des O_{V_i,P_i}, donc, pour un i_0, $R \subset O_{V_{i_0},P_{i_0}}$.

Maintenant, j'applique l'algorithme du théorème 1.6 à V' et aux points P'_j centres de ν, ces P'_j sont réguliers, $O_{V_{i_0},P_{i_0}}$ est essentiellement de type fini sur k, donc, pour j assez grand, $S = O_{V'_j,P'_j}$ domine $O_{V_{i_0},P_{i_0}}$. □

Théorème de principalisation. *Avec les notations précédentes, désignons par $\mathfrak{M}$ l'idéal maximal de $O_{V_{i_0},P_{i_0}}$, alors, $\mathfrak{M}S$ est un idéal inversible de S.*

Remarque 2.3. Le théorème de principalisation implique le résultat annoncé dans la remarque 2.1 et donc implique le théorème de Zariski-Lipman.

En effet, soit $O_{V_{i_0},P_{i_0}}$ est régulier et il n'y a rien à prouver, soit il n'est pas régulier, mais alors, $\mathfrak{M}S$ est un idéal inversible de $S = O_{V'_j,P'_j}$, l'éclaté de V_{i_0} en P_{i_0} est dominé par S dans un voisinage du centre de ν, mais, comme V'_j est normale, la normalisation V_{i_0+1} de cet éclaté est aussi dominée par S au voisinage de P_{i_0+1} centre de ν. Donc $O_{V_{i_0+1},P_{i_0+1}}$ satisfait aux hypothèses du théorème de principalisation, bref, si le résultat annoncé dans la remarque 2.1 est faux, pour tout entier n,

$$R \subset O_{V_{i_0+n},P_{i_0+n}} \subset S.$$

Mais, une fois de plus, cela contredit le théorème 1.6 car $S = O_{V'_j,P'_j}$ est essentiellement de type fini sur k.

Remarque 2.4. Un peu d'histoire. Le théorème de principalisation ci-dessus se trouve dans [L3] avec les hypothèses les plus minimales qui soient (V surface excellente), encore une fois, nous n'avons pas jugé utile de recopier ici la preuve, très courte, qui utilise la cohomologie et la suite spectrale de Leray [L3, 3.1]. La démonstration de [Z2] utilise l'hypothèse k algébriquement clos, le lecteur attentif s'apercevra qu'elle n'utilise aucune hypothèse sur la caractéristique de k. La preuve ci-dessous est dûe à l'auteur très fortement inspiré par Zariski. Elle ne suppose pas que k est algébriquement clos.

Hypothèses et notations. On note très abusivement $V = V_{i_0}$ et $P = P_{i_0}$. Nous avons donc trois anneaux locaux emboités, chacun dominant le précédent:

$$R = O_{V',P'} \subset O_{V,P} \subset S = O_{W,Q}.$$

On passe de R à S par une suite d'éclatements normalisés de points fermés suivis de localisation. D'autre part, quitte à remplacer V par son joint avec V' puis W par son joint avec V, nous pouvons supposer que les morphismes $V \to V'$ et $W \to V$ sont définis partout. Enfin, nous désignons par Z la surface de Riemann-Zariski de V, V', W. L'anneau $O_{V,P}$ est le localisé d'une R-algèbre de type fini. On a donc, avec des notations évidentes:

$$O_{V,P} = R[\frac{\Phi_1}{\Phi_0}, \dots, \frac{\Phi_n}{\Phi_0}]_{\mathfrak{M} \cap R[\dots]},$$

où $\Phi_i \in R$, $0 \leq i \leq n$, nous supposons qu'on peut extraire de $\frac{\Phi_1}{\Phi_0}, \dots, \frac{\Phi_n}{\Phi_0}$ un système de générateurs de l'idéal $\mathfrak{M} \subset O_{V,P}$ définissant P.

Nous allons supposer que *l'éclatement normalisé de V' en P' localisé en le centre de ν n'est pas dominé par P.*

Désignons par V_1 l'éclaté de V' le long de P'. La condition ci-dessus implique que *le fermé des points de V_1 en correspondance avec P est le diviseur exceptionnel E de V_1*. En effet, d'après le théorème principal de Zariski, ce fermé est connexe, s'il se réduit à un point P_1, on a $O_{V_1,P_1} \subset \cap_{\mu \in \mathbf{V}} R_\mu = O_{V,P}$, où $\mathbf{V}$ est l'ensemble des valuations de Z qui dominent P. Donc P dominerait P_1, ce qui contredirait notre hypothèse. Bien sûr, ce fermé se projette sur P': c'est le diviseur exceptionnel de V_1. Comme W domine V, le transformé strict du point générique de E dans W domine P, donc *le point générique de E domine P.*

Remarque 2.5. Si l'on suppose $\frac{\Phi_i}{\Phi_0}$ irréductible pour tout i, $1 \leq i \leq n$, les idéaux

$$J = (\Phi_1, \dots, \Phi_n) \subset I = (\Phi_0, \Phi_1, \dots, \Phi_n) \subset R,$$

sont M'-primaires, où M' désigne l'idéal maximal de R.

Pour I, c'est évident: R est factoriel, comme R est régulier de dimension 2, un idéal premier minimal appartenant à I serait engendré par un élément irréductible qui diviserait tous les Φ_i, $0 \leq i \leq n$, mais, les Φ_i sont premiers entre eux, il y a une contradiction. Supposons que J ne soit pas M'-primaire. Alors il existerait un irréductible $a \in R$ qui diviserait les Φ_i, $1 \leq i \leq n$. Posons $\Phi_i = a\phi_i$, $1 \leq i \leq n$. Tous les éléments de l'idéal I' de $O_{V,P}$ engendré par les $\frac{\Phi_1}{\Phi_0}, \dots, \frac{\Phi_n}{\Phi_0}$ sont de valuation a-adique positives, mais alors, I' étant l'idéal maximal de $O_{V,P}$, il est divisible par a qui, n'étant pas inversible dans R ne l'est pas dans $O_{V,P}$, ce qui est absurde.

Notations. (Les idéaux $\mathcal{I}$ et $\mathcal{J}$) On note $\mathbf{V}$ l'ensemble des valuations de Z qui dominent P. On pose

$$\mathcal{I} = \{x \in R; \mu(x) \geq \mu(\Phi_0),\ \mu \in \mathbf{V}\} \qquad \mathcal{J} = \{x \in R; \mu(x) > \mu(\Phi_0),\ \mu \in \mathbf{V}\}.$$

Définition. Soit $\mathcal{A}$ un idéal d'un anneau local régulier R, $\mathbf{V}$ un ensemble de valuations dominant R, on dit que $\mathcal{A}$ est $\mathbf{V}$-complet si on a

$$\mathcal{A} = \cap_{\mu \in \mathbf{V}} \mathcal{A} R_\mu \cap R,$$

où R_μ désigne l'anneau de valuation de μ.

Définition. (Zariski) Si R est un anneau local régulier de dimension 2, un idéal $\mathcal{A} \subset R$ est *complet* si $\mathcal{A} = \cap_{\mu \in Z} \mathcal{A}R_\mu \cap R$, où Z est la variété de Riemann de V'.

Exercices. On rappelle que, pour tout $\mu \in \mathbf{V}$, l'ensemble $\mu(O_{V,P} \setminus \{0\})$ est bien ordonné [V, 9.1], cela permet de définir $\mu(\mathcal{A})$ comme le minimum des valeurs des éléments de $\mathcal{A}$.

(1) Montrer que l'idéal $\mathcal{A}R_\mu$ est principal, en déduire: $\mathcal{A}R_\mu = \{x \in R_\mu | \mu(x) \geq \mu(\mathcal{A})\}$, et, $\mathcal{A}$ $\mathbf{V}$-complet $\Leftrightarrow \mathcal{A} = \{x \in R | \mu(x) \geq \mu(\mathcal{A})\}$, pour tout $\mu \in \mathbf{V}$.
(2) Montrer qu'une intersection d'idéaux $\mathbf{V}$-complets est un idéal $\mathbf{V}$-complet.
(3) Montrer que, si $\mathcal{A}$ est $\mathbf{V}$-complet et si I est un idéal quelconque de R, alors le transporteur $\mathcal{A} : I$ est égal à $\{x \in R |$pour tout $\mu \in \mathbf{V},\ \mu(x) + \mu(I) \geq \mu(\mathcal{A})\}$, en déduire que l'idéal $\mathcal{A} : I$ est $\mathbf{V}$-complet.
(4) Prouver qu'un idéal $\mathbf{V}$-complet est complet.
(5) Prouver que complet est équivalent à intégralement clos.

Proposition 2.6. *Les idéaux $\mathcal{I}$ et $\mathcal{J}$ sont complets, $\mathbf{V}$-complets (et M'-primaires si on suppose de plus que $\frac{\Phi_i}{\Phi_0}$ irréductible pour tout i, $1 \leq i \leq n$).*

Preuve. Il est clair que ces idéaux sont complets et $\mathbf{V}$-complets. Pour la deuxième assertion, remarquons d'abord que Φ_0 n'est pas inversible dans R, en effet, sinon, on aurait $R = O_{V,P}$, avec R régulier et $O_{V,P}$ non régulier. Donc, pour toute valuation μ de centre P', en particulier, pour toute valuation de $\mathbf{V}$, on a $\mu(\Phi_0) > 0$, on en déduit que $I \subset \mathcal{I} \subset M'$. Il est facile de voir que: $\mathcal{I}^2 \subset \mathcal{J} \subset \mathcal{I}$. Ce qui termine la preuve. □

Désormais, on suppose de plus que $\frac{\Phi_i}{\Phi_0}$ irréductible pour tout i, $1 \leq i \leq n$, c'est à dire $\mathcal{I}$ et $\mathcal{J}$ sont M'-primaires.

Proposition 2.7. *Soit a un élément de $\mathcal{I}$, on a équivalence entre les assertions suivantes:*

(i) *pour un $\mu \in \mathbf{V}$, $\mu(a) > \mu(\Phi_0)$,*
(ii) $\frac{a}{\Phi_0} \in \mathfrak{M}$,
(iii) *pour tout $\mu \in \mathbf{V}$, $\mu(a) > \mu(\Phi_0)$.*

De plus, si ces trois conditions équivalentes sont vérifiées, alors, pour toute valuation μ de $K(V)$ avec $O_{V,P} \subset R_\mu$, on a $\mu(a) \geq \mu(\Phi_0)$.

Preuve. (i) $\Rightarrow$ (ii). Comme $O_{V,P}$ est normal, $O_{V,P}$ est intersection des anneaux de valuations qui le dominent, on a $O_{V,P} = \cap_{\mu \in \mathbf{V}} R_\mu$, donc $\frac{a}{\Phi_0}$ est dans $O_{V,P}$, maintenant, $\frac{a}{\Phi_0}$ n'est pas inversible dans tous les R_μ, il n'est donc pas inversible dans $O_{V,P}$. Les implications (ii) $\Rightarrow$ (iii) et (iii) $\Rightarrow$ (i) sont triviales. Enfin, $\frac{a}{\Phi_0} \in \mathfrak{M}O_{V,P}$ implique la dernière assertion. □

Corollaire 2.8. *Soit $\mathcal{A}$ un idéal complet pour le système $\mathbf{V}$ avec $\mathcal{J} \subset \mathcal{A} \subset \mathcal{I}$, alors $\mathcal{A} = \mathcal{I}$ ou $\mathcal{A} = \mathcal{J}$.*

Preuve. Supposons que $\mathcal{A} \neq \mathcal{J}$, alors, d'après la proposition 2.7, il existe dans $\mathfrak{A}$ un élément a avec $\mu(a) = \mu(\phi_0)$ et donc, $\mathcal{A}R_\mu = \mathcal{I}R_\mu$, pour tout $\mu \in \mathbf{V}$ ce qui implique $\mathcal{I} = \mathcal{A}$. □

Proposition 2.9. *Soient a et b deux éléments de $\mathcal{I}$, $b \notin \mathfrak{J}$, alors $\frac{a}{b} \in O_{V,P}$, de plus, $\frac{a}{b} \in \mathfrak{M}$ si et seulement si $a \in \mathcal{J}$.*

Preuve. En effet, d'après la proposition 2.7, pour tout $\mu \in \mathbf{V}$, on a $\mu(a) \geq \mu(\Phi_0)$ et $\mu(b) = \mu(\Phi_0)$. Donc, $\mu(\frac{a}{b}) \geq 0$ pour tout $\mu \in \mathbf{V}$, comme $O_{V,P} = \cap_{\mu \in \mathbf{V}} R_\mu$, on a $\frac{a}{b} \in O_{V,P}$. D'autre part, $\frac{a}{b} \in \mathfrak{M}O_{V,P}$ si et seulement si pour tout $\mu \in \mathbf{V}$, on a $\mu(a) > \mu(b)$, c'est à dire, si et seulement si $a \in \mathcal{J}$. □

Proposition 2.10. *Soit V'' l'éclaté de V' le long de $\mathcal{I}$ et soit P'' le centre sur V'' pour une valuation $\mu \in \mathbf{V}$. Alors, $O_{V,P} = O_{V'',P''}$. C'est à dire que, dans un voisinage de P, V coïncide avec V''.*

Preuve. Il est clair que $\mathcal{I}O_{V,P} = \Phi_0 O_{V,P}$. Donc, par la propriété universelle de l'éclatement, on a:

$$O_{V'',P''} \subset O_{V,P}.$$

Mais, comme on a

$$O_{V,P} = R[\frac{\Phi_1}{\Phi_0}, \dots, \frac{\Phi_n}{\Phi_0}]_{\mathfrak{M} \cap R[\dots]},$$

les éléments $\frac{\Phi_1}{\Phi_0}, \dots, \frac{\Phi_n}{\Phi_0}$ sont de valeurs positives pour toute valuation $\mu \in \mathbf{V}$. Donc tous ces éléments sont dans $O_{V'',P''}$ et donc

$$O_{V,P} \subset O_{V'',P''},$$

d'où le résultat. □

Proposition 2.11. *On a $\mathcal{J}\, O_{V,P} = \Phi_0 \mathfrak{M} = \mathfrak{M}\mathcal{I}\, O_{V,P}$.*

Preuve. D'après la construction de Φ_0, on a: $\mathfrak{M} = (\frac{\Phi_1}{\Phi_0}, \dots, \frac{\Phi_n}{\Phi_0})$. D'où $\mathcal{J}\, O_{V,P} \supset \Phi_0 \mathfrak{M}$. D'autre part, on a $\mathcal{J} \subset \mathcal{I}$ et donc $\mathcal{J}\, O_{V,P} \subset \mathcal{I}O_{V,P} = \Phi_0\, O_{V,P}$, d'où le résultat. □

Théorème 2.12. *$\mathcal{J}O_{W,Q}$ est inversible.*

Remarque 2.13. Le théorème 2.12 prouve le théorème de principalisation. En effet, en utilisant la proposition 2.11, on a:

$$\mathcal{J}\, O_{W,Q} = \Phi_0 \mathfrak{M}\, O_{W,Q}.$$

Comme $O_{W,Q}$ est factoriel, un générateur de $\mathcal{J}\, O_{W,Q}$ divisé par Φ_0 engendre $\mathfrak{M}\, O_{W,Q}$.

Preuve. Pour prouver le théorème nous aurons besoin de trois lemmes.

Lemme 2.14. *On a* $\operatorname{ord}_{M'}(\mathcal{J}) = \operatorname{ord}_{M'}(\mathcal{I}) + 1$.

Preuve. On remarque que l'on a les inclusions: $\mathcal{I} \supset \mathcal{J} \supset M'\mathcal{I}$. Ce qui implique $\mathrm{ord}_{M'}(\mathcal{I}) \leq \mathrm{ord}_{M'}(\mathcal{J}) \leq \mathrm{ord}_{M'}(\mathcal{I})+1$. Comme la valuation $\mathrm{ord}_{M'}$ (cf. hypothèses et notations) est dans $\mathbf{V}$, les définitions de $(\mathcal{I})$ et $(\mathcal{J})$ impliquent $\mathrm{ord}_{M'}(\mathcal{I}) \neq \mathrm{ord}_{M'}(\mathcal{J})$. □

Lemme 2.15. *L'idéal M' n'apparait pas dans la décomposition de $\mathcal{I}$ en produits d'idéaux complets simples, mais apparait dans celle de $\mathcal{J}$, on définit alors l'idéal complet $\mathcal{J}_1$ par $\mathcal{J} = \mathcal{J}_1 M'$, on a:*

$$\mathcal{J}_1 = \mathcal{I} : M'.$$

Preuve. Soit $\mathcal{I} = \mathcal{I}_1 \ldots \mathcal{I}_n$ la décomposition de $\mathcal{I}$ en produits d'idéaux complets simples. Nous allons montrer que $\mathcal{I}_i O_{V,P}$ est principal, pour tout i, $1 \leq i \leq n$, comme P ne domine pas un point de l'éclaté de M', $M'O_{V,P}$ n'est pas principal et donc M' n'apparait pas parmi les $\mathcal{I}_i$. On peut trouver un élément $a \in \mathcal{I} - \mathcal{J}$ de la forme $a = a_1 \ldots a_n$ où $a_i \in \mathcal{I}_i$, $1 \leq i \leq n$. D'après la proposition 2.7, pour toute valuation $\mu \in \mathbf{V}$, $\mu(a) = \mu(\Phi_0)$, donc, $\mu(a_i) = \mu(\mathcal{I}_i)$, $1 \leq i \leq n$, comme $O_{V,P} = \cap_{\mu \in \mathbf{V}} R_\mu$ et les $\mathcal{I}_i$ sont complets, on en déduit que $\mathcal{I}_i O_{V,P} = a_i O_{V,P}$.

Nous avons vu (hypothèses et notations) que $\mathrm{ord}_{M'} \in \mathbf{V}$, on en déduit facilement que l'idéal $M'^{\rho+1}$ est $\mathbf{V}$-complet. Posons $\mathrm{ord}_{M'}(\mathcal{I}) = \rho$, d'après les exercices avant la proposition 2.6, l'idéal $\mathcal{I} \cap M'^{\rho+1}$ est $\mathbf{V}$-complet, cet idéal est strictement inclus dans $\mathcal{I}$ et contient $\mathcal{J}$, d'après le corollaire 2.8, $\mathcal{I} \cap M'^{\rho+1} = \mathcal{J}$. Il est clair que $M'\mathcal{I} \subset \mathcal{J}$, d'après le lemme 2.14, la valuation M'-adique de ces deux idéaux est $\rho+1$. Donc l'idéal $\mathrm{cl}_{M'}^{\rho+1}(\mathcal{J}) \subset \mathrm{Gr}_{M'}(O_{V',P'})$ n'est pas principal, donc M' apparait dans la décomposition de $\mathcal{J}$ en produits d'idéaux complets.

Pour terminer, montrons que $\mathcal{J}_1 = \mathcal{I} : M'$. L'inclusion $\mathcal{J}_1 \subset \mathcal{I} : M'$ est claire. Prouvons l'inclusion inverse. Notons $\mathcal{B} = \mathcal{I} : M'$. On a $\mathrm{ord}_{M'}(\mathcal{I}) = \rho \geq \mathrm{ord}_{M'}(\mathcal{B}) \geq \rho - 1 = \mathrm{ord}_{M'}(\mathcal{J}_1)$. Comme $M'\mathcal{B} \subset \mathcal{I}$, et que l'idéal initial de $\mathcal{I}$ est principal, on a $\rho + 1 = \mathrm{ord}_{M'}(M'\mathcal{B}) > \mathrm{ord}_{M'}(\mathcal{I})$. Donc $\mathrm{ord}_{M'}(\mathcal{B}) = \rho$ et $M'\mathcal{B} \subset M'^{\rho+1}$. Ce qui donne:

$$M'\mathcal{B} \subset M'^{\rho+1} \cap \mathcal{I} = \mathcal{J} = M'\mathcal{J}_1 \subset M'\mathcal{B}.$$

Ainsi, $M'\mathcal{B} = M'\mathcal{J}_1$, comme $\mathcal{B}$ et $\mathcal{J}_1$ sont complets, on a $\mathcal{B} = \mathcal{J}_1$. □

Lemme 2.16. *Si $\mathcal{A}$ est un idéal complet d'un anneau local R régulier de dimension 2 essentiellement de type fini sur un corps k et $\mathfrak{b}$ un idéal quelconque de cet anneau, alors tout facteur simple de $\mathcal{B} = \mathcal{A} : \mathfrak{b}$ est un prédecesseur ou coïncide avec un facteur simple de $\mathcal{A}$.*

Nous rappelons qu'un idéal complet simple est soit principal irréductible, soit M-primaire, M désignant le maximal de R, si cet idéal est M-primaire, son arbre de principalisation est unibranche. On dit qu'un idéal complet simple J est un *prédécesseur* de l'idéal complet simple I si l'arbre de principalisation de J est un sous-arbre de celui de I. Admettons ce lemme pour le moment et montrons qu'il prouve le théorème 2.12. En effet, comme nous l'avons vu dans la proposition 17, l'idéal $\mathcal{I}$ est principalisé dans $O_{V,P}$ et donc dans $O_{W,Q}$, donc, d'après le lemme 2.16, $\mathcal{J}_1 = I : M'$ aussi, donc, $\mathfrak{J} = M'\mathcal{J}_1$ aussi.

Prouvons le lemme 2.16. Soit R' l'anneau local d'un point de l'éclaté de R en M. On a: $\mathcal{B} = \{x \in R | \mu(x) + \mu(\mathfrak{b}) \geq \mu(\mathfrak{A}), R \subset R_\mu\}$, c'est un idéal complet. Donc [SZ2, appendice 5], $\mathcal{B}R'$ est complet, $\mathcal{B}R' = \{x' \in R' | \mu(x') + \mu(\mathfrak{b}) \geq \mu(\mathcal{A}), R' \subset R_\mu\}$, donc $\mathcal{B}R' = \mathcal{A}R' : \mathfrak{b}R'$. Si $\mathcal{A}R'$ est principal, $\mathcal{A}R' = x_1^{a_1} \cdots x_r^{a_r} R'$ et $\mathfrak{b}R' = x_1^{b_1} \cdots x_r^{b_r} B'$ avec $B' = R'$ ou B' est un idéal M'-primaire, on vérifie que $\mathcal{B}R' = x_1^{c_1} \cdots x_r^{c_r}$ où $c_i = a_i - b_i$ si $a_i \geq b_i$, $c_i = 0$ sinon, $1 \leq i \leq r$. Le lemme est donc vrai si $\mathcal{A}R'$ est principal, sinon, une récurrence sur le max des longueurs des arbres de principalisation des facteurs simples de $\mathcal{A}$ donne le résultat. □

References

[A1] S.S. Abhyankar: *On the valuations centered in a local domain*, Amer. J. Math. 78 (1956), 321–348.

[A2] S.S. Abhyankar: *Resolution of singularities of arithmetical surfaces*, dans: Arithmetical algebraic geometry (éd. O.F.G. Schilling), New York, Harper and Row 1965, 111–152.

[AM] M.F. Atiyah, I.G. Mac Donald: *Introduction to commutative algebra*. Addison-Wesley 1966.

[CGL] A. Campillo, G. Gonzalez-Springberg, M. Lejeune-Jalabert: *Clusters of infinitely near points*. Math. Ann. 306 (1996), 169–194.

[C1] V. Cossart: *Sur le polyèdre caractéristique d'une singularité*. Bull. Soc, Math. de France, 103 (1975), 13–19.

[C2] V. Cossart: *Desingularization of embedded excellent surfaces*. Tohoku Math. Jour. 33, n°1 (1981), 25–33.

[CGO] V. Cossart, J. Giraud, U. Orbanz: *Resolution of Surfaces Singularities*. Lecture Notes in Mathematics 1101, Springer 1984.

[H] R. Hartshorne: *Algebraic Geometry*. Graduate Texts in Mathematics n°52, Springer-Verlag 1960.

[Ha] H. Hauser: *Excellent surfaces and their taut resolution*, dans ce volume.

[LT] M. Lejeune-Jalabert, B. Teissier: *Cloture intégrale des idéaux et équisingularité*. Séminaire Lejeune-Teissier, Centre de Mathématiques de l'École Polytechnique, Publication de l'Université de Grenoble 1974.

[LJ] M. Lejeune-Jalabert: *Linear systems with infinitely near base conditions and complete ideals in dimension two*. Dans: Singularity theory (éds. Lê, Saito, Teissier), World Scientific 1995.

[L1] J. Lipman: *Desingularization of two-dimensional schemes*. Ann. Math. 107 (1978), 151–207.

[L2] J. Lipman: *On Complete Ideals in Regular Local Rings*. Algebraic Geometry and Commutative Algebra in Honor of Masayoshi Nagata. Kinokuniya, Tokyo 1988, 203–231.

[L3] J. Lipman: *Rational singularities with applications to algebraic surfaces and unique factorization*. Publ. IHES. 36 (1969), 195–279.

[M] T.T. Moh: *On a Newton polygon approach to the uniformization of singularities of characteristic p*. In: Algebraic Geometry and Singularities (éds. A. Campillo, L. Narváez). Proc. Conf. on Singularities La Rábida. Birkhäuser 1996.

[N] M. Nagata: *Local rings*, Interscience Publishers, John Wiley & Sons, New-York 1962.

[SZ1] P. Samuel, O. Zariski: *Commutative algebra*, vol.1, Graduate Texts in Mathematics n°28, Springer 1975.

[SZ2] P. Samuel, O. Zariski: *Commutative algebra*, vol.2, Graduate Texts in Mathematics n°29, Springer 1975.

[S] M. Spivakovsky: *Valuations in function fields of surfaces*, Am. Jour. of Math. 112 (1990), 107–156.

[V] M. Vaquié: *Valuations*, dans ce volume.

[Z1] O. Zariski: *Some results in the arithmetic theory of algebraic varieties*, Amer. J. Math. 61 (1939), 249–294.

[Z2] O. Zariski: *The reduction of the singularities of an algebraic surface*, Ann. of Math. 40 (1939), 639–689.

[Z3] O. Zariski: *Local uniformization on algebraic varieties*, Ann. of Math. 41 (1940), 852–896.

[Z4] O. Zariski: *A simplified proof for the resolution of singularities of an algebraic surface*, Ann. of Math. 43 (1942), 583–593.

[Z5] O. Zariski: *Reduction of the singularities of algebraic three dimensional varieties*, Ann. of Math. 45, No. 3 (1944), 472–542.

[Z6] O. Zariski: *The compactness of the Riemann manifold of an abstract field of algebraic functions*, Bull. Amer. Math. Soc. 45 (1944), 683–691.

Laboratoire de mathématiques, LAMA,
Université de Versailles,
45 rue des États-Unis
78035 Versailles, France
cossart@math.uvsq.fr

Progress in Mathematics, Vol. 181,

Toric Varieties and Toric Resolutions

David A. Cox

Abstract. This paper is an introduction to toric varieties and toric resolutions. We begin with basic definitions and examples, and then cover standard topics in toric geometry, including fans, support functions, and ampleness criteria. The paper also explores alternate constructions of toric varieties and nonnormal toric varieties. Then we turn our attention to singularities. We will discuss blowing-up in the toric context and resolution of singularities for toric varieties.

1. Introduction

Given an algebraically closed field k, we define a toric variety as follows. One starts with the algebraic torus $T \simeq (\mathbb{G}_m)^n = (k^*)^n$, and then a *toric variety* X over k is a normal variety containing T as a dense Zariski open subset such that the natural action of T on itself extends to an action of T on X.

The most basic examples of toric varieties are affine space $\mathbb{A}^n$ and projective space $\mathbb{P}^n$. In the context of this volume, however, a more relevant example is given by the variety $X \subset \mathbb{A}^4$ defined by

$$X = \{(x, y, z, w) \in \mathbb{A}^4 : xy = zw\}. \tag{1}$$

The origin is the unique singular point of X, and this variety is interesting because it *fails* to have a unique minimal desingularization. As we will see, this is easy to understand from the toric point of view. To prove that X is a toric variety, consider the subset $T \subset X$ defined by

$$T = \{(x, y, z, w) \in X : xyzw \neq 0\}.$$

The map $(\mathbb{G}_m)^3 \simeq T$ sending $(t, u, v) \in (\mathbb{G}_m)^3$ to $(t, u, v, tuv^{-1}) \in X$ is an isomorphism of varieties, and the action of T on itself clearly extends to an action on X. Since X is known to be normal, it follows that X is an affine toric variety. As we will learn in Section 2, this implies that X is Cohen-Macaulay and has at worst rational singularities.

In the next section, we will show that every toric variety is uniquely determined by a combinatorial structure called a *fan*. By way of motivation, let's see what this means for the example just discussed. For concreteness, we will work over the complex numbers $\mathbb{C}$. If N is the lattice of 1-parameter subgroups of the torus T, then each $u \in N$ is a homomorphism $u : \mathbb{C}^* \to T$. For which $u \in N$

does the limit $\lim_{t\to 0} u(t)$ exist in X? This question has a surprisingly interesting answer.

Under the isomorphism $(\mathbb{G}_m)^3 \simeq T$, we have $\mathbb{Z}^3 \simeq N$, where a triple $(a, b, c) \in \mathbb{Z}^3$ corresponds to the 1-parameter subgroup $u \in N$ defined by

$$u(t) = (t^a, t^b, t^c, t^{a+b-c}).$$

Regarded as a map into X, it is clear that $\lim_{t\to 0} u(t)$ exists if and only if

$$a \geq 0,\ b \geq 0,\ c \geq 0,\ a + b - c \geq 0. \tag{2}$$

These inequalities define a cone $\sigma \subset \mathbb{R}^3$ pictured in Figure 1.

FIGURE 1. The cone σ

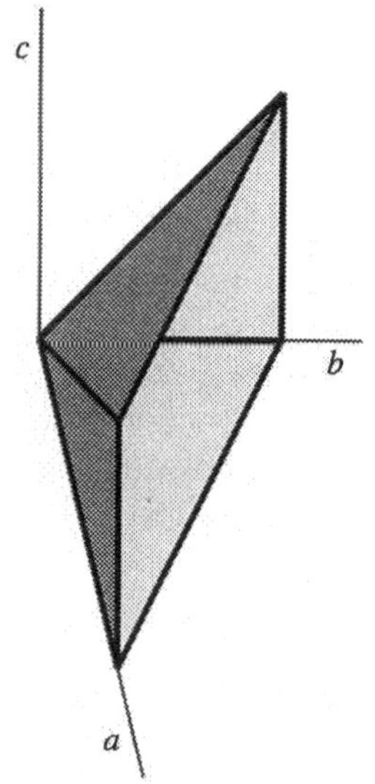

Given an integer point $(a, b, c) \in \sigma$, $\lim_{t\to 0} u(t)$ exists, but what point of X do we get? First observe that if we are in the *interior* of σ, then the inequalities in (2) are strict, which means that the limit is $(0, 0, 0, 0)$, the unique fixed point of the torus action. On the other hand, consider the bottom face of σ, where $c = 0$. A point in the relative interior of this face is $(a, b, 0)$ with $a, b > 0$, and the limit is clearly the point $(0, 0, 1, 0)$, which lies on a different orbit of the torus action.

The limit for the other faces of σ can be worked out equally easily, and one discovers one-to-one correspondences between the following interesting sets:

- The *limits* $\lim_{t\to 0} u(t)$ for $u \in N$.
- The *faces* of the cone $\sigma \subset \mathbb{R}^3 \simeq N \otimes_{\mathbb{Z}} \mathbb{R}$.
- The *orbits* of the torus action on X.

Furthermore, one can check that the dimension of each orbit is the codimension of the corresponding face, and whether one orbit lies in the closure of another can also be determined by looking at the corresponding faces. Thus the combinatorial structure provided by σ and its faces gives us a lot of information about the toric variety X.

Our next task is to develop the general theory of toric varieties.

2. Cones, fans, and toric varieties

Two classic references for toric varieties are Danilov's survey [Dan] and Oda's book [Oda1]. There are also the more recent books by Fulton [Ful] and Ewald [Ewa]. With a few exceptions, we will follow the notation of Fulton and Oda. We will state the basic definitions and properties of toric varieties and refer the reader to these references for proofs.

We will always work over an algebraically closed field k. To begin, suppose we have a torus $T \simeq (\mathbb{G}_m)^n$ over k. This gives the following objects:

- The characters $T \to \mathbb{G}_m$ form a lattice $M \simeq \mathbb{Z}^n$. We use additive notation for M. In terms of functions, we write $m \in M$ as $\chi^m : T \to \mathbb{G}_m$, and we multiply characters via $\chi^m \cdot \chi^{m'} = \chi^{m+m'}$.
- The 1-parameter subgroups $\mathbb{G}_m \to T$ form a lattice $N \simeq \mathbb{Z}^n$. We use additive notation for N. In terms of functions, we write $u \in N$ as $\lambda^u : \mathbb{G}_m \to T$, and we multiply 1-parameter subgroups via $\lambda^u \cdot \lambda^{u'} = \lambda^{u+u'}$.
- The composition of a character with a 1-parameter subgroup is an element of $\mathrm{Hom}(\mathbb{G}_m, \mathbb{G}_m) \simeq \mathbb{Z}$. This gives a pairing $\langle m, u\rangle \in \mathbb{Z}$ for $m \in M$, $u \in N$. In terms of the above notation, $\chi^m \circ \lambda^u(t) = t^{\langle m,u\rangle}$ for $t \in \mathbb{G}_m$.

We also let $M_{\mathbb{R}} = M \otimes_{\mathbb{Z}} \mathbb{R}$ and $N_{\mathbb{R}} = N \otimes_{\mathbb{Z}} \mathbb{R}$ denote the real vector spaces obtained from M and N.

We next define cones. A *rational polyhedral cone* $\sigma \subset N_{\mathbb{R}}$ is a cone generated by finitely many elements of N. In other words, there are $u_1, \ldots, u_s \in N$ such that

$$\sigma = \{\lambda_1 u_1 + \cdots + \lambda_s u_s \in N_{\mathbb{R}} : \lambda_1, \ldots, \lambda_s \geq 0\}.$$

We say that σ is *strongly convex* if $\sigma \cap (-\sigma) = \{0\}$. This is equivalent to saying that σ contains no subspaces of positive dimension. The cone shown in Figure 1 is strongly convex.

The *dimension* of a cone σ is the dimension of the smallest subspace containing σ. One also defines a *face* of σ to be the intersection $\{\ell = 0\} \cap \sigma$, where ℓ is a linear form which is nonnegative on σ. The set of faces of σ of dimension r is denoted $\sigma(r)$. For us, the most interesting faces of σ are the following:

- The *edges* of σ are the 1-dimensional faces $\rho \in \sigma(1)$. The *primitive element* n_ρ of $\rho \in \sigma(1)$ is the unique generator of the semigroup $\rho \cap N$. One can show that the primitive elements n_ρ, $\rho \in \sigma(1)$, generate the cone σ.

- The *facets* of σ are the codimension-1 faces. When $\dim \sigma = n$, each facet has an inward pointing normal which is naturally an element of $M_{\mathbb{R}}$. We get a unique inward normal by requiring that it is in M and has minimal length.

A cone $\sigma \subset N_{\mathbb{R}}$ has a *dual cone* $\sigma^\vee \subset M_{\mathbb{R}}$ defined by

$$\sigma^\vee = \{m \in M_{\mathbb{R}} : \langle m, u\rangle \geq 0 \text{ for all } u \in \sigma\}.$$

One can show that $\sigma^\vee$ is again a rational polyhedral cone. When σ has dimension n, $\sigma^\vee$ is generated by the inward pointing normals of σ. *Gordan's Lemma* [KKMS, §I.1] states that the semigroup $\sigma^\vee \cap M$ is finitely generated.

Definition. Given a strongly convex rational polyhedral cone $\sigma \subset N_{\mathbb{R}}$, let

$$X_\sigma = \mathrm{Spec}(k[\sigma^\vee \cap M]),$$

where $k[\sigma^\vee \cap M]$ is the semigroup algebra consisting of linear combinations of characters χ^m, with multiplication given by $\chi^m \cdot \chi^{m'} = \chi^{m+m'}$.

By Gordan's Lemma, $k[\sigma^\vee \cap M]$ is a finitely generated algebra over k, so that X_σ is a variety over k. Here are some examples of this definition.

Example. First consider an n-dimensional cone σ generated by a basis of N. The basis gives an isomorphism $N \simeq \mathbb{Z}^n$ which takes σ to the "first quadrant" where all coordinates are nonnegative. This coincides with $\sigma^\vee$, and it follows that $k[\sigma^\vee \cap M]$ can be identified with the usual polynomial ring $k[t_1, \ldots, t_n]$. Hence the associated toric variety is affine space $\mathbb{A}^n$.

Example. Next suppose that $\sigma = \{0\}$ is the trivial cone. Then $\sigma^\vee = M_{\mathbb{R}}$, so that $X_\sigma = \mathrm{Spec}(k[M])$. Since each $m \in M$ gives the function χ^m on T, $k[M]$ is the affine coordinate ring of T. Thus $T = \mathrm{Spec}(k[M])$ is the toric variety of $\sigma = \{0\}$.

Example. Consider the cone $\sigma \subset \mathbb{R}^3$ from Figure 1. The inward pointing normals of the facets of σ are

$$m_1 = (1,0,0),\ m_2 = (0,1,0),\ m_3 = (0,0,1),\ m_4 = (1,1,-1). \tag{3}$$

These generate the dual cone $\sigma^\vee$ and in this case also generate the semigroup $\sigma^\vee \cap M$. Under the ring homomorphism $k[x,y,z,w] \to k[\sigma^\vee \cap M]$ defined by

$$x \mapsto \chi^{m_1},\ y \mapsto \chi^{m_2},\ z \mapsto \chi^{m_3},\ w \mapsto \chi^{m_4},$$

one sees that $xy - zw \mapsto 0$ since $m_1 + m_2 = m_3 + m_4$. It follows easily that

$$k[x,y,z,w]/\langle xy - zw\rangle \simeq k[\sigma^\vee \cap M].$$

This shows that the cone of Figure 1 does indeed determine the toric variety defined in equation (1).

Example. Here is a nice example from [Oda1]. Let $\ell \geq 1$ be an integer, and consider the cone $\sigma \subset \mathbb{Z}^2$ generated by $(1,0)$ and $(1,\ell+1)$. We leave it as an exercise for the reader to show that $\sigma^\vee \cap M$ is generated by $m_1 = (0,1)$, $m_2 = (\ell+1,-1)$, $m_3 =$

$(1,0)$. Under the map $x_i \mapsto \chi^{m_i}$, proceeding as in the previous example shows that X_σ is the affine variety defined by

$$x_3^{\ell+1} - x_1 x_2 = 0. \tag{4}$$

Thus X_σ has a unique singular point which is a rational double point of type A_ℓ.

We next show that X_σ is a toric variety. Since $k[M]$ is the affine coordinate ring of T, applying Spec to the natural inclusion $k[\sigma^\vee \cap M] \subset k[M]$ induces

$$T \to X_\sigma.$$

To see that this is the inclusion of a Zariski open set, first note that σ strongly convex implies that $\sigma^\vee$ has dimension n. Then fix m_0 in the interior of $\sigma^\vee$. Given any $m \in M$, we have $m + dm_0 \in \sigma^\vee$ for $d \gg 0$, which implies

$$\chi^m = \frac{\chi^{m+dm_0}}{(\chi^{m_0})^d}.$$

It follows that $k[M]$ is the localization of $k[\sigma^\vee \cap M]$ at χ^{m_0}. Geometrically, this says that T is the Zariski open subset where $\chi^{m_0} \neq 0$, which makes sense since a character takes values in k^* on the torus. We will omit the proof that X_σ is normal (see [Oda1, Prop. 1.2]), and the action of T on X_σ will be explained in Section 4.

The inclusions $T \subset X_\sigma$ and $k[\sigma^\vee \cap M] \subset k[M]$ show that the affine toric variety X_σ is specified by saying which characters on the torus are allowed to extend to functions defined on all of X_σ.

Our next step is to create more general toric varieties by gluing together affine toric varieties containing the same torus T. This brings us to the concept of a *fan*, which is defined to be a finite collection Σ of cones in $N_\mathbb{R}$ with the following three properties:

1. Each $\sigma \in \Sigma$ is a strongly convex rational polyhedral cone.
2. If $\sigma \in \Sigma$ and τ is a face of σ, then $\tau \in \Sigma$.
3. If $\sigma, \tau \in \Sigma$, then $\sigma \cap \tau$ is a face of each.

Each $\sigma \in \Sigma$ gives an affine toric variety X_σ, and one can show that if τ is a face of σ, then there is a natural embedding $X_\tau \to X_\sigma$ as a Zariski open subset (above, we did the special case where $\{0\} \subset \sigma$ gives $X_{\{0\}} = T \subset X_\sigma$). This allows us to make the following definition.

Definition. Given a fan Σ in $N_\mathbb{R}$, X_Σ is the variety obtained from the affine varieties X_σ, $\sigma \in \Sigma$, by gluing together X_σ and X_τ along their common open subset $X_{\sigma\cap\tau}$ for all $\sigma, \tau \in \Sigma$.

Since the inclusions $T \subset X_\sigma$ are compatible with the identifications made in creating X_Σ, we see that X_Σ contains the torus T as a Zariski open set. One can also check that the maps $X_\tau \to X_\sigma$ are T-equivariant, which implies that T acts on X_Σ. Finally, the diagonal map $X_{\sigma\cap\tau} \to X_\sigma \times X_\tau$ is a closed embedding [Ful, Sect. 1.4], which shows that X_Σ is separated. It follows immediately that X_Σ is a toric variety as defined in Section 1. In [KKMS, §I.2, Thm. 5], it is shown that *all* toric varieties arise in this way, i.e., every toric variety is determined by a fan.

We next consider some examples of toric varieties.

Example. Given $\sigma \subset N_{\mathbb{R}}$, we get a fan by taking all faces of σ (including σ). One easily sees that the toric variety of this fan is the affine toric variety X_σ.

Example. The fans for $\mathbb{P}^2$ and $\mathbb{P}^1 \times \mathbb{P}^1$ are as follows:

FIGURE 2. The fans for $\mathbb{P}^2$ (left) and $\mathbb{P}^1 \times \mathbb{P}^1$ (right)

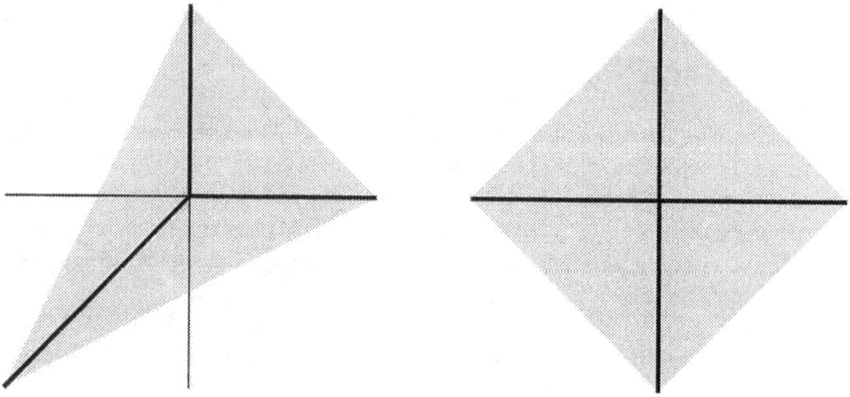

In this figure, 1-dimensional cones are indicated with thick lines, and 2-dimensional cones (which extend to infinity) are shaded. Thus the fan for $\mathbb{P}^2$ has three 2-dimensional cones, while the fan for $\mathbb{P}^1 \times \mathbb{P}^1$ has four such cones.

To see that the fan on the left gives $\mathbb{P}^2$, observe that the three 2-dimensional cones $\sigma_1, \sigma_2, \sigma_3$ are generated by bases of $\mathbb{Z}^2$. As noted in an earlier example, this implies that the affine toric varieties X_{σ_i} are copies of $\mathbb{A}^2$. By checking how these fit together along $X_{\sigma_i \cap \sigma_j}$, one gets the usual way of constructing $\mathbb{P}^2$ by gluing together three copies of $\mathbb{A}^2$. A similar argument shows that the fan on the right gives $\mathbb{P}^1 \times \mathbb{P}^1$.

Example. Let $e_1, \dots, e_n$ be a basis of $N = \mathbb{Z}^n$, and set $e_0 = -e_1 - \cdots - e_n$. Then we get a fan by taking the cones generated by all proper subsets of $\{e_0, e_1, \dots, e_n\}$. We leave it as an exercise for the reader to show that the associated toric variety is $\mathbb{P}^n$. When $n = 2$, this gives the first fan of Figure 2.

There are *many* other nice examples of toric varieties, including products of projective spaces, weighted projective spaces, and Hirzebruch surfaces. We will see in the next section that every rational convex polytope in $M_{\mathbb{R}}$ determines a projective toric variety.

We next consider the combinatorial structure of toric varieties. The basic idea is that the relations between faces, orbits and limits discussed in Section 1 generalize. More precisely, there are one-to-one correspondences between the following sets of objects:

- The limits $\lim_{t\to 0} \lambda^u(t)$ for $u \in |\Sigma| = \bigcup_{\sigma\in\Sigma} \sigma$ ($|\Sigma|$ is the *support* of Σ).
- The cones $\sigma \in \Sigma$.
- The orbits O of T on X_Σ.

The correspondences work as follows: an orbit O corresponds to a cone σ if and only if $\lim_{t\to 0} \lambda^u(t)$ exists and lies in O for all u in the relative interior of σ. If $\mathrm{orb}(\sigma)$ is the orbit corresponding to σ, then one has

- $\dim \sigma + \dim \mathrm{orb}(\sigma) = n$.
- $\mathrm{orb}(\sigma) \subset \overline{\mathrm{orb}(\tau)}$ if and only if $\tau \subset \sigma$.

In particular, fixed points of the torus action correspond to n-dimensional cones in the fan.

If we work with the orbit closures $\overline{\mathrm{orb}(\sigma)}$, we get similar correspondences. In addition, each orbit closure is a toric variety, and its fan can be described as follows. Given σ, let $N_\sigma = \mathbb{R}\sigma \cap N$, where $\mathbb{R}\sigma$ is the smallest subspace of $N_{\mathbb{R}}$ containing σ. Then N/N_σ is naturally dual to $\sigma^\perp \cap M$. Further, any cone $\tau \in \Sigma$ containing σ gives the cone

$$\overline{\tau} = (\tau + \mathbb{R}\sigma)/\mathbb{R}\sigma = (\tau + (N_\sigma)_{\mathbb{R}})/(N_\sigma)_{\mathbb{R}} \subset (N/N_\sigma)_{\mathbb{R}}.$$

These cones form a fan in $(N/N_\sigma)_{\mathbb{R}}$ which gives the toric variety $\overline{\mathrm{orb}(\sigma)}$. Geometrically, this fan is obtained from the *star* of σ (= all cones of Σ containing σ) by collapsing σ to a point in order to create a new fan in $(N/N_\sigma)_{\mathbb{R}}$.

Example. Let's apply the concepts just discussed to $\mathbb{P}^2$. The torus $T \subset \mathbb{P}^2$ is $(\mathbb{G}_m)^2 \simeq T$ via the map $(u, v) \in (\mathbb{G}_m)^2 \mapsto (u, v, 1) \in \mathbb{P}^2$.

We will first show how to construct the fan of $\mathbb{P}^2$ using limits. Given $u = (a, b)$, we get the 1-parameter subgroup $\lambda^u(t) = (t^a, t^b, 1)$. If we ask when $u = (a, b)$ satisfies $\lim_{t\to 0} \lambda^u(t) = (0, 0, 1)$, the answer is clearly $a, b > 0$. This defines the interior of one of the cones in the first fan of Figure 2. Similarly, if we want the limit to be $(0, 1, 0)$, the equality $(t^a, t^b, 1) = (t^{a-b}, 1, t^{-b})$ (these are homogeneous coordinates) shows that we must have $a - b, -b > 0$, which defines the interior of another cone from Figure 2. By considering the other possible limits, one gets the entire fan in this way. A good exercise for the reader would be to derive the fans for $\mathbb{P}^1 \times \mathbb{P}^1$ and $\mathbb{P}^n$ by the same method.

We can also describe the orbit closures corresponding to the cones in the fan of $\mathbb{P}^2$. The 2-dimensional cones correspond to fixed points. What about the 1-dimensional cones? Looking at the fan, we can figure this out as follows. Given an edge, the star containing it consists of the two 2-dimensional cones sharing the edge. When we collapse the edge to a point, we get a fan in $\mathbb{R}$ consisting of two rays emanating from the origin. This is the fan of $\mathbb{P}^1$. Hence $\mathbb{P}^1$ is the orbit closure associated to each edge of the fan of $\mathbb{P}^2$, which is what we would expect.

We conclude this section with some basic properties of toric varieties. Proofs can be found in [Oda1, Thm. 1.11, Thm. 1.10 and Cor. 3.9].

Theorem 2.1. *Let X_Σ be the toric variety determined by a fan Σ in $N_\mathbb{R}$. Then:*

1. *X_Σ is complete $\iff$ the support $|\Sigma| = \bigcup_{\sigma\in\Sigma}\sigma$ is all of $N_\mathbb{R}$.*
2. *X_Σ is smooth $\iff$ every $\sigma \in \Sigma$ is generated by a subset of a basis of N.*
3. *X_Σ is Cohen-Macaulay.*

Being Cohen-Macaulay is useful, for it means that Grothendieck duality is especially nice on toric varieties. See [Oda1, Section 3.2] for a careful explanation. In Section 3, we will give criteria for X_Σ to be projective, and in Section 5, we will show that X_Σ has at worst rational singularities.

Remark. Toric varieties are sometimes called *torus embeddings*, and [Ful] and [Oda1] call the fan Δ. Also, the toric variety determined by Σ is variously denoted X_Σ, $X(\Sigma)$, $Z(\Sigma)$, and $T_N\mathrm{emb}(\Sigma)$. Furthermore, polytopes (which we will encounter in Sections 3 and 4 below) are denoted P, $\Box$, and (just to confuse matters more) Δ. The lack of uniform notation is unfortunate, so that the reader of a paper using toric methods needs to look carefully at the notation.

We now turn our attention to divisors and line bundles on toric varieties.

3. Divisors and support functions

Our discussion of divisors begins with the observation that every toric variety X_Σ comes equipped with finitely many torus-invariant divisors, namely the closures of the orbits corresponding to 1-dimensional cones of the fan. If $\Sigma(1)$ is the set of all 1-dimensional cones of Σ, then we will let D_ρ denote the divisor corresponding to $\rho \in \Sigma(1)$. Note that $T = X_\Sigma \setminus \bigcup_\rho D_\rho$ (we use $\bigcup_\rho$, $\sum_\rho$, etc. to denote union, summation, etc. over all $\rho \in \Sigma(1)$).

The divisors D_ρ play an important role with respect to the toric variety X_Σ. For example, being Cohen-Macaulay, X_Σ has a dualizing sheaf ω_{X_Σ}, and [Oda1, Remark following Cor. 3.3] shows that

$$\omega_{X_\Sigma} \simeq \mathcal{O}_{X_\Sigma}(-\textstyle\sum_\rho D_\rho), \tag{5}$$

which generalizes the classic formula $\omega_{\mathbb{P}^n} \simeq \mathcal{O}_{\mathbb{P}^n}(-n-1)$. One subtlety is that $\sum_\rho D_\rho$ might be only a Weil divisor, so that one has to interpret (5) as an equality of reflexive sheaves of rank one (see [Rei] for a discussion of the sheaves coming from Weil divisors on normal varieties).

There is also a nice relation between the D_ρ and the characters χ^m coming from $m \in M$. Since χ^m maps T to $\mathbb{G}_m$, we can regard χ^m as a rational function on

X_Σ which is nonvanishing on T. Hence the divisor of χ^m is supported on $\bigcup_\rho D_\rho$, and by [Ful, Sect. 3.3], we have

$$\operatorname{div}(\chi^m) = \textstyle\sum_\rho \langle m, n_\rho\rangle D_\rho, \tag{6}$$

where, as in Section 2, n_ρ is the unique generator of $\rho \cap N$ for $\rho \in \Sigma(1)$.

Another useful fact proved in [Ful, Sect. 3.4] is that the torus invariant divisors generate the Chow group $A_{n-1}(X_\Sigma)$ of Weil divisors modulo rational equivalence. In fact, we get an exact sequence

$$M \xrightarrow{\alpha} \textstyle\bigoplus_\rho \mathbb{Z}D_\rho \xrightarrow{\beta} A_{n-1}(X_\Sigma) \longrightarrow 0, \tag{7}$$

where α is defined by (6) and β is the map taking a Weil divisor to its divisor class in the Chow group. Furthermore, if the elements of $\Sigma(1)$ span $N_\mathbb{R}$, then α is injective, so that (7) becomes a short exact sequence in this case.

Instead of Weil divisors and reflexive sheaves of rank one, we can also consider Cartier divisors and invertible sheaves. This is where *support functions* enter the picture. Recall that the support of Σ is $|\Sigma| = \bigcup_{\sigma\in\Sigma} \sigma$.

Definition. Given a fan Σ, a **support function** φ is a function $\varphi : |\Sigma| \to \mathbb{R}$ which is linear on each $\sigma \in \Sigma$ and takes integer values on $|\Sigma| \cap N$. The $\mathbb{Z}$-module of all support functions for Σ is denoted $\mathrm{SF}(N, \Sigma)$.

Equivalently, φ is a support function if for each $\sigma \in \Sigma$, we can find $m(\sigma) \in M$ such that

$$\varphi(u) = \langle m(\sigma), u\rangle \quad \text{for all } u \in \sigma. \tag{8}$$

From the support function φ, we get the divisor

$$D_\varphi = -\textstyle\sum_\rho \varphi(n_\rho) D_\rho. \tag{9}$$

(The minus sign is a standard convention.) The discussion in [Ful, Sect. 3.4] shows that D_φ is a Cartier divisor and that *all* T-invariant Cartier divisors arise in this way. The first assertion is easy to see, for if we apply (6) and (8) to the affine toric variety X_σ, we obtain $D_\varphi|_{X_\sigma} = \operatorname{div}(\chi^{-m(\sigma)})|_{X_\sigma}$, which shows that D_φ is Cartier. Using this, one easily obtains an exact sequence

$$M \longrightarrow \mathrm{SF}(N,\Sigma) \longrightarrow \mathrm{Pic}(X_\Sigma) \longrightarrow 0$$

which is compatible with (7) and is short exact when $\Sigma(1)$ spans $N_\mathbb{R}$. Furthermore, one can prove that $\mathrm{Pic}(X_\Sigma)$ is free whenever X_Σ is complete (see [Kaj]). (Early printings of [Ful] incorrectly stated that $\mathrm{Pic}(X_\Sigma)$ was always free.) Note also that when X_Σ is smooth, we have $\mathrm{SF}(N,\Sigma) \simeq \bigoplus_\rho \mathbb{Z}D_\rho$ and $A_{n-1}(X_\Sigma) \simeq \mathrm{Pic}(X_\Sigma)$.

Example. When is a given toric variety X_Σ Gorenstein? Since X_Σ is Cohen-Macaulay, being Gorenstein is equivalent to invertibility of the dualizing sheaf. By (5), ω_{X_Σ} is a line bundle if and only if $-\sum_\rho D_\rho$ comes from a support function φ, which by (9) means that for each $\sigma \in \Sigma$, there is $m(\sigma) \in M$ such that

$$\langle m(\sigma), n_\rho\rangle = 1 \quad \text{for all } \rho \in \sigma(1).$$

This makes it straightforward to check whether X_Σ is Gorenstein.

Example. In equation (4), we studied the affine toric variety X_σ, where σ is generated by $(1,0)$ and $(1,\ell+1)$. Let the corresponding toric divisors be D_1 and D_2. Then the reader can check that $a_1D_1 + a_2D_2$ is Cartier if and only if $a_1 - a_2$ is divisible by $\ell+1$. This shows that $A_1(X_\sigma) = \mathbb{Z}/(\ell+1)\mathbb{Z}$. Also note that $-D_1 - D_2$ is Cartier, so that X_σ is Gorenstein by the previous example.

When Σ is complete (meaning $|\Sigma| = N_\mathbb{R}$), we know that X_Σ is complete. In this case, we are interested in the following two types of support functions.

Definition. Let Σ be a complete fan, and let $\varphi : N_\mathbb{R} \to \mathbb{R}$ be in $\mathrm{SF}(N, \Sigma)$. Then:

1. φ is **convex** (sometimes called **upper convex**) if $\varphi(u+v) \geq \varphi(u) + \varphi(v)$ for all $u, v \in N_\mathbb{R}$.
2. φ is **strictly convex** if φ is convex and in addition $m(\sigma) \neq m(\tau)$ whenever σ, τ are distinct n-dimensional cones of Σ (recall that $m(\sigma)$ is determined by $\varphi(u) = \langle m(\sigma), u\rangle$ for $u \in \sigma$).

Convexity and strict convexity relate to the line bundle $\mathcal{O}_{X_\Sigma}(D_\varphi)$ as follows. Proofs can be found in [Oda1, Thm. 2.7 and Cor. 2.14].

Theorem 3.1. *Given a complete fan Σ, let D_φ be the divisor determined by a support function $\varphi \in \mathrm{SF}(N, \Sigma)$. Then:*

1. *$\mathcal{O}_{X_\Sigma}(D_\varphi)$ is generated by global sections $\iff$ φ is convex.*
2. *$\mathcal{O}_{X_\Sigma}(D_\varphi)$ is ample $\iff$ φ is strictly convex.*

It is possible to construct smooth complete fans which have *no* support functions other than the linear ones. The above theorem shows that such varieties cannot be projective. [KKMS, Ch. III] shows how to extend the definition of strictly convex to certain non-complete fans.

Example. As a simple example of the concepts introduced in this section, consider line bundles on $\mathbb{P}^1 \times \mathbb{P}^1$. Using the second fan in Figure 2, we see that the n_ρ's are $n_1 = (1,0), n_2 = (-1,0), n_3 = (0,1), n_4 = (0,-1)$. If the corresponding divisors are D_1, D_2, D_3, D_4, then the exact sequence (7) becomes

$$0 \longrightarrow \mathbb{Z}^2 \xrightarrow{\alpha} \bigoplus_{i=1}^4 \mathbb{Z}D_i \xrightarrow{\beta} \mathbb{Z}^2 \longrightarrow 0,$$

where

$$\begin{aligned} \alpha(a,b) &= aD_1 - aD_2 + bD_3 - bD_4, \\ \beta(a_1D_1 + \cdots + a_4D_4) &= (a_1 + a_2, a_3 + a_4). \end{aligned} \tag{10}$$

Furthermore, if φ is the corresponding support function, the reader should check that if $D = a_1D_1 + \cdots + a_4D_4$, then

$$\begin{aligned} D \text{ is ample} &\iff \varphi \text{ is strictly convex} \\ &\iff a_1 + a_2 > 0,\ a_3 + a_4 > 0 \\ &\iff \varphi(n_1 + n_2) > \varphi(n_1) + \varphi(n_2),\ \varphi(n_3 + n_4) > \varphi(n_3) + \varphi(n_4). \end{aligned}$$

The characterization of strictly convex given in the last line above has been generalized by Batyrev. He defines *primitive collections* and uses them to characterize strict convexity for toric varieties coming from *simplicial* fans (meaning that every $\sigma \in \Sigma$ is generated by linearly independent vectors). See [Bat1] or [Cox2] for precise statements.

There are also results which tell us when an ample line bundle on a toric variety is very ample.

Theorem 3.2. *Let D be an ample divisor on X_Σ. Then:*

1. *If X_Σ is smooth, then D is very ample.*
2. *If X_Σ has dimension $n \geq 2$, then $(n-1)D$ is very ample.*

The first assertion of this theorem is a classic result of Demazure [Dem], while the second is due to Ewald and Wessels [EW].

There is also a nice relationship between polytopes and line bundles on toric varieties. Given a torus invariant divisor $D = \sum_\rho a_\rho D_\rho$, consider the polyhedron in $M_\mathbb{R}$ defined by

$$P_D = \{m \in M_\mathbb{R} : \langle m, n_\rho \rangle \geq -a_\rho \text{ for all } \rho \in \Sigma(1)\}.$$

If $D = D_\varphi$ comes from the support function φ, one can show that P_D is given by

$$P_D = \{m \in M_\mathbb{R} : \langle m, u \rangle \geq \varphi(u) \text{ for all } u \in |\Sigma|\}.$$

The integral points in P_D determine the global sections of $\mathcal{O}_{X_\Sigma}(D)$. More precisely, if $m \in M$, then (6) tells us that

$$\operatorname{div}(\chi^m) + D = \textstyle\sum_\rho \langle m, n_\rho \rangle D_\rho + \sum_\rho a_\rho D_\rho.$$

This divisor is ≥ 0 if and only if $m \in P_D \cap M$, which implies

$$H^0(X_\Sigma, \mathcal{O}_{X_\Sigma}(D)) = \textstyle\bigoplus_{m \in P_D \cap M} k\chi^m. \tag{11}$$

In general, P_D might be unbounded, but when Σ is complete, P_D is bounded, i.e., it is a polytope.

The nicest case is when D is ample. By [Oda1, Lem. 2.12], this implies that P_D is the convex hull of $P_D \cap M$, has dimension n, and is combinatorially dual to Σ. The latter means that there is a one-to-one inclusion reversing correspondence

$$\sigma \in \Sigma \longleftrightarrow \operatorname{face}(\sigma) \subset P_D$$

between cones of Σ and faces of P_D (provided we count P_D as a face) such that

$$\dim(\sigma) + \dim(\operatorname{face}(\sigma)) = n$$

for all $\sigma \in \Sigma$. If we combine this with the correspondence between cones of Σ and torus orbits (or their closures) from Section 2, we get a one-to-one dimension preserving correspondence between faces of P_D and torus orbits. Thus P_D determines the combinatorics of the toric variety X_Σ.

Example. Using the notation for $\mathbb{P}^1 \times \mathbb{P}^1$ introduced earlier in this section, let $D = a_1 D_1 + \cdots + a_4 D_4$ be an ample divisor. The reader should verify that the polytope P_D is a rectangle in the plane. How does this relate to the torus orbits?

We can also go the other way. Given an n-dimensional polytope $P \subset M_{\mathbb{R}}$ with vertices in M, we can recover a toric variety and line bundle as follows. The *normal fan* of P consists of cones σ_F, where $F \subset P$ is a face, defined by

$$\sigma_F = \big\{u \in N_{\mathbb{R}} : \langle P, u\rangle \geq 0,\ \langle F, u\rangle = 0\big\},$$

where $\langle P, u\rangle \geq 0$ is short for $\langle m, u\rangle \geq 0$ for all $m \in P$, and similarly for $\langle F, u\rangle = 0$. From the normal fan of P, we get a toric variety X_P, and we also get a support function $\varphi : N_{\mathbb{R}} \to \mathbb{R}$ defined by

$$\varphi(u) = \min_{m \in P \cap M} \langle m, u\rangle. \tag{12}$$

Theorem 2.22 of [Oda1] shows that φ is strictly convex, so that $D = D_\varphi$ is ample on X_P.

The relation between toric varieties and polytopes is very deep. For example, if P is a simple polytope (meaning that the facet normals at each vertex are linearly independent), then Poincaré duality for the simplicial toric variety X_P implies the Dehn-Sommerville equations, which generalize Euler's formula $V - E + F = 2$ to higher dimensions. This and other applications can be found in [Ful, Chapter 5]. The reader may also wish to consult [Ewa] and [Stu2] for more on polytopes and toric varieties.

We next discuss new ways of creating toric varieties.

4. Other constructions of toric varieties

In addition to the classic method of defining toric varieties by gluing together affine toric varieties (as done in Section 2), three other techniques for constructing toric varieties have been given in recent years. These methods are:

- Proj of a graded ring.
- Generalized homogeneous coordinates.
- Toric ideals.

We begin with the first item. It is well known that if $S = k[x_0, \ldots, x_n]$ is a polynomial ring with the usual grading, then $\mathbb{P}^n = \mathrm{Proj}(S)$. This gives not only $\mathbb{P}^n$ but also the ample line bundle $\mathcal{O}_{\mathbb{P}^n}(1)$. From what we discussed at the end of Section 3, this indicates a connection with polytopes.

Suppose that $P \subset M_{\mathbb{R}}$ is an n-dimensional polytope with vertices in M. Then, using an auxiliary variable t_0, consider the ring S_P defined by

$$S_P = k\big[t_0^\ell \chi^m : m \in \ell P,\ \ell \geq 0\big].$$

This ring is graded by setting $\deg(t_0^\ell \chi^m) = \ell$, and then Batyrev [Bat2] noted that the toric variety X_P determined by P is given by

$$X_P = \operatorname{Proj}(S_P).$$

Furthermore, Batyrev also shows that the ample line bundle $\mathcal{O}_{X_P}(1)$ is precisely the line bundle coming from P (as determined by the support function (12)).

Example. Let $P \subset \mathbb{R}^n$ be the standard simplex defined by $t_i \geq 0$, $\sum_{i=1}^n t_i \leq 1$. What is the ring S_P? To work this out, first note that relative to the coordinates $t_1, \ldots, t_n$ of $\mathbb{R}^n$, a character χ^m is simply the monomial $t^m = t_1^{a_1} \cdots t_n^{a_n}$ for $m = (a_1, \ldots, a_n) \in \mathbb{Z}^n$. Then $m \in \ell P$ gives the monomial

$$t_0^\ell \chi^m = t_0^\ell t_1^{a_1} \cdots t_n^{a_n}, \quad a_i \geq 0,\ a_1 + \cdots + a_n \leq \ell.$$

Under the mapping which sends $t_0 \mapsto x_0$, $t_1 \mapsto x_1/x_0, \ldots, t_n \mapsto x_n/x_0$, this becomes the monomial

$$x_0^{\ell-(a_1+\cdots+a_n)} x_1^{a_1} \cdots x_n^{a_n}$$

of degree ℓ, and it follows that $S_P \simeq k[x_0, \ldots, x_n]$ as graded rings. Hence, taking Proj, we get $\mathbb{P}^n$. We invite the reader to double-check this by computing the normal fan of the simplex P.

Another way to view S_P is to consider the cone $\sigma \subset N_\mathbb{R} \times \mathbb{R}$ which is dual to the cone over $P \times \{1\} \subset M_\mathbb{R} \times \mathbb{R}$. One sees easily that

$$S_P = k\big[\sigma^\vee \cap (M \times \mathbb{Z})\big].$$

Thus S_P is Cohen-Macaulay. Since S_P is the coordinate ring of X_P under its projective embedding, we see that X_P is arithmetically Cohen-Macaulay.

We now turn to the second construction on our list, which concerns homogeneous coordinates for toric varieties. Returning to our basic example of $\mathbb{P}^n$, the usual homogeneous coordinates give not only the graded ring $k[x_0, \ldots, x_n]$ but also the quotient construction $\mathbb{P}^n \simeq (\mathbb{A}^{n+1} \setminus \{0\})/\mathbb{G}_m$. Given an arbitrary toric variety X_Σ, we can generalize this as follows. For each $\rho \in \Sigma(1)$, introduce a variable x_ρ, which gives the polynomial ring

$$S = k\big[x_\rho : \rho \in \Sigma(1)\big].$$

To grade this ring, note that a monomial $\Pi_\rho x_\rho^{a_\rho}$ gives a divisor $D = \sum_\rho a_\rho D_\rho$. If we write the monomial as x^D, its degree is defined to be $\deg(x^D) = \beta(D) \in A_{n-1}(X_\Sigma)$, where β is the map from (7) which takes a divisor to its class in the Chow group. The ring S with this grading is called the *homogeneous coordinate ring* of X_Σ. Note that when X_Σ is smooth, the grading is by $\operatorname{Pic}(X_\Sigma)$.

Example. We studied $\mathbb{P}^1 \times \mathbb{P}^1$ in Section 3, where the divisors corresponding to elements of $\Sigma(1)$ were denoted D_1, D_2, D_3, D_4. If the corresponding variables are x_1, x_2, x_3, x_4, then we get the ring $S = k[x_1, x_2, x_3, x_4]$. This is graded by the Picard group, which is $\mathbb{Z}^2$ in this case. Using (10), we see that

$$\deg(x_1^{a_1} x_2^{a_2} x_3^{a_3} x_4^{a_4}) = (a_1 + a_2, a_3 + a_4),$$

which is precisely the usual bigrading on $k[x_1, x_2; x_3, x_4]$, where each graded piece consists of bihomogeneous polynomials in x_1, x_2 and x_3, x_4.

One can also check that for $\mathbb{P}^n$, this construction gives $S = k[x_0, \dots, x_n]$ with the usual grading, and for $\mathbb{P}^n \times \mathbb{P}^m$, this gives $S = k[x_0, \dots, x_n; y_0, \dots, y_m]$ with the usual bigrading.

We can also use the variables x_ρ to give coordinates on X_Σ. To do this, we need an analog of the "irrelevant" ideal $\langle x_0, \dots, x_n \rangle \subset k[x_0, \dots, x_n]$. For each cone $\sigma \in \Sigma$, let $x^{\hat{\sigma}}$ be the monomial

$$x^{\hat{\sigma}} = \textstyle\prod_{\rho \in \Sigma(1) \setminus \sigma(1)} x_\rho,$$

and then define the ideal $B \subset S$ to be

$$B = \langle x^{\hat{\sigma}} : \sigma \in \Sigma \rangle.$$

When Σ is the fan giving $\mathbb{P}^n$, the reader should check that $B = \langle x_0, \dots, x_n \rangle$.

The basic idea is that X_Σ should be a quotient of $\mathbb{A}^{\Sigma(1)} \setminus \mathbf{V}(B)$, where $\mathbf{V}(B) \subset \mathbb{A}^{\Sigma(1)}$ is the variety defined by the ideal B. The quotient is by the group G, where

$$G = \mathrm{Hom}_{\mathbb{Z}}(A_{n-1}(X_\Sigma), \mathbb{G}_m).$$

Note that applying $\mathrm{Hom}_{\mathbb{Z}}(-, \mathbb{G}_m)$ to (7) gives the exact sequence

$$1 \longrightarrow G \longrightarrow (\mathbb{G}_m)^{\Sigma(1)} \longrightarrow T. \tag{13}$$

This shows that G acts naturally on $\mathbb{A}^{\Sigma(1)}$ and leaves $\mathbf{V}(B)$ invariant since this subvariety consists of coordinate subspaces.

The following representation of X_Σ was discovered independently by a variety of people (see [Cox1] for references and a proof).

Theorem 4.1. *Assume that X_Σ is a toric variety such that $\Sigma(1)$ spans $N_{\mathbb{R}}$ and that k has characteristic 0. Then:*

1. *X_Σ is the universal categorical quotient $(\mathbb{A}^{\Sigma(1)} \setminus \mathbf{V}(B))/G$.*
2. *X_Σ is a geometric quotient $(\mathbb{A}^{\Sigma(1)} \setminus \mathbf{V}(B))/G$ if and only if Σ is simplicial.*

Under the hypotheses of the theorem, one can *define* X_Σ to be the quotient $(\mathbb{A}^{\Sigma(1)} \setminus \mathbf{V}(B))/G$. To see why, note that (13) is short exact in this case, so that T is the quotient $(\mathbb{G}_m)^{\Sigma(1)}/G$. Thus

$$T = (\mathbb{G}_m)^{\Sigma(1)}/G \subset (\mathbb{A}^{\Sigma(1)} \setminus \mathbf{V}(B))/G.$$

Furthermore, since the "big" torus $(\mathbb{G}_m)^{\Sigma(1)}$ acts naturally on $\mathbb{A}^{\Sigma(1)} \setminus \mathbf{V}(B)$, it follows that T acts on X_Σ. Quotients preserve normality, so that all of the requirements of being a toric variety are satisfied by the quotient in Theorem 4.1.

Example. Continuing our example of $\mathbb{P}^1 \times \mathbb{P}^1$, the reader should check that $B = \langle x_1x_3, x_1x_4, x_2x_3, x_2x_4 \rangle$. Then, thinking of $\mathbb{A}^{\Sigma(1)}$ as $\mathbb{A}^2 \times \mathbb{A}^2$, one has

$$\mathbf{V}(B) = \{0\} \times \mathbb{A}^2 \cup \mathbb{A}^2 \times \{0\}.$$

One can also check that $G \simeq (\mathbb{G}_m)^2$ acts on $\mathbb{A}^2 \times \mathbb{A}^2$ via

$$(\lambda, \mu) \cdot (x_1, x_2, x_3, x_4) = (\lambda x_1, \lambda x_2, \mu x_3, \mu x_4).$$

Hence the quotient of Theorem 4.1 becomes

$$\left(\mathbb{A}^2 \times \mathbb{A}^2 \setminus (\{0\} \times \mathbb{A}^2 \cup \mathbb{A}^2 \times \{0\})\right)/(\mathbb{G}_m)^2,$$

which is exactly the way one usually represents $\mathbb{P}^1 \times \mathbb{P}^1$ as a quotient.

Example. Let $e_1, \dots, e_n$ be a basis of $N = \mathbb{Z}^n$, and let $\sigma = \mathbb{R}^n_+$ be the cone they generate. The resulting affine toric variety is $\mathbb{A}^n$. The goal of this example is to construct global coordinates for the blow-up of $0 \in \mathbb{A}^n$. A first observation is that if $\partial\sigma$ is the fan consisting of all proper faces of σ, then $X_{\partial\sigma} = \mathbb{A}^n \setminus \{0\}$ since the n-dimensional cone σ corresponds to the fixed point 0.

Now let $e_0 = e_1 + \cdots + e_n$ and consider the fan Σ whose cones are generated by all proper subsets of $\{e_0, \dots, e_n\}$, excluding $\{e_1, \dots, e_n\}$. Let's first argue that X_Σ is the blow-up of $0 \in \mathbb{A}^n$. In Σ, consider the edge ρ_0 generated by e_0. This corresponds to a divisor $D_0 \subset X_\Sigma$. We can describe D_0 using the methods of Section 2. The star of ρ_0 consists of all cones of Σ containing e_0. If we collapse ρ_0 to a point, we get a fan in an $(n-1)$-dimensional quotient of $\mathbb{R}^n$, which is easily seen to be the fan of $\mathbb{P}^{n-1}$. Thus $D_0 \simeq \mathbb{P}^{n-1}$. Furthermore, if we remove the star of ρ_0 from Σ, we are left with the fan $\partial\sigma$ from the previous paragraph. It follows that $X_\Sigma \setminus D_0 = \mathbb{A}^n \setminus \{0\}$. This makes it clear that we have the desired blow-up.

If x_i corresponds to the edge generated by e_i, then the reader should show that the homogeneous coordinate ring of X_Σ is $k[x_0, \dots, x_n]$ where $\deg(x_0) = -1$ and $\deg(x_i) = +1$ for $1 \le i \le n$. Furthermore, $\mathbf{V}(B) = \mathbb{A}^1 \times \{0, \dots, 0\}$ and $G = \mathbb{G}_m$ acts on $\mathbb{A}^{\Sigma(1)} = \mathbb{A}^1 \times \mathbb{A}^n$ by $\mu \cdot (x_0, \mathbf{x}) = (\mu^{-1}x_0, \mu\mathbf{x})$. Then, given $(x_0, \mathbf{x}) \in \mathbb{A}^1 \times \mathbb{A}^n \setminus \mathbf{V}(B)$, we can act on this point using G to obtain

$$\begin{array}{ll} (x_0, \mathbf{x}) \sim_G (1, x_0\mathbf{x}) & \text{if } x_0 \neq 0 \\ (0, \mathbf{x}) \sim_G (0, \mu\mathbf{x}) & \text{if } \mu \neq 0. \end{array}$$

It is now easy to see that $X_\Sigma = \left(\mathbb{A}^1 \times \mathbb{A}^n \setminus \mathbf{V}(B)\right)/G$ is the blow-up of $0 \in \mathbb{A}^n$. This approach gives global coordinates $x_0, x_1, \dots, x_n$ for the blow-up (subject to the action of $G = \mathbb{G}_m$). In terms of these coordinates, the blow-up map $X_\Sigma \to \mathbb{A}^n$ is given by $(x_0, x_1, \dots, x_n) \mapsto (x_0x_1, \dots, x_0x_n)$ (note that x_0x_i has degree 0 and hence is invariant under the group action).

Finally, we come to our third construction, which involves toric ideals. Let's begin with the affine case. Let $\sigma \subset N_\mathbb{R}$ be a cone, and suppose that the set $\mathcal{A} = \{m_1, \dots, m_s\}$ generates the semigroup $\sigma^\vee \cap M$. The map sending $y_i \mapsto \chi^{m_i}$ gives a surjective homomorphism $k[y_1, \dots, y_s] \to k[\sigma^\vee \cap M]$, and the kernel $I_\mathcal{A}$ is an example of a *toric ideal*.

A key observation is that $I_\mathcal{A}$ is generated by *binomials* (a binomial is a difference to two monomials). To state this precisely, note that each $\alpha = (a_1, \dots, a_s) \in \mathbb{Z}^s$ can be uniquely written $\alpha = \alpha^+ - \alpha^-$, where α^+ and α^- have nonnegative entries and disjoint support. Then one can prove that the toric ideal $I_\mathcal{A} \subset$

$k[y_1, \ldots, y_s]$ is given by

$$I_{\mathcal{A}} = \langle y^{\alpha^+} - y^{\alpha^-} : \alpha = (a_1, \ldots, a_s) \in \mathbb{Z}^s,\ \textstyle\sum_{i=1}^s a_i m_i = 0\rangle. \tag{14}$$

A proof of this assertion can be found in [Stu2, Cor. 4.3].

By (14), we can regard X_σ as the subvariety of $\mathbb{A}^s$ defined by the equations $y^{\alpha^+} = y^{\alpha^-}$. Given $t \in T$ and $y = (y_1, \ldots, y_s) \in X_\sigma$, what is $t \cdot y$? In Section 2, we deferred this question, but we can now give an explicit answer. We identify $t \in T$ with the point $u = (u_1, \ldots, u_s)$, where $u_i = \chi^{m_i}(t)$. The point u satisfies $u^{\alpha^+} = u^{\alpha^-}$, and combining this with the equations for y gives $(uy)^{\alpha^+} = (uy)^{\alpha^-}$. Hence we have a point of X_σ, so that

$$t \cdot y = (u_1 y_1, \ldots, u_s y_s) = (\chi^{m_1}(t) y_1, \ldots, \chi^{m_s}(t) y_s). \tag{15}$$

Viewed inside $\mathbb{A}^s$, the action of T on X_σ is now obvious.

In practice, toric ideals are defined in much greater generality and are closely related to nonnormal toric varieties. To set this up, let $\mathcal{A} = \{m_1, \ldots, m_s\}$ be *any* finite subset of M. Then we define the *toric ideal* $I_{\mathcal{A}}$ using equation (14). Toric ideals are easy to characterize: an ideal in $k[y_1, \ldots, y_s]$ is a toric ideal $I_{\mathcal{A}}$ if and only if it is a prime ideal and is generated by binomials (see [Stu1]).

Thinking geometrically, the ideal $I_{\mathcal{A}}$ defines a subvariety $X_{\mathcal{A}} \subset \mathbb{A}^s$. One can show that $X_{\mathcal{A}}$ is the Zariski closure of the image of the map $T \to \mathbb{A}^s$ defined by

$$t \mapsto (\chi^{m_1}(t), \ldots, \chi^{m_s}(t)). \tag{16}$$

Note also that $X_{\mathcal{A}}$ contains a torus (the image of T under the map (16)) which acts on all of $X_{\mathcal{A}}$ via (15). Hence $X_{\mathcal{A}}$ satisfies all of the criteria for being a toric variety, except possibly normality. For this reason, we call $X_{\mathcal{A}}$ a *generalized affine toric variety*. Basic references for toric ideals and generalized toric varieties are [GKZ1] and [Stu2].

According to [Stu2, Prop. 13.5], $X_{\mathcal{A}}$ is a toric variety in the usual sense (i.e., is normal) if and only if

$$\mathbb{N}\mathcal{A} = \mathrm{Cone}(\mathcal{A}) \cap \mathbb{Z}\mathcal{A},$$

where $\mathrm{Cone}(\mathcal{A})$ is the cone generated by $\mathcal{A}$, and $\mathbb{Z}\mathcal{A}$ (resp. $\mathbb{N}\mathcal{A}$) is the set of all integer (resp. nonnegative integer) combinations of elements of $\mathcal{A}$. More generally, the normalization of $X_{\mathcal{A}}$ is an affine toric variety X_σ, where $\sigma \subset N_{\mathbb{R}}$ is the cone dual to $\mathrm{Cone}(\mathcal{A})$ and N is the dual of $\mathbb{Z}\mathcal{A}$.

Example. Our first example of a toric variety was equation (1), which described the ideal $\langle xy - zw\rangle$. This is the toric ideal $I_{\mathcal{A}}$ for $\mathcal{A} = \{m_1, m_2, m_3, m_4\}$, where the m_i are given in (3).

Example. Given exponents $\mathcal{A} = \{\beta_1, \ldots, \beta_s\} \subset \mathbb{Z}$, we get a *monomial curve* in $\mathbb{A}^s$ parametrized by

$$t \mapsto (t^{\beta_1}, \ldots, t^{\beta_s}).$$

Since t^{β_i} is a character on $\mathbb{G}_m$, this is the generalized affine toric variety $X_{\mathcal{A}}$. It is nonnormal precisely when $X_{\mathcal{A}}$ fails to be smooth. The simplest example is the cusp parametrized by $t \mapsto (t^2, t^3)$. Here, the corresponding toric ideal is generated by the binomial $y^2 - x^3$.

Example. Consider the surface in $\mathbb{A}^4$ parametrized by

$$(t, u) \mapsto (t^4, t^3u, tu^3, u^4).$$

These monomials are characters on the torus $(\mathbb{G}_m)^2$, so that the image is $X_{\mathcal{A}}$ for $\mathcal{A} = \{(4,0), (3,1), (1,3), (0,4)\} \subset \mathbb{Z}^2$. By Exercise 3.18 of [Har, Chapter I], $X_{\mathcal{A}}$ is not normal. One can also show that $X_{\mathcal{A}}$ is not Cohen-Macaulay (see [Stu1]).

Projectively, the above parametrization defines a twisted quartic curve $C \subset \mathbb{P}^3$ which is normal. Hence C is normal but not projectively normal. A surprising number of basic examples in algebraic geometry are toric varieties in disguise.

Besides the generalized affine toric variety $X_{\mathcal{A}} \subset \mathbb{A}^s$, we also get a projective variety $Y_{\mathcal{A}} \subset \mathbb{P}^{s-1}$ by regarding (16) as a map $T \to \mathbb{P}^{s-1}$. More precisely, the *generalized projective toric variety* $Y_{\mathcal{A}}$ is defined to be the Zariski closure of the image of this map. One can easily show that:

- The affine cone of $Y_{\mathcal{A}}$ in $\mathbb{A}^s$ is the variety $X_{\mathcal{A}\times\{1\}}$. In this case, the toric ideal $I_{\mathcal{A}\times\{1\}}$ is homogeneous and defines the projective variety $Y_{\mathcal{A}}$.
- The normalization of $Y_{\mathcal{A}}$ is the toric variety X_P, where $P \subset M_{\mathbb{R}}$ is the convex hull of $\mathcal{A}$ and N is the dual of $\mathbb{Z}\mathcal{A}$.

Generalized projective toric varieties arise naturally in many different contexts.

Example. Let G be a connected semisimple algebraic group over k, and let T be a maximal torus. If P is a parabolic subgroup of G containing T, then T acts naturally on the flag variety G/P. The closure of a generic T-orbit is normal by [Dab, Thm. 3.2], so that a generic orbit closure is a toric variety. But in general, it is unknown whether arbitrary orbit closures are normal. Hence generalized projective toric varieties provide a natural language for discussing these varieties. See [Stu1, Sect. 2] for further discussion and references.

Example. Suppose that $\mathcal{A} = \{m_1, \dots, m_s\} \subset \mathbb{Z}^n$ and that $\mathcal{A}$ generates $\mathbb{Z}^n$. Then let $L(\mathcal{A})$ be the set of Laurent polynomials with exponent vectors in $\mathcal{A}$, i.e.,

$$L(\mathcal{A}) = \{a_1 t^{m_1} + \cdots + a_s t^{m_s} : a_i \in k\},$$

where $t^m = t_1^{a_1} \cdots t_n^{a_n}$ for $m = (a_1, \dots, a_n) \in \mathbb{Z}^n$. Given $n+1$ Laurent polynomials $f_0, \dots, f_n \in L(\mathcal{A})$, their *$\mathcal{A}$-resultant*

$$\mathrm{Res}_{\mathcal{A}}(f_0, \dots, f_n)$$

is a polynomial in the coefficients of the f_i whose vanishing is necessary and sufficient for the equations $f_0 = \cdots = f_n = 0$ to have a solution (see [GKZ1, Prop. 2.1]). However, one must be careful where the solution lies. The f_i are defined initially on the torus $(\mathbb{G}_m)^n$, but the definition of generalized projective toric variety shows that the equation $f_i = 0$ makes sense on $Y_{\mathcal{A}}$. Then one can prove that

$$\mathrm{Res}_{\mathcal{A}}(f_0, \dots, f_n) = 0 \iff f_1 = \cdots = f_n = 0 \text{ have a solution in } Y_{\mathcal{A}}.$$

Generalized toric varieties and toric ideals occur when studying hypergeometric equations of Gelfand, Kapranov and Zelevinski [GKZ2]. Also, some applications to combinatorics can be found in [Stu2, Ch .14].

Finally, we should mention that over $\mathbb{C}$, there is a fourth construction of toric varieties, via symplectic reduction. A brief discussion can be found in [Cox2].

It is time to turn our attention to the singularities of toric varieties.

5. Toric blow-ups and resolution of singularities

Before studying toric resolutions, we need to discuss maps between toric varieties. Suppose we have toric varieties $X_{\Sigma'}$ and X_Σ coming from fans $\Sigma' \subset N'_{\mathbb{R}}$ and $\Sigma \subset N_{\mathbb{R}}$. Let $f : N' \to N$ be a homomorphism such that for every cone $\sigma' \in \Sigma'$, there is $\sigma \in \Sigma$ satisfying $f_{\mathbb{R}}(\sigma') \subset \sigma$, where $f_{\mathbb{R}} : N'_{\mathbb{R}} \to N_{\mathbb{R}}$ is induced by f. This data determines a morphism

$$f_* : X_{\Sigma'} \longrightarrow X_\Sigma$$

such that on the torus $T' = N' \otimes_{\mathbb{Z}} \mathbb{G}_m \subset X_{\Sigma'}$, the restriction $f_*|_{T'}$ is the map given by

$$f \otimes 1 : N' \otimes_{\mathbb{Z}} \mathbb{G}_m \to N \otimes_{\mathbb{Z}} \mathbb{G}_m.$$

Furthermore, f_* is equivariant with respect to $f \otimes 1$. We call f_* a *toric morphism*. (As usual, we refer the reader to [Ful] and [Oda1] for proofs of the assertions made in this section.)

Example. In the fan for $\mathbb{P}^1 \times \mathbb{P}^1$ shown in Figure 2, projection onto the horizontal axis takes us to the fan for $\mathbb{P}^1$. The corresponding toric morphism is the projection map $\mathbb{P}^1 \times \mathbb{P}^1 \to \mathbb{P}^1$ onto one of the factors. We get the projection onto the other factor by projecting the fan onto the vertical axis in Figure 2.

Example. If $N' \subset N$ is a submodule of finite index and $\Sigma' = \Sigma$, then the inclusion $i : N' \hookrightarrow N$ gives a toric morphism

$$i_* : X_{\Sigma,N'} \longrightarrow X_{\Sigma,N},$$

where we have modified our usual notation to take the lattice into account. In this situation, the finite group N/N' acts naturally on $X_{\Sigma,N'}$ with quotient $X_{\Sigma,N}$, and i_* is the quotient map. See [Oda1, Cor. 1.16] for a proof.

One can show that a toric morphism $f_* : X_{\Sigma'} \to X_\Sigma$ is proper if and only if $f_{\mathbb{R}}^{-1}(|\Sigma|) = |\Sigma'|$. This is proved in [Oda1, Thm. 1.15]. If we apply this criterion to the zero map $f : N' \to \{0\}$, we leave it to the reader to show that this implies our earlier result that $X_{\Sigma'}$ is complete if and only if $|\Sigma'| = N_{\mathbb{R}}$.

We say that a fan Σ' *refines* or *subdivides* a fan Σ if $|\Sigma'| = |\Sigma|$ and every cone of Σ' is contained in a cone of Σ. In this situation, the identity map $1 : N \to N$ induces a toric morphism $1_* : X_{\Sigma'} \to X_\Sigma$ which is proper (by the previous paragraph) and birational (since it is the identity on T). This is the heart of how toric blow-ups and toric resolutions work.

Example. The most basic example of a toric blow-up was given in Section 4, where we used a basis $e_1, \dots, e_n$ of N to generate a cone $\sigma = \mathbb{R}^n_+$ such that $X_\sigma = \mathbb{A}^n$. Then, using $e_0 = e_1 + \cdots + e_n$, we subdivided σ into n cones of dimension n to get a fan Σ subdividing σ, and we showed that $X_\Sigma \to X_\sigma = \mathbb{A}^n$ was the blow-up of $0 \in \mathbb{A}^n$.

To generalize this example, suppose that X_Σ is a smooth toric variety, and let $Y \subset X_\Sigma$ be the closure of the torus orbit corresponding to a fixed cone $\sigma \in \Sigma$. Using the description from Section 2 of Y as a toric variety, one easily sees that Y is smooth, so that the blow-up $B_Y(X_\Sigma)$ of X_Σ along Y is also smooth. We claim that $B_Y(X_\Sigma)$ is toric and the blow-up map $B_Y(X_\Sigma) \to X_\Sigma$ is a toric morphism.

To prove this, we must construct a refinement Σ' of Σ which gives $B_Y(X_\Sigma)$. We will use the *star* of σ, which was introduced in Section 2 and consists of all cones $\tau \in \Sigma$ containing σ. The idea is to subdivide the star of σ and leave the other cones of Σ alone. We begin by subdividing σ in a manner similar to the previous example. If the primitive generators of σ are $n_1, \dots, n_d$ (where $d = \dim(\sigma)$), then set $n_0 = n_1 + \cdots + n_d$. For each i between 1 and d, define σ_i to be the cone generated by n_0 and $n_1, \dots, n_{i-1}, n_{i+1}, \dots, n_d$.

Now let τ be in the star of σ. We can write τ uniquely as

$$\tau = \sigma + \sigma',$$

where $\sigma' \in \Sigma$ satisfies $\sigma \cap \sigma' = \{0\}$ (this decomposition exists because τ is generated by a subset of a basis of N). Then τ gets replaced by the cones $\sigma_i + \sigma'$, $1 \le i \le d$. Doing this for all τ in the star of σ (and leaving the other cones of Σ unchanged) gives a refinement Σ' of Σ called the *star subdivision* of Σ relative to σ. Since X_Σ is smooth, [Oda1, Prop. 1.26] shows that $X_{\Sigma'} \to X_\Sigma$ is the desired blow-up.

If we turn to singular toric varieties, then one can use refinements to resolve singularities. We begin with two examples.

Example. In Section 2, we showed that the cone $\sigma \subset \mathbb{R}^2$ generated by $(1,0)$ and $(1, \ell+1)$ gave a 2-dimensional toric variety X_σ with a singularity of type A_ℓ. To resolve this singularity, we need to refine σ into smooth cones. The simplest way of doing this is pictured in Figure 3. In this picture, the original cone σ is shaded, and the refinement adds ℓ edges through the points $(1,i)$ for $1 \le i \le \ell$.

One easily checks that the fan Σ in Figure 3 is smooth. To see what the resolution X_Σ looks like, note that we removed σ from the fan of X_σ and replaced it by ℓ edges and $\ell+1$ 2-dimensional cones. By the methods of Section 2, each new edge gives a $\mathbb{P}^1$, and the cone generated by two consecutive new edges corresponds to the point of intersection of the two $\mathbb{P}^1$'s. Geometrically, this means that on X_Σ, the singular point of X_σ has been replaced by a chain of ℓ copies of $\mathbb{P}^1$, where successive curves intersect transversely. Furthermore, the methods for computing intersection numbers on smooth toric surfaces (see [Ful, Section 2.5] or [Oda1, Section 1.6]) show that each $\mathbb{P}^1$ has self-intersection -2. Hence we have constructed the minimal resolution of an A_ℓ singularity by purely toric means.

FIGURE 3. The fan Σ resolving a singularity of type A_ℓ

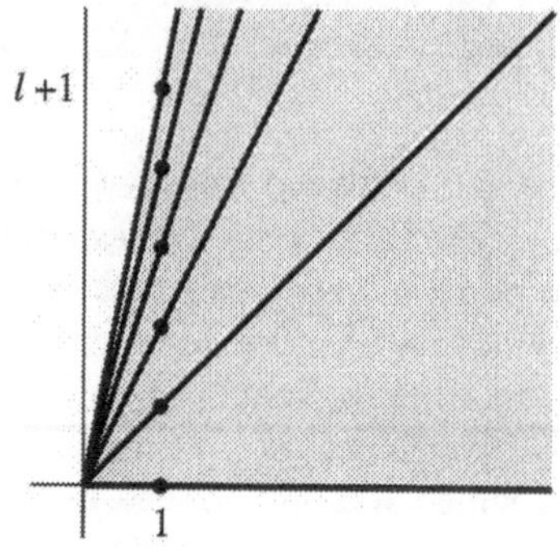

FIGURE 4. Fans Σ_1 and Σ_2 resolving the singularity $xy = zw$

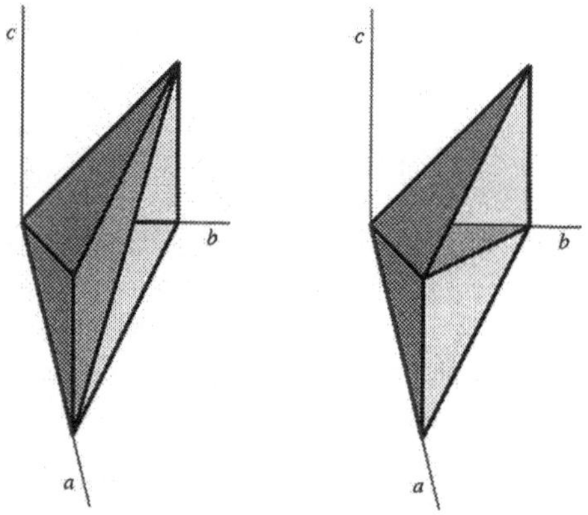

Example. We next resolve the singularity $xy = zw$ considered in equation (1). This is the toric variety X_σ, where σ is the cone pictured in Figure 1. Note that X_σ is clearly singular since σ has four generators—not a basis of $\mathbb{Z}^3$. The simplest way to refine σ to a smooth fan is to subdivide σ into two 3-dimensional cones. There are two ways of doing this, as shown in Figure 4.

The reader can check that these two refinements are smooth. In each case, σ is replaced by two new 3-dimensional cones which meet along a common 2-dimensional face. If we call these fans Σ_1 and Σ_2, then the common 2-dimensional face corresponds to a $\mathbb{P}^1$ in the resolution. This follows easily by considering the star of the 2-dimensional face and collapsing the face to a point. Thus we get two distinct small resolutions of the 3-dimensional variety X_σ, where in each case the singular point is replaced by a $\mathbb{P}^1$.

To compare these two resolutions, we take the common refinement of Σ_1 and Σ_2, which is illustrated in Figure 5. If we call this fan Σ_3, then we get the following commutative diagram:

$$\begin{array}{ccc} & X_{\Sigma_3} & \\ \swarrow & & \searrow \\ X_{\Sigma_1} & & X_{\Sigma_2} \\ \searrow & & \swarrow \\ & X_\sigma & \end{array} \tag{17}$$

Looking at Figure 5, we see that we've added a new edge (drawn extra long in the figure). This edge corresponds to a divisor on X_{Σ_3}, and by considering the star of the edge and collapsing the edge to a point, we see that the fan for this divisor is precisely the second fan in Figure 2. Hence the exceptional divisor of $X_{\Sigma_3} \to X_\sigma$ is $\mathbb{P}^1 \times \mathbb{P}^1$. Furthermore, by understanding how these fans fit together, one sees that over the singular point of X_σ, the diagram (17) is just the projection of $\mathbb{P}^1 \times \mathbb{P}^1$ onto its two factors. Hence we get a complete explanation of this classic example by means of toric geometry.

These examples can be generalized as follows.

Theorem 5.1. *If X_Σ is the toric variety coming from a fan Σ, then Σ has a refinement Σ' such that the toric morphism $X_{\Sigma'} \to X_\Sigma$ is a resolution of singularities.*

Proof. First observe that the smooth cones of Σ (i.e., those generated by subsets of $\mathbb{Z}$-bases) form a subfan $\Sigma^\circ \subset \Sigma$. Then the toric variety X_{Σ° is a smooth open subset of X_Σ. We claim that X_{Σ° is the smooth locus of X_Σ.

FIGURE 5. The common refinement of Σ_1 and Σ_2

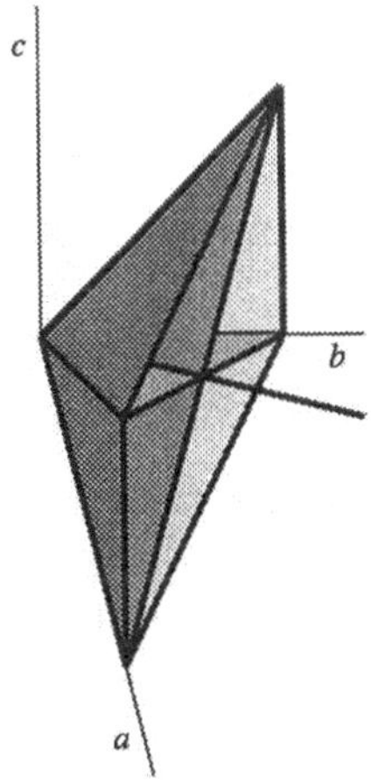

To prove this, it suffices to show that for a nonsmooth cone $\sigma \in \Sigma$, every point of the corresponding orbit $\mathrm{orb}(\sigma)$ is a singular point of X_Σ. We will use the following two facts, proved in [Ful, Section 2.1]:

- If a nonsmooth cone $\sigma \in \Sigma$ has dimension n, then the corresponding orbit $\mathrm{orb}(\sigma)$ consists of a single point which is a singular point of X_σ. Recall also that X_σ is Zariski open in X_Σ.
- If a cone $\sigma \in \Sigma$ has dimension $d < n$, let $N_\sigma = \mathbb{R}\sigma \cap N$, where $\mathbb{R}\sigma$ is the smallest subspace of $N_\mathbb{R}$ containing σ. Then σ can be regarded as a top-dimensional cone $\bar{\sigma} \subset (N_\sigma)_\mathbb{R}$, and there is a (noncanonical) isomorphism
$$X_\sigma \simeq X_{\bar{\sigma}} \times (\mathbb{G}_m)^{n-d}$$
such that
$$\mathrm{orb}(\sigma) \simeq \mathrm{orb}(\bar{\sigma}) \times (\mathbb{G}_m)^{n-d}.$$

If σ is a nonsmooth cone, the first assertion applied to $\bar{\sigma}$ shows that $\mathrm{orb}(\bar{\sigma})$ is a singular point of $X_{\bar{\sigma}}$, and then the second assertion shows that $\mathrm{orb}(\sigma)$ lies in the singular locus of X_Σ, as claimed.

Suppose we can find a smooth refinement Σ' of Σ which leaves the cones of Σ° unchanged. Then $X_{\Sigma'}$ is smooth and the induced toric morphism $X_{\Sigma'} \to X_\Sigma$ (we will call it ϕ) is proper. Furthermore, since X_{Σ° is the smooth locus of X_Σ and Σ° is a subfan of both Σ and Σ', it follows that ϕ is an isomorphism away from the singular locus of X_Σ. Hence ϕ is a resolution of singularities.

We will construct the refinement Σ' in two stages. In the first stage, the goal is to find a simplicial refinement Σ'' of Σ which doesn't change Σ°. (Recall that simplicial means that every cone in Σ'' is generated by linearly independent vectors.) We prove this by induction on the number of cones in Σ. Thus, suppose that σ is a maximal cone of Σ (i.e., not properly contained in any cone of Σ). Then $\Sigma \setminus \{\sigma\}$ is a fan, which by our induction hypothesis has a simplical refinement Σ_1'' that doesn't change smooth cones.

If σ is smooth, then every proper face $\tau \subset \sigma$ is smooth and hence, by our induction hypothesis, lies in Σ_1''. It follows easily that $\Sigma'' = \Sigma_1'' \cup \{\sigma\}$ is a fan with the desired properties.

On the other hand, if σ is not smooth, then let $d = \dim(\sigma)$ and pick $u_0 \in M$ in the relative interior of σ. If ρ_0 is the edge generated by u_0, then consider the set of cones
$$\Sigma'' = \Sigma_1'' \cup \{\tau + \rho_0 : \tau \in \Sigma_1'' \text{ and } \tau \subset \sigma\} \cup \{\rho_0\}.$$
The cones τ such that $\tau \in \Sigma_1''$ and $\tau \subset \sigma$ refine the boundary of σ. Each cone $\tau + \rho_0$ is easily seen to be simplicial (since τ is) and the union $\bigcup_\tau (\tau + \rho_0)$ over all such τ's is clearly equal to σ. It follows easily that Σ'' has the desired property.

In the second stage of the proof, we can assume that Σ is simplicial, and we seek a smooth refinement which doesn't affect the cones in Σ°. We begin by defining a measure of how close a simplical cone σ is to being smooth. Suppose σ

is generated by primitive elements $n_1, \ldots, n_d$. As above, let $N_\sigma = \mathbb{R}\sigma \cap N$. Then the *multiplicity* of σ is the index

$$\operatorname{mult}(\sigma) = \big[N_\sigma : \mathbb{Z}n_1 + \cdots + \mathbb{Z}n_d\big].$$

One easily checks that σ is smooth if and only if $\operatorname{mult}(\sigma) = 1$. As noted by Danilov [Dan, §8.2], the multiplicity has the following two properties:

1. $\operatorname{mult}(\sigma)$ is the number of points in $P_\sigma \cap N$, where
$$P_\sigma = \{\textstyle\sum_{i=1}^d \alpha_i n_i : 0 \le \alpha_i < 1\}.$$
2. $\operatorname{mult}(\sigma)$ is the normalized volume of $P_\sigma \subset \mathbb{R}\sigma$, where "normalized" means that the parallelepiped determined by a basis of $N_\sigma \subset \mathbb{R}\sigma$ has volume 1.

Then, to measure how close the whole fan Σ is to being smooth, we define

$$\operatorname{mult}(\Sigma) = \max_{\sigma \in \Sigma} \operatorname{mult}(\sigma).$$

Note that $\operatorname{mult}(\Sigma) = 1$ if and only if Σ is a smooth fan. We will show that whenever $\operatorname{mult}(\Sigma) > 1$, we can find a simplicial refinement Σ_1 which doesn't change Σ° and satisfies either

$$\begin{aligned} &\operatorname{mult}(\Sigma_1) < \operatorname{mult}(\Sigma), \text{ or} \\ &\operatorname{mult}(\Sigma_1) = \operatorname{mult}(\Sigma) \text{ and } \Sigma_1 \text{ has fewer cones of this multiplicity.} \end{aligned} \tag{18}$$

Once this is done, the theorem will follow immediately.

Suppose that $\operatorname{mult}(\Sigma) > 1$, and let $\sigma \in \Sigma$ be a nonsmooth cone of minimal dimension. Since $0 \in P_\sigma$ and $\operatorname{mult}(\sigma) > 1$, the first property of $\operatorname{mult}(\sigma)$ listed above shows that we can find a nonzero $u \in N$ of the form

$$u = \alpha_1 n_1 + \cdots + \alpha_d n_d, \quad 0 \le \alpha_i < 1,$$

where $n_1, \ldots, n_d$ are the primitive generators of σ. We can assume that u is primitive. Furthermore, since all proper faces of σ are smooth by minimality, we must have $\alpha_i > 0$ for all i (otherwise u would lie in a proper smooth face of σ, which would contradict the first property of $\operatorname{mult}(\sigma)$ listed above).

Now let $\tau \in \Sigma$ be a cone in the star of σ. We can write τ uniquely as

$$\tau = \sigma + \tau', \quad \tau' \in \Sigma,\ \tau' \cap \sigma = \{0\}.$$

This decomposition exists because τ is simplicial. For each $1 \le i \le d$, define the simplicial cone τ_i to be

$$\tau_i = \operatorname{Cone}(u, n_1, \ldots, n_{i-1}, n_{i+1}, \ldots, n_d) + \tau'. \tag{19}$$

Comparing the normalized volumes of P_τ and P_{τ_i}, one easily obtains

$$\operatorname{vol}(P_{\tau_i}) = \alpha_i \operatorname{vol}(P_\tau),$$

so that by the second bullet above, we have

$$\operatorname{mult}(\tau_i) = \alpha_i \operatorname{mult}(\tau) < \operatorname{mult}(\tau). \tag{20}$$

We can now prove the existence of a fan Σ_1 satisfying condition (18). Let $\tau_0 \in \Sigma$ be a cone of maximal multiplicity, and let σ be a nonsmooth face of τ_0

of minimal dimension. Let the star of σ be denoted Star(σ). Then, applying the construction (19) to each $\tau \in \mathrm{Star}(\sigma)$, we can define the set of cones

$$\Sigma_1 = \Sigma \setminus \mathrm{Star}(\sigma) \cup \bigcup_{\tau \in \mathrm{Star}(\sigma)} \{\tau_1, \dots, \tau_d\}.$$

It is straightforward to show that Σ_1 is a simplicial fan. By (20), this fan replaces the cones in Star(σ) with cones of strictly smaller multiplicity. Hence we lose at least one cone of maximum multiplicity (namely, τ_0) and create no new cones of this multiplicity. It follows that Σ_1 satisfies (18). Furthermore, since σ is nonsmooth, no smooth cone of Σ can lie in Star(σ), which shows that Σ_1 doesn't change the smooth cones of Σ. Hence Σ_1 has all of the desired properties, and the proof of the theorem is complete. □

By refining the argument of Theorem 5.1, one can prove the existence of a toric resolution $\phi : X_{\Sigma'} \to X_\Sigma$ such that ϕ is a *projective* morphism.

Our final result shows that toric varieties have especially mild singularities.

Theorem 5.2. *A toric variety X_Σ has at worst rational singularities.*

Proof. By Theorem 5.1, we can find a toric resolution $\phi : X_{\Sigma'} \to X_\Sigma$. To prove that X_Σ has rational singularities, we must show that

$$\phi_* \mathcal{O}_{X_{\Sigma'}} \simeq \mathcal{O}_{X_\Sigma} \quad \text{and} \quad R^i\phi_* \mathcal{O}_{X_{\Sigma'}} = 0 \text{ for } i > 0.$$

This is a local assertion on X_Σ, so that we can reduce to the affine case where $\phi : X_{\Sigma'} \to X_\sigma$ and Σ' refines the fan consisting of a cone σ and its faces. Then it suffices to prove

$$H^0(X_{\Sigma'}, \mathcal{O}_{X_{\Sigma'}}) \simeq H^0(X_\sigma, \mathcal{O}_{X_\sigma}) \quad \text{and} \quad H^i(X_{\Sigma'}, \mathcal{O}_{X_{\Sigma'}}) = 0 \text{ for } i > 0.$$

We first consider global sections. Since $\mathcal{O}_{X_{\Sigma'}}$ is the line bundle of the zero divisor, equation (11) tells us that

$$H^0(X_{\Sigma'}, \mathcal{O}_{X_{\Sigma'}}) = \bigoplus\nolimits_{m \in P_0 \cap M} k\chi^m,$$

where P_0 is the polytope

$$P_0 = \{m \in M_{\mathbb{R}} : \langle m, n_\rho \rangle \geq 0 \text{ for all } \rho \in \Sigma'(1)\}.$$

However, Σ' being a refinement implies that $|\Sigma'| = \sigma$. From this, one easily sees that $P_0 = \sigma^\vee$, and it follows that

$$H^0(X_{\Sigma'}, \mathcal{O}_{X_{\Sigma'}}) = \bigoplus\nolimits_{m \in \sigma^\vee \cap M} k\chi^m = k[\sigma^\vee \cap M] = H^0(X_\sigma, \mathcal{O}_{X_\sigma}),$$

as desired.

Finally, $H^i(X_{\Sigma'}, \mathcal{O}_{X_{\Sigma'}}) = 0$ for $i > 0$ since $\mathcal{O}_{X_{\Sigma'}}$ is generated by its global sections. This is a general fact which is true for *any* toric variety. A proof can be found in [Ful, Sect. 3.5] or [Oda1, Cor. 2.9]. The proof of the theorem is now complete. □

6. Conclusion

This brings us to the end of our discussion of toric varieties and toric resolutions. However, it is important to realize that a lot more can be done with toric varieties, with applications to many areas of mathematics. Even when we restrict to singularities and resolutions, there are many topics related to toric varieties not covered in this introduction. In particular, we would like to draw the reader's attention to the following:

- There have been numerous recent studies of the singularities of toric varieties. These are surveyed (very briefly) in [Cox2, Section 11].
- The closely related notions of *toroidal embeddings* [KKMS] and *toroidal varieties* [Dan] allow one to study more general varieties which locally analytically look like toric varieties. These are used in [KKMS] to prove the semi-stable reduction theorem.
- By using infinite fans invariant under a discrete group action, one can resolve some interesting singularities. Two-dimensional Hilbert cusp singularities are discussed in [Oda1, Section 4.1], and the general case can be found in [AMRT].
- Recent applications of toric geometry to resolution of singularities and singularities of curves can be found in the articles by Abramovich–Oort and Teissier–Goldin in this volume.

Toric varieties are an interesting, useful, and unusually accessible part of algebraic geometry. We hope the introduction given here will encourage the reader to learn more about this beautiful subject.

7. Acknowledgements

The author is very grateful to the reviewers for their useful suggestions and careful reading of the manuscript.

References

[AMRT] Ash, A., Mumford, D., Rapoport, M., and Tai, Y.: Smooth Compactifications of Locally Symmetric Varieties. Lie Groups: History, Frontiers and Applications IV, Math. Sci. Press, Brookline, MA, 1975.

[Bat1] Batyrev, V.: Quantum cohomology of toric manifolds. In: Journées de Géométrie Algébrique d'Orsay (Juillet 1992), Astérisque **218**, Société Math. France, Paris, 1993, 9–34.

[Bat2] Batyrev, V.: Variations of the mixed Hodge structure of affine hypersurfaces in algebraic tori. Duke Math. J. **69** (1993), 349–409.

[Cox1] Cox, D.: The homogeneous coordinate ring of a toric variety. J. Algebraic Geom. **4** (1995), 17–50.

[Cox2] Cox, D.: Recent developments in toric geometry. In: Algebraic Geometry – Santa Cruz 1995, Volume 2. J. Kollár, R. Lazarsfeld and D. Morrison, editors, AMS, Providence, RI, 1997, 389–436.

[Dab] Dabrowski, R.: On normality of the closure of a generic torus orbit in G/P. Pacific J. Math. **172** (1996), 321–330.

[Dan] Danilov, V.: The geometry of toric varieties. Russian Math. Surveys **33**:2 (1978), 97–154.

[Dem] Demazure, M.: Sous-groupes algébriques de rang maximum du groupe de Cremona. Ann. Sci. École Norm. Sup. **3** (1970), 507–588.

[Ewa] Ewald, G.: Combinatorial Convexity and Algebraic Geometry. Springer-Verlag, New York Berlin Heidelberg, 1996.

[EW] Ewald, G., Wessels, U.: On the ampleness of invertible sheaves in complete toric varieties. Results Math. **19** (1991), 275–278.

[Ful] Fulton, W.: Introduction to Toric Varieties. Princeton Univ. Press, Princeton, NJ, 1993.

[GKZ1] Gelfand, I., Kapranov, M., Zelevinski, A.: Discriminants, Resultants, and Multidimensional Determinants. Birkhäuser, Boston Basel Berlin, 1994.

[GKZ2] Gelfand, I., Kapranov, M., Zelevinski, A.: Hypergeometric functions and toral manifolds. Funct. Anal. Appl. **23** (1989), 94–106. Corrections in Funct. Anal. Appl. **27** (1995), 295.

[Har] Hartshorne, R.: Algebraic Geometry. Springer-Verlag, New York Berlin Heidelberg, 1977.

[Kaj] Kajiwara, T.: The functor of a toric variety with enough invariant effective Cartier divisors. Tôhoku Math J. **50** (1998), 139–157.

[KKMS] Kempf, G., Knudsen, F., Mumford, D., Saint-Donat, B.: Toroidal Embeddings, I. Lecture Notes in Math. **339**, Springer-Verlag, New York Berlin Heidelberg, 1973.

[Oda1] Oda, T.: Convex Bodies and Algebraic Geometry. Springer-Verlag, New York Berlin Heidelberg, 1988.

[Oda2] Oda, T.: Lectures on Torus Embeddings and Applications (Based on joint work with Katsuya Miyake). Tata Inst. Fund. Research **58**, Springer-Verlag, New York Berlin Heidelberg, 1978.

[Rei] Reid, M.: Canonical 3-folds. In: Journées de Géometrie Algébrique d'Angers. A. Beauville, editor, Sijthoff & Noordhoff, Alphen aan den Rijn, 1980, 273–310.

[Stu1] Sturmfels, B.: Equations defining toric varieties. In: Algebraic Geometry – Santa Cruz 1995, Volume 2. J. Kollár, R. Lazarsfeld and D. Morrison, editors, AMS, Providence, RI, 1997, 437–449.

[Stu2] Sturmfels, B.: Gröbner Bases and Convex Polytopes. University Lecture Series **8**, AMS, Providence, RI, 1996.

Dept. of Mathematics and Computer Science
Amherst College
Amherst, MA 01002 USA
`dac@cs.amherst.edu`

Progress in Mathematics, Vol. 181, © 2000 Birkhäuser Verlag Basel/Switzerland

A Compactification of a Fine Moduli Space of Curves

Bert van Geemen and Frans Oort

Introduction

In [4] Deligne and Mumford define stable curves and prove that the moduli space $\overline{\mathcal{M}_g}$ of stable curves of genus g is a "compactification" of the moduli space $\mathcal{M}_g$ of smooth curves. For any given integer $m \geq 3$ (invertible on some base scheme) Mumford has constructed a fine moduli scheme $\mathcal{M}_{g,m}$ of curves of genus g with level-m-structure; moreover $\mathcal{M}_{g,m} \to \mathcal{M}_g$ is a Galois covering. It is useful to have a compactification of $\mathcal{M}_{g,m}$,

$$\begin{array}{ccc} \mathcal{M}_{g,m} & \hookrightarrow & ? \\ \downarrow & & \downarrow \\ \mathcal{M}_g & \hookrightarrow & \overline{\mathcal{M}_g}. \end{array}$$

We find definitions and properties of such a compactification in [4], p. 106, in [12], Lecture 10, and in [1], § 2. With the convenient definitions, the results are not so difficult to find, and in this note we put these properties together. The main results are:

- *A compactification $\overline{\mathcal{M}_{g,m}}$, with a tautological family $\mathcal{D} \to \overline{\mathcal{M}_{g,m}}$ exists* (Theorem 2.1).
- *This space is not constructed as a coarse or a fine moduli scheme associated with a moduli functor.*
- *The compactification $\overline{\mathcal{M}_{g,m}}$ is a normal space, singular for* $g \geq 3$; for the description of its local structure, see Theorem 3.1.

The results of this note were written up in preprint 301 (August 1983) of the Mathematics Department of the University of Utrecht. Our results were partly contained in [9], and we never published this preprint.

We thank J.-L. Brylinski and J. Steenbrink for stimulating discussions.

1. Notation

Let us first introduce level structures. We fix an integer $m \in \mathbb{Z}_{\geq 1}$ (and soon we shall suppose $m \geq 3$). In [10], p. 129, a level structure on an abelian variety X of dimension g is defined as an isomorphism $X[m] \cong (\mathbb{Z}/m)^{2g}$. Here we adopt a slightly different concept.

Let S be a base scheme, and $m \in \mathbb{Z}_{\geq 1}$. Note that $((\mathbb{Z}/m)_S)^D \cong \mu_{m,S}$ (here superscript D refers to Cartier duality of finite group schemes). The natural bi-homomorphism

$$e : ((\mathbb{Z}/m)_S)^g \times (\mu_{m,S})^g) \times ((\mathbb{Z}/m)_S)^g \times (\mu_{m,S})^g) \longrightarrow \mu_{m,S}$$

defined by Cartier duality is called the symplectic pairing.

Let $X \to S$ be an abelian scheme of relative dimension g. Suppose m is invertible on S; i.e. there is a canonical morphism $S \to \mathrm{Spec}\ \mathbb{Z}[\frac{1}{m}]$. A *symplectic level m-structure on X/S* is an isomorphism

$$\phi : X[m] \xrightarrow{\sim} ((\mathbb{Z}/m)_S)^g \times (\mu_{m,S})^g$$

which identifies the Weil pairing $e_X : X[m] \times X[m] \to \mu_{m,S}$ with the symplectic pairing. Let $C \to S$ be a smooth and proper curve over S. A *symplectic level m-structure on C/S* is a symplectic level m-structure on $J := \mathrm{Pic}^0_C$.

Remark. Suppose m is invertible on S, and suppose a choice of a primitive m-th root of unity $\zeta_m \in \Gamma(S, \mathcal{O}_S)$ is possible, and has been made. Then we obtain an identification $(\mathbb{Z}/m)_S \cong \mu_{m,S}$, and the notion of a symplectic level-m-structure just given is the same as the one given in [10], p. 129. We could work with schemes over $T = \mathrm{Spec}\ \mathbb{Z}[\zeta_m, \frac{1}{m}]$, and define a level-$m$-structure using the identification $(\mathbb{Z}/n)_T \cong \mu_{m,T}$.

Remark. The definition given above can be generalized as follows. Let m be invertible on S and suppose given a finite flat group scheme $\mathcal{H} \to S$ such that every geometric fiber is isomorphic to $(\mathbb{Z}/m)^{2g}$ with a skew pairing $e : \mathcal{H} \times \mathcal{H} \to \mu_{m,S}$. Use this to define an e-symplectic pairing. The advantage of this is shown in the following example. Choose an elliptic curve E, say over $\mathbb{Q}$, and let $\mathcal{H} := E[m]$ equipped with its Weil-pairing. Call it e. The modular curve representing full level m-structure with this e-symplectic pairing is representable (say for $m \geq 3$), and it has a $\mathbb{Q}$-rational point, given by the existence of E.

If you feel that all these fine points are too fancy, just stick to a symplectic structure on $(\mathbb{Z}/m)^{2g}$. If you work with base schemes over $T = \mathrm{Spec}\ \mathbb{Z}[\zeta_m, \frac{1}{m}]$, there will be no difference.

In the sequel we shall fix

- an integer $g \geq 2$ (the genus),
- an integer $m \geq 3$ (the level),
- and an integer $\nu \geq 5$ (used in multi-canonical embeddings),

- and $N := (2\nu - 1)(g-1) - 1$; note that the ν-multi-canonical map gives $\Phi_{\nu \cdot K} : C \hookrightarrow \mathbb{P}$ with $\mathbb{P} = \mathbb{P}^N$ the projective space of dimension N. We write PGL for PGL(N) (= Aut($\mathbb{P}$)) and set $S_n = \text{Spec } (\mathbb{Z}[\frac{1}{n}])$.

By $\mathcal{M}_g \to S_1 := \text{Spec } (\mathbb{Z})$ we denote the (coarse) moduli scheme of curves of genus g as defined and constructed in [4]. We write

$$\begin{array}{ccc} \mathcal{H}_g^0 & \hookrightarrow & \mathcal{H}_g \\ & \searrow & \downarrow \\ & & S \end{array}$$

for the Hilbert schemes of smooth (respectively stable) ν-canonically embedded curves of genus g. Note that

$$\mathcal{M}_g = \text{PGL}\backslash\mathcal{H}_g^0 \quad \text{and} \quad \overline{\mathcal{M}_g} = \text{PGL}\backslash\mathcal{H}_g.$$

The existence theorems for $\mathcal{M}_g$ and for $\mathcal{M}_{g,m}$ are contained in [10]. By [7] we conclude that PGL is geometrically reductive, and by [11], Th. 5.1, we know that its action on $\mathcal{H}_g$ is stable, hence the required geometric quotient exists.

By $\mathcal{M}_{g,m} \to S_m$ we denote the (fine) moduli scheme of curves with a simplectic level-m-structure. Note that this is a fine moduli scheme, i.e. there exists a universal curve

$$\mathcal{D}^0 \to \mathcal{M}_{g,m}$$

with a symplectic level-m-structure representing the functor of smooth curves of genus g with such levels. The level-m-structures will be symplectic, thus the covering $\tau : \mathcal{M}_{g,m} \to \mathcal{M}_g$ is Galois with group $\Gamma = \text{Sp}(2g, \mathbb{Z}/m)$ (cf. [10], 7.3). Note that $\mathcal{M}_{g,m} \to S_m$ is smooth (because $m \geq 3$); in particular $\mathcal{M}_{g,m}$ is a normal space. Note that for every field k of characteristic not dividing m the fiber $\mathcal{M}_{g,m} \otimes \text{Spec } (K)$ is a regular variety.

We define

$$\overline{\mathcal{M}_{g,m}} \longrightarrow S_m := \text{Spec } \mathbb{Z}[\frac{1}{m}]$$

as the normalization of $\overline{\mathcal{M}_g}$ in the function field of $\mathcal{M}_{g,m}$; thus $\mathcal{M}_{g,m} \hookrightarrow \overline{\mathcal{M}_{g,m}}$. Compare with [1], p. 307; [4], p. 106. Note that for every field k of characteristic not dividing m the fiber $\overline{\mathcal{M}_{g,m}} \otimes \text{Spec } (K)$ is a normal variety, see [4], p. 106, Th. 5.9.

Remark. It would be much more natural to define a moduli functor of "stable curves with level structures" first, and then try to have a coarse or fine moduli scheme, thus arriving at a definition of $\overline{\mathcal{M}_{g,m}}$. However we do not know of such a representable moduli functor defining $\overline{\mathcal{M}_{g,m}}$. Once the local structure is studied, see Section 3, it will be clear that no "easily defined" functor will do.

To define tautological curves, let T be a scheme, and $f : T \to M$ a morphism, where M is a moduli space of curves. Consider a curve $\mathcal{C} \to T$ (equipped with extra structure ...). We say this curve is a *tautological curve* if it defines f. In particular, in this case, and given a geometric point $t \in T$, the moduli point of the fiber $\mathcal{C}_t$ equals $f(t)$:

$$[\mathcal{C}_t] \quad = f(t) \in M.$$

Sometimes this is also called a "universal curve". This terminology can be misleading! However, if we have a fine moduli scheme, the universal curve is tautological.

2. Construction of a tautological family

Theorem 2.1. *Let $g \in \mathbb{Z}_{\geq 2}$ and $m \in \mathbb{Z}_{\geq 3}$. There is a unique*

$$\mathcal{D} \to \overline{\mathcal{M}_{g,m}}$$

which is tautological, and a symplectic level-m-structure on

$$\mathcal{D}|_{\mathcal{M}_{g,m}} =: \mathcal{D}^0 \to \mathcal{M}_{g,m}$$

representing the moduli functor of smooth curves with level structure.

One could also formulate the theorem for $g = 0$. Though pedantic, it would be useful for later purposes in the case of moduli spaces of pointed curves; in that case the first claim of the theorem is still valid. For $g = 1$, considering curves of genus $g = 1$ with one base point (called elliptic curves) the theorem is not so difficult and well known. From now on we suppose $g \geq 2$.

Suppose given an open set U in a scheme T and a stable curve over U. If U is dense in T with T normal, and if this curve extends to a stable curve over T, this extension to a stable curve is unique once this is possible (this follows using [4], Th. 1.11, and Zariski's Main Theorem). The uniqueness in the theorem is not the surprising part, but existence will require some work.

We write $\mathcal{H}^0_{g,m} \longrightarrow S$ for the Hilbert scheme of ν-canonically embedded smooth curves of genus g with symplectic level-m-structure.
Note that $\mathrm{PGL}\backslash\mathcal{H}^0_{g,m} = \mathcal{M}_{g,m}$.

We define (cf. [12], p. 137; this definition differs from the one given in [1], p. 307)

$$\mathcal{H}_{g,m} \longrightarrow S$$

as the normalization of $\mathcal{H}_g$ in the function field of $\mathcal{H}^0_{g,m}$,

$$\begin{array}{ccc} \mathcal{H}^0_{g,m} & \hookrightarrow & \mathcal{H}_{g,m} \\ \downarrow & & \downarrow \\ \mathcal{H}^0_{g} & \hookrightarrow & \mathcal{H}_{g}. \end{array}$$

Note that $\mathcal{H}_g$ and $\mathcal{H}_{g,m}$ are non-complete varieties. This is the reason why we write $\mathcal{H}_{g,m}$ instead of $\overline{\mathcal{H}^0_{g,m}}$.

A point $y \in \mathcal{H}_g$ corresponds to a ν-canonically embedded curve $C_y \subset \mathbb{P}$. A point $x \in \mathcal{H}^0_{g,m}$ corresponds to a pair $x = (C_y \subset \mathbb{P}, \phi)$ where

$$\phi : J(C_y)[m] \xrightarrow{\sim} (\mathbb{Z}/m)^{2g}$$

is a symplectic isomorphism. For $a \in \mathrm{PGL}$ we define $ax = (C_{ay} \subset \mathbb{P}, \phi \circ (a_{|C_y})^*)$ where

$$\begin{array}{ccc} C_y & \longrightarrow & \mathbb{P} \\ {\scriptstyle a_{|C_y}}\downarrow & & \downarrow{\scriptstyle a} \\ a(C_y) = C_{ay} & \longrightarrow & \mathbb{P}. \end{array}$$

The Picard aspect of the Jacobian variety gives isomorphisms

$$J(C_{ay})[m] \xrightarrow{(a_{|C_y})^*} J(C_y)[m] \xrightarrow{\phi} (\mathbb{Z}/m)^{2g}.$$

This gives an action $\mathrm{PGL} \times \mathcal{H}^0_{g,m} \longrightarrow \mathcal{H}^0_{g,m}$ which extends uniquely to an action of PGL on $\mathcal{H}_{g,m}$ (by [15], Lemma 6.1). For $h \in \Gamma = \mathrm{Sp}(2g, \mathbb{Z}/m)$ and $x \in \mathcal{H}_{g,m}$ we define $h \cdot x$ in the natural way by

$$h \cdot x = h \cdot (C_y \subset \mathbb{P}, \phi) := (C_y \subset \mathbb{P}, h \circ \phi).$$

This action commutes with the action of PGL

$$h \cdot ax = (C_{ay} \subset \mathbb{P}, h \circ \phi \circ (a_{|C_y})^*) = a(h \cdot x).$$

The action of Γ on $\mathcal{H}^0_{g,m}$ extends to an action on $\mathcal{H}_{g,m}$ and we obtain an action

$$(\mathrm{PGL} \times \Gamma) \times \mathcal{H}_{g,m} \longrightarrow \mathcal{H}_{g,m}.$$

Note that PGL also acts on the universal families $\mathcal{C} \to \mathcal{H}_{g,m}$ and $\mathcal{E} \to \mathcal{H}_g$. We conclude that in the diagram

$$\begin{array}{ccccccccc} \mathcal{H}_{g,m} & \supset & \mathcal{H}^0_{g,m} & \longrightarrow & \mathrm{PGL}\backslash\mathcal{H}^0_{g,m} = & \mathcal{M}_{g,m} & \subset & \overline{\mathcal{M}_{g,m}} & \\ {\scriptstyle \pi}\downarrow & & \downarrow & & & \downarrow & & \downarrow{\scriptstyle \tau} & \\ \mathcal{H}_g & \supset & \mathcal{H}^0_g & \longrightarrow & \mathrm{PGL}\backslash\mathcal{H}^0_g = & \mathcal{M}_g & \subset & \overline{\mathcal{M}_g} & = \mathrm{PGL}\backslash\mathcal{H}_g \end{array}$$

the vertical arrows are Galois coverings, all with Galois group $\Gamma = \mathrm{Sp}(2g, \mathbb{Z}/m)$.

Claim 2.2. *The action of* PGL *on $\mathcal{H}_{g,m}$ has no fixed points.*

The claim implies the theorem, see the end of this section. To prove the claim will require some preparations and auxiliary results. We will see then that $\mathrm{PGL}\backslash\mathcal{H}_{g,m} = \overline{\mathcal{M}_{g,m}}$, and that the universal family of curves $\mathcal{C} \to \mathcal{H}_{g,m}$ descends to a family of stable curves

$$\begin{array}{ccc} \mathcal{C} & \longrightarrow & \mathrm{PGL}\backslash\mathcal{C} = \mathcal{D} \\ \downarrow & & \downarrow \\ \mathcal{H}_{g,m} & \longrightarrow & \mathrm{PGL}\backslash\mathcal{H}_{g,m} = \overline{\mathcal{M}_{g,m}}. \end{array}$$

Before we give proofs we introduce some further notation. Let $\mathcal{E} \to \mathcal{H}_g$ be the universal family; by results of Raynaud, cf. [13], it can be proved that

$$J := \mathrm{Jac}(\mathcal{E}/\mathcal{H}_g) := \mathbf{Pic}^0(\mathcal{E}/\mathcal{H}_g)$$

exists and that this formation commutes with base change (cf. [4], Th. 2.5, for precise arguments). Denote by $J[m] \subset J \to S$ the scheme of m-torsion points. Note that the morphism $J[m] \to S$ is étale and quasi-finite. Let $x \in \mathcal{H}_{g,m}$, $y = \pi(x) \in \mathcal{H}_g$, let $U_y \subset \mathcal{H}_g$ be the formal neighborhood of y in $\mathcal{H}_g$, and let $\Delta = \Delta_y \subset U_y$

denote the locus of points over which $\mathcal{E} \to \mathcal{H}_g$ is not smooth. By [4], 1.9, we know that Δ is a divisor with normal crossings. We denote by v the generic point of U_y, and by $\pi_1(U_y \setminus \Delta, v)$ the algebraic fundamental group prime to the characteristic of $k(y)$. Note that

$$J[m]_{|(U_y \setminus \Delta)} \longrightarrow U_y \setminus \Delta$$

is an étale covering. Thus we obtain a representation

$$R : \pi_1(U_y \setminus \Delta, v) \longrightarrow \mathrm{Sp}(J_v[m]).$$

The image of this "monodromy" R is denoted by

$$I_y := R(\pi_1(U_y \setminus \Delta, v)) \subset \mathrm{Sp}(J_v[m]),$$

where $J_v[m]$ is considered as the abstract group $J_v[m](\overline{k(v)})$.

Lemma 2.3. *The fiber $\pi^{-1}(y) \subset \mathcal{H}_{g,m}$ is given by*

$$\pi^{-1}(y) = I_y \backslash \{\phi | \, \phi : J_v[m] \xrightarrow{\sim} (\mathbb{Z}/m)^{2g}, \ \phi \text{ symplectic}\}.$$

Proof. Consider

$$\begin{array}{ccccccc} X = & \pi^{-1}(Y) & \hookrightarrow & \pi^{-1}(U_y) & \supset & \pi^{-1}(y) \\ \downarrow \pi^0 & & & \pi \downarrow & & \\ Y = & U_y \setminus \Delta & \hookrightarrow & U_y & \supset & \{y\}. \end{array}$$

As we work with stable curves, the monodromy around each component of Δ is unipotent, so of order dividing n, so not divisible by $\mathrm{char}(k(y))$ (we work over $S_n = \mathrm{Spec}\ (\mathbb{Z}[n^{-1}])$). The covering $\pi^0 : X \to Y$ is étale so that each component is determined by the monodromy R. Using that R is tame, Δ a divisor with normal crossings, and $\mathcal{H}_{g,m}$ normal we see that π is a generalized Kummer covering in the sense of [6], p. 12 (use [6], p. 39, Corollary 2.3.4). For Kummer coverings one easily proves that the fiber over a point (as a set) is canonically isomorphic with the orbits of the inertia group in the Galois group. Hence the same result follows for generalized Kummer coverings. By definition of $\mathcal{H}_{g,m}$ the fiber $\pi^{-1}(v)$ corresponds to the set of all ϕ as indicated. Thus the lemma follows. □

Lemma 2.4. *We have $J_y[m] = (J_v[m])^{I_y}$.*

Proof. First we note that $J[m] \to \mathcal{H}_g$ is an étale morphism, hence a point of $J_y[m]$ extends to a section of $J[m]_{|U_y} \to U_y$, cf. [5], 18.5.17; thus

$$J_y[m] \hookrightarrow (J_v[m])^{I_y}.$$

Let k be the number of singular points of the curve $C = \mathcal{E}_y$. By [4], Theorem 1.6, we can choose local coordinates $\{t_i\}$ $(1 \le i \le N)$ in $y \in U_y \subset \mathcal{H}_g$ so that $\Delta \subset U_y$ is given by the union of the divisors defined by $t_j = 0$, $1 \le j \le k$; moreover the singularities of $\mathcal{E}_y$ are locally given inside $\mathcal{E}$ by equations of the form $a_j b_j - t_j = 0$. Define $V_i \subset U_y$ by the equations $t_i = t_{i+1} = \cdots = t_N = 0$. Thus $V_1 = \{y\}$, $V_i \subset V_{i+1}$ and $V_{N+1} = U_y$. Let $v_i \in V_i$ be the generic point. In each step $V_i \subset V_{i+1}$, $1 \le i \le N$, we can apply local monodromy on one parameter, cf. [16], p. 495, Lemma 2: let $\mathcal{A}$ be the Néron minimal model of $J_{v_{i+1}}$ over $\mathrm{Spf}(k(v_i)[[t_i]])$

with special fiber $A := \mathcal{A}(t_i \mapsto 0)$. Then the invariants under the monodromy group in $J_{v_{i+1}}[m]$ are precisely $A[m]$. As the singularities are ordinary quadratic singularities, by [2], p. 192, Proposition 3.3.5, we conclude that the relevant part of the monodromy is either trivial or equals

$$\begin{pmatrix} 1 & 1 \\ 0 & 1 \end{pmatrix}.$$

Therefore $A[m] \cong (\mathbb{Z}/m)^{m-1}$ when $J_{v_{i+1}}[m] \cong (\mathbb{Z}/m)^m$ in the first case and $A[m] \cong J_{v_{i+1}}[m]$ in the second case. We conclude that $J_y[m] \supset (J_v[m])^{I_y}$.

Note that the term "relevant part of the monodromy" has the following meaning. Consider a stable curve C_0 with singularities $P_1, \dots, P_d$. The universal deformation space D of C_0 contains $\Delta \subset D$, the discriminant locus, the closed part over which the universal curve is non-smooth. This is a union of divisors $\Delta = \cup_{1 \le i \le d}\ H_i$, such that "$P_i$ stays singular" over H_i. The monodromy around H_i is trivial if and only if deleting P_i gives a disconnected curve $C_0 \setminus \{P_i\}$. In this case the generic point of H_i parameterizes a "curve of compact type", i.e. a curve where the Jacobian variety is an abelian variety. The monodromy around H_i is non-trivial, and in fact is as indicated above, if and only if the curve $C_0 - \{P_i\}$ is connected. This is the case if the generic point of H_i parameterizes an irreducible (singular) curve, and in this case its Jacobian variety is not an abelian variety. □

Proposition 2.5. *In the diagram*

$$\begin{array}{ccc} F := \{\phi \mid \phi : J_v[m] \xrightarrow{\sim} (\mathbb{Z}/m)^{2g},\ \phi \text{ symplectic}\} & \longrightarrow & I_y\backslash F \\ \downarrow & \swarrow \eta & \\ F' := \{\rho \mid \rho : J_y[m] \hookrightarrow (\mathbb{Z}/m)^{2g},\ \rho \text{ symplectic}\} & & \end{array}$$

the natural map η is surjective, but in general not injective.

Proof. From the preceding lemma we have $J_y[m] \hookrightarrow (J_v[m])^{I_y}$, and this shows the existence of η. The symplectic structure on $J_y[m]$ (which is degenerate if and only if J_y is not an abelian variety) is induced by the symplectic structure on $J_v[m]$, and it is not difficult to see that a given ρ extends

$$\begin{array}{ccc} J_v[m] & \xrightarrow{\exists} & (\mathbb{Z}/m)^{2g} \\ \cup & & \rho \uparrow \\ (J_v[m])^{I_y} & \supset & J_y[m]. \end{array}$$

We choose an irreducible curve C with two ordinary double points. We can choose a symplectic base (intersection form in standard form)

$$\{\alpha_1, \dots, \alpha_g, \beta_1, \dots, \beta_g\} \quad \text{for} \quad J_v[m]$$

such that I_y acts on $J_v[m]$ by

$$\begin{cases} \{\alpha_1, \dots, \alpha_{g-2}, \beta_1, \dots, \beta_g\} \subset (J_v[m])^{I_y} = J_y[m], \\ \alpha_{g-1} \longmapsto \alpha_{g-1} + k_1\beta_{g-1}, \\ \alpha_g \longmapsto \alpha_g + k_2\beta_g, \end{cases}$$

with $k_1, k_2 \in \mathbb{Z}/m$ depending on which element of I_y is acting. This means that I_y acts via matrices of the form

$$\left(\begin{array}{c|ccc} & 0 & \vdots & 0 \\ 1 & \cdots & & \cdots \\ & 0 & \vdots & B \\ \hline 0 & 1 & & \end{array}\right) \qquad B = \begin{pmatrix} k_1 & 0 \\ 0 & k_2 \end{pmatrix},$$

where 1 stands for a diagonal-1-matrix and 0 for a zero matrix. This determines the action of I_y, and $\{\alpha_1, \dots, \alpha_{g-2}, \beta_1, \dots, \beta_g\}$ is a base for $J_y[m]$. Take a ρ and extend it to some ϕ. Note that the matrix

$$N = \left(\begin{array}{c|ccc} & 0 & \vdots & 0 \\ 1 & \cdots & & \cdots \\ & 0 & \vdots & A \\ \hline 0 & 1 & & \end{array}\right) \qquad A = \begin{pmatrix} 0 & 1 \\ 1 & 0 \end{pmatrix}$$

is symplectic; hence $\psi := \phi \circ N$ is symplectic. Note that $N \notin I_y$, thus ϕ and ψ define different elements of $I_y \backslash F$, but they restrict to the same $\rho \in \eta(I_y \backslash F)$. This proves the proposition. □

Remark. Note that this shows that [12], p. 137, Theorem 10.6, is incorrect.

One would like to have an a priori definition of the functor which is represented by the scheme $\overline{\mathcal{M}_{g,m}}$ (and the same for $\mathcal{H}_{g,m}$); we were unable to do so, and the previous proposition indicates why we have chosen our definition.

Proof of the claim 2.2. We denote by $\mathbb{P} \supset \mathcal{C} \longrightarrow \mathcal{H}_{g,m}$ the universal family, by $J = \mathrm{Jac}(\mathcal{C}/\mathcal{H}_{g,m})$ its relative Jacobi scheme (which is the pull-back by $\mathcal{H}_{g,m} \to \mathcal{H}_g$ of $\mathrm{Jac}(\mathcal{E}/\mathcal{H}_g)$). Let

$$\mathcal{N} = J[m] \supset \mathcal{N}^0 = \mathcal{N}_{|\mathcal{H}^0_{g,m}}$$

be endowed with the universal symplectic level-m-structure

$$\Phi : \mathcal{N}^0 \xrightarrow{\sim} (\mathbb{Z}/n\mathbb{Z})^{2g} \times \mathcal{H}^0_{g,m}.$$

Let $a \in \mathrm{PGL}$, $x \in \mathcal{H}_{g,m}$ and $a(x) = x$. By the universal property of Hilbert schemes there exists a unique morphism A such that

$$\begin{array}{ccccc} \mathcal{E} & \xrightarrow{A} & a^{-1}\mathcal{E} & \longrightarrow & \mathcal{E} \\ & \searrow & \downarrow & & \downarrow \\ & & \mathcal{H}_g & \xrightarrow{a} & \mathcal{H}_g \end{array}$$

commutes. As $a(x) = x$, for $y = \pi(x) \in \mathcal{H}_g$ we have $a(y) = y$, thus

$$\begin{array}{ccc} \mathcal{E}_y & \hookrightarrow & \mathbb{P} \\ {\scriptstyle A_y}\big\downarrow & & \big\downarrow{\scriptstyle a} \\ (a^{-1}\mathcal{E}) = \mathcal{E}_y & \hookrightarrow & \mathbb{P}. \end{array}$$

This implies $A_y \in \mathrm{Aut}(\mathcal{E}_y)$. If we show that $A_y = \mathrm{id}$, it follows that $a = \mathrm{id}$ (because $\mathcal{E}_y \hookrightarrow \mathbb{P}$ is ν-canonical). Consider

$$F' := \{\rho |\, \rho : J_y[m] \hookrightarrow (\mathbb{Z}/m)^{2g},\ \rho \text{ symplectic}\},$$

and let $a_* : F' \longrightarrow F'$ and $a_*(\rho) := \rho \circ (A_y)^*$. Note that $a \in \mathrm{PGL}$ acts on $\mathcal{N}^0 \to \mathcal{H}^0_{g,m}$; it follows that

$$\begin{array}{ccccc} (\mathbb{Z}/m)^{2g} \times \mathcal{H}^0_{g,m} & & & & (\mathbb{Z}/m)^{2g} \times \mathcal{H}^0_{g,m} \\ {\scriptstyle \Phi}\big\uparrow & & & a^{-1}(\Phi) \nearrow & \big\uparrow{\scriptstyle \Phi} \\ \mathcal{N}^0 & \xleftarrow{a^*} & a^{-1}(\mathcal{N}^0) & \longrightarrow & \mathcal{N}^0 \\ & \searrow & \downarrow & & \downarrow \\ & & \mathcal{H}^0_{g,m} & \xrightarrow{a} & \mathcal{H}^0_{g,m} \end{array}$$

has the property $\Phi \circ a^* = a^{-1}(\Phi)$. If $ay = y$ then a maps $\pi^{-1}(y)$ to itself. Now observe, using the notation of Lemma 2.3, that the diagram

$$\begin{array}{ccc} \pi^{-1}(y) & \xrightarrow{a} & \pi^{-1}(y) \\ \| & & \| \\ I_y\backslash F & \xrightarrow{a} & I_y\backslash F \end{array}$$

gives rise to a map a' such that

$$\begin{array}{ccc} I_y\backslash F & \xrightarrow{a} & I_y\backslash F \\ {\scriptstyle \eta}\big\downarrow & & \big\downarrow{\scriptstyle \eta} \\ F' & \xrightarrow{a'} & F' \end{array}$$

and $a' = a_* : F' \longrightarrow F'$ as constructed above. Indeed, we have seen that $\Phi \circ a^* = a^{-1}(\Phi)$ on $\mathcal{N}^0$. From $a(y) = y$ we get $\phi \circ a^* = a^{-1}(\phi)$ where $\phi = \Phi_{\mathcal{N}_y}$.

Recall from [1], Lemma 4:

Lemma 2.6 (Serre's lemma). *Let C be a stable curve and let $A \in \mathrm{Aut}(C)$ be such that*

$$A^* : \mathrm{Jac}(C)[m] \longrightarrow \mathrm{Jac}(C)[m]$$

equals the identity (where $m \geq 3$ is an integer, prime to the characteristic). Then A equals the identity.

End of proof of Claim 2.2. We have assumed that $ax = x$, so for $y = \pi x$ we can apply the assertions proved above. There exists $\rho \in F'$, with $a'(\rho) = \rho$. This implies

$$((A_y)^* : J_y[m] \longrightarrow J_y[m]) = \mathrm{id}_{J_y[m]}$$

(because ρ is injective). We see that the diagrams above and Serre's lemma prove Claim 2.2. □

Proof that Claim 2.2 implies Thm.2.1. Note that $\overline{\mathcal{M}_{g,m}}$ is a geometric quotient of a Hilbert scheme. For $\nu \geq 5$ the points of $\mathcal{H}_g$ are stable with respect to the action of of PGL (cf. [11], Theorem 5.1), so the same holds for points of $\mathcal{H}_{g,m}$. Hence the quotient $\mathrm{PGL}\backslash\mathcal{H}_{g,m}$ exists (same arguments as in [11]); it is a normal variety (cf. [10], p. 5, and use that $\mathcal{H}_{g,m}$ is integral and normal). We get the diagram

$$\begin{array}{ccccc} & & \mathcal{M}_{g,m} = \mathrm{PGL}\backslash\mathcal{H}_{g,m} & & \\ & i \swarrow & \downarrow & \searrow i' & \\ \overline{\mathcal{M}_{g,m}} & & M & & \mathrm{PGL}\backslash\mathcal{H}_g. \\ & \tau \searrow & \downarrow & \swarrow \tau' & \\ & & \overline{\mathcal{M}_g} & & \end{array}$$

Here i, i' are inclusions and τ, τ' are finite maps from normal spaces which coincide on the set $\mathcal{M}_{g,m}$ which is dense in both; by uniqueness of the normalization we conclude (and may identify)

$$\overline{\mathcal{M}_{g,m}} = \mathrm{PGL}\backslash\mathcal{H}_{g,m}.$$

The action of PGL on $\mathcal{H}_{g,m}$ extends naturally to the universal curve $\mathcal{C} \longrightarrow \mathcal{H}_{g,m}$, and by the Claim 2.2 we conclude that this action has no fixed points on $\mathcal{C}$. Thus

$$\mathrm{PGL}\backslash\mathcal{C} =: \mathcal{D} \longrightarrow \overline{\mathcal{M}_{g,m}} = \mathrm{PGL}\backslash\mathcal{H}_{g,m}$$

is a family of stable curves (which extends the universal family over $\mathcal{M}_{g,m}$). This proves Theorem 2.1. □

3. Local structure of the moduli space

We introduce some notation, needed below. We choose $x \in \mathcal{H}_{g,m}$ and denote by C the corresponding curve. The images of this point are denoted as follows

$$\begin{array}{rcccl} x \in & \overline{\mathcal{H}^0_{g,m}} & \longrightarrow & \overline{\mathcal{M}_{g,m}} & \ni z \\ & \downarrow \pi & & \tau \downarrow & \\ \pi(x) = y \in & \mathcal{H}_g & \longrightarrow & \overline{\mathcal{M}_g} & \ni w = \tau(z). \end{array}$$

We denote by

$$I = I_x = \mathrm{Inertia}(x \in \overline{\mathcal{H}^0_{g,m}} \xrightarrow{\pi} \mathcal{H}_g) \subset \Gamma$$

the inertia group at x of the covering π, and analogously

$$G = G_z = \mathrm{Inertia}(z \in \overline{\mathcal{M}_{g,m}} \xrightarrow{\tau} \overline{\mathcal{M}_g}) \subset \Gamma;$$

note that we identified the Galois groups

$$\mathrm{Gal}(\overline{\mathcal{H}^0_{g,m}} \longrightarrow \mathcal{H}_g) = \Gamma = \mathrm{Gal}(\overline{\mathcal{M}_{g,m}} \longrightarrow \mathcal{M}_{g,m}).$$

A choice of an embedding $k(v) \subset k(U_x) \subset \overline{k(v)}$ induces an isomorphism $I_y \cong I_x$ (which explains the notation).

Let $k = \overline{k(w)}$ be an algebraically closed field. We write $W = k$ if $\mathrm{char}(k) = 0$, and $W = W_\infty(k)$ if $\mathrm{char}(k) = p > 0$, where $W_\infty(k)$ denotes the ring of infinite Witt vectors with coordinates in k. Note that the universal deformation space

$$\mathcal{X} \longrightarrow \mathrm{Spf}(W[[t_1, \dots, t_{3g-3}]]) = D$$

exists (cf. [4], pp. 81–83).

Theorem 3.1. (1) *There is an exact sequence of groups*

$$0 \to I_x \longrightarrow G_z \xrightarrow{\beta} \mathrm{Aut}(C) \to 0.$$

(2) *The fibers of* π *and* τ *are* $\pi^{-1}(y) \cong F/I_y$ *and* $\tau^{-1}(w) \cong F/G_z$ *with* F *as in Proposition 2.5.*
(3) *We have the following commutative diagram with isomorphisms as indicated*

$$\begin{array}{ccccc}
U_z & & \hookrightarrow & & \overline{\mathcal{M}_{g,m}} \\
\downarrow & & & & \\
I\backslash U_z & \cong D & & & \downarrow \\
\downarrow & \downarrow & & & \\
G\backslash U_z & \cong \mathrm{Aut}(C)\backslash D & \cong U_w \subset & & \overline{\mathcal{M}_g}.
\end{array}$$

(4) *For* $g \geq 3$ *the schemes* $\mathcal{H}_{g,m}$ *and* $\overline{\mathcal{M}_{g,m}}$ *have singularities in codimension 2.*

Remark. Suppose that $w \in \mathcal{M}_g$, i.e. that the curve C is nonsingular. Then $I_x = \{1\}$, and we have $G_z = \mathrm{Aut}(C)$, and $D \cong U_z$ (note that $m \geq 3$). Locally on $\mathcal{M}_g$ we have $U_w \cong \mathrm{Aut}(C)\backslash D \cong \mathrm{Aut}(C)\backslash U_z$. This is well known.

Proof. We have isomorphisms $C := \mathcal{E}_y \cong \mathcal{C}_x \cong \mathcal{D}_z$. Let $h \in G_z \subset \Gamma$. Then $h \in \Gamma \cong \mathrm{Gal}(\mathcal{H}_{g,m} \to \mathcal{H}_g)$ operates on $\mathcal{H}_{g,m}$, and from $h \in G_z$ it follows that z, x and hx are congruent modulo PGL. Hence there exists an element $a \in \mathrm{PGL}$, unique by 2.2, such that $hx = ax$. Then $ay = y$, and thus

$$\begin{array}{ccc}
C & \xrightarrow{y} & \mathbf{P} \\
\downarrow a_{|C} & & \downarrow a \\
C & \xrightarrow{y} & \mathbf{P}.
\end{array}$$

We define $\beta_x(h) = a_{|C} \in \mathrm{Aut}(C)$. Clearly β_x is a group homomorphism. If $\beta_x(h) = \mathrm{id}_C$, then $a = \mathrm{id}_C$, so $h \in I_x$; if $h \in I_x$ then $a = \mathrm{id}$, thus $\beta_x(h) = \mathrm{id}_C$. Equivalently $\mathrm{Ker}(\beta_x) = I_x$. By Lemma 2.3 we conclude $\sharp\pi^{-1}(y) = \sharp\Gamma/\sharp I_x$. The group $\mathrm{Aut}(C)$ acts faithfully on the fiber $\pi^{-1}(y)$, it acts via PGL, so every $\mathrm{Aut}(C)$-orbit is mapped to a point in $\overline{\mathcal{M}_{g,m}}$, thus $\sharp\pi^{-1}(w) \leq \sharp\pi^{-1}(y)/\sharp\mathrm{Aut}(C)$. Moreover $\sharp\pi^{-1}(w) = \sharp\Gamma/\sharp G$. We conclude that $\sharp G = \sharp\mathrm{Aut}(C) \cdot \sharp I_x$, proving the exactness of the sequence in 3.1 (1).

Note that 3.1 (2) has already been proved (cf. Lemma 2.3 plus the identification of I_x and I_y).

From the isomorphism $C \cong \mathcal{D}_z$ and the universality of the deformation space D of the curve C we obtain canonically a commutative diagram

$$\begin{array}{ccccccc} & & U_x & \longrightarrow & U_z & & \\ & & \downarrow & & \downarrow & & \\ I_x\backslash U_x = & & U_y & \longrightarrow & D & \longrightarrow \operatorname{Aut}(C)\backslash D & \longrightarrow U_w, \end{array}$$

with $C \cong \mathcal{D}_z$. Note that $U_x \to D$ is surjective (any point of D can be lifted to U_y by taking a base for the sections in the ν-canonical sheaf, and $U_x \to U_y$ is surjective), hence $U_z \to D$ is surjective, and we conclude that $U_z \to D$ factors as follows

$$U_z \longrightarrow I_x\backslash U_z \longrightarrow D.$$

Moreover, since $\overline{\mathcal{M}_{g,m}} \to \overline{\mathcal{M}_g}$ is a Galois covering and $\overline{\mathcal{M}_g}$ is normal we get $G_z\backslash U_z \xrightarrow{\sim} U_w$. The resulting commutative diagram

$$\begin{array}{ccc} & & U_z \\ & \swarrow & \downarrow \\ I_x\backslash U_z & \longrightarrow & D \\ \downarrow & & \downarrow \\ G_z\backslash U_z & \longrightarrow & \operatorname{Aut}(C)\backslash D \\ & \cong\searrow & \downarrow \\ & & U_w \end{array}$$

implies $\operatorname{Aut}(C)\backslash D \xrightarrow{\sim} U_w$ (this seems to be known, cf. [8], §1). Using (1) we conclude the proof of 3.1 (3).

To prove 3.1 (4), let $g \geq 3$ and let C be a stable curve obtained by choosing regular curves of genus

$$g(C'') = i \geq 1, \qquad g(C') = g - i - 1 \geq 1,$$

which intersect transversally in 2 different points $\{P, Q\} = C' \cap C''$. The universal deformation family $\mathcal{X} \to D$ is smooth over $D \setminus \Delta$, and $\Delta = \Delta_P \cup \Delta_Q$ consists of two divisors intersecting transversally. Let $d \in \Delta_P$, $d \notin \Delta_Q$ and $e \in \Delta_Q$, $e \notin \Delta_P$ and let $\mathcal{X}_d$, $\mathcal{X}_e$ denote the fibers of $\mathcal{X}$ over d and e respectively. Note that

$$\operatorname{Jac}(\mathcal{X}_d)[m] \cong \operatorname{Jac}(C)[m] \cong \operatorname{Jac}(\mathcal{X}_e)[m] \quad (\cong (\mathbb{Z}/m)^{2g-1}),$$

and one easily sees that

$$\mathbb{Z} \times \mathbb{Z} \cong \pi_1(D - \Delta, v) \xrightarrow{R} \operatorname{Aut}(\operatorname{Jac}(\mathcal{X}_v)[m]),$$

where the isomorphism can be taken such that

$$\operatorname{Ker}(R) = \langle (m,0),\ (0,m),\ (1,-1) \rangle, \qquad R(\pi_1(D - \Delta), v) \cong \mathbb{Z}/m.$$

Moreover, the *unique* normal cover $U \to D$ which is étale outside Δ and which is given by this representation R has a local description

$$U = \operatorname{Spf}(W[[t_1, \ldots, t_{3g-3}]][T]/(T^m - t_1 t_2));$$

this is a singularity of "type A_{m-1}". To see this, observe that U is normal, and has outside $t_1 = 0 = t_2$ the correct structure, hence it is the one we are looking for.

By what has been proved we know that $U \cong U_z$; thus for the choice of C we made, any point $z \in \tau^{-1}([C])$ is singular on $\overline{\mathcal{M}_{g,m}}$; clearly this gives a closed subset in codimension two as $\overline{\mathcal{M}_{g,m}}$ is normal, and hence non-singular in codimension one. If we take

$$\mathcal{E}_{|U_y} \longrightarrow U_y$$

the same arguments apply, and we conclude that any point in $\mathcal{H}_{g,m}$ above the $[C] \in \overline{\mathcal{M}_g}$ chosen above is singular on $\mathcal{H}_{g,m}$. This concludes the proof of Theorem 3.1. □

Remark. It is easy to check that $\overline{\mathcal{M}_{g,m}} \to S_m$ is smooth if $g = 2$ and $m \geq 3$. This can be proved by a direct (easy) verification. We can also use [9], p. 91, Satz I.

References

[1] J.-L. Brylinski, *Propriétés de ramification à l'infini du groupe modulaire de Teichmüller.* Ann. Sci. École Norm. Sup. (4) **12** (1979), 295–333.

[2] P. Deligne, *La formule de Picard-Lefschetz.* Exp. XV in: SGA 7 II, Groupes de monodromie en géométrie algébrique. LNM 340, Springer-Verlag 1973.

[3] P. Deligne, *Le lemme de Gabber.* In: Sém. sur les pinceaux arithmétiques: la conjecture de Mordell (Ed. L. Szpiro). Astérisque **127** (1985), 131–150.

[4] P. Deligne, D. Mumford, *The irreducibility of the space of curves of a given genus.* Publ. Math. IHES **36** (1969), 75–109.

[5] A. Grothendieck, J. Dieudonné, *Éléments de géométrie algébrique IV^4: Étude locale des schémas et des morphisms de schémas.* Publ. Math. IHES **32** (1967).

[6] A. Grothendieck, J. P. Murre, *The tame fundamental group of a formal neighbourhood of a divisor with normal crossings on a scheme.* Lecture Notes in Math. 208, Springer-Verlag 1971.

[7] W. J. Haboush, *Reductive groups are geometrically reductive.* Ann. Math. **102** (1975), 67–83.

[8] J. Harris, D. Mumford, *On the Kodaira dimension of the moduli space of curves.* Invent. Math. **67** (1982), 23–86.

[9] S. M. Mostafa, *Die Singularitäten der Modulmannigfaltigkeit $\overline{M}_g(n)$ der stabilen Kurven vom Geschlecht $g \geq 2$ mit n-Teilungspunktstruktur.* Journ. Reine Angew. Math. **343** (1983), 81–98.

[10] D. Mumford, *Geometric invariant theory.* Ergebnisse der Mathematik, Neue F. Vol. 34, Springer-Verlag 1965. 3rd enlarged edition: Mumford, D., Fogarty, J., Kirwan, F., Springer-Verlag 1993.

[11] D. Mumford, *Stability of projective varieties.* L'Enseignement Math. **23** (1977), 39–110.

[12] H. Popp, *Moduli theory and classification theory of algebraic varieties.* Lecture Notes in Math. 620, Springer-Verlag 1977.

[13] M. Raynaud, *Spécialisation du foncteur de Picard.* Publ. Math. IHES **38** (1970), 27–76.

[14] A. Grothendieck (with M. Raynaud and D. S. Rim), *Groupes de monodromie en géométrie algébrique (SGA 7 I).* Séminaire de géométrie algébrique du Bois-Marie. Lecture Notes in Math. 288, Springer-Verlag 1972.

[15] C. S. Seshadri, *Quotient spaces modulo reductive algebraic groups.* Ann. Math. **95** (1972), 511–556.

[16] J-P. Serre, J. Tate, *Good reduction of abelian varieties.* Ann. Math. **88** (1968), 492–517.

Dipartimento di Matematica
Via Ferrata 1
27100 Pavia Italy
`geemen@dragon.ian.pv.cnr.it`

Mathematisch Instituut
Budapestlaan 6
3508 TA Utrecht The Netherlands
`oort@math.uu.nl`

Progress in Mathematics, Vol. 181, © 2000 Birkhäuser Verlag Basel/Switzerland

Applications of de Jong's Theorem on Alterations

Thomas Geisser

1. Introduction

The purpose of this article is to give a survey of some of the applications of de Jong's theorem on alterations [dJ]. Most of the applications fall into one of the following two categories:

The first type of application deals with contravariant functors $\mathcal{F}$ from some subcategory of the category of schemes to the category of $\mathbb{Q}$-vector spaces with extra structure (e.g. Galois action), which are equipped with a trace map for finite étale morphisms. In a situation like this, for a given scheme X, $\mathcal{F}(X)$ will be a direct summand of $\mathcal{F}(X')$ for an alteration X' of X. This allows to deduce properties of $\mathcal{F}(X)$ for general X if one only knows the same property for smooth schemes. The following are examples of this kind of application: The independence of l in Grothendieck's monodromy theorem, the p-adic monodromy theorem, finiteness of rigid cohomology, and a (conditional) vanishing theorem for motivic cohomology. All but the last application were already discussed by Berthelot in his Bourbaki talk, and we follow his exposition.

The second type of application is more direct. There are certain Grothendieck topologies which admit proper surjective maps as coverings. For such a topology, de Jong's theorem tells us that any variety is locally smooth. The two main examples we are giving here are Deligne's topology of universal cohomological descent, and the h-topology of Suslin and Voevodsky. In the first case, we get a generalization to characteristic p of Deligne's theorem that any scheme admits a proper hypercovering. In the second case, a theorem of Suslin and Voevodsky comparing their singular cohomology of varieties to étale cohomology, and a theorem of Suslin comparing Bloch's higher Chow groups to étale cohomology, generalize to characteristic p.

We must point out that none of the work presented here is original, and that our exposition follows other papers closely in parts.

2. de Jong's theorem and Serre's conjecture

In this section, we explain the notation, give de Jong's theorem, and describe Serre's conjecture on intersection multiplicities.

Let X be an integral Noetherian scheme. An *alteration* X' of X is an integral scheme X' together with a proper dominant morphism $\varphi : X' \to X$ which is finite over a non-empty open subset of X.

There are two versions of de Jong's theorem; the first one deals with varieties over fields k, and the second one with varieties over complete discrete valuation rings. We denote by k-variety an integral scheme which is separated and of finite type over k.

Theorem 2.1. [dJ, Theorem 4.1] *Let X be a variety over a field k and $Z \subseteq X$ a proper closed subset. Then there exists an alteration $\varphi : X' \to X$ and an open immersion of X' into a regular, projective k-variety $\bar{X}'$, such that the closed subset $\varphi^{-1}(Z) \cup (\bar{X}' - X')$ is the support of a strict normal crossing divisor on $\bar{X}'$.*

There is a finite extension k' of k such that the structure morphism $\bar{X}' \to k$ factors through k' and that $\bar{X}'$ is geometrically irreducible and smooth over k'. If k is perfect, then $\bar{X}'$ is smooth over k, and φ can be chosen generically étale.

Let $S = \operatorname{Spec} A$ be the spectrum of a complete discrete valuation ring A with generic point η and closed point s. An *S-variety* is an integral scheme X, separated, flat and of finite type over S. Let X be an S-variety whose closed fiber X_s has irreducible components X_i, $i \in I$. For $J \subseteq I$, let $X_J = \cap_{j \in J} X_j$. Then X is *strictly semi-stable* over S, if

- X_η is smooth over $k(\eta)$
- X_s is reduced
- each X_i is a divisor on X
- for each $J \subseteq I$, X_J is smooth over $k(s)$ of codimension $\#J$ in X.

Let (X, Z) be a pair consisting of an S-variety X together with a closed subset $Z \subseteq X$, viewed as a reduced closed subscheme. Write $Z = Z_f \cup Z'$, where $Z_f \to S$ is flat and $Z' \subseteq X_s$. Let Z_i, $i \in I$, be the irreducible components of Z_f, and $Z_J = \cap_{j \in J} Z_j$. The pair (X, Z) is a *strictly semi-stable pair* if

- X is strictly semi-stable over S
- Z is divisor with normal crossings on X
- for each $J \subseteq I$, Z_J is a union of strictly semi-stable S-varieties.

Theorem 2.2. [dJ, Theorem 6.5] *Let X be an S-variety and $Z \subseteq X$ a proper closed subset containing the closed fiber. Then there exists a discrete valuation ring A', finite over A, a variety X' over $S' = \operatorname{Spec} A'$, an alteration $\varphi : X' \to X$ over S, and an open immersion $j : X' \to \bar{X}'$ of S'-varieties, such that $\bar{X}'$ is projective over S' with geometrically irreducible generic fiber, and $(\bar{X}', \varphi^{-1}(Z)_{\mathrm{red}} \cup (\bar{X}' - X'))$ is a strictly semi-stable pair.*

2.1. Serre's conjecture

The first application we give concerns intersection multiplicities, see [R2] for an overview. We give it here because it does not fit into one of the other categories mentioned in the introduction.

Let A be a regular local ring of finite Krull dimension with maximal ideal $\mathfrak{m}$ and residue field k. Let M and N be two finitely generated A-modules such that $M \otimes_A N$ is of finite length. This implies $\dim_A M + \dim_A N \leq \dim A$, where $\dim_A M$ is the Krull dimension of the ring $A/\operatorname{Ann}_A(M)$. Geometrically, if M and N are ideals of A defining subvarieties, one would like to define the multiplicity of intersection of these subvarieties in the point given by the maximal ideal of A. Of course, one wants this multiplicity to be non-negative, and to be zero if the subvarieties do not meet. Serre [Se] proposed to define the intersection multiplicity of M and N as

$$\chi_A(M,N) = \sum_{i\geq 0}(-1)^i \operatorname{lg}_A \operatorname{Tor}_i^A(M,N),$$

and conjectured the following properties:

Theorem 2.3. *Under the above assumptions, we have*
$\chi_A(M,N) \geq 0$ *(Positivity), and*
$\chi_A(M,N) = 0$ *if* $\dim_A M + \dim_A N < \dim A$ *(Annihilation).*

Serre proved this for A of equal characteristic, and for A of unequal characteristic and non-ramified. The annihilation conjecture was proved by Gillet-Soulé [GS] and Roberts [R1]. Finally, using de Jong's theorem, Gabber [GA] proved the positivity conjecture, see [B1] for more details.

3. Grothendieck topologies for which alterations are coverings

In this section, we give some applications which use Grothendieck topologies admitting proper surjective maps as coverings. De Jong's theorem implies that every variety is locally smooth for such a topology.

3.1. Proper hypercoverings

The reader may consult Deligne [D, Section 5] for more details on this subject. Let Δ be the category with objects finite ordered sets $[n] := \{0,\dots,n\}$ and morphisms maps respecting the ordering; let Δ_t be the full subcategory of sets $[n]$ with $n \leq t$. Recall that a simplicial object (respectively a t-truncated simplicial object) in the category $\mathcal{C}$ is a contravariant functor $U_. : \Delta \to \mathcal{C}$ (respectively $U_. : \Delta_t \to \mathcal{C}$). One usually writes X_n for $X_.([n])$. The restriction functor sk_t (t-skeleton) from simplicial objects to t-truncated simplicial objects has a right adjoint functor cosk_t (t-coskeleton) such that $\operatorname{sk}_t = \operatorname{sk}_t \operatorname{cosk}_t \operatorname{sk}_t$. The notion of a simplicial scheme generalizes the notion of a scheme by taking $X_n = X$ for all n, and all simplicial maps the identity.

A sheaf $\mathcal{F}^\cdot$ on a simplicial topological space $X_.$ is a family of sheaves $\mathcal{F}^n$ on X_n together with morphisms of sheaves on X_m, $f^*\mathcal{F}^n \to \mathcal{F}^m$, for each map

$f : [n] \to [m]$ satisfying obvious compatibilities. A sheaf on $X_.$ can be viewed as a functor on pairs (n, U) with $U \subseteq X_n$, satisfying certain compatibility conditions. In particular, the sheaves on $X_.$ can be viewed as the category of sheaves on a site. The global sections of the sheaf $\mathcal{F}^\cdot$ are

$$\Gamma(X_., \mathcal{F}^\cdot) = \ker\left(\Gamma(X_0, \mathcal{F}^0) \to \Gamma(X_1, \mathcal{F}^1)\right),$$

where the map is the difference of the maps induced by the two maps ∂_0, ∂_1 from $[0]$ to $[1]$. Let $H^i(X_., \mathcal{F}^\cdot)$ be the ith derived functor of the global section functor. Looking at an acyclic resolution (for example the Godement resolution), one sees that there is a spectral sequence

$$E_1^{pq} = H^q(X_p, \mathcal{F}^p) \Rightarrow H^{p+q}(X_., \mathcal{F}^\cdot).$$

Let $a : X_. \to S$ be an augmented simplicial scheme, i.e. a simplicial scheme together with a map $X_0 \to S$. This induces a (unique) map $a_n : X_n \to S$ for each n, and a functor a^* from sheaves on S to sheaves on $X_.$, sending $\mathcal{F}$ to the sheaf $a_n^*\mathcal{F}$ on X_n. The functor a^* has a right adjoint a_*, explicitly,

$$a_*\mathcal{F}^\cdot = \ker\left(a_{0*}\mathcal{F}^0 \xrightarrow{\partial_0^* - \partial_1^*} a_{1*}\mathcal{F}^1\right).$$

This can be derived to give a functor

$$Ra_* : D^+(X_.) \to D^+(S)$$

from the derived category of bounded above complexes of sheaves of abelian groups on $X_.$ to the corresponding category on S. Let $\varphi : \mathrm{id} \to Ra_*a^*$ be the associated adjunction morphism. Then a is said to be of *cohomological descent* if φ is an isomorphism. Since $\Gamma(X_., \mathcal{F}^\cdot) = \Gamma(S, Ra_*\mathcal{F}^\cdot)$ for any sheaf on $X_.$, cohomological descent implies

$$H^i(S, \mathcal{F}) \cong H^i(S, Ra_*a^*\mathcal{F}) \cong H^i(X_., a^*\mathcal{F}^\cdot),$$

and the spectral sequence above reads

$$E_1^{p,q} = H^q(X_p, a_p^*\mathcal{F}) \Rightarrow H^{p+q}(S, \mathcal{F}).$$

If we can find for a given S an $a : X_. \to S$ of cohomological descent such that all the schemes X_n are smooth, then this formalism allows us to study the cohomology groups of singular schemes in terms of the cohomology groups of smooth schemes.

We have the following basic example of morphisms of cohomological descent [SGA 4, V bis]: A t-truncated simplicial scheme $X_.$ is called a t-truncated proper hypercovering if the adjoint maps

$$\varphi_{n+1} : X_{n+1} \to (\operatorname{cosk}_n \operatorname{sk}_n X_.)_{n+1} \tag{1}$$

are proper and surjective for all $n \leq t-1$. In this case, the map $\operatorname{cosk}_t X_. \to S$ is of cohomological descent. This construction is used to prove the following theorem:

Theorem 3.1. *Let S be a variety over a perfect field k. Then there exists a simplicial scheme $\bar{X}_.$, projective and smooth over k, a strict normal crossing divisor $D_.$ in $\bar{X}_.$ with open complement $X_. = \bar{X}_. - D_.$, and an augmentation $a : X_. \to S$ which is a proper hypercovering of S.*

For the proof one constructs inductively, using de Jong's theorem, t-truncated simplicial schemes ${}_tX_.$ over S with compactification ${}_t\bar{X}_.$, such that ${}_tX_.$ satisfies the condition (1). The limit of these t-truncated schemes then satisfies the statement of the theorem, see Deligne [D, 6.2.5].

3.2. Singular cohomology of varieties

Suslin and Voevodsky define in [SV] singular homology $H_*^{\text{sing}}(X, A)$ and cohomology groups $H^*_{\text{sing}}(X, A)$ for any scheme X of finite type over a field k, and any abelian group A. For $k = \mathbb{C}$, and $A = \mathbb{Z}/n$, these groups generalize the usual singular homology groups. We give a short outline of the construction, see Levine [L2] for another survey.

Let $\mathcal{F}$ be a presheaf of abelian groups on Sch$/k$, the category of schemes of finite type over a field k of exponential characteristic p. We define presheaves $\mathcal{F}_q$ by

$$\mathcal{F}_q(X) := \mathcal{F}(X \times \Delta_q). \tag{2}$$

Here

$$\Delta_q = \operatorname{Spec} k[t_0, \dots, t_q]/(\sum t_i = 1)$$

is the algebraic q-simplex. As in topology, $\Delta_.$ is a cosimplicial scheme, hence every presheaf $\mathcal{F}$ on Sch$/k$ gives rise to a simplicial presheaf $\mathcal{F}_.$ on Sch$/k$ via (2). By the Dold-Kan equivalence, this corresponds to a complex of presheaves $\mathcal{F}_*$ on Sch$/k$. We let

$$C_*(\mathcal{F}) = \mathcal{F}_*(k)$$

be the global sections over k of this complex of presheaves. Note that in order to define $C_*(\mathcal{F})$, we only need to know the values of $\mathcal{F}$ on the algebraic q-simplices, for example it suffices for $\mathcal{F}$ to be defined on smooth schemes over k.

Let $c_0(X)$ and $z_0(X)$ be the presheaves which associate to every smooth connected k-scheme S the free abelian group generated by the closed integral subschemes $Z \subseteq X \times S$ which are finite and surjective over S and quasi-finite over S, respectively. Note that if X is proper, then $c_0(X) = z_0(X)$. For an abelian group A one defines

$$H_*^{\text{sing}}(X, A) = \operatorname{Tor}_*^{\text{Ab}}(C_*(c_0(X)), A),$$
$$H^*_{\text{sing}}(X, A) = \operatorname{Ext}^*_{\text{Ab}}(C_*(c_0(X)), A).$$

This generalizes singular (co)homology; for X a scheme of finite type over $\mathbb{C}$, one has the following natural isomorphisms [SV, Theorem 8.3]:

$$H_*^{\text{sing}}(X, \mathbb{Z}/m) \xrightarrow{\sim} H_*(X(\mathbb{C}), \mathbb{Z}/m)$$
$$H^*_{\text{sing}}(X, \mathbb{Z}/m) \xleftarrow{\sim} H^*(X(\mathbb{C}), \mathbb{Z}/m).$$

The right-hand side is the ordinary (co)homology of the $\mathbb{C}$-valued points of X.

Suslin and Voevodsky also show that for X separated of finite type over an algebraically closed field of characteristic 0, their singular cohomology groups agree with étale cohomology groups. Using de Jong's theorem on alterations, one can show that the last hypothesis is spurious:

Theorem 3.2. [SV, Corollary 7.8] *Let X be a separated scheme of finite type over an algebraically closed field k, and let m be prime to the characteristic of k. Then*

$$H^*_{\mathrm{sing}}(X,\mathbb{Z}/m) \cong H^*_{\text{ét}}(X,\mathbb{Z}/m).$$

The rest of this subsection is devoted to give a sketch of the proof of this theorem. We introduce a Grothendieck topology on Sch/k, the h-topology. An h-cover of a scheme X is a finite family of morphisms of finite type $\{p_i : X_i \to X\}$ such that the map $\coprod X_i \to X$ is a universal topological epimorphism. The h-topology is the Grothendieck topology generated by all h-coverings. The h-topology is finer than the étale topology. In our context it is important to note that an alteration is an h-covering; in particular every separated and integral scheme of finite type over k is locally smooth for the h-topology by Theorem 2.1. For any presheaf $\mathcal{F}$ on Sch/k we denote the associated sheaf for the h-topology by $\mathcal{F}_{\mathrm{h}}$.

We use a collection of theorems in [SV] to express étale cohomology in terms of the h-topology: The presheaf $c_0(X)$ can be extended to a presheaf on normal integral schemes with the same definition on objects; however one has to invert the characteristic of k for functoriality [SV, Section 5]. This presheaf can be further extended to a presheaf on all schemes of finite type over k [SV, Section 6], which we will again denote by $c_0(X)$. The analogous statements for $z_0(X)$ hold.

By [SV, Theorem 6.7], after inverting the characteristic of k, the h-sheaf $c_0(X)_{\mathrm{h}}$ is isomorphic to the free sheaf $\mathbb{Z}(X)_{\mathrm{h}}$ generated by X, i.e. the sheaf associated to the presheaf which sends U to the free abelian group generated by $\mathrm{Hom}(U,X)$. This implies the following

Lemma 3.3. *Let X be a separated scheme over k and m prime to the characteristic of k. Then we have isomorphisms*

$$\mathrm{Ext}^*_{\mathrm{h}}(c_0(X)_{\mathrm{h}},\mathbb{Z}/m) \cong H^*_{\text{ét}}(X,\mathbb{Z}/m)$$
$$\mathrm{Ext}^*_{\mathrm{h}}(z_0(X)_{\mathrm{h}},\mathbb{Z}/m) \cong H^*_{\text{ét},c}(X,\mathbb{Z}/m).$$

Proof. For an étale sheaf $\mathcal{F}$, we get by comparing to an intermediate Grothendieck topology, the qfh-topology, the isomorphisms [SV, Corollary 10.10]

$$\mathrm{Ext}^*_{\text{ét}}(\mathcal{F},\mathbb{Z}/m) \cong \mathrm{Ext}^*_{\mathrm{qfh}}(\mathcal{F}_{\mathrm{qfh}},\mathbb{Z}/m) \cong \mathrm{Ext}^*_{\mathrm{h}}(\mathcal{F}_{\mathrm{h}},\mathbb{Z}/m).$$

Hence we have

$$\mathrm{Ext}^*_{\mathrm{h}}(c_0(X)_{\mathrm{h}},\mathbb{Z}/m) \cong \mathrm{Ext}^*_{\mathrm{h}}(\mathbb{Z}(X)_{\mathrm{h}},\mathbb{Z}/m) \cong \mathrm{Ext}^*_{\text{ét}}(\mathbb{Z}(X),\mathbb{Z}/m) \cong H^*_{\text{ét}}(X,\mathbb{Z}/m).$$

To prove the second isomorphism, we choose an open embedding $j : X \to \bar{X}$ into a complete separated scheme $\bar{X}$. Let $i : Y \to \bar{X}$ be the closed embedding of the complement. There is an exact sequence of h-sheaves [Su]:

$$0 \to z_0(Y)_{\mathrm{h}} \xrightarrow{i_*} z_0(X)_{\mathrm{h}} \xrightarrow{j^*} z_0(U)_{\mathrm{h}} \to 0. \tag{3}$$

Comparing the associated long exact $\mathrm{Ext}^*_{\mathrm{h}}(-,\mathbb{Z}/m)$-sequence to the long exact Gysin sequence for étale cohomology with compact supports, the second statement follows from the first. □

To formulate the following theorem, we need another definition. A presheaf $\mathcal{F}$ is a *homotopy invariant presheaf with transfers* if the projection induces an isomorphism $\mathcal{F}(X) \xrightarrow{\sim} \mathcal{F}(X \times \mathbb{A}^1)$, and if every element of $c_0(X)(Y)$ induces a map $\mathcal{F}(X) \to \mathcal{F}(Y)$. As an example, any sheaf for the qfh-topology can be equipped with transfers [SV, Section 6]. On the other hand, taking the homology groups of the complex $\mathcal{F}_*$ is a functorial way of making $\mathcal{F}$ homotopy invariant [SV, Corollary 7.5]. In particular, for any qfh-sheaf the homology presheaves $\mathcal{H}_q(\mathcal{F}_*)$ are homotopy invariant presheaves with transfers.

Theorem 3.4. *(Rigidity Theorem* [SV, Theorem 4.5]*) Let k be an algebraically closed field, $\mathcal{F}$ a homotopy invariant presheaf with transfers on Sch/k, and let m be prime to the characteristic of k. Then there are canonical isomorphisms*

$$\mathrm{Ext}^*_{\mathrm{h}}(\mathcal{F}_{\mathrm{h}},\mathbb{Z}/m) \cong \mathrm{Ext}^*_{\mathrm{Ab}}(\mathcal{F}(k),\mathbb{Z}/m).$$

Proof. We only give an outline of the argument. Let $\mathcal{F}_0$ be the constant presheaf $\mathcal{F}(\mathrm{Spec}\, k)$. Let $\mathcal{F}'$ be the cokernel of the natural map $\mathcal{F}_0 \to \mathcal{F}$, which is an inclusion because k is algebraically closed, hence every scheme of finite type has a k-rational point.

An explicit calculation [SV, Theorem 4.4] shows that for a homotopy invariant m-torsion presheaf with transfer $\mathcal{G}$, and X^h_x the henselization of the smooth scheme X at a closed point x, $\mathcal{G}(X^h_x) \cong \mathcal{G}(k)$. Since alterations and étale covers are h-covers, every scheme is locally smooth for the h-topology by Theorem 2.1.

Applying this to the presheaves $\mathcal{F}'/m$ and ${}_m\mathcal{F}'$ (cokernel and kernel of multiplication by m of $\mathcal{F}'$), which are again homotopy invariant presheaves with transfers, we see that $\mathcal{F}'_{\mathrm{h}}$ is uniquely m-divisible, hence the natural map $(\mathcal{F}_0)_{\mathrm{h}} \to \mathcal{F}_{\mathrm{h}}$ induces the isomorphism of the theorem, noting

$$\mathrm{Ext}^*_h((\mathcal{F}_0)_{\mathrm{h}},\mathbb{Z}/m) \cong \mathrm{Ext}^*_{\mathrm{Ab}}(\mathcal{F}(k),\mathbb{Z}/m).$$

□

Corollary. *For any homotopy invariant presheaf with transfers $\mathcal{F}$ we have*

$$\mathrm{Ext}^*_{\mathrm{h}}(\mathcal{F}_{\mathrm{h}},\mathbb{Z}/m) \cong \mathrm{Ext}^*_{\mathrm{h}}((\mathcal{F}_*)_{\mathrm{h}},\mathbb{Z}/m) \cong \mathrm{Ext}^*_{\mathrm{Ab}}(C_*(\mathcal{F}),\mathbb{Z}/m).$$

Proof. To prove the first isomorphism, one uses the first hypercohomology spectral sequence

$$E_1^{pq} = \mathrm{Ext}^q_{\mathrm{h}}((\mathcal{F}_p)_{\mathrm{h}},\mathbb{Z}/m) \Rightarrow \mathrm{Ext}^{p+q}_{\mathrm{h}}((\mathcal{F}_*)_{\mathrm{h}},\mathbb{Z}/m),$$

where $(\mathcal{F}_*)_{\mathrm{h}}$ is the sheafification of the simplicial presheaf $\mathcal{F}_*$ on Sch/k. This sequence collapses at E_2 to the isomorphism $\mathrm{Ext}^q_{\mathrm{h}}(\mathcal{F}_{\mathrm{h}},\mathbb{Z}/m) \cong \mathrm{Ext}^q_{\mathrm{h}}((\mathcal{F}_*)_{\mathrm{h}},\mathbb{Z}/m)$, [SV, Corollary 7.3].

To prove the second isomorphism, Suslin and Voevodsky employ the second hypercohomology spectral sequence

$$E_2^{pq} = \mathrm{Ext}^p_{\mathrm{h}}(\mathcal{H}_q((\mathcal{F}_*)_{\mathrm{h}}),\mathbb{Z}/m) \Rightarrow \mathrm{Ext}^{p+q}_{\mathrm{h}}((\mathcal{F}_*)_{\mathrm{h}},\mathbb{Z}/m).$$

Since sheafification is exact, $\mathcal{H}_q((\mathcal{F}_*)_{\mathrm{h}}) \cong \mathcal{H}_q(\mathcal{F}_*)_{\mathrm{h}}$, and by definition $\mathcal{H}_q(\mathcal{F}_*)(k) = H_q(\mathcal{F}_*(k)) = H_q(C_*(\mathcal{F}))$. The rigidity theorem now shows that

$$\begin{aligned}\operatorname{Ext}^*_{\mathrm{h}}(\mathcal{H}_q((\mathcal{F}_*)_{\mathrm{h}}), \mathbb{Z}/m) &\cong \operatorname{Ext}^*_{\mathrm{h}}(\mathcal{H}_q(\mathcal{F}_*)_{\mathrm{h}}, \mathbb{Z}/m)\\ &\cong \operatorname{Ext}^*_{\mathrm{Ab}}(\mathcal{H}_q(\mathcal{F}_*)(k), \mathbb{Z}/m) \cong \operatorname{Ext}^*_{\mathrm{Ab}}(H_q(C_*(\mathcal{F})), \mathbb{Z}/m).\end{aligned}$$

Hence the natural map induced by taking the associated constant sheaf from the spectral sequence

$$E_2^{pq} = \operatorname{Ext}^p_{\mathrm{Ab}}(H_q(C_*(\mathcal{F})), \mathbb{Z}/m) \Rightarrow \operatorname{Ext}^{p+q}_{\mathrm{Ab}}(C_*(\mathcal{F}), \mathbb{Z}/m)$$

is an isomorphism on E_2-terms, and gives an isomorphism of the abutments. □

To finish the proof of Theorem 3.2, we apply the Corollary to $\mathcal{F} = c_0(X)$, noting that the presheaf $c_0(X)$ is actually a qfh-sheaf, hence admits transfers. We get

$$H^*_{\mathrm{sing}}(X, \mathbb{Z}/m) = \operatorname{Ext}^*_{\mathrm{Ab}}(C_*(c_0(X)), \mathbb{Z}/m) \cong \operatorname{Ext}^*_{\mathrm{h}}(c_0(X)_{\mathrm{h}}, \mathbb{Z}/m),$$

and conclude with Lemma 3.3.

3.3. Higher Chow groups and étale cohomology

Let $z^i(X, -)$ be Bloch's cycle complex, i.e. $z^i(X, n)$ is the free abelian group generated by the closed irreducible subschemes of codimension i of $X \times \Delta^n_k$ which intersect all faces properly; see [Bl] for the basic properties. Then for an abelian group A, higher Chow groups with A-coefficients are defined as

$$\mathrm{CH}^i(X, n, A) = H_n(z^i(X, -) \otimes A). \tag{4}$$

Suslin proves in [Su] that for an equidimensional scheme over an algebraically closed field k of characteristic 0, higher Chow groups are dual to étale cohomology with compact support. Again, the hypothesis that the base field has characteristic 0 is spurious:

Theorem 3.5. *Let X be an equidimensional quasi-projective scheme over an algebraically closed field k, and let $i \geq d = \dim X$. Then for any m prime to the characteristic of k,*

$$\mathrm{CH}^i(X, n, \mathbb{Z}/m) \cong H^{2(d-i)+n}_{\text{ét},c}(X, \mathbb{Z}/m(d-i))^\vee.$$

In particular, if X is smooth, we have

$$\mathrm{CH}^i(X, n, \mathbb{Z}/m) \cong H^{2i-n}_{\text{ét}}(X, \mathbb{Z}/m(i)).$$

Proof. Suslin shows that for X an *affine* equidimensional scheme of dimension d, the canonical injection of complexes

$$C_*(z_0(X)) \to z^d(X, -)$$

is a quasi-isomorphism [Su, Theorem 2.1]. We show how one can get the general statement from this. We proceed by induction on the dimension of X. For a quasi-projective scheme X, one can find an effective Cartier divisor $Y \subset X$ such that the open complement U is affine. It follows from Corollary 3.2 that we have for a complex of presheaves $\mathcal{F}^\cdot$

$$H_*(C_*(\mathcal{F}^\cdot) \otimes^{\mathbb{L}} \mathbb{Z}/m)^\vee \cong H^*(\mathbb{R}\operatorname{Hom}(C_*(\mathcal{F}^\cdot), \mathbb{Z}/m)) = \operatorname{Ext}^*_{\mathrm{Ab}}(C_*(\mathcal{F}^\cdot), \mathbb{Z}/m) \cong \operatorname{Ext}^*_{\mathrm{h}}(\mathcal{F}^\cdot_{\mathrm{h}}, \mathbb{Z}/m).$$

In particular, the complex $C_*(\mathcal{F}^\cdot) \otimes^{\mathbb{L}} \mathbb{Z}/m$ is acyclic if the complex of sheaves $\mathcal{F}^\cdot_{\mathrm{h}}$ is exact. Applying this to the exact sequence (3), we get the upper short exact sequence in the following commutative diagram of complexes of abelian groups

$$\begin{array}{ccccc} C_*(z_0(Y)) \otimes^{\mathbb{L}} \mathbb{Z}/m & \to & C_*(z_0(X)) \otimes^{\mathbb{L}} \mathbb{Z}/m & \to & C_*(z_0(U)) \otimes^{\mathbb{L}} \mathbb{Z}/m \\ \downarrow & & \downarrow & & \downarrow \\ z^{d-1}(Y,-) \otimes^{\mathbb{L}} \mathbb{Z}/m & \to & z^d(X,-) \otimes^{\mathbb{L}} \mathbb{Z}/m & \to & z^d(U,-) \otimes^{\mathbb{L}} \mathbb{Z}/m. \end{array}$$

The lower row is an exact triangle in the derived category by [Bl]. Since the outer vertical maps are quasi-isomorphisms by induction and the affine case, the same holds for the middle vertical map. The theorem follows now for $i = d$ because by Lemma 3.3 and Corollary 3.2

$$\begin{aligned} \mathrm{CH}^d(X, n, \mathbb{Z}/m) &= H_n(z^d(X,-) \otimes^{\mathbb{L}} \mathbb{Z}/m) \cong H_n(C_*(z_0(X)) \otimes^{\mathbb{L}} \mathbb{Z}/m) \\ &\cong \operatorname{Ext}^n_{\mathrm{Ab}}(C_*(z_0(X)), \mathbb{Z}/m)^\vee \cong \operatorname{Ext}^n_{\mathrm{h}}(z_0(X)_{\mathrm{h}}, \mathbb{Z}/m)^\vee \cong H^n_{\text{ét},c}(X, \mathbb{Z}/m)^\vee. \end{aligned}$$

The general case can be derived by applying this to $X \times \mathbb{A}^{i-d}$, and using homotopy invariance. □

4. Applications using trace maps

There are two main mechanisms how de Jong's theorem is used to prove properties of cohomology groups. Since the mechanism of the proof is the most important point and has to be adapted to the specific situation, we are somewhat vague in the formulation:

Lemma 4.1. *Let $\mathcal{P}$ be a property of cohomology groups of varieties over finite extensions of a perfect field K. Suppose that $\mathcal{P}$ is*

1. *preserved by extensions,*
2. *holds for the cohomology of smooth projective varieties.*

Assume that the cohomology theory

1. *has a long exact localization sequence,*
2. *has trace maps for finite étale maps.*

Then $\mathcal{P}$ holds for the cohomology groups of all varieties.

Proof. Using the localization sequence and induction, one sees that it is equivalent to prove property $\mathcal{P}$ for the cohomology of a scheme X or of some open subscheme U of X. On the other hand, we can use Theorem 2.1 to show that for a given X, there is an X', an alteration $\varphi : X' \to X$ and an open embedding of X' into a smooth projective scheme $\bar{X}'$. Let U be a sufficiently small smooth open subscheme of X, then the morphism $U' = U \times_X X' \to U$ is a finite map between smooth schemes. Using the trace map, we see that the cohomology of U is a direct summand of the cohomology of U'. Since $\mathcal{P}$ holds for the cohomology of $\bar{X}'$, using the localization sequence it holds for the cohomology of U', hence for U and finally for X. □

Lemma 4.2. *Let $\mathcal{P}$ be a property of cohomology groups of varieties over finite extensions of the field of fractions K of a Henselian discrete valuation ring A. Suppose that property $\mathcal{P}$*

1. *holds for the cohomology groups of the generic fibers of semi-stable schemes over A,*
2. *is inherited by direct summands of cohomology groups*

Suppose that the cohomology theory admits a trace map for finite étale maps. Then property $\mathcal{P}$ holds for all cohomology groups of smooth and proper varieties over K.

Proof. Let X be a scheme which is smooth and proper over K. By Theorem 2.2 we can find a finite extension K' of K, a strictly semi-stable scheme $\mathcal{X}'$ over the ring of integers of K' with generic fiber X', and a K-alteration $\varphi : X' \to X$. Property $\mathcal{P}$ holds for X' by hypothesis, and the cohomology group of X is a direct summand of the cohomology group of X' using the trace map. □

The following examples use these two methods with only minor modifications. The first two examples were discussed by Berthelot in [B1].

4.1. Monodromy, $l \neq p$

Let A be a Henselian discrete valuation ring with field of fractions K, and X be a K-scheme of finite type. Let $\bar{K}$ be the algebraic closure of K, $X_{\bar{K}} = X \times_K \bar{K}$, and denote by $I \subseteq G = \mathrm{Gal}(\bar{K}/K)$ the inertia subgroup of the Galois group. We fix a prime l different from the characteristic of K. Then the étale cohomology groups $H^i_{\text{ét}}(X_{\bar{K}}, \mathbb{Q}_l)$ and the étale cohomology groups with compact support $H^i_{\text{ét},c}(X_{\bar{K}}, \mathbb{Q}_l)$ are equipped with an action of the Galois group G, giving an l-adic representation of I.

If l is also different from the residue characteristic of A, then the monodromy theorem of Grothendieck [SGA 7, Th. 2.2] states that the l-adic representation $H^i_{\text{ét}}(X_{\bar{K}}, \mathbb{Q}_l)$ of I is quasi-unipotent, i.e. there is a subgroup of finite index $I' \subseteq I$ such that $g - \mathrm{id}$ acts nilpotently for each $g \in I'$. It has been observed by Deligne that de Jong's theorem implies that such an I' can be chosen independently of l:

Theorem 4.3. *There exists a subgroup $I' \subseteq I$ of finite index such that the action of I' on $H^i_{\text{ét}}(X_{\bar{K}}, \mathbb{Q}_l)$ and on $H^i_{\text{ét},c}(X_{\bar{K}}, \mathbb{Q}_l)$ is unipotent for each $l \neq p$.*

Proof. We give the main idea, for more details, see Berthelot [B1]. If X is the generic fiber of a semi-stable scheme, then the action of I is seen to be unipotent by using the vanishing cycle spectral sequence. Using the method of Lemma 4.2, we see that, after a finite extension of K (which amounts to replacing I by a subgroup of finite index), the theorem holds for all smooth and proper schemes over K.

If X is separated and of finite type over K, one can use the method of Lemma 4.1 to show that the cohomology with compact support $H^i_{\text{ét},c}(X_{\bar{K}}, \mathbb{Q}_l)$ has the property of the theorem. For X smooth and separated over K, the statement about $H^i_{\text{ét}}(X_{\bar{K}}, \mathbb{Q}_l)$ follows from the previous case by Poincaré duality.

In the general case one uses the spectral sequence for a Cech-covering to reduce to the case X affine, and then reduces to the case X integral. We can apply Theorem 3.1 to construct a proper hypercovering $X'_\bullet \xrightarrow{\varphi} X$ such that all X_i are smooth and such that the adjunction map $\mathbb{Q}_{l,X} \to R\varphi_*\mathbb{Q}_{l,X'_\bullet}$ is an isomorphism. The hypercohomology spectral sequence and the result for $H^t_{\text{ét}}(X'_{s,\bar{K}}, \mathbb{Q}_l)$ proves the result for $H^i_{\text{ét}}(X_{\bar{K}}, \mathbb{Q}_l)$. □

4.2. Monodromy, $l = p$

The above techniques also allow to study the p-adic representations $H^*_{\text{ét}}(X_{\bar{K}}, \mathbb{Q}_p)$ if K is of characteristic 0 with residue characteristic p. For simplicity we assume that K is a finite extension of $\mathbb{Q}_p$. Let K_0 be the maximal unramified subextension of K, with Frobenius endomorphism σ, and let G be the absolute Galois group of K.

Let us recall some basic properties of Fontaine's rings [Fo]

$$B_{\text{crys}} \subsetneq B_{\text{st}} \subsetneq B_{dR}; \quad B_{HT}.$$

These rings carry a structure of a G-module, and for the invariants one has

$$B_{HT}^G \cong B_{dR}^G \cong K;$$
$$B_{\text{crys}}^G \cong B_{\text{st}}^G \cong K_0.$$

The ring B_{dR} is a complete discrete valuation field with residue field $\bar{K}^\wedge$, the completion of the algebraic closure of K. The algebra B_{HT} is the graded algebra associated to the filtration given by the valuation of B_{dR}, and

$$B_{HT} = \operatorname{gr} B_{dR} = \bigoplus_{i \in \mathbb{Z}} \bar{K}^\wedge(i).$$

The K_0-algebra B_{crys} is equipped with a σ-semilinear automorphism φ, and a G-equivariant injective homomorphism $B_{\text{crys}} \otimes_{K_0} K \to B_{dR}$ which induces a filtration on $B_{\text{crys}} \otimes_{K_0} K$. The associated map of graded algebras is an isomorphism.

Finally, B_{st} is a G-invariant polynomial extension in one variable u of B_{crys} inside B_{dR}; we extend φ to B_{st} by setting $\varphi(u) = pu$. The monodromy operator $N : B_{\text{st}} \to B_{\text{st}}$ is the unique B_{crys}-derivation such that $Nu = 1$; it follows that $N\varphi = p\varphi N$ and we can recover B_{crys} as the kernel of N. The natural injection $B_{\text{st}} \otimes_{K_0} K \to B_{dR}$ induces a filtration on $B_{\text{st}} \otimes_{K_0} K$, and $u \in \text{Fil}^1$.

For a p-adic representation E of $G = \text{Gal}(\bar{K}/K)$ and $*$ one of the symbols crys, st, dR and HT, Fontaine defines

$$D_*(E) = (B_* \otimes_{\mathbb{Q}_p} E)^G.$$

Then E is said to be crystalline, semi-stable, de Rham or Hodge-Tate, if the canonical injection

$$\alpha_* : B_* \otimes_{B_*^G} D_*(E) \to B_* \otimes_{\mathbb{Q}_p} E \tag{5}$$

is an isomorphism, for $*$ the corresponding symbol. It is easy to see the following implications:

$$\text{crystalline} \Rightarrow \text{semi-stable} \Rightarrow \text{de Rham} \Rightarrow \text{Hodge-Tate}.$$

Of special interest is the case of the representation $H^*_{\text{ét}}(X_{\bar{K}}, \mathbb{Q}_p)$, in this case $D_*(H^*_{\text{ét}}(X_{\bar{K}}, \mathbb{Q}_p))$ can sometimes be identified with other cohomology theories, so that by (5) this cohomology theory and étale cohomology determine each other.

The following conjectures of Fontaine have been proved by Faltings [Fa] and Tsuji [T1] based on the work of a number of people (Fontaine, Hyodo, Kato, Messing...).

1. (Faltings) Let X be smooth and proper over K. Let $H^*_{dR}(X/K)$ be the de Rham cohomology of X, equipped with its Hodge filtration. Then the representation $H^*_{\text{ét}}(X_{\bar{K}}, \mathbb{Q}_p)$ is de Rham, and

$$B_{dR} \otimes_K H^*_{DR}(X/K) \cong B_{dR} \otimes_{\mathbb{Q}_p} H^*_{\text{ét}}(X_{\bar{K}}, \mathbb{Q}_p),$$

as filtered Galois-modules.

2. (Tsuji) Let X be the generic fiber of a proper, semi-stable scheme over A. Let $H^*_{\text{st}}(X/W(k))$ be the logarithmic crystalline cohomology of Hyodo and Kato, equipped with a σ-linear endomorphism φ and a monodromy operator N satisfying $N\varphi = p\varphi N$. After extending scalars to K, it is isomorphic to de Rham cohomology, hence inherits the Hodge filtration. Then $H^*_{\text{ét}}(X_{\bar{K}}, \mathbb{Q}_p)$ is semi-stable and

$$B_{\text{st}} \otimes_{W(k)} H^*_{\text{st}}(X/W(k)) \cong B_{\text{st}} \otimes_{\mathbb{Q}_p} H^*_{\text{ét}}(X_{\bar{K}}, \mathbb{Q}_p),$$

compatible with Galois action, σ-semilinear endomorphism, monodromy operator, and filtration after extension of scalars to K.

Note that if X is the generic fiber of a smooth and proper scheme $\mathcal{X}$ over A, then logarithmic crystalline cohomology agrees with the usual crystalline cohomology, and the monodromy operator N is zero. In this situation, the above result has been proved by Faltings, and yields that $H^*_{\text{ét}}(X_{\bar{K}}, \mathbb{Q}_p)$ is crystalline, and

$$B_{\text{crys}} \otimes_{W(k)} H^*_{\text{crys}}(\mathcal{X}/W(k)) \cong B_{\text{crys}} \otimes_{\mathbb{Q}_p} H^*_{\text{ét}}(X_{\bar{K}}, \mathbb{Q}_p),$$

as Galois modules with σ-semilinear endomorphism and filtration after extending scalars to K.

A representation E is called *potentially semi-stable*, if its restriction to an open subgroup of finite index of the Galois group is semi-stable. This is the closest analogy to the monodromy theorem of Grothendieck in the p-adic situation. Obviously, semi-stable representations are potentially semi-stable, and one can show that potentially semi-stable representations are de Rham. Using de Jong's and Tsuji's theorem, we get the following strengthening and alternate proof of (1):

Proposition 4.4. *For a smooth and proper scheme X over K, $H^*_{\acute{e}t}(X_{\bar{K}}, \mathbb{Q}_p)$ is potentially semi-stable.*

Proof. By Tsuji's theorem, the theorem holds for the generic fiber of a proper semi-stable scheme. Since a subrepresentation of a semi-stable representation is again semi-stable, we see using the method of Lemma 4.2 that $H^*_{\acute{e}t}(X_{\bar{K}}, \mathbb{Q}_p)$ is semi-stable as a $\mathrm{Gal}(\bar{K}/K')$-module, hence potentially semi-stable as a $\mathrm{Gal}(\bar{K}/K)$-module. □

Note that extensions of semi-stable representations need not be semi-stable, so that Lemma 4.1 does not apply to prove Proposition 4.4 for all K-varieties X. However, in a recent paper Tsuji uses proper hypercoverings to extend his method to prove the following generalization:

Theorem 4.5. [T2, Corollary 2.2.3] *Let X be a proper scheme over K. Then $H^*_{\acute{e}t}(X_{\bar{K}}, \mathbb{Q}_p)$ is potentially semi-stable.*

4.3. Finiteness of rigid cohomology

Let k be a field of characteristic $p > 0$, W a Cohen-ring for k, and K its field of fractions. For a smooth, affine scheme X over k, Monsky and Washnitzer defined the cohomology $H^*_{MW}(X/K)$, which is a K-vector space. Not much is known about these groups. On the other hand, for X smooth and proper over k, Grothendieck defined crystalline cohomology groups $H^*_{crys}(X/W)$, which are finitely generated W-modules. De Jong's theorem allows to prove finite generation of $H^*_{MW}(X/K)$, starting from the corresponding statement for crystalline cohomology.

The bridge between the two theories is Berthelot's rigid cohomology. We refer the reader to his paper [B2] for a proper definition of $H^*_{rig}(X/K)$ for X a separated scheme of finite type over k, and rigid cohomology with support $H^*_{Z,rig}(X/K)$ for Z a closed subscheme of X. We will only need the following properties:

Trace If X' is étale over X, then there exists a trace map $H^*_{MW}(X'/K) \to H^*_{MW}(X/K)$.

Proper If X is proper and smooth over k, then there is an isomorphism

$$H^*_{rig}(X/K) \xrightarrow{\sim} H^*_{crys}(X/W) \otimes_W K.$$

Affine If X is affine and smooth over k, then there is an isomorphism

$$H^*_{rig}(X/K) \xrightarrow{\sim} H^*_{MW}(X/K).$$

Gysin Let $Y \to X$ be a closed immersion of codimension r between two smooth schemes over k which can be lifted to characteristic 0. If Z a closed subscheme of Y then there exists a Gysin isomorphism

$$H^*_{Z,rig}(Y/K) \xrightarrow{\sim} H^{*+2r}_{Z,rig}(X/K).$$

Excision For $T \subseteq Z \subseteq X$, there is a long exact sequence

$$\ldots \to H^i_{T,rig}(X/K) \to H^i_{Z,rig}(X/K) \to H^i_{Z-T,rig}(X - T/K) \to \ldots.$$

Theorem 4.6. *(Berthelot,* [B2, Théorème 3.1]*) Let X be a smooth, separated scheme over k and $Z \subseteq X$ a closed subscheme. Then the groups $H^*_{Z,\mathrm{rig}}(X/K)$ are finite dimensional K-vector spaces. In particular, if X is a smooth affine scheme over k, the groups $H^*_{MW}(X/K)$ are finite dimensional K-vector spaces.*

Proof. (cf. Berthelot [B1]) The proof is an induction over n on the following two assertions for each field k of characteristic p and all smooth and separated schemes X over k:

$(a)_n$ $H^*_{\mathrm{rig}}(X/K)$ is finite dimensional for X of dimension at most n.
$(b)_n$ $H^*_{Z,\mathrm{rig}}(X/K)$ is finite dimensional for each closed subscheme Z of dimension at most n.

To prove $(a)_n$ from (b_{n-1}) one applies the method of Lemma 4.1. The statement follows for smooth and proper schemes by comparison to crystalline cohomology. To get a trace map, one finds an alteration over the algebraic closure of the base field (which is then generically étale), and observes that this alteration is already defined over a finite extension of the base field. The proof of (b_n) from (a_n) does not require de Jong's theorem, so we omit it. □

4.4. Rational motivic cohomology in characteristic p

We give an example of how de Jong's theorem can be used to study motivic cohomology of fields and varieties in characteristic p. For X a smooth variety over a field k, define motivic cohomology with coefficients in an abelian group A to be Bloch's higher Chow groups (4):

$$H^i(X, A(n)) = \mathrm{CH}^n(X, 2n-i, A).$$

By [L1], rationally motivic cohomology agrees with the weight n-part of the algebraic K-theory of X: $H^i(X,\mathbb{Q}(n)) \cong K_{2n-i}(X)^{(n)}_{\mathbb{Q}}$. A conjecture of Parshin states that if X is smooth and projective over a finite field, then $H^i(X,\mathbb{Q}(n)) = 0$ unless $i = 2n$. This is motivated by the idea that motivic cohomology should be the Ext-groups in a category of mixed motives, and that the category of mixed motives over a finite field should be semi-simple, hence the Ext-groups vanish. In order to convince algebraic geometers of the validity of Parshin's conjecture, we note that it is a consequence of Tate's conjecture on algebraic cycles and the conjecture that rational and numerical equivalence agrees for smooth, projective varieties over finite fields up to torsion [Ge, Theorem 3.3].

Using de Jong's theorem, we can show that Parshin's conjecture gives bounds for rational motivic cohomology of fields and of varieties in characteristic p:

Theorem 4.7. *Let k be a field of characteristic $p > 0$ of transcendence degree e (possibly infinite) over $\mathbb{F}_p$. Assume Parshin's conjecture and let X be a variety of dimension d over k. Then*
i) *$H^i(k,\mathbb{Q}(n)) = 0$ unless $i = n \le e$.*
ii) *$H^i(X,\mathbb{Q}(n)) = 0$ unless $n \le i \le \min\{n+d, e+d\}$.*

Proof. i) See [Ge, Theorem 3.4] It follows from the definition that $H^i(k, \mathbb{Q}(n)) = 0$ for $i > n$. By induction, we can assume that if $n' < n$, then for any field k of characteristic p, $H^i(k, \mathbb{Q}(n')) = 0$ for $i \neq n'$ and for $n' > e$. Since motivic cohomology commutes with direct limits, we can assume that e is finite. By de Jong, we can find a smooth projective variety X over $\mathbb{F}_p$ such that the function field $k(X)$ of X is a finite extension of k. Since the composition of the inclusion and transfer map $H^i(k, \mathbb{Q}(n)) \to H^i(k(X), \mathbb{Q}(n)) \to H^i(k, \mathbb{Q}(n))$ is multiplication by the degree of the extension, we can assume $k = k(X)$. Now consider the coniveau spectral sequence for motivic cohomology [Bl]

$$E_1^{s,t} = \bigoplus_{x \in X^{(s)}} H^{t-s}(k(x), \mathbb{Q}(n-s)) \Rightarrow H^{s+t}(X, \mathbb{Q}(n)).$$

We have $H^i(k, \mathbb{Q}(n)) = E_1^{0,i}$, and the differentials leaving $E_r^{0,i}$ end in $E_r^{r,i-r+1}$, which is a subquotient of a sum of groups $H^{i+1-2r}(k(x), \mathbb{Q}(n-r))$ for various fields $k(x)$ of transcendence degree $e - r$. If $i < n$, then $i + 1 - 2r < n - r$, and if $n > e$, then $e - r > n - r$. Hence, by induction all differentials leaving $E_r^{0,i}$ are zero and $H^i(k, \mathbb{Q}(n)) = E_\infty^{0,i}$. This is a quotient of $H^i(X, \mathbb{Q}(n))$, which is trivial by Parshin's conjecture.

ii) See [Ge, Corollary 3.5] The bound $i \leq n + d$ follows from the definition of higher Chow groups. For the other bounds, we use the coniveau spectral sequence and note that by (i) the E_1-terms vanish outside the specified bounds. □

References

[B1] P. Berthelot, Altérations de variétés algébriques, Séminaire Bourbaki, exp. 815 (1995/96), Astérisque 241 (1997), 273–311.

[B2] P. Berthelot, Finitude et pureté cohomologique en cohomologie rigide, Invent. Math. (1997), 329–377.

[Bl] S. Bloch, Algebraic Cycles and Higher K-theory, Adv. in Math. 61 (1986), 267–304.

[D] P. Deligne, Théorie de Hodge III, Publ. Math. IHES 44 (1975), 5–77.

[Fa] G. Faltings, Crystalline cohomology and p-adic Galois representations, in: Algebraic Analysis, Geometry and Number Theory, Johns Hopkins Univ. Press (1989), 25–80.

[Fo] J.M. Fontaine, Sur certains types de représentations p-adiques du groupe de Galois d'un corps local, construction d'un anneau de Barsotti-Tate, Ann. of Math. 115 (1982), 529–577.

[GA] O. Gabber, Non negativity of Serre's intersection multiplicity, lecture at the IHES 1995.

[Ge] T. Geisser, Tate's conjecture, algebraic cycles and rational K-theory in characteristic p, K-theory 13 (1998), 109–122.

[GS] H. Gillet, C. Soulé, Intersection theory using Adams operations, Invent. Math. 90 (1987), 243–277.

[I] L. Illusie, Cohomologie de de Rham et cohomologie étale p-adique, Séminaire Bourbaki 726 (1990), Astérisque 223 (1994), 9–57.

[dJ] A.J. de Jong, Smoothness, semi-stability and alterations, Publ. Math. IHES 83 (1996), 51–93.

[L1] M. Levine, Bloch's higher Chow groups revisited, in K-theory, Strasbourg 1992, Asterisque 226 (1994), 235–320.

[L2] M. Levine, Homology of algebraic varieties: an introduction to recent results of Suslin and Voevodsky, Bull. AMS 34 (1997), 293–312.

[MW] P. Monsky, G. Washnitzer, Formal cohomology: I, Annals of Math. 88 (1968), 181–217.

[M] P. Monsky, Formal cohomology II: The cohomology sequence of a pair, Annals of Math. 88 (1968), 218–238.

[R1] P. Roberts, The vanishing of intersection multiplicities of perfect complexes, Bull. AMS 13 (1985), 127–130.

[R2] P. Roberts, Intersection theory and the Homological Conjectures in Commutative Algebra, Proc. Int. Cong. Math., Kyoto 1990 (1991), 361–368.

[Se] J.P. Serre, Algèbre locale, Multiplicités, Lecture Notes in Math. 11, Springer 1975.

[Su] A. Suslin, Higher Chow groups of affine varieties and étale cohomology, Preprint 1993.

[SV] A. Suslin, V. Voevodsky, Singular homology of abstract algebraic varieties, Invent. Math. 123 (1996), 61–94.

[T1] T. Tsuji, p-adic étale cohomology and crystalline cohomology in the semi-stable reduction case, to appear in Invent. Math.

[T2] T. Tsuji, p-adic Hodge theory in the semi-stable reduction case, Documenta Mathematica, Extra Volume ICM 1998 II (1998), 207–216.

Department of Mathematics
University of Illinois at Urbana-Champaign
1409 West Green St.
Urbana, IL 61801 USA
geisser@math.uiuc.edu

Progress in Mathematics, Vol. 181,

Resolving Singularities of Plane Analytic Branches with one Toric Morphism

Rebecca Goldin and Bernard Teissier

Abstract. Let $(C,0)$ be an irreducible germ of complex plane curve. Let $\Gamma \subset \mathbb{N}$ be the semigroup associated to it and $C^\Gamma \subset \mathbb{C}^{g+1}$ the corresponding monomial curve, where g is the number of Puiseux exponents of $(C,0)$. We show, using the specialization of $(C,0)$ to $(C^\Gamma,0)$, that the same toric morphisms $Z(\Sigma) \to \mathbb{C}^{g+1}$ which induce an embedded resolution of singularities of $(C^\Gamma,0)$ also resolve the singularities of $(C,0) \subset (\mathbb{C}^{g+1},0)$, the embedding being defined by elements of the analytic algebra $\mathcal{O}_{C,0}$ whose valuations generate the semigroup Γ.

To Heisuke Hironaka

1. Introduction

In the last few years Mark Spivakovsky has proposed a program to prove the resolution of singularities of excellent schemes. A part of this program is a new look at Zariski's local uniformisation theorem for arbitrary valuations. A fundamental object of study in this approach is the graded ring associated to the filtration of a local domain R naturally provided by a valuation of R.

In the special case where R is the local analytic algebra $\mathcal{O}$ of a plane branch C, with its unique valuation, this graded ring was studied in detail by Monique Lejeune-Jalabert and the second author (see [L-T], [T1]), as a special case of the $\overline{\mathrm{gr}}_I R$ appearing in their study of the "$\overline{\nu}_I$-filtration" of a ring (roughly, by the integral closures of the fractional powers of an ideal I of R).
Recall that one calls *branch* a germ of analytically irreducible excellent curve. We will in this paper deal only with complex analytic branches.

The following facts which appeared at that time in that special case may be relevant to the understanding of the role of the graded ring in the general case:

- The graded ring is the ring of the monomial curve C^Γ with the same semigroup Γ as the given plane branch (see definitions below).
- The generators of the semigroup are the intersection numbers with the given plane branch C of a transversal non-singular germ $x = 0$ and of plane branches $f_j(x,y) = 0$ with a smaller number j , $0 \le j \le g-1$, of Puiseux exponents and having with C maximal contact in the sense defined by M. Lejeune-Jalabert (see [Z], pp. 16–17, [L-J]). In particular the initial forms

in the graded ring of the images in the algebra $\mathcal{O}$ of the equations of these branches generate it as a $\mathbb{C}$-algebra.

- There exists a one parameter deformation of C^Γ having all its fibers except the special one isomorphic to C, and this deformation is equisingular in the sense that it has a simulateous resolution of singularities by normalization; the normalization of the total space of the family is non-singular and induces normalization for each fiber (see [T1]).

We may therefore hope that a process of resolution for C^Γ will induce resolution for C. In this paper we show that the curve C^Γ can be resolved by a single toric modification of its ambient space $\mathbb{C}^{g+1}$ (where g is the number of characteristic Puiseux exponents of the plane branch C) and that this toric modification also resolves the curve C if we view it as embedded in $\mathbb{C}^{g+1}$ by $g+1$ elements of its maximal ideal whose valuations generate the semigroup Γ. This is shown by a generalization of the usual non degeneracy argument, which can be found also in [O2], [O3]. We need more details than in these papers because we want to show that the toric modification resolves not only C^Γ but also C.

From this follows the rather interesting fact that *any plane branch with g Puiseux exponents can be embedded in $\mathbb{C}^{g+1}$ in such a way as to be resolved by one single toric modification, i.e. it is in some sense non-degenerate with respect to its Newton polyhedron.* Moreover, if we denote by $\pi\colon Z \to \mathbb{C}^{g+1}$ a toric map resolving C, and by $\pi|S'\colon S' \to S$ the strict transform by this toric map of a non-singular surface S containing our plane branch C, we can identify $\pi|S'$ to the composition of g toric modifications of non-singular surfaces, thus recovering the known fact that a plane branch can be resolved by a composition of g toric morphisms. The first non trivial example, the simplest branch with two characteristic exponents, is computed in an Appendix.

The description of this composition of toric maps of surfaces obtained in this way has the advantage over the "static" one of [O1] of showing explicitly its analytic dependence on the coefficients of the equation or the parametrization of the curve, in view of the results of [T1], and also of giving a geometric vision of the relationship with the resolution process of the "singular curves with maximal contact" of M. Lejeune-Jalabert, and probably also of the approximate roots à la Abhyankar. Indeed, the embedding of the plane branch C in $\mathbb{C}^{g+1}$ is obtained by adding to the coordinates x, y the images in $\mathcal{O}$ of the equations of plane branches with $< g$ characteristic exponents having maximal contact with C. The disadvantage is that the construction given here is for the time being restricted to irreducible germs.

It is tempting to ask whether in general, *given any germ of an algebraic or analytic space and a valuation of its local algebra, the germ can be embedded in an affine space in such a way as to be non-degenerate with respect to its Newton polyhedron and to the given valuation, in the sense that its strict transform is non-singular at the point specified by the valuation in a toric modification of the ambient space subordinate to the Newton polyhedron.*

This would be an effective local uniformization theorem and the results of this paper indicate that perhaps the specialization to the graded algebra is the key to such a result.

In the general case the graded algebra associated with a valuation is not even Nœtherian, and as the reader will see, the proofs in this paper rely on the extensive knowledge we have of the structure of the semigroup of a plane branch, a knowledge essentially lacking in dimension ≥ 3 (the case of dimension 2 is currently under study). In the paper "Valuations, Deformations, and toric Geometry" following this one the second author begins to prepare the way for a proof of the local uniformization theorem along these lines. It is worth noting that one of the results of [L-T] is that the graded algebra associated with the $\overline{\nu}_I$ filtration, i.e. $\overline{\mathrm{gr}}_I R$, is an R-algebra of finite type. Some very interesting finiteness results on the graded algebra associated with a valuation and its relation with Abhyankar's inequality have recently been obtained by O. Piltant ([P]).

This text is based on a lecture given at the singularities Seminar by the second author in January 1994, where examples largely stood in for proofs. The ideas are presented in the framework of complex geometry for simplicity, but the interested reader will see how to transform this into a characteristic-blind resolution for one-dimensional excellent henselian equicharacteristic local integral domains with an algebraically closed residue field.

We are grateful to Tadao Oda and Michel Vaquié for pointing out some errors and imprecisions in preliminary versions of this text.

2. Puiseux expansion and the semigroup of a curve

Definition. Let $(C,0) \subset (\mathbb{C}^2,0)$ be a germ of an irreducible analytic curve (a plane branch) defined by the equation $f(X,Y)=0, \quad f \in \mathbb{C}\{X,Y\}$. We will call $\mathcal{O} = \mathbb{C}\{X,Y\}/(f)$ the *local algebra of the curve* C, and denote by $\mathbf{m}_{\mathcal{O}}$ its maximal ideal. Note that $\mathcal{O} = \mathbb{C}\{x,y\}$ where x and y are the residue classes in $\mathbb{C}\{X,Y\}/(f)$ of X and Y respectively.

According to Newton and Puiseux, there is a parametric representation of C, called the *Newton-Puiseux expansion*, i.e an injection of the algebra $\mathcal{O}$ into $\mathbb{C}\{t\}$ described in suitable coordinates X,Y,t by:

$$\begin{array}{rcll} X &=& x(t) = t^n & (*) \\ Y &=& y(t) = \sum_{i \geq m} a_i t^i & \text{with m > n.} \end{array}$$

The Puiseux expansion of x and y; since the two series converge for small enough $\|t\|$, say $\|t\| < \epsilon$. Denoting by $\mathbb{D}(0,\epsilon)$ the disk with center 0 and radius ϵ in $\mathbb{C}$, the image of the map $\mathbb{D}(0,\epsilon) \to \mathbb{C}^2$ defined by $x(t), y(t)$ is a representative of the germ $(C,0) \subset (\mathbb{C}^2,0)$. From the integer n and the exponents appearing in the expansion of $y(t)$ one can extract the Puiseux characteristic exponents, which characterize the topological type of a small representative of our branch (see [Z]). In particular, we may choose the coordinates X,Y in such a way that the exponent m appearing in the expansion of $y(t)$ is not divisible by the multiplicity n. It is then generally denoted by β_1. We shall do so from now on. More generally, any analytically irreducible germ of a curve has a normalization isomorphic to $\mathbb{C}\{t\}$, and its algebra is an analytic subalgebra of $\mathbb{C}\{t\}$.

Given this parametrization, certain properties of an analytic branch C, plane or not, can be described using the valuation on the algebra $\mathcal{O}$ induced via the injection $\mathcal{O} \hookrightarrow \mathbb{C}\{t\}$ by the t-adic valuation of $\mathbb{C}\{t\}$.

Definition. Given a branch $(C, 0)$ and its algebra $\mathcal{O} \hookrightarrow \mathbb{C}\{t\}$, let ν be the t-adic valuation on $\mathbb{C}\{t\}$. We define the *semigroup* Γ of $\mathcal{O}$ to be

$$\Gamma = \{\nu(\xi) : \ \xi \in \mathcal{O}\}.$$

We follow here the common usage, which is to denote by the same letter Γ the semigroup deprived of its zero element, i.e $\{\nu(\xi) : \ \xi \in \mathbf{m}_{\mathcal{O}}\}$. We say that the branch $(C, 0)$ *has the semigroup* Γ.

The semigroup Γ is finitely generated and has finite complement in $\mathbb{N}$ (see [Zariski, II.1] for the case of plane branches), and if $(C, 0)$ is a plane branch, its minimal set of generators $\{\overline{\beta}_0, \dots, \overline{\beta}_g\}$ can be uniquely determined by:

(a) $\overline{\beta}_0 = n$
(b) $\overline{\beta}_i = \min\{z \in \Gamma \mid \ z \notin \langle\overline{\beta}_0, \dots, \overline{\beta}_{i-1}\rangle\}$
where $\langle\overline{\beta}_0, \dots, \overline{\beta}_{i-1}\rangle \subset \mathbb{N}$ is the semigroup generated by $\{\overline{\beta}_0, \dots, \overline{\beta}_{i-1}\}$ (in particular, $\overline{\beta}_1 = \beta_1$). We then have that

1. $\{\overline{\beta}_0, \dots, \overline{\beta}_g\}$ generates Γ,
2. $\overline{\beta}_0 < \dots < \overline{\beta}_g$, and
3. $\gcd(\overline{\beta}_0, \dots, \overline{\beta}_g) = 1$

We call the vector $(\overline{\beta}_0, \dots, \overline{\beta}_g) \subset \mathbb{Z}^{g+1}$ the *weight vector*. Of course the datum of its semigroup Γ does *not* determine a branch up to analytic isomorphism. In fact, the datum of the semigroup Γ of a plane branch is equivalent to the datum of its Puiseux characteristic exponents (see [Z]). Among all the branches, plane or not, having a given semigroup, one may single out the affine curve $C^\Gamma \subset \mathbb{C}^{g+1}$ defined by the parametrization:

$$C^\Gamma : \ u_i = t^{\overline{\beta}_i} \quad 0 \leq i \leq g. \tag{1}$$

Consider $(C^\Gamma, 0)$ as the germ of a curve. Its algebra

$$\mathcal{O}_{C^\Gamma,0} = \mathbb{C}\{t^{\overline{\beta}_0}, \dots, t^{\overline{\beta}_g}\} \subset \mathbb{C}\{t\}$$

clearly has semigroup Γ. It is shown in [T1] that all branches having a given semigroup Γ are complex analytic deformations of $(C^\Gamma, 0)$ and these deformations are in some sense *equisingular*.

Definition. Given a branch $(C,0)$ and its semigroup Γ, we call the curve C^Γ described above by the parametrization (1) the *monomial curve* associated to C. If $(C,0)$ is a plane germ, the monomial curve C^Γ is a complete intersection in $\mathbb{C}^{g+1}$, where g is the number of Puiseux characteristic exponents of $(C,0)$.

3. Deforming curves

The curve C^Γ mentioned above, as we shall see, is the center of a finite-dimensional flat family of branches which contains (up to complex analytic isomorphism of germs) every germ of analytically irreducible curve (or branch) with semigroup Γ.

We begin by showing (see [T1]) that given any branch $(C,0)$, one can construct explicitly a one-parameter analytic deformation of $(C^\Gamma,0)$, whose general fibres are isomorphic to the curve $(C,0)$. Let us take for simplicity the case of a plane branch. Consider the parametric representation of the curve C as in $(*)$. After a change of coordinates we may assume that it has the form:

$$\begin{aligned} x(t) &= t^{\overline{\beta}_0} \\ y(t) &= t^{\overline{\beta}_1} + \textstyle\sum_{j>\overline{\beta}_1} c_j^{(1)} t^j . \end{aligned}$$

Note that $\overline{\beta}_0$ is the multiplicity n of C. By the definition of the $\overline{\beta}_i'$s, there exist elements $\xi_i(t) \in \mathcal{O} = \mathbb{C}\{x,y\}$, $2 \le i \le g$ such that

$$\xi_i(t) = t^{\overline{\beta}_i} + \sum_{j>\overline{\beta}_i} c_j^{(i)} t^j .$$

where $\xi_0 = x(t)$. Now consider the family, parametrized by v, of curves parametrized by t, in $\mathbb{C}^{g+1} \times \mathbb{C}$:

$$\mathfrak{X}(v) \begin{cases} x = t^{\overline{\beta}_0} \\ y = t^{\overline{\beta}_1} + \sum_{j>\overline{\beta}_1} c_j^{(1)} v^{j-\overline{\beta}_1} t^j \\ u_2 = t^{\overline{\beta}_2} + \sum_{j>\overline{\beta}_2} c_j^{(2)} v^{j-\overline{\beta}_2} t^j \\ \cdot = \cdot \\ \cdot = \cdot \\ \cdot = \cdot \\ u_g = t^{\overline{\beta}_g} + \sum_{j>\overline{\beta}_g} c_j^{(g)} v^{j-\overline{\beta}_g} t^j . \end{cases}$$

Proposition 3.1. *The branch $(\mathfrak{X}(0),0)$ is isomorphic to $(C^\Gamma,0)$, and for any $v_0 \neq 0$, the branch $(\mathfrak{X}(v_0),0)$ is isomorphic to $(C,0)$ by the isomorphism $u_i(t) \mapsto v_0^{-\overline{\beta}_i} u_i(v_0 t)$ (where we set $x = u_0,\ y = u_1$).*

Proof. For $v = 0$ it is obvious, and for $v \neq 0$, since $\mathcal{O}$ is generated as an analytic subalgebra of $\mathbb{C}\{t\}$ by $x(t), y(t)$, it follows from the fact that the $\xi_i(t)$ are in $\mathcal{O}$ (see [T1]). □

The semigroups of plane branches are characterized by the following arithmetical properties (see [T1]); let us set $e_i = \gcd(\overline{\beta}_0, \overline{\beta}_1, \dots, \overline{\beta}_i), e_{i-1} = n_i e_i$. Then

$$\begin{aligned} n_i\overline{\beta}_i \in &\quad \langle \overline{\beta}_0, \dots, \overline{\beta}_{i-1} \rangle \\ n_i\overline{\beta}_i < &\quad \overline{\beta}_{i+1} \end{aligned}$$

The first relation implies the existence of integers $\ell_j^{(i)}$, $1 \le i \le g$, $0 \le j \le i-1$, such that

$$n_i\overline{\beta}_i = \ell_0^{(i)}\overline{\beta}_0 + \dots + \ell_{i-1}^{(i)}\overline{\beta}_{i-1}$$

So that on the curve C^Γ as parametrized, we have equations:

$$\begin{array}{lllll} f_1 & = & u_1^{n_1} & - & u_0^{\ell_0^{(1)}} & = & 0 \\ f_2 & = & u_2^{n_2} & - & u_0^{\ell_0^{(2)}} u_1^{\ell_1^{(2)}} & = & 0 \\ \cdot & & \cdot & & & & \cdot \\ \cdot & & \cdot & & & & \cdot \\ \cdot & & \cdot & & & & \cdot \\ f_g & = & u_g^{n_g} & - & u_0^{\ell_0^{(g)}} \cdots u_{g-1}^{\ell_{g-1}^{(g)}} & = & 0. \end{array} \qquad (**)$$

It is shown in [T1] that these equations define the curve C^Γ, which is therefore a complete intersection in $\mathbb{C}^{g+1}$. Now using the theory of miniversal deformations (see [T3]), we have:

Theorem 3.2. **[Teissier]** *There exists a germ of a flat morphism*

$$p\colon\ (\mathcal{X}_u, 0) \to (\mathbb{C}^{\tau_-}, 0)$$

endowed with a section σ such that $(p^{-1}(0), 0)$ is analytically isomorphic to $(C^\Gamma, 0)$, and for any representative of the germ of the morphism p and any branch $(C, 0)$ with semigroup Γ, there exists $v_C \in \mathbb{C}^{\tau_-}$ such that $(p^{-1}(v_C), \sigma(v_C))$ is analytically isomorphic to $(C, 0)$. Moreover, $(\mathcal{X}_u, 0)$ is embedded in $(\mathbb{C}^{g+1} \times \mathbb{C}^{\tau_-}, 0)$ in such a way that p is induced by the second projection.

Proof. See [T1]; this is the miniversal constant semigroup deformation of C^Γ, and it corresponds to the part of the base of the miniversal (equivariant) deformation of C^Γ which is spanned by the coordinates with negative weight. It is shown in [T1], 4.4 that $\tau_- \le \#(\mathbb{N} \setminus \Gamma)$.

More precisely the miniversal constant semigroup deformation of C^Γ is a map $p\colon \mathcal{X}_u \to \mathbb{C}^{\tau_-}$, where $\mathcal{X}_u$ is embedded in $\mathbb{C}^{g+1} \times \mathbb{C}^{\tau_-}$ in such a way that p is induced by the second projection, and defined by the equations

$$\begin{array}{lllll} F_1 & = & f_1 + \sum_{r=1}^{\tau_-} v_r \phi_{r,1}(u_0, \dots, u_g) & = & 0 \\ \cdot & & \cdot & = & \cdot \\ \cdot & & \cdot & \cdot & \cdot \\ \cdot & & \cdot & \cdot & \cdot \\ F_g & = & f_g + \sum_{r=1}^{\tau_-} v_r \phi_{r,g}(u_0, \dots, u_g) & = & 0 \end{array}$$

where the $\phi_{r,j}$ are polynomials and each monomial in $(u_0, \dots, u_g)$ appearing in $\phi_{r,j}$ is of weight $> n_j\overline{\beta}_j$ when each u_k is given the weight $\overline{\beta}_k$; the equation f_j

is then homogeneous of weight $n_j\overline{\beta}_j$. In fact one may choose the vectors ϕ_r to have only one nonzero coordinate which is a monomial $\phi_{r,j}$. Moreover, the images vectors $\phi_1, \dots, \phi_{\tau_-}$ in $\mathbb{C}[C^\Gamma]^{g+1}/N$, where N is a certain jacobian submodule of $\mathbb{C}[C^\Gamma]^{g+1}$, are linearly independant over $\mathbb{C}$.

The basic property of the miniversal deformation is that for any deformation $d\colon (\mathcal{Y}, y) \to (S, 0)$ of the germ $(C^\Gamma, 0)$ such that all the fibers have a singular point with semigroup Γ there exists an analytic map germ $h\colon (S, 0) \to (\mathbb{C}^{\tau_-}, 0)$ such that d is isomorphic to the deformation obtained from p by pull-back by h. Moreover, since C^Γ is quasi-homogeneous, it is also the case for p in the sense that there are weights on the v_i such that the equations F_j are homogeneous of degree $n_j\overline{\beta}_j$. These weights are negative, which is the reason for the notation $\mathbb{C}^{\tau_-}$; in fact the weight w_r of v_r is $n_j\overline{\beta}_j$ minus the weight of $\phi_{r,j}$, this difference being independent of j.

The recipe to find vectors ϕ_r making the deformation miniversal is due to J. Mather; it is the following (see [T1]): Consider the ring of algebraic functions $\mathbb{C}[C^\Gamma]$ on the monomial curve, and the submodule N of $\mathbb{C}[C^\Gamma]^g$ generated by the vectors $\partial_{u_i} f$ where $f = (f_1, \dots, f_g)$. The quotient $\mathbb{C}[C^\Gamma]^g/N$ is a finite dimensional vector space over $\mathbb{C}$, and so we may choose vectors $\psi_i \in \mathbb{C}[u_0, \dots, u_g]^g$, each having a single nonzero coordinate which is a monomial, such that the natural images of the ψ_i in $\mathbb{C}[C^\Gamma]^g/N$ form a basis of this vector space. The ϕ_r are those among the $\psi_i = (0, \dots, 0, u^{q_{i,j}}, 0, \dots, 0)$ such that if $u^{q_{i,j}}$ is at the j-th line, the weight of $u^{q_{i,j}}$ is $> n_j\overline{\beta}_j$, so that in the corresponding deformation the only equation modified, $f_j + v_i u^{q_{i,j}}$ is modified by a term of weight greater than that of f_j.

In particular, the one-parameter specialization of a branch C to C^Γ which we have seen above can be obtained in this way, up to isomorphism, by a map $h_1\colon (\mathbb{C}, 0) \to (\mathbb{C}^{\tau_-}, 0)$. The map p is equivariant with respect to the action of the group $\mathbb{C}^*$ on $\mathfrak{X}_u$ (resp $\mathbb{C}^{\tau_-}$) which is described by $u_k \mapsto \lambda^{\overline{\beta}_j} u_k$, $v_r \to \lambda^{w_r} v_r$ (resp. only the action on the v_r). If $v_C \in \mathbb{C}^{\tau_-}$ is a point corresponding to a branch isomorphic to $(C, 0)$, the image of h_1 is contained in the orbit of v_C under this action of $\mathbb{C}^*$. The action of $\mathbb{C}^*$ on $\mathbb{C}^{\tau_-}$ ensures that any branch with semigroup Γ appears up to isomorphism as a fiber in any representative of the germ $p\colon (\mathfrak{X}_u, 0) \to (\mathbb{C}^{\tau_-}, 0)$. □

4. Resolution using toric morphisms

We produce *toric morphisms* $\pi(\Sigma)\colon Z(\Sigma) \to \mathbb{C}^{g+1}$ to resolve the singularities of $C^\Gamma \subset \mathbb{C}^{g+1}$, and then show that the same morphisms which resolves C^Γ resolve any fibre of the miniversal constant semigroup deformation $p\colon \mathfrak{X}_u \to \mathbb{C}^{\tau_-}$. In particular, since all curves with semigroup Γ are represented as fibers of this deformation, Theorem 3.2 implies that $\pi(\Sigma)$ resolves $(C, 0)$. The toric resolution of the curve C^Γ will be a consequence of a generalization of the work of Varchenko [V] (expounded also in [M]) for "convenient" or "commode" functions, whose Newton polyhedron intersects each axis and which are non-degenerate.

4.1. The Toric Morphism

For the notions and basic results of toric geometry used in this section, we refer to [C]. A toric morphism is locally described by monomial maps $\pi(a)\colon \mathbb{C}^{g+1} \to \mathbb{C}^{g+1}$ where $a = (a^0, \dots, a^g)$ with $a^j \in \mathbb{N}^{g+1}$ for all j and $\mathrm{span}\{a^0, \dots, a^g\} = \mathbb{R}^{g+1}$:

$$\begin{array}{rcl} u_0 & = & {y_0}^{a_0^0} \cdot \dots \cdot {y_g}^{a_0^g} \\ u_1 & = & {y_0}^{a_1^0} \cdot \dots \cdot {y_g}^{a_1^g} \\ \cdot & & \cdot \\ \cdot & & \cdot \\ \cdot & & \cdot \\ u_g & = & {y_0}^{a_g^0} \cdot \dots \cdot {y_g}^{a_g^g} \end{array}$$

More precisely, a fan (see [Oda], Chap. 1) Σ with support $\mathbb{R}_+^{g+1}$ is a decomposition of the positive quadrant $\mathbb{R}_+^{g+1}$ into rational simplicial cones σ_α with the properties that any face of such a cone is also a part of the fan and that the intersection of two of them is a face of each. A fan is *non-singular* if the primitive integral vectors of the 1-skeleton of each cone σ of dimension $g+1$ form a basis of the integral lattice $\mathbb{N}^{g+1}$.

To a fan Σ are associated a toric variety $Z(\Sigma)$ and a toric (equivariant) map $\pi(\Sigma)\colon Z(\Sigma) \to \mathbb{C}^{g+1}$. The variety $Z(\Sigma)$ is obtained by glueing up affine varieties $Z(\sigma)$ corresponding to each cone of maximum dimension, and if the fan is non-singular, so is $Z(\Sigma)$; in this case, for each cone σ of maximal dimension $Z(\Sigma)$ is isomorphic to $\mathbb{C}^{g+1}$ and the map $\pi(\sigma)$ induced by $\pi(\Sigma)$ is equal to $\pi(a^0, \dots, a^g)$ where the a^j are the primitive integral vectors of the 1-skeleton of the cone σ. An upper convex map (see [C], [Oda] A.3 p. 182-185) $m\colon \mathbb{R}_+^{g+1} \to \mathbb{R}_+$ taking integral values on the integral points and linear in each cone $\sigma \in \Sigma$ determines a divisor on $Z(\Sigma)$, which in the chart $\pi(a^0, \dots, a^g)\colon Z(\Sigma) \to \mathbb{C}^{g+1}$ has the equation

$$y_0^{m(a^0)} \dots y_g^{m(a^g)} = 0.$$

We will call it the exceptional divisor of the toric map $\pi(\Sigma)$ corresponding to the upper convex function m. Just like in the absolute case, the datum of such a map is equivalent, in the case where the function m is *strictly convex,* to the datum of an embedding $Z(\Sigma) \subset \mathbb{C}^{g+1} \times \mathbb{P}^N$ such that the exceptional divisor is the pull back of a hyperplane section of $\mathbb{P}^N$. Such an upper convex map is called a *support function.*

The *Newton polyhedron* of a function $f\colon \mathbb{C}^{g+1} \to \mathbb{C}$ can be used to define fans and a support function.

Definition. For a function $f = \sum_{p \in \mathbb{N}^{g+1}} f_p u^p$, let $\mathrm{supp}\, f = \{p \in \mathbb{N}^{g+1} : f_p \neq 0\}$ and $\mathcal{N}_+(f) =$ *boundary of the convex hull of* $(\{\mathrm{supp}\, f\} + \mathbb{R}^{g+1})$ in $\mathbb{R}^{g+1}$. We call $\mathcal{N}_+(f)$ the *Newton polyhedron* of the function f. Note that $\mathcal{N}_+(f)$ has finitely many compact faces and that its non-compact faces of dimension $\leq g$ are parallel to coordinate hyperplanes.

Define the function m by:

$$m(q) = \inf_{p \in \mathcal{N}_+(f)} \langle q, p \rangle$$

and define an equivalence relation: two vectors are equivalent if and only if the same elements of $\mathcal{N}_+(f)$ minimize the inner product with each vector. In other words,

$$\begin{array}{ccc} q & \sim & q' \\ & \Updownarrow & \\ \{p \in \mathcal{N}_+(f) \mid \langle q, p \rangle = m(q)\} & = & \{p \in \mathcal{N}_+(f) \mid \langle q', p \rangle = m(q')\} \end{array} \qquad (***)$$

The function m is homogenous and piecewise linear, thus the equivalence classes $(***)$ form a fan; to each class (and corresponding set of $p \in \mathcal{N}_+(f)$), we associate a cone σ which consists of those vectors whose inner product is minimized by (and only by) these p. We obtain a convex rational fan Σ_0 in $\mathbb{R}_+^{g+1}$ with vertex 0. By construction, the function m is linear on each cone $\sigma \in \Sigma$, and it is easy to verify that it is strictly upper convex. The non-compact faces of the boundary $\mathcal{N}(f)$ of $\mathcal{N}_+(f)$ correspond to cones σ in the fan which are contained in a coordinate hyperplane of $\mathbb{R}^{g+1}$. Moreover, given a subset $J \subset \{0, \dots, g\}$, we can consider the projection of the Newton polyhedron to the subspace $\mathbb{R}^J = \{p \in \mathbb{R}^{g+1} \mid p_k = 0 \text{ for } k \notin J\}$ of $\mathbb{R}^{g+1}$. If that projection contains the origin, then for all $a \in \check{\mathbb{R}}^{g+1} \mid a_k = 0$ for $k \notin J$, we have $m(a) = 0$, and therefore we may assume that the cone $\mathbb{R}_+^J$ is in the fan, so that the basis vectors of $\mathbb{R}_+^J$ are in the 1-skeleton of the fan.

By a theorem of Kempf, Mumford, et al. (see [TE], pp. 32–35, [Oda], p. 23), any fan can be refined into a non-singular fan, still containing as faces the spaces $\mathbb{R}_+^J$ such that $m(a) = 0$ for $a \in \mathbb{R}_+^J$. The function m is of course linear in each cone of the finer fan and upper convex (but not strictly so in general), and we are in the situation described above.

If we now consider several functions $f_j; 1 \leq j \leq k$, to each of them corresponds a Newton polyhedron, and therefore a fan $\Sigma_0^{(j)}$ and a support function m_j.

Definition. Let Σ_0 be the fan consisting of the intersections of the cones of the fans $\Sigma_0^{(j)}$. The fan Σ_0 is the least fine common refinement of all the $\Sigma_0^{(j)}$.

It is also the fan associated in the manner we have just seen to a single Newton polyhedron, which is the *Minkowski sum* of the Newton polyhedra of the f_j and is also the Newton polyhedron of the product $f_1 \dots f_k$.

Recall that the Minkowski sum $\mathcal{N}_1 + \mathcal{N}_2$ of two Newton polyhedra $\mathcal{N}_1$, $\mathcal{N}_2$ is the convex domain spanned by vector sums $\{p_1 + p_2 \mid p_1 \in \mathcal{N}_1,\ p_2 \in \mathcal{N}_2\}$. It is a commutative and associative operation. If $\mathcal{N}_k$ is the Newton polyhedron of f_k, $\mathcal{N}_1 + \mathcal{N}_2$ is the Newton polyhedron of $f_1 f_2$.

All the functions m_j are linear in each cone of Σ_0. Taking a non-singular refinement Σ of Σ_0 as described above, we finally have a non-singular fan with support functions m_j and thus, since this fan is also a refinement of the "trivial" fan of $\mathbb{R}_+^{g+1}$, we get a proper toric map $\pi(\Sigma)\colon Z(\Sigma) \to \mathbb{C}^{g+1}$ where $Z(\Sigma)$ is a non singular toric variety, and each function $f_j \circ \pi(\Sigma)$ defines a map from $Z(\Sigma)$ to $\mathbb{C}$.

To study the effect of the modification $\pi(\Sigma)$ on the functions f_j, we restrict ourselves to a chart $Z(\sigma)$ of $Z(\Sigma)$ corresponding to a cone σ whose primitive integral vectors are denoted by $a = (a^0, \ldots, a^g)$. We assume that they form a basis of the integral lattice, and we shall write $\sigma = \langle a^0, \ldots, a^g\rangle$ for the convex cone spanned by the vectors $(a^0, \ldots, a^g)$. If we write the function

$$f_j = \sum_{p \in \mathbb{N}^{g+1}} f_p^{(j)} u^p \ ,$$

where $u^p = u_0^{p_0} u_1^{p_1} \cdots u_g^{p_g}$, then the composition

$$f_j \circ \pi(a^0, \ldots, a^g) = \sum_{p \in \mathbb{N}^{g+1}} f_p^{(j)} y_0^{\langle a^0, p\rangle} \cdots y_g^{\langle a^g, p\rangle}, \tag{2}$$

where $\langle , \rangle$ is the standard inner product, can be written

$$\begin{aligned} f_j \circ \pi(a) &= y_0^{m_j(a^0)} \cdots y_g^{m_j(a^g)} \textstyle\sum_{p \in \mathbb{N}^{g+1}} f_p^{(j)} y_0^{\langle a^0, p\rangle - m_j(a^0)} \cdots y_g^{\langle a^g, p\rangle - m_j(a^g)} \\ &= y_0^{m_j(a^0)} \cdots y_g^{m_j(a^g)} \tilde{f}_j(y_0, \ldots, y_g) \ . \end{aligned}$$

The function $\tilde{f}_j$ is called the *strict transform* of f_j; it satisfies $\tilde{f}_j(0) \neq 0$.

We use the following fact, which is obvious from the definitions:
For each j, $1 \leq j \leq k$, there is a unique $p^{(j)} \in \mathcal{N}_+(f_j)$ such that $\langle a^i, p^{(j)}\rangle = m_j(a^i)$, $0 \leq i \leq g$, and the strict transforms take the form

$$\tilde{f}_j = f_{p^{(j)}}^{(j)} + \sum_{p \in \mathcal{N}_+(f_j) \setminus \{p^{(j)}\}} f_p^{(j)} y_0^{\langle a^0, p\rangle - m_j(a^0)} \cdots y_g^{\langle a^g, p\rangle - m_j(a^g)} .$$

Moreover we can compute the critical locus of $\mathrm{crit}(\pi(\Sigma))$ of $\pi(\Sigma)$, which locally is the critical locus of $\pi(\sigma)$ for each $\sigma = \langle a^0, \ldots, a^g\rangle \in \Sigma$.
By direct inspection we find the following relation for the jacobian matrix $\mathrm{jac}\pi(\sigma)$ of $\pi(\sigma)$:

$$y_0 \cdots y_g \ \mathrm{jac}\pi(\sigma) = u_0 \ldots u_g \ \det(a^0, \ldots, a^g),$$

setting $\alpha_j = \left(\sum_{i=0}^g a_i^j\right) - 1$ this means since $\det(a^0, \ldots, a^g) = \pm 1$,

$$\mathrm{jac}\pi(\sigma) = \pm y_0^{\alpha_0} \ldots y_g^{\alpha_g}.$$

So the divisor $y_j = 0$ is contained in the critical locus of $\pi(\sigma)$ if and only if a^j is not a coordinate vector.

Note that the divisor $y_j = 0$ is contained in $\pi(\sigma)^{-1}(0)$ if and only if the vector a^j has all its coordinates different from zero.

5. Existence of a toric resolution

In this section, we study a germ at 0 of a complete intersection $X \subset \mathbb{C}^{g+1}$ defined by a set of equations $\{f_1 = f_2 = \cdots = f_k = 0\}$ with $f_j \in \mathbb{C}\{u_0, \ldots, u_g\}$, $j = 1, \ldots k$. We assume that the series f_j have no constant term. We use the notations introduced above and consider a toric map of non-singular spaces $\pi(\Sigma)\colon Z(\Sigma) \to \mathbb{C}^{g+1}$. We make our computations in a chart corresponding to a regular simplicial cone $\sigma = \langle a^0, \ldots, a^g \rangle$ of Σ.

Definition. Given a family of functions $\{f_j\}_{1\le j\le k}$ as above, and a toric morphism associated to a regular fan Σ such that all the support functions $(m_j)_{1\le j\le k}$ are linear in each cone σ of Σ, for each map $\pi(\sigma)\colon \mathbb{C}^{g+1}(\sigma) \to \mathbb{C}^{g+1}$, we will call the divisor $y_0^{m(a^0)} \ldots y_g^{m(a^g)}$, where $m(a) = \sum_{j=1}^k m_j(a)$, the *toric exceptional divisor of $\pi(\sigma)$ associated to the support functions m_j*. This definition globalizes to $\pi(\Sigma)$.

From what we have seen one deduces for each $\pi(\sigma)$ with $\sigma = \langle a^0, \ldots, a^g \rangle$ the following statements:

Proposition 5.1. *The fiber $\pi(\sigma)^{-1}(0) \subset \mathbb{C}^{g+1}$ is the union of the intersections $y_{i_1} = \cdots = y_{i_t} = 0$ over all minimal families $J = (i_1, \ldots, i_t)$ of indices such that for each ℓ , $0 \le \ell \le g$, there is an $i \in J$ such that the ℓ-th coordinate of a^i is $\neq 0$. In particular the divisorial part of $\pi(\sigma)^{-1}(0)$ is the union of the divisors $y_i = 0$ for those i such that the vector a^i has no zero component.*
The critical locus crit$(\pi(\sigma))$ *is the union of the divisors $y_s = 0$ for those s such that a^s is not a coordinate vector of $\mathbb{Z}^{g+1}$.*

For any set of functions $\{f_j\}$, we label the compact faces of $\mathcal{N}_+(f_j)$ by γ_j. Any compact face of the Newton polyhedron $\mathcal{N}_+ = \sum_{j=1}^k \mathcal{N}_+(f_j)$ which is the Minkowski sum of the Newton polyhedra of the $\{f_j\}$ will be of the form $\gamma = \gamma_1 + \cdots + \gamma_k$. Each face γ_j in turn is of the form

$$\gamma_j = \gamma_j(I) = \{p \in \mathcal{N}_+(f_j) \mid \langle a^h, p\rangle = m_j(a^h) \text{ for } h \in I\}\ ,$$

where I is a subset of $\{0, \ldots, g\}$. If the face $\gamma_j(I)$ is not compact, it contains a line parallel to a coordinate axis, say the ℓth, so that there is an ℓ, $0 \le \ell \le g$, such that for all $h \in I$ we have $a_\ell^h = 0$.

Definition. We call the set of functions $\{f_j\}_{1\le j\le k}$ *non-degenerate* if for all compact faces $\gamma = \gamma_1 + \cdots + \gamma_k$ of $\mathcal{N}_+ = \sum_{j=1}^k \mathcal{N}_+(f_j)$, denoting by $f_j|_{\gamma_j}$, the sum

$$f_j|_{\gamma_j} = \sum_{p\in\gamma_j} f_p^{(j)} u^p,$$

the $k \times (g+1)$ matrix

$$\begin{pmatrix} \partial_{u_0} f_1|_{\gamma_1} & \cdot & \cdot & \cdot & \partial_{u_g} f_1|_{\gamma_1} \\ \partial_{u_0} f_2|_{\gamma_2} & \cdot & \cdot & \cdot & \partial_{u_g} f_2|_{\gamma_2} \\ \cdot & & & & \cdot \\ \cdot & & & & \cdot \\ \cdot & & & & \cdot \\ \partial_{u_0} f_k|_{\gamma_k} & \cdot & \cdot & \cdot & \partial_{u_g} f_k|_{\gamma_k} \end{pmatrix}$$

has maximal rank k on $(\mathbb{C}^*)^{g+1}$. This means that the equations

$$f_1|_{\gamma_1} = \cdots = f_k|_{\gamma_k} = 0$$

define a non singular complete intersection in $(\mathbb{C}^*)^{g+1}$. For example a single function f is non-degenerate if for each compact face γ of $\mathcal{N}_+(f)$, the $(g+1)$-vector $(\partial_{u_i} f|_\gamma)_i$ does not vanish in $(\mathbb{C}^*)^{g+1}$; this is the original definition of Varchenko [V].

Definition. The morphism $\pi : \tilde{X} \longrightarrow X$ is a *resolution* if the following conditions hold:

- π is a proper morphism,
- $\tilde{X}$ is non-singular, and
- $\tilde{X} \setminus \pi^{-1}(\mathrm{Sing}X) \longrightarrow X \setminus \mathrm{Sing}X$ is an isomorphism.

If the morphism π is a toric morphism $\pi(\Sigma)$, the last condition is a consequence of the inclusion

$$\tilde{X} \cap \mathrm{crit}(\pi(\Sigma)) \subset \pi(\Sigma)^{-1}(\mathrm{Sing}X).$$

However, this condition is difficult to check if X is not itself a toric subvariety, which motivates the following definition of a toric pseudo-resolution, after recalling that the strict transform $\tilde{X}$ of $X \subset \mathbb{C}^{g+1}$ by $\pi(\Sigma)$ is the closure in $Z(\Sigma)$ of $\pi(\Sigma)^{-1}(X) \setminus \mathrm{crit}(\pi(\Sigma))$.

Definition. A toric morphism $\pi(\Sigma)\colon Z(\Sigma) \to \mathbb{C}^{g+1}$ is a *toric (embedded) pseudo-resolution* of a subvariety $X \subset \mathbb{C}^{g+1}$ if the strict transform $\tilde{X}$ of X by $\pi(\Sigma)$ is smooth and transversal to a stratification of the critical locus of $\pi(\Sigma)$.

Note that a toric pseudo-resolution is not necessarily a resolution of singularities in the usual sense since it may not induce an isomorphism

$$\tilde{X} \setminus \pi(\Sigma)^{-1}(\mathrm{Sing}X) \to X \setminus \mathrm{Sing}X.$$

A toric pseudo-resolution only induces an isomorphism

$$\tilde{X} \setminus \mathrm{crit}(\pi(\Sigma)) \to X \setminus \mathrm{disc}(\pi(\Sigma)),$$

where $\mathrm{disc}(\pi(\Sigma))$ is the image by $\pi(\Sigma)$ of $\mathrm{crit}(\pi(\Sigma))$, and since $\tilde{X}$ is non singular, the inclusion $\mathrm{Sing}X \subset \mathrm{disc}(\pi(\Sigma))$ holds. In the case of a single function f, Varchenko introduced in [V] the condition of being *commode* which is equivalent to asking that if a primitive vector a^i of a cone σ (of a fan compatible with the Newton polyhedron of f) is contained in a hyperplane (which means that if in the chart $\mathbb{C}^{g+1}(\sigma)$ the divisor $y_i = 0$ is not contained in $\pi(\sigma)^{-1}(0)$), then we have $m(a^i) = 0$; it follows then from the conditions satisfied by our fans that a^i must in fact be a basis vector, so that $y_i = 0$ is not in the critical locus. For com-mode functions, in fact, the toric exceptional divisor and the critical locus of $\pi(\Sigma)$ both coincide set theoretically with $\pi(\Sigma)^{-1}(0)$ so that a toric pseudo-resolution of $f(u_0, \dots, u_g) = 0$ is also a resolution in the usual sense.

For complete intersections, Oka introduced in [O3] a notion of *convenient* generalizing Varchenko's definition. It is much too strong for our purposes. Our monomial curve is only 1-convenient in the sense of [O3], and the product of its equations is far from being commode in the sense of [V].

However, in the case of a curve, a toric embedded pseudo-resolution is an embedded resolution in the usual sense unless the curve is contained in a coordinate hyperplane.

5.1. The inverse image of X by the morphism $\pi(\Sigma)$

Theorem 5.2. *If the set of functions $\{f_j\}_{1\le j\le k}$ defining the complete intersection $X \subset \mathbb{C}^{g+1}$ is non-degenerate at the origin, there exists a neighborhood U of 0 in $\mathbb{C}^{g+1}$ such that in $\pi(\Sigma)^{-1}(U)$ the strict transform of X by $\pi(\Sigma)$ is non-singular and transversal in $Z(\Sigma)$ to the strata of a stratification of the divisor $\pi(\Sigma)^{-1}(0)$.*

Proof. We consider for each $\pi(\sigma)$ a natural stratification of $\mathbb{C}^{g+1}(\sigma)$ such that $\pi(\sigma)^{-1}(0)$ is a union of strata, as follows: For each $I \subset \{0, \dots, g\}$, define S_I to be the constructible subset of $\mathbb{C}^{g+1}(\sigma)$ defined by $y_i = 0$ for $i \in I$, $y_i \neq 0$ for $i \notin I$. The sets S_I form a partition of $\mathbb{C}^{g+1}(\sigma)$ by non-singular varieties.

If the subset I is such that for each ℓ ,$0 \le \ell \le g$, there is a $j \in I$ such that the ℓ-th coordinate of a^j is $\neq 0$ (see 5.1), the stratum S_I is contained in $\pi(\sigma)^{-1}(0)$, and conversely, so that $\pi(\sigma)^{-1}(0)$ is a union of strata. This stratification of each $\mathbb{C}^{g+1}(\sigma)$ is compatible with the chart decomposition of $Z(\Sigma)$ and so gives a stratification of $Z(\Sigma)$. Moreover, this stratification satisfies the Whitney conditions, so that to check the transversality of a non-singular subspace to every stratum in a neighborhood it is sufficient to check transversality to the strata contained in $\pi(\sigma)^{-1}(0)$; as we have seen, these strata correspond to compact faces of the Newton polyhedra. Now we can compute in each chart $\mathbb{C}^{g+1}(\sigma)$ the restriction of each $\tilde{f}_j$ to S_I:

$$\tilde{f}_j|_{S_I} = \sum_{p|\langle a^i,p\rangle = m_j(a^i) \text{ for } i\in I} f_p^{(j)} y_{j_1}^{\langle a^{j_1},p\rangle - m_j(a^{j_1})} \cdots y_{j_\ell}^{\langle a^{j_\ell},p\rangle - m_j(a^{j_\ell})},$$

where $\{j_1, \dots, j_\ell\} = \{0, \dots, g\} \setminus I$.
The set $\{p \mid \langle a^i, p\rangle = m_j(a^i) \text{ for } i \in I\}$ is by definition a face $\gamma_{j,I}$ of the Newton polyhedron $\mathcal{N}_+(f_j)$, and this face is compact by the assumption on I.

The function $\tilde{f}_j|_{S_I}$ is the restriction to S_I of a function on $\mathbb{C}^{g+1}(\sigma)$ which is independent of the y_i ; $i \in I$. Moreover, this last function differs from the composition with $\pi(\sigma)$ of the function

$$f_{j,\gamma_{j,I}} = \sum_{p\in\gamma_{j,I}} f_p^{(j)} u^p$$

by multiplication by a monomial in $(y_0, \dots, y_g)$. Since on S_I the coordinates y_k; $k \notin I$ are $\neq 0$, the behavior of the functions $\tilde{f}_j|_{S_I}$ is determined by the behavior of the $f_{j,\gamma_{j,I}} \circ \pi(\sigma)$ at points where all the y_i are $\neq 0$. It follows that if the jacobian determinant of the functions $f_{1,\gamma_{1,I}}, \dots, f_{k,\gamma_{k,I}}$ has a $k \times k$ minor which does not vanish on $(\mathbb{C}^*)^{g+1}$ then the functions $\tilde{f}_1|_{S_I}, \dots, \tilde{f}_k|_{S_I}$ define a non-singular subspace of codimension k in S_I (in particular it is empty if $k > \dim S_I$). This implies that the strict transform of X by $\pi(\Sigma)$ is non-singular and transversal to all the strata S_I in a neighborhood of $\pi(\sigma)^{-1}(0)$; in particular it is transversal to the critical locus of $\pi(\sigma)$. □

Applying this to each chart of $\pi(\Sigma)$ gives:

Corollary. *If the set of functions $\{f_j\}_{1\leq j\leq k}$ defining the complete intersection $X \subset \mathbb{C}^{g+1}$ is non-degenerate at 0, there exists a neighborhood U of 0 in $\mathbb{C}^{g+1}$ such that the strict transform $\tilde{X} \to X$ of X by $\pi(\Sigma)$ induces in $\pi(\Sigma)^{-1}(U)$ an embedded toric pseudo-resolution of singularities of $X \cap U$.*

5.2. The toric resolution of the monomial curve

Let C^Γ be the monomial curve derived from C. In this section, we study the embedded resolution of the curve C^Γ by a toric morphism from two different viewpoints, which correspond to its parametric and equational presentations respectively.

First, the monomial curve is the closure in $\mathbb{C}^{g+1}$ of an orbit of $\mathbb{C}^*$ described by

$$t \mapsto (t^{\overline{\beta}_0}, t^{\overline{\beta}_1}, \dots, t^{\overline{\beta}_g}).$$

By the theory of toric varieties (see [O], Chap 1) this orbit is described combinatorially by the linear map $\mathbb{Z} \to \mathbb{Z}^{g+1}$ such that the image of 1 is the *weight vector*

$$w = (\overline{\beta}_0, \overline{\beta}_1, \dots, \overline{\beta}_g) \in \mathbb{Z}^{g+1}.$$

This means that the corresponding morphism of semigroup algebras

$$\mathbb{C}[u_0, \dots, u_g] \to \mathbb{C}[t]$$

obtained from the dual map $\check{\mathbb{Z}}^{g+1} \to \check{\mathbb{Z}}$ by restriction to $\mathbb{N}^{g+1}$ is the morphism of algebras which corresponds to our parametrization of the monomial curve.

We use the notations of §4. Now if we take a regular fan Σ in $\mathbb{R}_+^{g+1}$ which is compatible with the weight vector w, in the sense that w is contained in an edge of any simplicial cone it meets outside 0, we obtain a toric map

$$\pi(\Sigma)\colon Z(\Sigma) \to \mathbb{C}^{g+1}.$$

To compute the strict transform of C^Γ by $\pi(\Sigma)$, we need to find out first which charts $Z(\sigma)$ of $Z(\Sigma)$, corresponding to $(g+1)$-dimensional simplicial cones σ of Σ, have an image which contains C^Γ. Seeking solutions of the form $y_i(t) = c_i t^{\alpha_i} + \cdots$

to the equations $y_0^{a_i^0}\cdot\ldots\cdot y_g^{a_i^g} = t^{\overline{\beta}_i}$, we see that the vector $(\alpha_0,\ldots,\alpha_g)$ must satisfy $M.\alpha = w$ where $M = (a_i^j)$ is the matrix of the column vectors $(a^0,\ldots,a^g)$ which are the generators of the simplicial cone σ. Since we want the α's to be positive, this implies that σ contains w, so that by our assumption that the fan is compatible with w, the vector w must be equal to one of the a^j. We see that we need to consider only those simplicial cones $\sigma = \langle a^0,\ldots,a^g\rangle$, where up to reordering we have $w = a^0$. Then, w is the image of the coordinate vector $(1,0,\ldots,0)$ by the linear map $p(\sigma)\colon \mathbb{Z}^{g+1} \to \mathbb{Z}^{g+1}$ described by the matrix M, i.e sending the i-th basis vector to a^i. The strict transform of C^Γ by the map $\pi(\sigma)$ is again a monomial curve, but its weight vector is the pull back of w by $p(\sigma)$, which is the basis vector $(1,0,\ldots,0)$; in fact, the simplicial cone σ is the image by $p(\sigma)$ of the cone generated by the basis vectors. So the strict transform of C^Γ has the parametric representation

$$y_0 = t, y_1 = \cdots = y_g = 1,$$

and it is indeed non singular and transversal to the exceptional divisor.

Remark.

1) The regular simplicial fan compatible with w is by no means unique; however, there are, at least in the case of plane branches, algorithms which produce such fans. Their description is beyond the intent of this text, but the reader can get some idea of the problem by looking at the example which concludes it.
2) This resolution process works for any monomial curve: the embedded toric resolution reduces entirely to the combinatorial problem of finding a regular simplicial fan Σ in $\mathbb{R}_+^{g+1}$ compatible with the weight vector. We can summarize this in:

Theorem 5.3. *Let $C^\Gamma \subset \mathbb{C}^{g+1}$ be a monomial curve with weight vector $w \in \mathbb{N}^{g+1}$. For any regular simplicial fan Σ in $\mathbb{R}_+^{g+1}$ compatible with w, the toric map $\pi(\Sigma)\colon Z(\Sigma) \to \mathbb{C}^{g+1}$ has the property that the strict transform of C^Γ is non singular and transverse to the exceptional divisor. In fact, it appears only in the charts corresponding to simplicial cones containing w and there it is defined parametrically, up to reordering of the coordinates, by $y_0 = t, y_1 = \cdots = y_g = 1$.* □

This construction of the resolution is simpler than the following one, but not as easily adaptable to the study of the effect of the toric map $\pi(\Sigma)$ on deformations of C^Γ such as our curve $C \subset \mathbb{C}^{g+1}$. The other viewpoint is to explore the implications of the previous section on C^Γ. We reprove the result above from this viewpoint. As specified in Equations (**), C^Γ is defined by equations $\{f_j = 0\}_{1\le j\le g}$, where each $f_j = u_j^\alpha - M_j$ and M_j is a monomial in $u_0,\ldots,u_{j-1}$. Note that $\text{supp} f_j$ has exactly two elements, corresponding to the exponents of the two monomials in f_j, and each Newton polyhedron $\mathcal{N}_+(f_j)$ has only one compact face, which is a segment. For each function f_j, let $\Sigma_0^{(j)}$ be the fan representing the equivalence classes derived from the function m_j. Let Σ be a non-singular fan obtained by refining $\Sigma_0 = \bigcap_j \Sigma_0^{(j)}$, and $\mathcal{N}_+$ be the corresponding Newton polyhedron. We refer to the compact face of the Newton polyhedron $\mathcal{N}_+(f_j)$ as γ_j.

Proposition 5.4. *The curve defined by $\{\tilde{f}_1 = \tilde{f}_2 = \cdots = \tilde{f}_g = 0\}$ intersects the divisor $\pi(\sigma)^{-1}(0)$ only in the charts $\mathbb{C}^{g+1}(\sigma)$, $\sigma = \langle a^0, \ldots, a^g\rangle$, in which for some b the vector a^b is the weight vector $(\overline{\beta}_0, \ldots, \overline{\beta}_g)$. If a^b is the weight vector, then $(\Pi_{i=0}^g y_i = 0) \bigcap \{\tilde{f}_1 = \tilde{f}_2 = \cdots = \tilde{f}_k = 0\}$ is contained in $\{y_b = 0\}$ but not in any other hyperplane $y_i = 0$ for $i \neq b$.*

Proof. Each strict transform $\tilde{f}_j$ has only two terms

$$\tilde{f}_j = f^{(j)}_{p_0^{(j)}} + f^{(j)}_{p_1^{(j)}} y_0^{\langle a^0, p_1^{(j)}\rangle - m_j(a^0)} \cdots y_g^{\langle a^g, p_1^{(j)}\rangle - m_j(a^g)}$$

since each equation f_j has only two terms. The only possibility for the g equations $\tilde{f}_j = 0$ to have a common root with $y_i = 0$ for some i is that $\langle a^i, p_1^{(j)}\rangle = m_j(a^i) = \langle a^i, p_0^{(j)}\rangle$. But this implies that a^i is constant on the Minkowski sum of the g segments constituting the Newton polyhedra of the f_j. By the structure of the equations, this sum is of dimension g; a^i is uniquely determined as the primitive normal vector of this face, and it has to be the weight vector a^b.
Indeed, the monomial map $\pi(\sigma)\colon\ Z(\sigma) \to \mathbb{C}^{g+1}$ maps the strict transform $\tilde{C}^\Gamma$ to C^Γ; if the coordinate y_b corresponds to the weight vector, we see that y_b must be a coordinate on the strict transform of C^Γ, and all other coordinates equal to 1. In other words, the equations in $Z(\sigma)$ of the desingularization of C^Γ are $(y_i = 1,$ for $i \neq b)$; they are the strict transforms of the equations of C^Γ, which have only ± 1's as coefficients. □

Theorem 5.5. *Let C^Γ be a monomial curve defined by $\{f_j\}_{1\le j\le g}$ as above. The closure of the inverse image of $\{f_1 = \cdots = f_g = 0\}$ by the morphism $\pi(\Sigma)$ is non-singular and transversal in $Z(\Sigma)$ to the strata of $\pi(\Sigma)^{-1}(0)$.*

Proof. This follows directly from Theorem 5.2 if we can prove the fact that the family of functions $\{f_j\}_{1\le j\le g}$ is non-degenerate.

But there is only one compact face for the Newton polyhedron $\sum_{j=1}^g \mathcal{N}_+(f_j)$; it is the (Minkowski) sum $\sum_{j=1}^g \gamma_j$. By Proposition 5.4, this compact face corresponds to the subset $I = \{j\} \subset \{0, \ldots, g\}$ such that a^j is the weight vector. Moreover, for each j, the function f_j is equal to f_{j,γ_j}, so that we only have to check that the equations of the monomial curve define a non-singular complete intersection in $(\mathbb{C}^*)^{g+1}$ and we have

$$df_1 \wedge \cdots \wedge df_g = n_1 \ldots n_g u_1^{n_1 - 1} \ldots u_g^{n_g - 1} du_1 \wedge \cdots \wedge du_g + \cdots$$

so the differential form does not vanish outside of the coordinate hyperplanes: the set of equations $f_1, \ldots, f_g$ is non-degenerate.
Note that unlike the first proof, this one encounters a (minor) difficulty if we work over a field of characteristic dividing one of the n_i. □

6. Simultaneous resolution

We will now show that the toric morphism not only resolves C^{Γ}, but simultaneously resolves all curves in the miniversal deformation with constant semigroup $\mathcal{X}_u$ of C^{Γ} which we saw in section 3.

Definition. Let $f : (X,0) \to (Y,0)$ be a flat map with reduced fibres and Y reduced. We say (see [T2]) that f admits a *very weak simultaneous resolution* if, for all sufficiently small representatives, there exists a proper morphism $\pi : \tilde{X} \to X$ such that:

1. The composition $q = f \circ \pi : \tilde{X} \to Y$ is an analytic submersion, i.e. q is flat, and for all $y \in Y$, the fiber $\tilde{X}(y) = q^{-1}(y)$ is non-singular.
2. For all $y \in Y$, the induced morphism $\tilde{X}(y) \to X(y)$ is a resolution of the singularities of $X(y)$. Let Y_1 be the image of a section $\sigma : Y \to X$.

We say that f admits a *weak simultaneous resolution along* Y_1 if it also satisfies the condition:

1. The morphism $q_{Y_1} : (\pi^{-1}(Y_1))_{\text{red}} \to Y$ induced by q is locally topologically a fibration, in the sense that every point $\tilde{x} \in \pi^{-1}(Y_1)$, has an open neighborhood $U \in (\pi^{-1}(Y_1))_{\text{red}}$ such that $U_{Y_1} \simeq V \times \tilde{Y}_0$, where V is an open set in Y, $\tilde{Y}_0$ is an open neighbourhood of $\tilde{x}$ in the fibre of q_{Y_1} passing through $\tilde{x}$, and $\simeq$ is a Y-homeomorphism.

We say that f admits a *strong simultaneous resolution along* Y_1 if it also satisfies the condition:
The morphism $q_{Y_1} : (\pi^{-1}(Y_1)) \to Y$ induced by q is locally analytically trivial (before reduction), in the sense that every point $\tilde{x} \in \pi^{-1}(Y_1)$, has an open neighborhood $U \in (\pi^{-1}(Y_1))$ such that $U_{Y_1} \simeq V \times \tilde{Y}_0$, where V is an open set in Y, $\tilde{Y}_0$ is an open neighbourhood of $\tilde{x}$ in the fibre of q_{Y_1} passing through $\tilde{x}$, and $\simeq$ is a Y-isomorphism.

An *embedded resolution* for $\mathcal{X} \subset \mathbb{C}^N$ is a bimeromorphic map $\pi\colon Z \to \mathbb{C}^N$ where Z is smooth, the exceptional locus E of π is a divisor with normal crossings in Z and the strict transform of $\mathcal{X}$ is smooth and transversal to the canonical stratification of E. As usual, $\mathbb{C}^N$ stands for an open subset of $\mathbb{C}^N$.

Recall from Section 3 that the miniversal deformation of the curve C^{Γ} yields a family of curves, $p\colon \mathcal{X}_u \to \mathbb{C}^{\tau_-}$, embedded in $\mathbb{C}^{g+1} \times \mathbb{C}^{\tau_-}$ in such a way that p is induced by the second projection, and where $\mathcal{X}_u$ is defined by the equations

$$
\begin{array}{rcccl}
F_1 & = & f_1 + \sum_{r=1}^{\tau_-} v_r \phi_{r,1}(u_0, \dots, u_g) & = & 0 \\
\vdots & & \vdots & & \vdots \\
F_g & = & f_g + \sum_{r=1}^{\tau_-} v_r \phi_{r,g}(u_0, \dots, u_g) & = & 0
\end{array}
$$

where the $\phi_{r,j}$ are polynomials and each monomial in $(u_0, \dots, u_g)$ appearing in $\phi_{r,j}$ is of weight $> n_j\overline{\beta}_j$ when each u_k is given the weight $\overline{\beta}_k$; the equation f_j is then homogeneous of weight $n_j\overline{\beta}_j$. We have $\mathfrak{X}_u(0) \simeq C^\Gamma$. Let $\pi(\Sigma)\colon Z(\Sigma) \to \mathbb{C}^{g+1}$ be the toric modification associated to a regular fan in $\mathbb{R}_+^{g+1}$ compatible with the sum of the Newton polyhedra of the equations f_j of C^Γ. By construction, C^Γ is resolved by the morphism $\pi(\Sigma)$.

Theorem 6.1. *If Σ is a regular fan compatible with the Newton polyhedra of the equations $(**)$ of C^Γ, the morphism*

$$\Pi(\Sigma) = \pi(\Sigma) \times \mathrm{Id}_{\mathbb{C}^{\tau_-}} : Z(\Sigma) \times \mathbb{C}^{\tau_-} \to \mathbb{C}^{g+1} \times \mathbb{C}^{\tau_-}$$

induces by restriction to the strict transform $\tilde{\mathfrak{X}}_u$ a resolution of singularities of $\mathfrak{X}_u$, which is a strong simultaneous resolution for $p\colon \mathfrak{X}_u \to \mathbb{C}^{\tau_-}$ with respect to the subspace $Y_1 = \{0\} \times \mathbb{C}^{\tau_-}$. In addition, it is an embedded resolution for $\mathfrak{X} \subset \mathbb{C}^{g+1} \times \mathbb{C}^{\tau_-}$.

Proof. Consider in a chart $\mathbb{C}^{g+1}(\sigma) \times \mathbb{C}^{\tau_-}$ the composition

$$F_j \circ \Pi(\sigma) = f_j \circ \Pi(\sigma) + \sum_{r=1}^{\tau_-} v_r(\phi_{r,j} \circ \pi(\sigma)).$$

Since for each support function $m(a)$ and fixed q the function $\langle a, q\rangle / m(a)$ is a continuous function of a, and since the $\phi_{j,r}$ are polynomials, after possibly refining the fan Σ we may assume that for each cone $\sigma = \langle a^b, a^1, \dots, a^g\rangle$ containing the weight vector a^b, for each j, $1 \le j \le g$, and for each monomial appearing in one of the polynomials $\phi_{j,r}$, the inequality $\langle a^b, q\rangle > m_j(a^b)$, which is equivalent to the weight inequalities mentioned above, implies the inequalities $\langle a^s, q\rangle \ge m_j(a^s)$ for $1 \le s \le g$. We can therefore write the composition $F_j \circ \Pi(\sigma)$ as follows:

$$F_j \circ \Pi(\sigma) = y_0^{m_j(a^0)} \dots y_g^{m_j(a^g)}(\tilde{f}_j + \sum_{r=1}^{\tau_-} v_r\tilde{\phi}_{r,j}),$$

where $\tilde{\phi}_{r,j} = \sum_q \tilde{\phi}_{r,j,q} y_0^{\langle a^b,q\rangle - m_j(a^b)} \dots y_g^{\langle a^g,q\rangle - m_j(a^g)}$. From this follows, since it is true for the $\tilde{f}_j$, that at least for small $\|v\|$ the $\tilde{F}_j = \tilde{f}_j + \sum_{r=1}^{\tau_-} v_r\tilde{\phi}_{r,j}$ define a non-singular complete intersection $\tilde{\mathfrak{X}}_u$ in $Z(\sigma) \times \mathbb{C}^{\tau_-}$: the strict transform of $\mathfrak{X}_u$; moreover, each fiber over a point $v \in \mathbb{C}^{\tau_-}$ for sufficiently small $\|v\|$ is the resolution of the corresponding fiber of $\mathfrak{X}_u(v)$. Let us now consider $\tilde{Y}_1 = (\pi(\sigma) \times \mathrm{Id}_{\mathbb{C}^{\tau_-}}|_{\tilde{\mathfrak{X}}_u})^{-1}(Y_1)$. Its equations are $y_b = 0, \tilde{F}_j = 0$; by the Implicit Function theorem, it is non-singular and admits at least for small $\|v\|$ the coordinates $v_1, \dots, v_{\tau_-}$; the morphism $\tilde{Y}_1 \to Y_1$ is not only a homeomorphism, but a local analytic isomorphism. Since there is an equivariant action of $\mathbb{C}^*$, all this is in fact true for all v.

Let us now consider a cone σ not adjacent to the weight vector. Let u^q be a monomial appearing as a $\phi_{r,j}$, set $G_j = f_j + v_r u^q$, and let $K_j \subset \{0, \dots g\}$ be the set of those indices s such that $\langle a^s, q\rangle < m_j(a^s)$. Set $n_j(a^s) = \langle a^s, q\rangle$ for $s \in K_j$. Then $G_j \circ \Pi(\sigma)$ equals

$$\Pi_{s\notin K_j} y_s^{m_j(a^s)} \Pi_{s\in K_j} y_s^{n_j(a^s)} \big(\tilde f_j \Pi_{s\in K_j} y_s^{m_j(a^s)-n_j(a^s)} + v_r \Pi_{s\notin K_j} y_s^{\langle a^s,q\rangle - m_j(a^s)}\big).$$

By our choice of σ we know that the $\tilde f_j$ do not vanish together on the exceptional divisor. The common zeroes of the strict transforms

$$\tilde G_j = \tilde f_j \Pi_{s\in K_j} y_s^{m_j(a^s)-n_j(a^s)} + v_r \Pi_{s\notin K_j} y_s^{\langle a^s,q\rangle - m_j(a^s)}$$

constitute in the chart $\mathbb{C}^{g+1}(\sigma) \times \mathbb{C}^{\tau-}$ the strict transform $\tilde{\mathcal{X}}_u$ of the space $\mathcal{X}_u$, therefore this strict transform meets the exceptional divisor only for $v_r = 0$ and then it coincides with the union of some components in $\mathbb{C}^{g+1}(\sigma) \times \{0\}$ of the exceptional divisor of $\pi(\sigma)$. However, these components do not meet the strict transform (see 5.4) of the curve C^Γ, and therefore the inverse image of $\{0\}$ in $\tilde{\mathcal{X}}_u$ is not connected. Since $\mathcal{X}_u$ is the total space of an equisingular deformation of C^Γ it is analytically irreducible at the origin and by Zariski's main theorem the inverse image of $\{0\}$ must be connected. We therefore obtain a contradiction, which shows that for each cone $\sigma = \langle a^0, \dots, a^g\rangle \in \Sigma$ all the monomials u^q appearing in the miniversal equisingular deformation of C^Γ satisfy the inequalities $\langle a^s, q\rangle \geq m_j(a^s)$ for $0 \leq s \leq g$. We are in the situation described at the beginning of the proof, and this shows that the map $\tilde{\mathcal{X}}_u \to \mathcal{X}_u$ induced by the toric modification $Z(\Sigma) \times \mathbb{C}^{\tau-} \to \mathbb{C}^{g+1} \times \mathbb{C}^{\tau-}$ is a weak embedded simultaneous resolution along Y_1.

Finally we remark that a refinement Σ' of our original regular fan corresponds to a birational toric map $Z(\Sigma') \to Z(\Sigma)$ which is an isomorphism outside the exceptional divisor. Moreover, in a chart corresponding to a cone σ adjacent to the weight vector, it is an isomorphism outside $\Pi_{s=1}^g y_s = 0$; this open set contains the strict transform of C^Γ since we saw that it meets only $y_0 = 0$ (see 5.4). It follows from this that if the strict transform of $\mathcal{X}_u$ in $Z(\Sigma') \times \mathbb{C}^{\tau-}$ is a simultaneous resolution of $\mathcal{X}_u$, such was already the case for the strict transform in $Z(\Sigma) \times \mathbb{C}^{\tau-}$. This concludes the proof. □

Corollary. *For a plane branch $(C, 0)$ with g characteristic exponents, with semi group $\Gamma = \langle \overline{\beta}_0, \dots, \overline{\beta}_g\rangle$ and given parametrically by $x = x(t)$, $y = y(t)$ with $\nu(x) = \overline{\beta}_0$, $\nu(y) = \overline{\beta}_1$, let $\xi_i(t) \in \mathcal{O}_{C,0} = \mathbb{C}\{x(t), y(t)\} \subset \mathbb{C}\{t\}$ for $2 \leq i \leq g$ be series such that*

$$(\nu(x), \nu(y), \nu(\xi_2), \dots, \nu(\xi_g)) = (\overline{\beta}_0, \dots, \overline{\beta}_g),$$

where ν is the t-adic valuation. Then the embedding $(C, 0) \to (\mathbb{C}^{g+1}, 0)$ given by

$$u_0 = x(t), u_1 = y(t), u_2 = \xi_2(t), \dots, u_g = \xi_g(t)$$

has the property that its image is resolved by one toric modification.

7. The transforms of plane curves

Let us return to the construction of the miniversal constant semigroup deformation in §3. It is easy to check that the vectors $\phi_j \in \mathbb{C}[C^\Gamma]^g$, $1 \leq j \leq g-1$, containing u_{j+1} at the j-th line and zero elsewhere, are independent modulo N with the notations of §3. In the miniversal constant semigroup deformation, the plane curves appear as those for which the coefficients of all these vectors are $\neq 0$ (see [T1]). For each point v_0 in the corresponding open set of $\mathbb{C}^{\tau-}$, the equation of the corresponding plane branch is obtained as follows, in a small neighborhood of 0:

One uses the first equation F_1 to express u_2 as a series in u_0, u_1, using the implicit function theorem. Then one substitutes this value in F_2 and uses this equation to express u_3 as a function of u_0, u_1, and so on. Finally the equation $F_{g-1} = 0$ allows us to express u_g as a series in u_0, u_1, and at this point we have the equations of a non-singular surface S_{v_0} containing our branch, thus explicitly shown to be planar.

The next step is to consider the strict transform $\tilde{S}_{v_0}$ of S_{v_0} in $\mathbb{C}^{g+1} \times \{v_0\}$ and the induced map $\tilde{S}_{v_0} \to S_{v_0}$ induced by $\Pi(\Sigma)$. One shows that, in a neighborhood of the point of the exceptional divisor picked by the strict transform of C, it factors through the toric map $S^1_{v_0} \to S_{v_0}$ which on S_{v_0} with the coordinates u_0, u_1 resolves the singularities of the branch $u_1^{n_1} - u_0^{\ell_1^{(0)}} = 0$. One then takes the strict transform of this equation as a new coordinate on $S^1_{v_0}$ and then the strict transform of C appears as a branch on $S^1_{v_0}$ with one less characteristic exponent; its equations are the strict transforms of $F_2, \ldots, F_g$ on $S^1_{v_0}$, in the new coordinates. Again one must look at the toric map (in the new coordinates) which resolves the first characteristic pair and show that it is dominated by $\tilde{S}_{v_0}$, and so on; finally we have factored the map $\tilde{S}_{v_0} \to S_{v_0}$ into the composition of g toric maps. This is similar to the process described by Spivakovsky in [S] of approximating a given valuation by a sequence of monomial valuations. However we see that if we allow a change in ambient space, the uniformization of a single monomial valuation, namely the t-adic valuation on the monomial curve, gives the uniformization of the t-adic valuation of C.

On the surface S, each of the coordinates u_i, $2 \leq i \leq g$ is expressed as a series in u_0, u_1. For each i, $2 \leq i \leq g$, the equation $u_i = 0$ on S defines a curve with i characteristic exponents having at the origin maximal contact with our original plane branch; this is a consequence of what we have seen in Corollary 6, and gives a geometric construction of curves (singular or not) having maximal contact in the sense of [L-J] with our plane branch; they are the curves on S_{v_0} defined by the equations $u_i = 0$, $i \geq 1$. Remark that for $i \geq 2$ they can also be written $f_{i-1}(u_0, \ldots, u_g) +$ higher degree, and with a little more work we find the structure of the approximate roots.

The construction just outlined needs to be detailed but we will not do it here since it is not part of the main purpose of this text. It is performed in the first non trivial special case (two characteristic exponents $(3/2, 7/4)$, i.e. semigroup $\langle 4, 6, 13 \rangle$) at the end of the example to which we now turn.

8. Appendix: An example

Example. Suppose we have a plane branch C with semigroup $\langle 4,6,13\rangle$. The corresponding monomial curve C^{Γ} is described (see [T]) by the polynomials

$$\begin{array}{rcccl} f_1 & = & {u_1}^2 - {u_0}^3 & = & 0 \\ f_2 & = & {u_2}^2 - {u_0}^5 u_1 & = & 0. \end{array}$$

As specified above, we construct $\mathcal{N}_+(f_1)$ and $\mathcal{N}_+(f_2)$, the Newton polygons, by the convex hull of the union of positive quadrants beginning at points corresponding to the exponents of the $\{u_i\}$'s . Thus,

$$\begin{array}{rcl} \mathcal{N}_+(f_1) & = & \text{convex hull of } \{\{(3,0,0),(0,2,0)\} + \mathbb{R}^3_+\} \\ \mathcal{N}_+(f_2) & = & \text{convex hull of } \{\{(5,1,0),(0,0,2)\} + \mathbb{R}^3_+\} \end{array}$$

and the corresponding fans are defined, respectively, by

$$\begin{array}{rcl} \Sigma_0^{(1)} & = & \text{the first quadrant of } \mathbb{R}^3 \text{ cut by the plane } 3x = 2y, \text{and} \\ \Sigma_0^{(2)} & = & \text{the first quadrant of } \mathbb{R}^3 \text{ cut by the plane } 5x + y = 2z. \end{array}$$

Then the equivalence class fan of f_1 and f_2 together is the intersection

$$\Sigma_0 = \Sigma_0^{(1)} \cap \Sigma_0^{(2)}$$

which has four maximal dimension cones:

1. $\langle (2,3,0),\ (0,1,0),\ (4,6,13),\ (0,2,1)\rangle$
2. $\langle (0,2,1),\ (0,0,1)\ (4,6,13)\rangle$
3. $\langle (2,0,5),\ (1,0,0),\ (2,3,0),\ (4,6,13)\rangle$
4. $\langle (0,0,1),\ (2,0,5),\ (4,6,13)\rangle$

We build here by an inductive method a refinement which resolves the fan $\Sigma_0 = \Sigma_0^{(1)} \cap \Sigma_0^{(2)}$. Notice that each maximal dimension cone spans $\mathbb{Z}^3$ (since the matrix of the spanning skeleton is unimodular). For each cone, we have listed the corresponding morphism $\pi(\sigma)$ in terms of the resulting $\{u_i\}$, the composition $f_i \circ \pi(\sigma)(y_0, y_1, y_2)$ for $i = 1,2$, and the exceptional divisor $\pi(\sigma)^{-1}(0)$. We show that the two surfaces are transverse to each other and to the exceptional divisor. We start from a refinement for the fan in two dimensions associated to the first equation $u_1^2 - u_0^3 = 0$. The weight vector is $(2,3)$ and from the geometric interpretation of the continued fraction expansion we find that a regular fan subdividing it is composed of the following four 2-dimensional cones and their faces:

1. $\sigma_1^{(2)} = \langle (1,0),\ (1,1)\rangle$
2. $\sigma_2^{(2)} = \langle (1,1),\ (2,3)\rangle$
3. $\sigma_3^{(2)} = \langle (2,3),\ (1,2)\rangle$
4. $\sigma_4^{(2)} = \langle (1,2),\ (0,1)\rangle$

Now we have to lift the vectors of the 1-skeleton to $\mathbb{R}^3$ in such a way that they form the 1-skeleton of a regular fan subdividing Σ_0. It suffices to show that in each cone of this fan both functions m_1 and m_2 are linear, i.e that all linear forms take their minimum at a vertex of the sum $\mathcal{N}$ of the Newton polyhedra of f_1 and f_2. This Newton polyhedron has four vertices: $(3,0,2),(8,1,0),(0,2,2),(5,3,0)$, so we are especially interested in finding four cones having the weight vector as one of their faces.

Let us begin by lifting the weight vector $(2,3)$:
We seek integral vectors of the form $(2k,3k,z)$ with $k>0,\ z>0$; we know that they take their minimum value on the compact segment of the Newton polyhedron of f_1. On $\mathcal{N}_+(f_2)$ they take their minimum at $((5,1,0)$ if $13k \leq 2z$, at $(0,0,2)$ if $13k \geq 2z$, at both if $13k=2z$. Remembering that we seek primitive vectors, we find three natural possibilities: $k=1, z=6$, giving the vector $a^1=(2,3,6)$, which takes its minimum at $(0,0,2)$, $k=1, z=7$ giving $a^2=(2,3,7)$, which takes its minimum at $(5,1,0)$, and of course the weight vector $a^0=(4,6,13)$ itself. We remark that $a^0=a^1+a^2$.
Let us now try to lift $(1,1)$; we seek vectors of the form (k,k,z); on $\mathcal{N}_+(f_1)$ they take their minimum value at $(0,2,0)$, and on $\mathcal{N}_+(f_2)$ at $(5,1,0)$ if $3k \leq z$, at $(0,0,2)$ if $3k \geq z$. In this case we can take $k=1, z=3$, which gives the vector $a^3=(1,1,3)$ which takes its minimum on the compact segment of $\mathcal{N}_+(f_2)$.
Finally take $(1,2)$; we seek vectors $(k,2k,z)$ and there is one taking its minimum on the compact face of $\mathcal{N}_+(f_2)$: it corresponds to $k=1, z=3$, giving the vector $a'^4=(1,2,3)$. We remark that $a^0=a^2+a^3+a'^4$, and it is therefore tempting to take a'^4 as our fourth vector. However, it is not on the hyperplane $5x+y=2z$ so that the support function m_2 will *not* be linear on a cone spanned by $\{a^0,a^2,a'^4\}$. A better choice is to take $a^4=a^0-a^3=(3,5,10)$.

Now we check that the four cones (which we have decorated with the points of the Newton polyhedra of f_1 and f_2 where the support functions $m_1(a),\ m_2(a)$ take their minimum for $a \in \sigma_i$):

1. $\sigma_1^{(3)}=\langle a^0,a^1,a^3\rangle;\ \ p_0^{(1)}=(0,2,0);\ p_0^{(2)}=(0,0,2)$
2. $\sigma_2^{(3)}=\langle a^0,a^2,a^3\rangle;\ \ p_0^{(1)}=(0,2,0);\ p_0^{(2)}=(5,1,0)$
3. $\sigma_3^{(3)}=\langle a^0,a^2,a^4\rangle;\ \ p_0^{(1)}=(3,0,0);\ p_0^{(2)}=(5,1,0)$
4. $\sigma_4^{(3)}=\langle a^0,a^1,a^4\rangle;\ \ p_0^{(1)}=(3,0,0);\ p_0^{(2)}=(0,0,2)$

are unimodular and such that their union is a neighborhood of $\mathbb{R}_+a^0$ in $\mathbb{R}^3$; the second fact is obvious since the weight vector is in the interior of the convex hull of the σ_i , $1 \leq i \leq 4$, and to check the first it suffices to check that one of them is unimodular. Now we can complete this subfan into a regular fan of $\mathbb{R}^3_+$, either by the same method or by invoking a general theorem, but we do not much care about it, since we have seen that the only charts where something interesting happens are those which correspond to a cone containing the weight vector in its 1-skeleton, and we have those above. Moreover, the projections in $\mathbb{R}^2$ of the σ_i form part of a regular subfan of the $\sigma_i^{(2)}$.

Let us now study the behavior of f_1, f_2 under the monomial maps corresponding to the $\sigma_i^{(3)}$, $1 \leq i \leq 4$. For economy we write σ_i for $\sigma_i^{(3)}$.

1. The cone σ_1 spanned by $\{(4,6,13),\ (2,3,6),\ (1,1,3)\}$

$$\begin{aligned}\pi(\sigma_1) : u_0 &= {y_0}^4{y_1}^2y_2\\ u_1 &= {y_0}^6{y_1}^3y_2\\ u_2 &= {y_0}^{13}{y_1}^6{y_2}^3\end{aligned}$$

Then,

$$\begin{aligned}f_1 \circ \pi(\sigma_1)(y_0, y_1, y_2) &= {y_0}^{12}{y_1}^6{y_2}^2(1-y_2)\\ f_2 \circ \pi(\sigma_1)(y_0, y_1, y_2) &= {y_0}^{26}{y_1}^{12}{y_2}^6(1-y_1)\end{aligned}$$

and

$$\pi(\sigma_1)^{-1}(0) = \{y_0 = 0\} \cup \{y_1 = 0\} \cup \{y_2 = 0\}.$$

Here the strict transforms are

$$\begin{aligned}\tilde{f}_1 &= 1 - y_2\\ \tilde{f}_2 &= 1 - y_1.\end{aligned}$$

It is clear that the equations $\tilde{f}_1 = \tilde{f}_2 = 0$ define a non-singular complete intersection meeting the exceptional divisor transversally at the point $y_0 = 0, y_1 = y_2 = 1$.

To make the same computation for the other charts $\pi(\sigma_i)$ is in fact superfluous, since by construction of the toric modification, we will only observe the same phenomenon in a different chart.

For verification's sake let us compute for $\pi(\sigma_3)$:

2. The cone σ_3 spanned by $\{(4,6,13),\ (2,3,7),\ (3,5,10)\}$

$$\begin{aligned}\pi(\sigma_3) : u_0 &= {y_0}^4{y_1}^2{y_2}^3\\ u_1 &= {y_0}^6{y_1}^3{y_2}^5\\ u_2 &= {y_0}^{13}{y_1}^7{y_2}^{10}\end{aligned}$$

Then,

$$\begin{aligned}f_1 \circ \pi(\sigma_3)(y_0, y_1, y_2) &= {y_0}^{12}{y_1}^6{y_2}^9(y_2 - 1)\\ f_2 \circ \pi(\sigma_3)(y_0, y_1, y_2) &= {y_0}^{26}{y_1}^{13}{y_2}^{20}(y_1 - 1)\end{aligned}$$

And indeed it is the same situation viewed in another chart.

The miniversal deformation with constant semigroup of C^Γ is computed in [T1]: here $\tau_- = 2$ and $\mathfrak{X}_u$ is defined in $\mathbb{C}^3 \times \mathbb{C}^2$ by the equations

$$\begin{aligned}F_1 &= u_1^2 - u_0^3 + v_1u_2 + v_2u_0u_2 = 0\\ F_2 &= u_2^2 - u_0^5u_1 = 0\end{aligned}$$

If we compose F_1 and F_2 with $\Pi(\sigma) = \pi(\sigma_1) \times \mathrm{Id}_{\mathbb{C}^2}\colon\ \mathbb{C}^{g+1}(\sigma) \times \mathbb{C}^2 \to \mathbb{C}^{g+1} \times \mathbb{C}^2$ we get

$$F_1 \circ \Pi(\sigma)(y_0, y_1, y_2) = {y_0}^{12}{y_1}^{6}{y_2}^{2}(y_2 - 1 + v_1 y_0 y_2 + v_2 y_0^5 y_1^2 y_2^2)$$
$$F_2 \circ \Pi(\sigma_1)(y_0, y_1, y_2) = {y_0}^{26}{y_1}^{13}{y_2}^{20}(y_1 - 1)$$

The strict transform of $\mathfrak{X}_u$ is defined in this chart by

$$\tilde{F}_1(y_0, y_1, y_2) = y_2 - 1 + v_1 y_0 y_2 + v_2 y_0^5 y_1^2 y_2^2 = 0$$
$$\tilde{F}_2(y_0, y_1, y_2) = y_1 - 1 = 0$$

It is indeed a simultaneous resolution for $\mathfrak{X}_u$. There remains to check that the strict transform of $\mathfrak{X}_u$ does not meet the charts corresponding to cones that are not adjacent to the weight vector. This depends on our construction of a regular fan and will not be done here.

Let us now consider the plane branch with equations in $\mathbb{C}^3$ (for $v = v_0 \neq 0$):

$$F_1(v_0; u_0, u_1, u_2) = u_1^2 - u_0^3 - v_0 u_2 = 0$$
$$F_2(u_0, u_1, u_2) = u_2^2 - u_0^5 u_1 = 0$$

It lies on the non-singular surface S_{v_0} with equation $F_1(v_0; u_0, u_1, u_2) = 0$. Let $\tilde{S}_{v_0}$ be the strict transform of S_{v_0} by the toric map $Z(\Sigma) \to \mathbb{C}^{g+1}$. As before we need to examine the situation only in a chart $Z(\sigma)$ where σ is adjacent to the weight vector. We take $a^0 = (4, 6, 13)$, $a^1 = (2, 3, 7)$, $a^2 = (3, 5, 10)$. So $\pi(\sigma)$ is described as:

$$\pi(\sigma) : u_0 = {y_0}^{4}{y_1}^{2}{y_2}^{3}$$
$$u_1 = {y_0}^{6}{y_1}^{3}{y_2}^{5}$$
$$u_2 = {y_0}^{13}{y_1}^{7}{y_2}^{10}$$

and we have

$$(u_1^2 - u_0^3 - v_0 u_2) \circ \pi(\sigma) = y_0^{12} y_1^6 y_2^9 (y_2 - 1 - v_0 y_0 y_1 y_2)$$

so that the equation of $\tilde{S}_{v_0}$ in this chart is

$$y_2(1 - v_0 y_0 y_1) = 1,$$

and the mapping $\tilde{S}_{v_0} \to S$ is described by

$$u_0 = {y_0}^{4}{y_1}^{2}\left(\frac{1}{1 - v_0 y_0 y_1}\right)^3$$
$$u_1 = {y_0}^{6}{y_1}^{3}\left(\frac{1}{1 - v_0 y_0 y_1}\right)^5$$

All computations are made in a neighborhood of the exceptional divisor, i.e for $|y_0|$ small.

Now the toric map $S^1_{v_0} \to S_{v_0}$ which resolves the plane branch with one characteristic pair $u_1^2 - u_0^3 = 0$ has a chart:

$$u_0 = x_0 x_1^2$$
$$u_1 = x_0 x_1^3$$

and we can check that there is a factorization $\tilde{S}_{v_0} \to S^1_{v_0}$ given as follows in the coordinates y_0, y_1 on $\tilde{S}_{v_0}$:

$$x_0 = 1 - v_0 y_0 y_1$$
$$x_1 = y_0^2 y_1 (\frac{1}{1 - v_0 y_0 y_1})^2$$

and we can view this factorization itself as composed of the maps

$$\begin{aligned} x_0 &= w_0 \\ x_1 &= w_0^2 w_1 \end{aligned} \quad \text{and} \quad \begin{aligned} w_0 &= 1 - v_0 y_0 y_1 \\ w_1 &= y_0^2 y_1 \end{aligned}$$

Now this last map is monomial after a change of the coordinates w_0, w_1 and moreover the substitution of the w's in the x's gives a monomial map $S^2_{v_0} \to S_{v_0}$ described as follows

$$u_0 = w_0^5 w_1^2$$
$$u_1 = w_0^7 w_1^3$$

which is again a monomial map, so that we have factorized our map $\tilde{S}_{v_0} \to S_{v_0}$ as a composition of two monomial maps, up to a change of variables (essentially a translation $w_0 \mapsto w_0 - 1$). Note the fact that it was convenient to refine part of the fan of the plane branch $u_1^2 - u_0^3 = 0$ from $(1, 1), (2, 3)$ to $(5, 7), (2, 3)$ before writing the translation, but that we could also have made a change in the variables y_0, y_1 (for $|y_0|$ small) to bring directly the map $\tilde{S}_{v_0} \to S^1_{v_0}$ into monomial form.

Finally, note that the strict transform on $\tilde{S}_v$ of our plane curve $C_{v_0} \subset S_{v_0}$ has equation $y_1 - 1 = 0$. In the general case, since we do not deform the last equation f_g of the monomial curve, the last equation of the strict transform of C_{v_0}, in the proper chart will still be of the form $y_k - 1 = 0$.

References

[C] D. Cox, *Toric varieties and toric resolutions.* This volume.

[F] W. Fulton, *Introduction to Toric Varieties.* Princeton University Press, 1993.

[TE] G. Kempf, F. Knudsen, D. Mumford, B. Saint-Donat, *Toroidal embeddings.* Lecture Notes in Math. 339, Springer-Verlag 1973.

[L-J] M. Lejeune-Jalabert, Thèse d'Etat, Université Paris 7, 1973.

[L-T] M. Lejeune-Jalabert, B. Teissier, *Cloture intégrale des idéaux et équisingularité.* In: Séminaire sur les singularités, Ecole Polytechnique, 1974–75. Institut Fourier, Université de Grenoble.

[M] M. Merle, *Polyèdre de Newton, Eventail et Désingularisation, d'après Varchenko.* In Séminaire sur les Singularités des Surfaces, Palaiseau, France, 1976–1977, Edited by M. Demazure, H. Pinkham, B. Teissier, Lecture Notes in Math. 777, Springer-Verlag 1980.

[Oda] T. Oda, *Convex bodies and algebraic geometry,* Ergeb. der Math., 3. Folge, Band 15, Springer-Verlag 1988.

[O1] M. Oka, *Geometry of plane curves via toroidal resolution.* In: Algebraic Geometry and Singularities, A. Campillo et al., editors, Progress in Math., No. 134, Birkhäuser, 95–121.

[O2] M. Oka, *Canonical stratification of non degenerate complete intersection varieties.* Jour. Math. Soc. Japan, 1990, **42**, 397–422.

[O3] M. Oka, *Non degenerate complete intersection singularities.* Actualités Mathématiques, Hermann, Paris, 1997.

[S] M. Spivakovsky, *Resolution of singularities.* Preprint, University of Toronto, 1994.

[T1] B. Teissier, Appendix in [Z].

[T2] B. Teissier, *Résolution simultanée II.* In: Séminaire sur les Singularités des Surfaces, Palaiseau, France, 1976–1977, Edited by M. Demazure, H. Pinkham, B. Teissier, Lecture Notes in Mathematics 777, Springer-Verlag, 1980.

[T3] B. Teissier, *The hunting of invariants in the geometry of discriminants.* In: Real and Complex Singularities, Nordic Summer School, Oslo 1976, Edited by Per Holm, Noordhoff and Sijthof, 1977.

[V] A.N. Varchenko, *Zeta function of the monodromy and Newton's diagram.* Invent. Math. **37** (1976), 253–262.

[Z] Oscar Zariski. *Le Problème des Modules pour les Branches Planes.* Hermann, Paris, 1986.

Dept. of Mathematics
M.I.T.
Cambridge, MA 02139, USA
goldin@math.mit.edu

Laboratoire de Mathématiques de lE.N.S.,
URA 762 du C.N.R.S.,
Ecole Normale Supérieure
75005 Paris, France
teissier@dmi.ens.fr

Progress in Mathematics, Vol. 181,

Excellent Surfaces and Their Taut Resolution

Herwig Hauser

1. Introduction

Purpose of the present paper is to reveal the beauty and subtlety of resolution of singularities in the case of excellent two-dimensional schemes embedded in three-space and defined over an algebraically closed field of arbitrary characteristic. The proof of strong embedded resolution we describe here combines arguments and techniques of O. Zariski, H. Hironaka, S. Abhyankar and the author.

Theorem 1.1. *Let W be an excellent regular three-dimensional ambient scheme over an algebraically closed field K of arbitrary characteristic. For every reduced hypersurface X in W there exists a sequence of blowups*

$$W^n \to \cdots \to W^0 = W$$

with closed centers Z^i inside the singular locus of the $(i-1)$-st strict transform X^{i-1} of X such that the last strict transform X^n is smooth and has normal crossings with the exceptional divisor E^n.

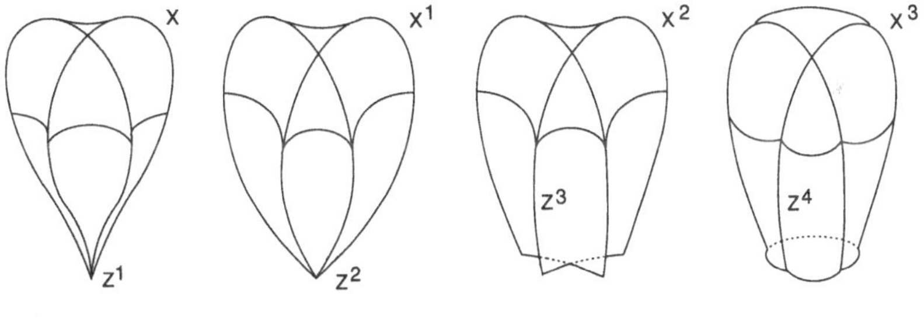

FIGURE 1. Blowups

The centers Z^i will be chosen inside the equimultiple locus of X^i. This is the subscheme of points where X has maximal multiplicity. We allow singular and non reduced centers as long as the intermediate ambient spaces W^i remain smooth. By scheme we understand a scheme of finite type over K, in particular locally noetherian and quasi-compact. The surface X is a closed reduced subscheme of W of codimension one.

The proof of the theorem splits into two parts: First, at each stage of the process the next center to be blown up has to be chosen suitably. Then it has to be shown that when passing to the strict transform X' of X the situation improves. This is done by exhibiting certain local invariants of singularities which have dropped.

For different proofs we refer to the work of Abhyankar, Lipman and Spivakovsky [Ab 2, Lp, Sp 3], as well as to the contributions of Cossart and Lê in this volume [Co 2, Le].

Definition of center. The resolution process described in this paper consists of two stages. First, the strict transform of X has to made smooth. Secondly, the total transform has to be modified so as to have normal crossings. The centers for the first stage are defined through the equimultiple locus S of X. For surfaces, it consists of a certain number of irreducible curves and isolated points. This simplifies the definition of the centers substantially compared to higher dimensions: as long as there are singular curves in S, blow up their singular points so as to make them smooth. Using resolution of curves and determining the equimultiple locus S' of the blown up scheme X' it is seen that this process yields in a finite number of steps a scheme whose equimultiple locus consists of isolated points and smooth curves having at most normal crossings. Thus it suffices to consider this situation. Here we blow up the whole equimultiple locus equipped with a suitable non reduced structure at the intersection points of its components. This choice of the center and the resulting blowup will be called *taut.* It would also be sufficient, though less canonical with respect to extensions to higher dimensions, to blow up S with the reduced structure, since the only singularities appearing on the ambient scheme W' are ordinary double points which lie outside the strict transform X' of X and which can be resolved easily by one point blowup.

It turns out that under the chosen blowup the highest multiplicity occurring on X either drops or remains constant when passing to the strict transform X' of X. At any point a' of X' with the same multiplicity, the equimultiple locus S' has again at most normal crossings, thus allowing to repeat the process. This defines a sequence of taut blowups. It will then be shown that finitely many steps yield a smooth strict transform of X.

For the second stage, we consider the total transform X^* of X. By the preceding blowups it can be assumed that X^* is a union of smooth surfaces. Under blowup, the strict transform of a smooth surface intersects the exceptional divisor transversally. Therefore, when applying stage one to X^*, finitely many taut blowups make the strict transforms of the components of X^* and the appearing new exceptional components intersect pairwise transversally. Suitable additional blowups are then necessary to guarantee that the union of all these components has normal crossings.

Improvement of invariant. The second part of the proof consists in showing that the preceding algorithm terminates, i.e. that after a finite number of steps the obtained scheme is smooth. This is proven by induction. At each closed point a

of X one defines a local invariant i_a. It belongs to a well ordered subset Γ of $\mathbb{Q}^4$. Its first component is the multiplicity, i.e. the order of the power series f defining locally X at a in W. The other three components are orders of certain coefficient ideals associated to f which can be expressed through the Newton polyhedron of f. It will be shown that if X is singular, i_a drops when passing from a point a of S to a point a' over it under the blowup as defined above. An additional argument shows that this local improvement implies a drop of the global multiplicity of X after finitely many blowups. By induction, a smooth surface is achieved.

Structure of the paper. After recalling basic properties of blowups we define in Section 3 the centers of blowup selected at each stage of the resolution process. In difference to the classical treatment we allow singular centers of mild type, namely normal crossing curves with embedded components at the intersection points. These components are chosen so that the blown up ambient scheme remains regular. The choice of such centers is natural because it preserves possibly existing local or global symmetries of the scheme which may permute the two components of the normal crossings, and moreover reduces the number of required induction invariants in comparison to the treatment of Bierstone-Milman and Encinas-Villamayor [BM, E-V 1, E-V 2]. Taking into account the history of the resolution process by distinguishing old and young exceptional divisors becomes superfluous.

As a variation, we indicate what happens when changing the structure of the embedded components at the intersection points of the center. For certain choices, the blown up ambient scheme is again smooth, but the induction invariants do not necessarily improve.

Section 4 follows Zariski's exposition [Za 1] in showing how to reduce the equimultiple locus to a normal crossings situation, and proving that this situation persists under taut blowup (Theorem 4.4). This is a prerequisite to make the induction work.

In Section 5 we introduce flags. By this we understand full flags of local regular schemes centered at each point of the equimultiple locus. In three dimensions, a flag at a point a consists of a smooth curve inside a smooth surface, both passing through a. Flags are very useful to put some ordering on the coordinates and to reduce the number of allowed coordinate choices. Local coordinates subordinate to the chosen flag are needed to define the induction invariants from the Newton polyhedron of the locally defining equation by imposing in $\mathbb{N}^3$ a hierarchy among the vertices of the polyhedron. This idea appears implicitly already in resolution of plane curves when using the Weierstrass form of the defining equation, and presents the basis of Hironaka's argument in [Hi 1] for surfaces. A key property of flags is their compatibility with blowup (in contrast to e.g. a collection of coordinate hyperplanes appearing as components of the exceptional divisor). A purely geometric argument allows to construct canonically from any flag at a and transversal to the center an induced flag at *any* point a' over a of the exceptional divisor (Theorem

5.1). Therefore subordinate coordinates are preserved under blowup and the induction invariants can be defined again in W' and compared with those in W. On the way, we will have to show that the transversality of the flag with respect to the equimultiple locus is preserved under blowup (Theorem 5.2). For this, and for the control of the invariants under blowup, it is shown that for any choice of a' above a, one can perform a subordinate local coordinate choice at a which makes the local blowup *monomial* in the resulting coordinates. In this sense, flags are sufficiently restrictive to prohibit permutations of the coordinates – these present one of the main difficulties of the topic –, and sufficiently flexible to render blowups combinatorial without quitting the local setting when passing from a to a'.

Section 6 is devoted to the construction of the induction invariants. We follow the suggestion of Hironaka [Hi 1], being aware that this choice of invariants is very specific to dimension three and has no evident extension to higher dimensions. The construction is done by introducing first a vector of numbers which belongs to a certain ordered set and which depends on the choice of the coordinates subordinate to the chosen flag. To make them to genuine invariants, i.e., independent of the subordinate coordinate choice (though dependent on the flag), it is natural and appropriate to define the invariant as the maximal value of the vector over all subordinate coordinate choices. This has been done by many authors in different contexts, e.g. [Ab 1, Hi 1, Mo], and reflects the observation that the finest information on the singularity can be extracted in most specific coordinates. And these turn out – according to the setting – to be maximizing coordinates. It has to be shown that the maximum exists, at least within the set of formal coordinate choices. This is done either directly or using the general argument of [Ha 1] based on the Artin Approximation Theorem. Obviously the resulting invariant does not depend on the coordinates. In characteristic zero, maximality is usually achieved by the existence of a hypersurface of maximal contact, and it is known that this hypersurface accompanies the resolution process along any sequence of points where the multiplicity remains constant. Moreover it allows to prove semi-continuity properties for the invariant. In arbitrary characteristic, this reasoning breaks down, and one has to show that maximality of the vector persists under blowup. Actually, it suffices to realize the maximum on the blown up scheme, see Section 8 for more details. It seems that semi-continuity properties can be dispensed with in the special case of surfaces.

The induction invariants proposed in [BM, E-V 1, E-V 2] in characteristic zero are inspired by Hironaka's paper on idealistic exponents [Hi 3] and are more conceptual than the ones described here. They cannot be used directly in positive characteristic, even for surfaces, due to the failure of maximal contact. There is some perspective to adapt them to the present setting (simplifying them at the same time by discarding their memorative aspect on the history) and to make them work for surfaces of arbitrary characteristic. This discussion could not be included in this article.

Section 7 establishes the induction step in the combinatorial situation. It is proven that for taut monomial blowups (and fixed coordinates) the vector of invariants drops in the lexicographic order (Theorem 7.1). This is done by explicit case by case calculations. Experimentation shows that there is not much freedom in changing the invariant and still having it drop.

The following section shows how to reduce an arbitrary local blowup to the combinatorial situation (Theorem 8.1). For any point a' of the exceptional divisor and sitting over a local subordinate coordinates are chosen at a moving a' to the origin of one chart and making the blowup monomial. Moreover this can be done so that the vector realizes in the induced coordinates at a' the maximal value, i.e., equals the actual invariant.

Section 9 combines Theorems 7.1 and 8.1 for proving the existence of resolution for surfaces (Theorem 1.1). Since the invariant is not obviously semicontinuous, the argument has to make a small detour to show that after finitely many blowups the resulting surface has smaller global multiplicity. By induction, a smooth surface will be obtained.

Many of the concepts presented in this article have analogues in higher dimensions, viz flags, coefficient ideals, maximality of invariants, see [Ha 1] and [E-V 2]. They lend themselves for approaching resolution in arbitrary dimension. Our exposition is occasionally more explicit than necessary in order to stress this aspect. Others like the invariants themselves or the reduction of the equimultiple locus to normal crossings have fatal drawbacks already for threefolds. Observe that in the algorithms of [BM, E-V 1, E-V 2] the stratification used is much finer than the one given by the multiplicity and that the smallest stratum defining the center is automatically regular.

Problem 1.2. Extend the present proof to fields which are not necessarily algebraically closed or to schemes defined over **Z**. The assumption algebraically closed is only used in Lemma 4.2 and in the proofs of Theorems 1.1, 7.1 and 8.1 where we neglect residually algebraic irrational points in the exceptional divisor. You may consult [Bn], [Co 1], Theorem 1, p. 218, of [Hi 2], Theorem 8 of [Ha 1] and the introduction of [E-V 2].

Problem 1.3. Extend the present proof to the case of surfaces embedded in a regular scheme W of arbitrary dimension (non-hypersurface case).

Problem 1.4. Given a reduced hypersurface X in a regular four dimensional ambient scheme W, assume that its equimultiple locus S consists of isolated points, smooth curves and possibly smooth surfaces, all of them meeting with normal crossings. Define a non reduced structure Z on S and local invariants of X such that blowing up Z in W gives a regular scheme W' and a strict transform X' all of whose invariants have dropped (see [P, R] for the first assertion). Observe that the normal crossings structure of the equimultiple locus may get lost under blowup, see [Ha 2, ex. 10].

In a first reading, it might to be desirable to proceed as follows: Start with the definition of the center of blowup given in Section 3 omitting the propositions given there. Taking into account Theorems 4.4 and 5.1 pass directly to the construction of the invariants in Section 6, followed by the study of their behaviour in Sections 7 and 8. Conclude by Section 9 proving Theorem 1.1.

The author has profited from discussions with many people, among them O. Villamayor, V. Cossart, S. Encinas, J. Lipman, A. Quirós and M. Spivakovsky. Various ideas and concepts presented here have their source in the existing literature, especially in the papers [Za 1, Hi 1, Hi 2, Ab 1, BM, CGO, E-V 1, E-V 2, Mo, Sp 2]. We are very indebted to the two referees for a careful reading of the manuscript and valuable suggestions. The work on this article has been supported in part by the scientific exchange program "Acciones Integradas" of the Austrian Ministery of Science.

2. Preliminaries

We collect several basic properties of blowups and multiplicities. For proofs and more details, see [Bn, Gi, Hi 1, Hi 2, Ha 1, Ha 2, O]. In this section, X and W may have arbitrary dimension, where X is reduced and closed in W regular. Let $\mathcal{I}$ be the defining ideal sheaf of X in W with stalks $\mathcal{I}_a$ and local rings $\mathcal{O}_{X,a} = \mathcal{O}_{W,a}/\mathcal{I}_a$. For a a closed point of X, let m_a denote the maximal ideal of $\mathcal{O}_{W,a}$ with residue field $\mathcal{O}_{W,a}/m_a = K$. Let

$$o_a = \max\,\{k \in \mathbb{N},\ \mathcal{I}_a \subseteq m_a^k\}$$

denote the order of X at a. It is invariant under completion of the local rings, upper-semicontinuous with respect to deformation and localization and takes only finitely many values (since X is noetherian). For a proof of this in characteristic zero, see [Hi 1], p. 106. In arbitrary characteristic, we refer to [Hi 2], Thm. 1, chap. III 3, p. 218, [Bn] and [E-V 2]. In particular, the maximum $o_X = \max_{a \in X} o_a$ exists and the *equimultiple locus* $S = \{a \in X,\ o_a = o_X\}$ is a closed reduced subscheme of X, strictly contained and non-empty in X if X is singular (by excellence), see [Za 2], Section 4. It is defined in W by the minimal ideal Q such that $I \subset Q^{o_X}$ and does not depend on the embedding of X in W. For non-hypersurfaces, the stratification given by the order can be refined by the Hilbert-Samuel stratification [Bn].

A scheme X is said to have normal crossings at a point a, if it can be defined locally at a in a regular ambient scheme W by an ideal of the local ring $\mathcal{O}_{W,a}$ generated by monomials in a regular parameter system of $\mathcal{O}_{W,a}$. For surfaces, this signifies that X consists locally at a of at most three smooth components meeting transversally like the three coordinate planes in affine three-space (i.e., having different tangent planes at a intersecting in one point).

For Z a closed subscheme of X, let $\pi : W' \to W$ denote the blowup of W in Z, and denote by X^*, respectively X^{st}, the total and the strict transform of X in W' under π. Then X^{st} equals the blowup X' of X with center Z. Objects associated to X' in analogy with X will be marked by a prime. Thus a' will denote a point in X', S' the equimultiple locus of X', etc. Let $E = \pi^{-1}Z$ be the (reduced) exceptional divisor in W' and let E_a denote the fibre $\pi^{-1}(a)$ of a point a of X.

When a and a' are fixed, we let $\mathbf{R} = \mathcal{O}_{W,a}$ and $\mathbf{R}' = \mathcal{O}_{W',a'}$ denote the local rings with completions $\overline{\mathbf{R}}$ and $\overline{\mathbf{R}'}$ with respect to the maximal ideals $\mathbf{M} = m_a$ and $\mathbf{M}' = m_{a'}$. Let $\mathbf{P}$ in $\mathbf{R}$ be the stalk at a of the ideal sheaf defining Z in W. For $a \in X$ and $a' \in E_a$ we call the induced map $\mathbf{R} \to \mathbf{R}'$ the local blowup of $\mathbf{R}$ with center $\mathbf{P}$ [Ha 2]. For $a \in Z$, $a' \in E_a$ and $Z \subseteq S$ one has $o_{a'} \leq o_a$, see e.g. the appendix to [Hi 1]. This implies that S' is contained in the total transform S^* of S if $o_{X'} = o_X$.

Assume that Z is smooth and let $a \in Z$. For any closed point a' over a there exist a regular system of parameters $x_1, \dots, x_n$ of $\mathbf{R}$, a subset J of $\{1, \dots, n\}$ and an element $j \in J$ such that

(a) x_i, $i \in J$, generate $\mathbf{P}$.
(b) W' is covered locally along E by the affine charts Spec $\mathbf{R}[\frac{x}{x_i}, k \in J]$ with $i \in J$.
(c) $y_1, \dots, y_n$ defined by $y_i = x_i/x_j$ for $i \in J \setminus j$ and $y_i = x_i$ for $i \notin J \setminus j$ form a regular system of parameters of $\mathbf{R}'$.
(d) E is defined in W' locally at a' by $y_j = 0$.

See [Hi 2, chap. III], [Ha 2] or [Bn] for more details and a description of the situation when K is not algebraically closed. We then say that a' is the origin of the x_j-chart of the blowup with respect to the coordinates $x_1, \dots, x_n$ and that $\mathbf{R} \to \mathbf{R}'$ is monomial with respect to $x_1, \dots, x_n$. Note that a regular system of parameters of $\mathbf{R}$ is also one for its completion. We often write $\mathbf{x}$ for short and speak of local coordinates of W at a. As the affine charts of W and W' at a and a' are isomorphic we shall write again $\mathbf{x}$ for the coordinates $\mathbf{y}$ at a' defined above.

Passing to completion is compatible with local blowup, i.e. the corresponding diagram commutes.

Let Z_1, Z_2 be two disjoint centers in W and denote by W'_{12} and W'_{21} the schemes obtained from W by blowing up first Z_1 and then the strict (= total) transform of Z_2, respectively inversely. Let W' be the scheme obtained by blowing up $Z_1 \cup Z_2$. Then W', W'_{12} and W'_{21} are canonically isomorphic.

Exercise 2.1. Let $a \in Z$ be a point and let X be a hypersurface in W defined locally at a by $f \in \mathbf{R}$. Let $\text{hom}_a f$ be the homogeneous polynomial of lowest degree appearing in the Taylor expansion of f at a. Then the points of E_a where the multiplicity has remained constant are contained in the intersection of E_a with the strict transform of the zero set of $\text{hom}_a f$.

Exercise 2.2. (Resolution of plane curves) Show that blowing up the singular points of a plane curve resolves the curve in a finite number of steps. To prove this, show that the pair $i_a = (o_a, s_a)$ drops in the lexicographic order when passing from singular points $a \in X$ to points $a' \in E_a$ of the strict transform X' of the curve. Here, s_a denotes the maximum over all coordinate choices of the slope of the first segment (from the left) of the Newton polygon of f (see Section 6 for how to prove that s_a is well defined). You may use Sections 7 and 8 to find a proof with slightly different induction invariants. See [Rg] for a detailed treatment in arbitrary characteristic.

3. Definition of the centers of blowup

From now on X denotes a singular reduced surface in a regular three-dimensional ambient scheme W. We shall determine a convenient center to be blown up such that the invariant defined later decreases when passing to the strict transform of X. As the order defines an upper semicontinuous function on X and X is singular, the equimultiple locus S of X consists of finitely many points and of finitely many irreducible curves. We say that S has at most normal crossings at $a \in S$, if a is either an isolated point of S, or a smooth point of a curve of S, or a normal crossings point of two or three local components of S at a. The last condition means that S looks locally at a like two or three coordinate axes in three-space.

Exercise 3.1. The set T of points where S does not have at most normal crossings is finite. In a normal crossings point there cannot pass three components of S (cf. with the proof of Proposition 4.5).

Define the center Z in X as follows.

(a) If T is not empty, let $Z = T$.

(b) If T is empty, let Z be the closed subscheme of X supported by S with embedded components at intersection points of S given in suitable local coordinates by the ideal $\mathbf{P} = (x, yz)(x, y)(x, z) = (x, y)^2 \cap (x, z)^2 \cap (x, y, z)^3$ with generators $x^3, x^2y, x^2z, xyz, y^2z^2$. Clearly, the definition does not depend on the choice of coordinates, and Z is reduced if there are no intersection points.

If the center is chosen in this way we call $\pi : W' \to W$ the taut blowup of X in Z ($\tau\grave{o}$ $\alpha\grave{\upsilon}\tau\acute{o}$, the same). We do not require in (b) that $\mathrm{ord}_{\mathbf{P}}X = \mathrm{ord}_a X$ for all $a \in Z$. As a variant we shall also discuss the case where the embedded components equal the cube of the maximal ideal of the intersection points. Both centers yield smooth ambient blown up schemes, but the first choice makes the chosen invariant drop, whereas the second does not.

The choice of the non-reduced structures on normal crossing centers was motivated by the theory of complete ideals and the studies of Pfeifle in [P], see also [R].

Exercise 3.2. Blowing up the reduced ideal (x, yz) in W gives a threefold W' with precisely one singular point which is an ordinary double point defined locally in four-space by the equation $xw - yz = 0$.

In the algorithms of [BM] and [E-V 1, E-V 2], when applied to surfaces, the centers are almost always points, and only at the very end of the process when the situation has become almost combinatorial, also smooth curves are blown up. In occurrence of normal crossings curves the ambiguity which component to choose for the center is solved by blowing up the intersection point, thus creating two old components, the strict transforms of the original components, and a new component, the new exceptional divisor. There then appear in W' two normal crossings, but the symmetry can now be untied by the "age" of the components, and the two old curves are blown up. This, of course, requires additional invariants as bookholders of the history. Compare this with Propositions 3.8 and 4.8 below, where the chosen non-reduced center encapsulates in one blowup this composition of blowups.

Zariski shows that if one chooses for the center instead of normal crossings one of the smooth components and then the strict transform of the other component, the resulting strict transform of X does not depend on the choice *provided that the multiplicity remains constant* [Za 1]. This fails in higher dimension [Ha 2, Sp 2].

There is another possibility to get rid of normal crossings by localizing X along one component and applying resolution of curves to make the localization smooth. This allows to reduce the equimultiple locus to a finite set of isolated points, but is not canonical since there is no natural indication which component to choose (imagine that there is a local but not global symmetry of X permuting the two components).

Proposition 3.3. *Let $S = S_y \cup S_z$ be a normal crossing at a with two smooth components S_y and S_z. Choose local coordinates x, y, z at a such that S_y and S_z are defined by $x = z = 0$ and $x = y = 0$ respectively. Let Z be the center with ideal $\mathbf{P} = (x, yz)(x, y)(x, z)$. The blowup W' of W in Z is smooth with five affine charts given as follows.*

x^3 :	$\operatorname{Spec} K[x, \frac{y}{x}, \frac{z}{x}]$	$(x,y,z) \to (x, xy, xz)$	$E = (x)$	$E_a = (x)$
x^2y :	$\operatorname{Spec} K[\frac{x}{y}, y, \frac{z}{x}]$	$(x,y,z) \to (xy, y, xyz)$	$E = (xy)$	$E_a = (y)$
x^2z :	$\operatorname{Spec} K[\frac{x}{z}, \frac{y}{x}, z]$	$(x,y,z) \to (xz, xyz, z)$	$E = (xz)$	$E_a = (z)$
xyz :	$\operatorname{Spec} K[\frac{yz}{x}, \frac{x}{z}, \frac{x}{y}]$	$(x,y,z) \to (xyz, xy, xz)$	$E = (xyz)$	$E_a = (xy, xz)$
y^2z^2 :	$\operatorname{Spec} K[\frac{x}{yz}, y, z]$	$(x,y,z) \to (xyz, y, z)$	$E = (yz)$	$E_a = (y, z)$

Proof. The assertions are checked by computation, using the fact that W' is covered locally along E by the charts $\operatorname{Spec} \mathbf{R}[\mathbf{P}a_i^{-1}]$ with a_i generators of $\mathbf{P}$. □

Exercise 3.4. Show that under taut blowup the order of the strict transform X' of X at a point a' over an intersection point a of S does not increase.

Proposition 3.5. *The taut blowup of W over an intersection point a of $S = S_y \cup S_z$ is the composition of the blowup of W in the reduced ideal (x, yz) of S followed by the blowup of the unique singular point s obtained by this blowup. The exceptional divisor E has three components E_y, E_z and E_s, where E_y and E_z are the closures of $\pi^{-1}(S_y \setminus a)$ and $\pi^{-1}(S_z \setminus a)$ in W' and where E_s is the fiber over s under the second blowup. The fiber E_a over $a = S_y \cap S_z$ consists of a one-dimensional component $E_t = E_y \cap E_z \cong \mathbb{P}^1$ and a two-dimensional component $E_s \cong \mathbb{P}^1 \times \mathbb{P}^1$ intersecting in the origin c of the xyz-chart.*

Proof. The blowup $\pi_1 : W^1 \to W$ of S with reduced ideal (x, yz) in W has two charts.

$yz : W^1 = \operatorname{Spec} K[x, y, z, \frac{x}{yz}] = \operatorname{Spec} K[\frac{x}{yz}, y, z]$ smooth , $(x, y, z) \to (xyz, y, z)$

$x :\ W^1 = \operatorname{Spec} K[x, y, z, \frac{yz}{x}] \cong \operatorname{Spec} K[x, y, z, w]/xw - yz$ singular.

To resolve the singularity of W^1, we embed W^1 into a four dimensional ambient space V with local coordinates x, y, z, w and blow up the singular point s, getting $\pi_2 : W^2 \to W^1$. Let $I^1 = (xw - yz)$ be the defining ideal of W^1 in V and I^2 its strict transform in W^2. In addition to the smooth yz-chart we get four charts in W^2 with smooth strict transforms $I^2 = (w - yz)$, $(xw - z)$, $(xw - y)$ and $(x - yz)$ and substitutions as follows.

$$x: \ \mathrm{Spec}\, K[x, \tfrac{y}{x}, \tfrac{z}{x}, \tfrac{w}{x}]/\tfrac{w}{x} - \tfrac{y}{x}\tfrac{z}{x} = \mathrm{Spec}\, K[x, \tfrac{y}{x}, \tfrac{z}{x}], \qquad (x,y,z) \to (x, xy, xz)$$
$$y: \ \mathrm{Spec}\, K[\tfrac{x}{y}, y, \tfrac{z}{y}, \tfrac{w}{y}]/\tfrac{x}{y}\tfrac{w}{y} - \tfrac{z}{y} = \mathrm{Spec}\, K[\tfrac{x}{y}, y, \tfrac{z}{x}], \qquad (x,y,z) \to (xy, y, xyz)$$
$$z: \ \mathrm{Spec}\, K[\tfrac{x}{z}, \tfrac{y}{z}, z, \tfrac{w}{z}]/\tfrac{x}{z}\tfrac{w}{z} - \tfrac{y}{z} = \mathrm{Spec}\, K[\tfrac{x}{z}, \tfrac{y}{x}, z], \qquad (x,y,z) \to (xz, xyz, z)$$
$$w: \ \mathrm{Spec}\, K[\tfrac{x}{w}, \tfrac{y}{w}, \tfrac{z}{w}, w]/\tfrac{x}{w} - \tfrac{y}{w}\tfrac{z}{w} = \mathrm{Spec}\, K[\tfrac{x}{z}, \tfrac{x}{y}, \tfrac{yz}{x}], \quad (x,y,z) \to (xyz, xy, xz)$$

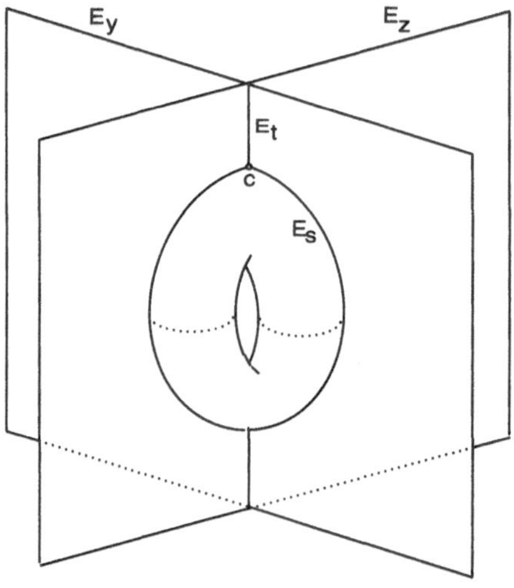

FIGURE 2. Normal crossings blowup

Exercise 3.6. Show that the resulting 5 charts glue in a way which gives the blowup W' of W in the ideal $\mathbf{P} = (x, yz)(x, y)(x, z)$.

Blowing up (x, yz) in W equals outside the origin the blowup of a smooth curve. Hence the inverse images $\pi_1^{-1}(S_y \setminus a)$ and $\pi_1^{-1}(S_z \setminus a)$ are isomorphic to $(S_y \setminus a) \times \mathbb{P}^1$ and $(S_z \setminus a) \times \mathbb{P}^1$. Let $E_y^1 \cup E_z^1$ be the closure in W^1 of $\pi_1^{-1}(S \setminus a)$. The singular point s of W^1 lies in the intersection $E_y^1 \cap E_z^1$ and is blown up via π_2. The fiber $E_s = \pi_2^{-1}(s)$ is isomorphic to $\mathbb{P}^2$. It follows that the exceptional divisor E of the blowup π has three components E_y, E_z and E_s, where the first two are the strict transforms under π_2 of E_y^1 and E_z^1. The fiber $E_a = (\pi_2\pi_1)^{-1}(a)$ has two components, namely $E_y \cap E_z$ and E_s. Four of the charts cover E_s, whereas the

yz-chart covers $(E_y \cap E_z) \setminus c$. It is checked that $E_y \cap E_z$ and E_s intersect in the origin of the xyz-chart. We shall see in Section 4 that if S is the equimultiple locus of X the strict transform X' meets E_a only in the yz-chart obtained already by the blowup π_1. □

Variation. There is an alternate non-reduced structure on normal crossings centers which yields under blowup a regular ambient scheme. It is less appropriate for resolution purposes than the taut structure since the invariants we use may increase and additional invariants are necessary to show that the situation improves, cf. with [BM] and [E-V 1]. We include a description thereof because this blowup is the composition of the blowup of the intersection point of the normal crossings followed by the blowup of the strict transforms of the two components. This is similar to the procedure used in loc. cit.

Proposition 3.7. *Let $Z = Z_y \cup Z_z$ be a normal crossings in W with embedded component the cube of the maximal ideal at the intersection point a of the two smooth curves, i.e. given locally by the ideal $(x,yz) \cap (x,y,z)^3$ with generators $x^3, xy^2, xz^2, y^2z, yz^2, x^2y, x^2z, xyz$. The blowup W' of W in Z is smooth with charts and substitutions as follows:*

$$\begin{array}{llll} x^3: & W' = \operatorname{Spec} K[x, \frac{y}{x}, \frac{z}{x}], & (x,y,z) \to (x, xy, xz), & E = (x), \quad E_a = (x) \\ xy^2: & W' = \operatorname{Spec} K[\frac{x}{y}, y, \frac{z}{x}], & (x,y,z) \to (xy, y, xyz), & E = (xy), \quad E_a = (y) \\ y^2z: & W' = \operatorname{Spec} K[\frac{x}{z}, y, \frac{z}{y}], & (x,y,z) \to (xyz, y, yz), & E = (yz), \quad E_a = (y) \end{array}$$

The charts xz^2 and yz^2 are symmetric to the charts xy^2 and y^2z respectively. The remaining charts are open subsets of the preceding ones.

$x^2y : W' = \operatorname{Spec} K[\frac{x}{y}, (\frac{x}{y})^{-1}, y, \frac{z}{x}]$ *contained in chart xy^2.*
$x^2z : W' = \operatorname{Spec} K[\frac{x}{z}, (\frac{x}{z})^{-1}, z, \frac{y}{x}]$ *symmetric to the chart x^2y*
$xyz : W' = \operatorname{Spec} K[\frac{x}{z}, (\frac{x}{z})^{-1}, \frac{y}{x}, (\frac{y}{x})^{-1}, z]$ *contained in chart xz^2*

Proof. This is verified by computation. □

Proposition 3.8. *The blowup $\pi : W' \to W$ defined in the preceding proposition is the composition of the point blowup of the intersection point a followed by the blowup of the strict transform of the two curves S_y and S_z. The exceptional divisor E of π has three components E_a, E_y and E_z, where $E_a \cong \mathbb{P}^2$ and where $E_y \cong S_y \times \mathbb{P}^1$, $E_z \cong S_z \times \mathbb{P}^1$ are the closures in W' of $\pi^{-1}(S_y \setminus a)$ and $\pi^{-1}(S_z \setminus a)$ respectively.*

Proof. The blow up of $\mathbf{P} = (x,y,z)$ has three charts. The strict transforms of the two curves S_y and S_z lie in two of them.

chart x: $W^1 = \operatorname{Spec} K[x, \frac{y}{x}, \frac{z}{x}]$ with substitution $(x,y,z) \to (x, xy, xz)$. The strict transforms of S_y and S_z do not meet this chart.
chart y: $W^1 = \operatorname{Spec} K[\frac{x}{y}, y, \frac{z}{y}]$ with substitution $(x,y,z) \to (xy, y, yz)$. Only the strict transforms of S_y lies in this chart.
chart z: symmetric to preceding chart.

We next blow up in the y-chart the strict transform S_y^1 of S_y and get two charts. Denote by $x' = xy$, $y' = y$, and $z' = zy$ the induced coordinates in this chart so that S_y^1 is defined by $x' = z' = 0$.

chart x': $W^2 = \operatorname{Spec} K[x', y', z', \frac{z'}{x'}] = \operatorname{Spec} K[\frac{x}{y}, y, \frac{z}{x}]$ with substitution $(x', y', z') \to (x', y', x'z')$, and $W^2 \to W$ given by $(x, y, z) \to (xy, y, xyz)$.
chart z': $W^2 = \operatorname{Spec} K[x', y', z', \frac{x'}{z'}] = \operatorname{Spec} K[\frac{x}{z}, y, \frac{z}{y}]$ with substitution $(x', y', z') \to (x'z', y', z')$, and $W^2 \to W$ given by $(x, y, z) \to (xyz, y, yz)$.

As a result we have five charts, namely x, (y, x'), (z, x'), (y, z'), (z, y'), with substitutions $(x, y, z) \to (x, xy, xz)$, (xy, y, xyz), (xz, xyz, z), (xyz, y, yz), (xyz, yz, z). These coincide with the ones obtained by blowing up $\mathbf{P} = (x, yz) \cap (x, y, z)^3$.

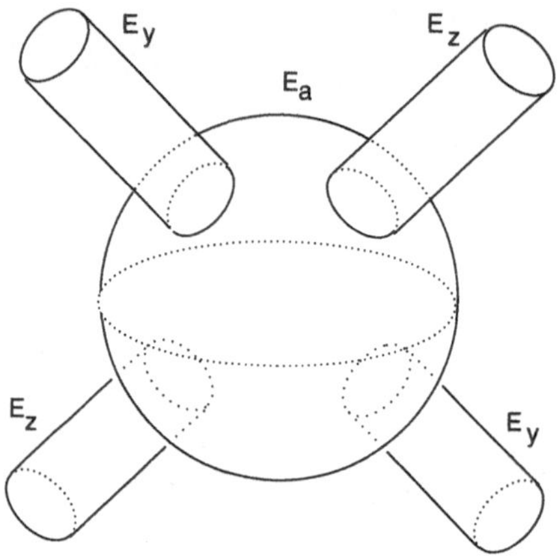

FIGURE 3. Variation normal crossings blowup

Exercise 3.9. Show that the charts glue as the charts from Proposition 3.7 and that the exceptional divisor decomposes as asserted. □

Exercise 3.10. Compute the strict transform of $x^2 + y^5z^5$ under the blowup of $\mathbf{P} = (x, yz)(x, y)(x, z)$ and $\mathbf{P} = (x, yz) \cap (x, y, z)^3$ respectively. Determine the charts where the order has remained constant and compare the degree of the monomials in y and z with respect to the two blowups (cf. with the last two paragraphs of Section 7).

4. Transformation of equimultiple locus under blowup

Let $\pi : W \to W'$ be the taut blowup with center Z. In this section, S need not have normal crossings. Assume that the maximal multiplicity has remained constant, $o_{X'} = o_X$. We describe possible configurations of the equimultiple locus S' of X' in terms of the strict and total transforms S^{st} and $S^* = \pi^{-1}S$ of S and the exceptional divisor $E = \pi^{-1}Z$.

We shall need an invariant τ which is used by various authors as induction invariant. Here it only plays an auxiliary role. Let $a \in X$ and $f \in \mathbf{R} = \mathcal{O}_{W,a}$ be a defining equation of X locally at a. Let $f = \sum_{ijk} c_{ijk}x^i y^j z^k$ be the expansion of f in the completion $\overline{\mathbf{R}} \cong K[[x,y,z]]$ of $\mathbf{R}$. The order $o = o_a$ of f at a will be the minimal value $i+j+k$ with non zero coefficient c_{ijk}. Let $\mathrm{hom}_a f = \sum c_{ijk}x^i y^j z^k$ be the homogeneous form of lowest degree of f, where the sum ranges over those indices for which $i+j+k = o_a$. It defines the tangent cone TC of X at a. Let τ_a be the minimal number of variables necessary to write $\mathrm{hom}_a f$ over all coordinate choices.

Exercise 4.1. Show that it suffices to consider only linear coordinate changes to realize τ_a. More explicitly, show the following: Let ε be the graded reverse monomial order on $\mathbb{N}^3$ and $\mathrm{in}_{\varepsilon\mathbf{x}} f$ the initial monomial of f with respect to ε and $\mathbf{x} = (x,y,z)$. Set $\max_\varepsilon f = \max_{\mathbf{x}} \mathrm{in}_{\varepsilon\mathbf{x}} f$. Then τ_a equals the index of the last variable appearing in $\max_\varepsilon f$, counting x, y, z in this order [Ha 1].

Lemma 4.2. *Assume that K is algebraically closed. Let a be a point of S where S has at most normal crossings. If $\tau_a = 3$ then a is an isolated point of S. If a is a smooth point of S, then $\tau_a \leq 2$, and $\tau_a = 2$ implies that there are coordinates in which* $\mathrm{hom}_a f \in (x,y)$ *and S is given locally at a by the ideal (x,y). If a is an intersection point of two smooth components of S then $\tau_a = 1$ and there exist local coordinates such that* $\mathrm{hom}_a f = x^o$ *and S is defined by the ideal (x,yz).*

Proof. Assume that a is not an isolated point of S. Choose coordinates at a such that S is defined by the ideal (x,y) or (x,yz). In the first case, $f \in (x,y)^o$ and therefore $\mathrm{hom}_a f \in (x,y)^o$, say $\tau_a \leq 2$. In the second case, we get $f \in (x,yz)^o$ and therefore $\mathrm{hom}_a f \in (x)^o$, say $\tau_a = 1$. □

Exercise 4.3. Assume K algebraically closed, and let a be an isolated point of S with $\tau_a = 3$. Blow up X in a, and let a' be a point above a in the strict transform X' of X. Show that $o_{a'} < o_a$ for all such a'. Compare this with the counterexample of Hironaka for K not algebraically closed [Hi 4, Thm. 3, p. 331, Ha 2, ex. 7].

Theorem 4.4. *Let $W' \to W$ be the taut blowup of X with center Z. Assume that the multiplicity has remained constant.*

(a) *If S has singular irreducible components or smooth components not intersecting transversally, S' is the strict transform of S augmented possibly by a curve isomorphic to $\mathbb{P}^1$. If this curve meets another smooth component of S' the intersection is transversal.*

(b) *If the equimultiple locus has at most normal crossings singularities S' has again at most normal crossings singularities.*

It follows from (a) and resolution of curves in three-space that by a finite sequence of point blowups the equimultiple locus can be transformed into a normal crossings curve as above. Then its components will be blown up until the highest occurring multiplicity drops (Theorem of Beppo Levi, see Section 6 and 7 and [Za 1, p. 522]).

The theorem will be proven through three propositions which describe more accurately the possible transformations of the equimultiple locus in each case. We suppose throughout to be in the situation of the theorem.

Proposition 4.5. *Assume that the center $Z = a$ is a closed point of S, S being an isolated point or an arbitrary curve. Let a' be a point over a where the multiplicity has remained constant. Then, locally at a', either $S' = S^{st}$ or $S' = S^{st} \cup (E \cap X')$ with $E \cap X' \cong \mathbb{P}^1$. In the second case, and if S has only smooth components, S^{st} and $E \cap X'$ meet transversally.*

The assertion does not hold in higher dimensions, see [Ha 2, ex. 10].

Proof. We follow [Za 1], proof of Thm. 1 and Lemma 3.2, p. 479. As we blow up a, $\pi : W \to W'$ is an isomorphism outside a and hence $X \setminus a \cong X' \setminus E$ and $S \setminus a \cong S^* \setminus E$. This implies that $S^* \setminus E \subseteq S'$. As S' is closed, the strict transform S^{st} of S is formed by certain components of S'. Therefore S' is contained in the union of S^{st} and $E \cap X'$. Observe that $E \cong \mathbb{P}^2$ and that $E \cap X'$ could a priori be a reducible and singular curve. We distinguish two cases. If $\tau_a \geq 2$ only finitely many points of $X' \cap E$ can have order o, namely, by exercise 4.1, the intersections with E of the strict transforms of the zero set of $\mathrm{hom}_a f$ (if $\tau_a = 3$, the multiplicity drops in all points of E.) This implies that $S' = S^{st}$.

If $\tau_a = 1$ we have for suitable local coordinates x, y, z in W at a that $f \equiv x^o$ modulo $(x, y, z)^{o+1}$. At the origin of the x-chart the multiplicity drops (and also at the other points of this chart, by an argument as in the proof of the next proposition). Similarly, in the y-chart it drops at all points of E except those where $x = 0$, since $X' \cap E$ is given by $(y^{-o} f(xy, y, zy), y) = (x^o, y)$. The situation in the z-chart is symmmetric to the preceding one. Hence $S' = S^{st} \cup (X' \cap E) \cong S^{st} \cup \mathbb{P}^1$. Assume that S has only smooth components. Choose one. It is given in suitable coordinates of $\mathbf{R}$ by $x = z = 0$. The strict transform S^{st} only appears in the y-chart and has the same equations $x = z = 0$. Therefore it is transversal to $X' \cap E$. □

Proposition 4.6. *Assume that $Z = S$ is a smooth curve. Let a' be a point over a where the multiplicity has remained constant. Then $S' \subseteq E \cap X'$. If S' has a one dimensional component, $S' = E \cap X' \cong S$ under π.*

For reduced but possibly singular centers see [Za 1], Thm. 2, p. 484, and its Corollary, p. 485. See also [Hi 1, p. 109].

Proof. Let $a \in Z$. Observe that $\tau_a \leq 2$ by the lemma. If $\tau_a = 2$ the multiplicity drops at all points of E. So we may assume $\tau_a = 1$, say $f = x^o + g(x, y, z)$

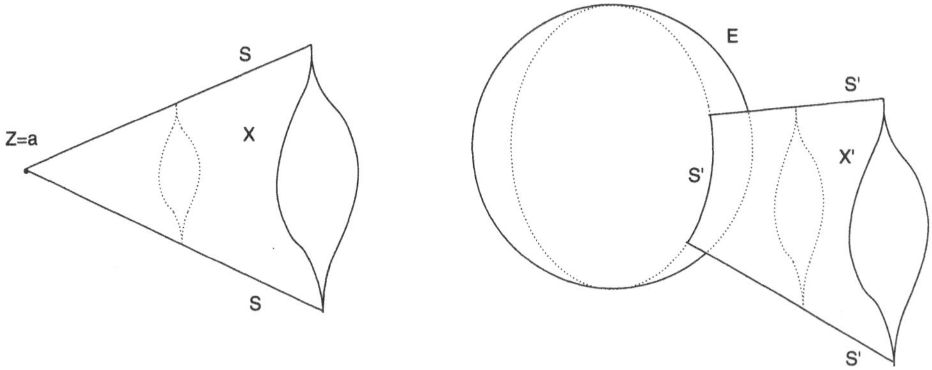

FIGURE 4. Equimultiple locus and point blowup

with $g \in (x, y, z)^{o+1}$. In the x-chart the total transform f^* of f is of form $f^* = x^o \cdot (1 + \sum c_{ijk} x^{i+j-o} y^j z^k) = x^o \cdot f'$ with $i+j > o$. It follows that f' does not vanish at the origin. The translation $x, y+s, z+t$ preserves the unit since the x-exponents in the sum remain ≥ 1. Hence f' does not vanish inside E_a in the x-chart. In the y-chart $X' \cap E$ is given by the ideal $(y^{-o} f(xy, y, z), y) = (x^o + y^{-o} g(xy, y, z), y)$ where $f = x^o + g(x, y, z)$. This intersection is only o-fold if the variety defined by $x^o + y^{-o} g(xy, y, z)$ in $y = 0$ is o-fold, i.e. an o-th power $(x + h(x, z))^o$ for some h. The curve in W' given by the ideal $(x + h(x, z), y)$ is isomorphic to Z given by (x, y) in W. □

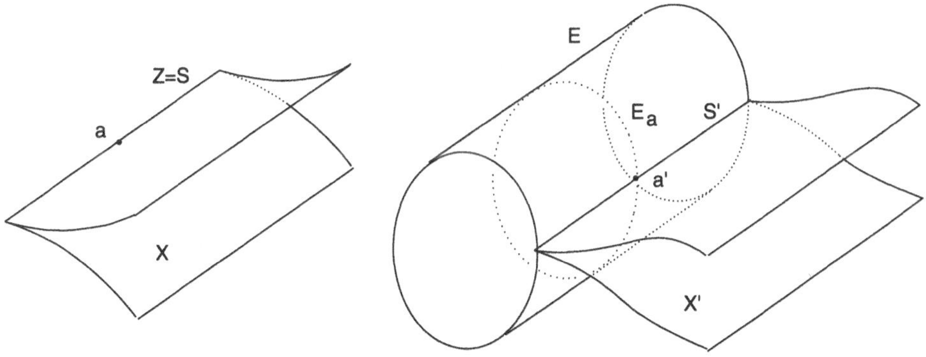

FIGURE 5. Equimultiple locus and curve blowup

Proposition 4.7. *Let S have normal crossing at a. Then $X' \cap E$ lies inside the y^2z^2-chart. If the multiplicity has remained constant every one-dimensional component of S' is isomorphic to a component of S under π.*

Proof. The lemma implies $\tau_a = 1$. Choose local coordinates x, y, z such that Z is defined by the ideal $\mathbf{P} = (x, yz)(x, y)(x, z)$ locally at a. We write

$$f = x^o + \sum c_{ijk} x^i y^j z^k$$

where the sum runs over all triples i, j, k satisfying $i+j \geq o$, $i+k \geq o$, $i+j+k > o$ and $0 \leq i < o$. We consider the strict transform f' of f at a' in the various charts covering E.

chart x^3: We have $f' = 1+\sum c_{ijk} x^{i+j+k-o} y^j z^k$ at the origin of this chart, and the sum remains in the ideal generated by x under the translations $(x, y+s, z+t)$ of this chart. Hence f' is invertible in E and $X' \cap E_a$ does not meet this chart.
chart x^2y: We have $f' = 1+\sum c_{ijk} x^{i+k-o} y^{i+j+k-o} z^k$ at the origin of this chart, and the sum remains in the ideal generated by y under the translations $(x+s, y, z+t)$ of this chart. Hence f' is invertible everywhere and $X' \cap E_a$ does not meet this chart.
chart x^2z: symmetric to preceding chart.
chart xyz: We have $f' = 1 + \sum c_{ijk} x^{i+j+k-o} y^{j+i-o} z^{k+i-o}$ at the origin of this chart, and the sum remains in the ideal generated by x under the translations $(x, y+s, z+t)$ of this chart. The other translation $(x+s, y, z)$ requires the following argument. The sum lies in the ideal generated by y and z except possibly if $j+i = o$ and $k+i = o$, hence $j = k = o - i$. This implies $f' = 1 + \sum c_{ijk} x^{o-i}$ and X' may intersect E in this chart in a point of the x-axis off the origin. Such points lie in the y^2z^2-chart and will be treated there.
chart y^2z^2: We have $f' = x^o + \sum c_{ijk} x^i y^{j+i-o} z^{k+i-o}$ at the origin of this chart. Translations are $(x+s, y, z)$. It follows that $X' \cap E_a$ intersects the x-axis in isolated points given as solutions of $x^o + \sum_{i=0}^{o-1} c_{i,o-i,o-i} x^i = 0$. The component E_y of E is given by $z = 0$, which gives $f'_{|E_y} = x^o \cdot \sum_{ij} c_{i,j,o-i} x^i y^{j+i-o}$. To have there an o-fold curve requires that $f'_{|E_y}$ equals a power $(x + a(y))^o$ for some series $a(y)$. Therefore this curve is isomorphic to S_y. This shows that each component of S' is isomorphic to a component of S, and two components can only meet on the x-axis of this chart. □

Proof of Theorem 4.4. The first proposition proves (a) of the theorem. The other two show that if the multiplicity remains constant, S' will consist of smooth irreducible curves intersecting transversally and isomorphic to certain components of S, isolated points and possibly a $\mathbb{P}^1$. If C' is a one dimensional component of S' then either C' is the strict transform of some component C of S, or $C' \cong S$ and S is smooth or $C' = E \cap X' \cong \mathbb{P}^1$. This concludes the proof of the theorem. □

Variation. We describe the transformation of the equimultiple locus when we change the embedded components of the normal crossing center Z at the intersection points as in Proposition 3.7.

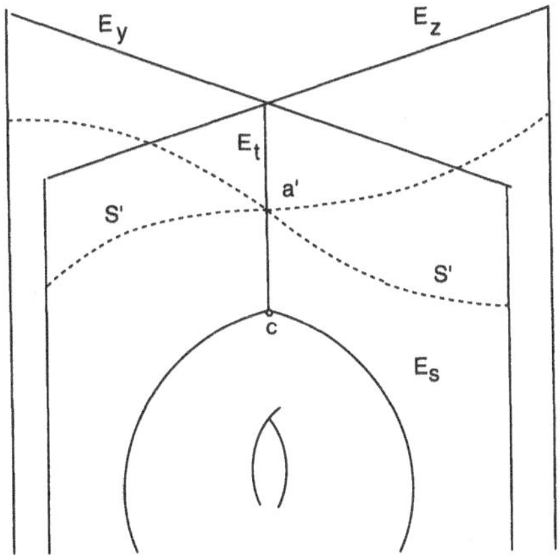

FIGURE 6. Equimultiple locus and crossings blowup

Proposition 4.8. *Let Z be a normal crossings center with embedded components the cubes of the maximal ideal at the intersection points. Assume that $o_{X'} = o_X$. If a one dimensional component of S' lies inside $E \cap X'$ then $E \cap X'$ is irreducible, equal to this component and isomorphic to a component of S under π. Moreover, S' meets E only in the y^2z- or yz^2-chart.*

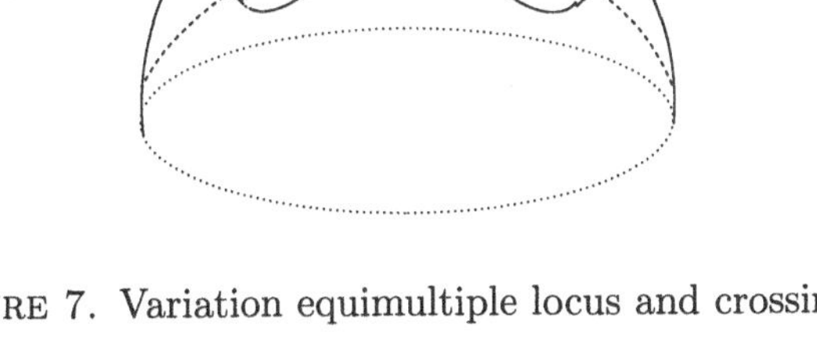

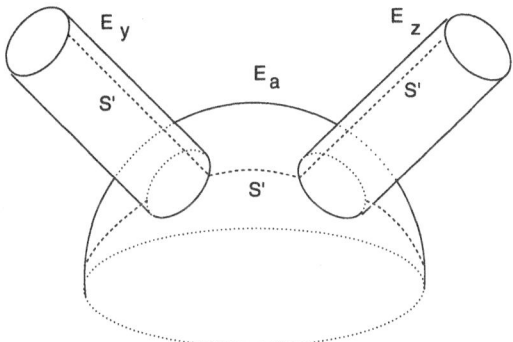

FIGURE 7. Variation equimultiple locus and crossings blowup

Proof. Let us place at an intersection point $a \in Z$. Choosing local coordinates x, y, z we may assume $a = 0$ and Z defined by the ideal $\mathbf{P} = (x, yz) \cap (x, y, z)^3$ locally at a. By the lemma, $\tau_a = 1$, and we can write

$$f = x^o + \sum c_{ijk} x^i y^j z^k$$

where the sum runs over all triples i, j, k satisfying $i + j \geq o$, $i + k \geq o$ and $0 \leq i < o$. We consider the total transform f^* in the various charts.

chart x^3: The substitution is (x, xy, xz), hence $f^* = x^o + \sum c_{ijk} x^{i+j+k} y^j z^k = x^o \cdot (1 + \sum c_{ijk} x^{i+j+k-o} y^j z^k) = x^o \cdot f'$. Possible translations are $(x, y+s, z+t)$. We may restrict by symmetry to translations $y+s$. This changes only those monomials of the expansion of f' where $j \geq 1$. As $i + k \geq o$ for all i, j, k these monomials have x-exponent $i + j + k - o \geq 1$. As the x-exponent remains unchanged under the translation, the resulting f' is again a unit. Hence $X' \cap E_a$ does not meet this chart.

chart xy^2: The substitution is (xy, y, xyz), hence $f^* = x^o y^o + \sum x^{i+k} y^{i+j+k} z^k = x^o y^o \cdot (1 + \sum c_{ijk} x^{i+k-o} y^{i+j+k-o} z^k) = x^o y^o \cdot f'$. Possible translations are $(x, y + s, z + t)$ or $(x + s, y, z + t)$. Translations $y + s$ affect only those monomials where $i + j + k - o \geq 1$. If $i + k - o \geq 1$ or $k \geq 1$ the resulting series f' remains a unit after the translation. So assume $i + k - o = 0$ and $k = 0$ so that these monomials are of form y^j with $j \geq 1$. This implies that in the expansion of f there appeared monomials $x^o y^j$. This case can be excluded by supposing f in Weierstrass form. Translations $z + t$ affect only those monomials where $k \geq 1$ and hence y-exponent $i + j + k - o \geq 1$, i.e. f' has no zeroes in this chart of E. Translations $x + s$ affect only those monomials where $i + k - o \geq 1$ and hence y-exponent $i + j + k - o \geq 1$, i.e. $f'|_{E_a}$ has no zeroes in this chart.

chart y^2z: The substitution is (xyz, y, yz), hence $f^* = x^o y^o z^o + \sum c_{ijk} x^i y^{i+j+k} z^{i+k} = y^o z^o \cdot (x^o + \sum c_{ijk} x^i y^{i+j+k-o} z^{i+k-o}) = y^o z^o \cdot f'$. Here E is given by $yz = 0$ and $X' \cap E$ by the ideal $(f', yz) = (x^o + \sum c_{ijk} x^i y^{i+j+k-o} z^{i+k-o}, yz) = (x^o + \sum_{ij} x^i y^j, yz)$. This is a curve in E. If it belongs to the equimultiple locus S' of X', the series $x^o + \sum_{ij} x^i y^j$ must have order o along this curve, which implies that $x^o + \sum_{ij} x^i y^j = (x + h(x, y))^o$ is an o-th power. Then $S' = S^{st} \cup (E \cap X')$ and $E \cap X'$ is a subscheme of X' locally isomorphic to S under π.

The charts xz^2 and yz^2 are symmetric to the preceding ones, the charts x^2z, x^2y and xyz are contained in the preceding charts as open subsets.

Summarizing, S' can only lie in the y^2z-chart with equation $x = y = 0$ or in the yz^2-chart with equation $x = z = 0$ or inside E_y or E_z. As we restrict to $a \in S$ the intersection point and hence $a' \in E_a$, only the first two cases will be relevant. □

The assertion of the last proposition can also be proven by interpreting the blowup as a composition of blowups of points and smooth curves and applying Propositions 4.5 and 4.6.

5. Transformation of flags under blowup

A flag in a regular ambient scheme W at a point a is a full chain of local regular subschemes F_i of dimension i of W at a. Flags will be needed to define the induction invariant. Let X be a surface in three-space W and let a be a closed point of the equimultiple locus S of X. We assume that S has at most normal crossings at a. The flag $\mathcal{F}$ at a consists of a smooth curve F_1 contained in a smooth surface F_2 in W, both passing through a. It is called transversal to S if one of the following cases occurs.

If a is an isolated point of S, F_1 and F_2 can be arbitrary. If a is on precisely one smooth curve of S, either F_1 and F_2 are both transversal to this curve, or F_1 is transversal to S and F_2 contains S. If a is the intersection point of two smooth curves of S meeting transversally at a, F_1 is transveral to both curves and F_2 contains one curve and is transversal to the other. We don't intend the choice of $\mathcal{F}$ to be canonical.

Local coordinates x, y, z at a are called subordinate to the flag $\mathcal{F}$ if F_1 and F_2 are defined by $y = z = 0$ and $z = 0$ respectively. In the construction of the invariants we will work in the completion $\overline{\mathbf{R}}$ and choose coordinates there. Subordinate coordinate changes are of type $(x + a, y + b, z + c)$ where $a \in \overline{\mathbf{R}}$ is arbitrary, b belongs to the ideal (y, z) and c to the ideal (z) of $\overline{\mathbf{R}}$. Without loss of generality it will suffice to consider only coordinate changes where $a \in K[[y, z]]$, $b \in K[[z]]$ and $c = 0$, compare with the Gauss-Bruhat decomposition of Aut $\overline{\mathbf{R}}$ in [Ha 1].

The flag is called partially transversal to the tangent cone TC of X if F_1 is transversal to the maximal linear space along which TC is a product [Hi 4]. This is a void condition if $\tau_a = 3$. In subordinate coordinates, it signifies that the variable x appears in at least one monomial of $\mathrm{hom}_a f$.

Theorem 5.1. *Assume that S has at most normal crossings at a. Let $\pi : W' \to W$ be the taut blowup of X with center Z and let a' be a point over a. Assume that either S is smooth at a or, in case where S has normal crossings at a, that a' belongs to the one dimensional component $E_y \cap E_z$ of E_a. For any flag $\mathcal{F}$ in W at a transversal to S and partially transversal to TC there exist formal subordinate local coordinates $\mathbf{x}$ of W at a such that π is monomial at a' with respect to $\mathbf{x}$ and Z is defined locally at a by the ideals (x, y, z), (x, y), (x, z) or $(x, yz)(x, y)(x, z)$.*

Here, the blowup is called monomial with respect to $\mathbf{x}$ if the affine charts are given as in Proposition 3.3. For normal crossings centers, it seems impossible to achieve monomiality at all points $a' \in E_a$. Variation: If Z has embedded components of form $(x, yz) \cap (x, y, z)^3$ at the intersection points, this is possible, see the proof below, but destroys the form of $\mathbf{P} = (x, yz) \cap (x, y, z)^3$.

Proof. For blowups of smooth centers in regular schemes of arbitrary dimension this has been proven in [Ha 1]. We adapt the proof to dimension three, complementing it by the case of normal crossings centers.

Start with any formal subordinate coordinates $\mathbf{x}$ in $\overline{\mathbf{R}}$. We first simplify by subordinate coordinate changes the ideal $\mathbf{P}$ defining Z. If S is a point, $\mathbf{P} = (x, y, z)$. If S is a smooth curve, partial transversality of $\mathcal{F}$ with respect to TC implies that $\mathbf{P}$ can be written $(x + b(y,z), y + cz + d(y,z))$ or $(x + b(y,z), z + d(y))$ with series b and d of order ≥ 2 and a constant $c \in K$. The obvious subordinate coordinate change transforms $\mathbf{P}$ to (x, y) or $(x, z + d(y))$. Transversality of $\mathcal{F}$ with S forces in the second case $d = 0$. If S is a normal crossings, at least one component of S is defined by an ideal of form $(x + b(y,z), y + cz + d(y,z))$, hence, after subordinate coordinate change, of form (x, y). The other component is defined by an ideal of form $(x, cy + z + d(y,z))$, and transversality of $\mathcal{F}$ with S implies $c = d = 0$. The asserted form of $\mathbf{P}$ follows in all cases.

(1) *Point blowup.* Decompose $E_a \cong \mathbb{P}^2$ into the origin of the x-chart, the x-axis of the y-chart and the affine xy-plane of the z-chart. If a' is the origin of the x-chart, the local blowup is already monomial at a'. If a' lies in the x-axis of the y-chart, apply a translation $v = (x - s, y, z)$ in E_a to move a' to the origin of the y-chart. This translation is induced by the local subordinate coordinate change $u = (x - sy, y, z)$ of W at a. Clearly, u preserves the ideal $\mathbf{P}$, and π has become monomial at a'. For a' in the z-chart, apply a translation $v = (x - s, y - t, z)$ in E_a to move a' to the origin of the z-chart. This translation is induced by the local subordinate coordinate change $u = (x - sz, y - tz, z)$ of W at a. Clearly, u preserves the ideal $\mathbf{P}$, and π has become monomial at a'.

(2) *Curve blowup.* If S is a smooth curve defined by (x, y) or (x, z), a similar argument applies. Here, $E_a \cong \mathbb{P}^1$ decomposes into the origin of the x-chart and the x-axis in the y-, respectively z-chart. We leave the details as exercise.

(3) *Crossings blowup.* Let S be defined by the ideal (x, yz). We have seen earlier that E_a has two components $E_y \cap E_z \cong \mathbb{P}^1$ and $E_s \cong \mathbb{P}^1 \times \mathbb{P}^1$ intersecting at the origin of the xyz-chart. Let $a' \in E_y \cap E_z$. If it equals this origin, π is monomial at a'. Else a' lies in the y^2z^2 chart. As $E_y \cap E_z$ is the x-axis in this chart, there is a translation in W' of type $(x - s, y, z)$ which moves a' to the origin. This translation is induced from the local subordinate coordinate change $(x - syz, y, z)$ in W at a. It preserves $\mathbf{P}$. This proves the assertion. $\square$

Variation: Let $\mathbf{P} = (x, yz) \cap (x, y, z)^3$. Here, $E_a \cong \mathbb{P}^2$ will be decomposed into the affine plane $z = 0$ in the yz^2-chart (in which E_a is given by $z = 0$), the x-axis in the xz^2-chart (which is the projective line at infinity of $\mathbb{P}^2$ of the preceding chart) and the origin of the x^3-chart (which is the point at infinity of the preceding curve). In the first two charts translations $(x - s, y - t, z)$ and $(x + s, y, z)$ in E_a move a' to the origin of the chart. They correspond, by the substitution formulas for π, to local subordinate coordinate changes $(x - sy, y - sz, z)$ and $(x - sz, y, z)$ in W at a. The local blowup has become monomial. Note that the coordinate changes in W at a affect the equations of Z, yielding the ideals $(x - sy, yz - sz^2) \cap (x, y, z)^3$, respectively $(x - sz, yz) \cap (x, y, z)^3$.

Theorem 5.2. *Assume that S has at most normal crossings and let $\pi : W' \to W$ be the taut blowup of X with center Z. Let $\mathcal{F}$ be a flag at a transversal to S and partially transversal to the tangent cone TC of X. Let a' be a closed point over a. If a is an intersection point of S we suppose that a' lies in the one dimensional component $E_y \cap E_z$ of E_a, or equals one of the origins of the remaining charts. There exists a canonically defined induced flag $\mathcal{F}'$ at a'. Subordinate coordinates at a induce canonically subordinate coordinates at a'. For $a' \in S'$ with $o_{a'} = o_a$, the flag $\mathcal{F}'$ is transversal to the equimultiple locus S' and partially transversal to the tangent cone TC' of X' in a'.*

Canonical means that the flag $\mathcal{F}'$ is invariant under local isomorphisms of W' induced by automorphisms of W at a preserving $\mathcal{F}$. The assertion of the theorem holds for blowups of smooth centers in any dimension [Ha 1].

Proof. (1) *Point blowup.* Define $\mathcal{F}'$ at a' as follows. Let F_1^{st} and F_2^{st} be the strict transforms of F_1 and F_2 under π. They intersect the exceptional divisor E transversally in a point $F_1^{st} \cap E$ and a smooth curve $F_2^{st} \cap E$. If $a' = F_1^{st} \cap E$ is the intersection point (which is the origin of the x-chart), let $F_1' = F_1^{st}$ and $F_2' = F_2^{st}$. If a' lies on $F_2^{st} \cap E$ but not on F_1^{st} (i.e., in the y-chart), let $F_1' = F_2^{st} \cap E$ and $F_2' = F_2^{st}$. If $a' \notin F_2^{st}$ (i.e., in the z-chart), set $F_2' = E$ and let F_1' be the projective line in $E = \mathbb{P}^2$ connecting a' with $F_1^{st} \cap E$.

Exercise 5.3. In each case, letting x, y, z denote the coordinates at a' induced by the usual formulas, show that F_1' and F_2' will be defined by $y = z = 0$ and $z = 0$.

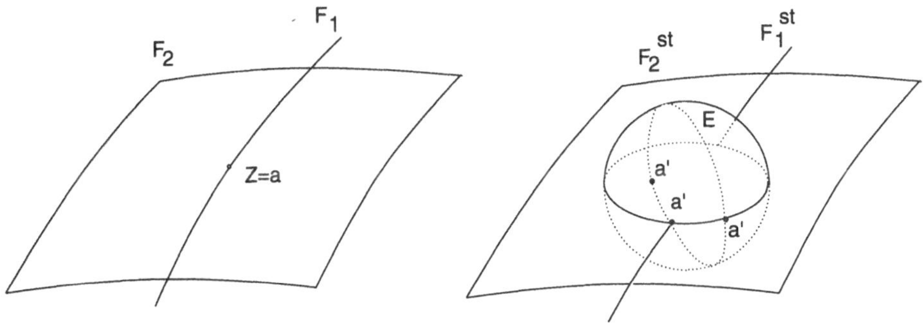

FIGURE 8. Flag and point blowup

By Proposition 4.5, we know that S' equals the strict transform of S, possibly augmented by $E \cap X' \cong \mathbb{P}^1$. In the latter case, by the same proposition, S' does not meet the x-chart, and has equation $x = z = 0$ in the y-chart, respectively $x = y = 0$ in the z-chart. Hence F_1' and S' are always transversal, and F_2' and S' are either transversal or $S' \subseteq F_2'$.

(2) *Curve blowup.* (a) If F_1 and F_2 are both transversal to S, choose coordinates x, y, z at a such that S is defined by $x = y = 0$. Define $\mathcal{F}'$ as follows. If $a' = F_1^{st} \cap E_a$

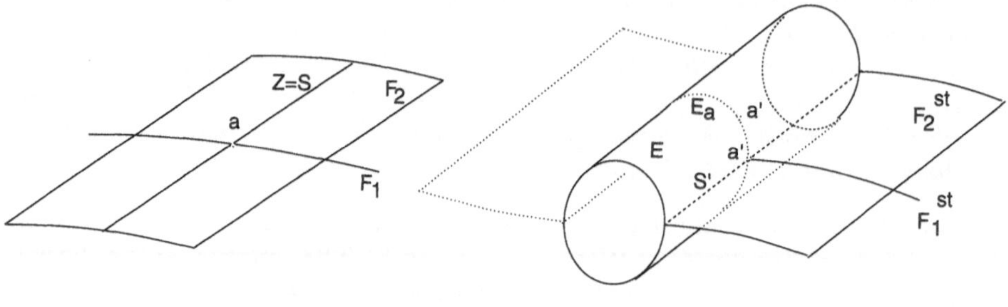

FIGURE 9. Flag and curve blowup

(which is the origin of the x-chart) let $F_1' = F_1^{st}$ and $F_2' = F_2^{st}$. If $a' \in E_a \setminus F_1^{st} \cap E_a$ (i.e., a' lies in the y-chart), set $F_1' = E_a$ and $F_2' = F_2^{st}$. Observe that E_a is a curve. In both cases F_1' and F_2' are defined by $y = z = 0$ and $z = 0$ respectively.

Consider now S'. If S' is a point, nothing is to show. If S' is a curve it must lie inside $E \cap X'$ and by Proposition 4.6, $S' = E \cap X'$ and $S \cong S'$. Moreover S' lies in the y-chart and is given there by $x + h(z) = y = 0$ for some h. As the flag $\mathcal{F}'$ is given by $y = z = 0$ and $z = 0$ it is transversal to $F_1' = E_a$ and F_2'.

(b) If F_1 is transversal to S and $S \subseteq F_2$ choose coordinates such that S is defined by $x = z = 0$. If $a' \in F_1^{st} \cap E_a$ (which is the origin of the x-chart) let $F_1' = F_1^{st}$ and $F_2' = F_2^{st}$. If $a' \in E_a \setminus F_1^{st} \cap E_a$ (i.e., a' lies in the y-chart), set $F_1' = E_a$ and $F_2' = E$. Again E_a is a curve. In both cases F_1' and F_2' are defined by $y = z = 0$ and $z = 0$.

By Proposition 4.6, $S' = E \cap X' \cong S$ lies in the z-chart and is defined there by $x + h(y) = z = 0$. The flag $\mathcal{F}'$ being given by $y = z = 0$ and $z = 0$ we have S' transversal to F_1' and contained in F_2'.

(3) *Crossings blowup.* Let $S = S_y \cup S_z$ at a. Let $a' \in E_a$ be a point of $E_y \cap E_z$ or an origin of the other charts. By Theorem 3 there is a subordinate coordinate change in W at a such that S is defined by the ideal (x, yz) and such that the local blowup $\mathbf{R} \to \mathbf{R}'$ is monomial. In particular, a' moves by the induced translation in $E_a \subseteq W'$ to the origin of one of the charts.

With this prior choice of coordinates it suffices to define $\mathcal{F}'$ at the origins of the charts. In the x^3-chart, set $F_1' = F_1^{st}$ and $F_2' = F_2^{st}$.

Exercise 5.4. Check that $F_1^{st} \cap E_a$ is the origin of this chart.

In the x^2y-chart, set $F_1' = E_a \cap F_2^{st}$ and $F_2' = F_2^{st}$.

Exercise 5.5. Check that F_2^{st} contains the origin of that chart.

In the x^2z-chart, the component of E_a containing the origin is isomorphic to $\mathbb{P}^1 \times \mathbb{P}^1$. Let F_1' be the (unique) line in $\mathbb{P}^1 \times \mathbb{P}^1$ through the origin of this chart and the origin of the x^3-chart, and set $F_2' = E_a$. In the xyz-chart and in the y^2z^2-chart, set $F_1' = E_y \cap E_z$ and $F_2' = E_y$.

Exercise 5.6. Check that the substitutions of the coordinates associated to the blowup and described in Proposition 3.5 define coordinates at a' which are subordinate to $\mathcal{F}'$.

Exercise 5.7. Prove that if the multiplicity remains constant, $\mathcal{F}'$ is partially transversal to the tangent cone TC' of X'.

This completes the proof of Theorem 5.2. □

Variation: Let $\mathbf{P} = (x, yz) \cap (x, y, z)^3$ define Z where S is a normal crossings of two smooth curves S_y and S_z at a. For any a' over a, the flag $\mathcal{F}'$ is defined as follows. Choose subordinate coordinates such that S is given by (x, yz). If $a' = F_1^{st} \cap E_a$ (which is the origin of the x^3-chart), set $F_1' = F_1^{st}$ and $F_2' = F_2^{st}$. If $a' \in E_a \cap F_2^{st} \setminus F_1^{st}$ (i.e., a' lies in the xy^2-chart) set $F_1' = F_2^{st} \cap E_a$ and $F_2' = F_2^{st}$.

If $a' \in E_a \setminus F_2^{st}$ we distinguish several cases. Let $\tau : W^1 \to W$ denote the blowup of W in the intersection point a of S, and denote by S_z^1 the strict transform of S_z. There is a unique line C^1 in the exceptional divisor $E_a^1 = \tau^{-1}a \cong \mathbb{P}^2$ going through $F_1^1 \cap E_a^1$ and $S_z^1 \cap E_a^1$. Let C be the inverse image of C^1 under the blowup of S_z^1 in W^1. This is a curve in E_a which goes through the origins of the x^3-chart and the xz^2-chart.

(a) If $a' \in E_a \cap E_z \setminus C$ (i.e., a' lies in the yz^2-chart) set $F_1' = E_a \cap E_z$ and $F_2' = E_a$.

(b) If $a' \in E_a \setminus (E_z \cup C)$ (i.e., a' lies in the x^3-chart) let F_1' be the unique line in $E_a \setminus (E_y \cup E_z) \cong E_a^1 \setminus (S_y^1 \cup S_z^1) \cong \mathbb{P}^2 \setminus \{\text{two points}\}$ going through a' and $F_1^{st} \cap E_a$ and set $F_2' = E_a$.

(c) If $a' \in C \cap E_z$ (i.e., a' is the origin of the xz^2-chart) set $F_1' = C$ and $F_2' = E_a$.

It is checked that in all cases F_1' and F_2' are defined in the induced coordinates by $y = z = 0$ and $z = 0$.

The equimultiple locus S' does not appear in the charts x^3, xy^2 and xz^2, since at any point there the multiplicity has dropped, see Proposition 4.8. In the two remaining charts y^2z and yz^2, we may assume that S' is a curve. By the same proposition, S' is defined in both charts by $x + h(y) = yz = 0$ for some series $h(y)$. Hence it meets $\mathcal{F}'$ as prescribed. □

Exercise 5.8. Replace the construction of the flag $\mathcal{F}'$ for normal crossings of the variation by interpreting the blowup as a composition of blowups in smooth centers.

6. Construction of the induction invariant

This section is largely inspired by Hironaka's definition of invariants used in [Hi 1]. We fix again a hypersurface X in a regular three dimensional scheme W whose equimultiple locus S has at most normal crossings.

The local invariant i_a we shall associate to a closed point $a \in X$ will be defined by first constructing a quadruple $i_{a\mathbf{x}} \in \mathbb{Q}^4$ which depends on the choice of coordinates $\mathbf{x}$ at a and by then specifying a set of formal coordinates for which $i_{a\mathbf{x}}$ takes the same value. We will then set $i_a = i_{a\mathbf{x}}$ for $\mathbf{x}$ in this set of coordinates. All points considered will be closed.

Coordinate dependent definition. Let $f \in \mathbf{R} = \mathcal{O}_{W,a}$ be a local equation of X in W at a. For coordinates $\mathbf{x} = (x, y, z)$ in the completion $\overline{\mathbf{R}}$ of $\mathbf{R}$ let $f = \sum c_{ijk} x^i y^j z^k$ be the expansion of f with respect to $\mathbf{x}$ at a. We set

$$i_{a\mathbf{x}} = (o_a, \beta_y, s_{\beta\gamma}, |\beta|)$$

with $o_a = \mathrm{ord}_a f$ the order of f at a. The components β_y, $s_{\beta\gamma}$ and $|\beta| = \beta_y + \beta_z$ are defined as follows. Let $NP_{yz} \subseteq \mathbb{Q}^2_+$ be the projection from $(o_a, 0, 0)$ of the

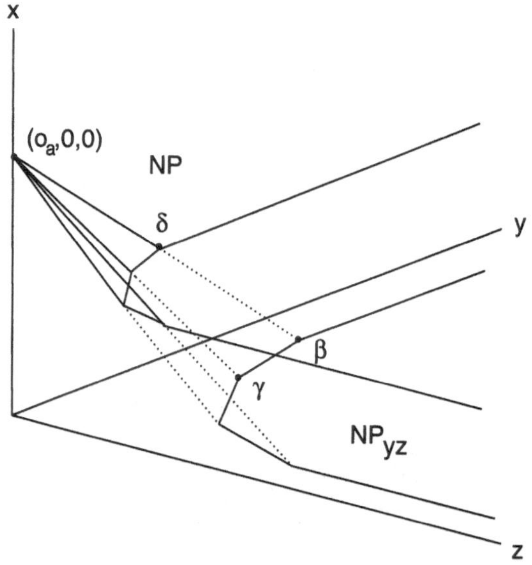

FIGURE 10. Projection of Newton Polyhedron

Newton polyhedron NP of f, neglecting the portion of NP in $(o_a, 0, 0) + \mathbb{Q}^3_+$. A point $(\delta_x, \delta_y, \delta_z)$ with $\delta_x < o_a$ is sent to $(\frac{o_a \delta_y}{o_a - \delta_x}, \frac{o_a \delta_z}{o_a - \delta_x})$. Let β be the vertex of NP_{yz} which is closest to the y-axis, i.e., with minimal second component β_z (i.e. the point of NP_{yz} which is minimal with respect to the inverse lexicographic order in $\mathbb{Q}^2_+$). If β does not exist, f equals in the given coordinates the monomial x^{o_a}

times a unit of $\overline{\mathbf{R}}$ and nothing is to prove. So we may discard this case. Let γ be the vertex of the segment of NP_{yz} adjacent to β, i.e., the vertex of NP_{yz} with second component γ_z minimal among the vertices of $NP_{yz} \setminus \beta$ (γ exists if and only if NP_{yz} is not a quadrant).

With these choices, β_y will denote the first component of β and $s_{\beta\gamma}$ the slope of the segment from β to γ, i.e., $s_{\beta\gamma} = \frac{\beta_y - \gamma_y}{\beta_z - \gamma_z} \in \mathbb{Q}_-$ (we draw the y-axis vertically). We set $s_{\beta\gamma} = -\infty$ if γ does not exist. We may equally take instead of β_y the slope of the segment in $\mathbb{Q}^2_+$ between $(o_a, 0)$ and $(0, \beta_y)$. Observe that $\beta_y + \beta_z \geq o$ and $0 > s_{\beta\gamma} \geq -\infty$.

This ad-hoc definition of the invariant i_a may seem little natural. It depends highly on the ordering chosen among x, y, z and leaves rather unclear how it could be extended to higher dimensions. However, without taking into account exceptional divisors when defining the invariant for the strict transforms, experimentation shows that there is not too much flexibility how to choose i_a such that it drops under taut blowup. Observe that β and $s_{\beta\gamma}$ may also serve as invariants for resolution of plane curves, see exercise 2.2. Projecting the Newton polyhedron from $(o_a, 0, 0)$ to the yz-plane explains why in many expositions the case where f has form $f = x^{o_a} + g(y, z)$ is considered as representative for the problem.

Exercise 6.1. Show that the invariant is independent of the choice of the defining equation f of X at a.

As the slopes $s_{\beta\gamma}$ are negative and may become arbitrarily small it is not immediate that the invariant belongs to a well ordered set. Let Γ be the set of quadruples (a, b, c, d) with $a, b, d \in \mathbb{N}$ and c of form $c = -\frac{p}{q}$ with p, q in $\mathbb{N}$ and $p \leq b$ (allowing $c = -\infty$). By construction, $i_{a\mathbf{x}} \in \Gamma$.

Exercise 6.2. Show that Γ is well ordered.

Example 6.3. Let $f = x^3 + y^4 z + y^2 z^2 + y z^7 + z^{14}$. Then $i_{a\mathbf{x}} = (3, 4, -\frac{1}{2}, 5)$ and $\beta = (4, 1)$, $\gamma = (2, 2)$.

A more conceptual definition of the invariant through coefficient ideals goes as follows [E-V 2, Ha 3]. Set $o = o_a$. For any ideal I of $\mathbf{R}$, let

$$I_{yz} = \sum_i (a_i(y, z),\ f \in I)^{\frac{o!}{o-i}}$$

be the coefficient ideal of I with respect to $\mathbf{x}$ and o. Here $a_i(y, z)$ denote the coefficients of elements $f \in I$ given by the expansion $f \equiv \sum_{i=0}^{o-1} a_i(y, z) x^{o-i}$ modulo $x^o \cdot (x, y, z)$. Observe that I_{yz} is compatible with coordinate changes in y, z but not in x. For a principal ideal $I = (f)$ with f expanded as above it is in general not true that $I_{yz} = (a_i(y, z)^{\frac{o!}{o-i}}, i = 0, \dots, o-1)$. This can be seen in the example of S. Encinas where $f = x^3 + xy^4 + z^5$ with $I_{yz} = (y^{12}, z^{10})$. Multiplying f by the unit $1 - x$ of $\mathbf{R}$ gives $I_{yz} = (y^{12} + y^8 z^5, z^{10})$. Consider now the Newton polygon of I_{yz}

$$NP_{yz} = \bigcup_{g \in I_{yz}} \bigcup_{\alpha \in \operatorname{supp} g} \alpha + \mathbb{Q}^2_+.$$

Then β equals the inverse lexicographically minimal vertex of $\frac{1}{(o-1)!} \cdot NP_{yz}$. In [E-V 1, E-V 2] the second component of the invariant is defined as the order of I_{yz}. This order generally increases under blowup. It is necessary to factor suitable powers of the exceptional divisor from the transform $(I')_{yz}$ to get the weak transform of I_{yz} with non-increasing order. There arise, however, serious problems when making the reasoning coordinate independent, since the exceptional divisors loose their monomial form when changing coordinates.

Example 6.4. Let $f = x^2 + y^{10} + y^3z^3 + z^{10}$ with isolated singular point at zero. Hence $S = 0$ and $I_y = (y^{10} + y^3z^3 + z^{10})$ has order 6. Blowing up the origin yields in the y-chart a strict transform $f' = x^2 + y^8 + y^4z^3 + y^8z^{10}$ with $(I')_y = (y^8+y^4z^3+y^8z^{10})$ of order 7. However $\beta = (10,0)$ has transformed into $\beta' = (8,0)$ and the first component has dropped.

Coordinate free definition. We specify the coordinates which we shall select to define i_a. The coordinates will have to be subordinate to a well chosen flag, and, secondly, have to maximize $i_{a\mathbf{x}}$ lexicographically over all formal subordinate coordinate choices at a (restricted slightly by maximizing also β_z). Hironaka calls such coordinates well and very well prepared [Hi 1].

We first choose and then fix forever a flag $\mathcal{F}$ in W at a, transversal to S and partially transversal to the tangent cone TC of X (Section 5). Given such a flag, we will only consider coordinates x, y, z in the completion $\overline{\mathbf{R}}$ which are subordinate to $\mathcal{F}$. Hence F_1 and F_2 are defined in $\overline{\mathbf{R}}$ by $y = z = 0$ and $z = 0$ respectively. Among all subordinate coordinates, consider those for which the vector $(\beta_z, \beta_y, s_{\beta\gamma}, |\beta|) \in \mathbb{Q}^4$ becomes maximal with respect to the lexicographic order on $\mathbb{Q}^4$ [instead of $s_{\beta\gamma}$ we may also maximize the projection of γ to $\mathbb{Q}_+$, and $|\beta|$ is maximized automatically through (β_z, β_y)]. Observe that we first maximize β_z which does not appear as a component of $i_{a\mathbf{x}}$. It can be seen, either directly as in [Hi 1, p. 115] or via the Artin Approximation Theorem as in [Ha 1], that the maximum actually exists and is achieved by some formal coordinates. In the algebraic setting it is convenient to work in the completion $\overline{\mathbf{R}}$. For analytic spaces the maximum also exists in the analytic category, see [Ha 1]. In all cases, $\beta = (\beta_y, \beta_z)$ and $s_{\beta\gamma}$ are the same for such coordinates. Then set

$$i_a = i_{a\mathbf{x}} = (o_a, \beta_y, s_{\beta\gamma}, |\beta|)$$

where $\mathbf{x}$ are coordinates in $\overline{\mathbf{R}}$ subordinate to $\mathcal{F}$ and maximizing $(\beta_z, \beta_y, s_{\beta\gamma}, |\beta|)$ lexicographically. Note that instead of the slope $s_{\beta\gamma}$ one could as well project N_{yz} from β to the z-axis $\mathbb{Q}_+$ and take the slope of this line. However, the (unique) vertex of the projection may increase under blowup if one does not factor an appropriate power of the exceptional divisor.

7. Transformation of $i_{a\mathbf{x}}$ under monomial blowup

In the situation of the last paragraph, let $\pi : W' \to W$ be the taut blowup associated to the singular surface X with center Z. We assume that the equimultiple locus S has at most normal crossings. Let $a \in X$ be a closed point inside S. We shall show in this and the next section that for all closed points $a' \in X'$ above a one has $i_{a'} < i_a$ in the lexicographic order on $\mathbb{Q}^4$. As the set Γ where i_a can vary is well ordered, there cannot occur an infinite sequence of points a', a'', ... over a along which the invariant does not stabilize. The stage at which it stabilizes depends on the choice of a, and a priori moving a in S may produce strictly decreasing sequences of arbitrary length. This does not guarantee resolution of X. We will have to show in Section 9 that all sequences stabilize uniformly at a certain stage independent of a. Observe that it is not clear whether i_a takes only finitely many values on S.

The proof that i_a decreases under blowup splits into a combinatorial argument on monomial blowups and a reduction argument which reduces the general case to the combinatorial situation.

Theorem 7.1. *Let $\mathbf{R} \to \mathbf{R}'$ be a taut local monomial blowup with respect to X and formal coordinates $\mathbf{x}$ of $\overline{\mathbf{R}}$ subordinate to a chosen flag $\mathcal{F}$ at a. Assume that S has at most normal crossings and that $\mathcal{F}$ is transversal to S and partially transversal to TC. Denote by $\mathbf{x}$ also the induced coordinates at a'. Then $i_{a'\mathbf{x}} < i_{a\mathbf{x}}$.*

Proof. The assumption implies that the point a' sits in one of the origins of the various affine charts induced in W' by the coordinates $\mathbf{x}$. We denote by S', β', γ', $\mathcal{F}'$ etc. all objects associated to the strict transform X' of X as was done before for X. An asterisque * will denote objects in W' obtained from below by applying the monomial substitution of the variables and factoring the exceptional divisor o_a-times. For instance, β^* will equal in the z-chart of a point blowup the vector $(\beta_y, \beta_z + \beta_y - o_a)$ which may equal β' but may as well have moved to the interior of the projected Newton polygon NP'_{yz} of f'. If the latter occurs, a new vertex of NP'_{yz} assumes the role of β', i.e, is the inverse lexicographically minimal vertex of NP'_{yz}. In each case we will precise in the concrete setting what is meant. The proof is purely computational and goes case by case.

(1) *Point blowup.* This only occurs when S has an isolated point at a. The argument has inductive character with respect to the components of $i_{a\mathbf{x}}$ and the three charts. We shall show that in all charts $i_{a'\mathbf{x}} < i_{a\mathbf{x}}$ lexicographically.

(a) Behaviour of o_a. First consider the x-chart. A vertex $\alpha = (\alpha_x, \alpha_y, \alpha_z)$ of the Newton polygon NP of f transforms into the point $\alpha^* = (\alpha_x+\alpha_y+\alpha_z-o_a, \alpha_y, \alpha_z)$ which may be a vertex of NP' or not. Take for α an exponent of the tangent cone $\mathrm{hom}_a f$ of f, viz satisfying $|\alpha| = \alpha_x + \alpha_y + \alpha_z = o_a$. Since the coordinates are subordinate to the flag $\mathcal{F}$ which is partially transversal to TC, it follows that $\alpha_x > 0$ for at least one such α. This implies $\alpha_y + \alpha_z < o_a$ for this α and hence $|\alpha^*| < o_a$, say $o_{a'} < o_a$ and $i_{a'\mathbf{x}} < i_{a\mathbf{x}}$ in this chart.

Exercise 7.2. Show that $o_{a'} \leq o_a$ in the y-chart and z-chart. Determine when equality occurs.

(b) Behaviour of β_y. We may assume $o_{a'} = o_a$, which reduces by (a) to the y- and z-chart. In the y-chart we argue as follows. As S has an isolated point at a there exists a vertex δ of NP satisfying $\delta_x + \delta_z < o_a$ (else the curve defined by $x = z = 0$ would lie in S). Projecting δ from $(o_a, 0, 0)$ to $\mathbb{Q}^2_+$ yields the point $(\frac{o_a\delta_y}{o_a-\delta_x}, \frac{o_a\delta_z}{o_a-\delta_x})$ in NP_{yz}. This shows that $\beta = (\beta_y, \beta_z)$ has second component $< o_a$. A short computation gives $\beta^* = (\beta_y + \beta_z - o_a, \beta_z)$, hence β_z remains constant and β^* has again minimal second component among all vertices of the projected polyhedron NP'_{yz}. Therefore $\beta^* = \beta'$. Moreover, since $\beta_z < o_a$, we get $\beta'_y = \beta_y + \beta_z - o_a < \beta_y$. We have shown that $i_{a'\mathbf{x}} < i_{a\mathbf{x}}$ in this chart.

Now consider the z-chart. Here, $\beta^* = (\beta_y, \beta_z + \beta_y - o_a)$. If β^* lies in the interior of the projected polyhedron, it is replaced by a vertex β' with smaller first component because β_y was maximal among all vertices of the projected polyhedron (recall that β_z was minimal). In this case $\beta'_y < \beta_y$. If β^* remains a vertex, it equals necessarily β', and its first component β'_y has remained constant, say $\beta'_y = \beta_y$. In both cases we have $\beta'_y \leq \beta_y$.

(c) Behaviour of $s_{\beta\gamma}$ and $|\beta|$. We may assume that $o_{a'} = o_a$ by (a) and that $\beta^* = \beta'$ by (b). Hence we are left with the z-chart. By definition, the vertex $\gamma = (\gamma_y, \gamma_z)$ satisfies, if it exists, that $\beta_y > \gamma_y$ and $\beta_z < \gamma_z$. The slope $s_{\beta\gamma}$ equals $\frac{\beta_y - \gamma_y}{\beta_z - \gamma_z}$. After blowing up it becomes $\frac{\beta_y - \gamma_y}{\beta_z - \gamma_z + \beta_y - \gamma_y}$. If the denominator is negative the slope has decreased since $\beta_y - \gamma_y > 0$. If it is ≥ 0 the vertex $\gamma^* = (\gamma_y, \gamma_z + \gamma_y - o_a)$ has second component less or equal $\beta^* = (\beta_y, \beta_z + \beta_y - o_a)$ which contradicts $\beta^* = \beta'$. Therefore the slope has dropped. If γ does not exist, $NP_{yz} = \beta + \mathbb{Q}^2_+$ is a quadrant and S can only be an isolated point if both β_z and β_y are $< o_a$. Clearly, NP'_{yz} is then again a quadrant and $|\beta'| = |\beta^*| = \beta_y + \beta^*_z = \beta_y + \beta_z + \beta_y - o_a < |\beta|$. This gives $i_{a'\mathbf{x}} < i_{a\mathbf{x}}$ also in this chart.

(2) *Curve blowup.* Lemma 4.2 implies $\tau_a = 1$. If $\mathbf{P}$ equals (x, y), the x-chart is irrelevant since $X' \cap E_a$ does not pass there. In the y-chart a vertex $(\delta_x, \delta_y, \delta_z)$ transforms into $(\delta_x, \delta_y + \delta_x - o_a, \delta_z)$ which shows that $o_{a'} \leq o_a$. If $o_{a'} = o_a$, the vertex $\beta = (\beta_y, \beta_z)$ moves to $\beta^* = (\beta_y - o_a, \beta_z)$. As the second component remains constant, $\beta^* = \beta'$ and β_y has dropped. If $\mathbf{P}$ equals (x, z) we may restrict, similarly as before, to the z-chart and $o_{a'} = o_a$. Here β moves to $\beta^* = (\beta_y, \beta_z - o_a)$. If $\beta^* = \beta'$ we get $\beta'_y = \beta_y$ and $s_{\beta\gamma} < s_{\beta\gamma}$, provided γ exists. If it does not exist we have $|\beta'| < |\beta|$. If $\beta^* \neq \beta'$, then $\beta'_y < \beta^*_y = \beta_y$. In all cases we conclude by $i_{a'\mathbf{x}} < i_{a\mathbf{x}}$.

(3) *Crossings blowup.* Let $\mathbf{P} = (x, yz)(x, y)(x, z)$. Again we have by Lemma 4.8 that $\tau_a = 1$. In Proposition 3 of Section 4 it was shown that X' meets E_a only in the chart y^2z^2. There the blowup is given by $(x, y, z) \to (xyz, y, z)$ with E and E_a defined by $yz = 0$, respectively $y = z = 0$. Therefore the strict transform of $f = x^{o_a} + \sum c_{ijk}x^iy^jz^k$ is given by $f' = x^{o_a} + \sum c_{ijk}x^iy^{j+i-o_a}z^{k+i-o_a}$. Hence

$o_{a'} \leq o_a$. Let $(\delta_x, \delta_y, \delta_z)$ project to β. By definition of NP_{yz} we have $\delta_x < o_a$. Then $(\delta_x, \delta_y + \delta_x - o_a, \delta_z + \delta_x - o_a)$ projects to β', hence $\beta'_y = \frac{o_a}{o_a - \delta_x} \cdot (\delta_y + \delta_x - o_a) < \frac{o_a}{o_a - \delta_x} \cdot \delta_y = \beta_y$. This gives $i_{a'\mathbf{x}} < i_{a\mathbf{x}}$. Only the first two components of $i_{a\mathbf{x}}$ were used. □

Variation: (3′) Let $\mathbf{P} = (x, yz) \cap (x, y, z)^3$. By Proposition 4.8, only the y^2z- and yz^2-charts are relevant. We show that $i_{a\mathbf{x}}$ may increase. In the y^2z-chart the vertices of NP move according to $(\delta_x, \delta_y, \delta_z) \to (\delta_x, \delta_y + \delta_x + \delta_z - o_a, \delta_z + \delta_x - o_a)$ and if β_z remains minimal, the component β_y may increase to $\beta'_y = \beta_y + \delta_x + \delta_z - o_a$. Observe here that $\delta_x + \delta_z \geq o_a$ since S contains the curve defined by $x = z = 0$. Hence i_a may increase in this chart.

In the yz^2-chart, nevertheless, the invariant decreases. Here a vertex $(\delta_x, \delta_y, \delta_z)$ moves to $(\delta_x, \delta_y + \delta_x - o_a, \delta_z + \delta_x + \delta_y - o_a)$. Consider $\beta = (\beta_y, \beta_z)$ moving to $(\beta_y, \beta_z + \beta_y - o_a)$. If β survives, say $\beta^* = \beta'$, then $\beta_z + \beta_y - o_a$ must again be minimal, and we have $\beta'_y = \beta^*_y = \beta_y - o_a < \beta_y$, hence β_y drops. If β moves to the interior of NP'_{yz} and the new vertex is $\beta' = \delta^*$ for some vertex δ of NP_{yz}, then $\beta^*_z = \beta_y + \beta_z - o_a > \delta^*_z = \delta_y + \delta_z - o_a$ and $\beta_z < \delta_z$. Hence $\beta'_y = \delta^*_y = \delta_y - o_a < \beta_y - o_a < \beta_y$ and β_y drops. □

8. Reduction to monomial blowup

We show that for local blowups $\mathbf{R} \to \mathbf{R}'$ the improvement $i_{a'} < i_a$ follows from the improvement of $i_{a\mathbf{x}}$ under monomial blowup. We work as in Section 6 with the completions of the local rings. We assume that S has at most normal crossings.

Theorem 8.1. *Given $\pi : W' \to W$ the taut blowup of X and $a \in S$ a closed point, let $a' \in W'$ be a closed point over a where the multiplicity $o_{a'}$ of X' has remained constant. Assume chosen a flag $\mathcal{F}$ at a transversal to S and partially transversal to the tangent cone TC of X. There exist formal subordinate coordinates $\mathbf{x}$ at $a \in W$ such that $\overline{\mathbf{R}} \to \overline{\mathbf{R}'}$ is monomial with respect to $\mathbf{x}$, β_z is maximal and such that the induced coordinates in $\overline{\mathbf{R}'}$ maximize β'_z and realize $i_{a'}$ up to the relevant component.*

The various cases of the proof of Theorem 7.1 show that it suffices for the induction on i_a – according to the blowup and the chart considered – to maximize $i_{a'\mathbf{x}}$ only up to the component which drops under the corresponding monomial blowup. This is the meaning of the relevant component in Theorem 8.1. If $o_{a'}$ has dropped, a new flag transversal to the S' and TC' has to be chosen at a' in W'.

Proof. Start with any subordinate coordinates in $\overline{\mathbf{R}}$. By Theorem 5.1, we may apply a subordinate coordinate change in $\overline{\mathbf{R}}$ which makes the blowup monomial and hence moves a' to the origin of one of the charts. For crossings blowup, we have seen already that $X' \cap E$ lies in the y^2z^2-chart, so Theorem 5.1 does apply in this case as well. Denote by $\mathbf{x}$ also the induced coordinates in $\overline{\mathbf{R}'}$.

By Theorem 5.2 and since $o_a = o_{a'}$, the induced flag $\mathcal{F}'$ in $\overline{\mathbf{R}'}$ is again transversal to S'.

We now maximize $(\beta'_z, \beta'_y, s'_{\beta\gamma}, |\beta'|)$ by subordinate coordinate changes in $\overline{\mathbf{R}'}$. By the Gauss-Bruhat decomposition, see [Ha 1], we only have to consider subordinate coordinate changes in $\overline{\mathbf{R}'}$ of type $(x + a(y,z), y + b(z), z)$ to maximize β'_z and $i_{a'\mathbf{x}}$. We shall show that there exists such a change in $\overline{\mathbf{R}'}$ which is induced from subordinate coordinate changes in $\overline{\mathbf{R}}$ and leaves $\overline{\mathbf{R}} \to \overline{\mathbf{R}'}$ monomial. Simultaneously, β_z will be maximized in $\overline{\mathbf{R}}$.

Exercise 8.2. Show that in the z-chart of point blowup the change $v = (x + y^k z^{k-2}, y, z)$ with $k \geq 2$ in $\mathbf{R}'$ is not induced by a change in $\mathbf{R}$, though v fixes the exceptional divisor in this chart.

(1) *Point blowup.* It suffices to consider the y- and z-chart which cover all of $E = E_a$ except the origin of the x-chart where we already know that the multiplicity drops. In the y-chart we will maximize (β'_z, β'_y) lexicographically and can discard the remaining components by the proof of Theorem 7.1. It is immediate from the definition that only changes $(x + a(y,z), y, z)$ can alter β'. Now observe that in the induced coordinates in $\mathbf{R}'$, NP'_{yz} is contained in the cone $\{(j,k) \in \mathbb{Q}^2_+,\ j \geq k - o_a\}$ because points (j,k) of NP_{yz} move under monomial blowup in this chart to $(j + k - o_a, k)$. Set $a(y,z) = \sum a_{jk} y^j z^k$. If $v = (x + a(y,z), y, z)$ maximizes (β'_z, β'_y) the sum in $a(y,z)$ can be chosen over pairs (j,k) for which $j \geq k - 1$, since we wish to eliminate in f' only monomials whose projection lies in NP'_{yz}. This can be seen inductively, first eliminating all monomials in the expansion of f' whose projection gives β', and then starting again with the new β'.

Exercise 8.3. Write down this argument in all details and compare it with Lemma 3(b) of [Ha 1]. Note that it is characteristic independent.

Once we can restrict to changes of this form, it is immediate to see that v is induced from the subordinate change $u = (x + \sum_{j \geq k-1} a_{jk} y^{j-(k-1)} z^k, y, z)$ in $\overline{\mathbf{R}}$ and that this change preserves the monomiality of the blowup (use the constancy of multiplicity and $j + k \geq 1$. If a change $(x + y, y, z)$ would be needed to maximize β_z, o_a would have dropped at a').

Exercise 8.4. Show that since v maximizes β'_z, u maximizes β_z.

We next treat the z-chart, where all components of i_a are relevant. To maximize (β'_z, β'_y) in $\overline{\mathbf{R}'}$ and β_z in $\overline{\mathbf{R}}$ the argument is the same as before with the role of y and z interchanged. To maximize $s'_{\beta\gamma}$ in $\overline{\mathbf{R}'}$ we will need more general changes of type $v = (x + a(y,z), y + b(z), z)$. These are products of changes of type $(x + a(y,z), y, z)$ and $(x, y + b(z), z)$. For the first, the preceding reasoning applies again, since we may require that v does not create new monomials outside NP'_{yz}. For the second, it suffices to observe that changes $(x, y + b(z), z)$ are induced under point blowup in the z-chart from changes in $\overline{\mathbf{R}}$ which have the same coordinate expression.

(2) *Curve blowup.* For $\mathbf{P} = (x, y)$ and $\mathbf{P} = (x, z)$ it suffices to consider the y- and z-chart respectively, and to restrict to maximizing (β'_z, β'_y). It is then straightforward

to see that any change of type $(x+a(y,z),y,z)$ in $\overline{\mathbf{R}}'$ is induced from a subordinate change in $\overline{\mathbf{R}}$ which preserves monomiality of the blowup and that β_z is maximal if β'_z is maximal.

(3) *Crossings blowup.* Only the y^2z^2-chart and β' are relevant. The monomial substitution $\overline{\mathbf{R}} \to \overline{\mathbf{R}}'$ is given by $(x,y,z) \to (xyz,y,z)$. Therefore any change of type $(x+a(y,z),y,z)$ in $\overline{\mathbf{R}}'$ is induced from a subordinate change in $\overline{\mathbf{R}}$ which preserves monomiality of the blowup. Check again that β_z is maximal if β'_z is maximal. This concludes the proof of the theorem. □

9. Proof of Theorem 1.1

Assume given a singular reduced surface X in a regular three-dimensional ambient scheme W. We show first that finitely many taut blowups provide a smooth strict transform of X. Further blowups will then make the intersection with the exceptional divisor transversal, i.e. will produce a total transform of X with normal crossings.

By Proposition 4.5 we may assume, applying a finite number of point blowups, that the equimultiple locus S of X consists of finitely many isolated points and finitely many smooth curves which have at most normal crossings. Let $\pi : W' \to W$ be the taut blowup of X with center Z as defined in stage 1 of Section 3. It yields the strict transform X' of X and an exceptional divisor E in W'. Choose at each point a of S a flag $\mathcal{F}$ transversal to S and partially transversal to the tangent cone TC of X.

Since it is not clear whether i_a takes finitely many values on S we cannot argue by stratifying S according to the invariant. Stratifications only work well for invariants which are semi-continuous.

Instead, we show first that any smooth component C of S dissolves after a finite number of taut blowups into finitely many isolated points (possibly none) with the same multiplicity. By Propositions 4.6 and 4.7 we know that if the component persists after one blowup with the same multiplicity, no new component has appeared and the components C and C' are isomorphic under π. Assume that this happens infinitely many times, producing an infinite sequence of components C, C', C'', ... of the equimultiple loci of the strict transforms of X. Choose in each component $C^{(k)}$ a point $a^{(k)}$ over a where the invariant takes its minimal value on $C^{(k)}$. Such points exist because i_a varies in a well ordered set. By Theorems 7.1 and 8.1 the sequence $i_{a^{(k)}}$ strictly decreases yielding the contradiction.

Consider now an isolated point a of S. Its fibre $E_a \cap X'$ under π may be a component of S', necessarily isomorphic to $\mathbb{P}^1$. It is disjoint from the other components, cf. Proposition 4.5. By the preceding observation, finitely many further blowups decompose it into finitely many points of the same multiplicity as a. Using again Theorems 7.1 and 8.1 we know that for each of these points a' we have $i_{a'} < i_a$.

It follows that after finitely many blowups, the multiplicity must drop at each point a' over any point a of S. Hence the maximal multiplicity occuring on the surface has dropped and finitely many taut blowups yield as *strict transform* of X a smooth surface.

To achieve normal crossings of the *total transform* of X, we may consider from the beginning a finite union $X = \cup_{i=0}^k X_i$ of smooth surfaces such that $Y = \cup_{i=1}^k X_i$ has normal crossings. Here, Y corresponds to the exceptional divisor accumulated so far by the prior blowups, and X_0 to the smooth strict transform of the singular scheme we started with. A finite sequence of taut blowups will yield a smooth strict transform of $\cup_{i=0}^k X_i$. As the X_i were smooth, its components will meet each exceptional component transversally. However, the total transform need not have normal crossings yet. To achieve normal crossings by further blowups is now a combinatorial problem which will be left to the reader in form of the following

Exercise 9.1. Draw real pictures of the five singularities defined in three-space by $xy(x+y)$, $x(x+y)(y+z^2)$, $xy(x+y)z$, $x(x+y)(y+z^2)z$ and $xyz(x+y+z)$. Determine for each of them a center of blowup such that the total transform has normal crossings. Use this to show that the total transform of $\cup_{i=0}^k X_i$ from above can be made into normal crossings.

Combining the two stages of taut blowups as described above establishes the embedded resolution of our original singular scheme X in W. This completes the proof of Theorem 1.1 and concludes the paper. □

References

[Ab 1] Abhyankar, S.: Desingularization of plane curves. Proc. Symp. Pure Appl. Math. **40**, Amer. Math. Soc. 1983.

[Ab 2] Abhyankar, S.: *Resolution of singularities of embedded algebraic surfaces.* Acad. Press 1966.

[Bn] Bennett, B.-M.: On the characteristic function of a local ring. Ann. Math. **91** (1970), 25–87.

[BM 1] Bierstone, E., Milman, P.: Canonical desingularization in characteristic zero by blowing up the maximum strata of a local invariant. Invent. Math. **128** (1997), 207–302.

[CGO] Cossart, V., Giraud, J., Orbanz, U.: *Resolution of surface singularities.* Springer Lecture Notes in Math. vol 1101.

[Co 1] Cossart, V.: Desingularization of embedded excellent surfaces. Tôhoku Math. J. **33** (1981), 25–33.

[Co 2] Cossart, V.: Uniformisation et désingularisation des surfaces d'après Zariski. This volume.

[E-V 1] Encinas, S., Villamayor, O.: Good points and constructive resolution of singularities. Acta Math. **181** (1998), 109–158.

[E-V 2] Encinas, S., Villamayor, O.: A course on constructive desingularization and equivariance. This volume.

[Gi] Giraud, J.: Etude locale des singularités. Cours de 3e Cycle, Orsay 1971/72.

[Ha 1] Hauser, H.: Resolution techniques. Preprint 1999.

[Ha 2] Hauser, H.: Seventeen obstacles for resolution of singularities. In: *Singularities*, The Brieskorn Anniversary Volume (eds: V. I. Arnold, G.-M. Greuel, J. Steenbrink), Birkhäuser 1998.

[Ha 3] Hauser, H.: Tetrahedra, prismas and triangles. Preprint 1998.

[Hi 1] Hironaka, H.: Desingularization of excellent surfaces. Notes by B. Bennett at the Conference on Algebraic Geometry, Bowdoin 1967. Reprinted in: Cossart, V., Giraud, J., Orbanz, U.: *Resolution of surface singularities.* Lecture Notes in Math. 1101, Springer 1984.

[Hi 2] Hironaka, H.: Resolution of singularities of an algebraic variety over a field of characteristic zero. Ann. of Math. **79** (1964), 109–326.

[Hi 3] Hironaka, H.: Idealistic exponents of singularity. In: *Algebraic Geometry*, the Johns Hopkins Centennial Lectures. Johns Hopkins University Press 1977.

[Hi 4] Hironaka, H.: Additive groups associated with points of a projective space. Ann. Math. **92** (1970), 327–334.

[Le] Lê, D.T.: Les singularités sandwich. This volume.

[Lp] Lipman, J.: Desingularization of 2-dimensional schemes. Ann. Math. **107** (1978), 151–207.

[Mo] Moh, T.T.: On a Newton polygon approach to the uniformization of singularities in characteristic p. In: *Algebraic Geometry and Singularities* (eds.: A. Campillo, L. Narváez). Proc. Conf. on Singularities La Rábida. Birkhäuser 1996.

[O] Orbanz, U.: Embedded resolution of algebraic surfaces after Abhyankar (characteristic 0). In: Cossart, V., Giraud, J., Orbanz, U.: *Resolution of surface singularities.* Lecture Notes in Math. 1101, Springer 1984.

[P] Pfeifle, J.: Das Aufblasen algebraischer Varietäten und monomialer Ideale. Masters thesis, Innsbruck 1997.

[Rg] Regensburger, G.: Die Auflösung von ebenen Kurvensingularitäten. Masters theses, Innsbruck 1999.

[R] Rosenberg, J.: Blowing up nonreduced toric subschemes of $\mathbb{A}^n$. Preprint 1998.

[Sp 1] Spivakovsky, M.: A counterexample to the theorem of Beppo Levi in three dimensions. Invent. Math. **96** (1989), 181–183.

[Sp 2] Spivakovsky, M.: Resolution of singularities. Preprint 1997.

[Za 1] Zariski, O.: Reduction of singularities of algebraic three dimensional varieties. Ann. Math. **45** (1944), 472–542.

[Za 2] Zariski, O.: The concept of a simple point of an abstract algebraic variety. Trans. Amer. Math. Soc. **62** (1947), 1–52.

Institut für Mathematik
Universität Innsbruck
A-6020 Austria
herwig.hauser@uibk.ac.at

Progress in Mathematics, Vol. 181, © 2000 Birkhäuser Verlag Basel/Switzerland

An Application of Alterations to Dieudonné Modules

A. Johan de Jong

Introduction

In this note we indicate a non-cohomological application of the results of [2] and [3]. The application concerns the crystalline Dieudonné module functor.

Let S be a scheme of characteristic $p > 0$. A *Barsotti-Tate group,* or *p-divisible group G over S* is a sheaf of abelian groups on the $fppf$-site of S which is p-divisible, p-power torsion, and such that $G[p]$ is representable by a finite locally free group scheme over S. The basic example is the p^∞-torsion of an abelian scheme over S, i.e., $G = A[p^\infty]$. It is very analogous to the ℓ-adic Tate module of A, which can be seen as a $\mathbb{Z}_\ell$-sheaf on the étale site of S. The category of p-divisible groups over S is denoted BT_S. For more information on p-divisible groups, see [6], [7], and [8].

The crystalline Dieudonné module functor transforms a p-divisible group over S into a *Dieudonné crystal over S*. Let us explain what a crystal over $S = \operatorname{Spec} \mathbb{F}_p[X]$ amounts to: It is given by a quadruple (M, ∇, F, V), where

- M is a finite free $\mathbb{Z}_p\{X\}$-module, where $\mathbb{Z}_p\{X\}$ is the p-adic completion of $\mathbb{Z}_p[X]$,
- ∇ is a connection $M \to M \otimes \widehat{\Omega}^1_{\mathbb{Z}_p\{X\}}$ (continuous differentials),
- F is a σ-linear map $M \to M$, i.e., a linear map $M \otimes_\sigma \mathbb{Z}_p\{X\}$, where $\sigma : \mathbb{Z}_p\{X\} \to \mathbb{Z}_p\{X\}$ is the ring map such that $X \mapsto X^p$, and
- V is a linear map $M \to M \otimes_\sigma \mathbb{Z}_p\{X\}$.

There are conditions, expressing that F anf V are compatible with ∇, and that $FV = p$ and $VF = p$. For general schemes S, the category of Dieudonné crystals can be described in a similar, but more involved manner, see [9], [10] and [4, 2.3.4] for example. The category of Dieudonné crystals over S is denoted DC_S. An important point is that, even though Dieudonné crystals are hard to describe and hard to get hold of, they are easier to deal with than p-divisible groups. Also, to follow the gist of this note, the reader does not need to understand precisely what such a crystal amounts to.

We recall the crystalline Dieudonné module functor

$$\mathbb{D} : \mathrm{BT}_S \longrightarrow \mathrm{DC}_S$$

from the category of Barsotti-Tate groups to the category of Dieudonné crystals over S. The relation between Barsotti-Tate groups and modules with connections evidenced by $\mathbb{D}$ is analogous to the Riemann-Hilbert correspondence between perverse sheaves and $\mathcal{D}$-modules in characteristic 0. We do not give the definition of $\mathbb{D}$ here, which is delicate and not easy to explain. For references we suggest the original works [7], [8]; here Grothedieck raises several questions concerning the faithfulness and surjectivity of $\mathbb{D}$. For a definition of $\mathbb{D}$ that is currently in vogue we refer to [11]. There has been a lot of work on this functor, we refer to the introduction of [4] for some history and results. See [12] for an application of $\mathbb{D}$ to homomorphisms of p-divisible groups and Abelian varieties.

1. The application of alterations

Let k be a field with a finite p-basis and let $S \to \operatorname{Spec} k$ be a morphism of finite type. In this case it was proved in [4] that

(i) $\mathbb{D}$ is an equivalence if S is regular [4, Main Theorem 1], and

(ii) $\mathbb{D}$ is fully faithful up to isogeny for any S [4, Main Theorem 2].

The proof of (ii) in [4] was quite complicated and necessitated proving some non-trivial results on formal and rigid geometry. In this note we will sketch how to deduce (ii) from (i) using the alterations theorem [2, Theorem 4.1]. The proof given below still uses some of the rigid analytic techniques of [4, Chapters 5, 6 & 7], but they will be isolated in the proof of the lemma below. In addition, at the end of this note we discuss the question whether the functor $\mathbb{D}$ might even be essentially surjective up to isogeny.

2. Full faithfulness up to isogeny

Let S of finite type over k as above be given. Let $G_1, G_2 \in \mathrm{BT}_S$ and let $\psi : \mathbb{D}(G_2) \to \mathbb{D}(G_1)$ be a morphism of the associated Dieudonné crystals. We have to show that there is an integer $n \in \mathbb{N}$ and a morphism $\varphi : G_1 \to G_2$ such that $\mathbb{D}(\varphi) - p^n\psi$ is torsion in $\operatorname{Hom}(\mathbb{D}(G_2), \mathbb{D}(G_1))$.

Lemma 2.1. *Suppose we have n and $\varphi : G_1 \to G_2$ such that for all points s of S the map $\mathbb{D}(\varphi_s) - p^n\psi|_{\mathrm{CRIS}(s/\Sigma)}$ is zero. Then $\mathbb{D}(\varphi) - p^n\psi$ is torsion.*

Proof. (Sketch.) The idea is that the assumption implies that the induced map $\mathbb{D}(\varphi)^{rig} - p^n\psi^{rig}$ of convergent F-isocrystals is zero. It is known that this implies that $\mathbb{D}(\varphi) - p^n\psi$ is torsion, see [1, Theorem 2.4.2]. See also [4, Section 5.5] where a similar argument is given. ○

Remark. A more general lemma will be proved in the forthcoming paper [5]. In [5] it is proven that the hypothesis on k (finite p-basis) is actually superfluous for the results of [4].

We return to the problem of finding n and φ. We remark that $\mathbb{D}$ is faithful over any reasonable scheme. We also remark that we may replace S by its reduction; the category of p-divisible groups up to isogeny is unchanged by this, and similar for the category of Dieudonné crystals (see [4, proof of 5.1.2]).

Thus we may assume that S is reduced. Let $\pi : S' \to S$ be the normalization mapping (i.e., S' is the disjoint union of the normalizations of the irreducible components of S). Let $i : T \to S$ be a proper closed subscheme of S and let $i' : T' \to S'$ be the closed immersion of $T' = \pi^{-1}(T)$ into S'. There exists a T such that the following diagram of sheaves on S is cartesian (i.e., it is a fibre product diagram):

$$\begin{array}{ccc} \mathcal{O}_S & \longrightarrow & i_*\mathcal{O}_T \\ \downarrow & & \downarrow \\ \pi_*\mathcal{O}_{S'} & \longrightarrow & \pi_*i'_*\mathcal{O}_{T'} \end{array}$$

Exercise. Construct such a T.

Assume, for the moment, that (ii) is proved in the case of normal schemes of finite type over k. Then we can find n' and $\varphi' : G_1 \times_S S' \to G_2 \times_S S'$ such that $\mathbb{D}(\varphi') - p^{n'}\psi|_{\mathrm{CRIS}(S'/\Sigma)}$ is torsion. By Noetherian induction we have n_T and $\varphi_T : G_{1,T} \to G_{2,T}$ adapted to $\psi|_{\mathrm{CRIS}(T/\Sigma)}$. Let n be the maximum of n' and n_T. By faithfulness of $\mathbb{D}$ we get that $p^{n-n'}\varphi'|_{T'} = p^{n-n_T}\varphi_T|_{T'}$.

The cartesian diagram above will imply (see below) that $p^{n-n'}\varphi'$ and $p^{n-n_T}\varphi_T$ "glue" to a morphism $\varphi : G_1 \to G_2$. Once this is established, it follows from the lemma that $p^n\psi - \mathbb{D}(\varphi)$ is torsion.

Let us write $\mathcal{A}_{i,N}$ for the locally free sheaf of $\mathcal{O}_S$-modules (or Hopf algebras) corresponding to the finite flat group scheme $G_i[p^N]$ over S. There are maps

$$\mathcal{A}_{2,N} \otimes \mathcal{O}_T \longrightarrow \mathcal{A}_{1,N} \otimes \mathcal{O}_T$$

induced by $p^{n-n_T}\varphi_T$ acting on $G_{1,T}[p^N]$ and maps

$$\mathcal{A}_{2,N} \otimes \mathcal{O}_{S'} \longrightarrow \mathcal{A}_{1,N} \otimes \mathcal{O}_{S'}$$

induced by $p^{n-n'}\varphi$ acting on $G_{1,S'}[p^N]$. These agree as maps on $\mathcal{A}_{2,N} \otimes \mathcal{O}_{T'}$ and by the cartesian diagram (which stays cartesian after tensoring with the locally free sheaf $\mathcal{A}_{i,N}$) this induces maps $\mathcal{A}_{2,N} \to \mathcal{A}_{1,N}$. This is a compatible system of maps of Hopf algebras giving a morphism $\varphi : G_1 \to G_2$ as desired.

In this manner we have reduced our problem to the case of integral normal schemes S of finite type over k. Let $a : S' \to S$ be an alteration with S' regular [2, Theorem 4.1]. By (i) [4, Main Theorem 1] we get a morphism $\varphi' : G_1 \times_S S' \to G_2 \times_S S'$ corresponding to $\psi|_{\mathrm{CRIS}(S'/\Sigma)}$. On the other hand, let $U \subset S$ denote the open dense regular locus of the scheme S. Again by (i) there is a morphism $\varphi_U : G_{1,U} \to G_{2,U}$ corresponding to the restriction of ψ to U. By functoriality we have $a^*\varphi_U = \varphi'|_{a^{-1}(U)}$.

We claim that this implies that there is a unique $\varphi : G_1 \to G_2$ over S that restricts back to φ_U and φ'. This is proven in exactly the same way as above using the following cartesian diagram of sheaves on S:

$$\begin{array}{ccc} \mathcal{O}_S & \longrightarrow & j_*\mathcal{O}_U \\ \downarrow & & \downarrow \\ a_*\mathcal{O}_{S'} & \longrightarrow & a_*j'_*\mathcal{O}_{a^{-1}(U)} \end{array}$$

Notations used: $U \xrightarrow{j} S$ inclusion, $a^{-1}(U) \xrightarrow{j'} S'$ inclusion.

Exercise. Prove that the diagram is cartesian. (Hint: use that S is normal.)

3. Essential surjectivity up to isogeny

We can try to prove more. One can ask whether $\mathbb{D}$ is essentially surjective up to isogeny for certain S.

A more precise question is the following: given an object $\mathcal{E} \in \mathrm{DC}_S$ does there exist an $n \in \mathbb{N}$ and a $G \in \mathrm{BT}_S$ such that $\mathbb{D}(G) \cong \mathcal{E}^{(n)}$? Here $\mathcal{E}^{(n)}$ is the pull back of $\mathcal{E}$ under the nth iterate of the absolute Frobenius morphism of S. This formulation is suggested by the following two facts: (1) any Dieudonné crystal $\mathcal{E}$ is isogenous to $\mathcal{E}^{(1)}$, and (2) the question has a positive answer if S is the spectrum of an Artinian ring. The result (2) is a consequence of the fact that if $(A, \mathfrak{m}, k)$ is Artinian local of characteristic p, then for some n the nth iterate of Frobenius factors as $A \to k \to A$ (and the fact that $\mathbb{D}$ is an equivalence over k, see [4, Main Theorem 1] and the remark following the lemma above).

Let us consider an integral normal scheme S separated and of finite type over the field k with a finite p-basis. Let $U \subset S$ be the dense open regular subscheme of S and let $a : S' \to S$ be an alteration with S' regular [2, Theorem 4.1]. Let $\mathcal{E}$ be a Dieudonné crystal over S. Using (i) we deduce the existence of $G_U \in \mathrm{BT}_U$ corresponding to $\mathcal{E}|_U$ and $G' \in \mathrm{BT}_{S'}$ corresponding to $\mathcal{E}|'_S$ which have isomorphic restrictions over $a^{-1}(U)$. Furthermore, for every point $s \in S$ the restriction of G' to the reduced fibre $T := a^{-1}(s)_{red}$ is constant. This follows from [4, 5.1.1] and the fact that $\mathcal{E}|_{\mathrm{CRIS}(T/\Sigma)}$ is constant (being the pull back of $\mathcal{E}_s = \mathcal{E}|_{\mathrm{CRIS}(s/\Sigma)}$). We want to prove that this implies that $(G')^{(n)}$ descends to a p-divisible group G over S for some n. If we consider the sheaves of Hopf algebras associated to $G'[p^N]$ and $G_U[p^N]$ then we see that we get a triple as described below. Thus we want to know for which special situations $S' \to S \supset U$ the answer to the question below is positive.

Let S be an integral scheme of characteristic p, let $U \subset S$ be open dense, and let $a : S' \to S$ be an alteration. We consider triples $(\mathcal{F}', \mathcal{F}_U, \alpha)$, where

(a) $\mathcal{F}'$ is a locally free $\mathcal{O}_{S'}$-module,

(b) $\mathcal{F}_U$ is a locally free $\mathcal{O}_U$-module,

(c) $\alpha : a^*\mathcal{F}_U \to \mathcal{F}|_{a^{-1}(U)}$ is an isomorphism, and

(d) for all $s \in S$ the restriction of $\mathcal{F}'$ to the scheme $T := a^{-1}(s)_{red}$ is free (i.e., $\mathcal{F}'|_T \cong \mathcal{O}_T^m$ for some m).

In the following we use F to denote the Frobenius morphism of both S and S'.

Question. Under what conditions on S, $a : S' \to S$ and U does there exist an $n \in \mathbb{N}$ with the following property: for every triple $(\mathcal{F}', \mathcal{F}_U, \alpha)$ as above there exists a locally free sheaf $\mathcal{F}$ on S such that

$$a^*\mathcal{F} \cong (F^n)^*(\mathcal{F}') \text{ and } \mathcal{F}|_U \cong (F^n)^*\mathcal{F}_U \text{ ?}$$

Remark.

(a) In general we cannot find such an integer n. For example, when $p \neq 2$, take $a : S' = \mathrm{Spec}(k[x,y]) \to S = \mathrm{Spec}(k[x^2, xy, y^2])$ and $U = S \setminus \{(x^2, xy, y^2)\}$ to get a counter example. Indeed, take $\mathcal{F}' = \mathcal{O}_{S'}$, let $\mathcal{G}$ be the coherent sheaf on S corresponding to the ideal (x^2, xy) in the ring $k[x^2, xy, y^2]$, and let $\mathcal{F}_U = \mathcal{G}|_U$. Then $\mathcal{F}'$ and $\mathcal{F}_U$ are locally free of rank 1, and we leave it as an exercise to show the existence of a map α. We also leave it as an exercise to show that $\mathcal{F}_U^{\otimes n} \cong \mathcal{F}_U$ is n is odd: the Picard group of U has order 2 and $\mathcal{F}_U$ is a generator. Finally, note that $\mathcal{G} = j_*\mathcal{F}_U$. Hence for every extension of $(F^n)^*(\mathcal{F}_U) = \mathcal{F}_U^{\otimes p^n} \cong \mathcal{F}_U$ to a locally free sheaf $\mathcal{F}$ on S we have $\mathcal{F} \cong j_*(\mathcal{F}|_U) \cong j_*(\mathcal{F}_U) \cong \mathcal{G}$. As $\mathcal{G}$ is not free, $\mathcal{F}$ is not free, a contradiction.

(b) The answer to the question is yes if $S' \to S$ is the resolution of singularities of a normal surface in characteristic p, this the reader can prove by working through the set of exercises at the end of this note. Thus it seems quite likely that the above considerations will lead to a proof of essential surjectivity up to isogeny for the functor $\mathbb{D}$ in the case of normal surfaces of finite type over k.

(c) Using (b) we can try to bootstrap (e.g. by trying to use (b) to prove the existence of a G outside of codimension 3 and then working on the question above in the case where $S \setminus U$ has codimension at least 3). We do not want to get into this here, but this demonstrates a general principle in trying to apply the alteration theorems: some work remains relating upstairs to downstairs. This work usually involves a trace map (in the case of cohomological applications) or some kind of descent question (like the question above).

Exercise. Let $a : S' \to S$ be the resolution of singularities of a normal surface of finite type over a field of characeristic p. Let $U = S^{\mathrm{reg}}$ be the regular locus of S. Prove that there exists an $n \in \mathbb{N}$ as in the question by completing the following exercises:

(i) Show that it suffices to prove this when S has a unique singular point $s \in S$.

(ii) Let $E \subset S'$ be the reduced exceptional fibre, and let $I \subset \mathcal{O}_{S'}$ be its ideal sheaf. Show that there exists a natural number $n \in \mathbb{N}$ such that for all $m \geq n$ we have $H^1(E, I^m/I^{m+1}) = (0)$.

(iii) Let E_n be the nth infinitesimal neigbourhood of E in S', i.e., the closed subscheme defined by I^{n+1}. Show that the nth iterate of the Frobenius morphism of E_n factorizes as $E_n \to E \to E_n$, where $E \to E_n$ is the natural closed immersion.

(iv) Let n be as in (ii) and $\mathcal{F}'$ be a locally free sheaf of rank r over S' such that $\mathcal{F}'|_{E_n} \cong \mathcal{O}^r_{E_n}$. Show that $a_*\mathcal{F}'$ is locally free, and that $\mathcal{F}' \cong a^*a_*\mathcal{F}'$.

(v) Show that the integer n found in (ii) works. Hint: apply (iv) to the sheaf $(F^n)^*(\mathcal{F}')$.

References

[1] P. Berthelot, *Cohomologie rigide et cohomologie rigide a support propre*, Version provisoire du 9-08-1991.

[2] A.J. de Jong, *Smoothness, semi-stability and alterations*, Publ. Math. I.H.E.S. **83**, 51–93 (1996).

[3] A.J. de Jong, *Families of curves and alterations,* Ann. de l'Institut Fourier **47**, 599–621 (1997).

[4] A.J. de Jong, *Crystalline Dieudonné module theory via formal and rigid geometry,* Publ. Math. I.H.E.S. **82**, 5–96 (1995).

[5] A.J. de Jong and W. Messing, *Crystalline Dieudonné Theory over excellent schemes,* Bull. Soc. math. France **127**, 1999, pp. 333–348.

[6] J. Tate, *p-divisible groups*, Proceedings of a conference on local fields, Driebergen (1966), Springer-Verlag.

[7] A. Grothendieck, *Groupes de Barsotti-Tate et cristaux*, Actes Congrès Intern. Math. 1970. Tome 1, pp. 431–436.

[8] A. Grothendieck, *Groupes de Barsotti-Tate et cristaux de Dieudonné*, Séminaire de mathématiques supérieures, Université de Montréal, Les Presses de l'Université de Montréal, 1974.

[9] P. Berthelot, *Cohomologie cristalline des schémas de charactéristique* $p > 0$, Lecture Notes in Math. **407**, Springer 1974.

[10] P. Berthelot and A. Ogus, *Notes on crystalline cohomology,* Princeton University Press, 1978.

[11] P. Berthelot, L. Breen, W. Messing, *Théorie de Dieudonné cristalline II*, Lecture Notes in Math. **930**, Springer 1982.

[12] A.J. de Jong, *Homomorphisms of Barsotti-Tate groups and crystals in positive characteristic*, Inventiones Mathematicae **134**, 301–333 (1998).

Department of Mathematics
Massachusetts Institute of Technology
77 Massachusetts Avenue
Cambridge, MA 02139-4307 USA

Progress in Mathematics, Vol. 181,

Valuation Theoretic and Model Theoretic Aspects of Local Uniformization

Franz-Viktor Kuhlmann

1. Introduction

In this paper, I will take you on an excursion from Algebraic Geometry through Valuation Theory to Model Theoretic Algebra, and back. If our destination sounds too exotic for you, you may jump off at the Old World (Valuation Theory) and divert yourself with problems and examples until you catch our plane back home.

As a preparation for the foreign countries of the Old World, I recommend to read the first sections of the paper [V]. You may also look at the basic facts about valuations in the books [ZS], [R1], [EN], [WA2], [JA]. All other equipment will be distributed on the excursion. I do not recommend that you try to read about the exotic countries of model theory. An introduction will be given on the excursion. If you want to read more, you may ask me for a copy of the chapter "Introduction to model theoretic algebra" of the forthcoming book [K2], and later you may look at books like [CK], [CH1], [HO], [SA].

Since I want our excursion (and myself) to be as relaxed as possible, all varieties we meet will be assumed to be affine and irreducible (wherever it matters). But certainly, we do not assume them to be non-singular, nor do we assume that the base fields be algebraically closed or the characteristic be 0.

2. What does local uniformization mean?

Here is the bitter truth of mankind: In most cases we average human beings are too stupid to solve our problems globally. So we try to solve them locally. And if we are clever enough (and truly interested), we then may think of patching the local solutions together to obtain a global solution.

What is the problem we are considering here? It is the fact that an algebraic variety has singularities, and we want to get rid of them. That is, we are looking for a second variety having the same function field, and having no singularities. This would be the global solution of our problem. As we are too stupid for it, we are first looking for a local solution. Naively speaking, "local" means something like "at a point of the variety". So local solution would mean that we get rid of one singular point. We are looking for a new variety where our point becomes non-singular. But wait, this was nonsense. Because what is our old, singular point on the new variety? We cannot talk of the same points of two different varieties, unless we deal with subvarieties. But passing from varieties to subvarieties or vice versa will in general not provide the solution we are looking for. So do we have to forget about local solutions of our problem?

The answer is: no. Let us have a closer look at our notion of "point". Assume our variety V is given by polynomials $f_1, \ldots, f_n \in K[X_1, \ldots, X_\ell]$. Naively, by a point of V we then mean an ℓ-tupel $(a_1, \ldots, a_\ell)$ of elements in an arbitrary extension field L of K such that $f_i(a_1, \ldots, a_\ell) = 0$ for $1 \leq i \leq n$. This means that the kernel of the "evaluation homomorphism" $K[X_1, \ldots, X_\ell] \to L$ defined by $X_i \mapsto a_i$ contains the ideal $(f_1, \ldots, f_n)$. So it induces a homomorphism η from the coordinate ring $K[V] = K[X_1, \ldots, X_\ell]/(f_1, \ldots, f_n)$ into L over K. (The latter means that it leaves the elements of K fixed.) However, if $a'_1, \ldots, a'_\ell \in L'$ are such that $a_i \mapsto a'_i$ induces an isomorphism from $K(a_1, \ldots, a_\ell)$ onto $K(a'_1, \ldots, a'_\ell)$, then we would like to consider $(a_1, \ldots, a_\ell)$ and $(a'_1, \ldots, a'_\ell)$ as the same point of V. That is, we are only interested in η up to composition $\sigma \circ \eta$ with isomorphisms σ. This we can get by considering the kernel of η instead of η. This leads us to the modern approach: to view a point as a prime ideal of the coordinate ring.

But I wouldn't have told you all this if I intended to follow this modern approach. Instead, I want to build on the picture of homomorphisms. So I ask you to accept temporarily the convention that a point of V is a homomorphism of $K[V]$ over K (i.e., leaving K elementwise invariant), modulo composition with isomorphisms. Recall that $K[V] = K[x_1, \ldots, x_\ell]$, where x_i is the image of X_i under the canonical epimorphism $K[X_1, \ldots, X_\ell] \to K[X_1, \ldots, X_\ell]/(f_1, \ldots, f_n) = K[V]$. The function field $K(V)$ of V is the quotient field $K(x_1, \ldots, x_\ell)$ of $K[V]$. It is generated by $x_1, \ldots, x_\ell$ over K, hence it is finitely generated. Every finite extension of a field K of transcendence degree at least 1 is called an *algebraic function field* (over K), and it is in fact the function field of a suitable variety defined over K. When we talk of function fields in this paper, we will always mean algebraic function fields.

Now recall what it means to look for another variety V' having the same function field $F := K(V)$ as V (i.e., being birationally equivalent to V). It just means to look for another set of generators $y_1, \ldots, y_k$ of F over K. Now the points of V' are the homomorphisms of $K[y_1, \ldots, y_k]$ over K, modulo composition with isomorphisms. But in general, $y_1, \ldots, y_k$ will not lie in $K[x_1, \ldots, x_\ell]$, hence we do not see how a given homomorphism of $K[x_1, \ldots, x_\ell]$ could determine a homomorphism of $K[y_1, \ldots, y_k]$. But if we could extend the homomorphism of $K[x_1, \ldots, x_\ell]$ to all of $K(x_1, \ldots, x_\ell)$, then this extension would assign values to every element of $K[y_1, \ldots, y_k]$. Let us give a very simple example.

Example 1. Consider the coordinate ring $K[x]$ of $V = \mathbb{A}^1_K$. That is, x is transcendental over K, and the function field $K(V)$ is just the rational function field $K(x)$ over K. A homomorphism of the polynomial ring $K[V] = K[x]$ is just given by "evaluating" every polynomial $g(x)$ at $x = a$. I have seen many people who suffered in school from the fact that one can also try to evaluate rational functions $g(x)/h(x)$. The obstruction is that a could be a zero of h, and what do we get then by evaluating $1/h(x)$ at a? (In fact, if our homomorphism is not an embedding, i.e., if a is not transcendental over K, then there will always be a polynomial h over K having a as a root.) So we have to accept that the evaluation will not only render elements in $K(a)$, but also the element ∞, in which case we say that the evaluated rational function has a pole at a. So we can extend our homomorphism to a map P on all of $K(x)$, taking into the bargain that it may not always render finite values. But on the subring $\mathcal{O}_P = \{g(x)/h(x) \mid h(a) \neq 0\}$ of $K(x)$ on which P is finite, it is still a homomorphism.

What we have in front of our eyes in this example is one of the two basic classical examples for the concept of a *place*. (The other one, the p-adic place, comes from number theory.) Traditionally, the application of a place P is written in the form $g \mapsto gP$, where instead of gP also $g(P)$ was used in the beginning, reminding of the fact that P originated from an evaluation homomorphism. If you translate the German "g an der Stelle a auswerten" literally, you get "evaluate g at the place a", which explains the origin of the word "place".

Associated to a place P is its *valuation ring* $\mathcal{O}_P$, the maximal subring on which P is finite, and a valuation v_P. In our case, the value $v_P(g/h)$ is determined by computing the zero or pole order of g/h (pole orders taken to be negative integers). In this way, we obtain values in $\mathbb{Z}$, which is the value group of v_P. In general, given a field L with place P and associated valuation v_P, the valuation ring $\mathcal{O}_P = \{b \in L \mid bP \neq \infty\} = \{b \in L \mid v_P b \geq 0\}$ has a unique maximal ideal $\mathcal{M}_P = \{b \in L \mid bP = 0\} = \{b \in L \mid v_P b > 0\}$. The *residue field* is $LP := \mathcal{O}_P/\mathcal{M}_P$ so that P restricted to $\mathcal{O}_P$ is just the canonical epimorphism $\mathcal{O}_P \to LP$. The characteristic of LP is called the *residue characteristic* of (L, P). If P is the identity on $K \subseteq L$, then $K \subseteq LP$ canonically. The valuation v_P can be defined to be the homomorphism $L^\times \to L^\times/\mathcal{O}_P^\times$. The latter is an ordered abelian group, the *value group* of (L, v_P). We denote it by $v_P L$ and write it additively. Note that $bP \neq \infty \Leftrightarrow b \in \mathcal{O}_P \Leftrightarrow v_P b \geq 0$, and $bP = 0 \Leftrightarrow b \in \mathcal{M}_P \Leftrightarrow v_P b > 0$.

Instead of (L, P), we will often write (L, v) if we talk of valued fields in general. Then we will write av and Lv instead of aP and LP. If we talk of an *extension of valued fields* and write $(L|K, v)$ then we mean that v is a valuation on L and K is endowed with its restriction. If we only have to consider a single extension of v from K to L, then we will use the symbol v for both the valuation on K and that on L. Similarly, we use "$(L|K, P)$".

Observe that in Example 1, P is uniquely determined by the homomorphism on $K[x]$. Indeed, we can always write g/h in a form such that a is not a zero of both g and h. If then a is not a zero of h, we have that $(g/h)P = g(a)/h(a) \in K(a)$. If a is a zero of h, we have that $(g/h)P = \infty$. Thus, the residue field of P is $K(a)$, and the value group is $\mathbb{Z}$. On the other hand, we have the same non-uniqueness for places as we had for homomorphisms: also places can be composed with isomorphisms. If P, Q are places of an arbitrary field L and there is an isomorphism $\sigma : LP \to LQ$

such that $\sigma(bP) = bQ$ for all $b \in \mathcal{O}_P$, then we call P and Q *equivalent places.* In fact, P and Q are equivalent if and only if $\mathcal{O}_P = \mathcal{O}_Q$. Nevertheless, it is often more convenient to work with places than with valuation rings, and we will just identify equivalent places wherever this causes no problems.

Two valuations v and w are called *equivalent valuations* if they only differ by an isomorphism of the value groups; this holds if and only if v and w have the same valuation ring. As for places, we will identify equivalent valuations wherever this causes no problems, and we will also identify the isomorphic value groups.

At this point, we shall introduce a useful notion. Given a function field $F|K$, we will call P a *place of* $F|K$ if it is a place of F whose restriction to K is the identity. We say that P is *trivial* on K if it induces an isomorphism on K. But then, composing P with the inverse of this isomorphism, we find that P is equivalent to a place of F whose restriction to K is the identity. Note that a place P of F is trivial on K if and only if v_P is *trivial* on K, i.e., $v_P K = \{0\}$. This is also equivalent to $K \subset \mathcal{O}_P$. A place P of $F|K$ is said to be a *rational place* if $FP = K$. The *dimension* of P, denoted by $\dim P$, is the transcendence degree of $FP|K$. Hence, P is *zero-dimensional* if and only if $FP|K$ is algebraic.

Let's get back to our problem. The first thing we learn from our example is the following. Clearly, we would like to extend our homomorphism of $K[V]$ to a place of $K(V)$ because then, it will induce a map on $K[V']$. Then we have the chance to say that the point we have to look at on the new variety (e.g., in order to see whether this one is simple) is the point given by this map on $K[V']$. But this only makes sense if this map is a homomorphism of $K[V']$. So we have to require:

$$y_1, \ldots, y_k \in \mathcal{O}_P$$

(since then, $K[y_1, \ldots, y_k] \subseteq \mathcal{O}_P$, which implies that P is a homomorphism on $K[y_1, \ldots, y_k]$).

This being granted, the next question coming to our mind is whether to every point there corresponds exactly one place (up to equivalence), as it was the case in Example 1. To destroy this hope, I give again a very simple example. It will also serve to introduce several types of places and their invariants.

Example 2. Consider the coordinate ring $K[x_1, x_2]$ of $V = \mathbb{A}^2_K$. That is, x_1 and x_2 are algebraically independent over K, and the function field $K(V) = K(x_1, x_2)$ is just the rational function field in two variables over K. A homomorphism of the polynomial ring $K[V] = K[x_1, x_2]$ is given by "evaluating" every polynomial $g(x_1, x_2)$ at $x_1 = a_1$, $x_2 = a_2$. For example, let us take $a_1 = a_2 = 0$ and try to extend the corresponding homomorphism of $K[x_1, x_2]$ to $K(x_1, x_2)$. It is clear that $1/x_1$ and $1/x_2$ go to ∞. But what about x_1/x_2 or even x_1^m/x_2^n? Do they go to 0, ∞ or some non-zero element in K? The answer is: all that is possible, and there are infinitely many ways to extend our homomorphism to a place of $K(x_1, x_2)$.

There is one way, however, which seems to be the most well-behaved. It is to construct a *place of maximal rank*; we will explain this notion later in full generality. The idea is to learn from Example 1 where we replace K by $K(x_2)$ and x by x_1, and extend the homomorphism defined on $K(x_2)[x_1]$ by $x_1 \mapsto 0$ to a unique place Q of $K(x_1,x_2)$. Its residue field is $K(x_2)$ since $x_1Q = 0 \in K(x_2)$, and its value group is $\mathbb{Z}$. Now we do the same for $K(x_2)$, extending the homomorphism given on $K[x_2]$ by $x_2 \mapsto 0$ to a unique place $\overline{Q}$ of $K(x_2)$ with residue field K and value group $\mathbb{Z}$. We compose the two places, in the following way. Take $b \in K(x_1,x_2)$. If $bQ = \infty$, then we set $bQ\overline{Q} = \infty$. If $bQ \neq \infty$, then $bQ \in K(x_2)$, and we know what $bQ\overline{Q} = (bQ)\overline{Q}$ is. In this way, we obtain a place $P = Q\overline{Q}$ on $K(x_1,x_2)$ with residue field K. We observe that for every $g \in K[x_1,x_2]$, we have that $g(x_1,x_2)Q\overline{Q} = g(0,x_2)\overline{Q} = g(0,0)$, so our place P indeed extends the given homomorphism of $K[x_1,x_2]$. Now what happens to our critical fractions? Clearly, $(1/x_1)P = (1/x_1)Q\overline{Q} = (\infty)\overline{Q} = \infty$, and $(1/x_2)P = (1/x_2)Q\overline{Q} = (1/x_2)\overline{Q} = \infty$. But what interests us most is that for all $m > 0$ and $n \geq 0$, $(x_1^m/x_2^n)P = (x_1^m/x_2^n)Q\overline{Q} = 0\overline{Q} = 0$. We see that "$x_1$ goes more strongly to 0 than every x_2^n". We have achieved this by sending first x_1 to 0, and only afterwards x_2 to 0. We have arranged our action "lexicographically".

What is the associated value group? General valuation theory (cf. [V], §3 and §4, or [ZS]) tells us that for every composition $P = Q\overline{Q}$, the value group $v_{\overline{Q}}(FQ)$ of the place $\overline{Q}$ on FQ is a convex subgroup of the value group v_PF, and that the value group v_QF of P is isomorphic to $v_PF/v_{\overline{Q}}(FQ)$. If the subgroup $v_{\overline{Q}}(FQ)$ is a direct summand of v_PF (as it is the case in our example), then v_PF is the lexicographically ordered direct product $v_QF \times v_{\overline{Q}}(FQ)$. Hence in our case, $v_PK(x_1,x_2) = \mathbb{Z} \times \mathbb{Z}$, ordered lexicographically. The *rank of an abelian ordered group* G is the number of proper convex subgroups of G (or rather the order type of the chain of convex subgroups, ordered by inclusion, if this is not finite). The *rank of* (F,P) is defined to be the rank of v_PF. See under the name "hauteur" in [V]. In our case, the rank is 2. We will see in Section 7 that if P is a place of $F|K$, then the rank cannot exceed the transcendence degree of $F|K$. So our place $P = Q\overline{Q}$ has maximal possible rank.

There are other places of maximal rank which extend our given homomorphism, but there is also an abundance of places of smaller rank. In our case, "smaller rank" can only mean rank 1, i.e., there is only one proper convex subgroup of the value group, namely $\{0\}$. For an ordered abelian group G, having rank 1 is equivalent to being archimedean ordered and to being embeddable in the ordered additive group of $\mathbb{R}$. Which subgroups of $\mathbb{R}$ can we get as value groups? To determine them, we look at the *rational rank* of an ordered abelian group G. It is $\mathrm{rr}\, G := \dim_{\mathbb{Q}} \mathbb{Q} \otimes_{\mathbb{Z}} G$ (note that $\mathbb{Q} \otimes_{\mathbb{Z}} G$ is the *divisible hull* of G). This is the maximal number of rationally independent elements in G. We will see in Section 7 that for every place P of $F|K$ we have that

$$\mathrm{rr}\, v_PF \;\leq\; \mathrm{trdeg}\, F|K\;. \tag{1}$$

Hence in our case, also the rational rank of P can be at most 2. The subgroups of $\mathbb{R}$ of rank 2 are well known: they are the groups of the form $r\mathbb{Z} + s\mathbb{Z}$ where $r > 0$ and $s > 0$ are rationally independent real numbers. Moreover, through multiplication by $1/r$, the group is order isomorphic to $\mathbb{Z} + \frac{s}{r}\mathbb{Z}$. As we identify equivalent valuations, we can assume all rational rank 2 value groups (of a rank 1 place) to be of the form $\mathbb{Z}+r\mathbb{Z}$ with $0<r\in\mathbb{R}\setminus\mathbb{Q}$. To construct a place P with this value group on $K(x_1,x_2)$, we proceed as follows. We want that $v_Px_1=1$ and $v_Px_2=r$; then it will follow that $v_PK(x_1,x_2)=\mathbb{Z}+r\mathbb{Z}$ (cf. Theorem 7.1 below). We observe that for such P, $v_P(x_1^m/x_2^n)=m-nr$, which is >0 if $m/n>r$, and <0 if $m/n<r$. Hence, $(x_1^m/x_2^n)P=0$ if $m/n>r$, and $(x_1^m/x_2^n)P=\infty$ if $m/n<r$. I leave it to you as an exercise to verify that this defines a unique place P of $K(x_1,x_2)|K$ with the desired value group and extending our given homomorphism.

Observe that so far every value group was finitely generated, namely by two elements. Now we come to the groups of rational rank 1. If such a group is finitely generated, then it is simply isomorphic to $\mathbb{Z}$. How do we get places P on $K(x_1,x_2)$ with value group $\mathbb{Z}$? A place with value group $\mathbb{Z}$ is called a *discrete place*. The idea is to first construct a place on the subfield $K(x_1)$. We know from Example 1 that every place of $K(x_1)|K$ (if it is not trivial on $K(x_1)$) will have value group $\mathbb{Z}$ (cf. Theorem 7.1). Then we can try to extend this place from $K(x_1)$ to $K(x_1,x_2)$ in such a way that the value group doesn't change.

There are many different ways how this can be done. One possibility is to send the fraction x_1/x_2 to an element z which is transcendental over K. You may verify that there is a unique place which does this and extends the given homomorphism; it has value group $\mathbb{Z}$ and residue field $K(z)$. If, as in this case, a place P of $F|K$ has the property that $\operatorname{trdeg} FP|K=\operatorname{trdeg} F|K-1$, then P is called a *prime divisor* and v_P is called a *divisorial valuation.* The places $Q,\overline{Q}$ were prime divisors, one of F, the other one of FQ.

But maybe we don't want a residue field which is transcendental over K? Maybe we even insist on having K as a residue field? Well, then we can employ another approach. Having already constructed our place P on $K(x_1)$ with residue field K, we can consider the completion of $(K(x_1),P)$. The *completion* of an arbitrary valued field (L,v) is the completion of L with respect to the topology induced by v. Both v and the associated place P extend canonically to this completion, whereby value group and residue field remain unchanged. Let us give a more concrete representation of this completion.

Let t be any transcendental element over K. We consider the unique place P of $F|K$ with $tP=0$. The associated valuation is called the *t-adic valuation*, denoted by v_t. It is the unique valuation v on $K(t)$ (up to equivalence) which is trivial on K and satisfies that $vt>0$. We want to write down the completion of $(K(t),v_t)$. We define the *field of formal Laurent series* (I prefer *power series field*) over K. It is denoted by $K((t))$ and consists of all formal sums of the form

$$\sum_{i=n}^{\infty} c_it^i \quad \text{with } n\in\mathbb{Z} \text{ and } c_i\in K\,. \tag{2}$$

I suppose I don't have to tell you in which way the set $K((t))$ can be made into a field. But I tell you how v_t extends from $K(t)$ to $K((t))$: we set

$$v_t \sum_{i=n}^{\infty} c_i t^i = n \quad \text{if } c_n \neq 0 \,. \tag{3}$$

One sees immediately that $v_t K((t)) = v_t K(t) = \mathbb{Z}$. For $b = \sum_{i=n}^{\infty} c_i t^i$ with $c_n \neq 0$, we have that $bv_t = \infty$ if $m < 0$, $bv_t = 0$ if $m > 0$, and $bv_t = c_0 \in K$ if $m = 0$. So we see that $K((t))v_t = K(t)v_t = K$. General valuation theory shows that $(K((t)), v_t)$ is indeed the completion of $(K(t), v_t)$.

It is also known that the transcendence degree of $K((t))|K(t)$ is uncountable. If K is countable, this follows directly from the fact that $K((t))$ then has the cardinality of the continuum. But it is quite easy to show that the transcendence degree is at least one, and already this suffices for our purposes here. The idea is to take any $y \in K((t))$, transcendental over $K(t)$; then $x_1 \mapsto t$, $x_2 \mapsto y$ induces an isomorphism $K(x_1, x_2) \to K(t, y)$. We take the restriction of v_t to $K(t, y)$ and pull it back to $K(x_1, x_2)$ through the isomorphism. What we obtain on $K(x_1, x_2)$ is a valuation v which extends our valuation v_P of $K(x_1)$. As is true for v_t, also this extension still has value group $\mathbb{Z} = v_P K(x_1)$ and residue field $K = K(x_1)P$. The desired place of $K(x_1, x_2)$ is the place associated with this valuation v.

We have now constructed essentially all places on $K(x_1, x_2)$ which extend the given homomorphism of $K[x_1, x_2]$ and have a finitely generated value group (up to certain variants, like exchanging the role of x_1 and x_2). The somewhat shocking experience to every "newcomer" is that on this rather simple rational function field, there are also places extending the given homomorphism and having a value group which is not finitely generated. For instance, the value group can be $\mathbb{Q}$. (In fact, it can be any subgroup of $\mathbb{Q}$.) We postpone the construction of such a place till Section 18.

After we have become acquainted with places and how one obtains them from homomorphisms of coordinate rings, it is time to formulate our problem of local desingularization. Instead of looking for a desingularization "at a given point" of our variety V, we will look for a desingularization at a given place P of the function field $F|K$ (we forget about the variety from which F originates). Suppose we have any V such that $K(V) = F$, that is, we have generators $x_1, \ldots, x_\ell$ of $F|K$ and the coordinate ring $K[x_1, \ldots, x_\ell]$ of V. If we talk about the *center of P on V*, we always tacitly assume that $x_1, \ldots, x_\ell \in \mathcal{O}_P$, so that the restriction of P is a homomorphism on $K[x_1, \ldots, x_\ell]$. With this provision, the center of P on V is the point $(x_1 P, \ldots, x_\ell P)$ (or, if we so want, the induced homomorphism). We also say that P is *centered on V at $(x_1 P, \ldots, x_\ell P)$*. If V is a variety defined over K with function field F, then we call V a *model of $F|K$*. Our problem now reads:

(LU) *Take any function field $F|K$ and a place P of $F|K$. Does there exist a model of $F|K$ on which P is centered at a simple point?*

This was answered in the positive by Oscar Zariski in [Z] for the case of K having characteristic 0. Instead of "local desingularization", he called this principle *local uniformization.*

3. Local uniformization and the Implicit Function Theorem

Let's think about what we mean by "simple point". I don't really have to tell you, so let me pick the most valuation theoretic definition, which will show us our way on our excursion. It is the Jacobi criterion: Given our variety V defined by $f_1, \ldots, f_n \in K[X_1, \ldots, X_\ell]$ and having function field F, then a point $a = (a_1, \ldots, a_\ell)$ of V is called *simple* (or *smooth*) if $\operatorname{trdeg} F|K = \ell - r$, where r is the rank of the Jacobi matrix

$$\left(\partial_{X_j} f_i(a)\right)_{1 \leq i \leq n, 1 \leq j \leq \ell}$$

But wait — I have seen the Jacobi matrix long before I learned anything about algebraic geometry. Now I remember: I saw it in my first year calculus course in connection with the *Implicit Function Theorem*. Let's have a closer look. First, let us assume that we don't have too many f_i's. Indeed, when looking for a local uniformization we will construct varieties V defined by $\ell - \operatorname{trdeg} F|K$ many polynomial relations, whence $n = \ell - \operatorname{trdeg} F|K$. In this situation, if a is a simple point, then n is equal to r and after a suitable renumbering we can assume that for $k := \ell - n = \operatorname{trdeg} F|K$, the submatrix

$$\left(\partial_{X_j} f_i(a)\right)_{1 \leq i \leq n, k+1 \leq j \leq \ell}$$

is invertible. Then, assuming that we are working over the reals, the Implicit Function Theorem tells us that for every $(a'_1, \ldots, a'_k)$ in a suitably small neighborhood of $(a_1, \ldots, a_k)$ there is a unique $(a'_{k+1}, \ldots, a'_\ell)$ such that $(a'_1, \ldots, a'_\ell)$ is a point of V. Working in the reals, the existence is certainly interesting, but for us here, the main assertion is the uniqueness. Let's look at a very simple example.

Example 3. I leave it to you to draw the graph of the function $y^2 = x^3$ in $\mathbb{R}^2$. It only exists for $x \geq 0$. Starting from the origin, it has two branches, one positive, one negative. Now assume that we are sitting on one of these branches at a point (x, y), away from the origin. If somebody starts to manipulate x then we know exactly which way we have to run (depending on whether x increases or decreases). But if we are sitting at the origin and somebody increases x, then we have the freedom of choice into which of the two branches we want to run. So we see that everywhere but at the origin, y is an implicit function of x in a sufficiently small neighborhood. Indeed, with $f(x, y) = x^3 - y^2$, we have that $\frac{\partial f}{\partial x}(x, y) = 3x^2$. If $x \neq 0$, then this is non-zero, whence $r = 1$ while $\operatorname{trdeg} F|K = 1$ and $\ell = 2$, so for $x \neq 0$, (x, y) is a simple point. On the other hand, $\frac{\partial f}{\partial x}(0, 0) = 0$ and $\frac{\partial f}{\partial y}(0, 0) = 0$, so $(0, 0)$ is singular.

We have now seen the connection between simple points and the Implicit Function Theorem. "Wait!" you will interrupt me. "You have used the topology of $\mathbb{R}$. What if we don't have such a topology at hand? What then do you mean by 'neighborhood'?" Good question. So let's look for a topology. My luck, that the Implicit Function Theorem is also known in valuation theory. Indeed, we have already remarked in connection with the notion "completion of a valued field" that every valuation induces a topology. And since we have our place on F, we have the topology right at hand. That is why I said that the Jacobi criterion renders the most valuation theoretical definition of "simple".

But now this makes me think: haven't I seen the Jacobi matrix in connection with an even more famous valuation theoretical theorem, one of central importance in valuation theory? Indeed: it appears in the so-called "multidimensional version" of *Hensel's Lemma*. This brings us to our next sightseeing attraction on our excursion.

4. Hensel's Lemma

Hensel's Lemma is originally a lemma proved by Kurt Wilhelm Sebastian Hensel for the field of p-adic numbers $\mathbb{Q}_p$. It was then extended to all complete discrete valued fields and later to all maximal fields (see Corollary 3 below). A valued field (L, v) is called *maximal* (or *maximally complete*) if it has no proper extensions for which value group and residue field don't change. A complete field is not necessarily maximal, and if it is not of rank 1 (i.e., its value group is not archimedean), then it also does not necessarily satisfy Hensel's Lemma. However, complete discrete valued fields are maximal. In particular, $(K((t)), v_t)$ is maximal.

In modern valuation theory (and its model theory), Hensel's Lemma is rather understood to be a property of a valued field. The nice thing is that, in contrast to "complete" or "maximal", it is an elementary property (I will tell you in Section 12 what this means). We call a valued field *henselian* if it satisfies Hensel's Lemma. Here is one version of Hensel's Lemma for a valued field with valuation ring $\mathcal{O}_v$:

Hensel's Lemma. *For every polynomial $f \in \mathcal{O}_v[X]$ the following holds: if $b \in \mathcal{O}_v$ satisfies*

$$vf(b) > 0 \quad \text{and} \quad vf'(b) = 0\,, \tag{4}$$

then f admits a root $a \in \mathcal{O}_v$ such that $v(a-b) > 0$.

Here, f' denotes the derivative of f. Note that a more classical version of Hensel's Lemma talks only about monic polynomials.

For the multidimensional version, we introduce some notation. For polynomials $f_1, \dots, f_n$ in variables $X_1, \dots, X_n$, we write $f = (f_1, \dots, f_n)$ and denote by J_f the Jacobian matrix $\left(\partial_{X_j} f_i\right)_{i,j}$. For $a \in L^n$, $J_f(a) = \left(\partial_{X_j} f_i(a)\right)_{i,j}$.

Multidimensional Hensel's Lemma. *Let $f = (f_1, \dots, f_n)$ be a system of polynomials in the variables $X = (X_1, \dots, X_n)$ and with coefficients in $\mathcal{O}_v$. Assume that there exists $b = (b_1, \dots, b_n) \in \mathcal{O}_v^n$ such that*

$$vf_i(b) > 0 \text{ for } 1 \le i \le n \quad \text{and} \quad v \det J_f(b) = 0\,. \tag{5}$$

Then there exists a unique $a = (a_1, \dots, a_n) \in \mathcal{O}_v^n$ such that $f_i(a) = 0$ and that $v(a_i - b_i) > 0$ for all i.

And here is the valuation theoretical Implicit Function Theorem:

Implicit Function Theorem: *Take $f_1,\dots,f_n \in L[X_1,\dots,X_\ell]$ with $n<\ell$. Set*

$$\tilde{J} := \begin{pmatrix} \frac{\partial f_1}{\partial X_{\ell-n+1}} & \cdots & \frac{\partial f_1}{\partial X_\ell} \\ \vdots & & \vdots \\ \frac{\partial f_n}{\partial X_{\ell-n+1}} & \cdots & \frac{\partial f_n}{\partial X_\ell} \end{pmatrix} . \tag{6}$$

Assume that $f_1,\dots,f_n$ admit a common zero $a=(a_1,\dots,a_\ell)\in L^\ell$ and that $\det\tilde{J}(a)\neq 0$. *Then there is some $\alpha\in vL$ such that for all $(a'_1,\dots,a'_{\ell-n})\in L^{\ell-n}$ with $v(a_i-a'_i)>2\alpha$, $1\leq i\leq \ell-n$, there exists a unique $(a'_{\ell-n+1},\dots,a'_\ell)\in L^n$ such that $(a'_1,\dots,a'_\ell)$ is a common zero of $f_1,\dots,f_n$, and $v(a_i-a'_i)>\alpha$ for $\ell-n<i\leq\ell$.*

It is (not all too well) known that Hensel's Lemma holds in (L,v) if and only if the Multidimensional Hensel's Lemma holds in (L,v), and this in turn is true if and only if the Implicit Function Theorem holds in (L,v). For a proof, see [K2] or [PZ]. The latter paper is particularly interesting since it shows the connection between the Implicit Function Theorem in henselian fields and the "real" Implicit Function Theorem in $\mathbb{R}$.

There are many more versions of Hensel's Lemma which all are equivalent to the above (the classical Hensel's Lemma for monic polynomials, Krasner's Lemma, Newton's Lemma, Hensel-Rychlik, ...). See [R2] or [K2] for a listing of them. It is indeed often very useful to have the different versions at hand. One particularly important is given in the following lemma:

Lemma 4.1. *A valued field (L,v) is henselian if and only if the extension of v to the algebraic closure $\tilde{L}$ of L is unique.*

Since any valuation of any field can always be extended to any extension field (cf. [V], §5), the following is an easy consequence of this lemma: *(L,v) is henselian if and only if v admits a unique extension to every algebraic extension field.* Also, we immediately obtain:

Corollary 1. *Every algebraic extension of a henselian field is again henselian.*

This is hard to prove if you use Hensel's Lemma instead of the unique extension property in the proof. On the other hand, the next lemma is hard to prove using the unique extension property, while it is immediate if you use Hensel's Lemma:

Lemma 4.2. *Take a henselian field (L,v) and a relatively algebraically closed subfield L' of L. Then also (L',v) is henselian.*

Let us take a short break to see how Hensel's Lemma can be applied. The following two examples will later have important applications.

Example 4. Assume that $\mathrm{char}\, L=p>0$. A polynomial $f(X)=X^p-X-c$ with $c\in L$ is called an *Artin-Schreier polynomial* (over L). If ϑ is a root of f in some extension of L, then $\vartheta,\vartheta+1,\dots,\vartheta+p-1$ are the distinct roots of f.

Hence if f is irreducible over L, then $L(\vartheta)|L$ is a Galois extension of degree p. It is called an *Artin-Schreier extension.* Conversely, *every* Galois extension of degree p in characteristic p is generated by a root of a suitable Artin-Schreier polynomial, i.e., is an Artin-Schreier extension (see [K2] for a proof).

Let us prove our assertion about the roots of f. We note that in characteristic $p > 0$, the map $x \mapsto x^p$ is a ring homomorphism (the *Frobenius*). Therefore, the polynomial $\wp(X) := X^p - X$ is an *additive polynomial.* A polynomial g is called additive if $g(a+b) = g(a) + g(b)$ for all a, b (for details, cf. [L2], VIII, §11). Thus, if $i \in \mathbb{F}_p$, then $f(\vartheta + i) = \wp(\vartheta + i) - c = \wp(\vartheta) - c + \wp(i) = 0 + i^p - i = i - i = 0$ since $i^p = i$ for every $i \in \mathbb{F}_p$.

Now assume that (L, v) is henselian. Suppose first that $vc > 0$. Take $b = 0 \in \mathcal{O}_v$. Then $vf(b) = vc > 0$. On the other hand, $f'(X) = pX^{p-1} - 1 = -1$ since $p = 0$ in characteristic p. Hence, $vf'(b) = v(-1) = 0$. Therefore, Hensel's Lemma shows that f admits a root in L, which by our above observation about the roots of f means that f splits completely over L.

Suppose next that $vc = 0$. Then for $b \in \mathcal{O}_v$ we have that $v(b^p - b - c) > 0$ if and only if $0 = (b^p - b - c)v = (bv)^p - bv - cv$. Hence, $v(b^p - b - c) > 0$ if and only if bv is a root of the Artin-Schreier polynomial $X^p - X - cv \in Lv[X]$. If $cv = 0$, which is our previous case where $vc > 0$, then 0 is a root of $X^p - X - cv = X^p - X$ and we can choose $b = 0$. But in our present case, $cv \neq 0$, and everything depends on whether $X^p - X - cv$ has a root in Lv or not. If it has a root η in Lv, then we choose $b \in \mathcal{O}_v$ such that $bv = \eta$. We obtain that $(b^p - b - c)v = (bv)^p - bv - cv = \eta^p - \eta - cv = 0$, hence $vf(b) > 0$. Then by Hensel's Lemma, $X^p - X - c$ has a root $a \in \mathcal{O}_v$ with $v(a - b) > 0$, hence $av = bv = \eta$. Conversely, if $X^p - X - c$ has a root a in L, then one easily shows that $a \in \mathcal{O}_v$, and $0 = 0v = (a^p - a - c)v = (av)^p - av - cv$ yields that $X^p - X - cv$ has a root in Lv.

The only remaining case is that of $vc < 0$. In this case, $X^p - X - c \notin \mathcal{O}_v[X]$, so Hensel's Lemma doesn't give us any immediate information about whether f has a root in L or not.

Example 5. Take a field K of characteristic $p > 0$. In the field $(K((t)), v_t)$ (which is henselian, cf. Corollary 3 below), the Artin-Schreier polynomial

$$X^p - X - t \tag{7}$$

has the root

$$a = \sum_{i=0}^{\infty} (-t)^{p^i} \tag{8}$$

since

$$a^p - a = \sum_{i=0}^{\infty} (-t)^{p^{i+1}} - \sum_{i=0}^{\infty} (-t)^{p^i} = \sum_{i=1}^{\infty} (-t)^{p^i} - \sum_{i=0}^{\infty} (-t)^{p^i} = t\,.$$

Take any polynomial $f \in \mathcal{O}_v[X]$. By fv we mean the *reduction of the polynomial f modulo v*, that is, the polynomial we obtain from f by replacing every coefficient c_i of f by its residue $c_i v$. As the residue map is a homomorphism on

$\mathcal{O}_v$, we have that $f'v = (fv)'$. Suppose there is some $b \in L$ such that $vf(b) > 0$ and $vf'(b) = 0$. This is equivalent to $f(b)v = 0$ and $f'(b)v \neq 0$. But $f(b)v = fv(bv)$ and $f'(b)v = (fv)'(bv)$, so the latter is equivalent to bv being a simple root of fv. Conversely, if fv has a simple root ζ, find some b such that $bv = \zeta$, and you will have that $vf(b) > 0$ and $vf'(b) = 0$. Hence, Hensel's Lemma is also equivalent to the following version:

Hensel's Lemma, Simple Root Version. *For every polynomial $f \in \mathcal{O}_v[X]$ the following holds: if fv has a simple root ζ in Lv, then f admits a root $a \in \mathcal{O}_v$ such that $av = \zeta$.*

Example 6. Take a henselian valued field (L, v) and a relatively algebraically closed subfield L' of L. Assume there is an element ζ of the residue field Lv which is algebraic over $L'v$, and denote its minimal polynomial over $L'v$ by $h \in L'v[X]$. Find a monic polynomial $f \in (\mathcal{O}_v \cap L')[X]$ such that $fv = h$.

If ζ is separable over $L'v$, then ζ is a simple root of h. As $\zeta \in Lv$ and (L, v) is henselian by assumption, the Simple Root Version of Hensel's Lemma tells us then that there is some $a \in L$ such that $h(a) = 0$ and $av = \zeta$. But as a is algebraic over L' we have that $a \in L'$, so that $\zeta = av \in L'v$. If on the other hand ζ is not separable over $L'v$, then it is quite possible that $\eta \notin L'v$. But we have proved:

Lemma 4.3. *If (L, v) is henselian and L' is relatively algebraically closed in L, then $L'v$ is* relatively separable-algebraically closed *in Lv, i.e., every element of Lv already belongs to $L'v$ if it is separable-algebraic over $L'v$.*

Something similar can be shown for the value groups, provided that $Lv = L'v$. Pick an element $\delta \in vL$ such that for some $n > 0$, $n\delta \in vL'$. Choose some $d \in L$ such that $vd = \delta$. Hence, $vd^n = nvd \in vL'$ and we can choose some $d' \in L'$ such that $vd'd^n = 0$. Assuming that $Lv = L'v$, we can also pick some $d'' \in L'$ such that $(d'd''d^n)v = 1$.

An element u with $uv = 1$ is called a *1-unit.* We consider the polynomial $X^n - u$. Its reduction modulo v is simply the polynomial $X^n - 1$. Obviously, 1 is a root of that polynomial, but is it a simple root? The answer is: 1 is a simple root of $X^n - 1$ if and only if the characteristic of Lv does not divide n. Hence in that case, Hensel's Lemma shows that there is a root $a \in L$ of the polynomial $X^n - u$ such that $av = 1$. This proves:

Lemma 4.4. *Take a 1-unit u in the henselian field (L, v) and $n \in \mathbb{N}$ such that the characteristic of Lv does not divide n. Then there is a unique 1-unit $a \in L$ such that $a^n = u$.*

In our present case, this provides an element $a \in L$ such that $a^n = d'd''d^n$. We find that $(a/d)^n = d'd'' \in L'$. Since $a/d \in L$ and L' is relatively algebraically closed in L, this implies that $a/d \in L'$. On the other hand, $v(a/d)^n = vd'd'' = vd' = n\alpha$ so that $v(a/d) = \alpha$. This proves that $\alpha \in vL'$. We have proved:

Lemma 4.5. *If (L, v) is henselian and L' is relatively algebraically closed in L and $Lv = L'v$, then the torsion subgroup of vL/vL' is trivial if* $\operatorname{char} Lv = 0$, *and it is a p-group if* $\operatorname{char} Lv = p > 0$.

It can be shown that the assertion is in general not true without the assumption that $Lv = L'v$.

Let us return to our variety V which is defined by $f_1, \dots, f_n \in K[X_1, \dots, X_\ell]$ and has coordinate ring $K[x_1, \dots, x_\ell]$. We have seen that a point $a = (a_1, \dots, a_\ell)$ of V is simple if and only if after a suitable renumbering, the submatrix

$$\tilde{J} = \left(\partial_{X_j} f_i(a)\right)_{1\le i\le n, k+1\le j\le \ell} \tag{9}$$

of $J_f(a)$ is invertible, where $k := \ell - n = \operatorname{trdeg} F|K$. That means that $f_1, \dots, f_n$ and a satisfy the assumptions of the Implicit Function Theorem.

Since we are interested in the question whether the center of P on V is simple, we have to look at $a = (x_1P, \dots, x_\ell P)$. As P is a homomorphism on $\mathcal{O}_P$ and leaves the coefficients of the f_i invariant, we see that

$$\left(\partial_{X_j} f_i(x_1P, \dots, x_\ell P)\right) = \left(\partial_{X_j} f_i(x_1, \dots, x_\ell)\right) P\,. \tag{10}$$

We have omitted the indices since this holds for *every* submatrix of J_f. Again because P is a homomorphism, it commutes with taking determinants (since this operation remains inside the ring $\mathcal{O}_P$). Hence,

$$\det \tilde{J}(x_1P, \dots, x_\ell P) = (\det \tilde{J}(x_1, \dots, x_\ell))P\,. \tag{11}$$

Therefore, $\det \tilde{J}(x_1P, \dots, x_\ell P) \neq 0$ is equivalent to $v \det \tilde{J}(x_1, \dots, x_\ell) = 0$. This condition also appears in the Multidimensional Hensel's Lemma, but with J_f in the place of $\tilde{J}$. So we are led to the question: what is the connection? It is obvious that we have some variables too many for the case of the Multidimensional Hensel's Lemma. But they are exactly $\operatorname{trdeg} F|K$ too many, and on the other hand, at least the basic Hensel's Lemma obviously talks about algebraic elements (we will see that this is also true for the Multidimensional Hensel's Lemma). So why don't we just take $x_1, \dots, x_k$ as a transcendence basis of $F|K$ and view $f_1, \dots, f_n$ as polynomial relations defining the remaining $x_{k+1}, \dots, x_\ell$, which are algebraic over $K(x_1, \dots, x_k)$? But then, we should write every f_i as a polynomial $\tilde{f}_i$ in the variables $X_{k+1}, \dots, X_\ell$ with coefficients in $K(x_1, \dots, x_k)$, or actually, in $K[x_1, \dots, x_k]$. Then we have that

$$\begin{aligned} f_i(x_1, \dots, x_\ell)P &= f_i(x_1P, \dots, x_\ell P) \\ &= \tilde{f}_iP(x_{k+1}P, \dots, x_\ell P) = \tilde{f}_i(x_{k+1}, \dots, x_\ell)P\,. \end{aligned}$$

With $\tilde{f} := (\tilde{f}_1, \dots, \tilde{f}_n)$ and $\tilde{f}P := (\tilde{f}_1P, \dots, \tilde{f}_nP)$, it follows that

$$\det \tilde{J}(x_1P, \dots, x_\ell P) = \det J_{\tilde{f}P}(x_{k+1}P, \dots, x_\ell P) = (\det J_{\tilde{f}}(x_{k+1}, \dots, x_\ell))P\,. \tag{12}$$

Hence, $\det \tilde{J}(x_1P, \dots, x_\ell P) \neq 0$ is equivalent to $v \det J_{\tilde{f}}(x_{k+1}, \dots, x_\ell) = 0$, which means that the polynomials $\tilde{f}_1, \dots, \tilde{f}_n$ and the elements $x_{k+1}, \dots, x_\ell$ satisfy the assumption (5) of the Multidimensional Hensel's Lemma. Indeed, we have that $v\tilde{f}_i(x_{k+1}, \dots, x_\ell) = \infty > 0$ since $\tilde{f}_i(x_{k+1}, \dots, x_\ell) = 0$. So we see:

To find a model of $F|K$ on which P is centered at a simple point means to find generators $x_1, \dots, x_\ell \in \mathcal{O}_P$ such that $x_1, \dots, x_k$ form a transcendence basis of $F|K$ and $x_{k+1}, \dots, x_\ell$ together with the polynomials which define them over $K[x_1, \dots, x_k]$ satisfy the assumption of the Multidimensional Hensel's Lemma.

(Since f_i and $\tilde{f}_i$ are the same polynomial, just written in two different ways, we will later use "f_i" instead of "$\tilde{f}_i$"; cf. the definition of relative uniformization in Section 15. There, we will also prefer "$(\det J_{\tilde{f}}(x_{k+1}, \dots, x_\ell))P \neq 0$" over "$v \det J_{\tilde{f}}(x_{k+1}, \dots, x_\ell) = 0$".)

What we have derived now is still quite vague, and before we can make more out of it, I'm sorry, you have to go to a course again.

5. A crash course in ramification theory

Throughout, we assume that $L|K$ is an algebraic extension, not necessarily finite, and that v is a *non-trivial* valuation on K. If w is a valuation on L which extends v, then there is a natural embedding of the value group vK of v in the value group wL of w. Similarly, there is a natural embedding of the residue field Kv of v in the residue field Lw of w. If both embeddings are onto (which we just express by writing $vK = wL$ and $Kv = Lw$), then the extension $(L, w)|(K, v)$ is called **immediate**. WARNING: It may happen that $vK \simeq wL$ or $Kv \simeq Lw$ although the corresponding embedding is not onto and therefore, the extension is not immediate. For example, every finite extension of the p-adics $(\mathbb{Q}_p, v_p)$ will again have a value group isomorphic to $\mathbb{Z}$, but $v_p p$ may not be anymore the smallest positive element in this value group.

We choose an arbitrary extension of v to the algebraic closure $\tilde{K}$ of K. Then for every $\sigma \in \mathrm{Aut}\,(\tilde{K}|K)$, the map

$$\tilde{v}\sigma = \tilde{v} \circ \sigma : \; L \ni a \;\mapsto\; \tilde{v}(\sigma a) \in \tilde{v}\tilde{K} \tag{13}$$

is a valuation of L which extends v.

Theorem 5.1. *The set of all extensions of v from K to L is*

$$\{\tilde{v}\sigma \mid \sigma \text{ an embedding of } L \text{ in } \tilde{K} \text{ over } K\}\,.$$

(We say that "all extensions of v from K to L are *conjugate*".)

Corollary 2. *If $L|K$ is finite, then the number g of distinct extensions of v from K to L is smaller or equal to the extension degree $[L : K]$. More precisely, g is smaller or equal to the degree of the maximal separable subextension of $L|K$. In particular, if $L|K$ is purely inseparable, then v has a unique extension from K to L.*

Theorem 5.2. *Assume that $n := [L : K]$ is finite, and denote the extensions of v from K to L by $v_1, \dots, v_{\mathrm{g}}$. Then for every $i \in \{1, \dots, \mathrm{g}\}$, the* ramification index

$e_i = (v_iL : vK)$ *and the* inertia degree $f_i = [Lv_i : Kv]$ *are finite, and we have the* fundamental inequality

$$n \geq \sum_{i=1}^{g} e_i f_i \ . \tag{14}$$

From now on, let us assume that $L|K$ is normal. Hence, the set of all extensions of v from K to L is given by $\{\tilde{v}\sigma \mid \sigma \in \mathrm{Aut}\,(L|K)\}$. For simplicity, we denote the restriction of $\tilde{v}$ to L again by v. The valuation ring of v on L will be denoted by $\mathcal{O}_L$. We define distinguished subgroups of $G := \mathrm{Aut}\,(L|K)$. The subgroup

$$G^d := G^d(L|K, v) := \{\sigma \in G \mid \forall x \in \mathcal{O}_L : \ v\sigma x \geq 0\} \tag{15}$$

is called the *decomposition group* **of** $(L|K, v)$. It is easy to show that σ sends $\mathcal{O}_L$ into itself if and only if the valuations v and $v\sigma$ agree on L. Thus,

$$G^d = \{\sigma \in G \mid v\sigma = v \text{ on } L\} \ . \tag{16}$$

Further, the *inertia group* is defined to be

$$G^i := G^i(L|K, v) := \{\sigma \in \mathrm{Aut}\,(L|K) \mid \forall x \in \mathcal{O}_L : v(\sigma x - x) > 0\} \ , \tag{17}$$

and the *ramification group* is

$$G^r := G^r(L|K, v) := \{\sigma \in \mathrm{Aut}\,(L|K) \mid \forall x \in \mathcal{O}_L : v(\sigma x - x) > vx\} \ . \tag{18}$$

Let S denote the maximal separable extension of K in L (we call it the *separable closure of* K *in* L). The fixed fields of G^d, G^i and G^r in S are called *decomposition field, inertia field* and *ramification field* **of** $(L|K, v)$. For simplicity, let us abbreviate them by Z, T and V. (These letters refer to the german words "Zerlegungskörper", "Trägheitskörper" and "Verzweigungskörper".)

Remark. In contrast to the classical definition used by other authors, we take decomposition field, inertia field and ramification field to be the fixed fields of the respective groups *in the maximal separable subextension.* The reason for this will become clear in Section 8.

By our definition, V, T and Z are separable-algebraic extensions of K, and S|V, S|T, S|Z are (not necessarily finite) Galois extensions. Further,

$$1 \subset G^r \subset G^i \subset G^d \subset G \ \text{ and thus, } \ \mathrm{S} \supset \mathrm{V} \supset \mathrm{T} \supset \mathrm{Z} \supset K \ . \tag{19}$$

(For the inclusion $G^i \subset G^d$ note that $vx \geq 0$ and $v(\sigma x - x) > 0$ implies that $v\sigma x \geq 0$.)

Theorem 5.3. *G^i and G^r are normal subgroups of G^d, and G^r is a normal subgroup of G^i. Therefore,* T|Z, V|Z *and* V|T *are (not necessarily finite) Galois extensions.*

First, we consider the decomposition field Z. In some sense, it represents all extensions of v from K to L.

Theorem 5.4.

a) *$v\sigma = v\tau$ on L if and only if $\sigma\tau^{-1}$ is trivial on* Z.

b) *$v\sigma = v$ on* Z *if and only if σ is trivial on* Z.

c) *The extension of v from* Z *to L is unique.*

d) *The extension* $(\mathrm{Z}|K, v)$ *is immediate.*

WARNING: It is in general not true that $v\sigma \neq v\tau$ holds already on Z if it holds on L.

a) and b) are easy consequences of the definition of G^d. c) follows from b) by Theorem 5.1. For d), there is a simple proof using a trick which is mentioned in the paper [AX] by James Ax.

Now we turn to the inertia field T. Let $\mathcal{M}_L$ denote the valuation ideal of v on L (the unique maximal ideal of $\mathcal{O}_L$). For every $\sigma \in G^d(L|K, v)$ we have that $\sigma\mathcal{O}_L = \mathcal{O}_L$, and it follows that $\sigma\mathcal{M}_L = \mathcal{M}_L$. Hence, every such σ induces an automorphism $\overline{\sigma}$ of $\mathcal{O}_L/\mathcal{M}_L = Lv$ which satisfies $\overline{\sigma}\,\overline{a} = \overline{\sigma a}$. Since σ fixes K, it follows that $\overline{\sigma}$ fixes Kv.

Lemma 5.5. *Since $L|K$ is normal, the same is true for $Lv|Kv$. The map*

$$G^d(L|K, v) \ni \sigma \;\mapsto\; \overline{\sigma} \in \mathrm{Aut}\,(Lv|Kv) \tag{20}$$

is a group homomorphism.

Theorem 5.6.

a) *The homomorphism (20) is onto and induces an isomorphism*

$$\mathrm{Aut}\,(\mathrm{T}|\mathrm{Z}) \;=\; G^d/G^i \;\simeq\; \mathrm{Aut}\,(\mathrm{T}v|\mathrm{Z}v)\;. \tag{21}$$

b) *For every finite subextension $F|\mathrm{Z}$ of* $\mathrm{T}|\mathrm{Z}$,

$$[F:\mathrm{Z}] \;=\; [Fv:\mathrm{Z}v]\;. \tag{22}$$

c) *We have that $v\mathrm{T} = v\mathrm{Z} = vK$. Further, $\mathrm{T}v$ is the separable closure of Kv in Lv, and therefore,*

$$\mathrm{Aut}\,(\mathrm{T}v|\mathrm{Z}v) \;=\; \mathrm{Aut}\,(Lv|Kv)\;. \tag{23}$$

If $F|\mathrm{Z}$ is normal, then b) is an easy consequence of a). From this, the general assertion of b) follows by passing from F to the normal hull of the extension $F|\mathrm{Z}$ and then using the multiplicativity of the extension degree. c) follows from b) by use of the fundamental inequality.

We set $p := \mathrm{char}\, Kv$ if this is positive, and $p := 1$ if $\mathrm{char}\, Kv = 0$. Given any extension $\Delta \subset \Delta'$ of abelian groups, the *p'-divisible closure of Δ in Δ'* is defined to be the subgroup $\{\alpha \in \Delta' \mid \exists n \in \mathbb{N} :\ (p,n) = 1 \wedge n\alpha \in \Delta\}$ of all elements in Δ' whose order modulo Δ is prime to p.

Theorem 5.7.

a) *There is an isomorphism*

$$\mathrm{Aut}\,(\mathrm{V}|\mathrm{T}) \;=\; G^i/G^r \;\simeq\; \mathrm{Hom}\,(v\mathrm{V}/v\mathrm{T}\,,\,(\mathrm{T}v)^\times)\;, \tag{24}$$

where the character group on the right-hand side is the full character group of the abelian group $v\mathrm{V}/v\mathrm{T}$. *Since this group is abelian,* $\mathrm{V}|\mathrm{T}$ *is an abelian Galois extension.*

b) *For every finite subextension* $F|\mathrm{T}$ *of* $\mathrm{V}|\mathrm{T}$,

$$[F:\mathrm{T}] = (vF:v\mathrm{T}) . \tag{25}$$

c) $\mathrm{V}v = \mathrm{T}v$, *and* $v\mathrm{V}$ *is the* p'*-divisible closure of* vK *in* vL.

b) follows from a) since for a finite extension $F|\mathrm{T}$, the group $vF/v\mathrm{T}$ is finite and thus there exists an isomorphism of $vF/v\mathrm{T}$ onto its full character group. The equality $\mathrm{V}v = \mathrm{T}v$ follows from b) by the fundamental inequality. The second assertion of c) follows from the next theorem and the fact that the order of all elements in $(\mathrm{T}v)^\times$ and thus also of all elements in $\mathrm{Hom}\,(v\mathrm{V}/v\mathrm{T}\,,\,(\mathrm{T}v)^\times)$ is prime to p.

Theorem 5.8. *The ramification group* G^r *is a* p*-group and therefore,* $\mathrm{S}|\mathrm{V}$ *is a* p*-extension. Further,* $vL/v\mathrm{V}$ *is a* p*-group, and the residue field extension* $Lv|\mathrm{V}v$ *is purely inseparable. If* $\mathrm{char}\, Kv = 0$, *then* $\mathrm{V} = \mathrm{S} = L$.

We note:

Lemma 5.9. *Every* p*-extension is a tower of Galois extensions of degree* p. *In characteristic* p, *all of them are Artin-Schreier-extensions, as we have mentioned in Example 4.*

From Theorem 5.8 it follows that there is a canonical isomorphism

$$\mathrm{Hom}\left(v\mathrm{V}/v\mathrm{T}\,,\,(\mathrm{T}v)^\times\right) \simeq \mathrm{Hom}\left(vL/vK\,,\,(Lv)^\times\right) . \tag{26}$$

We summarize our main results in the table on the next page.

We state two more useful theorems from ramification theory. If we have two subfields K, L of a field M (in our case, we will have the situation that $L \subset \tilde{K}$) then $K.L$ will denote the smallest subfield of M which contains both K and L; it is called the *field compositum of* K *and* L.

Theorem 5.10. *If* $K \subseteq K' \subseteq L$, *then the decomposition field of the normal extension* $(L|K', v)$ *is* $\mathrm{Z}.K'$, *its inertia field is* $\mathrm{T}.K'$, *and its ramification field is* $\mathrm{V}.K'$.

Theorem 5.11. *If* $E|K$ *is a normal subextension of* $L|K$, *then the decomposition field of* $(E|K, v)$ *is* $\mathrm{Z}\cap E$, *its inertia field is* $\mathrm{T}\cap E$, *and its ramification field is* $\mathrm{V}\cap E$.

If we take for $L|K$ the normal extension $\tilde{K}|K$, then we speak of *absolute ramification theory.* The fixed fields K^d, K^i and K^r of $G^d(\tilde{K}|K, v)$, $G^i(\tilde{K}|K, v)$ and $G^r(\tilde{K}|K, v)$ in the separable-algebraic closure K^{sep} of K are called *absolute decomposition field, absolute inertia field* and *absolute ramification field* of (K, v) (with respect to the given extension of v from K to its algebraic closure $\tilde{K}$). If $\mathrm{char}\, Kv = 0$, then by Theorem 5.8, $K^r = K^{\mathrm{sep}} = \tilde{K}$.

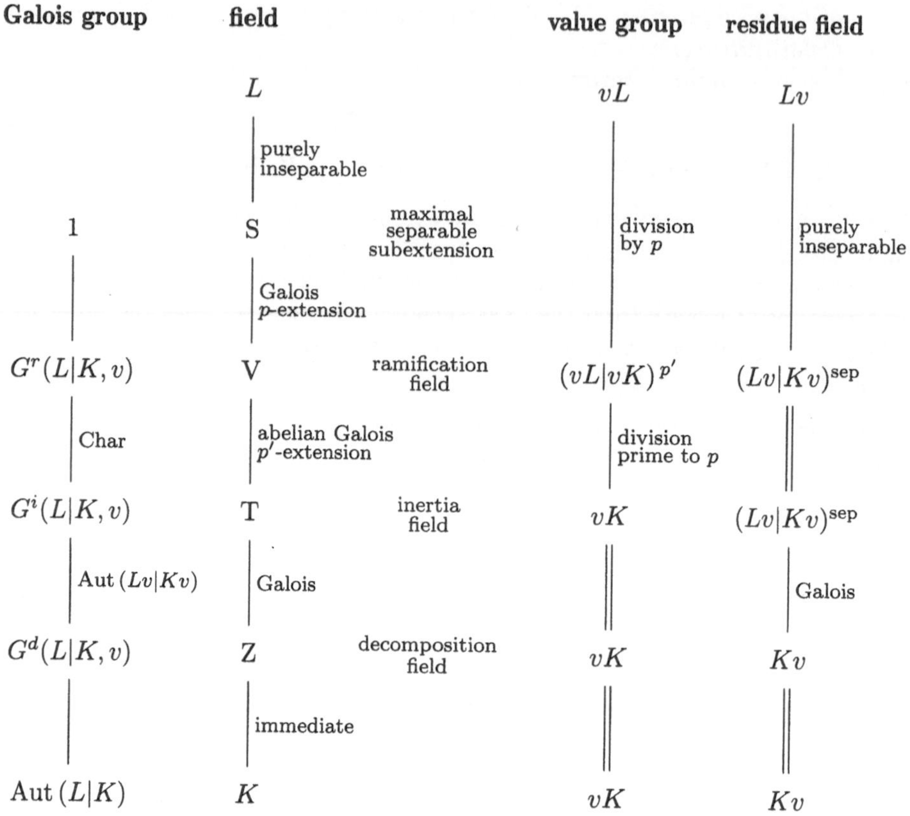

$(vL|vK)^{p'}$ denotes the p'-divisible closure of vK in vL, $(Lv|Kv)^{\mathrm{sep}}$ denotes the separable closure of Kv in Lv, and Char denotes the character group (26).

Lemma 5.12. *Fix an extension of v from K to $\tilde{K}$. Then the absolute inertia field of (K,v) is the unique maximal extension of (K,v) within the absolute ramification field having the same value group as K.*

Proof. Let $(L|K,v)$ be any extension within the absolute ramification field s.t. $vL = vK$. Then $vL^i = vL = vK = vK^i$. By Theorem 5.10, $L^i = L.K^i$. Further, $L \subseteq K^r$ yields that $L.K^i \subseteq K^r$. If the subextension $L^i|K^i$ of $K^r|K^i$ were proper, it contained a proper finite subextension $L_1|K^i$, and by part b) of Theorem 5.7 we had that $vK^i \subsetneqq vL_1 \subseteq vL^i$. As this contradicts the fact that $vL^i = vK^i$, we find that $L^i = K^i$, that is, $L \subseteq K^i$. □

From part c) of Theorem 5.4 we infer that the extension of v from K^d to $\tilde{K}$ is unique. On the other hand, if L is any extension field of K within K^d, then by Theorem 5.10, $K^d = L^d$. Thus, if $L \neq K^d$, then it follows from part b) of Theorem 5.4 that there are at least two distinct extensions of v from L to K^d and thus also to $\tilde{K} = \tilde{L}$. This proves that the absolute decomposition field K^d is a minimal algebraic extension of K admitting a unique extension of v to its algebraic closure. So it is the minimal algebraic extension of K which is henselian (cf. Lemma 4.1). We call it the *henselization of* (K, v) *in* $(\tilde{K}, v)$. Instead of K^d, we also write K^h. A valued field is henselian if and only if it is equal to its henselization. Henselizations have the following universal property:

Theorem 5.13. *Let (K, v) be an arbitrary valued field and (L, v) any henselian extension field of (K, v). Then there is a unique embedding of (K^h, v) in (L, v) over K.*

From the definition of the henselization as a decomposition field, together with part d) of Theorem 5.4, we obtain another very important property of the henselization:

Theorem 5.14. *The henselization (K^h, v) is an immediate extension of (K, v).*

Corollary 3. *Every maximal valued field is henselian. In particular, $(K((t)), v_t)$ is henselian.*

Finally, we employ Theorem 5.10 to obtain:

Theorem 5.15. *If $K'|K$ is an algebraic extension, then the henselization of K' is $K'.K^h$.*

6. A valuation theoretical interpretation of local uniformization

We return to where we stopped before entering the crash course in ramification theory. The first question is: what does it mean that $x_{k+1}, \dots, x_\ell$ together with the polynomials which define them over $K[x_1, \dots, x_k]$ satisfy the assumption of the Multidimensional Hensel's Lemma? First of all, general valuation theory tells us that a rational function field $K(x_1, \dots, x_k)$ is much too small to be henselian (unless the valuation is trivial). But we could pass to the henselization of $(K(x_1, \dots, x_k), P)$. So does it mean that $x_{k+1}, \dots, x_\ell$ lie in this henselization? If we look closely, there is something fishy in the way we have satisfied the assumption of the Multidimensional Hensel's Lemma. Instead of talking about a so-called "approximative root" $b = (b_1, \dots, b_n)$ which lies in the henselian field we wish to work in, we have talked already about the actual root, and we do not know where it lies. Let us modify our Example 3 a bit to see that it does not always lie in the henselization of $(K(x_1, \dots, x_k), v)$.

Example 7. Let us consider the function field $\mathbb{Q}(x, y)$ where $y^2 = x^3$. Take the place given by $xP = 2$, $yP = 2\sqrt{2}$. The minimal polynomial of y over $\mathbb{Q}(x)$ is $f(Y) = Y^2 - x^3$. As $f(y) = 0$, we have that $v_P f(y) = \infty > 0$. As $f'(Y) = 2Y$, we have that $v_P f'(y) = v_P 2y = 0$ (since $2yP = 4\sqrt{2} \neq 0$). Hence, f and y satisfy the assumption (4) of Hensel's Lemma. But y does not lie in the henselization of $(\mathbb{Q}(x), P)$. Indeed, P on $\mathbb{Q}(x)$ is just the place coming from the evaluation homomorphism given by $x \mapsto 2$; hence, $\mathbb{Q}(x)^h P = \mathbb{Q}(x)P = \mathbb{Q}$. But $\mathbb{Q}(x,y)P \neq \mathbb{Q}$ since $yP = 2\sqrt{2} \notin \mathbb{Q}$.

So we see that extensions of the residue field can play a role. We could try to suppress them by requiring that K be algebraically closed. This works for those P for which $FP|K$ is algebraic, but if this is not the case, then we have no chance to avoid them. At least, we can show that they are the only reason why $x_{k+1}, \ldots, x_\ell$ may not lie in the henselization of $(K(x_1, \ldots, x_k), P)$.

Theorem 6.1. *If $x_{k+1}, \ldots, x_\ell$ together with the polynomials f_i which define them over $K[x_1, \ldots, x_k]$ satisfy the assumption (5) of the Multidimensional Hensel's Lemma, then $x_{k+1}, \ldots, x_\ell$ lie in the absolute inertia field of $(K(x_1, \ldots, x_k), P)$, and the extension $FP|K(x_1P, \ldots, x_kP)$ is separable-algebraic. If in addition P is a rational place, then $x_{k+1}, \ldots, x_\ell$ lie in the henselization of $(K(x_1, \ldots, x_k), P)$.*

Proof. Denote by (L, P) the absolute inertia field of $(K(x_1, \ldots, x_k), P)$. First,

$$\det J_{\tilde{f}P}(x_{k+1}P, \ldots, x_\ell P) = \det J_{\tilde{f}}(x_{k+1}, \ldots, x_\ell)P \neq 0 \tag{27}$$

and the fact that the f_iP are polynomials over $K(x_1P, \ldots, x_kP)$ imply that $x_{k+1}P, \ldots, x_\ell P$ are separable algebraic over $K(x_1P, \ldots, x_kP)$ (cf. [L2], Chapter X, §7, Proposition 8). On the other hand, LP is the separable-algebraic closure of $K(x_1, \ldots, x_k)P$. Therefore, there are elements $b_1, \ldots, b_n$ in L such that $b_iP = x_{k+i}P$. Since (L, P) is henselian, the Multidimensional Hensel's Lemma now shows the existence of a common root $(b'_1, \ldots, b'_n) \in L^n$ of the f_i such that $b'_iP = b_iP = x_{k+i}P$. But the uniqueness assertion of the Multidimensional Hensel's Lemma also holds in the algebraic closure $\tilde{L}$ of L (which is also henselian). So we find that $(b'_1, \ldots, b'_n) = (x_{k+1}, \ldots, x_\ell)$. Hence, $x_{k+1}, \ldots, x_\ell$ are elements of L.

If we have in addition that P is a rational place, then $x_{k+1}P, \ldots, x_\ell P \in K$. In this case, we can choose $b_1, \ldots, b_n$ and $b'_1, \ldots, b'_n$ already in the henselization of $(K(x_1, \ldots, x_k), P)$, which implies that also $x_{k+1}, \ldots, x_\ell$ lie in this henselization. □

Since the absolute inertia field is a separable-algebraic extension and every rational function field is separable, we obtain:

Corollary 4. *If the place P of $F|K$ admits local uniformization, then $F|K$ is separable.*

We see that we are slowly entering the *structure theory of valued function fields*, that is, the algebraic theory of function fields $F|K$ equipped with a valuation (which may or may not be trivial on K). Later, we will see some main results from this theory (Theorems 14.1 and 17.4).

Given a place P of F, not necessarily trivial on K, we will say that $(F|K, P)$ is *inertially generated* if there is a transcendence basis T of $F|K$ such that (F, P) lies in the absolute inertia field of $(K(T), P)$. Similarly, $(F|K, P)$ is *henselian generated* if there is a transcendence basis T of $F|K$ such that (F, P) lies in henselization of $(K(T), P)$. Now we see a valuation theoretical interpretation of local uniformization:

Theorem 6.2. *If the place P of $F|K$ admits local uniformization, then $(F|K, P)$ is inertially generated. If in addition $FP = K$, then $(F|K, P)$ is henselian generated.*

So if local uniformization holds in arbitrary characteristic for every $F|K$ with perfect K, then for every place P of $F|K$, the valued function field $(F|K, P)$ is inertially generated. In the context of valuation theory, at least to me, this is a quite surprising assertion. Here is our first open problem:

Open Problem 1: Is the converse also true, i.e., if $(F|K, P)$ is inertially generated, does it then admit local uniformization?

I will discuss this question in Section 15. A partial answer to this question is given in the papers [K5] and [K6]. What we see is that in order to get local uniformization, one has to avoid ramification. Indeed, ramification is the valuation theoretical symptom of branching, the violation of the Implicit Function Theorem at a point of the variety. Let us look again at our simple Example 3:

Example 8. Consider the function field $\mathbb{R}(x, y)$ where $y^2 = x^3$. Take the place given by $xP = 0 = yP$. As P on $K(x)$ originates from the evaluation homomorphism given by $x \mapsto 0$, we have that $v_P K(x) = \mathbb{Z}$, with $v_P x = 1$ the smallest positive element in the value group. Now compute $v_P y$. We have that $y^2 = x^3$, whence $2vy = vy^2 = vx^3 = 3$. It follows that $vy = 3/2 \notin \mathbb{Z}$, that is, the extension $(K(x, y)|K(x), P)$ is *ramified*, or in other words, $(K(x, y), P)$ does not lie in the absolute inertia field of $(K(x), P)$. We see that we have ramification at the singular point $(0, 0)$. As an exercise, you may check that $(K(x, y), Q)$ lies in the absolute inertia field of $(K(x), Q)$ whenever $xQ \neq 0$.

7. Inertial generation and Abhyankar places

We may now ask ourselves: How could we show that for a given place P of $F|K$, the valued function field $(F|K, P)$ is inertially generated?

Example 9. Let us start with the most simple case, where $\operatorname{trdeg} F|K = 1$. Assuming that P is not trivial on F (if it is trivial, then local uniformization is trivial if $F|K$ is separable), we pick some $z \in F$ such that $zP = 0$. As we have seen in Example 1, $v_P K(z) = \mathbb{Z}$. Since $z \notin K$ and $\operatorname{trdeg} F|K = 1$, we know that $F|K(z)$ is algebraic; since $F|K$ is finitely generated, it follows that $F|K$ is finite. From Theorem 5.2 we infer that the ramification index $(v_P F : v_P K(z))$ is finite. Therefore, $v_P F$ is again isomorphic to $\mathbb{Z}$ and we can pick some $x \in F$ such that $x \in \mathcal{O}_P$ and $v_P F = \mathbb{Z} v_P x$.

We have achieved that $v_P F = v_P K(x)$. If $\operatorname{char} FP = \operatorname{char} K$ is 0, then we know from Lemma 5.12 that the absolute inertia field $K(x)^i$ is the unique maximal extension still having the same value group as $K(x)$. In this case, we find that F

must lie in this absolute inertia field, and we have proved that $(F|K, P)$ is inertially generated. But we are lost, it seems, if the characteristic is $p > 0$, since in this case, the absolute inertia field is not necessarily the maximal algebraic extension of $K(x)$ having the same value group. To solve this case, we yet have to learn some additional tools.

In this example, the fact that $v_P F$ was finitely generated played a crucial role. As we have shown, this is always the case if $\operatorname{trdeg} F|K = 1$. But in general, we can't expect this to hold. We will give counterexamples in Section 18. But prior to the negative, we want to start with the positive, i.e., criteria for the value group to be finitely generated.

The following theorem has turned out in the last years to be amazingly universal in many different applications of valuation theory. It plays an important role in algebraic geometry as well as in the model theory of valued fields, in real algebraic geometry, or in the structure theory of exponential Hardy fields (= nonarchimedean ordered fields which encode the asymptotic behaviour of real-valued functions including exp and log, cf. [KK]). For more details and the easy proof of the theorem, see [V], Theorem 5.5, or [B], Chapter VI, §10.3, Theorem 1, or [K2].

Theorem 7.1. *Let $(L|K, P)$ be an extension of valued fields. Take $x_i, y_j \in L$, $i \in I$, $j \in J$, such that the values $v_P x_i$, $i \in I$, are rationally independent over $v_P K$, and the residues $y_j P$, $i \in J$, are algebraically independent over KP. Then the elements x_i, y_j, $i \in I$, $j \in J$, are algebraically independent over K, the value of each polynomial in $K[x_i, y_j \mid i \in I, j \in J]$ is equal to the least of the values of its monomials, and*

$$v_P K(x_i, y_j \mid i \in I, j \in J) = v_P K \oplus \bigoplus_{i\in I} \mathbb{Z} v_P x_i \qquad (28)$$

$$K(x_i, y_j \mid i \in I, j \in J)P = KP(y_j P \mid j \in J)\,. \qquad (29)$$

Moreover, the valuation v_P on $K(x_i, y_j \mid i \in I, j \in J)$ is uniquely determined by its restriction to K, the values $v_P x_i$ and the residues $y_j P$.

For the proof of the following corollary, see [V] or [K2].

Corollary 5. *Let $(L|K, P)$ be an extension of valued fields of finite transcendence degree. Then*

$$\operatorname{trdeg} L|K \geq \operatorname{trdeg} LP|KP + \operatorname{rr}(v_P L / v_P K)\,. \qquad (30)$$

If in addition $L|K$ is a function field, and if equality holds in (30), then the extensions $v_P L | v_P K$ and $LP|KP$ are finitely generated. In particular, if P is trivial on K, then $v_P L$ is a product of finitely many (namely, $\operatorname{rr} v_P L$) copies of $\mathbb{Z}$, and LP is again a function field over K.

If P is a place of $F|K$, then (30) reads as follows:

$$\operatorname{trdeg} F|K \geq \operatorname{trdeg} FP|K + \operatorname{rr} v_P F\,. \qquad (31)$$

The famous *Abhyankar inequality* is a generalization of this inequality to the case of noetherian local rings (see [V]). We call P an *Abhyankar place* if equality holds in (31).

The rank of an ordered abelian group is always smaller or equal to its rational rank. This is seen as follows. If G_1 is a subgroup of G, then its divisible hull $\mathbb{Q}\otimes G_1$ lies in the convex hull of G_1 in $\mathbb{Q}\otimes G$. Hence if G_1 is a proper convex subgroup of G, then $\mathbb{Q}\otimes G_1$ is a proper convex subgroup of $\mathbb{Q}\otimes G$ and thus, $\dim_{\mathbb{Q}}\mathbb{Q}\otimes G_1 < \dim_{\mathbb{Q}}\mathbb{Q}\otimes G$. It follows that if $\{0\} = G_0 \subsetneqq G_1 \subsetneqq \cdots \subsetneqq G_n = G$ is a chain of convex subgroups of G, then $\mathrm{rr}\, G = \dim_{\mathbb{Q}}\mathbb{Q}\otimes G \geq n$. In view of (31), this proves that the rank of a place P of a function field $F|K$ cannot exceed $\mathrm{trdeg}\, F|K$ and thus is finite. We say that P is of *maximal rank* if the rank is equal to $\mathrm{trdeg}\, F|K$.

If $\mathrm{trdeg}\, F|K = 1$, then every place P of $F|K$ is an Abhyankar place. It is of maximal rank if and only if it is non-trivial. Indeed, if $v_P F$ is not trivial, then $\mathrm{rr}\, v_P F \geq 1$, and it follows from (31) that $\mathrm{trdeg}\, F|K = 1 = \mathrm{rr}\, v_P F$. Then also the rank is 1 since a group of rational rank 1 is a non-trivial subgroup of $\mathbb{Q}$. If on the other hand $v_P F$ is trivial, then P is an isomorphism on F so that $\mathrm{trdeg}\, F|K = 1 = \mathrm{trdeg}\, FP|K$.

Using Corollary 5, we can now generalize our construction given in Example 9. Let P be an arbitrary place of $F|K$. We set $\rho = \mathrm{rr}\, v_P F$ and $\tau = \mathrm{trdeg}\, FP|K$. We take elements $x_1,\ldots,x_\rho \in F$ such that $v_P x_1,\ldots,v_P x_\rho$ are rationally independent elements in $v_P F$. Further, we take elements $y_1,\ldots,y_\tau \in F$ such that $y_1P,\ldots,y_\tau P$ are algebraically independent over K. Then by Theorem 7.1, $x_1,\ldots,x_\rho,y_1,\ldots,y_\tau$ are algebraically independent over K. The restriction of P to $K(x_1,\ldots,x_\rho,y_1,\ldots,y_\tau)$ is an Abhyankar place. We fix this situation for later use. We call

$$\left.\begin{array}{l} F_0 := K(x_1,\ldots,x_\rho,y_1,\ldots,y_\tau) \text{ with} \\ \rho = \mathrm{rr}\, v_P F \text{ and } \tau = \mathrm{trdeg}\, FP|K, \\ v_P x_1,\ldots,v_P x_\rho \text{ rationally independent in } v_P F, \text{ and} \\ y_1P,\ldots,y_\tau P \text{ algebraically independent over } K \end{array}\right\} \qquad (32)$$

an *Abhyankar subfunction field* of $(F|K,P)$.

Now let us assume in addition that P is an Abhyankar place of $F|K$. That is, $\rho+\tau = \mathrm{trdeg}\, F|K$. It follows that $x_1,\ldots,x_\rho,y_1,\ldots,y_\tau$ is a transcendence basis of $F|K$. We refine our choice of these elements as follows. From Corollary 5 we know that $v_P F$ is product of ρ copies of $\mathbb{Z}$. So we can choose $x_1,\ldots,x_\rho \in \mathcal{O}_P$ in such a way that $v_P F = \mathbb{Z}v_P x_1 \oplus \ldots \oplus \mathbb{Z}v_P x_\rho$, which implies that $v_P F = v_P K(x_1,\ldots,x_\rho)$. From Corollary 5 we also know that $FP|K$ is finitely generated. We shall also assume that $FP|K$ is separable. Then it follows that there is a separating transcendence basis for $FP|K$. We choose $y_1,\ldots,y_\tau \in \mathcal{O}_P$ in such a way that $y_1P,\ldots,y_\tau P$ is such a separating transcendence basis. Now we can choose some $a \in FP$ such that $FP = K(y_1P,\ldots,y_\tau P,a)$. We take a monic polynomial f with coefficients in the valuation ring of (F_0,P) such that its reduction fv_P

is the minimal polynomial of a over $F_0P = K(y_1P,\dots,y_\tau P)$. Since $a \in FP$ is separable-algebraic over $K(y_1P,\dots,y_\tau P)$, by Hensel's Lemma (Simple Root Version) there exists a root η of f in the henselization of (F,P) such that $\eta P = a$. Take $\sigma \in \mathrm{Aut}(\tilde{F}_0|F_0)$ such that $v(\sigma x - x) > 0$ for all x in the valuation ring of P on $\tilde{F}$. Then in particular, $v(\sigma\eta - \eta) > 0$. But if $\sigma\eta \neq \eta$, then it follows from $\deg(f) = \deg(fv_P)$ that $(\sigma\eta)P \neq \eta P$, i.e., $v(\sigma\eta - \eta) = 0$. Hence, $\sigma\eta = \eta$, which shows that η lies in the absolute inertia field of F_0.

Now the field $F_0(\eta)$ has the same value group and residue field as F, and it is contained in the henselization of F. Hence by Theorem 5.14,

$$(F^h|F_0(\eta)^h, P) \tag{33}$$

is an immediate algebraic extension. As η lies in the absolute inertia field of F_0 and this field is henselian, we have that $F_0(\eta)^h$ is a subfield of this absolute inertia field. If we could show that $F^h = F_0(\eta)^h$, then F itself would lie in this absolute inertia field, which would prove that $(F|K,P)$ is inertially generated. If the residue characteristic $\mathrm{char}\, FP = \mathrm{char}\, K$ is 0, then again Lemma 5.12 tells us that the absolute inertia field of (F_0,P) is the unique maximal extension having the same value group as F_0; so F^h must be a subfield of it. Hence in characteristic 0 we have now shown that $(F|K,P)$ is inertially generated. But what happens in positive characteristic? Can the extension (33) be non-trivial? To answer this question, we have to take a closer look at the main problem of valuation theory in positive characteristic.

8. The defect

Assume that (K,v) is henselian and $(L|K,v)$ is a finite extension of degree n. Then we have to deal only with a single ramification index e and a single inertia degree f. Hence, the fundamental inequality now reads as

$$n \geq \mathrm{e}\,\mathrm{f}\,. \tag{34}$$

If L is contained in K^i then by Theorem 5.6, $n = \mathrm{f}$. If $K = K^i$ and L is contained in K^r, then by Theorem 5.7, $n = \mathrm{e}$. Putting these observations together (using Theorems 5.10 and 5.11 and the fact that extension degree, ramification index and inertia degree are multiplicative), one finds:

Lemma 8.1. *If (K,v) is henselian and $L|K$ is a finite subextension of $K^r|K$, then it satisfies the* fundamental equality

$$n = \mathrm{e}\,\mathrm{f}\,. \tag{35}$$

Hence, an inequality can only result from some part of the extension which lies beyond the absolute ramification field. So Theorem 5.8 shows that the missing factor can only be a power of p. In this way, one proves the important *Lemma of Ostrowski:*

Theorem 8.2. *Set* $p := \operatorname{char} Kv$ *if this is positive, and* $p := 1$ *if* $\operatorname{char} Kv = 0$. *If* (K, v) *is henselian and* $L|K$ *is of degree* n, *then*

$$n = \mathrm{d\,e\,f}\,, \tag{36}$$

where d *is a power of* p. *In particular, if* $\operatorname{char} Kv = 0$, *then we always have the fundamental equality* $n = \mathrm{e\,f}$.

The integer $\mathrm{d} \geq 1$ is called the *defect* of the extension $(L|K, v)$. This can also be taken as a definition for the defect if (K, v) is not henselian, but the extension of v from K to L is unique. We note:

Corollary 6. *If* (K, v) *is henselian and* $(L|K, v)$ *is a finite immediate extension, then the defect of* $(L|K, v)$ *is equal to* $[L : K]$.

A henselian field is called *defectless* if every finite extension has trivial defect $\mathrm{d} = 1$. In rigid analysis, this is also called *stable.* A not necessarily henselian field is called defectless if for every finite extension of it, equality holds in the fundamental inequality (14) (if the field is henselian, this coincides with our first definition). A proof of the next theorem can be found in [K2] and, partially, also in [E].

Theorem 8.3. *A valued field is a defectless field if and only if its henselization is.*

We also note the following fact, which is easy to prove:

Lemma 8.4. *Every finite extension field of a defectless field is again a defectless field.*

The following are examples of defectless fields:

(DF1) All valued fields with residue characteristic 0. This is a direct consequence of the Lemma of Ostrowski.

(DF2) Every discretely valued field of characteristic 0. An easy argument shows that every finite extension with non-trivial defect of a discretely valued field must be inseparable. In particular, the field $(\mathbb{Q}_p, v_p)$ of p-adic numbers with its p-adic valuation, and all of its subfields, are defectless fields.

(DF3) All maximal fields (and hence also all power series fields, see Section 9) are defectless fields. For the proof, see [R1] or [K2].

We have seen that the extensions beyond the absolute ramification field are responsible for non-trivial defects. To get this picture, we have chosen a modified approach to ramification theory (cf. our remark in Section 5). We have shifted the purely inseparable extensions to the top (cf. our table). In fact, that is where the purely inseparable extensions belong, because from the ramification theoretical point of view, they can be nasty, and in this respect, they have much in common with the extension $\mathrm{S}|\mathrm{V}$.

The defect, appearing only for positive residue characteristic, is essentially the cause of the problems that we have in algebraic geometry as well as in the model theory of valued fields, in positive characteristic. Therefore, it is very important that you get a feeling for what the defect is. Let us look at three main examples. The first one is the most basic and was probably already known to most of the early valuation theorists. But it seems reasonable to attribute it to F. K. Schmidt.

Example 10. We consider the power series field $K((t))$ with its t-adic valuation $v = v_t$. We have already remarked in Example 2 that $K((t))|K(t)$ is transcendental. So we can choose an element $s \in K((t))$ which is transcendental over $K(t)$. Since $vK((t)) = \mathbb{Z} = vK(t)$ and $K((t))v = K = K(t)v$, the extension $(K((t))|K(t), v)$ is immediate. The same must be true for the subextension $(K(t,s)|K(t), v)$ and thus also for $(K(t,s)|K(t,s^p), v)$. The latter extension is purely inseparable of degree p (since s,t are algebraically independent over K, the extension $K(s)|K(s^p)$ is linearly disjoint from $K(t,s^p)|K(s^p)$). Hence, Corollary 2 shows that there is only one extension of the valuation v from $K(t,s^p)$ to $K(t,s)$. Consequently, its defect is p.

To give an example of a henselian field which is not defectless, we build on the foregoing example.

Example 11. By Theorem 5.13, there is a henselization $(K(t,s),v)^h$ of the field $(K(t,s),v)$ in the henselian field $K((t))$ and a henselization $(K(t,s^p),v)^h$ of the field $(K(t,s^p),v)$ in $(K(t,s),v)^h$. We find that the extension $K(t,s)^h|K(t,s^p)^h$ is again purely inseparable of degree p. Indeed, $K(t,s)|K(t,s^p)$ is linearly disjoint from the separable extension $K(t,s^p)^h|K(t,s^p)$, and by virtue of Corollary 5.15, $K(t,s)^h = K(t,s).K(t,s^p)^h$. Also for this extension, we have that $\mathrm{e} = \mathrm{f} = 1$ and again, the defect is p. Note that by what we have said earlier, an extension of degree p with non-trivial defect over a discretely valued field like $(K(t,s^p),v)^h$ can only be purely inseparable.

Now we will give an example of a finite *separable* extension with non-trivial defect. It seems to be the generic example for our purposes since its importance is also known in algebraic geometry.

Example 12. Take an arbitrary field K of characteristic $p > 0$, and t to be transcendental over K. On $K(t)$ we take the t-adic valuation $v = v_t$. We set $L := K(t^{1/p^i} \mid i \in \mathbb{N})$. This is a purely inseparable extension of $K(t)$; if K is perfect, then it is the perfect hull of $K(t)$. By Corollary 2, v has a unique extension to L. We set $L_k := K(t^{1/p^k})$ for every $k \in \mathbb{N}$; so $L = \bigcup_{k\in\mathbb{N}} L_k$. We observe that $1/p^k = vt^{1/p^k} \in vL_k$, so $(vL_k : vK(t)) \geq p^k$. Now the fundamental inequality shows that $(vL_k : vK(t)) = p^k$ and that $L_k v = K(t)v = K$. The former shows that $vL_k = \frac{1}{p^k}\mathbb{Z}$. We obtain that $vL = \bigcup_{k\in\mathbb{N}} vL_k = \frac{1}{p^\infty}\mathbb{Z}$ and that $Lv = \bigcup_{k\in\mathbb{N}} L_k v = K$. We consider the extension $L(\vartheta)|L$ generated by a root ϑ of the Artin-Schreier-polynomial $X^p - X - \frac{1}{t}$. We set

$$\vartheta_k := \sum_{i=1}^{k} t^{-1/p^i} \qquad (37)$$

and compute

$$\begin{aligned}\vartheta_k^p - \vartheta_k - \frac{1}{t} &= \sum_{i=1}^{k} t^{-1/p^{i-1}} - \sum_{i=1}^{k} t^{-1/p^i} - t^{-1} \\ &= \sum_{i=0}^{k-1} t^{-1/p^i} - \sum_{i=1}^{k} t^{-1/p^i} - t^{-1} = -t^{-1/p^k} .\end{aligned}$$

Therefore,

$$\begin{aligned}(\vartheta - \vartheta_k)^p - (\vartheta - \vartheta_k) &= \vartheta^p - \vartheta - \frac{1}{t} - \left(\vartheta_k^p - \vartheta_k - \frac{1}{t}\right) \\ &= 0 + t^{-1/p^k} = t^{-1/p^k} .\end{aligned}$$

If we have the equation $b^p - b = c$ and know that $vc < 0$, then we can conclude that $vb < 0$ since otherwise, $v(b^p - b) \geq 0 > vc$, a contradiction. But since $vb < 0$, we have that $vb^p = pvb < vb$, which implies that $vc = v(b^p - b) = pvb$. Consequently, $vb = \frac{vc}{p}$. In our case, we obtain that

$$v(\vartheta - \vartheta_k) = \frac{vt^{-1/p^k}}{p} = -\frac{1}{p^{k+1}} . \tag{38}$$

We see that $1/p^{k+1} \in vL_k(\vartheta)$, so that $(vL_k(\vartheta) : vL_k) \geq p$. Since $[L_k(\vartheta) : L_k] \leq p$, the fundamental inequality shows that $[L_k(\vartheta) : L_k] = p$, $vL_k(\vartheta) = 1/p^{k+1}\mathbb{Z}$, $L_k(\vartheta)v = L_k v = K$ and that the extension of the valuation v from L_k to $L_k(\vartheta)$ is unique. As $L(\vartheta) = \bigcup_{k\in\mathbb{N}} L_k(\vartheta)$, we obtain that $vL(\vartheta) = \bigcup_{k\in\mathbb{N}} 1/p^{k+1}\mathbb{Z} = \frac{1}{p^\infty}\mathbb{Z} = vL$ and that $L(\vartheta)v = \bigcup_{k\in\mathbb{N}} L_k v = K = Lv$. We have thus proved that the extension $(L(\vartheta)|L, v)$ is immediate. Since ϑ has degree p over every L_k, it must also have degree p over their union L. Further, as the extension of v from L_k to $L_k(\vartheta)$ is unique for every k, also the extension of v from L to $L(\vartheta)$ is unique. So we have $n = p$, $\mathrm{e} = \mathrm{f} = \mathrm{g} = 1$, and we find that the defect of $(L(\vartheta)|L, v)$ is p.

To obtain a defect extension of a henselian field, we show that L can be replaced by its henselization. By Theorem 5.15, $L^h(\vartheta) = L^h.L(\vartheta) = L(\vartheta)^h$. By Theorem 5.14, $vL(\vartheta)^h = vL(\vartheta) = vL = vL^h$ and $L(\vartheta)^h v = L(\vartheta)v = Lv = L^h v$. Hence, also the extension $(L^h(\vartheta)|L^h, v)$ is immediate. We only have to show that it is of degree p. This follows from the general valuation theoretical fact that if an extension $L'|L$ admits a unique extension of the valuation v from L to L', then $L'|L$ is linearly disjoint from $L^h|L$. But we can also give a direct proof. Again by Theorem 5.15, $L_k^h(\vartheta) = L_k(\vartheta)^h$, and by Theorem 5.14, $vL_k(\vartheta)^h = vL_k(\vartheta)$ and $vL_k^h = vL_k$. Therefore, $(vL_k^h(\vartheta) : vL_k^h) = (vL_k(\vartheta) : vL_k) = p$, showing that $[L_k^h(\vartheta) : L_k^h] = p$ for every k. Again by Theorem 5.15, $L^h = L.L_1^h = (\bigcup_{k\in\mathbb{N}} L_k).L_1^h = \bigcup_{k\in\mathbb{N}} L_k.L_1^h = \bigcup_{k\in\mathbb{N}} L_k^h$. By the same argument as before, it follows that $[L^h(\vartheta) : L^h] = p$.

Hence, we have found an immediate Artin-Schreier extension of degree p and defect p of a henselian field which is only of transcendence degree 1 over K.

A valued field (K, v) is called *algebraically maximal* if it admits no proper immediate algebraic extension, and it is called *separable-algebraically maximal* if it admits no proper immediate separable-algebraic extension. Since the henselization is an immediate separable-algebraic extension by Theorem 5.14, every separable-algebraically maximal field is henselian. The converse is not true, since the field (L^h, v) of our foregoing example is henselian but not separable-algebraically maximal. Corollary 6 shows that every henselian defectless field is algebraically maximal. The converse is not true, as was shown by Françoise Delon [DEL] (cf. also [K2]).

Example 13. In the foregoing example, we may replace $K(t)$ by $K((t))$, taking L to be the field $K((t))(t^{1/p^k} \mid k \in \mathbb{N})$. It is not hard to show (by splitting up the power series in a suitable way) that $K((t^{1/p^k})) = K((t))[t^{1/p^k}]$, which is algebraic over $K((t))$. Hence, $L = \bigcup_{k\in\mathbb{N}} K((t^{1/p^k}))$, a union of an ascending chain of power series fields. By Lemma 12.2 below, (L, v) is henselian, and $(L(\vartheta)|L, v)$ gives an instant example of an immediate extension of a henselian field. But this L is "very large": it is of infinite transcendence degree over K. On the other hand, this version of our example shows that an infinite algebraic extension of a maximal field (or a union over an ascending chain of maximal fields) is in general not even defectless (and hence also not maximal). The example can also be transformed into the p-adic situation, showing that there are infinite extensions of $(\mathbb{Q}_p, v_p)$ which are not defectless fields (cf. [K2]).

9. Maximal immediate extensions

Based on our examples, we can observe another obstruction in positive characteristic. In many applications of valuation theory, one is interested in the embedding of a given valued field in a power series field which, if possible, should have the same value group and residue field. (We give an example relevant for algebraic geometry in the next section.) Then this power series field would be an immediate extension of our field, and since every power series field is maximal, it would be a *maximal immediate extension* of our field. So we see that we are led to the problem of determining maximal immediate extensions, in particular, whether maximal immediate extensions of a given valued field are unique up to valuation preserving isomorphism. It was shown by Wolfgang Krull [KR] that maximal immediate extensions exist for every valued field. The proof uses Zorn's Lemma in combination with an upper bound for the cardinality of valued fields with prescribed value group and residue field. Krull's deduction of this upper bound is hard to read; later, Kenneth A. H. Gravett [GRA] gave a nice and simple proof.

The uniqueness problem for maximal immediate extensions was considered by Irving Kaplansky in his important paper [KA1]. He showed that if the so-called *hypothesis A* holds, then the field has a unique maximal immediate extension (up to valuation preserving isomorphism). For a Galois theoretic interpretation of hypothesis A and more information about the uniqueness problem, see [KPR]. Let us mention a problem which was only partially solved in [KPR] and in [WA1]:

Open Problem 2: If a valued field does not satisfy Kaplansky's hypothesis A, does it then admit two non-isomorphic maximal immediate extensions?

We can give a quick example of a valued field with two non-isomorphic maximal immediate extensions.

Example 14. In the setting of Example 13, suppose that K is not *Artin-Schreier closed*, that is, there is an element $c \in K$ such that $X^p - X - c$ is irreducible over K. Take ϑ_c to be a root of $X^p - X - (\frac{1}{t} + c)$; note that $v(\frac{1}{t} + c) = v\frac{1}{t} < 0$ since $vc = 0 > v\frac{1}{t}$. Then in exactly the same way as for ϑ, one shows that the extension $(L(\vartheta_c)|L, v)$ is immediate of degree p and defect p. So we have two distinct immediate extensions of L. We take (M_1, v) to be a maximal immediate extension of $L(\vartheta)$, and (M_2, v) to be a maximal immediate extension of $L(\vartheta_c)$. Then (M_1, v) and (M_2, v) are also maximal immediate extensions of (L, v). If they were isomorphic over L, then M_1 would also contain a root of $X^p - X - (\frac{1}{t} + c)$; w.l.o.g., we can assume that it is the one called ϑ_c. Now we compute:

$$(\vartheta_c - \vartheta)^p - (\vartheta_c - \vartheta) \;=\; \vartheta_c^p - \vartheta_c - (\vartheta^p - \vartheta) \;=\; \frac{1}{t} + c - \frac{1}{t} \;=\; c\,.$$

Since $vc = 0$, we also have that $v(\vartheta_c - \vartheta) = 0$ (you may prove this along the lines of an argument given earlier). Applying the residue map to $\vartheta_c - \vartheta$, we thus obtain a root of $X^p - X - c$. But by our assumption, this root is not contained in $K = Lv$. Consequently, $M_1 v \neq Lv$, contradicting the fact that (M_1, v) was an immediate extension of (L, v). This proves that (M_1, v) and (M_2, v) cannot be isomorphic over (L, v).

We have used that K is not Artin-Schreier closed. And in fact, one of the consequences of hypothesis A for a valued field (L, v) is that its residue field be Artin-Schreier closed (see Section 11).

We will need a generalization of the field of formal Laurent series, called *(generalized) power series field.* Take any field K and any ordered abelian group G. We take $K((G))$ to be the set of all maps μ from G to K with well-ordered *support* $\{g \in G \mid \mu(g) \neq 0\}$. You can visualize the elements of $K((G))$ as formal power series $\sum_{g\in G} c_g t^g$ for which the support $\{g \in G \mid c_g \neq 0\}$ is well ordered. Using this condition one shows that $K((G))$ is a field in a similar way as it is done for $K((t))$. Also, one uses it to introduce the valuation:

$$v \sum_{g\in G} c_g t^g \;=\; \min\{g \in G \mid c_g \neq 0\} \tag{39}$$

(the minimum exists because the support is well ordered). This valuation is often called the *canonical valuation of* $K((G))$, and sometimes called the *minimum support valuation.* With this valuation, $K((G))$ is a maximal field.

The fields L constructed in Examples 12 and 13 are subfields of $K((\mathbb{Q}))$ in a canonical way. It is interesting to note that the element ϑ is an element of $K((\mathbb{Q}))$:

$$\vartheta = \sum_{i\in\mathbb{N}} t^{-1/p^i} = t^{-1/p} + t^{-1/p^2} + \ldots + t^{-1/p^i} + \ldots . \tag{40}$$

Indeed,

$$\begin{aligned}\vartheta^p - \vartheta - \frac{1}{t} &= \sum_{i\in\mathbb{N}} t^{-1/p^{i-1}} - \sum_{i\in\mathbb{N}} t^{-1/p^i} - t^{-1}\\ &= \sum_{i=0}^{\infty} t^{-1/p^i} - \sum_{i=1}^{\infty} t^{-1/p^i} - t^{-1} = 0 .\end{aligned}$$

Note that the values $vt^{-1/p^n} = -1/p^n$ converge from below to 0. Therefore, ϑ does not even lie in the completion of L. In fact, there cannot be a root of $X^p - X - 1/t$ in the completion; if a would be such a root, then there would be some $b \in L$ such that $v(a-b) > 0$. We would have that

$$(a-b)^p - (a-b) = a^p - a - (b^p - b) = \frac{1}{t} - (b^p - b) . \tag{41}$$

Because of $v(a-b) > 0$, the left-hand side and consequently also the right-hand side has value > 0. But as we have seen in Example 4, the polynomial $X^p - X - c$ splits over every henselian field containing c if $vc > 0$. Hence, in the cases where L is henselian, there exists a root $a' \in L$ of $X^p - X - 1/t + b^p - b$. It follows that $(a'+b)^p - (a'+b) - 1/t = 1/t - (b^p - b) + b^p - b - 1/t = 0$. As $a' + b \in L$, this would imply that $X^p - X - 1/t$ splits over L, a contradiction.

Let us illustrate the influence of the defect by considering an object which is well known in algebraic geometry.

10. A quick look at Puiseux series fields

Recall that $K((\mathbb{Q}))$ is the field of all formal sums $\sum_{q\in\mathbb{Q}} c_q t^q$ with $c_q \in K$ and well-ordered support. The subset

$$\mathrm{P}(K) := \left\{ \sum_{i=n}^{\infty} c_i t^{i/k} \mid c_i \in K,\, n \in \mathbb{Z},\, k \in \mathbb{N} \right\} = \bigcup_{k\in\mathbb{N}} K((t^{1/k})) \subset K((\mathbb{Q})) \tag{42}$$

is itself a field, called the *Puiseux series field over* K. Here, the valuation v on $K((t^{1/k}))$ is again the minimum support valuation, in particular, we have that $vt^{1/k} = 1/k$. In this way, the valuation v on every $K((t^{1/k}))$ is an extension of the t-adic valuation v_t of $K((t))$ and of the valuation of every subfield $K((t^{1/m}))$ where m divides k.

$\mathrm{P}(K)$ can also be written as a union of an ascending chain of power series fields in the following way. We take p_i to be the i-th prime number and set $m_k :=$

$\prod_{i=1}^{k} p_i^k$. Then $m_k | m_{k+1}$ and thus $K((t^{1/m_k})) \subset K((t^{1/m_{k+1}}))$ for every $k \in \mathbb{N}$, and every natural number will divide m_k for large enough k. Therefore,

$$\mathrm{P}(K) = \bigcup_{k\in\mathbb{N}} K((t^{1/m_k}))\,. \tag{43}$$

If one does not want to work in the power series field $K((\mathbb{Q}))$, then one simply has to choose a compatible system of k-th roots $t^{1/k}$ of t (that is, for $k = \ell m$ we must have $(t^{1/k})^\ell = t^{1/m}$; this is automatic for the elements $t^{1/k} \in K((\mathbb{Q}))$ by definition of the multiplication in this field). Then (42) can serve as a definition for the Puiseux series field over K.

Lemma 10.1. *The Puiseux series field $P(K)$ is an algebraic extension of $K((t))$, and it is henselian with respect to its canonical valuation v. Its residue field is K and its value group is $\mathbb{Q}$.*

Proof. For every $k \in \mathbb{N}$, the element $t^{1/k}$ is algebraic over $K((t))$. Similarly as in Example 13, we have that $K((t^{1/k})) = K((t))[t^{1/k}]$, which is algebraic over $K((t))$. Consequently, also the union $\mathrm{P}(K)$ of the $K((t^{1/k}))$ is algebraic over $K((t))$. By Corollary 3, $K((t))$ is henselian w.r.t. its canonical valuation v_t. As the canonical valuation v of $\mathrm{P}(K)$ is an extension of v_t, Corollary 1 yields that $(\mathrm{P}(K), v)$ is henselian.

The value group of every $(K((t^{1/k})), v)$ is $\mathbb{Z}vt^{1/k} = \mathbb{Z}\frac{vt}{k} = \frac{1}{k}\mathbb{Z}$, so the union over all $K((t^{1/k}))$ has value group $\bigcup_{k\in\mathbb{N}} \frac{1}{k}\mathbb{Z} = \mathbb{Q}$. The residue field of every $(K((t^{1/k})), v)$ is K, hence also the residue field of their union is K. □

Theorem 10.2. *Let K be a field of characteristic* 0. *Then* $(\mathrm{P}(K), v)$ *is a defectless field. Further, $P(K)$ is the algebraic closure of $K((t))$ if and only if K is algebraically closed.*

Proof. The residue field of $(\mathrm{P}(K), v)$ is K, hence if $\mathrm{char}\, K = 0$, then $(\mathrm{P}(K), v)$ is a defectless field by **(DF1)** in Section 8.

For the second assertion, we use the following valuation theoretical fact (try to prove it, it is not hard):

Let (L, v) be a valued field and choose any extension of v to the algebraic closure $\tilde{L}$. Then $v\tilde{L}$ is the divisible hull of vL, and $\tilde{L}v$ is the algebraic closure of Lv.

Hence, $v\widetilde{K((t))} = \mathbb{Q} = v\mathrm{P}(K)$ and $\widetilde{K((t))}P = \tilde{K}$. Thus if $\widetilde{K((t))} = \mathrm{P}(K)$, then $\tilde{K} = \mathrm{P}(K) = K$, which shows that K must be algebraically closed. For the converse, note that by the foregoing lemma, $\mathrm{P}(K) \subseteq \widetilde{K((t))}$. Assume that $\tilde{K} = K$. Then the extension $(\widetilde{K((t))}|\mathrm{P}(K), v)$ is immediate. But since $(\mathrm{P}(K), v)$ is henselian (by the foregoing lemma) and defectless, every finite subextension must be trivial by Theorem 8.2. This proves that $\widetilde{K((t))} = \mathrm{P}(K)$, i.e., $\mathrm{P}(K)$ is algebraically closed. □

The assertion of this theorem does not hold if K has positive characteristic:

Example 15. In Example 12, we can replace L_k by $K((t^{1/k}))$ for every $k \in \mathbb{N}$ (as opposed to $K((t^{1/p^k}))$, which we used in Example 13). Still, everything works the same, producing the henselian Puiseux series field $L = \mathrm{P}(K)$ with an immediate Artin-Schreier extension $(L(\vartheta)|L, v)$ of degree p and defect p.

By construction, $\mathrm{P}(K)$ is a subfield of $K((\mathbb{Q}))$. Hence, the arguments at the end of the last section show that there is no root of $X^p - X - 1/t$ in the completion of $\mathrm{P}(K)$. The arguments of Example 14 show that $\mathrm{P}(K)$ has non-isomorphic maximal immediate extensions if K is not Artin-Schreier closed.

Our example proves:

Theorem 10.3. *Let K be a field of characteristic $p > 0$. Then $(\mathrm{P}(K), v)$ is not defectless. In particular, $\mathrm{P}(K)$ is not algebraically closed, even if K is algebraically closed. Not even the completion of $\mathrm{P}(K)$ is algebraically closed.*

There is always a henselian defectless field extending $K((t))$ and having residue field K and divisible value group, even if K has positive characteristic. We just have to take the power series field $K((\mathbb{Q}))$. But in contrast to the Puiseux series field, this field is "very large": it has uncountable transcendence degree over $K((t))$. Nevertheless, having serious problems with the Puiseux series field in positive characteristic, we tend to replace it by $K((\mathbb{Q}))$. But this seems problematic since it might not be the unique maximal immediate extension of the Puiseux series field. However, if K is perfect and does not admit a finite extension whose degree is divisible by p (and in particular if K is algebraically closed), then Kaplansky's uniqueness result shows that the maximal immediate extension is unique. On the other hand, our example shows that the assumption "K is perfect" alone is not sufficient, since there are perfect fields which are not Artin-Schreier closed.

11. The tame and the wild valuation theory

Before we carry on, let us describe some advanced ramification theory based on the material of Sections 5 and 8. Throughout, we let (K, v) be a henselian non-trivially valued field. We set $p = \mathrm{char}\, Kv$ if this is positive, and $p = 1$ otherwise. If $(L|K, v)$ is an algebraic extension, then we call $(L|K, v)$ a *tame extension* if for every finite subextension $L'|K$ of $L|K$,

1) $(vL' : vK)$ is not divisible by the residue characteristic $\mathrm{char}\, Kv$,
2) $L'v|Kv$ is separable,
3) $[L' : K] = (vL' : vK)[L'v : Kv]$, i.e., $(L'|K, v)$ has trivial defect.

From the ramification theoretical facts presented in Section 5, one derives:

Theorem 11.1. *If (K, v) is henselian, then its absolute ramification field (K^r, v) is the unique maximal tame extension of (K, v), and its absolute inertia field (K^i, v) is the unique maximal tame extension of (K, v) having the same value group as K.*

An extension $(L|K,v)$ is called *purely wild* if $L|K$ is linearly disjoint from $K^r|K$. An ordered group G is called p-divisible if for every $\alpha \in G$ and $n \in \mathbb{N}$ there is $\beta \in G$ such that $p^n\beta = \alpha$. The *p-divisible hull* of G, denoted by $\frac{1}{p^\infty}G$, is the smallest subgroup of the divisible hull $\mathbb{Q}\otimes G$ which contains G and is p-divisible; it can be written as $\{\alpha/p^n \mid \alpha \in G\,,\, n \in \mathbb{N}\}$. The following was proved by Matthias Pank (cf. [KPR]):

Theorem 11.2. *If (K,v) is henselian, then there exists a field complement W to K^r over K, that is, $W.K^r = \tilde{K}$ and $W \cap K^r = K$. The degree of every finite subextension of $W|K$ is a power of p. Further, vW is the p-divisible hull $\frac{1}{p^\infty}vK$ of vK, and Wv is the perfect hull of Kv.*

So (W,v) is a maximal purely wild extension of (K,v). It was shown by Pank and is shown in [KPR] via Galois theory that W is unique up to isomorphism over K if Kv does not admit finite separable extensions whose degree are divisible by p. On the other hand, if vK is p-divisible and Kv is perfect, then $(W|K,v)$ is an immediate extension, and since every subextension of $K^r|K$ has trivial defect, it follows that the field complements W of K^r over K are precisely the maximal immediate algebraic extensions of (K,v).

It was shown by George Whaples [WH2] and by Françoise Delon [D] that Kaplansky's original hypothesis A consists of the following three conditions:

1) Kv does not admit finite separable extensions whose degree are divisible by p,
2) vK is p-divisible,
3) Kv is perfect.

So if (K,v) satisfies Kaplansky's hypothesis A, then it follows from what we said above that the maximal immediate algebraic extensions of (K,v) are unique up to isomorphism over K. But this is the kernel of the uniqueness problem for the maximal immediate extensions: using Theorem 2 of [KA1], one can easily show that the maximal immediate extensions are unique as soon as the maximal immediate algebraic extensions are.

Since all finite tame extensions have trivial defect, the defect is located in the purely wild extensions $(W|K,v)$. So we are interested in their structure. Here is one amazing result, due to Florian Pop (for the proof, see [K2], and for the notion of "additive polynomial", see Example 4):

Theorem 11.3. *Let $(L|K,v)$ be a minimal purely wild extension, i.e., there is no subextension $L'|K$ of $L|K$ such that $L \neq L' \neq K$. Then there is an additive polynomial $\mathcal{A} \in K[X]$ and some $c \in K$ such that $L|K$ is generated by a root of $\mathcal{A}(X)+c$.*

The degree of $\mathcal{A}$ is a power of p (since it is additive), and in general it may be larger than p.

Now we shall quickly develop the theory of tame fields. The henselian field (K,v) is said to be a *tame field* if all of its algebraic extensions are tame extensions. By Theorem 11.2, this holds if and only if K^r is algebraically closed. Similarly,

(K, v) is said to be a *separably tame field* if all of its separable-algebraic extensions are tame extensions. This holds if and only if K^r is separable-algebraically closed.

By Theorem 5.10, $\tilde{K} = K^r.W$ is the absolute ramification field of W. If $W'|K$ is a proper subextension, then $\tilde{K} \neq K^r.W'$. This proves:

Lemma 11.4. *Every maximal purely wild extension W is a tame field. No proper subextension of $W|K$ is a tame field. The maximal separable subextension is a separably tame field.*

By Theorem 5.8, $K^{\mathrm{sep}}|K^r$ is a p-extension. Hence if $\mathrm{char}\, Kv = 0$, then this extension is trivial. Since then also $\mathrm{char}\, K = 0$, it follows that $K^r = K^{\mathrm{sep}} = \tilde{K}$. Therefore,

Lemma 11.5. *Every henselian field of residue characteristic 0 is a tame field.*

Suppose that $K_1|K$ is an algebraic extension. Then $K^r \subseteq K_1^r$ by Theorem 5.10. Hence if K^r is algebraically closed, then so is K_1^r, and if K^r is separable-algebraically closed, then so is K_1^r. This proves:

Lemma 11.6. *Every algebraic extension of a tame field is again a tame field. Every algebraic extension of a separably tame field is again a separably tame field.*

If $K^r = \tilde{K}$, then every finite extension of (K, v) is a tame extension and thus has trivial defect, which shows that (K, v) is a defectless field. If $K^r = K^{\mathrm{sep}}$, then every finite separable extension has trivial defect. So we note:

Lemma 11.7. *Every tame field is henselian defectless and perfect. Every separably tame field is henselian and all of its finite separable extensions have trivial defect.*

We give a characterization of tame and separably tame fields (for the proof, see [K2]):

Lemma 11.8. *The valued field (K, v) is tame if and only if it is algebraically maximal, vK is p-divisible and Kv is perfect. If* $\mathrm{char}\, K = \mathrm{char}\, Kv$ *then (K, v) is tame if and only if it is algebraically maximal and perfect.*

Further, a non-trivially valued field (K, v) is separably tame if and only if it is separable-algebraically maximal, vK is p-divisible and Kv is perfect.

This lemma together with Lemma 11.7 shows that for perfect valued fields (K, v) with $\mathrm{char}\, K = \mathrm{char}\, Kv$, the two properties "algebraically maximal" and "henselian and defectless" are equivalent.

Corollary 7. *Assume that* $\mathrm{char}\, K = \mathrm{char}\, Kv$. *Then every maximal immediate algebraic extension of the perfect hull of (K, v) is a tame field (and no proper subextension of it is a tame field). If* $\mathrm{char}\, Kv = 0$ *then already the henselization (K^h, v) is a tame field.*

The following is a crucial lemma in the theory of tame fields. For its proof, see [K1] or [K2].

Lemma 11.9. *Let (L, v) be a tame field and $K \subset L$ a relatively algebraically closed subfield. If in addition $Lv|Kv$ is an algebraic extension, then (K, v) is also a tame field and moreover, vK is pure in vL and $Kv = Lv$. The same holds for "separably tame" in the place of "tame".*

The break we took for the development of the theory of tame fields is at the same time a good occasion to do some model theoretic preparation for later sections.

12. Some notions and tools from model theoretic algebra

The basic idea of model theoretic algebra is to analyze the assertions that an algebraist wants to prove, and to apply principles that are valid for certain types of assertions. Such principles prove once and for all facts that otherwise are proved over and over again in different settings (as a little example, see Lemma 12.1 below). To state and apply such principles, it is necessary to make it precise what it is that we are talking about, and in which mathematical language we are talking. The reader may interpose that mathematicians are talking about mathematical structures, which are fixed by definitions, that is, by axioms. If for instance we are talking about a group, then we talk about a set of elements G and a binary function $G \times G \rightarrow G$ which associates with every two elements a third one. So besides the underlying set, we are using a *function symbol* for this function, which is $+$ if we write the group additively, and $\cdot$ if we write it multiplicatively. But a group also has a unit element, for which we use the *constant symbol* 0 in the additive and 1 in the multiplicative case. If we talk about ordered groups, then for expressing the ordering we need a further symbol, which might be $<$ or $\leq$. Although it is also binary like $+$ or $\cdot$, it is not a function symbol. Since the ordering is a relation between the elements of the group, it is called a *relation symbol.* The description of a mathematical object may need more than one constant, function or relation symbol. For a field, we need two binary functions, $+$ and $\cdot$, and two constants, 0 and 1. Further, we may need function symbols or relation symbols of any (fixed) number of entries.

A *language* is defined to be

$$\mathcal{L} = \mathcal{F} \cup \mathcal{C} \cup \mathcal{R}$$

where

- $\mathcal{F}$ is a set of function symbols,
- $\mathcal{C}$ is a set of constant symbols,
- $\mathcal{R}$ is a set of relation symbols.

For example,

$$\mathcal{L}_{\mathrm{G}} := \{+, -, 0\}$$

is the *language of groups* (where $-$ is a function symbol with one entry), $\mathcal{L}_{\mathrm{OG}} := \{+,-,0,<\}$ is the *language of ordered groups*,

$$\mathcal{L}_{\mathrm{F}} := \{+,\cdot,-,^{-1},0,1\}$$

is the *language of fields*, and

$$\mathcal{L}_{\mathrm{VF}} := \{+,\cdot,-,^{-1},0,1,\mathcal{O}\}$$

is the *language of valued fields*, where $\mathcal{O}$ is a relation symbol with one entry.

For a given language $\mathcal{L}$, an *$\mathcal{L}$-structure* is a quadruple

$$\mathfrak{A} = (A, \mathcal{R}_{\mathfrak{A}}, \mathcal{F}_{\mathfrak{A}}, \mathcal{C}_{\mathfrak{A}})$$

where

- A is a set, called the *universe* **of** $\mathfrak{A}$,
- $\mathcal{F}_{\mathfrak{A}} = \{f_{\mathfrak{A}} \mid f \in \mathcal{F}\}$ such that every $f_{\mathfrak{A}}$ is a function on A of the same arity as the function symbol f,
- $\mathcal{C}_{\mathfrak{A}} = \{c_{\mathfrak{A}} \mid c \in \mathcal{C}\}$ such that every $c_{\mathfrak{A}}$ is an element of A (called a *constant*),
- $\mathcal{R}_{\mathfrak{A}} = \{R_{\mathfrak{A}} \mid R \in \mathcal{R}\}$ such that every $R_{\mathfrak{A}}$ is a relation on A of the same arity as the relation symbol R.

We call $R_{\mathfrak{A}}$ the *interpretation* **of** R **on** A, and similarly for the functions $f_{\mathfrak{A}}$ and the constants $c_{\mathfrak{A}}$. Let $\mathfrak{A}$ and $\mathfrak{B}$ be two $\mathcal{L}$-structures. Then we will call $\mathfrak{A}$ a *substructure* of $\mathfrak{B}$ if the universe A of $\mathfrak{A}$ is a subset of the universe B of $\mathfrak{B}$ and the restrictions to A of the interpretations of the relation and function symbols and the constant symbols on B coincide with the interpretations of the same relation, function and constant symbols on A.

Any set A with a distinguished element 0, a binary function $+ : A \times A \to A$ and a unary function $- : A \to A$ will be a structure for the language $\mathcal{L}_{\mathrm{G}} = \{+,-,0\}$ of groups. This does not say anything about the behaviour of the functions $+$ and $-$ and the element 0. For instance, 0 may not at all behave as a neutral element for $+$. Such properties of structures cannot be fixed by the language. They have to be described by axioms.

By an *elementary $\mathcal{L}$-formula* we mean a syntactically correct string built up using the symbols of the language $\mathcal{L}$, variables, $=$, and the logical symbols $\forall$, $\exists$, $\neg$, $\wedge$, $\vee$, $\to$, $\leftrightarrow$. An elementary $\mathcal{L}$-formula is called an *elementary $\mathcal{L}$-sentence* if every variable is bound by some quantifier. An elementary $\mathcal{L}$-sentence is called *existential* if it is of the form $\exists X_1 \ldots \exists X_n \varphi(X_1,\ldots,X_n)$ where $\varphi(X_1,\ldots,X_n)$ is a quantifier free $\mathcal{L}$-formula and $X_1,\ldots,X_n$ are the only variables appearing in φ. Hence an existential sentence is a sentence which only talks about the existence of certain elements. An elementary $\mathcal{L}$-sentence is called *universal existential* if it is of the form $\forall X_1 \ldots \forall X_k \exists X_{k+1} \ldots \exists X_n \varphi(X_1,\ldots,X_n)$ where φ is as above ($k=0$ or $k=n$ are admissible, so existential is also universal existential).

For example, the usual sentences expressing associativity, commutativity, the fact that 0 is a neutral element, and the existence of inverses are universal existential elementary $\mathcal{L}_{\mathrm{G}}$-sentences. They form an elementary axiom system for the class of abelian groups. Similarly, we have elementary axiom systems for the classes of ordered abelian groups, fields, valued fields. It is not necessary that an axiom system consists of only finitely many axioms. For instance, properties like "algebraically closed" or "real closed" can be axiomatized by an infinite scheme of universal existential elementary axioms. One can quantify over all possible polynomials of fixed degree n by quantifying over their $n+1$ coefficients. But in order to express that *all* polynomials have a root in a field K we need countably many axioms talking about polynomials of increasing degree. In a similar way, the property "henselian" is axiomatized by an infinite scheme of universal existential elementary axioms in the language of valued fields. In contrast to this, properties like "complete" or "maximal" have no elementary axiomatization in $\mathcal{L}_{\mathrm{VF}}$; we would have to quantify over subsets of the universe, which is impossible in elementary sentences. This shows that it makes sense to replace "complete" or "maximal" by "henselian", wherever possible.

The following lemma expresses (once and for all) a fact that is (intuitively) known to every good mathematician.

Lemma 12.1. *The union over an ascending chain of $\mathcal{L}$-structures $\mathfrak{A}_i$, $i \in \mathbb{N}$, satisfies all universal existential elementary sentences which are satisfied in all of the $\mathfrak{A}_i$.*

This proves, for instance:

Lemma 12.2. *The union of an ascending chain of henselian valued fields (L_i, v), $i \in \mathbb{N}$, is again a henselian valued field.*

Any set $\mathcal{T}$ of elementary $\mathcal{L}$-sentences is called an *elementary axiom system* (or an *$\mathcal{L}$-theory*). If an $\mathcal{L}$-structure satisfies all axioms in $\mathcal{T}$, then we call it a *model of* $\mathcal{T}$. So if we have an $\mathcal{L}_{\mathrm{G}}$-structure which is a model of the axiom system of groups, then we know that $+, -, 0$ behave as we expect them to do. The axiom system for valued fields expresses that $\{x \mid \mathcal{O}(x) \text{ holds}\}$ is a valuation ring. Since v is not a function from the field into itself, we cannot simply take a function symbol for v into the language. However, we can express "$vx \leq vy$" by the elementary sentence "$\mathcal{O}(yx^{-1}) \vee x = y = 0$".

We say that two $\mathcal{L}$-structures $\mathfrak{A}, \mathfrak{B}$ are *elementarily equivalent*, denoted by $\mathfrak{A} \equiv \mathfrak{B}$, if $\mathfrak{A}$ and $\mathfrak{B}$ satisfy the same elementary $\mathcal{L}$-sentences. An axiom system is *complete* if and only if all of its models are elementarily equivalent. Syntactically, that means that every elementary $\mathcal{L}$-sentence or its negation can be deduced from that axiom system. For example, the axiom system of divisible ordered abelian groups and the axiom system of algebraically closed fields of fixed characteristic are complete. It was shown by Abraham Robinson [RO] that the axiom system of algebraically closed non-trivially valued fields of fixed characteristic and fixed residue characteristic is complete.

The completeness of an axiom system yields a *Transfer Principle* for the class axiomatized by it. For example, the completeness of the axiom system for algebraically closed fields of fixed characteristic tells us that every elementary $\mathcal{L}_{\mathrm{F}}$-sentence which holds in one algebraically closed field will also hold in all other algebraically closed fields of the same characteristic. This reminds of the *Lefschetz Principle* which was stated and partially proved by Weil and Lefschetz and says (roughly speaking) that algebraic geometry over all algebraically closed fields of a fixed characteristic is the same ("there is but one algebraic geometry in characteristic p"). As for the elementary $\mathcal{L}_{\mathrm{F}}$-sentences of algebraic geometry, this indeed follows from the completeness. However, Lefschetz and Weil had in mind more than just the elementary sentences. That is why Weil worked with so-called "universal domains" which are algebraically closed and of infinite transcendence degree over their prime field. So the assertion was that there is but one algebraic geometry over universal domains of characteristic p. A satisfactory formalization and model theoretic proof by use of an *infinitary language* is due to Paul Eklof [EK]. Infinitary languages admit the conjunction of infinitely many elementary sentences. With such infinitary sentences, one can also express the fact that a field has infinite transcendence degree over its prime field. This cannot be done by elementary sentences. Indeed, algebraically closed fields are elementarily equivalent to the algebraic closure of their prime field, even if they have infinite transcendence degree.

In the theory of valued fields, we consider two important invariants: value groups and residue fields. If two valued fields are equivalent (in the language of valued fields), then so are their value groups (in the language of ordered groups) and their residue fields (in the language of fields). We are interested in a converse of this implication: under which additional assumptions do we have the so-called *Ax-Kochen-Ershov principle*:

$$vK \equiv vL \text{ and } Kv \equiv Lv \;\Rightarrow\; (K,v) \equiv (L,v) \tag{44}$$

It follows from the fact that the henselization is an immediate extension that this principle can only hold for henselian fields. And indeed, it was shown by James Ax and Simon Kochen [AK1] and independently by Yuri Ershov [ER2] that this principle holds for all henselian fields of residue characteristic 0. This is the famous *Ax-Kochen-Ershov Theorem.* Ax and Kochen proved it in order to deduce a correct version of *Artin's conjecture* about the fields $\mathbb{Q}_p$ of p-adic numbers ([AK1]; cf. [CK] or [K2]). From the Ax-Kochen-Ershov Theorem, one obtains an equivalence of two ultraproducts:

$$\prod_{p \text{ prime}} \mathbb{Q}_p \,/\mathcal{D} \equiv \prod_{p \text{ prime}} \mathbb{F}_p((t)) \,/\mathcal{D} \tag{45}$$

because both fields carry a canonical henselian valuation of residue characteristic 0 and have equal value groups and residue fields. Since a product like $\prod_p \mathbb{Q}_p$ would not even be a field, one has to take the product modulo a non-principal ultrafilter $\mathcal{D}$ on the set of all primes. Let us quickly give the main facts about ultraproducts.

A filter $\mathcal{D}$ on a set I is called an *ultrafilter* if

$$J \subset I \wedge J \notin \mathcal{D} \implies I \setminus J \in \mathcal{D}\,, \tag{46}$$

and it is called *non-principal* if it is not of the form $\{J \subset I \mid i \in J\}$, for no $i \in I$. If $\mathfrak{A}_i$, $i \in I$ are $\mathcal{L}$-structures, then the *ultraproduct* $\prod_{i\in I} \mathfrak{A}_i / \mathcal{D}$ is defined by setting, for all $(a_i), (b_i) \in \prod_{i\in I} \mathfrak{A}_i$,

$$(a_i) \equiv (b_i) \text{ modulo } \mathcal{D} \iff \{i \in I \mid a_i = b_i\} \in \mathcal{D}\,. \tag{47}$$

The following theorem is due to J. Łos. For a proof, see [K2] or [CK].

Theorem 12.3. (Fundamental Theorem of Ultraproducts)
For every $\mathcal{L}$-sentence φ, $\prod_{i\in I} \mathfrak{A}_i / \mathcal{D}$ satisfies φ if and only if

$$\{i \in I \mid \mathfrak{A}_i \text{ satisfies } \varphi\} \in \mathcal{D}\,. \tag{48}$$

Thus, elementary sentences which are true for all $\mathbb{F}_p((t))$ can be transferred to $\prod_p \mathbb{F}_p((t))/\mathcal{D}$, from there via (45) to $\prod_p \mathbb{Q}_p/\mathcal{D}$, and from there, by varying over all possible ultrafilters on the set of primes, to almost all $\mathbb{Q}_p$. The elementary sentences we are interested in are deduced from the analogue of Artin's conjecture which holds for all $\mathbb{F}_p((t))$, as proved by Serge Lang [L1].

Since the proof of the Ax-Kochen-Ershov Theorem, the Ax-Kochen-Ershov principle (44) has also been proved for other classes of valued fields, like p-adically closed fields, or algebraically maximal fields satisfying Kaplansky's hypothesis A ("algebraically maximal" is stronger than "henselian" if the residue characteristic is positive, as we have seen in Example 11). The proofs used, more or less explicitly, that those fields have unique maximal immediate extensions. But this is not necessary for the validity of the Ax-Kochen-Ershov principle (44). Using instead the Generalized Stability Theorem (Theorem 14.1 below) and the Henselian Rationality of Immediate Function Fields (Theorem 17.4 below), I proved that Ax-Kochen-Ershov principle (44) also holds for all tame fields ([K1], [K2]).

There is another version of (44) which will bring us closer to applications in algebraic geometry. There is a notion which in the past years has turned out to be more basic and flexible than that of elementary equivalence. Through general tools of model theory (like Theorem 13.5 below), notions like elementary equivalence can often be reduced to it. Take $\mathfrak{B}$ to be an $\mathfrak{L}$-structure, and $\mathfrak{A}$ a substructure of $\mathfrak{B}$. We form a language $\mathcal{L}(A)$ by adjoining the universe A of $\mathfrak{A}$ to the language $\mathcal{L}$. That is, in the language $\mathcal{L}(A)$ we have a constant symbol for every element of the structure $\mathfrak{A}$, so we can talk about every single element. We say that $\mathfrak{A}$ *is existentially closed in* $\mathfrak{B}$, denoted by $\mathfrak{A} \prec_\exists \mathfrak{B}$, if every existential $\mathcal{L}(A)$-sentence holds already in $\mathfrak{A}$ if it holds in $\mathfrak{B}$. (The other direction is trivial: if something exists in $\mathfrak{A}$, then it also exists in $\mathfrak{B}$.) Let us illustrate the use of this notion by three important examples.

Example 16. Take a field extension $L|K$. If $K \prec_\exists L$ in the language of fields, then K is relatively algebraically closed in L. To see this, take $a \in L$ algebraic over K. Take $f = X^n + c_{n-1}X^{n-1} + \cdots + c_0 \in K[X]$ to be the minimal polynomial of a over K. Since $f(a) = 0$, we know that the existential $\mathcal{L}_{\mathrm{F}}(K)$-sentence

$$\exists X\; X^n + c_{n-1}X^{n-1} + \cdots + c_0 = 0$$

holds in L. ("X^n" is an abbreviation for "$X \cdot \ldots \cdot X$" where X appears n times. Observe that we need the constants from K since we use the coefficients c_i in our sentence.) Since $K \prec_\exists L$, it must also hold in K. That means, that f also has a root in K. But as a minimal polynomial, f is irreducible over K. This shows that f must be linear, i.e., $a \in K$. One can also show that $L|K$ must be separable.

Similarly, let $G \subset H$ be an extension of abelian groups such that $G \prec_\exists H$ in the language of groups. Take $\alpha \in H$ such that $n\alpha \in G$ for some integer $n > 0$. Set $\beta = n\alpha$. Then the existential $\mathcal{L}_{\rm G}(G)$-sentence "$\exists X\, nX = \beta$" holds in H. (Here, "nX" is just an abbreviation for "$X + \cdots + X$" where X appears n times.) Hence, it must also hold in G. That is, $\alpha = \beta/n \in G$. Hence, we have:

Lemma 12.4. *If $L|K$ is an extension of fields such that $K \prec_\exists L$ in the language of fields, then K is relatively algebraically closed in L and $L|K$ is separable. If $G \subset H$ is an extension of abelian groups such that $G \prec_\exists H$ in the language of groups, then H/G is torsion free.*

Example 17. Take a function field $F|K$ such that $K \prec_\exists F$ in the language of fields. Since $F|K$ is separable by Lemma 12.4, we can choose a separating transcendence basis $t_1, \ldots, t_k$ of $F|K$ and an element $z \in F$ such that $F = K(t_1, \ldots, t_k, z)$ with z separable-algebraic over $K(t_1, \ldots, t_k)$. Take $f \in K[X_1, \ldots, X_k, Z]$ to be the irreducible polynomial of z over $K[t_1, \ldots, t_k]$ (obtained from the minimal polynomial by multiplication with the common denominator of the coefficients from $K(t_1, \ldots, t_k)$). We have that $f(t_1, \ldots, t_k, z) = 0$. Since z is a simple root of f, we also have that $\frac{\partial f}{\partial Z}(t_1, \ldots, t_k, z) \neq 0$. Further, we take n arbitrary non-zero elements $z_1, \ldots, z_n \in F$ which we write as $g_1/h_1, \ldots, g_n/h_n$ with non-zero elements $g_i, h_i \in K[t_1, \ldots, t_k, z]$. Now the existential $\mathcal{L}_{\rm F}(K)$-sentence

$$\begin{aligned} \text{“}\, \exists Y_1 \cdots \exists Y_k \exists Y\; f(Y_1, \ldots, Y_k, Y) = 0 \,\wedge\, \frac{\partial f}{\partial Z}(Y_1, \ldots, Y_k, Y) \neq 0 \,\wedge \\ \wedge\, g_1(Y_1, \ldots, Y_k, Y) \neq 0 \,\wedge \cdots \wedge\, g_n(Y_1, \ldots, Y_k, Y) \neq 0 \,\wedge \\ \wedge\, h_1(Y_1, \ldots, Y_k, Y) \neq 0 \,\wedge \cdots \wedge\, h_n(Y_1, \ldots, Y_k, Y) \neq 0\, \text{”} \end{aligned}$$

holds in F. Hence it must also hold in K, that is, there are $c_1, \ldots, c_k, d \in K$ such that $f(c_1, \ldots, c_k, b) = 0$, $\frac{\partial f}{\partial Z}(c_1, \ldots, c_k, d) \neq 0$ and

$$g_i(c_1, \ldots, c_k, b)/h_i(c_1, \ldots, c_k, d) \neq 0, \infty\,.$$

On $K[t_1, \ldots, t_k]$, we have the homomorphism given by $t_i \mapsto c_i$, or equivalently, by $t_i - c_i \mapsto 0$. As in Example 2, we can construct a place P of maximal rank of $K(t_1, \ldots, t_k) = K(t_1 - c_1, \ldots, t_k - c_k)$ which extends this homomorphism. Its residue field is K. Now we consider the polynomial $g(Z) = f(t_1, \ldots, t_k, Z)$. Its reduction modulo P is the polynomial $f(c_1, \ldots, c_k, Z)$ which admits d as a simple root. Hence by Hensel's Lemma, $g(Z)$ has a root z' in the henselization $(K(t_1, \ldots, t_k)^h, P)$ of $(K(t_1, \ldots, t_k), P)$. Thus, the assignment $z \mapsto z'$ defines an embedding of F over $K(t_1, \ldots, t_k)$ in $K(t_1, \ldots, t_k)^h$, and pulling the place P from the image of this embedding back to F, we obtain on F a place P with residue field K and having maximal rank. In addition, $z_iP \neq 0, \infty$. We have proved:

Lemma 12.5. *Take a function field $F|K$ such that $K \prec_\exists F$ in the language of fields. Take non-zero elements $z_1, \ldots, z_n \in F$. Then there exists a rational place*

P of $F|K$ of maximal rank and such that $z_i P \neq 0, \infty$ for $1 \leq i \leq n$, and a model of $F|K$ on which P is centered at a smooth point.

Note that it was crucial for our proof that $F|K$ is finitely generated (because elementary sentences can only talk about finitely many elements). If a field extension $L|K$ is not finitely generated, then there may not exist a place P of L such that $LP = K$, even if $K \prec_\exists L$.

Example 18. A field K is called a *large field* (cf. [POP]) if every smooth curve over K has infinitely many K-rational points, provided it has at least one K-rational point. For the proof of the following theorem, see [K2] or [K3].

Theorem 12.6. *A field K is large if and only if $K \prec_\exists K((t))$ (in the language of fields).*

The use of "$\prec_\exists$" gives us another version of the Ax-Kochen-Ershov principle:

$$vK \prec_\exists vL \text{ and } Kv \prec_\exists Lv \;\Rightarrow\; (K,v) \prec_\exists (L,v)\,. \tag{49}$$

This principle also holds for the classes of valued fields that we mentioned above: henselian fields of residue characteristic 0, p-adically closed fields, algebraically maximal fields satisfying Kaplansky's hypothesis A, tame fields. A short proof for the case of henselian fields of residue characteristic 0 is given in the appendix of [KP]. This form of the Ax-Kochen-Ershov principle is applied in the proof of Theorem 20.1 below.

13. Saturation and embedding lemmas

How can a principle like (49) be proved? In fact, nice model theoretic results often just represent a good algebraic structure theory. Indeed, using a very useful model theoretic tool, we can easily transfer "$\prec_\exists$" to an algebraic fact. The tool is that of a *κ-saturated model* (where κ is a cardinal number). Saturation is a property which is not elementary, quite similar to "complete" or "maximal", but still different (in fact, "maximal" and "saturated" are to some extent mutually exclusive). Before defining "κ-saturated", I want to illustrate its meaning by an example which plays a remarkable role in the theory of ordered structures. We take fields, but the same can be done for ordered abelian groups and other ordered structures.

Example 19. Take any ordinal number α and an ordered field $(K,<)$. For $A, B \subset K$ we will write $A < B$ if every element of A is smaller than every element of B. Now $(K,<)$ is said to be an *η_α-field* if for every two subsets $A, B \subset K$ of cardinality less than $\aleph_\alpha$ (= the α-th cardinal number) such that $A < B$, there is some $c \in K$ such that $A < \{c\} < B$, i.e., c lies between A and B. Note that because of the restriction of the cardinality of A and B, this does not mean that $(K,<)$ is cut-complete (in fact, the only cut-complete field is $\mathbb{R}$, while there is an abundance of η_α-fields).

Given $A, B \subset K$ such that $A < B$, we consider the collection of elementary sentences (in the language of ordered fields with constants from K) "$a < X$",

$a \in A$, and "$X < b$", $b \in B$. It is clear that if we take any finite subset of these, then there is some element in K that we can insert for X so that all of these finitely many sentences hold. That is, our collection of sentences is *finitely realizable* in $(K, <)$. Now if $(K, <)$ is κ-saturated, then this tells us the following: if the cardinality of $A \cup B$ is smaller than κ, then there is an element $c \in K$ which simultaneously satisfies *all* of our above sentences (we say that c *realizes* the above set of elementary sentences). But this means that $A < \{c\} < B$. So we see that every $\aleph_\alpha$-saturated ordered field is an η_α-field.

Let us extract a definition from our example. An $\mathcal{L}$-structure $\mathfrak{A}$ will be called *κ-saturated* if for every subset S of its universe A of cardinality less than κ, every set of elementary $\mathcal{L}(S)$-sentences is realizable in $\mathfrak{A}$, provided that it is finitely realizable in $\mathfrak{A}$. To express the fact that there are enough κ-saturated $\mathcal{L}$-structures, we need one further notion. Given an $\mathcal{L}$-structure $\mathfrak{B}$ with substructure $\mathfrak{A}$, we say that $\mathfrak{B}$ *is an elementary extension of* $\mathfrak{A}$ and write $\mathfrak{A} \prec \mathfrak{B}$ if *every* $\mathcal{L}(A)$-sentence holds in $\mathfrak{A}$ if and only if it holds in $\mathfrak{B}$. (So in contrast to "existentially closed", here we do not restrict the scope to existential sentences.) For example, an algebraically closed field K is existentially closed in every extension field, and every algebraically closed extension field of K is an elementary extension of K. If $K \prec L$, then K is existentially closed in every intermediate field.

We are going to state the theorem which provides us with sufficiently many κ-saturated structures. It is a consequence of one of the basic theorems of model theory:

Theorem 13.1. **(Compactness Theorem)** *A set of elementary $\mathcal{L}$-sentences has a model if and only if each of its finite subsets has a model.*

For the proof of the next theorems, see [CK] or [K2].

Theorem 13.2. *For every infinite $\mathcal{L}$-structure $\mathfrak{A}$ and every large enough κ there exists a κ-saturated elementary extension of $\mathfrak{A}$.*

Here "large enough" means: larger than the cardinality of the language (which, if infinite, will in most cases be the cardinality of the set of constants appearing in the language), and larger than the cardinality of the universe of $\mathfrak{A}$. Now the reduction of "$\prec_\exists$" to an algebraic statement is done as follows:

Theorem 13.3. *Take an $\mathcal{L}$-structure $\mathfrak{B}$ with substructure $\mathfrak{A}$. Take κ larger than the cardinality of $\mathcal{L}$ and the cardinality of the universe of $\mathfrak{B}$. Further, choose a κ-saturated elementary extension $\mathfrak{A}^*$ of $\mathfrak{A}$. Then $\mathfrak{A} \prec_\exists \mathfrak{B}$ holds if and only if there is an embedding of $\mathfrak{B}$ over $\mathfrak{A}$ in $\mathfrak{A}^*$.*

If there is an embedding of $\mathfrak{B}$ over $\mathfrak{A}$ in $\mathfrak{A}^*$, then every existential sentence holding in $\mathfrak{B}$ will carry over to the image of $\mathfrak{B}$ in $\mathfrak{A}^*$, from where it goes up to $\mathfrak{A}^*$. Since $\mathfrak{A} \prec \mathfrak{A}^*$, it then also holds in $\mathfrak{A}$.

A nice additional feature of saturation is the following:

Theorem 13.4. *There is an embedding of $\mathfrak{B}$ over $\mathfrak{A}$ in $\mathfrak{A}^*$ already if for every finitely generated subextension $\mathfrak{A} \subset \mathfrak{B}'$ of $\mathfrak{A} \subset \mathfrak{B}$ there is an embedding of $\mathfrak{B}'$ over $\mathfrak{A}$ in $\mathfrak{A}^*$.*

So if we have a field extension $L|K$ and want to prove that $K \prec_\exists L$, we take a κ-saturated elementary extension K^* of K, for κ larger than the cardinality of L, and seek to embed L over K in K^*. By the last theorem, we only have to show that every finitely generated subextension $F|K$ of $L|K$ embeds in K^*. But a finitely generated extension is a function field (in view of Lemma 12.4 we can exclude the case where $F|K$ is algebraic).

If K is an algebraically closed field, then so is K^* because it is an elementary extension of K. The assumption that K^* is κ-saturated with κ larger than the cardinality of L implies that the transcendence degree of $K^*|K$ is at least as large as that of $L|K$. So we see that L embeds over K in K^* (even without employing the last theorem). This proves that every algebraically closed field is existentially closed in every extension field.

Let's see how we can prove a principle like (49) with the above tools. We take a κ-saturated elementary extension $(K,v)^* = (K^*,v^*)$ of (K,v) (with respect to the language of valued fields). Then it is easy to prove that v^*K^* is a κ-saturated elementary extension of vK (with respect to the language of ordered groups) and that K^*v^* is a κ-saturated elementary extension of Kv (with respect to the language of fields). Thus, we see from Theorem 13.3 that $vK \prec_\exists vL$ implies that vL embeds over vK in v^*K^*, and that $Kv \prec_\exists Lv$ implies that Lv embeds over Kv in K^*v^*. So Theorem 13.3 shows that we can prove (49) by an *embedding lemma* of the form: *If vL embeds over vK in v^*K^* and Lv embeds over Kv in K^*v^* and* (additional assumptions) *then (L,v) embeds over K in $(K,v)^*$ (as a valued field).* See Example 21 and Example 23 below for two different cases and a sketch of the proof of (49) for tame fields.

To conclude this section, let us come back to elementary extensions. An $\mathcal{L}$-theory $\mathcal{T}$ is called *model complete* if for every two models $\mathfrak{A}$ and $\mathfrak{B}$ of $\mathcal{T}$ such that $\mathfrak{A}$ is a substructure of $\mathfrak{B}$ we have that $\mathfrak{A} \prec \mathfrak{B}$. This is closely connected to the relation $\mathfrak{A} \prec_\exists \mathfrak{B}$ through the following important criterion (cf. [K2] or [CK]):

Theorem 13.5. (Robinson's Test)
Assume that for every two models $\mathfrak{A}$ and $\mathfrak{B}$ of $\mathcal{T}$ such that $\mathfrak{A}$ is a substructure of $\mathfrak{B}$ we have that $\mathfrak{A} \prec_\exists \mathfrak{B}$. Then $\mathcal{T}$ is model complete.

For example, this theorem together with the fact that every algebraically closed field is existentially closed in every extension field shows that the axiom system of algebraically closed fields is model complete. Furthermore, with this theorem together with Theorem 13.3, Theorem 7.1, Theorem 2 of [KA1] and Theorem 5.1, it is not hard to prove the following theorem of Abraham Robinson:

Theorem 13.6. *The elementary axiom system of non-trivially valued algebraically closed fields is model complete.*

Observe that we do not need "side conditions" about the value groups and the residue fields here (because they are divisible and algebraically closed, respectively). But there is also an Ax-Kochen-Ershov principle with $\prec$ in the place of $\prec_\exists$ that again holds for the classes of valued fields which I mentioned above.

Example 20. Another simple but useful example for a fact proved by an embedding lemma is the following:

Theorem 13.7. *If (K,v) is henselian and (L,v) is a separable extension of (K,v) within its completion, then $(K,v) \prec_\exists (L,v)$.*

This fact can be seen as the (much simpler) "field version of Artin Approximation". It was observed in the 1960s by Yuri Ershov; for a proof, see [K2]. Together with Theorem 12.6, a modification of Theorem 14.3 and the transitivity of $\prec_\exists$, this theorem (applied to $(K(t),v)^h$) can be used to prove (cf. [K2], [K3]):

Theorem 13.8. *If the field K admits a henselian valuation, then $K \prec_\exists K((t))$, i.e., K is a large field.*

14. The Generalized Grauert-Remmert Stability Theorem

Let us return to our problem of inertial generation as considered at the end of Section 7. Our problem was to show that the finite immediate extension (33) of henselian fields is trivial. If it is not, then by Corollary 6 it has non-trivial defect (which then is equal to its degree). So we would like to show that the field $F_0(\eta)^h$ is a defectless field. The reason for this would have to lie in the special way we have constructed this field.

At this point, let us invoke a deep and important theorem from the theory of valued function fields ([K1], [K2]). For historical reasons, I call it the *Generalized Grauert-Remmert Stability Theorem* although I do not like the notion "stable". It is one of those words in mathematics which is very often used in different contexts, but in most cases does not reflect its meaning. I replace it by "defectless".

If $(F|K,v)$ is an extension of valued fields of finite transcendence degree, then by inequality (30) of Corollary 5, $\mathrm{trdeg}\,F|K - \mathrm{trdeg}\,Fv|Kv - \mathrm{rr}\,(vF/vK)$ is a non-negative integer. We call it the *transcendence defect* of $(F|K,v)$. We say that $(F|K,v)$ is *without transcendence defect* if the transcendence defect is 0.

Theorem 14.1. *Let $(F|K,v)$ be a valued function field without transcendence defect. If (K,v) is a defectless field, then also (F,v) is a defectless field.*

This theorem has a long and interesting history. Hans Grauert and Reinhold Remmert [GR] first proved it in a very restricted case, where (K,v) is a complete discrete valued field and (F,v) is discrete too. There are generalizations by Laurent Gruson [GRU], Michel Matignon, and Jack Ohm [OH]. All of these generalizations are restricted to the case $\mathrm{trdeg}\,F|K = \mathrm{trdeg}\,Fv|Kv$, the case of *constant reduction.* The classical origin of it is the study of curves over number fields and the idea to reduce them modulo a p-adic valuation. Certainly, the reduction should again render a curve, this time over a finite field. This is guaranteed by the condition $\operatorname{trdeg} F|K = \operatorname{trdeg} Fv|Kv$, where F is the function field of the curve and Fv will be the function field of its reduction. Naturally, one seeks to relate the genus of $F|K$ to that of $Fv|Kv$. Several authors proved *genus inequalities*. To illustrate the use of the defect, we cite an inequality proved by Barry Green, Michel Matignon and Florian Pop in [GMP1]. Let $F|K$ be a function field of transcendence degree 1 and assume that K coincides with the constant field of $F|K$ (the relative algebraic closure of K in F). Let $v_1, \ldots, v_s$ be distinct constant reductions of $F|K$ which have a common restriction to K. Then:

$$1 - g_F \leq 1 - s + \sum_{i=1}^{s} \delta_i e_i r_i (1 - g_i) \tag{50}$$

where g_F is the genus of $F|K$ and g_i the genus of $Fv_i|Kv_i$, r_i is the degree of the constant field of $Fv_i|Kv_i$ over Kv_i, δ_i is the defect of $(F^{h(v_i)}|K^{h(v_i)}, v_i)$ where "$.^{h(v_i)}$" denotes the henselization with respect to v_i, and $e_i = (v_i F : v_i K)$ (which is always finite in the constant reduction case by virtue of Corollary 5). It follows that constant reductions v_1, v_2 with common restriction to K and $g_1 = g_2 = g_F \geq 1$ must be equal. In other words, for a fixed valuation on K there is at most one extension v to F which is a *good reduction*, that is, (i) $g_F = g_{Fv}$, (ii) there exists $f \in F$ such that $vf = 0$ and $[F : K(f)] = [Fv : Kv(fv)]$, (iii) Kv is the constant field of $Fv|Kv$. An element f as in (ii) is called a *regular function*.

More generally, f is said to have the *uniqueness property* if fv is transcendental over Kv and the restriction of v to $K(f)$ has a unique extension to F. In this case, $[F : K(f)] = \delta\, e\, [Fv : Kv(fv)]$ where δ is the defect of $(F^h|K^h, v)$ and $e = (vF : vK(f)) = (vF : vK)$. If K is algebraically closed, then $e = 1$, and it follows from the Stability Theorem that $\delta = 1$; hence in this case, every element with the uniqueness property is regular.

It was proved in [GMP2] that F has an element with the uniqueness property already if the restriction of v to K is henselian. The proof uses Theorem 13.6 and ultraproducts of function fields. Elements with the uniqueness property also exist if vF is a subgroup of $\mathbb{Q}$ and Kv is algebraic over a finite field. This follows from work in [GMP4] where the uniqueness property is related to the *local Skolem property* which gives a criterion for the existence of algebraic v-adic integral solutions on geometrically integral varieties.

As an application to rigid analytic spaces, the Stability Theorem is used to prove that the quotient field of the free Tate algebra $T_n(K)$ is a defectless field, provided that K is. This in turn is used to deduce the *Grauert-Remmert Finiteness Theorem*, in a generalized version due to Gruson; see [BGR].

Surprisingly, it was not before the model theory of valued fields developed in positive characteristic that an interest in a generalized version of the Stability Theorem arose. But a criterion like Robinson's Test (Theorem 13.5) forces us to deal with arbitrary extensions of arbitrarily large valued fields. For instance, it is virtually impossible to restrict oneself to rank 1 in order to prove model completeness or completeness of a class of valued fields. And the extensions $(L|K,v)$ in question won't obey a restriction like "vL/vK is a torsion group". Therefore, I had to prove the above Generalized Stability Theorem. At that time, I had not heard of the Grauert-Remmert Theorem, so I gave a purely valuation theoretic proof ([K1], [K2]), not based on the original proofs of Grauert-Remmert or Gruson like the other cited generalizations.

Later, I was amazed to see that the Generalized Stability Theorem is also the suitable version for an application to the problem of local uniformization. (If your valuation v is trivial on the base field K and you ask that $\operatorname{trdeg} F|K = \operatorname{trdeg} Fv|Kv$, then $vL/\{0\}$ is torsion, so $vL = \{0\}$ and v is also trivial on F; this is not quite the case we are interested in.) So let's now describe this application. By our assumption at the end of Section 7, P is an Abhyankar place on F and hence also on $F_0(\eta)$. That is, $(F_0(\eta)|K,P)$ is a function field without transcendence defect. As P is trivial on K, also v_P is trivial on K. But a trivially valued field (K,v) is always a defectless field since for every finite extension $L|K$ we have that $[L:K] = [Lv:Kv]$. Hence by the Generalized Stability Theorem, $(F_0(\eta),v_P)$ is a defectless field. By Theorem 8.3, also $(F_0(\eta)^h,v_P)$ is a defectless field. Therefore, since $(F^h|F_0(\eta)^h,v_P)$ is an immediate extension, Corollary 6 shows that it must be trivial. We have proved that $F^h = F_0(\eta)^h$. By construction, $F_0(\eta)^h$ was a subfield of the absolute inertia field of (F_0,P). Hence also F is a subfield of that absolute inertia field, showing that (F,P) is inertially generated. We have thus proved the first part of the following theorem (I leave the rest of the proof to you as an exercise; cf. [K6]):

Theorem 14.2. *Assume that P is an Abhyankar place of $F|K$ and that $FP|K$ is a separable extension. Then $(F|K,P)$ is inertially generated. If in addition $FP = K$ or $FP|K$ is a rational function field, then $(F|K,P)$ is henselian generated. In all cases, if $v_PF = \mathbb{Z}v_Px_1 \oplus \ldots \oplus \mathbb{Z}v_Px_\rho$ and $y_1P,\ldots,y_\tau P$ is a separating transcendence basis of $FP|K$, then $\{x_1,\ldots,x_\rho,y_1,\ldots,y_\tau\}$ is a generating transcendence basis.*

Example 21. Let's now describe the model theoretic use of the Generalized Stability Theorem. Take $(L|K,v)$ of finite transcendence degree and without transcendence defect. Note that then for every function field $F|K$ contained in $L|K$, also the extension $(F|K,v)$ is without transcendence defect. Further, we assume that (K,v) is a non-trivially valued henselian defectless field and that $vK \prec_\exists vL$ and $Kv \prec_\exists Lv$. Making again essential use of the Generalized Stability, one proves a generalization of Theorem 14.2 to the case of non-trivially valued ground fields. Using also Theorem 7.1, Lemma 12.4 and Theorem 5.13, it is easy to show that the embeddings of vL and Lv (which induce embeddings of vF and Fv) lift to an embedding of (F,v) in $(K,v)^*$. This proves:

Theorem 14.3. *Take a non-trivially valued henselian defectless field (K,v) and an extension $(L|K,v)$ of finite transcendence degree, without transcendence defect. If $vK \prec_\exists vL$ and $Kv \prec_\exists Lv$, then $(K,v) \prec_\exists (L,v)$.*

This theorem shows the advantage of the notion "$\prec_\exists$" since there is no analogue for "$\prec$".

To conclude this section, we give a short sketch of a main part of the proof of the Generalized Stability Theorem. This is certainly interesting because very similar methods have been used by Shreeram Abhyankar for the proof of his results in positive characteristic (see, e.g., [A1], [A6], [A7]).

We have to prove that a certain henselian field (L, v) is a defectless field. We take an arbitrary finite extension $(L'|L, v)$ and have to show that it has trivial defect. We may assume that this extension is separable since the case of purely inseparable extensions can be considered separately and is much easier. Looking at $(L'|L, v)$, we are completely lost since we have not the slightest chance to develop a good structure theory. But we only have to deal with the defect, and we remember that a defect only appears if extensions beyond the absolute ramification field L^r are involved. So instead of $(L'|L, v)$ we consider the extension $(L'.L^r|L^r, v)$ which has the same defect as $(L'|L, v)$, although it will in general not have the same degree (the use of this fact reminds of Abhyankar's "Going Up" and "Coming Down"; cf. [A1]). Now we use the fact that by Theorem 5.8 the separable-algebraic closure of L^r is a p-extension. It follows that its subextension $L'.L^r|L^r$ is a tower of Artin-Schreier extensions (cf. Lemma 5.9). Since the defect is multiplicative, to prove that $(L'.L^r|L^r, v)$ has trivial defect it suffices to show that each of these Artin-Schreier extensions has trivial defect. So we take such an extension, generated by a root ϑ of an irreducible polynomial $X^p - X - c$ over some field L'' in the tower. By what we learned in Example 4, $vc \leq 0$. If $vc = 0$, the extension (if it is not trivial) would correspond to a proper separable extension of the residue field; but as we are working beyond the absolute ramification field, our residue field is already separable-algebraically closed. So we see that $vc < 0$. If $b \in L''$, then also the element $\vartheta - b$ generates the same extension. By the additivity of the polynomial $X^p - X$, $\vartheta - b$ is a root of the Artin-Schreier polynomial $X^p - X - (c - b^p + b)$. The idea now is to use this principle to deduce a "normal form" for c from which we can read off that the extension has trivial defect. Still, we are quite lost if we do not make some reductions beforehand. First, it is clear that one can proceed by induction on the transcendence degree; so we can reduce to the case of $\operatorname{trdeg} L|K = 1$. Second, as v may not be trivial on K, it may have a very large rank. By general valuation theory, one has to reduce first to finite rank and then to rank 1. This being done, one can show that c can be taken to be a polynomial $g \in K[x]$, where $x \in L''$ is transcendental over K. Now the idea is the following: if $k = p \cdot \ell$ and g contains a non-zero summand $c_k x^k$, then we replace it by $c_k^{1/p} x^\ell$. This is done by setting $b = c_k^{1/p} x^\ell$ in the above computation. In this way one eliminates all p-th powers in g, and the thus obtained normal form for c will show that the extension has trivial defect.

This method (which I call "Artin-Schreier surgery") seems to have several applications; I used it again to prove a quite different result (Theorem 17.4 below). It can also be found in the paper [EPP].

Let us note that the Artin-Schreier polynomials appear in Abhyankar's work in a somewhat disguised form. This is because the coefficients have to lie in the local ring he is working in. For example, if $vc < 0$, we would rather prefer to have a polynomial having coefficients in the valuation ring, defining the same extension as $X^p - X - c$. Setting $X = cY$, we find that if ϑ is a root of $X^p - X - c$, then ϑ/c is a root of $Y^p - c^{1-p}Y - c^{1-p}$ with $vc^{1-p} = (1-p)vc > 0$. Therefore, Abhyankar considers polynomials of the form $Z^p - c_1 Z - c_2$ (cf. e.g., [A1], page 515, [A6], Theorem (2.2), or [A7], page 34). In an extension obtained from L by adjoining a $(p-1)$th root of c_1 (if (L,v) is henselian, then such an extension is tame), this polynomial can be transformed back into an Artin-Schreier polynomial.

Having shown inertial generation for function fields with Abhyankar places, let us return to our problem of local uniformization.

15. Relative local uniformization

Throughout, I have stressed the function field aspect of local uniformization. I have shown that function fields with Abhyankar places can be generated in a nice way. I have talked about the algebraic elements satisfying the assumption of the Multidimensional Hensel's Lemma. The logical consequence of all this is to try to build up our function field step by step: first, choose a nice transcendence basis, according to Theorem 14.2, then find algebraic elements, one after the other, each of them satisfying the assumptions of Hensel's Lemma over the previously generated field. This is the origin of the following definition.

Take a finitely generated extension $F|K$, not necessarily transcendental, and a place P on F, not necessarily trivial on K. We write $\mathcal{O}_F$ for the valuation ring of P on F, and $\mathcal{O}_K$ for the valuation ring of its restriction to K. Further, take elements $\zeta_1, \ldots, \zeta_m \in \mathcal{O}_F$. We will say that $(F|K,P)$ *is uniformizable with respect to* $\zeta_1, \ldots, \zeta_m$ if there are

- a transcendence basis $T = \{t_1, \ldots, t_s\} \subset \mathcal{O}_F$ of $F|K$ (may be empty),
- elements $\eta_1, \ldots, \eta_n \in \mathcal{O}_F$, with $\zeta_1, \ldots, \zeta_m$ among them,
- polynomials $f_i(X_1, \ldots, X_n) \in \mathcal{O}_K[t_1, \ldots, t_s, X_1, \ldots, X_n]$, $1 \leq i \leq n$,

such that $F = K(t_1, \ldots, t_s, \eta_1, \ldots, \eta_n)$, and

(U1) for $i < j$, X_j does not occur in f_i,

(U2) $f_i(\eta_1, \ldots, \eta_n) = 0$ for $1 \leq i \leq n$,

(U3) $(\det J_f(\eta_1, \ldots, \eta_n))P \neq 0$.

Assertion **(U1)** implies that J_f is triangular. Assertion **(U3)** says that $f_1, \ldots, f_n$ and $\eta_1, \ldots, \eta_n$ satisfy the assumption (5) of the Multidimensional Hensel's Lemma. By the triangularity, this implies that for each i, f_i and η_i satisfy the hypothesis of Hensel's Lemma over the ground field $K(t_1, \ldots, t_s, \eta_1, \ldots, \eta_{i-1})$.

We say that $(F|K,P)$ *is uniformizable* if it is uniformizable with respect to *every* choice of finitely many elements in $\mathcal{O}_F$.

Now assume in addition that P is trivial on K. Then $\mathcal{O}_K = K$, and the P-residues of the coefficients are obtained by just replacing t_j by t_jP, for $1 \leq j \leq n$. Hence if we view the polynomials f_i as polynomials in the variables $Z_1, \ldots, Z_s, X_1, \ldots, X_n$, then assertion **(U3)** means that the Jacobian matrix at the point $(t_1P, \ldots, t_sP, \eta_1P, \ldots, \eta_nP)$ has maximal rank. This assertion says that on the variety defined over K by the f_i (having generic point $(t_1, \ldots, t_s, \eta_1, \ldots, \eta_n)$ and function field F), the place P is centered at the smooth point

$$(t_1P, \ldots, t_sP, \eta_1P, \ldots, \eta_nP) .$$

By uniformizing with respect to the ζ's, we obtain the following important information: if we have already a model V of $F|K$ with generic point $(z_1, \ldots, z_k)$, where $z_1, \ldots, z_k \in \mathcal{O}_F$, then we can choose our new model $\mathcal{V}$ of $F|K$ in such a way that the local ring of the center of P on $\mathcal{V}$ contains the local ring of the center $(z_1P, \ldots, z_kP)$ of P on V. For this, we only have to let $z_1, \ldots, z_k$ appear among the ζ's. In fact, Zariski proved the following *Local Uniformization Theorem* (cf. [Z]):

Theorem 15.1. *Suppose that $F|K$ is a function field of characteristic 0, P is a place of $F|K$, and $\zeta_1, \ldots, \zeta_m$ are elements of $\mathcal{O}_F$. Then $(F|K, P)$ is uniformizable with respect to $\zeta_1, \ldots, \zeta_m$.*

But the ζ's also play another role. Through their presence, the above property becomes transitive (see [K6]):

Theorem 15.2. (Transitivity of Relative Uniformization) *If $E|F$ is uniformizable with respect to $\zeta_1, \ldots, \zeta_m$ and $F|K$ is uniformizable with respect to certain finitely many elements derived from $E|F$ and $\zeta_1, \ldots, \zeta_m$, then $E|K$ is uniformizable with respect to $\zeta_1, \ldots, \zeta_m$. In particular, if $E|F$ and $F|K$ are uniformizable, then so is $E|K$.*

As we do not require that P is trivial on K, we call the property defined above *relative (local) uniformization.* Its transitivity enables us to build up our function field step by step by extensions which admit relative uniformization. I give examples of uniformizable extensions; for the proofs, see [K5], [K6], [K7].

I) We consider a function field $F|K$ and a place P of F such that v_PK is a convex subgroup of v_PF. The latter always holds if P is trivial on K since then, $v_PK = \{0\}$. We take elements $x_1, \ldots, x_\rho$ in F such that $v_Px_1, \ldots, v_Px_\rho$ form a maximal set of rationally independent elements in v_PF modulo v_PK. Further, we take elements $y_1, \ldots, y_\tau$ in F such that $y_1P, \ldots, y_\tau P$ form a transcendence basis of FP over KP.

Proposition 15.3. *In the described situation, $(K(x_1, \ldots, x_\rho, y_1, \ldots, y_\tau)|K, P)$ is uniformizable. More precisely, the transcendence basis T can be chosen of the form $\{x'_1, \ldots, x'_\rho, y_1, \ldots, y_\tau\}$, where $x'_1, \ldots, x'_\rho \in \mathcal{O}_{K(x_1,\ldots,x_\rho)}$ and for some $c \in \mathcal{O}_K$, $c \neq 0$, (with $c = 1$ if P is trivial on K), the elements $cx'_1, \ldots, cx'_\rho$ generate the same multiplicative subgroup of $K(x_1, \ldots, x_\rho)^\times$ as $x_1, \ldots, x_\rho$.*

The proof uses Theorem 7.1. It also uses the following lemma, which was proved (but not explicitly stated) by Zariski in [Z] for subgroups of $\mathbb{R}$, using the *Algorithm of Perron*. I leave it as an easy exercise to you to prove the general case by induction on the (finite!) rank of the ordered abelian group. An instant proof of the lemma can be found in [EL] (Theorem 2.2), and I am sure there are several more authors who reproved the lemma, not knowing about Zariski's application.

Lemma 15.4. *Let G be a finitely generated ordered abelian group. Take any positive elements $\alpha_1, \ldots, \alpha_\ell$ in G. Then there exist positive elements $\gamma_1, \ldots, \gamma_\rho \in G$ such that $G = \mathbb{Z}\gamma_1 \oplus \ldots \oplus \mathbb{Z}\gamma_\rho$ and every α_i can be written as a sum $\sum_j n_{ij}\gamma_j$ with non-negative integers n_{ij}.*

II) The next result is based on Kaplansky's work [KA1]. First, we need:

Lemma 15.5. *Let $(K(z)|K, P)$ be an immediate transcendental extension. Assume further that (K, P) is separable-algebraically maximal or that $(K(z), P)$ lies in the completion of (K, P). Then for every polynomial $f \in K[X]$, the value $v_P f(a)$ is fixed for all $a \in K$ sufficiently close to z. That is,*

$$\begin{gathered}\forall f \in K[X]\ \exists \alpha \in v_P K\ \exists \beta \in \{v_P(z-b) \mid b \in K\}\ \forall a \in K : \\ v_P(z-a) \geq \beta \Rightarrow v_P f(a) = \alpha\,. \end{gathered} \tag{51}$$

Kaplansky proves that if (51) does not hold, then there is a proper immediate algebraic extension of (K, P). If $(K(z), P)$ does not lie in the completion of (K, P), then this can be transformed into a proper immediate separable-algebraic extension ([K1], [K2]; the proof uses a variant of the Theorem on the Continuity of Roots). The existence of such an extension is excluded if (K, v) is separable-algebraically maximal. If on the other hand we assume that $(K(z), P)$ lies in the completion of (K, P), then one can show that if f does not satisfy (51), then $v_P f(z) = \infty$. But this means that $f(z) = 0$, contradicting the assumption that $K(z)|K$ is transcendental. Note that by Lemma 11.8, every separably tame and hence also every tame field (K, P) satisfies the assumption of Lemma 15.5.

The following proposition is somewhat complementary to Proposition 15.3.

Proposition 15.6. *Let $(K(z)|K, P)$ be an immediate transcendental extension. If z satisfies (51), then $(K(z)|K, P)$ is uniformizable.*

III) The proof of the following proposition is quite easy; for the transcendental part, it uses Lemma 15.5 and Proposition 15.6.

Proposition 15.7. *Every separable extension of a valued field within its completion is uniformizable.*

The henselization of a valued field (K, P) is always a separable-algebraic extension. If (K, P) has rank 1, then moreover, the henselization lies in the completion of (K, P) (since in this case the completion is henselian). Therefore, Proposition 15.7 yields:

Corollary 8. *Assume that (K,P) has rank 1. If $(L|K,P)$ is a finite subextension of the henselization of (K,P), then it is uniformizable.*

IV) The following proposition is of interest in view of the inertial generation of function fields with Abhyankar places. Its proof is again quite easy.

Proposition 15.8. *Let (K,P) be a henselian field and $(L|K,P)$ a finite extension within the absolute inertia field of (K,P). Then $(L|K,P)$ is uniformizable.*

After these positive results, I have to talk about a serious problem. Throughout my work in the model theory of valued fields, my experience was that henselizations are very nice extensions and do not harm at all. Because of their universal property (Theorem 5.13), they behave well in embedding lemmas. Unfortunately, for the problem of local uniformization, this seems to be entirely different. While we have relative uniformization if the henselization lies in the completion, we get problems if this is not the case. And for a rank greater than 1, we cannot expect in general that the henselization lies in the completion (since the completion will in general not be henselian). This leads to the following important open problem:

Open Problem 3: Prove or disprove: every finite subextension in the henselization of a valued field is uniformizable.

This is a special case of a slightly more general problem, however. Again in view of inertial generation, we would like to know the following.

Open Problem 4: Assume that (K,v) is a field which is *not* henselian. Prove or disprove: every finite subextension in the absolute ramification field of (K,v) is uniformizable.

The obstruction is the following. Assume that $(L|K,v)$ is a finite subextension in the absolute ramification field of (K,v). Suppose there is an intermediate field L' such that $Lv = L'v$ and $[L':K] = [L'v:Kv]$. The former yields that L lies in the henselization of L'. The latter yields that $(L'|K,v)$ admits relative local uniformization, which can be proved in exactly the same way as Proposition 15.8, although (K,v) need not be henselian. So by the transitivity of relative uniformization, the problem would be reduced to that of subextensions within the henselization. But such intermediate fields L' may not exist!

A closer look reveals that this problem is also the kernel of our problem about subextensions within the henselization. Indeed, if we have rank > 1, then P is the composition $P = Q\overline{Q}$ of two non-trivial places. Advanced ramification theory shows that if $(L|K,P)$ is a subextension of the henselization of (K,P), then $(L|K,Q)$ is a subextension of the absolute inertia field of (K,Q). If we could split everything up by intermediate fields, then we could reduce to extensions like the above $(L'|K,v)$, and extensions within completions; this would solve our problem. But the necessary intermediate fields may only exist after enlarging the extension $L|K$. Nevertheless, in [K6] I prove a weak form of relative uniformization for finite extensions within the absolute ramification field.

Let us see what we get and what we do not get from the above positive results.

16. Local uniformization for Abhyankar places

Using the transitivity of relative uniformization, we can combine Proposition 15.3 with Corollary 8 to obtain:

Theorem 16.1. *Take an Abhyankar place of $F|K$ such that (F,P) has rank 1. If $FP = K$ or $FP|K$ is a rational function field, then $(F|K,P)$ is uniformizable.*

This works since $(F|K,P)$ is henselian generated. But as soon as $FP|K$ is not a rational function field, $(F|K,P)$ will not be henselian generated. Then we may run into the problems described in the last section. For the time being, our only chance is to accept to extend F. Then we can prove ([K6], [K7]):

Theorem 16.2. *Assume that P is an Abhyankar place of $F|K$ and take elements $\zeta_1,\ldots,\zeta_m \in \mathcal{O}_F$. Then there is a finite purely inseparable extension $\mathcal{K}|K$, a finite separable extension $\mathcal{F}|F.\mathcal{K}$, and an extension of P from F to $\mathcal{F}$ such that $(\mathcal{F}|\mathcal{K},P)$ is uniformizable with respect to $\zeta_1,\ldots,\zeta_m$. In addition, we have:*

1) *If (F,P) has rank 1, then $\mathcal{F}$ can be obtained from $F.\mathcal{K}$ by a Galois extension.*
2) *If (F,P) has rank $r > 1$, then $\mathcal{F}$ can be obtained from $F.\mathcal{K}$ by a sequence of at most $r-1$ Galois extensions if $FP|K$ is algebraic, or at most r Galois extensions otherwise.*
3) *Alternatively, $\mathcal{F}$ can always be chosen to lie in the henselization of $(F.\mathcal{K},P)$.*
4) *If $FP|K$ is separable, then in all cases we can choose $\mathcal{K}=K$.*

Unfortunately, "Galois extension" and "lie in the henselization" are mutually incompatible. Indeed, the normal hull of a subextension of the henselization will in general not again lie in the henselization.

We could prove local uniformization for all Abhyankar places of $F|K$ with separable $FP|K$, if we would know a positive answer to the following problem:

Open Problem 5: Take any Abhyankar place of $F|K$. Is it possible to choose the generating transcendence basis T for the inertial generation of $(F|K,P)$ in such a way that the extension $(F|K(T),P)$ is uniformizable?

Here is an even more daring idea. Since we have seen that we have problems with the henselization, why don't we try to avoid it?

Open Problem 6: Take any Abhyankar place of $F|K$. Is it possible to choose the generating transcendence basis T for the inertial generation of $(F|K,P)$ in such a way that the extension of P from $K(T)$ to F is unique?

In that case, general valuation theory shows that the extension $F|K(T)$ is linearly disjoint from the extension $K(T)^h|K(T)$. If T has that property, we would say that *T has the uniqueness property for $F|K$* (this is a generalization of the definition given in Section 14). But I warn you: this problem seems to be very hard. It is already non-trivial to prove the existence of elements with the uniqueness property in case of transcendence degree 1. For arbitrary transcendence degree, the necessary algebraic geometry is not in sight. But perhaps there is some connection with local uniformization or resolution of singularities?

A weak form of local uniformization, without extending the function field, can be proved for all Abhyankar places in the case of perfect ground fields [K6]. Also T. Urabe [U], building on Abhyankar's original approach, has a comparable result for the special case of places of maximal rank.

17. Non-Abhyankar places and the Henselian Rationality of immediate function fields

What can we do if the place P of $F|K$ is *not* an Abhyankar place? Still, the place may be nice. Assume for instance that v_PF is finitely generated and $FP = K$. Then we can choose $x_1, \ldots, x_\rho$ such that $v_P = \mathbb{Z}v_Px_1 \oplus \ldots \oplus \mathbb{Z}v_Px_\rho$, and set $F_0 := K(x_1, \ldots, x_\rho)$. Consequently, $(F|F_0, v_P)$ is an immediate extension. If P is not an Abhyankar place, then this extension is not algebraic. But we have already stated two tools to treat transcendental immediate extensions, namely Proposition 15.6 and Proposition 15.7.

Let us first apply Proposition 15.7. If (F, v_P) is a separable extension within the completion of (F_0, v_P), then by this proposition, the transitivity and Proposition 15.3, we find that $(F|K, P)$ is uniformizable. Do we need the assumption that the extension be separable? The answer is: this is automatically true. Indeed, by the Generalized Stability Theorem 14.1, (F_0, v_P) is a defectless field; thus, our assertion follows from the following lemma:

Lemma 17.1. *If (L, v) is a defectless field and $(L'|L, v)$ is an immediate extension, then $L'|L$ is separable.*

Proof. We have to show that $L'|L$ is linearly disjoint from every purely inseparable finite extension $E|L$. As the extension of the valuation from L to E is unique by Corollary 2, this implies that $[E : L] = (vE : vL)[Ev : Lv]$. Now we consider the compositum $L'.E$ (with the unique extension of the valuation from L' to the purely inseparable extension $L'.E$). Since $vE \subseteq v(L'.E)$, $vL' = vL$, $Ev \subseteq (L'.E)v$ and $L'v = Lv$, we have that $(v(L'.E) : vL') \geq (vE : vL)$ and $[(L'.E)v : L'v] \geq [Ev : Lv]$. Hence,

$$\begin{aligned}[L'.E : L'] &\geq (v(L'.E) : vL')[(L'.E)v : L'v] \\ &\geq (vE : vL)[Ev : Lv] = [E : L] \geq [L'.E : L'] \,.\end{aligned}$$

Thus, equality holds everywhere, showing that $L'|L$ is linearly disjoint from $E|L$. □

In view of this result, Proposition 15.7 and the transitivity prove:

Theorem 17.2. *The assertions of Theorem 16.1 and Theorem 16.2 remain true if (F, P) is replaced by a finitely generated extension within its completion.*

If P is a discrete rational place of $F|K$, then we only have to choose $t \in F$ such that $v_P t$ is the smallest positive element in $v_P F \simeq \mathbb{Z}$; this will imply that $v_P F = \mathbb{Z} v_P t$. The immediate extension (F, P) of $(K(t), P)$ will automatically lie in the completion of $(K(t), P)$ (since the completion is of the form $K((t))$ and hence maximal, and one can easily show that the completion is unique up to isomorphism). So we obtain:

Theorem 17.3. *Every discrete rational place is uniformizable.*

Now let us turn to the general case. Not every immediate extension lies in the completion, not even in rank 1.

Example 22. Take the valuation v on the rational function field $K(x_1, x_2)$ such that $vK(x_1, x_2) = \mathbb{Z} + r\mathbb{Z}$ with $r \in \mathbb{R} \setminus \mathbb{Q}$. We can view $K(x_1, x_2)$ as a subfield of the power series field $K((\mathbb{Z} + r\mathbb{Z}))$, with $x_1 = t$ and $x_2 = t^r$. In $\mathbb{Z} + r\mathbb{Z}$, we can choose a monotonically increasing sequence r_i converging to 0 from below. Then we take the element $z = \sum_{i=1}^{\infty} t^{r_i} \in K((\mathbb{Z} + r\mathbb{Z}))$. It does not lie in the completion of $(K(x_1, x_2), v)$. Nevertheless, the extension $(K(x_1, x_2, z)|K(x_1, x_2), v)$ is immediate. I leave it to you to show that z is transcendental over $K(x_1, x_2)$.

So let us look at the case of an immediate extension $(F_0(z)|F_0, P)$ which does not lie in the completion of (F_0, P). Then to apply Proposition 15.6, we need to know that z satisfies (51). By Lemma 15.5, this would hold if (F_0, P) were separable-algebraically maximal. But as the rational function field F_0 is not henselian unless P is trivial, it will admit its henselization as a proper immediate separable-algebraic extension. The only way out at this point is to pass to the immediate extension $(F_0^h(z)|F_0^h, P)$. There, we can apply Proposition 15.6 since by the Generalized Stability Theorem, (F_0^h, P) is a defectless field and thus is algebraically maximal. But now we have extended our function field F. (In the end, we will only need a finite subextension since the statement of local uniformization only talks about finitely many elements.) We will return to this aspect below.

Beforehand, let us think about two further problems. First, if our extension $F|F_0$ is of transcendence degree > 1, how do we carry on by induction? As the Generalized Stability Theorem does not apply any more in this situation, we do not know whether $(F_0(z), P)$ is a defectless field (and in fact, in general it will not be). Then we have to enlarge $F_0(z)$ even more to achieve the next induction step; in positive characteristic, the henselization will be too small for this purpose.

Second, if for instance $F|F_0(z)$ is a proper algebraic extension, then does it have relative uniformization? The only answers we have apply to the case where (F, P) lies in the henselization of $(F_0(z), P)$ (and even in this case our discussion has shown a bunch of problems). If (F, P) does not lie in the henselization, then we know nothing. Observe that since the extension $(F|F_0(z), P)$ is finite and immediate, (F, P) does not lie in the henselization if and only if $(F^h|F_0(z)^h, P)$ has non-trivial defect (by Theorem 5.14 and Corollary 6).

So the question arises: how can we avoid the defect in the case of immediate extensions? The answer is a theorem that I proved in [K1] (cf. [K2]). As for the Generalized Stability Theorem, the proof uses ramification theory and the deduction of normal forms for Artin-Schreier extensions. It also uses significantly a theory of immediate extensions which builds on Kaplansky's paper [KA1].

Theorem 17.4. (Henselian Rationality of Immediate Function Fields) *Let (K, P) be a tame field and $(F|K, P)$ an immediate function field of transcendence degree 1. Then*

$$\text{there is } x \in F \text{ such that } (F^h, P) = (K(x)^h, P) , \tag{52}$$

that is, $(F|K, P)$ is henselian generated. The same holds over a separably tame field (K, P) if in addition $F|K$ is separable.

Since the assertion says that F^h is equal to the henselization of a rational function field, we also call F *henselian rational.* For valued fields of residue characteristic 0, the assertion is a direct consequence of the fact that every such field is defectless. Indeed, take any $x \in F \setminus K$. Then $K(x)|K$ cannot be algebraic since otherwise, $(K(x)|K, P)$ would be a proper immediate algebraic extension of the tame field (K, P), a contradiction to Lemma 11.8. Hence, $F|K(x)$ is algebraic and immediate. Therefore, $(F^h|K(x)^h, P)$ is algebraic and immediate too. But since it cannot have a non-trivial defect, it must be trivial. This proves that $(F, P) \subset (K(x)^h, P)$. In contrast to this, in the case of positive residue characteristic only a very carefully chosen $x \in F \setminus K$ will do the job.

To illustrate the use of Theorem 17.4 in the model theory of valued fields, we give an example which is "complementary" to Example 21, treating the case of immediate extensions:

Example 23. Suppose that (K, v) is a tame field and that $(L|K, v)$ is an immediate extension. Then the conditions $vK \prec_\exists vL$ and $Kv \prec_\exists Lv$ are trivially satisfied. So do we have that $(K, v) \prec_\exists (L, v)$? Using the theory of tame fields, in particular the crucial Lemma 11.9, one reduces the proof to the case of transcendence degree 1. So we have to prove an embedding lemma for immediate function fields $(F|K, v)$ of transcendence degree 1 over tame fields. If we take any $x \in F \setminus K$ then it will satisfy (51) because (K, v) is tame and thus also algebraically maximal. Then with the help of Theorem 2 of [KA1] and the saturation of $(K, v)^*$, we can find an embedding of $(K(x), v)$ in $(K, v)^*$. But how do we carry on? We know that $(F|K(x), v)$ is immediate, but this does not mean that (F, v) lies in the henselization of $(K(x), v)$. But if it does, we can just use the universal property of the henselization (Theorem 5.13). Indeed, being a tame field, (K, v) is henselian. Since $(K, v)^*$ is an elementary extension of (K, v) (in the language of valued fields), it is also henselian. Hence if $(K(x), v)$ embeds in $(K, v)^*$, then this embedding can be extended to an embedding of $(K(x)^h, v)$ in $(K, v)^*$. This induces the desired embedding of F.

If F does not lie in the henselization of $(K(x), v)$, then $(F^h|K(x)^h, v)$ has non-trivial defect, and we have no clue how the embedding of $K(x)^h$ could be extended to an embedding of F^h. Again, our enemy is the defect, and we have to avoid it. Now this can be done by Theorem 17.4. It tells us that there is some $x \in F$ such that (F,v) lies in the henselization of $(K(x),v)$. So we have proved:

Theorem 17.5. *Suppose that (K,v) is a tame field and that $(L|K,v)$ is an immediate extension. Then $(K,v) \prec_\exists (L,v)$.*

Given an extension of tame fields of finite transcendence degree, then by use of Lemma 11.9, one can separate it into an extension without transcendence defect and an immediate extension. Both can be treated separately by Theorem 14.3 and Theorem 17.5. As $\prec_\exists$ is transitive, this proves (cf. [K1], [K2]):

Theorem 17.6. *The Ax-Kochen-Ershov principle (49) holds for every extension $(L|K,v)$ of tame fields.*

Let us return to our problem of local uniformization. So far, we have worked with the assumption that we can find a subfunction field F_0 in F such that the restriction of P to F_0 is an Abhyankar place and $(F|F_0,P)$ is immediate. But it is not always possible to achieve the latter. For example, take F to be the rational function field $K(x_1,x_2)$ and P such that $FP = K$ and $v_P F$ is a not finitely generated subgroup of $\mathbb{Q}$; we will construct such a place P in the next section. But for any F_0 on which P is an Abhyankar place, $v_P F_0$ is finitely generated, so we will always have that $v_P F \neq v_P F_0$.

In this situation, passing to henselizations may help again. Given an arbitrary place P of $F|K$, we choose an Abhyankar subfunction field as in (32). We have that $v_P F/v_P F_0$ is a torsion group and that $FP|F_0P$ is algebraic. Take F_1 to be the relative algebraic closure of F_0 in F^h. If $\mathrm{char}\, FP = \mathrm{char}\, K = 0$, then by Lemma 4.3 and Lemma 4.5, $(F^h|F_1,P)$ is an immediate extension; so we succeeded again in reducing to the case of immediate extensions. But if $\mathrm{char}\, FP = \mathrm{char}\, K = p > 0$, then we only know that $v_P F^h/v_P F_1$ is a p-group and that $F^hP|F_1P$ is purely inseparable. So in this case, passing to henselizations is not enough. But we obtain an immediate extension if we replace F^h and F_1 by their perfect hulls. In fact, to make all of our tools work, we have to take even bigger extensions. Namely, we have to pass to smallest algebraic extensions which are tame or at least separably tame fields. But these extensions still have nice properties; we will talk about them in Section 23.

With this approach, one can deduce the following theorems from the Generalized Stability Theorem, the Henselian Rationality of Immediate Function Fields, the results described in Section 15, and the transitivity of relative uniformization:

Theorem 17.7. *Let $F|K$ be a function field of arbitrary characteristic and P a place of $F|K$. Take any elements $\zeta_1,\ldots,\zeta_m \in \mathcal{O}_P$. Then there exist a finite extension $\mathcal{F}$ of F, an extension of P to $\mathcal{F}$, and a finite purely inseparable extension $\mathcal{K}$ of K within $\mathcal{F}$ such that $(\mathcal{F}|\mathcal{K},P)$ is uniformizable with respect to $\zeta_1,\ldots,\zeta_m$.*

Theorem 17.8. *The extension $\mathcal{F}|F$ can always be chosen to be normal.*

See [K6] for the proof. These theorems also follow from the results of Johan de Jong ([dJ]; cf. also [AO]). So this should be an interesting question:

Open Problem 7: Is it possible to recognize counterparts of the Generalized Stability Theorem and the Henselian Rationality of Immediate Function Fields in the theory of semi-stable reduction, or in any other part of de Jong's proof of desingularization by alteration?

The advantage of proving the above theorems by the described valuation theoretical approach is that we get additional information about the extension $\mathcal{F}|F$. We have already seen in this and the previous section that in certain cases we do not need an extension, i.e., we have local uniformization already for $(F|K, P)$. In other cases, we can obtain $\mathcal{F}$ from F by one or a tower of Galois extensions. We will see in Section 23 that in general, we can choose $\mathcal{F}|F.\mathcal{K}$ to be a separable (but not Galois) extension with additional information about the related extensions of value group and residue field.

A little bit of horror makes an excursion even more interesting. So let's watch out for bad places.

18. Bad places

In this section we will show that there are places of function fields $F|K$ whose value group or residue field are not finitely generated. By combining the methods you can construct examples where both is the case. The following two examples can already be found in [ZS], Chapter VI, §15. But our approach (using Hensel's Lemma) is somewhat easier and more conceptual.

Example 24. We construct a place on the rational function field $K(x_1, x_2)|K$ whose value group $G \subset \mathbb{Q}$ is not finitely generated, assuming that the order of every element in $G/\mathbb{Z}$ is prime to $\operatorname{char} K$. To this end, we just find a suitable embedding of $K(x_1, x_2)$ in $K((G))$. We do this by setting $S := \{n \in \mathbb{N} \mid 1/n \in G\}$ and

$$x_1 := t \quad \text{and} \quad x_2 := \sum_{n \in S} t^{-1/n} . \tag{53}$$

Further, take the valuation v on $K(x_1, x_2)$ to be the restriction of the canonical valuation v of $K((G))$. We wish to show that $1/S \subseteq vK(x_1, x_2)$, so that $G \subset vK(x_1, x_2)$. Since $(K(x_1, x_2), v) \subset (K((G)), v)$, it follows that $G = vK(x_1, x_2)$. If x_2 were algebraic over $K(x_1)$, we would know by Corollary 5 that $vK(x_1, x_2)$ is finitely generated. Hence if it is not, then x_2 must be transcendental over $K(x_1)$, so that $K(x_1, x_2)$ is indeed the rational function field over K in two variables.

Suppose that $\operatorname{char} K = 0$; then we can get $G = \mathbb{Q}$. Also in positive characteristic one can define the valuation in such a way that the value group becomes $\mathbb{Q}$; since then we have to deal with inseparability, our construction has to be refined slightly, which we will not do here.

Now let us prove our assertion. We take (L, v) to be the henselization of $(K(x_1, x_2), v)$. We are going to show that $t^{1/n} \in L$ for all $n \in S$. Suppose we have shown this for all $n < k$, where $k \in S$ (we can assume that $k > 1$). Then also

$s_k := \sum_{n\in S, n<k} t^{-1/n} \in L$. We write

$$x_2 - s_k = \sum_{n\in S, n\geq k} t^{-1/n} = t^{-1/k}(1+c) \tag{54}$$

where $c \in K((G))$ with $vc > 0$. Hence, $1+c$ is a 1-unit. We have that $(1+c)^k = t(x_2 - s_k)^k \in L$. On the other hand, $(1+c)^k v = ((1+c)v)^k = 1^k = 1$, which shows that $(1+c)^k$ is again a 1-unit. Since $k \in S$ we know that $\mathrm{char}\, LP = \mathrm{char}\, K$ does not divide k. Hence by Lemma 4.4, $1+c \in L$. This proves that $t^{1/k} = (1+c)(x_2 - s_k)^{-1} \in L$.

We have now proved that $t^{1/k} \in L$ for all $k \in S$. Hence, $1/k = vt^{1/k} \in vL$ for all $k \in S$. But since the henselization is an immediate extension, we know that $vL = vK(x_1, x_2)$, so we have proved that $1/S \subset vK(x_1, x_2)$.

Example 25. We take a field K for which the separable-algebraic closure K^{sep} is an infinite extension (i.e., K is neither separable-algebraically closed nor real closed). We construct a place of the rational function field $K(x_1, x_2)|K$ whose residue field is not finitely generated. We choose a sequence a_n , $n \in \mathbb{N}$ of elements which are separable-algebraic over K of degree at least n. We define an embedding of $K(x_1, x_2)$ in $K^{\mathrm{sep}}((t))$ by setting

$$x_1 := t \quad \text{and} \quad x_2 := \sum_{n\in\mathbb{N}} a_n t^n \,. \tag{55}$$

Further, we take the valuation v on $K(x_1, x_2)$ to be the restriction of the valuation of $K^{\mathrm{sep}}((t))$. We wish to show that $a_n \in K(x_1, x_2)v$ for all $n \in \mathbb{N}$, so that $K(x_1, x_2)v|K$ cannot be finitely generated. If x_2 were algebraic over $K(x_1)$, we would know by Corollary 5 that $K(x_1, x_2)v|K$ is finitely generated. So if it is not, then x_2 must be transcendental over $K(x_1)$, so that $K(x_1, x_2)$ is indeed the rational function field over K in two variables. By a modification of the construction, one can also generate infinite inseparable extensions of K. If K is countable, one can generate every algebraic extension of K as a residue field of $K(x_1, x_2)$.

We take again (L, v) to be the henselization of $(K(x_1, x_2), v)$. We are going to show that $a_n \in L$ for all $n \in \mathbb{N}$. Suppose we have shown this for all $n < k$, where $k \in \mathbb{N}$. Then also $s_k := \sum_{n=1}^{k-1} a_n t^n \in L$. We write

$$\frac{x_2 - s_k}{t^k} = \frac{1}{t^k} \sum_{n=k}^{\infty} a_k t^k = a_k(1+c) \tag{56}$$

where $c \in K^{\mathrm{sep}}((t))$ with $vc > 0$. Take $f \in K[X]$ to be the minimal polynomial of a_k over K and note that $f = fv$. Since $a_k \in K^{\mathrm{sep}}$, we know that a_k is a simple root of f. On the other hand, $a_k = a_k(1+c)v \in Lv$. Hence by Hensel's Lemma (Simple Root Version) there is a root a of f in L such that $av = a_k$. As we may assume that the place associated with v is the identity on K, this will give us that $a = a_k$; so $a_k \in L$.

We have now proved that $\forall n \in \mathbb{N}\; a_n \in L$. Hence, $\forall n \in \mathbb{N}\; a_n \in Lv = K(x_1, x_2)v$.

19. The role of the transcendence basis and the dimension

In our approach described in Section 17, we have obtained the subfunction field F_0 on which the restriction of P is an Abhyankar place by choosing the elements $x_1, \ldots, x_\rho, y_1, \ldots, y_\tau$. But then we have made no effort to improve our choice. With this "stiff" approach (which in fact gives additional information), one can prove Theorem 17.7, but it can be shown that in general one cannot get local uniformization without an extension of the function field. I want to show why not. The following example is particularly interesting since it is also a key example in the model theory of valued fields of positive characteristic (cf. Section 21).

Example 26. We denote by $\mathbb{F}_p$ the field with p elements. We consider the following function field of transcendence degree 3 over $\mathbb{F}_p$:

$$F = \mathbb{F}_p(x_1, x_2, y, z) \quad \text{with } z^p - z = x_1 - x_2 y^p . \tag{57}$$

Since $x_1 \in \mathbb{F}_p(x_2, y, z)$, F is a rational function field. However, in [K1] (cf. also [K2], [K4]) we have shown that there is a rational place P of $F|\mathbb{F}_p$ such that $v_P F = \mathbb{Z}v_P x_1 \times \mathbb{Z}v_P x_2$ (ordered lexicographically) and that the valued function field $(F|\mathbb{F}_p(x_1, x_2), P)$ is not henselian generated. It follows that F cannot lie in the henselization of $(\mathbb{F}_p(t_1, t_2, t_3), P)$ if $t_1, t_2 \in F$ are algebraic over $\mathbb{F}_p(x_1, x_2)$ (and hence lie in $\mathbb{F}_p(x_1, x_2)$ since this is relatively algebraically closed in F). Therefore, Theorem 6.2 shows that $(F|\mathbb{F}_p, P)$ admits no local uniformization with t_1, t_2 algebraic over $K(x_1, x_2)$. A function field having such a local uniformization must have degree at least p over F. And indeed, degree p suffices, as

$$(F(x_2^{1/p})|\mathbb{F}_p(x_1, x_2^{1/p}), P)$$

is a rational function field. (It is an interesting fact that there is also a Galois extension of $\mathbb{F}_p(x_1, x_2)$ of degree p such that the function field becomes henselian generated.)

The proof that $(F|\mathbb{F}_p(x_1, x_2), P)$ is not henselian generated is based on showing that $(\mathbb{F}_p(x_1, x_2), P)$ is not existentially closed in (F, P). I have not found an algebraic proof.

Open Problem 8: Develop a method to prove algebraically that a given valued function field $(F|K, v)$ (v not necessarily trivial on K) is *not* henselian generated or inertially generated.

A variant of the example (cf. [K7]) shows: *There are immediate transcendental extensions of valued fields which are not uniformizable.* The example teaches us that relative uniformization will not always hold without an extension of the function field. Hence, in general we will have to optimize our choice of the transcendence basis for F_0 or even for F in order to obtain local uniformization for $(F|K, P)$. Given a transcendence basis T of F, it is easy to measure how far F is from lying in the absolute inertia field $K(T)^i$ of $(K(T), P)$: we just have to take the degree

$$\mathrm{ig}(F, T) := [F.K(T)^i : K(T)^i] . \tag{58}$$

This raises the problem:

Open Problem 9: Develop a method to change T in such a way that $\mathrm{ig}(F,T)$ decreases.

In our above example, this is very easy since $F|K$ is actually a rational function field. But one can modify the example in such a way that $F|K$ is not rational. Instead of doing this for the above example, let us look at a slightly simpler example, which will also show that already a valued rational function field can have an immediate extension of degree p with defect p.

Example 27. We take an arbitrary field K of characteristic $p > 0$ and work in the power series field $K((\frac{1}{p^\infty}\mathbb{Z}))$ with its canonical valuation v. Recall that $\frac{1}{p^\infty}\mathbb{Z}$ is the p-divisible hull $\{m/p^n \mid m \in \mathbb{Z},\, n \in \mathbb{N}\}$ of $\mathbb{Z}$. We have that $K((t)) \subset K((\frac{1}{p^\infty}\mathbb{Z}))$. For every $i \in \mathbb{N}$, we set $\nu_i := \sum_{j=1}^{i} j$, and we define:

$$z := \sum_{i=1}^{\infty} t^{p^{\nu_i} - p^{-\nu_i}} \in K\left(\left(\frac{1}{p^\infty}\mathbb{Z}\right)\right) . \tag{59}$$

We show that $(K((\frac{1}{p^\infty}\mathbb{Z})), v)$ is an immediate extension of $(K(t,z), v)$. Since both fields have residue field K, we only have to show that $\frac{1}{p^\infty}\mathbb{Z} \subseteq vK(t,z)$. For $k \in \mathbb{N}$, we compute:

$$\begin{aligned} z^{p^{\nu_k}} - \sum_{i=1}^{k} t^{p^{\nu_k+\nu_i} - p^{\nu_k-\nu_i}} &= \sum_{i=1}^{\infty} \left(t^{p^{\nu_i} - p^{-\nu_i}}\right)^{p^{\nu_k}} - \sum_{i=1}^{k} \left(t^{p^{\nu_i} - p^{-\nu_i}}\right)^{p^{\nu_k}} \\ &= \sum_{i=k+1}^{\infty} t^{p^{\nu_k+\nu_i} - p^{\nu_k-\nu_i}} = t^{p^{\nu_k+\nu_{k+1}} - p^{\nu_k-\nu_{k+1}}} + \cdots \end{aligned}$$

So

$$v t^{p^{\nu_k+\nu_{k+1}}} \left(z^{p^{\nu_k}} - \sum_{i=1}^{k} t^{p^{\nu_k+\nu_i} - p^{\nu_k-\nu_i}}\right)^{-1} = p^{\nu_k-\nu_{k+1}} = \frac{1}{p^{k+1}} . \tag{60}$$

As the element on the left-hand side is in $K(t,z)$, this shows that $p^{-k}\mathbb{Z} \subset vK(t,z)$ for every $k \in \mathbb{N}$. Consequently, $\frac{1}{p^\infty}\mathbb{Z} \subseteq vK(t,z)$, as desired. This also proves that z is transcendental over $K(t)$ since otherwise, $(vK(t,z) : \mathbb{Z})$ would be finite.

From Section 8 we know that $\vartheta = \sum_{i\in\mathbb{N}} t^{-1/p^i} \in K((\mathbb{Q}))$ is a root of the Artin-Schreier polynomial $X^p - X - 1/t$. We see that this power series already lies in the subfield $K((\frac{1}{p^\infty}\mathbb{Z}))$ of $K((\mathbb{Q}))$. Hence, $(K(t,z,\vartheta)|K(t,z),v)$ is a subextension of $(K((\frac{1}{p^\infty}\mathbb{Z}))|K(t,z),v)$ and thus, it is immediate too. In order to show that it has non-trivial defect, we have to show that it has a unique extension of the valuation, or equivalently, that it is linearly disjoint from the henselization of $(K(t,z),v)$. Since it is a Galois extension of prime degree, it suffices to show that it does not lie in this henselization.

We take the subfield $K(t^{1/p^k} \mid k \in \mathbb{N})$ of $K((\frac{1}{p^\infty}\mathbb{Z}))$. By definition, z lies in the completion of $(K(t^{1/p^k} \mid k \in \mathbb{N}), v)$ (since the values of the summands form a sequence which is cofinal in the value group $\frac{1}{p^\infty}\mathbb{Z}$). Since this value group is archimedean, that is, v has rank 1, we know that the henselization of $(K(t,z),v)$ lies in the completion of $(K(t,z),v)$, which by what we have just shown lies in the completion of $(K(t^{1/p^k} \mid k \in \mathbb{N}), v)$. On the other hand, we have seen in Section 8 that ϑ does not lie in the completion of $(K(t^{1/p^k} \mid k \in \mathbb{N}), v)$. Hence, it does not lie in the henselization of $(K(t,z),v)$.

We have now shown that the function field $K(t,z,\vartheta)|K$ admits a place with a value group which is not finitely generated and such that the extension $(K(t,z,\vartheta)|K(t,z),v)$ is immediate of degree p and defect p. Now you will point out that our function field $K(t,z,\vartheta)$ is again rational: since $1/t = \vartheta^p - \vartheta$, we have that $K(t,z,\vartheta) = K(z,\vartheta)$. So let's change something. We take a polynomial $f(t) \in K[t]$ and note that $vf(t) \geq 0$. Now we replace ϑ by a root ϑ_f of the polynomial $X^p - X - (\frac{1}{t} + f(t))$. It can be shown that the new extension $(K(t,z,\vartheta_f)|K(t,z),v)$ will again be immediate of degree p and defect p. In fact, this is obvious if we choose f without constant term. In that case, $vf(t) > 0$ and we know from Example 4 that the polynomial $X^p - X - f(t)$ has a root in $K(t,z)^h$. By the additivity of $X^p - X$ it follows that the two polynomials $X^p - X - \frac{1}{t}$ and $X^p - X - (\frac{1}{t} + f(t))$ define the same extension of degree p and defect p over $K(t,z)^h$. Consequently, also $(K(t,z,\vartheta_f)|K(t,z),v)$ must be immediate of degree p and defect p.

Now we have that

$$\vartheta_f^p - \vartheta_f \;=\; \frac{1}{t} + f(t) \;=\; \frac{1 + tf(t)}{t}\,. \tag{61}$$

So the minimal polynomial of t over $K(z,\vartheta_f)$ will be

$$Xf(X) \,-\, (\vartheta_f^p - \vartheta_f)X \,+\, 1\,. \tag{62}$$

The transition from the representation $F = K(t,z,\vartheta_f)$ with ϑ_f algebraic over $K(t,z)$ to the representation $F = K(z,\vartheta_f,t)$ with t algebraic over $K(z,\vartheta_f)$ may be called *Artin-Schreier inversion.* With a suitable choice of f, the function field $F = K(t,z,\vartheta_f)|K$ will not be rational. However, whatever choice of f I computed, I found that after a little Artin-Schreier surgery on $X^p - X - (\frac{1}{t} + f(t))$ (which replaces f by a better polynomial), Artin-Schreier inversion will yield a tame extension $(K(z,\vartheta_f,t)|K(z,\vartheta_f),v)$. So at least we got rid of the defect, probably even of the ramification. After all, this is what we expect since we know from Abhyankar's work (cf. [A10]) that $(F|K,P)$ always admits local uniformization for trdeg $F|K$ up to 3 (with the possible exception of characteristic 2, 3, 5).

Getting rid of defect and ramification by Artin-Schreier inversion seems to be the algebraic kernel of local uniformization and, in particular, of Abhyankar's proofs. However, the following questions should be answered without a restriction of the dimension (i.e., the transcendence degree of $F|K$):

Open Problem 10: Prove (or disprove) that by Artin-Schreier inversion in connection with Artin-Schreier surgery one can always get rid of the defect. How about ramification?

The only case where I know that the answer is positive is the case of Theorem 17.4, the Henselian Rationality of Immediate Function Fields. There, it is the crucial part of the proof. There seems to be no reason why the answer to the above problem should depend on the transcendence degree of $F|K$ or on the particular value of the positive characteristic. Actually, what I am saying is not quite true since the proof of Theorem 17.4 so far only works under a strong assumption about the base field, and a generalization may more easily be achieved if the restriction of P to that base field is an Abhyankar place. That might indicate that there is more hope for function fields of transcendence defect at most 1 than for those with higher transcendence defect. In dimension 2 ($\mathrm{trdeg}\, F|K = 2$), every place P of $F|K$ will have transcendence defect at most 1 since there is always a subfunction field of transcendence degree at least 1 on which the restriction of P is an Abhyankar place. This seems to separate the case of dimension ≤ 2 from the case of dimension ≥ 3. But as Abhyankar was able to tackle with dimension 3 (where the transcendence defect may well be 2), there seems to be no reason why all this shouldn't work for even higher dimensions.

By looking at these crucial Artin-Schreier extensions, we have considered the kernel of the problem. But are we really sure that we can always reduce to this kernel (for instance, by passing to ramification fields as described in Section 14)?

Open Problem 11: Is it always (in all dimensions) possible to pull down local uniformization through tame extensions? That is, if $(\mathcal{F}|K, P)$ is uniformizable where $(\mathcal{F}^h|F^h, P)$ is a tame extension, will then also $(F|K, P)$ be uniformizable? Which additional assumptions do we possibly need? What answers can be extracted from Abhyankar's work? Is there some generalization of his "Going Up" and "Coming Down" techniques to all dimensions?

Again, there is no hint why the dimension should have an influence on this problem. Possibly it can be found in Abhyankar's work.

If we look at our examples, we see that bad places and defect extensions in the generation of a function field already appear in dimension 2, so from this point of view, the dividing line seems to lie between dimension 1 and dimension 2. Our consideration concerning the transcendence defect seems to suggest a dividing line between dimension 2 and dimension 3. Also Example 26 goes in this direction, although it is not clear what it actually means for local uniformization and whether there possibly is an analogue of transcendence degree 2.

There is, however, another point which we have not yet mentioned. If our place P has rank > 1, can we then always proceed by induction on the rank? If we are ready to extend our function field F, then the answer is: yes ([K5], [K6], [K7]). But what if we want to prove local uniformization without extension of F and we have $P = Q\overline{Q}$ such that $FQ|K$ is not finitely generated, hence not a function field?

In this situation, Q consumes already transcendence degree 2, and if we assume that also $\overline{Q}$ is non-trivial, then we have $\operatorname{trdeg} F|K \geq 3$. Analogously, if we find that critical things happen in dimension 3 but not in dimension 2, these things might only develop their destructive influence in connection with composition of places, which would lift the critical dimension up to 4. (But to be true, I do not believe that this could happen. I believe, if we have local uniformization in dimension 3, then ultimately there will be a proof which works for dimension 3 in the same way as for all higher dimensions.)

20. The space of all places of $F|K$

The set $S(F|K)$ of all places of $F|K$ (where equivalent places are identified) is called the *Zariski-Riemann manifold* or just the *Zariski space* of $F|K$. See [V] for the definition of the Zariski topology on $S(F|K)$, its compactness and other properties of this space. Here, we will consider yet another property. We have seen in Section 18 that the Zariski space even of very simple function fields can contain bad places. On the other hand, we have seen that there are good places (e.g., Abhyankar places) for which local uniformization is easier than in the general case. But do good places or Abhyankar places exist in every Zariski space?

An ad hoc method to prove that this is true is to construct places of maximal rank. Take a transcendence basis $t_1, \ldots, t_k$ of $F|K$ and set $K_{k+1} := K$ and $K_i := K(t_i, \ldots, t_k)$ for $1 \leq i < k$. Take P_1 to be any place of $F|K_2$ such that $t_1 P_1 = 0$. By Corollary 5, $FP_1|K_2$ is finite; hence, $FP_1|K_3$ is a function field. So we can choose a place P_2 of $FP_1|K_3$ such that $t_2 P_2 = 0$. Again, $FP_1 P_2|K_3$ is finite. By induction, we construct places (in fact, prime divisors) $P_1, \ldots, P_k$. Their composition $P = P_1 \ldots P_k$ is a place of $F|K$ of rank $k = \operatorname{trdeg} F|K$. Hence, P is a place of maximal rank and thus an Abhyankar place of $F|K$.

We wish to give a much more sophisticated method which shows that good places not only exist, but even are "very representative" in every Zariski space. The general result reads as follows (cf. [K3]; the special case of $\operatorname{char} K = 0$ was proved in [KP]):

Theorem 20.1. *Let $F|K$ be a function field in k variables. Let Q be a place of $F|K$ and $a_1, \ldots, a_m, b_1, \ldots, b_n \in F$. Then there exists a place P of $F|K$ such that:*

1) *$v_P F$ is a finitely generated group and extends the subgroup of $v_Q F$ generated by $v_Q b_1, \ldots, v_Q b_n$,*
2) *FP is finitely generated over K and extends $K(a_1 Q, \ldots, a_m Q)$,*
3) *the following holds:*

$$\begin{array}{rclll} a_i P &=& a_i Q & \text{for} & 1 \leq i \leq m\,, \\ v_P b_j &=& v_Q b_j & \text{for} & 1 \leq j \leq n\,. \end{array}$$

Moreover, P can be chosen such that $v_P F$ is a subgroup of the p-divisible hull $\frac{1}{p^\infty} v_Q F$ of $v_Q F$ if $\operatorname{char} K = p > 0$, *and $v_P F \subseteq v_Q F$ otherwise, and that FP is a* *subfield of the perfect hull of* FQ. *Alternatively, if* $r, d \in \mathbb{N} \cup \{0\}$ *satisfy*

$$\dim Q \leq d \leq k - \mathrm{rr}\, Q \,, \qquad \mathrm{rr}\, Q \leq r \leq k - d \,, \tag{63}$$

then P *may be chosen such that* $\dim P = d$ *and* $\mathrm{rr}\, P = r$. *(If* $r = k - d$*, then* P *is an Abhyankar place of* $F|K$*.)*

The theorem tells us that for every place Q of $F|K$ and every choice of finitely many elements in F there is an Abhyankar place P which agrees with Q on these elements.

For the proof of the above theorem, see [K3]. The main ideas are the following. First, we choose an Abhyankar subfunction field F_0 of $(F|K,Q)$ as in (32), where we replace P by Q. Using the model theory of tame fields, we "pull the situation down" into a finite extension F_1 of F_0. This is done as follows.

By our choice of F_0, $v_Q F/v_Q F_0$ is a torsion group and $FQ|F_0Q$ is algebraic. Now we take (L,Q) to be a maximal purely wild extension of the henselization of (F,Q); then by Lemma 11.4, (L,Q) is a tame field. Further, we take L' to be the relative algebraic closure of F_0 in L. Then by Lemma 11.9, (L',Q) is a tame field and $(L|L',Q)$ is an immediate extension. Hence, there are elements $a''_1,\dots,a''_m$ and $b''_1,\dots,b''_n$ in L' whose values or residues coincide with those of $a_1,\dots,a_m$ and $b_1,\dots,b_n$. Now we choose generators $t_1,\dots,t_k,z$ for the function field $F.L'|L'$, like for $F|K$ in Example 17. The elements a_i,b_i are rational functions in these generators. Similarly as in Example 17, we want to pull down these generators to L', preserving the values and residues of a_i,b_i. So the existential $\mathcal{L}_{\mathrm{VF}}(L')$-sentence we employ will now contain also the information that $v_Q a_i = v_Q a''_i$ and $b_iQ = b''_iQ$ (note that a''_i, b''_i are constants from L'). Since $(L',v_Q) \prec_\exists (L,v_Q)$ by Theorem 17.6 (because $v_Q K \prec_\exists v_Q L$ and $Kv_Q \prec_\exists Lv_Q$ trivially hold), we know that the existential sentence also holds in (L',v_Q). This gives us the elements $c_1,\dots,c_k,d \in L'$ and thus also the new elements a'_i,b'_i as rational functions in these new elements, satisfying that $v_Q a'_i = v_Q a''_i = v_Q a_i$ and $b'_iQ = b''_iQ = b_iQ$.

The elements $c_1,\dots,c_k,d$ generate a finite extension (F_1,Q) of (F_0,Q) inside of (L',Q). This extension will be responsible for the extension of the value group and the residue field. But as it lies in L, we can employ Theorem 11.2 to show that $v_Q F_1/v_Q F_0$ is a p-group and $F_1Q|F_0Q$ is purely inseparable, which yields the corresponding assertion in Theorem 20.1.

Adjoining enough transcendental elements and extending Q in a suitable way, we build up a function field (F_2,P) having dimension d and rational rank r and such that P is an Abhyankar place of $F_2|K$. Finally, using the Implicit Function Theorem, we embed F over K in the completion of (F_2,P) and pull back P through this embedding. We construct the embedding in such a way that the image of every a_i and every b_j is very close to a'_i and b'_j, respectively. This implies that $a_iP = a'_iP = a'_iQ = a_iQ$ and $v_P b_j = v_P b'_j = v_Q b'_j = v_Q b_j$ (recall that by construction, P and Q coincide on the field F_1 which contains the elements $a'_1,\dots,a'_m,b'_1,\dots,b'_n$).

In [K3], I prove several modifications of Theorem 20.1, which have various applications. Let me give an example.

Example 28. A modification of Theorem 20.1 (cf. [K3]) shows that one can replace Q by a discrete place P such that FP is a subfield of the perfect hull of FQ (again, one can preserve finitely many residues, but not values anymore). Hence if K is perfect and Q is a rational place, then also P will be rational. As (F, P) is discrete, F embeds over K in $K((t))$. If we assume that K is large, then by Theorem 12.6, $K \prec_\exists K((t))$. Since every existential elementary sentence holding in F will also hold in the bigger field $K((t))$, it follows that $K \prec_\exists F$. We have proved:

Theorem 20.2. *Assume that K is a large field. Assume further that K is perfect and that $F|K$ admits a rational place Q. Then K is existentially closed in F (in the language of fields).*

Reviewing the results on local uniformization that we have stated so far, we see that the best results can be obtained for zero-dimensional discrete or zero-dimensional Abhyankar places (and if K is assumed to be algebraically closed, then "zero-dimensional" is the same as "rational"). But the above theorem only renders places P with $\dim P \geq \dim Q$, so starting from a place which is not zero-dimensional, we will again get a place which is not zero-dimensional. This can be overcome by a modification like the one in the last example. For the formulation of the results we shall use the *Zariski patch topology* for which the basic open sets are the sets of the form

$$\{P \in S(F|K) \mid a_1P \neq 0, \ldots, a_kP \neq 0;\, b_1P = 0, \ldots, b_\ell P = 0\} \tag{64}$$

with $k, \ell \in \mathbb{N} \cup \{0\}$ and $a_1, \ldots, a_k, b_1, \ldots, b_\ell \in F \setminus \{0\}$. It is finer than the Zariski topology. But some proofs showing that the Zariski topology is compact actually show first that the Zariski patch topology is compact (cf. [SP]). Also, the compactness of the Zariski patch topology and the Zariski topology can easily be derived from the Compactness Theorem of model theory (Theorem 13.1); see [K2], [K3].

Theorem 20.3. *The following places lie dense in $S(F|K)$ with respect to the Zariski patch topology:*

a) *the zero-dimensional rank 1 Abhyankar places,*
b) *the zero-dimensional places of maximal rank,*
c) *the zero-dimensional discrete places,*
d) *the prime divisors.*

This can be proved by a combination of Theorem 20.1 and Lemma 12.5 (cf. [K3]). For the proof, one does not need the model theory of tame fields; an application of Theorem 13.6 will suffice. From Theorem 20.3 together with Theorem 17.3 or Theorem 16.1, we obtain:

Corollary 9. *If K is algebraically closed, then the uniformizable places P lie dense in $S(F|K)$ with respect to the Zariski patch topology.*

This result immediately generates the following questions:

Open Problem 12: If we have proved local uniformization for a set of places which lies dense in $S(F|K)$ with respect to the Zariski patch topology, can we patch the local solutions together to obtain the global resolution of singularities? If this doesn't work, how about finer topologies? What other properties of $S(F|K)$ can be deduced from dense subsets?

Certainly, it doesn't follow directly from Corollary 9 that all places in $S(F|K)$ are uniformizable. But it would follow if the next open problem had a positive answer. Observe that local uniformization is an open property, that is, if P is uniformizable, then there is an open neighborhood of P in which every place admits (the same) local uniformization.

Open Problem 13: Can we define something like a "radius" of these "local uniformization neighborhoods" and show that there is a lower bound for this radius?

After all, being a finitely generated field extension, a function field only contains "finite algebraic information". On the other hand, it should be clear from the examples of bad places that there are infinitely many ways of being bad ... So it would be nice if we could forget about bad places. However, the badness expresses itself already on a transcendence basis, and the lower bound for the radius might only depend on the algebraic extension above that transcendence basis.

Now let's have a look at the main open problem of the model theory of valued fields. It has been around since the work of Ax and Kochen, and several excellent model theorists have tried their luck on it, in vain.

21. $\mathbb{F}_p((t))$

May I introduce to you my dearest friend and scariest enemy: $\mathbb{F}_p((t))$. Recall that it appeared on the right-hand side of (45). On the left-hand side, there were the fields $\mathbb{Q}_p$ of p-adic numbers. In a second paper [AK2], Ax and Kochen gave a nice ("recursive") complete axiom system for $\mathbb{Q}_p$ with its p-adic valuation. This generated the problem to give a nice complete axiom system also for $\mathbb{F}_p((t))$ with its t-adic valuation. One can always give an axiom system by writing down *all* sentences which hold in a structure. But "writing down" is very optimistic: there are infinitely many such sentences, and we may not even have a procedure to generate them in some algorithmic way. In contrast to this, a finite axiom system causes no problem. Also schemes like we use for "algebraically closed" or "henselian" aren't problematic since increasingly large finite subsets of them can be produced by an algorithm. This is what "recursive" means. Now if we have a complete recursive axiom system then there is also an algorithm to decide whether a given elementary sentence holds in every model of that axiom system. This is what one means when asking the famous question:

Open Problem 14: Is the elementary theory of $(\mathbb{F}_p((t)), v_t)$ decidable? In other words, does $(\mathbb{F}_p((t)), v_t)$ admit a complete recursive axiomatization?

The complete recursive axiomatization for $(\mathbb{Q}_p, v_p)$ is not hard to state. It is essentially the following:

1) (K, v) is a valued field,
2) (K, v) is henselian,
3) $\operatorname{char} K = 0$,
4) $Kv = \mathbb{F}_p$,
5) vK is an ordered abelian group which is elementarily equivalent to $\mathbb{Z}$.

We can write $Kv = \mathbb{F}_p$ since every field which is elementarily equivalent to $\mathbb{F}_p$ is already equal to $\mathbb{F}_p$ (because $\mathbb{F}_p$ has finitely many elements, and their number can thus be expressed by an elementary sentence). In contrast to this, we cannot write $vK = \mathbb{Z}$; since $\mathbb{Z}$ has infinitely many elements, Theorem 13.2 implies that there are many other ordered abelian groups which are elementarily equivalent to $\mathbb{Z}$. These are called *$\mathbb{Z}$-groups.* An ordered abelian group is a $\mathbb{Z}$-group if and only if $\mathbb{Z}$ is a convex subgroup of G and $G/\mathbb{Z}$ is divisible.

Observe that $(\mathbb{F}_p((t), v_t)$ looks very much like $(\mathbb{Q}_p, v_p)$. Indeed, the only axiom that does not hold is axiom 3). So if we replace it by 3′): "$\operatorname{char} K = p$", will we get a complete axiom system (which by our above remarks would be recursive)? We have stated in Section 8 that every henselian discretely valued field of characteristic 0 is a defectless field. From this it follows that in the presence of the other axioms (including 3)!), axiom 2) implies that (K, v) is defectless. If we change 3) to "$\operatorname{char} K = p$", this is not any longer true, and we have to replace axiom 2) by 2′): "(K, v) is henselian and defectless".

For a long time, many model theorists believed that the axiom system 1), 2′), 3′), 4), 5) could be complete. But based on an observation by Lou van den Dries, I was able to show in [K1] that this is not the case (cf. [K2]). It is precisely Example 26 which proves the incompleteness. The point is that if K has positive characteristic, then we have non-linear additive polynomials which we can use to express additional elementary properties. In characteristic 0, the only additive polynomials are of the form cx, so there is nothing interesting about them. But in positive characteristic, the image $f(K)$ of an additive polynomial is a subgroup of the additive group of K. If $f_1, \cdots, f_n$ are additive polynomials, then one can consider the subgroup $f_1(K) + \ldots + f_n(K)$. For certain choices of the f's, these subgroups have nice elementary properties if K is elementarily equivalent to $\mathbb{F}_p((t))$. This implies that for certain choices, one can even show that $K = f_1(K) + \cdots + f_n(K)$. To some extent, this has the same flavour as Hensel's Lemma, but the incompleteness result shows that all this doesn't follow from Hensel's Lemma, or to be more precise, doesn't even follow from the axioms 1), 2′), 3′), 4), 5). See [K4] for details.

The subgroups of the form $f_1(K) + \cdots + f_n(K)$ are definable by an elementary sentence using constants from K (as coefficients of the polynomials f_i). This fact leads to the following questions:

Open Problem 15: Does the axiom system 1), 2′), 3′), 4), 5) become complete if we add the elementary properties of the subgroups of the form $f_1(K) + \cdots + f_n(K)$? What other subgroups of $\mathbb{F}_p((t))$ are elementarily definable, and what are their elementary properties?

As we have seen already that additive polynomials play a crucial role for local uniformization in positive characteristic, it makes sense to ask:

Open Problem 16: What is the relation between the elementary properties of $\mathbb{F}_p((t))$ expressible by use of additive polynomials and algebraic geometry in positive characteristic?

On the valuation theoretical side, I can say that work in progress indicates that these elementary properties have a crucial meaning for the structure theory of valued function fields. For example, it seems that the Henselian Rationality of Immediate Function Fields can be generalized to the case of base fields (K, v) which are not tame but have these properties. By the way, these properties don't play a role in the model theory of tame fields because all tame fields are perfect (like all other fields of positive characteristic for which we know that Ax-Kochen-Ershov principles hold). In contrast to this, $\mathbb{F}_p((t))$ is not perfect since t has no p-th root in $\mathbb{F}_p((t))$.

In comparison to the model theory of tame fields, that of $\mathbb{F}_p((t))$ is much more complex since some tools available for tame fields will not work anymore. As an example, we do not have an analogue of the crucial Lemma 11.9 which we used to separate extensions of tame fields into extensions without transcendence defect and immediate extensions. Thus, we cannot do this (in general) in the case of fields which are elementarily equivalent to $\mathbb{F}_p((t))$. We are also not able to "slice" immediate extension into extensions of transcendence degree 1. But then we would have to develop an analogue of Kaplansky's theory of immediate extensions for the case of higher transcendence degree, or even worse, simultaneously for all mixed extensions without a possibility of separation. However, this is more or less the generalization of the theory of approximate roots that is recently discussed.

22. Local uniformization vs. Ax-Kochen-Ershov

My first encounter with Zariski's Local Uniformization Theorem was when I studied the model theoretic proof of the p-adic Nullstellensatz by Moshe Jarden and Peter Roquette [JR]. At one point, they consider the following situation. They have an extension $L|K$ of p-adically closed fields, and a function field $F|K$ inside of $L|K$. Inside of F, they have a certain subring B containing K and an element $g \in B$ which is not a unit in B. Now they wish to show that there is a rational place P of $F|K$ such that $gP = 0$. They take a maximal ideal $\mathcal{M}$ in B such that $g \in \mathcal{M}$. By the existence theorem for places (see [V], Proposition 1.2), there is a place Q of $F|K$ such that $B \subseteq \mathcal{O}_Q$ and $\mathcal{M}_Q \cap B = \mathcal{M}$. It follows that $gQ = 0$. Since $K \subset B$ we know that $K \subseteq FQ$. But Q may not be the required place since it may not be rational. If $FQ|K$ is a function field, then one can proceed as follows. By the special choice of the ring B one knows that FQ is contained in some p-adically closed extension field L'. By the work of Ax-Kochen and Ershov, one knows that $K \prec L'$. It follows that $K \prec_{\exists} FQ$, hence by Lemma 12.5 there is a rational place $\overline{Q}$ of $FQ|K$. So the place $Q\overline{Q}$ is a rational place of $F|K$ which satisfies $gQ\overline{Q} = 0\overline{Q} = 0$, as desired.

But it may well happen that $FQ|K$ is not finitely generated. Then we can't apply Lemma 12.5. In this situation, Jarden and Roquette use Zariski's Local Uniformization (Theorem 15.1) to show that there exists a place P of $F|K$ such that $gP = 0$ and that $FP|K$ is finitely generated. More generally, they show:

Lemma 22.1. *Take a function field $F|K$ of characteristic 0, a place Q of $F|K$ and elements $y_1, \dots, y_n \in \mathcal{O}_Q$. Then there exists a place P of $F|K$ such that $y_iP = y_iQ$, $1 \le i \le n$, $FP \subseteq FQ$ and $FP|K$ is finitely generated.*

The proof works as follows. After adding elements if necessary, we can assume that $F = K(y_1, \dots, y_n)$. By Zariski's Local Uniformization Theorem, after adding further elements we can assume that $a = (y_1Q, \dots, y_nQ)$ is a simple point of the K-variety whose generic point is $(y_1, \dots, y_n)$. Hence, the local ring $\mathcal{O}_a$ is regular. Now Jarden and Roquette employ the following lemma from [A2]:

Lemma 22.2. *Suppose that R is a regular local ring with maximal ideal M and quotient field F. Then there exists a place P dominating R such that $FP = R/M$.*

(In fact, P is the place associated with the order valuation deduced from (R, M).)

Corollary 10. *Suppose that a is a simple point of V. Then there exists a place P of $F|K$ such that $a = (y_1P, \dots, y_nP)$ and $FP = K(a)$.*

Proof. The residue field $\mathcal{O}_a/\mathcal{M}_a$ of $\mathcal{O}_a$ is isomorphic to $K(a)$. We identify both fields, so that the residue map $\mathcal{O}_a \to \mathcal{O}_a/\mathcal{M}_a = K(a)$ maps every y_i to a_i. By applying the foregoing lemma to $R = \mathcal{O}_a$, we obtain a place P of $F|K$ dominating $\mathcal{O}_a$ and such that $FP = \mathcal{O}_a/\mathcal{M}_a = K(a)$. Since P dominates $\mathcal{O}_a$, it extends the residue map, whence $a = (y_1P, \dots, y_nP)$. □

This proves Lemma 22.1: by the definition of a we get that $(y_1P, \dots, y_nP) = (y_1Q, \dots, y_nQ)$ and $FP = K(y_1P, \dots, y_nP) = K(y_1Q, \dots, y_nQ) \subset FQ$, showing also that FP is a finitely generated extension of K.

Note that if $a \in K^n$ then the place P obtained from the foregoing corollary is rational. Hence we have (you may compare this with Lemma 12.5):

Corollary 11. *Suppose that a is a simple K-rational point of V. Then there exists a rational place P of $F|K$ such that $a = (y_1P, \dots, y_nP)$.*

In my Masters Thesis, I showed how to avoid the use of Zariski's Local Uniformization Theorem by constructing a place of maximal rank (which by Corollary 5 always has a finitely generated residue field). This trick was then used by Alexander Prestel and Peter Roquette in their book [PR] for the proof of the p-adic Nullstellensatz. It also provided the first idea for the paper [PK] in which we proved a version of Theorem 20.1 for function fields of characteristic 0 by using the Ax-Kochen-Ershov Theorem. This version has interesting applications to real algebra and real algebraic geometry ([PK], [P]). Surprisingly, a paper by Ludwig Bröcker and Heinz-Werner Schülting [BS] derives about the same results and applications, using resolution of singularities in characteristic 0 (Hironaka) in the place of the Ax-Kochen-Ershov Theorem. See also the survey paper [SCH]. The first question I have in this connection is:

Open Problem 17: To which extent is it (easily) possible to replace the use of resolution of singularities by local uniformization in real algebraic geometry?

Seeing that the Ax-Kochen-Ershov Theorem and Zariski's Local Uniformization Theorem can be used to deduce the same results, Roquette asked:

Open Problem 18: What is the relation between Ax-Kochen-Ershov Theorem and Zariski's Local Uniformization Theorem? Can one prove one from the other?

If that were true, then there would be some hope that a progress in positive characteristic made on one side could be transferred to the other side. For instance, as already mentioned, local uniformization or resolution of singularities in positive characteristic could possibly help to solve problems in the model theory of valued fields of positive characteristic.

In view of the details I have told you about, my own preliminary answer is that the relation between local uniformization and the model theory of valued fields lies in the facts and theorems from the structure theory of valued function fields which play a crucial role in both problems. Therefore, new insights in this structure theory will also be of importance for both problems.

On the other hand, there are ingredients on either side which do not appear on the other. For example, Lemma 15.4 does not seem to play any role on the model theoretic side. Further, the henselization causes a lot of serious problems for the local uniformization of places of rank > 1, whereas it is the best friend of model theorists. The need to optimize the choice of the transcendence basis (cf. Section 19) to avoid these problems and to avoid the defect has (so far) no analogue on the model theoretic side; however, this may change with a deeper insight in the theory of $\mathbb{F}_p((t))$. Conversely, model theory is forced to deal with extensions $(L|K, v)$ of valued fields where v is non-trivial on K. To some extent, we did the same when we considered relative uniformization. But in contrast to that situation, the valuations on K in the model theoretic case may be of arbitrary rank and arbitrarily nasty.

Our discussion would not be complete if we would not mention the following nice result, due to Jan Denef [DEN]. It says that "the existential theory of $\mathbb{F}_p((t))$ is decidable":

Theorem 22.3. *If resolution of singularities holds in positive characteristic, then there is an algorithm to decide whether a given existential elementary sentence in the language of valued fields holds in* $(\mathbb{F}_p((t)), v_t)$.

23. Back to local uniformization in positive characteristic

In the last section, we have seen that for certain applications it matters to know what the residue fields of our places are. In particular, when changing a bad place to a good place, we might wish to keep the residue field within a certain class of fields. For example, this could be the class of all fields in which the base field K is existentially closed. (Note that if K is perfect and existentially closed in L, then it will also be existentially closed in any purely inseparable algebraic extension of L.) If we take into the bargain an extension of the function field in order to obtain local uniformization, but require that this extension should be normal or even Galois, then we may not be able anymore to control the corresponding extension of the residue field. So it makes sense to ask for the minimal possible change of the residue field.

The key to this question is the fact that there are minimal algebraic extensions of (F, P) which are tame (or separably tame) fields; we just have to pass to the henselization of (F, P) and then choose a field W according to Theorem 11.2, cf. Lemma 11.4. (For the case of "tame", see also Corollary 7.) Working inside of such an extension, we can nicely apply the theory of tame and separably tame fields. In particular, we can use Lemma 11.9 and the transitivity of relative uniformization to reduce to extensions of the types discussed in **I)–IV)** in Section 15.

On the other hand, if (L, P) is such an extension of (F, P), then by Theorem 11.2, $v_P L/v_P F$ will be a p-torsion group if $\operatorname{char} K = p > 0$, and $LP|FP$ will be purely inseparable. If $\operatorname{char} K = 0$, then (L, P) is just the henselization of (F, P). As our extension $\mathcal{F}$ remains inside of L, we can show (cf. [K7]):

Theorem 23.1. *In addition to the assertion of Theorem 17.7, the finite extension $\mathcal{F}|F$ can be chosen to be separable and to satisfy:*

a) *if* $\operatorname{char} K = p > 0$, *then the finite group $v_P\mathcal{F}/v_P F$ is a p-torsion group, and the finite extension $\mathcal{F}P|FP$ is purely inseparable,*

b) *if* $\operatorname{char} K = 0$, *then $\mathcal{F}$ can be chosen to lie in the henselization of F.*

(Clearly, in the case of $\operatorname{char} K = 0$, our result is weaker than Zariski's Local Uniformization Theorem. However, it provides some more information since we can uniformize a finite extension of F within its henselization while keeping fixed the once chosen transcendence basis of the subfunction field $F_0|K$ on which P is an Abhyankar place.)

The following corollary illustrates the advantage of controlling the residue field extension. In fact, Theorem 20.2 can also be proved by use of this corollary.

Corollary 12. *Assume that K is perfect and that P is a rational place of $F|K$. Take $\zeta_1, \dots, \zeta_m \in \mathcal{O}_F$. Then there is a finite extension $\mathcal{F}|F$ and an extension of P to $\mathcal{F}$ such that $(\mathcal{F}|K, P)$ is uniformizable with respect to $\zeta_1, \dots, \zeta_m$ and P is still a rational place of $\mathcal{F}|K$.*

References

[A1] Abhyankar, S.: *Local uniformization on algebraic surfaces over ground fields of characteristic $p \neq 0$*, Annals of Math. **63** (1956), 491–526. Corrections: Annals of Math. **78** (1963), 202–203

[A2] Abhyankar, S.: *On the valuations centered in a local domain*, Amer. J. Math. **78** (1956), 321–348

[A3] Abhyankar, S.: *Simultaneous resolution for algebraic surfaces*, Amer. J. Math. **78** (1956), 761–790

[A4] Abhyankar, S.: *On the field of definition of a nonsingular birational transform of an algebraic surface*, Annals of Math. **65** (1957), 268–281

[A5] Abhyankar, S.: *Ramification theoretic methods in algebraic geometry*, Princeton University Press (1959)

[A6] Abhyankar, S.: *Uniformization in p-cyclic extensions of algebraic surfaces over ground fields of characteristic p*, Math. Annalen **153** (1964), 81–96

[A7] Abhyankar, S.: *Reduction to multiplicity less than p in a p-cyclic extension of a two dimensional regular local ring (p = characteristic of the residue field)*, Math. Annalen **154** (1964), 28–55

[A8] Abhyankar, S.: *Uniformization of Jungian local domains*, Math. Annalen **159** (1965), 1–43. Correction: Math. Annalen **160** (1965), 319–320

[A9] Abhyankar, S.: *Uniformization in p-cyclic extensions of a two dimensional regular local domain of residue field characteristic p*, Festschrift zur Gedächtnisfeier für Karl Weierstraß 1815–1965, Wissenschaftliche Abhandlungen des Landes Nordrhein-Westfalen **33** (1966), 243–317 (Westdeutscher Verlag, Köln und Opladen)

[A10] Abhyankar, S.: *Resolution of singularities of embedded algebraic surfaces*, Academic Press, New York (1966); 2nd enlarged edition: Springer, New York (1998)

[A11] Abhyankar, S.: *Nonsplitting of valuations in extensions of two dimensional regular local domains*, Math. Annalen **153** (1967), 87–144

[AO] Abramovich, D., Oort, F.: *Alterations and resolution of singularities*, this volume

[AK1] Ax, J., Kochen, S.: *Diophantine problems over local fields I*, Amer. Journ. Math. **87** (1965), 605–630.

[AK2] Ax, J., Kochen, S.: *Diophantine problems over local fields II*, Amer. Journ. Math. **87** (1965), 631–648.

[AX] Ax, J.: *A metamathematical approach to some problems in number theory*, Proc. Symp. Pure Math. **20**, Amer. Math. Soc. (1971), 161–190

[B] Bourbaki, N.: *Commutative algebra*, Paris (1972)

[BGR] Bosch, S., Güntzer, U., Remmert, R.: *Non-Archimedean Analysis*, Berlin (1984)

[BR] Brown, S. S.: *Bounds on transfer principles for algebraically closed and complete valued fields*, Memoirs Amer. Math. Soc. **15** No. 204 (1978)

[BS] Bröcker, L., Schülting, H. W.: *Valuations of function fields from the geometrical point of view*, J. reine angew. Math. **365** (1986), 12–32

[CH1] Cherlin, G.: *Model theoretic algebra*, Lecture Notes Math. **521** (1976)

[CH2] Cherlin, G.: *Model theoretic algebra*, J. Symb. Logic **41** (1976), 537–545

[CK] Chang, C. C., Keisler, H. J.: *Model Theory*, Amsterdam, London (1973)

[dJ] de Jong, A. J.: *Smoothness, semi-stability and alterations*, Publications Mathematiques I.H.E.S. **83** (1996), 51–93

[DEL] Delon, F.: *Quelques propriétés des corps valués en théories des modèles*, Thèse Paris VII (1981)

[DEN] Denef, J.: personal communication, Oberwolfach, October 1998

[DEU] Deuring, M.: *Reduktion algebraischer Funktionenkörper nach Primdivisoren des Konstantenkörpers*, Math. Z. **47** (1942), 643–654

[EK] Eklof, P.: *Lefschetz's Principle and local functors*, Proc. Amer. Math. Soc. **37** (1973), 333–339

[EL] Elliott, G. A.: *On totally ordered groups, and* K_0, in: Ring Theory Waterloo 1978, eds. D. Handelman and J. Lawrence, Lecture Notes Math. **734**, 1–49

[EN] Endler, O.: *Valuation theory*, Berlin (1972)

[EPP] Epp, H.: *Eliminating wild ramification*, Inventiones Math. **19** (1973), 235–249

[ER1] Ershov, Yu. L.: *On the elementary theory of maximal normed fields*, Dokl. Akad. Nauk SSSR **165** (1965), 21–23 English translation in: Sov. Math. Dokl. **6** (1965), 1390–1393.

[ER2] Ershov, Yu. L.: *On elementary theories of local fields*, Algebra i Logika **4**:2 (1965), 5–30

[ER3] Ershov, Yu. L.: *On the elementary theory of maximal valued fields I* (in Russian), Algebra i Logika **4**:3 (1965), 31–70

[ER4] Ershov, Yu. L.: *On the elementary theory of maximal valued fields II* (in Russian), Algebra i Logika **5**:1 (1966), 5–40

[ER5] Ershov, Yu. L.: *On the elementary theory of maximal valued fields III* (in Russian), Algebra i Logika **6**:3 (1967), 31–38

[GR] Grauert, H., Remmert, R.: *Über die Methode der diskret bewerteten Ringe in der nicht archimedischen Analysis*, Inventiones Math. **2** (1966), 87–133

[GRA] Gravett, K. A. H.: *Note on a result of Krull*, Cambridge Philos. Soc. Proc. **52** (1956), 379

[GRE1] Green, B.: *Recent results in the theory of constant reductions*, Séminaire de Théorie des Nombres, Bordeaux **3** (1991), 275–310

[GRE2] Green, B.: *The Relative Genus Inequality for Curves over Valuation Rings*, J. Alg. **181** (1996), 836–856

[GMP1] Green, B., Matignon, M., Pop, F.: *On valued function fields I*, manuscripta math. **65** (1989), 357–376

[GMP2] Green, B., Matignon, M., Pop, F.: *On valued function fields II*, J. reine angew. Math. **412** (1990), 128–149

[GMP3] Green, B., Matignon, M., Pop, F.: *On valued function fields III*, J. reine angew. Math. **432** (1992), 117–133

[GMP4] Green, B., Matignon, M., Pop, F.: *On the Local Skolem Property*, J. reine angew. Math. **458** (1995), 183–199

[GRB] Greenberg, M. J.: *Rational points in henselian discrete valuation rings*, Publ. Math. I.H.E.S. **31** (1967), 59–64

[GRL] Greenleaf, M.: *Irreducible subvarieties and rational points*, Amer. J. Math. **87** (1965), 25–31

[GRU2] Gruson, L.: *Fibrés vectoriels sur un polydisque ultramétrique*, Ann. Sci. Ec. Super., IV. Ser., **177** (1968), 45–89

[HO] Hodges, W.: *Model Theory*, Encyclopedia of mathematics and its applications **42**, Cambridge University Press (1993)

[JA] Jacobson, N.: *Lectures in abstract algebra, III. Theory of fields and Galois theory*, Springer Graduate Texts in Math., New York (1964)

[JR] Jarden, M., Roquette, P.: *The Nullstellensatz over $\wp$–adically closed fields*, J. Math. Soc. Japan **32** (1980), 425–460

[KA1] Kaplansky, I.: *Maximal fields with valuations I*, Duke Math. Journ. **9** (1942), 303–321

[KA2] Kaplansky, I.: *Maximal fields with valuations II*, Duke Math. Journ. **12** (1945), 243–248

[K1] Kuhlmann, F.-V.: *Henselian function fields and tame fields*, extended version of Ph.D. thesis, Heidelberg (1990)

[K2] Kuhlmann, F.-V.: *Valuation theory of fields, abelian groups and modules*, preprint, to appear in the "Algebra, Logic and Applications" series (Gordon and Breach), eds. A. Macintyre and R. Göbel

[K3] Kuhlmann, F.–V.: *Places of algebraic function fields in arbitrary characteristic*, submitted

[K4] Kuhlmann, F.–V.: *Elementary properties of power series fields over finite fields*, submitted; prepublication in: Structures Algébriques Ordonnées, Séminaire Paris VII (1997), and in: The Fields Institute Preprint Series, Toronto (1998)

[K5] Kuhlmann, F.–V.: *On local uniformization in arbitrary characteristic*, The Fields Institute Preprint Series, Toronto (1997)

[K6] Kuhlmann, F.–V.: *On local uniformization in arbitrary characteristic I*, submitted

[K7] Kuhlmann, F.–V.: *On local uniformization in arbitrary characteristic II*, submitted

[KK] Kuhlmann, F.-V., Kuhlmann, S.: *Valuation theory of exponential Hardy fields*, submitted

[KP] Kuhlmann, F.–V., Prestel, A.: *On places of algebraic function fields*, J. reine angew. Math. **353** (1984), 181–195

[KPR] Kuhlmann, F.-V., Pank, M., Roquette, P.: *Immediate and purely wild extensions of valued fields*, manuscripta math. **55** (1986), 39–67

[KR] Krull, W.: *Allgemeine Bewertungstheorie*, J. reine angew. Math. **167** (1931), 160–196

[L1] Lang, S.: *On quasi algebraic closure*, Ann. of Math. **55** (1952), 373–390

[L2] Lang, S.: *Algebra*, New York (1965)

[OH] Ohm, J.: *The henselian defect for valued function fields*, Proc. Amer. Math. Soc. **107** (1989)

[OR] Ore, O.: *On a special class of polynomials*, Trans. Amer. Math. Soc. **35** (1933), 559–584

[OS] Ostrowski, A.: *Untersuchungen zur arithmetischen Theorie der Körper*, Math. Z. **39** (1935), 269–404

[POP] Pop, F.: *Embedding problems over large fields*, Ann. of Math. **144** (1996), 1–34

[P] Prestel, A.: *Model theory of fields: an application to positive semidefinite polynomials*, Soc. Math. de France, mémoire no. **16** (1984), 53–64

[PR] Prestel, A., Roquette, P.: *Formally p-adic fields*, Lecture Notes Math. **1050**, Berlin–Heidelberg–New York–Tokyo (1984)

[PZ] Prestel, A., Ziegler, M.: *Model theoretic methods in the theory of topological fields*, J. reine angew. Math. **299/300** (1978), 318–341

[R1] Ribenboim, P.: *Théorie des valuations*, Les Presses de l'Université de Montréal (1964)

[R2] Ribenboim, P.: *Equivalent forms of Hensel's lemma*, Expo. Math. **3** (1985), 3–24

[RO] Robinson, A.: *Complete Theories*, Amsterdam (1956)

[RQ1] Roquette, P.: *Zur Theorie der Konstantenreduktion algebraischer Mannigfaltigkeiten*, J. reine angew. Math. **200** (1958), 1–44

[RQ2] Roquette, P.: *A criterion for rational places over local fields*, J. reine angew. Math. **292** (1977), 90–108

[RQ3] Roquette, P.: *On the Riemann p-space of a field: The p-adic analogue of Weierstrass' approximation theorem and related problems*, Abh. Math. Sem. Univ. Hamburg **47** (1978), 236–259

[RQ4] Roquette, P.: *p–adische und saturierte Körper. Neue Variationen zu einem alten Thema von Hasse*, Mitteilungen Math. Gesellschaft Hamburg **11** Heft 1 (1982), 25–45

[RQ5] Roquette, P.: *Some tendencies in contemporary algebra*, in: Perspectives in Mathematics, Anniversary of Oberwolfach 1984, Basel (1984), 393–422

[SA] Sacks, G. E.: *Saturated Model Theory*, Reading, Massachusetts (1972)

[SCH] Scheiderer, C.: *Real algebra and its applications to geometry in the last 10 years: some major developments and results*, in: Real algebraic geometry, Proc. Conf. Rennes 1991, eds. M. Coste and M. F. Roy, Lecture Notes in Math. **1524** (1992), 1–36 and 75–96

[SP] Spivakovsky, M.: *Resolution of singularities I: local uniformization*, manuscript (1996)

[U] Urabe, T.: *Resolution of singularities of germs in characteristic positive associated with valuation rings of iterated divisor type*, preprint 1999, 75 pages

[V] Vaquie, M.: *Valuations*, this volume

[WA1] Warner, S.: *Nonuniqueness of immediate maximal extensions of a valuation*, Math. Scand. **56** (1985), 191–202

[WA2] Warner, S.: *Topological fields*, Mathematics studies **157**, North Holland, Amsterdam (1989)

[WH1] Whaples, G.: *Additive polynomials*, Duke Math. Journ. **21** (1954), 55–65

[WH2] Whaples, G.: *Galois cohomology of additive polynomials and n-th power mappings of fields*, Duke Math. Journ. **24** (1957), 143–150

[Z] Zariski, O.: *Local uniformization on algebraic varieties*, Ann. Math. **41** (1940), 852–896

[ZS] Zariski, O., Samuel, P.: *Commutative Algebra*, Vol. II, New York–Heidelberg–Berlin (1960)

Department of Mathematics and Statistics
University of Saskatchewan
Saskatoon, Saskatchewan, Canada S7N 5E6
fvk@math.usask.ca
`http://math.usask.ca/~fvk/index.html`

Progress in Mathematics, Vol. 181,

Les Singularités Sandwich

Lê Dũng Tráng

Dédié à Pierre Samuel
pour son 77-ème anniversaire

1. Introduction

Dans cet article nous exposons et interprétons certains des résultats et concepts de M. Spivakovsky dans son article sur les singularités Sandwich [S], puis nous introduisons la notion de singularité Sandwich relative à une singularité rationnelle de surface. On ne considèrera que des espaces complexes, mais, comme il est signalé dans [S], on peut se placer sur un corps de caractéristique 0 algébriquement clos et, pour beaucoup de résultats, sur un corps de caractéristique quelconque.

Soit $X \subset U \subset \mathbf{C}^N$ une surface analytique complexe plongée dans un domaine ouvert U de $\mathbf{C}^N$. L'application de Gauss de X fait correspondre à un point non-singulier y de X le plan tangent à X au point y dans la Grassmannienne $G(2,N)$ des 2-plans dans $\mathbf{C}^N$. La fermeture $\tilde{X}$ du graphe de l'application de Gauss dans $U \times G(2,N)$ est un espace analytique. Le morphisme analytique complexe propre $\nu\colon \tilde{X} \to X$, induit par la projection de $U \times G(2,N)$ sur le premier facteur U, est *la modification de Nash* de X. La projection de $U \times G(2,N)$ sur le second facteur $G(2,N)$ induit une application analytique $\gamma\colon \tilde{X} \to G(2,N)$, qui est une extension de l'application de Gauss.

Nous étudions des surfaces normales X, c'est à dire des surfaces analytiques complexes dont tous les anneaux locaux $\mathcal{O}_{X,x}$ sont normaux, i.e. intègres et intégralement clos dans leur corps des fractions. On sait que X est une surface normale si et seulement si les singularités de X sont isolées et de Cohen-Macaulay. On supposera que X n'a qu'une seule singularité $\{x\}$. Dans ce cas une résolution de X est un morphisme analytique $\pi\colon Z \to X$ tel que: i) l'espace Z est non singulier; ii) le morphisme π est propre; iii) le morphisme π induit un isomorphisme analytique de $Z - \pi^{-1}(x)$ sur $X - \{x\}$; iv) l'espace $Z - \pi^{-1}(x)$ est dense dans Z. Un théorème de R. Walker et O. Zariski ([Z] §21) montre qu'on peut obtenir une résolution de X par la composition d'une suite finie d'éclatements de points suivis de normalisation.

D'après ([S] p. 412), au début des années 60, J. Nash aurait demandé à H. Hironaka s'il était possible de résoudre les singularités de surfaces en composant une suite finie de modifications de Nash. En 1975, A. Nobile a prouvé que, si la modification de Nash est un isomorphisme, alors X est non-singulier (cf. [No], voir aussi [Te] §2 Proposition p. 585). En particulier, ceci entraine qu'une courbe est résolue par une suite finie de modifications de Nash. Dans sa thèse doctorale [R], V. Rebassoo a montré que certaines hypersurfaces quasi-homogènes de $\mathbf{C}^3$ sont résolues par une suite finie de modifications de Nash. Dans [GS1] G. Gonzalez-Sprinberg a montré que, par une suite finie de modifications de Nash suivies de normalisation, on peut résoudre les singularités quotients cycliques, puis, dans [GS2], les points doubles rationnels de surfaces.

Il apparait alors que la modification de Nash suivie de la normalisation est une transformation de base qui permet de résoudre les singularités des surfaces complexes. Nous appellerons *modification de Nash normalisée* la composée de la modification de Nash et de la normalisation. Dans [H] H. Hironaka montre que, en composant une suite finie de modifications de Nash normalisées, on obtient une surface dont les singularités sont rationnelles au sens de M. Artin (cf. [A]) et qui domine birationnellement une surface non-singulière. Dans [S] M. Spivakovsky étudie les singularités qui proviennent de l'éclatement d'un idéal complet dans le plan. Il appelle *singularités Sandwich* ces singularités (voir [S] definition 1.1, Chap. II). M. Spivakovsky montre que ces singularités sont normales et que leurs anneaux locaux sont les anneaux locaux normaux qui dominent birationnellement les anneaux réguliers de dimension 2. Il établit qu'en composant une suite finie de modifications de Nash normalisées, on obtient des singularités Sandwich particulières appelées *singularités minimales.* Ces singularités sont caractérisées par le fait que le cône tangent est réduit. Enfin, il montre que ces singularités sont résolues après une suite finie de modifications de Nash normalisées.

Le lecteur pourra aussi lire un compte-rendu de la méthode de M. Spivakovsky dans le Séminaire Bourbaki de G. Gonzalez Sprinberg [GS3].

2. Normalisation et éclatements

Afin de rendre la lecture de cet article plus facile, nous rappelons les définitions des morphismes de normalisation et d'éclatements.

Soit $(X, \mathcal{O}_X)$ un espace analytique complexe réduit. Il existe sur X un faisceau cohérent $\mathcal{F}$ de $\mathcal{O}_X$-modules dont les fibres $\mathcal{F}_x$ en les points x de X sont les normalisations des anneaux locaux $\mathcal{O}_{X,x}$. On démontre (voir [Ho] Exp. 21 §B 4) que $\mathcal{F}$ est une $\mathcal{O}_X$-algèbre de présentation finie (qui est aussi un $\mathcal{O}_X$-module de type fini), ce qui fait qu'on lui associe un morphisme analytique

$$n \colon \overline{X} \to X$$

où $\overline{X}$ est l'espace analytique $Specan(\mathcal{F})$ ([Ho] Exp. 19 §1). Le morphisme analytique n est la normalisation de l'espace analytique X. C'est un morphisme fini (i.e. propre et à fibres finies) qui est un isomorphisme analytique au-dessus de la partie non-singulière et vérifie la propriété suivante (voir [Ho] Exp. 21 §B 4 Cor. 3):

Théorème 2.1. *Soit n la normalisation d'un espace analytique complexe réduit X. Soit $\varphi\colon Y \to X$ un morphisme analytique d'un espace normal dans X qui soit propre et induise un isomorphisme analytique au-dessus de la partie non-singulière de X, alors il existe un morphisme analytique $\tilde{\varphi}\colon Y \to \overline{X}$ et un seul tel que $\varphi = n \circ \tilde{\varphi}$.*

Soit $(X, \mathcal{O}_X)$ un espace analytique complexe réduit et $\mathcal{I}$ un faisceau cohérent d'idéaux de $\mathcal{O}_X$. On définit l'algèbre graduée

$$\mathcal{A} := \mathcal{O}_X \oplus \mathcal{I} \oplus \mathcal{I}^2 \oplus \cdots \oplus \mathcal{I}^n \oplus \cdots$$

Cette $\mathcal{O}_X$-algèbre est de présentation finie. Elle définit donc un espace analytique $Specan\mathcal{A}$ et un morphisme analytique de $Specan\mathcal{A}$ sur X dont les fibres sont des cônes. Le morphisme projectivisé associé

$$e_{\mathcal{I}}\colon Projan\mathcal{A} \to X$$

est propre et est un isomorphisme analytique au-dessus de l'ouvert $X - V(\mathcal{I})$ complément dans X du sous-espace analytique défini par $\mathcal{I}$ (voir pour le cas algébrique [Ha] Chap. II, §7 p. 160–164). On vérifie que l'image inverse $e_{\mathcal{I}}{}^*\mathcal{I}$ est un faisceau inversible d'idéaux de $Projan\mathcal{A}$. Le morphisme $e_{\mathcal{I}}$ est l'éclatement de X de centre $\mathcal{I}$. On notera souvent $X_{\mathcal{I}}$ l'espace $Projan\mathcal{A}$. L'espace $e_{\mathcal{I}}^{-1}(V(\mathcal{I}))$ est un diviseur de $X_{\mathcal{I}}$ appelé *diviseur exceptionnel* de l'éclatement $e_{\mathcal{I}}$. Le morphisme $e_{\mathcal{I}}$ vérifie la propriété universelle suivante:

Théorème 2.2. *Soit $(X, \mathcal{O}_X)$ un espace analytique complexe réduit et $\mathcal{I}$ un faisceau cohérent d'idéaux de $\mathcal{O}_X$. Pour tout morphisme analytique complexe propre $\psi\colon Y \to X$ tel que $\psi^*\mathcal{I}$ soit un faisceau inversible d'idéaux de Y, il existe un morphisme analytique $\tilde{\psi}\colon Y \to X_{\mathcal{I}}$ et un seul tel que $\psi = e_{\mathcal{I}} \circ \tilde{\psi}$.*

Dans le cas spécial où le faisceau d'idéaux $\mathcal{I}$ est engendré par des sections globales $g_1, \ldots, g_r$, on a la proposition suivante qui donne un moyen pratique de construire l'éclatement de centre $\mathcal{I}$.

Proposition 2.3. *Soit $(X, \mathcal{O}_X)$ un espace analytique complexe réduit et $\mathcal{I}$ un faisceau cohérent d'idéaux de $\mathcal{O}_X$ engendré par des sections globales $g_1, \ldots, g_r$. L'espace $X_{\mathcal{I}}$ est isomorphe à l'espace analytique complexe fermeture dans $X \times \mathbf{P}^{r-1}$ du graphe de l'application de $X - V(\mathcal{I})$ dans $\mathbf{P}^{r-1}$ qui à $z \in X - V(\mathcal{I})$ fait correspondre $(g_1(z) : \ldots : g_r(z)) \in \mathbf{P}^{r-1}$ et l'éclatement $e_{\mathcal{I}}$ est isomorphe à la projection de la fermeture de ce graphe sur X.*

Remarque 2.4. Dans cet article nous aurons à considérer l'éclatement de centre l'idéal maximal défini par un point sur un espace analytique. Cet éclatement est aussi appelé l'éclatement du point. En général l'éclatement d'un point est différent de l'éclatement de centre un idéal dont le support est le point.

Par exemple, l'éclatement de centre l'idéal engendré par les monômes x^2, y^3 dans le plan complexe $\mathbf{C}^2$ est singulier et correspond à la surface de $\mathbf{C}^2 \times \mathbf{P}^1$ d'équation $\lambda x^2 + \mu y^3 = 0$, avec $(x, y) \in \mathbf{C}^2$ et $(\mu : -\lambda) \in \mathbf{P}^1$, alors que l'éclatement du point dans $\mathbf{C}^2$ est non-singulier et correspond à la surface de $\mathbf{C}^2 \times \mathbf{P}^1$ d'équation

$\lambda x + \mu y = 0$. Dans ces deux cas, on remarque que la restriction aux surfaces de la projection sur $\mathbf{P}^1$ donne des familles de courbes qui sont des systèmes linéaires de courbes planes.

Nous aurons à considérer des éclatements suivis de normalisations. La raison en est la suivante. Soit I un idéal de l'anneau analytique local $\mathcal{O}$. On rappelle qu'un élément α de $\mathcal{O}$ est *entier sur l'idéal* I (cf. [SZ] Appendix 4, Definition 2 ou [JLT] §1, Définition 1.1) s'il existe une relation

$$\alpha^k + \sum_1^k u_j \alpha^{k-j} = 0$$

dans laquelle on a $u_j \in I^j$. L'ensemble des éléments de $\mathcal{O}$ entiers sur l'idéal I forme un idéal $\overline{I}$ de $\mathcal{O}$ appelé *clôture (ou fermeture) intégrale* de l'idéal I dans $\mathcal{O}$.

Supposons que $\mathcal{O}$ soit réduit. C'est l'anneau local en un point x d'un espace analytique réduit X. Quitte à choisir X assez petit, l'idéal I définit un faisceau cohérent d'idéaux $\mathcal{I}$ de $\mathcal{O}_X$. De même $\overline{I}$ définit un faisceau cohérent d'idéaux $\overline{\mathcal{I}}$ de $\mathcal{O}_X$. Avec les notations ci-dessus, on a (cf. [JLT] iv) du Th. 2.1):

Théorème 2.5. *Pour un choix convenable du représentant X du germe d'espace analytique (X, x), l'éclatement normalisé*

$$\overline{e}_{\mathcal{I}} \colon \overline{X}_{\mathcal{I}} \to X$$

de centre $\mathcal{I}$ dans X coincide avec l'éclatement normalisé de centre $\overline{\mathcal{I}}$ et

$$\mathcal{I}\mathcal{O}_{\overline{X}_{\mathcal{I}}} = \overline{\mathcal{I}}\mathcal{O}_{\overline{X}_{\mathcal{I}}}.$$

Pour éviter un langage trop lourd, l'éclatement de centre l'idéal cohérent $\mathcal{I}$ défini par un idéal I de $\mathcal{O}_{X,x}$ sur un représentant assez petit X du germe (X, x) sera appelé l'éclatement de centre I dans le germe d'espace analytique (X, x).

3. Singularités Sandwich

Soit I un idéal de l'anneau analytique local $\mathcal{O}_{\mathbf{C}^2,0}$ des germes de fonctions analytiques complexes en un point 0 du plan complexe $\mathbf{C}^2$. On suppose que l'idéal I est intégralement clos dans $\mathcal{O}_{\mathbf{C}^2,0}$, i.e. l'idéal I égale sa clôture intégrale $\overline{I}$: on dira aussi, comme O. Zariski, que c'est un *idéal complet* de $\mathcal{O}_{\mathbf{C}^2,0}$. L'éclatement de cet idéal dans $(\mathbf{C}^2, 0)$ donne une surface complexe Z. Par définition les singularités de Z sont des *singularités Sandwich.*

On a le résultat suivant très particulier:

Lemme 3.1. *Les singularités Sandwich sont normales.*

Preuve. En effet d'après [SZ] (Appendix 5 Theorem 2′), le produit d'idéaux complets de $\mathcal{O}_{\mathbf{C}^2,0}$ est un idéal complet, car $\mathcal{O}_{\mathbf{C}^2,0}$ est régulier. Il en résulte que, pour la clôture intégrale $\overline{I}$ d'un idéal I, la $\mathcal{O}_{\mathbf{C}^2,0}$-algèbre graduée de présentation finie

$$\overline{\mathcal{B}} = \mathcal{O}_{\mathbf{C}^2,0} \oplus \overline{I} \oplus \cdots \oplus \overline{I^n} \oplus \cdots$$

égale la $\mathcal{O}_{\mathbf{C}^2,0}$-algèbre graduée de présentation finie

$$\mathcal{B} = \mathcal{O}_{\mathbf{C}^2,0} \oplus \overline{I} \oplus \cdots \oplus \overline{I}^n \oplus \cdots$$

L'espace *Projan* de $\overline{\mathcal{B}}$ est normal car $\overline{\mathcal{B}}$ est la normalisation de $\mathcal{B}$, alors que l'espace *Projan* de $\mathcal{B}$ n'est autre que l'éclatement de $\overline{I} = I$ dans $(\mathbf{C}^2, 0)$. Ceci montre donc que l'éclatement de I dans $(\mathbf{C}^2, 0)$ donne un espace normal dont toutes les singularités sont normales. □

On peut définir les singularités Sandwich de la manière équivalente suivante:

Lemme 3.2. *Les singularités de l'éclatement normalisé d'un idéal primaire pour l'idéal maximal de $\mathcal{O}_{\mathbf{C}^2,0}$ sont des singularités Sandwich.*

Preuve. En fait ceci provient du fait général que les éclatements d'un idéal primaire pour l'idéal maximal d'une algèbre analytique locale et de sa clôture intégrale dans cette algèbre ont la même normalisation (voir Théorème 2.5 et [JLT] iv) du Th. 2.1). □

Ce dernier lemme permet d'observer le:

Corollaire 3.3. *Les singularités de l'éclatement normalisé d'un idéal (f, g) primaire pour l'idéal maximal de $\mathcal{O}_{\mathbf{C}^2,0}$ sont des singularités Sandwich.*

L'intérêt de ce corollaire provient du fait que l'éclatement de l'idéal (f, g) dans un voisinage ouvert U assez petit de 0 dans $\mathbf{C}^2$ n'est autre que la surface de déformation associée au pinceau linéaire engendré par f et g (voir la remarque 2.4), à savoir la surface d'équation

$$\mu f - \lambda g = 0$$

dans $U \times \mathbf{P}^1$, où les coordonnées de $\mathbf{P}^1$ sont $(\lambda : \mu)$. Le corollaire affirme donc que la normalisation de cette surface de déformation a des singularités Sandwich.

En fait toutes les singularités Sandwich sont obtenues comme dans le Corollaire 3.3, car on a une caractérisation intéressante et géométrique des singularités Sandwich (voir [LêW], Remarque sur le Théorème 4.5):

Proposition 3.4. *Toute singularité Sandwich est une singularité de l'éclatement normalisé d'un idéal (f, g) primaire pour l'idéal maximal de $\mathcal{O}_{\mathbf{C}^2,0}$.*

Preuve. En effet, si nous considérons un idéal complet I de $\mathcal{O}_{\mathbf{C}^2,0}$, on peut extraire une suite régulière (f, g) de I qui engendre un idéal dont la multiplicité égale celle de I (cf. [Na] Theorem 24.1). D'après un théorème de Rees (cf. [Re]), comme l'idéal (f, g) est contenu dans I et que ces idéaux ont même multiplicité, la fermeture intégrale de (f, g) égale celle de l'idéal I, donc égale I. L'idéal (f, g) et l'idéal I ont alors le même éclatement normalisé (cf. Théorème 2.5 et [JLT] iv) du Th. 2.1). Comme l'éclatement de I est déjà normal, le Corollaire 3.3 entraine la Proposition 3.4. □

Remarque 3.5. L'énoncé de la Proposition 3.4 signale qu'une singularité Sandwich peut ne pas être seule dans l'éclatement normalisé d'un idéal primaire pour l'idéal maximal de $\mathcal{O}_{\mathbf{C}^2,0}$ (voir aussi 5). Cependant, on peut choisir l'idéal complet à éclater pour qu'il n'y ait que la seule singularité Sandwich choisie dans l'éclatement (cf. Corollaire 4.18 ci-dessous).

En résumé, les singularités Sandwich sont les singularités d'une normalisation de surface de déformation d'un système linéaire de courbes planes engendré par deux éléments de $\mathcal{O}_{\mathbf{C}^2,0}$ sans composantes irréductibles communes.

Il est intéressant de remarquer que, d'après J. Lipman et B. Teissier (voir [LiT] §5), le produit d'idéaux intégralement clos dans un anneau analytique local d'une singularité rationnelle est aussi un idéal intégralement clos.

Rappelons que, d'après M. Artin (voir [A]), une singularité normale de surface (X,x) est rationnelle s'il existe une résolution des singularités de X (voir l'introduction) $\pi\colon X' \to X$ telle que le faisceau dérivé $\mathbf{R}^1\pi_*(\mathcal{O}_{X'})$ de l'image directe du faisceau structural $\mathcal{O}_{X'}$ par π ait sa fibre nulle en x, i.e.

$$\mathbf{R}^1\pi_*(\mathcal{O}_{X'})_x = H^1(\pi^{-1}(x), \mathcal{O}_{X'}) = 0.$$

Par une démonstration analogue à celle du Théorème 3.1, à l'aide du théorème de J. Lipman et B. Teissier cité ci-dessus nous pouvons montrer que l'éclatement d'un idéal intégralement clos de l'anneau local d'une singularité rationnelle de surface est une surface normale. A l'instar de M. Spivakovsky, pour une singularité rationnelle de surface (X,x) donnée nous appellerons *singularité Sandwich relative à la singularité rationnelle* (X,x) toute singularité de l'éclatement d'un idéal intégralement clos de l'anneau local $\mathcal{O}_{X,x}$.

Nous verrons plus loin que les singularités Sandwich relatives à une singularité rationnelle sont rationnelles. En particulier les singularités Sandwich sont rationnelles.

4. Approche combinatoire

Dans ce paragraphe nous aurons besoin de la notion d'intersection de courbes dans une surface non singulière et aussi de la notion d'auto-intersection.

Rappelons que dans une surface analytique complexe non singulière S, le nombre d'intersection locale $(C_1.C_2)_s$ de deux courbes distinctes C_1 et C_2 en $s \in S$ est donné par la longueur du quotient $\mathcal{O}_{S,s}/(f_1,f_2)$ de l'anneau local $\mathcal{O}_{S,s}$ par l'idéal (f_1,f_2) engendré par deux équations locales de C_1 et C_2 en s. Si C_1 et C_2 sont compactes, le nombre d'intersection $C_1.C_2$ de C_1 et C_2 dans S égale la somme des nombres d'intersection locaux en tous les points de $C_1 \cap C_2$. On démontre (cf. [Ha] Chap. V §1 ou [Be] Chap. I) que

$$C_1.C_2 = deg(\mathcal{O}_S(C_1) \otimes_{\mathcal{O}_S} \mathcal{O}_{C_2})$$

où $deg(\mathcal{O}_S(C_1) \otimes_{\mathcal{O}_S} \mathcal{O}_{C_2})$ est le degré de la restriction du faisceau $\mathcal{O}_S(C_1)$ des sections méromorphes φ, telles que $div(\varphi) + C_1 \geq 0$, à la courbe C_2 (Attention, le nombre d'intersection $C_1.C_2$ est évidemment supérieur (ou égal) aux nombres de points de $C_1 \cap C_2$).

Quand $C_1 = C_2 = C$, le degré

$$deg(\mathcal{O}_S(C) \otimes_{\mathcal{O}_S} \mathcal{O}_C)$$

est fini. On définit l'auto-intersection d'une courbe compacte C de S par

$$C.C := deg(\mathcal{O}_S(C) \otimes_{\mathcal{O}_S} \mathcal{O}_C).$$

Rappelons aussi que le diviseur exceptionnel de l'éclatement d'un point dans $\mathbf{C}^2$ est une courbe projective rationnelle non singulière d'auto-intersection égale à -1.

Considérons l'éclatement $e \colon U_I \to U$ d'un idéal complet I primaire pour l'idéal maximal d'un point 0 de l'ouvert U de $\mathbf{C}^2$. Soit $\pi \colon \tilde{U}_I \to U_I$ la résolution minimale de U_I, i.e. une résolution de U_I (morphisme propre pour lequel $\tilde{U}_I$ soit non-singulier et (i) qui soit un isomorphisme au-dessus de la partie non-singulière de U_I, (ii) pour laquelle l'image inverse par π de la partie non-singulière de U_I soit dense dans $\tilde{U}_I$) dans laquelle il n'existe aucune courbe rationnelle lisse d'auto-intersection -1 (voir [L] Theorem 5.9). La composée $e \circ \pi$ est propre et composée d'applications biméromorphes. On sait alors que c'est la composée d'éclatements de points (voir [L] Theorem 5.7). On a:

Lemme 4.1. *Soit U un voisinage ouvert d'un point 0 dans $\mathbf{C}^2$. Soit $f \colon \tilde{U} \to U$ une application biméromorphe propre d'un espace lisse $\tilde{U}$ dans U qui induise un isomorphisme analytique au-dessus de $U - \{0\}$. Alors f est une résolution plongée (non nécessairement minimale) d'une singularité de courbe plane dans laquelle les éclatements successifs sont toujours centrés en un point de la transformée stricte de la courbe au-dessus de la singularité.*

Preuve. On sait que f est une suite d'éclatements de points (cf. [L] Theorem 5.7). Comme f est un isomorphisme analytique au-dessus de $U - \{0\}$, ces éclatements sont centrés au-dessus de 0. Comme U est non singulier, le théorème principal de Zariski ([Ha] Chap. III, Corollary 11.5) donne que le diviseur exceptionnel $f^{-1}(0)$ de f est connexe. Les composantes de ce diviseur $f^{-1}(0)$ sont toutes des courbes rationnelles non-singulières. On considère les composantes D_i, $1 \leq i \leq k$, de ce diviseur exceptionnel dont l'auto-intersection est -1. Pour chacune de ces composantes D_i, $1 \leq i \leq k$, on considère un disque complexe Γ_i assez petit qui soit transverse à D_i en un point général. Le morphisme f est une résolution plongée du germe de courbe $\cup_{1 \leq i \leq k} f(\Gamma_i)$ en 0 dans laquelle les éclatements successifs sont toujours centrés en un point de la transformée stricte de la courbe au-dessus de la singularité, puisque l'image inverse de $\cup_{1 \leq i \leq k} f(\Gamma_i)$ par f est un diviseur à croisements normaux et que la transformée stricte de la courbe singulière n'intersecte que les composantes d'auto-intersection -1. On établit par récurrence sur le nombre d'éclatements de points qui composent f que les transformées strictes successives intersectent les diviseurs d'auto-intersection -1 qui apparaissent, ce qui signifie que les centres d'éclatements successifs sont en des points de la transformée stricte de la courbe au-dessus de 0. □

Remarque 4.2. Dans ce qui suit, afin d'éviter toute ambiguïté, quand nous parlerons de *résolution plongée* (non nécessairement minimale) d'une singularité de courbe plane, il s'agira d'une résolution dans laquelle les éclatements successifs sont toujours centrés en un point de la transformée stricte de la courbe au-dessus de la singularité.

Appliquons le lemme 4.1 au cas où $f := e \circ \pi$. Cette composée est donc la résolution plongée d'une courbe plane dans laquelle les éclatements successifs sont toujours centrés en un point de la transformée stricte de la courbe au-dessus de la singularité. Mais, attention, comme le rappelle le lemme, ce n'est pas nécessairement la résolution plongée minimale d'une courbe plane.

Remarquons que le graphe d'intersection du diviseur $f^{-1}(0)$, i.e. le graphe dont les sommets sont les composantes de la courbe $f^{-1}(0)$ et dont les arêtes correspondent au points d'intersection des courbes extrémités de ces arêtes, est un arbre *simplicial* (i.e. deux sommets de l'arbre sont joints par une arête au plus), car deux composantes ne se coupent au plus qu'en un point et qu'il n'y a pas de cycles dans le graphe d'intersection.

Pour chaque singularité Sandwich x_j, $1 \leq j \leq r$, qui apparait dans U_I, il résulte de ce qui précède que le graphe d'intersection du diviseur exceptionnel de la résolution minimale de x_j est un sous-arbre connexe $\mathcal{B}_j$ de l'arbre d'intersection du diviseur $f^{-1}(0)$ dont tous les sommets correspondent à des composantes d'auto-intersection ≤ -2.

Ainsi le diviseur exceptionnel de la résolution minimale d'une singularité Sandwich a toutes ses composantes qui sont des courbes rationnelles lisses et le graphe d'intersection de ces composantes est un arbre. Nous verrons ci-dessous qu'une singularité Sandwich est rationnelle, ce qui explique ces observations sur le diviseur exceptionnel de la résolution minimale (voir [B] Lemma 1.3).

Définition. Le graphe d'intersection des composantes du diviseur exceptionnel d'une résolution d'une singularité de surface normale, pondéré par les auto-intersections des composantes dans la surface de résolution, sera aussi appelé *graphe de la résolution* ou *graphe dual de la singularité* quand il s'agira de la résolution minimale.

Inversement on a le résultat suivant:

Théorème 4.3. *Soit C une courbe dont les composantes sont des courbes rationnelles non-singulières plongée dans une variété analytique complexe V de dimension 2. On pondère le graphe d'intersection des composantes de cette courbe par les auto-intersections et on suppose que le graphe ainsi pondéré soit un sous-arbre de l'arbre d'intersection pondéré de la résolution d'une courbe complexe plane, alors la singularité de la surface normale obtenue en contractant cette configuration de courbes est une singularité Sandwich.*

Preuve. La matrice d'incidence associée à l'arbre d'intersection pondéré des composantes du diviseur exceptionnel de la résolution d'une courbe plane (matrice d'intersection des composantes du diviseur exceptionnel) est définie négative (cf. [M]). Par conséquent la sous-matrice, matrice d'incidence de l'arbre d'intersection des composantes de la courbe de la proposition, est aussi définie négative. Un théorème de H. Grauert [G] montre alors que l'espace X obtenu à partir de V en *contractant* $\mathcal{C}$, i.e. en identifiant la courbe $\mathcal{C}$ à un point 0, a une structure d'espace analytique unique telle que l'application $p\colon V \to X$ soit analytique complexe et propre et que X soit un espace normal. Il nous faut montrer que la singularité $(X, 0)$ est Sandwich.

Considérons un voisinage tubulaire ouvert $T(\mathcal{C})$ de la courbe $\mathcal{C}$ dans V. On a supposé que l'arbre d'intersection pondéré $\mathcal{B}$ de la courbe $\mathcal{C}$ est sous-arbre de l'arbre d'intersection pondéré $\mathcal{A}$ de la résolution d'une courbe plane. Soit $\mathcal{B}'$ l'union des arbres obtenus de $\mathcal{A}$ en enlevant $\mathcal{B}$. On peut construire une courbe $\mathcal{C}'$ (non nécessairement connexe) plongée dans une surface analytique complexe V' dont l'union des arbres d'intersection pondérés des composantes connexes de $\mathcal{C}'$ soit $\mathcal{B}'$. En effet soit p un sommet de $\mathcal{B}'$ de poids $-k$. On va associer une courbe rationnelle lisse d'auto-intersection $-k$. On trouve une telle courbe rationnelle de la façon suivante. Pour chaque entier $k \geq 0$, on a une surface de Nagata-Hirzebruch (cf. [BPV] Chap. V §4), i.e. une surface rationnelle réglée qui est un fibré en courbes rationnelles lisses sur la courbe rationnelle lisse $\mathbf{P}^1$. Ce fibré a une section qui est une courbe rationnelle lisse d'auto-intersection $-k$. Notons T_p un voisinage tubulaire de cette section C_p dans la surface de Nagata-Hirzebruch. Ainsi, pour chaque sommet p de l'union $\mathcal{B}'$ des arbres d'intersection pondérés, nous associons un tel espace T_p et une courbe C_p d'auto-intersection $-k$. Si p et p' sont joints par une arête, nous collons les espaces T_p et $T_{p'}$ en deux polydisques centrés en des points sur les courbes C_p et $C_{p'}$. De proche en proche on recolle en un espace $\mathcal{T} := \cup T_p$ (non nécessairement connexe) les espaces T_p de telle sorte que la courbe $\cup C_p$ (non nécessairement connexe) ait $\mathcal{B}'$ pour union des graphes pondérés d'intersection (on dit que la variété $\mathcal{T}$ a été obtenue par *plombage* à partir de $\mathcal{B}'$).

Enfin, on recolle les composantes connexes de $\mathcal{T}$ à $T(\mathcal{C})$ pour obtenir une variété analytique complexe W de dimension 2. On peut vérifier que si les divers voisinages tubulaires sont choisis assez petits, W est un voisinage régulier de la courbe $\mathcal{D} := \mathcal{C} \cup_p C_p$. Dans cette variété W la courbe $\mathcal{D}$ a un graphe d'intersection qui est $\mathcal{A}$ et c'est donc l'arbre de la résolution d'une courbe complexe plane.

On utilise maintenant des résultats de M. Artin (cf. [A] Proposition 2). Dans le groupe abélien libre engendré par les composantes irréductibles de $\mathcal{D}$, le plus petit élément pour l'ordre produit dont l'intersection avec toutes les composantes irréductibles de $\mathcal{D}$ soit ≤ 0 (voir ci-dessous la Proposition 4.6) ne dépend que de l'arbre $\mathcal{A}$ et est appelé le cycle fondamental $Z_{\mathcal{A}}$ associé à $\mathcal{A}$. Comme l'arbre $\mathcal{A}$ est l'arbre d'intersection d'une résolution de $(\mathbf{C}^2, 0)$, ce cycle fondamental est donc de genre arithmétique nul (voir la définition ci-dessous) et son auto-intersection

$$Z_{\mathcal{A}}.Z_{\mathcal{A}} = -1.$$

D'après un théorème de M. Artin (critère de Castelnuovo généralisé, cf.[A] Corollary 7), si l'on contracte $\mathcal{D}$ dans W on trouve un espace analytique normal Z qui est non-singulier. En contractant $\mathcal{C}$ dans W on trouve un espace X_1 singulier et normal en un point 0

$$p\colon W \to X_1$$

et dans X_1, en contractant l'image $p(\mathcal{D})$, on trouve Z

$$q\colon X_1 \to Z.$$

Comme $T(\mathcal{C})$ est plongé dans W et la contraction de $\mathcal{C}$ dans $T(\mathcal{C})$ est un voisinage de la singularité de X dans X, cette singularité est 0 et les germes $(X,0)$ et $(X_1,0)$ coïncident.

Notons $z = q(p(\mathcal{D}))$. Pour chaque composante C du diviseur B de W correspondant aux sommets de $\mathcal{B}'$, considérons un disque Δ_C transverse en un point non singulier du diviseur D. L'image $\Gamma := q \circ p(\cup_C \Delta_C)$ est une courbe plane. Soit m_C la multiplicité de $(q \circ p)^{-1}(\Gamma)$ le long de C.

Lemme 4.4. *Les fonctions φ de $\mathcal{O}_{Z,z}$, pour lesquelles $val_C(\varphi) \geq m_C$, pour tout C, composante du diviseur B, forment un idéal complet I et l'image inverse q^*I est un faisceau inversible d'idéaux sur X_1.*

Preuve. Les propriétés des valuations val_C montrent que I est évidemment un idéal de $\mathcal{O}_{Z,x}$. Soit φ_Γ une équation de Γ. Par définition on a $\varphi_\Gamma \in I$. En choisissant des disques Δ'_C disjoints des Δ_C et aussi transverses aux composantes C de B en des points non-singuliers, on obtient une courbe Γ' image de $\cup_C \Delta_C$ par $q \circ p$. Si $\varphi_{\Gamma'}$ est une équation de Γ', on a aussi $\varphi_{\Gamma'} \in I$.

Si W est un voisinage régulier assez petit de $\mathcal{D}$, l'idéal $(\varphi_\Gamma, \varphi_{\Gamma'})\mathcal{O}_W$ est inversible. En effet, en tous les points w de $\mathcal{D}$ où ne passe pas la transformée stricte de $\varphi_\Gamma = 0$, la fonction $q \circ p \circ \varphi_\Gamma$ engendre $(\varphi_\Gamma, \varphi_{\Gamma'})\mathcal{O}_{W,w}$. Comme les points de $\mathcal{D}$ où passe la transformée stricte de $\varphi_\Gamma = 0$ sont disjoints des points de $\mathcal{D}$ où passe la transformée stricte de $\varphi'_\Gamma = 0$, on obtient notre assertion.

Comme W est non-singulier (donc normal), si W est un voisinage régulier assez petit de $\mathcal{D}$, le faisceau d'idéaux $(\varphi_\Gamma, \varphi_{\Gamma'})\mathcal{O}_W$ égale $I'\mathcal{O}_W$, où I' est la clôture intégrale de $(\varphi_\Gamma, \varphi_{\Gamma'})\mathcal{O}_{Z,z}$ dans $\mathcal{O}_{Z,z}$. Par ailleurs, comme φ_Γ et $\varphi_{\Gamma'}$ sont des fonctions constantes sur les composantes correspondant aux sommets de $\mathcal{C}$, le faisceau d'idéaux $(\varphi_\Gamma, \varphi_{\Gamma'})\mathcal{O}_{X_1}$ est aussi inversible. Les composantes C' du diviseur exceptionnel de q dans X_1 sont les images $p(C)$ des composantes C de B. La multiplicité de $I'\mathcal{O}_W$ le long de C' égale m_C. Comme l'idéal I' est intégralement clos, une fonction φ dans $\mathcal{O}_{Z,z}$ est dans I' si et seulement si

$$val_{C'}(\varphi) \geq m_C$$

(cf. [JLT] Conséquence de Théorème 2.1 iv)). Comme $val_{C'}(\varphi) = val_C(\varphi)$, les idéaux I' et I sont égaux. Ceci montre que I est complet.

En fait, pour établir que I est complet, on aurait pu aussi invoquer le résultat général selon lequel, dans un anneau local régulier de dimension 2 muni de valuations $(v_j)_{1 \leq j \leq k}$ centrées en l'idéal maximal, tout idéal formé des éléments x tels que $v_j(x) \geq n_j$, pour $1 \leq j \leq k$ et $(n_j)_{1 \leq j \leq k}$ entiers positifs, est complet. La démonstration ci-dessus en fournit une preuve.

Comme X_1 est normal, le faisceau d'idéaux $(\varphi_\Gamma, \varphi_{\Gamma'})\mathcal{O}_{X_1}$ égale

$$I'\mathcal{O}_{X_1} = I\mathcal{O}_{X_1}$$

(voir Théorème 2.5). Ce qui montre que $I\mathcal{O}_{X_1} = q^*I$ est inversible et termine la démonstration du lemme. □

Pour terminer la preuve du Théorème 4.3, nous allons montrer que q est isomorphe à l'éclatement normalisé de l'idéal complet I.

Il suffit de montrer que q est l'éclatement normalisé de l'idéal $(\varphi_\Gamma, \varphi_{\Gamma'})\mathcal{O}_{Z,z}$. En fait, comme $(\varphi_\Gamma, \varphi_{\Gamma'})\mathcal{O}_W$ est inversible, par définition l'espace W domine biméromorphiquement l'éclatement normalisé de l'idéal $(\varphi_\Gamma, \varphi_{\Gamma'})\mathcal{O}_{Z,z}$. On a donc une contraction de W dans l'éclatement normalisé de l'idéal $(\varphi_\Gamma, \varphi_{\Gamma'})\mathcal{O}_{Z,z}$ en contractant les composantes C_i du diviseur Z_I, défini par le faisceau inversible d'idéaux $(\varphi_\Gamma, \varphi_{\Gamma'})\mathcal{O}_W = I\mathcal{O}_W$, telles que $Z_I.C_i = 0$. Or le diviseur Z_I est la partie compacte du diviseur $div(\varphi_\Gamma \circ q \circ p)$ défini par la fonction $\varphi_\Gamma \circ q \circ p$. Comme la partie non-compacte de $div(\varphi_\Gamma \circ q \circ p)$ n'intersecte que les composantes dans B, les composantes à contracter sont les composantes du diviseur exceptionnel de $q \circ p$ qui ne sont pas dans B. Or cette contraction définit précisemment X_1, donc q est bien l'éclatement normalisé de l'idéal $(\varphi_\Gamma, \varphi_{\Gamma'})\mathcal{O}_{Z,z}$ et la singularité $(X_1, 0) = (X, 0)$ est Sandwich. □

A une résolution de la singularité d'un germe de surface normale on associe un graphe pondéré de la façon suivante. Les sommets du graphe sont les composantes du diviseur exceptionnel de la résolution considérée (ce sont des courbes projectives éventuellement singulières) et le nombre d'arêtes entre deux sommets égale le nombre d'intersection des deux courbes correspondant à ces sommets dans la surface non singulière de la résolution. Le théorème principal de O. Zariski (cf. [Ha] Chap. III, Corollary 11.5) implique que le graphe obtenu est effectivement connexe. On pondère ce graphe en donnant à chaque sommet le poids égal à l'auto-intersection de la composante correspondante. La matrice d'intersection des composantes du diviseur exceptionnel de la résolution considérée est une matrice symétrique. Comme nous l'avons déjà signalé dans la preuve du Théorème 4.3, un résultat de P. Du Val et D. Mumford montre que cette matrice symétrique définit une forme quadratique définie négative (voir [M]).

Quitte à éclater encore des points, on peut obtenir une résolution dans laquelle le diviseur exceptionnel est à croisements normaux, les composantes du diviseur exceptionnel sont non-singulières et deux composantes se coupent au plus en un point. Dans ce cas le graphe d'intersection est *simplicial*, i.e. deux sommets sont joints au plus par une arête et l'extrémité d'une arête ne coïncide avec son origine.

De façon plus générale, soit $\mathcal{A}$ un graphe simplicial pondéré par des poids entiers négatifs. On associe à ce graphe un espace vectoriel sur le corps des rationnels $\mathbf{Q}$, dont la base est donnée par les sommets du graphe que l'on a numérotés, et une forme quadratique donnée par la matrice symétrique $M = (m_{i,j})$ (appelée la *matrice d'incidence* du graphe pondéré) définie pour $i \neq j$ $m_{i,j} =$ le nombre d'arêtes entre les i-ème et j-ème sommets, i.e. 1 pour un graphe simplicial ou 0 si le i-ème sommet n'est pas lié au j-ème sommet par une arête et

$$m_{i,i} = \text{ le poids du i-ème sommet.}$$

Il est naturel d'appeler *graphe singulier* un graphe simplicial pondéré dont la matrice d'incidence symétrique associée est celle d'une forme quadratique définie négative. Dans ce cas cette matrice symétrique sera aussi appelée *matrice d'intersection du graphe singulier.* On appellera *forme d'intersection du graphe singulier* la forme bilinéaire associée à cette matrice. Le théorème de Grauert implique donc:

Théorème 4.5. *Tout graphe singulier est le graphe de résolution d'une surface normale.*

Comme ci-dessus, on obtient par plombage une surface non singulière et une courbe plongée dans cette surface dont le graphe pondéré d'intersection des composantes est le graphe singulier considéré. Par contraction de la courbe, on obtient une singularité normale dont le graphe de résolution est le graphe singulier considéré.

En fait, d'après un résultat de O. Zariski, à tout graphe singulier est attaché un cycle, appelé *cycle fondamental* (cf.[A] p.132) et défini par:

Proposition 4.6. *Soit $\mathcal{A}$ un graphe singulier de sommets E_i, $1 \leq i \leq k$ et de forme d'intersection associée I. Il existe une unique combinaison linéaire à coefficients entiers strictement positifs $Z = \sum_1^k a_i E_i$ telle que $I(Z, E_i) \leq 0$, pour tout i, $1 \leq i \leq k$, et pour toute combinaison linéaire non nulle à coefficients entiers $Z' = \sum_1^k a_i' E_i$ telle que $I(Z', E_i) \leq 0$, pour tout i, $1 \leq i \leq k$, on a $a_i' \geq a_i$, pour tout i, $1 \leq i \leq k$. On appelle Z le cycle fondamental de $\mathcal{A}$.*

Dans la suite nous noterons la forme d'intersection I associée à l'arbre $\mathcal{A}$ de la façon suivante:

$$I(E_i, E_j) := E_i.E_j \ (= 0 \text{ ou } 1$$

Pour tout graphe singulier $\mathcal{A}$ on appellera *cycle* $C = \sum_1^k c_i E_i$ toute combinaison linéaire à coefficients entiers dans l'espace vectoriel rationnel engendré par les sommets E_i. A un cycle C on associe

$$p(C) = \frac{C.C + \sum_1^k c_i(w_i - 2)}{2} + 1$$

où $w_i := -E_i.E_i$.

Définition. On appelle $p(C)$ le *genre arithmétique du cycle C.*

Ceci nous permet de définir:

Définition. On appelle *graphe rationnel* un graphe singulier dont le cycle fondamental a un genre arithmétique nul.

Dans [A] (Theorem 3), M. Artin montre:

Proposition 4.7. *Un graphe est rationnel si et seulement si c'est le graphe d'une résolution d'une singularité rationnelle. De plus un graphe rationnel est un arbre.*

Donc nous utiliserons le nom d'*arbre rationnel* au lieu de graphe rationnel. M. Artin montre aussi ([A] Proposition 1) que

Théorème 4.8. *Tout sous-arbre d'un arbre rationnel est rationnel.*

En particulier, on a le:

Corollaire 4.9. *L'arbre de résolution d'une singularité Sandwich est rationnel.*

Preuve. En effet, l'arbre de résolution plongée d'un germe de courbe plane est évidemment rationnel et on a vu (Proposition 4.3) que l'arbre de résolution d'une singularité Sandwich est un sous-arbre de l'arbre de résolution d'un germe de courbe plane. □

Suivant M. Spivakovsky ([S] Definition 1.9 (1)), il est commode de définir:

Définition. On appelle *arbre non-singulier* l'arbre de résolution plongée d'un germe de courbe plane. On appelle *arbre Sandwich* l'arbre de résolution d'une singularité Sandwich.

On résume les résultats précédents par (voir [S] Proposition 1.11):

Théorème 4.10. *Soit $\mathcal{A}$ un arbre simplicial pondéré. Les conditions suivantes sont équivalentes:*

i) *Tout germe de surface normale qui a une résolution des singularités dont les composantes du diviseur exceptionnel sont des courbes rationnelles lisses et, dont l'arbre d'intersection est $\mathcal{A}$, est Sandwich;*
ii) *Il y a une singularité Sandwich qui a une résolution des singularités dont l'arbre est $\mathcal{A}$;*
iii) *L'arbre $\mathcal{A}$ est Sandwich;*
iv) *L'arbre $\mathcal{A}$ est le sous-arbre de l'arbre d'une résolution de courbe plane.*

Remarque 4.11. Dans la définition d'un arbre rationnel et d'un arbre Sandwich on ne suppose pas que les poids soient ≤ -2. Si les poids sont ≤ -2, ces arbres sont les arbres d'intersection des composantes du diviseur exceptionnel des résolutions minimales des singularités correspondantes.

Une interprétation intéressante de l'auto-intersection du cycle fondamental d'un arbre rationnel est donnée par M. Artin (cf. [A] Corollary 6):

Lemme 4.12. *Le nombre d'auto-intersection $Z.Z$ du cycle fondamental Z d'un arbre rationnel $\mathcal{A}$ égale $-m(X,x)$, où $m(X,x)$ est la multiplicité d'une singularité rationnelle de surface (X,x) ayant une résolution dont l'arbre est $\mathcal{A}$. Le nombre $-Z.Z+1$ égale la dimension d'immersion de (X,x).*

La terminologie de M. Spivakovsky est justifiée par le théorème suivant:

Théorème 4.13. *Un anneau analytique local normal de dimension deux qui domine biméromorphiquement un anneau analytique local régulier de dimension deux est l'anneau local d'une singularité Sandwich.*

Preuve. On peut supposer que la singularité est algébrique, puisqu'elle est isolée et qu'elle domine birationnellement un anneau local régulier de dimension deux. On trouve alors une application birationnelle $\rho\colon X \to \mathbf{C}^2$ et un point x de $\rho^{-1}(0)$ tel que $\mathcal{O}_{X,x}$ soit l'anneau local considéré. Comme cet anneau est normal, on peut supposer que X est normal.

Soit $\pi\colon W \to X$ une résolution des singularités de X. On remarque que la courbe $\pi^{-1}(x)$ est contenue dans le diviseur de $\rho \circ \pi^{-1}(0)$. On obtient que X est obtenu en contractant une courbe contenue dans le diviseur exceptionnel de la résolution plongée d'une singularité de courbe plane. D'après le Théorème 4.3, il en résulte que (X, x) est une singularité Sandwich. □

On obtient aussi le résultat annoncé suivant:

Théorème 4.14. *Soit (X, x) une singularité rationnelle. Une singularité Sandwich relative à (X, x) est rationnelle.*

Preuve. Soit I un idéal complet de $\mathcal{O}_{X,x}$. Soit $e_I\colon X_I \to X$ l'éclatement normalisé associé à I dans un représentant assez petit de (X, x). Soit $\pi\colon W \to X_I$ une résolution de X_I. Comme W domine birationnellement la résolution minimale de (X, x), l'arbre d'intersection du diviseur exceptionnel de W est rationnel et la contraction de toute courbe connexe contenue dans ce diviseur exceptionnel donne une singularité rationnelle (cf. Théorème 4.10). Comme X_I est obtenu à partir de W en faisant de telles contractions, les singularités de X_I sont rationnelles. Par définition (voir §3), ce sont des singularités Sandwich relatives à (X, x). □

On a un critère combinatoire qui permet de caractériser une singularité Sandwich relative à une singularité rationnelle:

Théorème 4.15. *Une singularité de surface normale est une singularité Sandwich relative à une singularité rationnelle (X, x), si et seulement s'il existe une résolution $\pi\colon X' \to X$ de (X, x) et un cycle $Z \neq 0$ à support dans le diviseur exceptionnel $\pi^{-1}(x) = \cup_i C_i$, tel que, pour chaque composante C_i, on ait $Z.C_i \leq 0$, pour lesquels la singularité soit obtenue par contraction d'une composante de Tiourina-Spivakovsky de Z.*

Rappelons que les composantes de Tiourina-Spivakovsky de Z (cf. [S] Chap. III, Definition 3.1) sont les courbes connexes maximales dans $\pi^{-1}(x)$ formées de composantes C_i telle que $Z.C_i = 0$. Quand Z est le cycle fondamental de π, il s'agit des composantes de Tiourina (voir §6 ci-dessous).

Preuve. La condition est évidemment nécessaire, car, si une singularité est une singularité Sandwich relative à la singularité rationnelle (X, x), dans la résolution minimale de l'éclatement X_I de l'idéal complet I qui définit la singularité Sandwich relative, l'image inverse de cet idéal est inversible et définit un cycle Z qui vérifie les conditions du Théorème. Il est facile de voir que les singularités de X_I sont obtenues en contractant les composantes de Tiourina-Spivakovsky.

La réciproque provient du fait remarqué par J. Lipman (voir [Li] §18) que, comme (X,x) est une singularité rationnelle, tout cycle $Z \neq 0$ à support dans le diviseur exceptionnel $\pi^{-1}(x) = \cup_i C_i$, tel que, pour chaque composante C_i, on ait $Z.C_i \leq 0$, est défini par l'image inverse d'un idéal complet de (X,x) qui est inversible dans X'. □

Le Théorème 4.15 permet de donner une formulation aussi simple que celle du Théorème 4.3 pour caractériser combinatoirement les singularités Sandwich relative à une singularité rationnelle, grâce au lemme suivant:

Lemme 4.16. *Soit $\pi\colon X' \to X$ une résolution de singularité rationnelle de surface (X,x). Soient $(C_i)_{1\leq i\leq k}$ les composantes du diviseur exceptionnel $\pi^{-1}(x) = \cup_1^k C_i$. Soit $(C_i)_{1\leq i\leq r}$, avec $1 \leq r \leq k$, un ensemble de composantes du diviseur exceptionnel. Il existe un cycle $Z = \sum_1^k a_i C_i$ tel que*

$$Z.C_i < 0, \text{ pour } 1 \leq i \leq r,$$
$$Z.C_i = 0, \text{ pour } r < i.$$

Preuve. La matrice d'intersection dans X' des composantes $(C_i)_{1\leq i\leq k}$ du diviseur exceptionnel de π est définie négative (cf. [M]). Elle définit donc un automorphisme linéaire h de l'espace vectoriel sur le corps des rationnels $\mathbf{Q}$ dont la base est l'ensemble des composantes $(C_i)_{1\leq i\leq k}$. On considère des entiers positifs n_i, pour $1 \leq i \leq r$. Il existe un vecteur V unique tel que

$$h(V) = \sum_1^r (-n_i) C_i.$$

Quitte à multiplier V par un entier, on obtient un cycle $Z = \sum_1^k a_i C_i$, comme annoncé dans le lemme. □

Dans [A] (Proposition 2), on montre que, pour $1 \leq i \leq k$, on a $a_i > 0$. Ceci donne une généralisation du Théorème 4.3:

Corollaire 4.17. *Une singularité de surface normale est une singularité Sandwich relative à une singularité rationnelle (X,x), si et seulement si elle est obtenue en contractant une courbe connexe contenue dans le diviseur exceptionnel d'une résolution de (X,x).*

Preuve. Soit $p\colon X_1 \to X$ l'éclatement d'un idéal complet de l'anneau local $\mathcal{O}_{X,x}$. Soit $q\colon X' \to X_1$ une résolution des singularités de X_1. Les singularités de X_1 sont obtenues en contractant des courbes connexes contenues dans le diviseur exceptionnel de la résolution $p \circ q$ de (X,x). La condition du corollaire est donc nécessaire.

Elle est suffisante, si l'on montre que toute courbe connexe contenue dans le diviseur exceptionnel d'une résolution $\pi\colon X' \to X$ de (X,x) est la composante de Tiourina-Spivakovsky d'un cycle effectif Z.

Soit $\pi^{-1}(x) = \cup_1^k C_i$ le diviseur exceptionnel de la résolution π. Soit $\mathcal{C} := \cup_{r+1}^k C_i$, une courbe connexe contenue dans le diviseur exceptionnel de π. D'après le lemme précédent, il existe un cycle $Z = \sum_1^k a_i C_i$ tel que

$$Z.C_i < 0, \text{ pour } 1 \leq i \leq r,$$

$$Z.C_i = 0, \text{ pour } r < i.$$

On remarque que $\mathcal{C}$ est l'unique composante de Tiourina-Spivakovsky d'un tel cycle, ce qui démontre le corollaire. □

Un raisonnement analogue à celui de la preuve du corollaire 4.17 donne le:

Corollaire 4.18. *Pour toute singularité Sandwich (X, x) non régulière donnée, il existe un idéal complet de $\mathcal{O}_{\mathbf{C}^2,0}$ dont l'éclatement ne contient qu'une seule singularité effective isomorphe à (X, x).*

Preuve. En effet, soit $p\colon X \to \mathbf{C}^2$ l'éclatement d'un idéal complet I de $\mathcal{O}_{\mathbf{C}^2,0}$ dans lequel on a la singularité Sandwich (X, x). Soit $q\colon X' \to X$ une résolution des singularités de X. Soit $\mathcal{C}$ la courbe connexe contenue dans le diviseur exceptionnel de $p \circ q$ qui se contracte sur x. Comme dans la démonstration du corollaire précédent, il existe un cycle Z dont le support est le diviseur exceptionnel de $p \circ q$ et dont la seule composante de Tiourina-Spivakovsky soit $\mathcal{C}$. Le résultat de J. Lipman (voir [Li] §18) déjà invoqué montre qu'il existe un idéal complet I_1 de $\mathcal{O}_{\mathbf{C}^2,0}$ dont l'image inverse dans X' est inversible et qui définit le cycle Z.

On vérifie que l'éclatement de I_1 ne contient qu'une seule singularité effective isomorphe à (X, x). □

5. Singularités primitives, singularités minimales

Parmi les singularités Sandwich on distingue:

Définition. On dit qu'une singularité Sandwich est *primitive* si elle est isomorphe à celle obtenue par l'éclatement normalisé d'un idéal primaire pour l'idéal maximal de $\mathcal{O}_{\mathbf{C}^2,0}$ dont la fermeture intégrale dans $\mathcal{O}_{\mathbf{C}^2,0}$ est un idéal simple (voir [SZ] Appendix 5 p. 385).

On observe qu'un idéal complet est simple si et seulement si un élément général de l'idéal est l'équation d'une courbe analytiquement irréductible. On peut alors démontrer (voir [S] Chap. II §3):

Théorème 5.1. *Une singularité Sandwich est primitive si et seulement si son graphe dual est l'une des composantes connexes du complément dans le graphe de résolution plongée d'un germe de courbe analytiquement irréductible du sommet correspondant à l'unique composante d'auto-intersection -1.*

Preuve. Une résolution minimale de l'éclatement de l'idéal simple donne une résolution plongée d'un élément général de l'idéal (voir §4). Dans le diviseur exceptionnel de cette résolution, il n'y a donc qu'une seule composante d'auto-intersection -1. Les singularités de l'éclatement de l'idéal simple sont obtenues en contractant les parties compactes du diviseur exceptionnel qui correspondent au complémentaire de la composante d'auto-intersection -1. □

Remarque 5.2. Si la résolution minimale de l'éclatement de l'idéal simple donne la résolution plongée minimale d'un germe de courbe plane, on aura deux composantes connexes dans le complémentaire de la composante d'auto-intersection -1. Dans ce cas l'une des deux singularités Sandwich primitives obtenues ne peut être obtenue qu'en éclatant un idéal complet simple dont l'éclatement aura nécessairement deux points singuliers. Cependant d'après le Corollaire 4.18, il existe un idéal complet (qui n'est pas simple) dont l'éclatement ne contient qu'une seule singularité effective isomorphe à l'une ou l'autre des singularités précédentes.

Il est alors facile d'établir qu'une singularité dont le graphe dual est un *bambou* (i.e. un segment) avec des poids ≤ -2 est une singularité Sandwich primitive.

Mais si la résolution minimale de l'éclatement de l'idéal simple n'est pas la résolution plongée minimale d'un germe de courbe plane, l'éclatement de l'idéal simple ne contient qu'une seule singularité et le graphe dual de la singularité primitive est le sous-graphe d'un graphe non-singulier qui n'a qu'un seul sommet de plus.

Une autre classe importante de singularités Sandwich est

Définition. On dit qu'une singularité Sandwich est une singularité *minimale* si son cycle fondamental Z est réduit, i.e. si E_i désigne les composantes irréductibles de ce diviseur exceptionnel de la résolution minimale, $1 \leq i \leq k$, on a $Z = \sum_1^k E_i$.

En fait dans [K] (§3.4), J. Kollár montre que:

Proposition 5.3. *Une singularité rationnelle est minimale si et seulement si le cône tangent en la singularité est réduit.*

Caractérisation combinatoire des singularités minimales

Comme le remarque M. Spivakovsky ([S] Remark 2.3), le graphe dual d'une singularité minimale est très simple:

Théorème 5.4. *Un graphe pondéré est le graphe dual d'une singularité minimale si et seulement si c'est un arbre et, pour tout sommet x, la valeur absolue $w(x)$ du poids $-w(x)$ de ce sommet est au moins 2 et supérieur ou égal à la valence $v(x)$ de ce sommet, i.e. au nombre d'arêtes de l'arbre qui ont ce sommet comme extrémité*

$$w(x) \geq v(x).$$

Démonstration. La condition est évidemment nécessaire puisque le graphe dual d'une singularité minimale est un arbre pondéré qui est singulier. Il possède donc une forme d'intersection, qui définit une forme quadratique définie négative, et un cycle fondamental Z tel que $Z.x \leq 0$. Comme Z est réduit et que

$$w(x) = -x.x,$$

on en déduit l'inégalité

$$w(x) \geq v(x).$$

Montrons que la condition est suffisante. Considérons donc un arbre pondéré $\mathcal{A}$ tel que, pour tout sommet x, on ait $w(x) \geq v(x)$. On appelle $E_1, \ldots, E_k$ ses sommets.

On montre d'abord que l'arbre est singulier. En utilisant [A] (Proposition 2 (ii)), il suffit de montrer que le cycle $Z = \sum_1^k E_i$ vérifie, pour tout $1 \leq i \leq k$,

$$Z.E_i \leq 0 \text{ et } Z.Z < 0.$$

L'inégalité $w(E_i) \geq v(E_i)$ est équivalente à $Z.E_i \leq 0$. Il reste à démontrer que $Z.Z < 0$. Nous le montrons par récurrence sur k. En fait par récurrence nous montrons simultanément que $Z.Z \leq -2$ et que $\mathcal{A}$ est bien un arbre rationnel minimal.

Pour $k = 1$, c'est évident. Supposons $k \geq 2$ et le résultat vrai pour $k' < k$, i.e., pour tout k', $1 \leq k' < k$, un arbre pondéré ayant k' sommets pour lesquels on a $w(x) \geq v(x)$ est un arbre rationnel minimal. Soit E_1 une extrémité de $\mathcal{A}$. Appelons $\mathcal{A}_1$ le sous-arbre maximal de $\mathcal{A}$ qui contient tous les autres sommets de $\mathcal{A}$. Comme on suppose l'hypothèse de récurrence et que $\mathcal{A}_1$ est un arbre pondéré dans lequel tous les sommets vérifient l'inégalité considérée entre le poids et la valence, cette arbre est bien singulier et minimal. Il en est de même de l'arbre $\mathcal{A}_2$ constitué du seul point E_1 avec le poids $w(E_1)$. Le cycle fondamental de $\mathcal{A}_1$ est

$$Z_1 = \sum_2^k E_i$$

et celui de $\mathcal{A}_2$ est

$$Z_2 = E_1.$$

De plus, on a la nullité des genres arithmétiques

$$p(Z_1) = \frac{Z_1.Z_1 + \sum_2^k (w(E_i) - 2)}{2} + 1 = 0$$

$$p(Z_2) = \frac{Z_2.Z_2 + (w(E_1) - 2)}{2} + 1 = 0$$

Or

$$p(Z) = \frac{(Z_1 + Z_2).(Z_1 + Z_2) + \sum_1^k (w(E_i) - 2)}{2} + 1$$

$$(Z_1 + Z_2).(Z_1 + Z_2) = Z_1.Z_1 + Z_2.Z_2 + 2 = -\sum_1^k (w(E_i) - 2) - 2$$

car $Z_1.Z_2 = 1$, ce qui donne $p(Z) = 0$ et également $Z.Z \leq -2$, puisque, $w(E_i) \geq 2$, pour tout $1 \leq i \leq k$. Comme l'inégalité en chaque sommet entre le poids et la valence équivaut à $Z.E_i \leq 0$, $1 \leq i \leq k$, $Z = \sum_1^k E_i$ est bien le cycle fondamental, car il est clairement minimal.

Exemples 1. On a les exemples suivants que l'on peut facilement vérifier

i) Si l'arbre est un bambou, i.e. s'il est homéomorphe à un segment, il est rationnel minimal. En fait, dans ce cas il correspond à une singularité cyclique quotient (cf. [BPV] Chap. III §5). Nous avons vu (Remarque 5.2) que c'est une singularité Sandwich primitive.

ii) Parmi les points doubles rationnels, seuls les singularités du type de $\mathbf{A_n}$ sont Sandwich. En fait, elles sont aussi minimales.

iii) Tout sous-arbre d'un arbre rationnel minimal est rationnel minimal.

6. Courbes polaires des singularités rationnelles

Dans [LêT], B. Teissier et l'auteur ont introduit la notion de courbe polaire locale d'un espace analytique complexe. Dans le cas d'une surface normale on a:

Définition. Soit X une surface normale. Soit $0 \in X$. On appelle *courbe polaire* $\Gamma(\pi)$ d'une application $\pi\colon X \to \mathbf{C}^2$ finie en 0, la fermeture dans X de l'espace critique de la restriction de π à la partie non-singulière de X. Quand X est plongée localement en 0 dans un espace affine $\mathbf{C}^N$ et que l'application π est donnée par une projection linéaire sur $\mathbf{C}^2$ assez générale, on dit tout simplement que $\Gamma(\pi)$ est la courbe polaire de la singularité $(X,0)$ ou quelquefois la courbe polaire générique de la singularité $(X,0)$.

Il est possible que la courbe polaire soit vide, par exemple quand X est non singulier en 0.

En fait, dans ce cas où X est à singularité isolée, la courbe polaire de π n'est autre que le lieu critique de l'application finie π.

Composantes de Tiourina et courbe polaire

Dans [T], G. Tiourina a introduit des sous-graphes du graphe dual d'une résolution d'une singularité rationnelle qui sont les composantes connexes maximales dont les sommets sont les composantes du diviseur exceptionnel de la résolution minimale dont l'intersection avec le cycle fondamental de la résolution minimale est nulle. On appelle ces sous-graphes les *graphes de Tiourina de l'arbre rationnel.* Les composantes de ces sous-graphes sont appelées composantes de Tiourina de l'arbre rationnel.

G. Tiourina démontre alors que (cf. [T] Proposition 1.2):

Théorème 6.1. *Les composantes de Tiourina du graphe dual d'une singularité rationnelle sont les graphes duaux des singularités de l'espace obtenu en éclatant une fois la singularité rationnelle.*

Preuve. Soit $\pi\colon \tilde{X} \to X$ une résolution de la singularité $(X,0)$. Comme $(X,0)$ est rationnelle, l'image inverse par π de l'idéal maximal $\mathcal{M}_0$ qui définit 0 dans X est inversible dans $\tilde{X}$. La propriété universelle de l'éclatement implique que π factorise par $p\colon \tilde{X} \to X'$ à travers l'éclatement $e\colon X' \to X$ de $\mathcal{M}_0$. On a donc le triangle commutatif

$$\begin{array}{ccc} \tilde{X} & \xrightarrow{p} & X' \\ & \searrow & \downarrow e \\ & & X \end{array}$$

L'idéal $\mathcal{M}_0$ est un idéal complet de $\mathcal{O}_{X,0}$, donc le théorème de Lipman-Teissier (cf. §3) implique que X' est normal. Le morphisme p est une résolution des singularités de X'. Par ailleurs le cycle fondamental d'une singularité rationnelle coïncide avec son cycle maximal, i.e. le cycle de la résolution minimale de la singularité défini par l'image inverse de l'idéal maximal de la singularité (cf. [A] Theorem 4). Ainsi le cycle fondamental est obtenu de la façon suivante:

On considère la transformée totale de la section hyperplane générique de la surface et on ne retient que la partie compacte du diviseur contenue dans le diviseur exceptionnel de la résolution considérée. Comme la transformée stricte par e d'une section hyperplane générale de X en x ne passe pas par les points singuliers de X', d'après la caractérisation géométrique du diviseur fondamental que l'on vient de donner, ceci équivaut à dire que les intersections du cycle fondamental de π avec les composantes autres que les transformées strictes des composantes qui proviennent du cône tangent sont nulles. Par définition ces composantes sont les composantes irréductibles des composantes de Tiourina. L'application p est la contraction des composantes de Tiourina de la résolution π de X, car X' est normal. On en déduit que p est une résolution des singularités de X' et les composantes de Tiourina du cycle fondamental de la résolution π sont les diviseurs exceptionnels de ces singularités. On obtient le Théorème en considérant la résolution minimale de $(X,0)$. □

Un résultat intéressant, conséquence d'un travail récent de J. Snoussi (cf. [Sn] Théorème 11), relie les composantes de Tiourina et la courbe polaire générique:

Théorème 6.2. *La transformée stricte, par la résolution minimale d'une singularité rationnelle de surface, de la courbe polaire générique de la singularité intersecte chacune des composantes de Tiourina.*

Ce résultat, conjecturé par M. Spivakovsky ([S] III Remark 3.12), provient de ce que les points singuliers de l'éclatement de l'idéal maximal d'une singularité rationnelle de surface représentent des tangentes communes aux courbes polaires génériques (voir [Sn]).

7. Transformation de Nash des singularités minimales

Le point essentiel de la résolution des singularités d'une singularité rationnelle minimale par une suite de transformées de Nash normalisées est de comprendre comment se comporte le graphe dual de la transformée de Nash normalisée d'une singularité minimale. Pour cela on a d'abord besoin d'une caractérisation géométrique des transformées de Nash normalisées (cf. [GS2] §2 et [S] Theorem 1.2):

Théorème 7.1. *Soit $(X,0)$ un germe de surface normale. Soit $\nu\colon \tilde{X} \to X$ la transformée de Nash normalisée de X. Alors, un morphisme analytique propre surjectif μ d'une surface normale Z dans X factorise à travers ν si et seulement si les transformées strictes par μ des courbes polaires génériques de $(X,0)$ n'ont pas de point fixe.*

Rappelons qu'un point z de Z est un point fixe pour les transformées strictes par μ des courbes polaires génériques de $(X,0)$ si, en plongeant X localement en 0 dans un espace affine $\mathbf{C}^N$, il existe un ouvert dense de projections linéaires de $\mathbf{C}^N$ sur $\mathbf{C}^2$ qui induisent $\pi\colon X \to \mathbf{C}^2$ dont les courbes polaires $\Gamma(\pi)$ ont leurs transformées strictes par μ qui passent par z.

Pour résoudre les singularités rationelles minimales par une suite finie de modifications de Nash normalisées, le point crucial sera de démontrer le résultat suivant.

Théorème 7.2. *Soit $(X,0)$ un germe de singularité minimale. Soit k le nombre de sommets dans son graphe dual. Les singularités de la transformée de Nash normalisée de $(X,0)$ sont minimales et le nombre de sommets dans leur graphe dual est borné par $k/2$.*

Preuve. Tout d'abord on démontre le lemme suivant:

Lemme 7.3. *Soit $(X,0)$ un germe de singularité minimale. Les singularités de la transformée de Nash normalisée de $(X,0)$ sont minimales.*

Preuve. Dans [S] (Theorem 5.4), M. Spivakovsky donne une description explicite de la transformée stricte de la courbe polaire par la résolution minimale dans le cas où l'on considère une singularité minimale.

Pour cela on étiquette les sommets x du graphe de la résolution minimale par un entier s_x défini de la façon suivante. Si la composante D_x du diviseur exceptionnel qui correspond à x est dans la transformée stricte du projectivisé du cône tangent, i.e.

$$Z.D_x < 0$$

où Z est le cycle fondamental de la résolution minimale, alors on pose

$$s_x = 1.$$

Si $Z.D_x = 0$, on pose

$$s_x = d(x) + 1$$

où $d(x)$ désigne la distance dans le graphe de la résolution du sommet x à l'ensemble des sommets y pour lesquels on a $Z.D_y < 0$.

M. Spivakovsky définit alors les arêtes centrales comme des arêtes qui joignent des sommets x et y pour lesquels $s_x = s_y$ et les sommets centraux qui correspondent à des sommets x qui possèdent au moins deux sommets adjacents y et z pour lesquels

$$s_y = s_z = s_x - 1.$$

M. Spivakovsky montre le résultat suivant:

Proposition 7.4. *Avec les notations précédentes, la transformée stricte d'une courbe polaire générique n'intersecte que les composantes qui correspondent à un sommet central et, en leur point d'intersection, les composantes qui correspondent aux extrémités d'une arête centrale. De plus:*

i) *En la composante D_x qui correspond à un sommet central, le système linéaire des transformées strictes des courbes polaires génériques n'a pas de point fixe et D_x intersecte m_x composantes d'une transformée stricte, avec*

$$m_x = -s_x D_x.D_x - \sum_y s_y$$

où la somme s'étend à tous les sommets adjacents à x dans le graphe de la résolution minimale.

ii) *Les extrémités x et y d'une arête centrale correspondent à deux composantes D_x et D_y dont le point d'intersection est un point fixe simple du système linéaire des transformées strictes des courbes polaires génériques, i.e. on élimine ce point fixe après l'éclatement de ce point.*

Cette proposition montre que, dans la résolution $\pi := \pi_0 \circ q$ obtenue à partir de la résolution minimale π_0 d'une singularité rationnelle minimale en éclatant les points qui correspondent aux points associés aux arêtes centrales, le système linéaire des transformées strictes des courbes polaires génériques n'a pas de point fixe donc domine par un morphisme κ l'éclatement de Nash normalisé de la singularité minimale dont on est parti.

L'image par κ des composantes du diviseur exceptionnel de la résolution π qui sont intersectées par la transformée stricte de la courbe polaire générique sont des composantes du diviseur exceptionnel de la transformation de Nash normalisée.

En effet, l'image par κ de ces composantes sont évidemment les composantes intersectées par la transformée stricte de la courbe polaire générique par la transformation de Nash normalisée. Réciproquement, toute composante du diviseur exceptionnel de la transformation de Nash normalisée est intersectée par la transformée stricte de la courbe polaire générique, car, sinon on pourrait contracter cette composante et obtenir une surface normale strictement dominée biméromorphiquement par la transformation de Nash normalisée et sur laquelle le système linéaire des transformées strictes des courbes polaires génériques n'aurait aucun point fixe, ce qui contredirait la définition des transformées de Nash normalisées.

Dans la résolution π, le système linéaire des transformées strictes des courbes polaires n'a pas de point fixe, donc la résolution π domine la modification de Nash normalisée de la singularité rationnelle minimale. Les composantes du diviseur exceptionnel qui se projettent par κ sur les singularités de la modification de Nash normalisée de la singularité minimale forment donc des sous-arbres de la partie de l'arbre de la résolution π qui est la transformée stricte par q de l'arbre de résolution minimale de la singularité rationnelle minimale.

La transformée stricte par q de l'arbre de résolution minimale de la singularité rationnelle minimale a éventuellement plusieurs composantes connexes qui sont des arbres. Les sommets de ces composantes connexes ont des valences au plus égales aux valences de l'arbre de résolution minimale de la singularité rationnelle minimale dont on est parti et des poids au moins égaux aux poids originaux. Ceci montre que les sous-arbres de la transformée stricte par q de l'arbre de résolution minimale de la singularité rationnelle minimale sont des arbres de singularités minimales, d'après le Théorème 5.4. Les singularités de la modification de Nash normalisée d'une singularité rationnelle minimale sont donc aussi rationnelles minimales. □

On obtient aussi une estimation facile de la taille des arbres des résolutions minimales de ces singularités. Comme nous avons éclaté les points correspondants aux arêtes centrales le nombre de points dans l'arbre de résolution d'une singularité de la modification de Nash normalisée est majoré par $k/2$, si k est le nombre de composantes exceptionnelles de la résolution minimale de la singularité rationnelle minimale dont on est parti. Ceci démontre donc le Lemme 7.3 et le Théorème 7.2. □

Remarque 7.5. Une démonstration analogue à la précédente établit que les singularités de la transformation de Nash normalisée d'une singularité rationnelle sont rationnelles, mais on ne connait pas d'estimation de la complexité de ces singularités en fonction de celle de la singularité rationnelle considérée.

8. La résolution des singularités des surfaces via la transformée de Nash

Dans [H], Heisuke Hironaka a démontré:

Théorème 8.1. *Soit S une surface algébrique complexe normale. Soit*

$$\pi: X \to S$$

la résolution minimale de S. Après un nombre fini de modifications de Nash normalisées

$$\nu \circ \cdots \circ \nu_{i-1}: S^{(i)} \to S^{(i-1)} \to \cdots \to S^{(1)} \to S$$

l'application birationnelle de $S^{(i)}$ dans X est régulière.

Ce théorème implique, en particulier:

Corollaire 8.2. *Les singularités de $S^{(i)}$ sont Sandwich.*

Preuve. On peut appliquer le Théorème 4.13 ou bien faire le raisonnement suivant.

Soit $p\colon X^{(i)} \to S^{(i)}$ une résolution de $S^{(i)}$. La composée $\nu \circ \cdots \circ \nu_{i-1} \circ p$ est birationnelle et on a une application birationnelle régulière unique $P_i\colon X^{(i)} \to X$ telle que

$$\pi \circ P_i = \nu \circ \cdots \circ \nu_{i-1} \circ p.$$

Comme les espaces $X^{(i)}$ et X sont non singuliers, cette application P_i est une suite d'éclatements de points. Cette composée est donc la résolution plongée d'un nombre fini de germes de courbes holomorphes planes. On en déduit facilement que les arbres de résolution des singularités de $S^{(i)}$ sont alors sous-arbres des arbres de résolution de ces germes de courbes planes. D'après le iv) du théorème 4.10, on en déduit que les singularités de $S^{(i)}$ sont Sandwich. □

En fait, un premier résultat remarquable de Spivakovsky est:

Théorème 8.3. *Soit S une surface normale. Si l'entier i est assez grand, si l'on compose i modifications de Nash normalisées*

$$S^{(i)} \to S^{(i-1)} \to \cdots \to S^{(1)} \to S$$

les singularités de $S^{(i)}$ sont minimales.

Preuve. La preuve de Spivakovsky est basée sur le Lemme suivant.

Lemme 8.4. *Soit $\nu\colon \tilde{S} \to S$ la modification de Nash normalisée d'une surface holomorphe normale qui n'a qu'une seule singularité 0 qui est rationnelle. Soient $\pi\colon X \to S$ et $\tilde{\pi}\colon \tilde{X} \to \tilde{S}$ les résolutions minimales de S et $\tilde{S}$. On a alors une application birationnelle unique $N\colon \tilde{X} \to X$ telle que $\pi \circ N = \nu \circ \tilde{\pi}$. Dans ce cas une composante du diviseur exceptionnel dans $\tilde{X}$ d'une singularité de $\tilde{S}$ est mauvaise si elle est transformée stricte d'une composante du diviseur exceptionnel dans X qui est aussi mauvaise.*

Rappelons qu'on dit qu'une composante du diviseur exceptionnel dans la résolution minimale d'une singularité rationnelle est mauvaise, si la valence de cette composante égale son poids plus 1. On sait qu'une singularité rationnelle dont le diviseur exceptionnel dans sa résolution minimale n'a aucune composante mauvaise est minimale (voir Théorème 5.4).

Preuve. Nous n'allons donner qu'une esquisse de preuve du Lemme 8.4. On remarque que N n'est autre que la suite finie d'éclatements de points qui consiste à éliminer les points fixes du système des courbes polaires sur X. M. Spivakovsky remarque que ces points fixes ne sont autres que les points de X où le faisceau $\tilde{\Omega}$, quotient de $\pi^*\Omega_S^2$ par son idéal de torsion, n'est pas localement principal. De plus N est la modification minimale pour que le quotient de $(\pi \circ N)^*\Omega_S^2$ par son idéal de torsion soit inversible sur $\tilde{X}$. Pour trouver N, M. Spivakovsky montre que:

Théorème 8.5. *Soit α un point de X où $\tilde{\Omega}$ n'est pas inversible. Alors*

i) *Si α est un point non singulier du diviseur exceptionnel, il existe des coordonnées locales (u, v) en α, où $u = 0$ est une équation locale du diviseur exceptionnel, et un entier ℓ tels que*

$$\tilde{\Omega} \simeq (u^\ell, v)\mathcal{O}_X$$

localement en α;

ii) *Si α est un point du diviseur exceptionnel qui est l'intersection de deux composantes D_1 et D_2, il existe des coordonnées locales (u, v) en α, où $u = 0$ et $v = 0$ sont des équations locales de D_1 et D_2, telles que $\tilde{\Omega}$ est isomorphe à un idéal de $\mathcal{O}_X$ engendré par un nombre fini de monômes $u^a v^b$.*

Le cas i) donne une suite d'éclatements en ligne et le cas ii) introduit un bambou de composantes entre les transformées strictes de D_1 et D_2 dans $\tilde{X}$. □

Nous sommes alors en mesure de montrer le Théorème 8.3. Soit D une mauvaise composante dans le diviseur exceptionnel de la résolution minimale $\pi: X \to S$ de S. Soit i un entier assez grand pour que l'on ait une application régulière birationnelle Q_i de $S^{(i)}$ dans X. La transformée stricte de D par cette application birationnelle Q_i est une courbe dans $S^{(i)}$, donc la transformée stricte D_i de D par le morphisme P_i (défini dans la preuve du Corollaire 8.2) n'est donc pas une composante du diviseur exceptionnel de la résolution de $S^{(i)}$.

En appliquant le Lemme 8.4 par récurrence sur i, on montre qu'une mauvaise composante $D^{(i)}$ du diviseur exceptionnel de la résolution $p: X^{(i)} \to S^{(i)}$ est la transformée stricte par P_i d'une mauvaise composante D dans X de la résolution minimale π de S. Ceci contredit le fait observé ci-dessus que la transformée stricte $D_i = D^{(i)}$ de D par P_i n'est pas une composante du diviseur exceptionnel de la résolution de $S^{(i)}$. Donc le diviseur exceptionnel de la résolution p ne contient pas de mauvaise composante et les singularités de $S^{(i)}$ sont donc minimales. □

On est ramené à montrer qu'une singularité rationnelle qui est minimale peut être résolue par une suite finie de modification de Nash normalisée. Ce résultat est obtenu grâce au Théorème 7.2 qui montre que le nombre de composantes dans le diviseur de la résolution minimale d'une singularité rationnelle minimale diminue au moins de moitié après une modification de Nash normalisée.

References

[A] M. Artin, *On isolated rational singularities of surfaces*, Amer. J. Math. **88** (1966), 129–136.

[BPV] W. Barth, C. Peters, A. Van de Ven, Compact complex surfaces, Ergebnisse der Math. und ihrer Grenzgebiete 3. Folge, Band 4, Springer Verlag (1984).

[Be] A. Beauville, Surfaces algébriques complexes, Astérisque **54**, Soc. Math. Fr. (1984).

[B] E. Brieskorn, *Rationale Singularitäten komplexer Flächen*, Inv. Math. **4** (1968), 336–358.

[GS1] G. Gonzalez-Sprinberg, *Eventails en dimension deux et transformée de Nash*, Notes miméographiées Ec. Norm. Sup. Paris (1977), 67p.

[GS2] G. Gonzalez-Sprinberg, *Résolution de Nash des points doubles rationnels*, Ann. Inst. Fourier **32** (1982), 111–178.

[GS3] G. Gonzalez-Sprinberg, *Désingularisation des surfaces par des modifications de Nash normalisées*, Sém. Bourbaki, vol. 1985/1986, Astérisque **145–146** (1987), 187–207.

[G] H. Grauert, *Über Modifikationen und exzeptionnelle analytische Mengen*, Math. Ann. **146** (1962), 331–368.

[Ha] R. Hartshorne, Algebraic Geometry, GTM **52**, Springer Verlag (1977).

[H] H. Hironaka, *On Nash blowing-up*, in Arithmetic and Geometry ed. by M. Artin and J. Tate, in Prog. in Math. **36** (1983), 103–111, Birkhäuser.

[Ho] C. Houzel, *Géométrie analytique locale I, II, III, IV*, in Sém. H. Cartan, 1960–1961, Notes miméographiées de l'Inst. H. Poincaré (1962), Paris.

[JLT] M. Jalabert-Lejeune, B. Teissier, *Clôture intégrale des idéaux et équisingularité*, Séminaire à l'Ecole Polytechnique 1973–74, Publ. Inst. Fourier, Grenoble. A paraître chez Hermann, Coll. Actualités Mathématiques.

[K] J. Kollár, *Toward moduli of singular varieties*, Comp. Math. **56** (1985), 369–398.

[L] H. Laufer, Normal Surface Singularities, Ann. Math. Stud. **71**, Princeton University Press (1971), Princeton.

[LêT] Lê Dũng Tráng, B. Teissier, *Variétés polaires et classes de Chern des variétés singulières*, Ann. Math. **114** (1981), 457–491.

[LêW] Lê Dũng Tráng, C. Weber, *Equisingularité dans les pinceaux de germes de courbes planes et C^0 suffisance*, Ens. Math. **43** (1997), 355–380.

[Li] J. Lipman, *Rational singularities with applications to algebraic surfaces and unique factorization*, Publ. Math. IHES **36** (1969), 195–279.

[LiT] J. Lipman, B. Teissier, *Pseudo-rational rings and a theorem of Briançon-Skoda about integral closure of ideals*, Mich. J. Math. **28** (1981), 97–116.

[M] D. Mumford, *The topology of normal singularities of an algebraic surface and a criterion for simplicity*, Publ. Math. IHES **9** (1961).

[Na] M. Nagata, Local Rings, Interscience Tracts **13**, John Wiley & Sons (1962), New York.

[No] A. Nobile, *Some properties of Nash blowing-up*, Pacific J. Math. **60** (1975), 297–305.

[R] V. Rebassoo, Desingularization properties of the Nash blowing-up process, Thesis, University of Washington (1977).

[Re] D. Rees, *a-transformation of local rings and a theorem on multiplicities of ideals*, Proc. Camb. Phil. **57** (1961), 8–17.

[Sn] J. Snoussi, *Limites d'espaces tangents à une surface normale*, C. R. Acad. Sc. **327** (1998), 369–372.

[S] M. Spivakovsky, *Sandwiched singularities and desingularization of surfaces by normalized Nash transformations*, Ann. Math. **131** (1990), 411–491.

[SZ] P. Samuel, O. Zariski, Commutative Algebra, Van Nostrand 1960.

[Te] B. Teissier, *The hunting of invariants in the geometry of discriminants*, Nordic Summer School, Oslo 1976, ed. by Per Holm, Nordhoff & Sijthoff (1977), 565–677.

[T] G. N. Tiourina, *Absolute isolatedness of rational singularities and triple rational points*, Funk. Analiz i ievo Prilojenia **2** (1968), 70–81.

[Z] O. Zariski, *The reduction of singularities of an algebraic surface*, Ann. Math. **40** (1939), 639–689.

CMI-Université de Provence
39 Rue Joliot Curie,
13453 Marseille CEDEX 13, France
`ledt@gyptis.univ-mrs.fr`

Progress in Mathematics, Vol. 181,

Equisingularity and Simultaneous Resolution of Singularities

Joseph Lipman

Abstract. Zariski defined equisingularity on an n-dimensional hypersurface V via stratification by "dimensionality type," an integer associated to a point by means of a generic local projection to affine n-space. A possibly more intuitive concept of equisingularity can be based on stratification by simultaneous resolvability of singularities. The two approaches are known to be equivalent for families of plane curve singularities. In higher dimension we ask whether constancy of dimensionality type along a smooth subvariety W of V implies the existence of a simultaneous resolution of the singularities of V along W. (The converse is false.)

The underlying idea is to follow the classical inductive strategy of Jung – begin by desingularizing the discriminant of a generic projection – to reduce to asking if there is a canonical resolution process which when applied to quasi-ordinary singularities depends only on their characteristic monomials. This appears to be so in dimension 2. In higher dimensions the question is quite open.

1. Introduction – equisingular stratifications

The purpose, mainly expository and speculative, of this paper – an outgrowth of a survey lecture at the September 1997 Obergurgl working week – is to indicate some (not all) of the efforts that have been made to interpret *equisingularity*, and connections among them; and to suggest directions for further exploration.

The term "equisingularity" has various connotations. Common to these is the idea of stratifying an algebraic or analytic $\mathbb{C}$-variety V in such a way that along each stratum the points are, as singularities of V, equivalent in some pleasing sense, and somehow get worse as one passes from a stratum to its boundary. A stratification of V is among other things a partition into a locally finite family of submanifolds, the strata (see §2), so that whatever "equivalence" is taken to mean, there should be, locally on V, only finitely many equivalence classes of singularities.

Locally along any stratum V gives rise to families of singularities, realized as the fibers of a retraction onto the stratum. Accordingly one also thinks about equisingularity in terms of some kind of equivalence of the individual singularities in a family. Equivalence may be specified, for example, by equality of some numerical invariants.

Here are some simple suggestive examples, and a little more background.

Examples. (a) The surface V in $\mathbb{C}^3$ given by the equation $X^3 = TY^2$ has singular locus $L\colon X = Y = 0$ (the T-axis), along which it has multiplicity 2 except at the origin O, where it has multiplicity 3. So there is something quite special about O. Interpreting V along L as a family of curve germs in the (X, Y)-plane, with parameter T, makes it plausible that the equisingular strata ought to be $V \setminus L$, $L \setminus \{O\}$, and $\{O\}$. Changing coordinates changes the family, but not the generic member, and not the fact that something special happens at the origin.

(b) Replacing X^3 by X^2, we get multiplicity 2 everywhere along L, including O. The corresponding family of plane curve germs consists of a pair of intersecting lines degenerating to a double line; so there is still something markedly special about O – a feeling reinforced by the following picture.

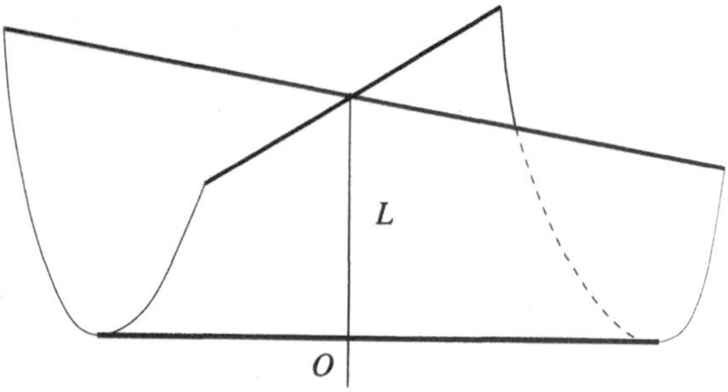

This indicates that equimultiplicity along L, while presumably *necessary*, is *not sufficient* for equisingularity.

(c) The surface $V\colon X(X + Y)(X - Y)(TX + Y) = 0$, along its singular locus $L\colon X = Y = 0$, can be regarded as a family of plane-curve germs all of which look topologically the same, namely four distinct lines through a point. Thus we have *topological equisingularity* along L. (In this example the homeomorphisms involved can be made piecewise linear, hence *bi-Lipschitz;* see e.g., [T2, pp. 349–361], [Z3, Introduction, §2, pp. 4–7], and [P] for more on Lipschitz equisingularity.) From this point of view, the equisingular strata should be $V \setminus L$ and L. However, from the analytic – or even differentiable – point of view, all these germs are distinct, since the cross-ratio of the four lines varies with T. Thus differential isomorphism is too stringent a condition for equisingularity – there are too many equivalence classes.

(d) The notion of *Whitney-equisingularity* (alias *differential equisingularity*) of V along L at a smooth point x of L is associated with the Whitney conditions $\mathfrak{W}(V, L)$ holding at x. These conditions signify that if $y \in L$ and $z \in V \setminus \mathrm{Sing}(V)$ (where $\mathrm{Sing}(V)$ is the singular locus of V) approach x in such a way that the tangent space $T_{V,z}$ and the line joining y to z both have limiting positions, then the limit of the tangent spaces contains the limit of the lines. (The conditions can be described intrinsically, i.e., as a condition on the prime ideal of L in the local ring $\mathcal{O}_{V,x}$, see proof of Theorem 3.2 below.)

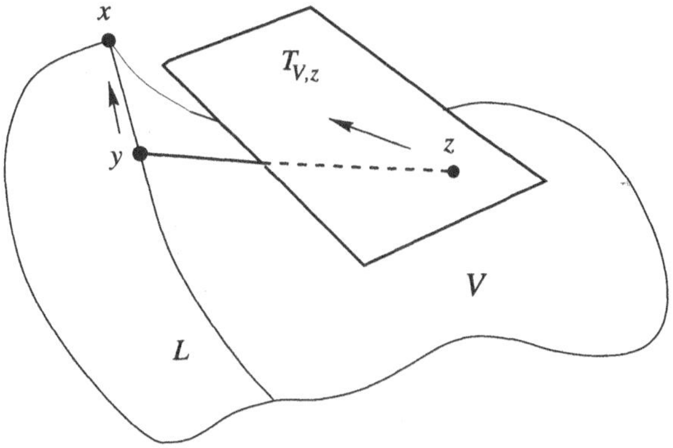

The conditions fail in example (b) above for $y = (X_y, Y_y, T_y) = (0, 0, t^{1/2})$ and $z = (X_z, Y_z, T_z) := (t^2, t, t^2)$, the limiting tangent space as $t \to 0$ being the plane $T = 2X$, while the limiting line is just L: $X = Y = 0$.

The conditions $\mathfrak{W}(V, L)$ by themselves are not enough: on the surface in $\mathbb{C}^3$ given by $X^2 = T^2Y^3$, $\mathfrak{W}(V, L_i)$ holds for $L_1: X = T = 0$ and $L_2: X = Y = 0$, but topological equisingularity fails at the origin along either of these components of the singular locus [Z1, p. 488]. One should require that a neighborhood of x in L be contained in a stratum of some Whitney stratification – a stratification such that $\mathfrak{W}(\overline{S_1}, S_2)$ holds everywhere along S_2 for any strata S_1, S_2 with $\overline{S_1} \supset S_2$.

Indeed, Thom and Mather showed that a Whitney stratification of V is topologically equisingular in the following sense: for any point x of a stratum S, the germ (V, x) together with its induced stratification is topologically the product of (S, x) with a stratified germ $(V_0, x) \subset (V, x)$ in which x is a stratum by itself (see [M1, p. 202, (2.7), and p. 220, Corollary (8.4)]).

(e) The obvious formulation of a converse to the Thom-Mather theorem is false: in [BS], Briançon and Speder showed that the family of surface germs

$$Z^5 + TY^6Z + Y^7X + X^{15} = 0$$

(each member, for small T, having an isolated singularity at the origin) is topologically equisingular, but not differentially so. On the other hand, there is a beautiful converse, due to Lê and Teissier, equating topological and Whitney-equisingularity of the totality of members of the family $(V \cap H, L, x)$ as H ranges over general smooth germs containing L at x (see [LT, §5], [T2, p. 480, §4]).

Though not the principal focus of this paper, the approach to equisingularity via $\mathfrak{W}$ is the most extensively explored one. Whitney stratifications are of basic importance, for example, in the classification theory of differentiable maps ([M1], [M2], [GL]), in D-module theory and the solution of the Riemann-Hilbert problem ([LM]), and in the Goresky-Macpherson theory of intersection homology ([Mc], [GMc]). (See also [DM], for the existence of Whitney stratifications in a general class of geometric categories.)

Hironaka proved that a Whitney stratification is *equimultiple:* for any two strata S_1, S_2, the closure $\overline{S_1}$ has the same multiplicity (possibly 0) at every point of S_2 [H1, p. 137, 6.2]. Work of Lê and Teissier in the early 1980's led to characterizations of Whitney-equisingularity by the constancy of a finite sequence of "polar multiplicities" [T2], [T3]. In the 1990's, work by Gaffney and others in the context of subfamilies of families of complete intersection singularities has resulted in characterizations of Whitney-equisingularity in families via constancy of sequences of Milnor numbers of the fibers, or constancy of multiplicities of Jacobian modules. (For this, and much more, see, e.g., [GM], [K]).

It is therefore hard to envision an acceptable definition of equisingularity which does not at least *imply* Whitney-equisingularity. But one may still wish to think about conditions which reflect the analytic – not just the differential – structure of V. Indeed, §§3–5 of this paper are devoted to describing two plausible formulations of "analytic equisingularity" and to comparing them with each other and with Whitney-equisingularity.

In brief – and informally – we regard equisingularity theory as being *the study of conditions which give rise to a satisfying notion of "equisingular stratification," and of connections among them.*

In practice, such conditions on a variety-germ V often reflect local constancy of some numerical characters of singularities in a family whose total space is V.

2. Stratifying conditions

To introduce more precision into the preceding vague indications about equisingularity (see e.g., 2.8), we spend a few pages on basic remarks about stratifications. Knowledgeable readers may prefer going directly to §3.

There is a good summary of the origins of stratification theory in pp. 33–44 of [GMc]. What follows consists mostly of variants of material in [T2, pp. 382–402], and is straightforward to verify. The main result, Proposition 2.7, generalizes [T2, pp. 478–480] (which treats the case of Whitney stratifications).

2.1. We work either in the category of reduced complex-analytic spaces or in its subcategory of reduced finite-type algebraic $\mathbb{C}$-schemes; in either case we refer to objects V simply as "varieties." For any such V the set $\mathrm{Sing}(V)$ of singular (non-smooth) points is a closed subvariety of V. A *locally closed subvariety of* V is a subset W each of whose points has an open neighborhood $U \subset V$ whose intersection with W is a closed subvariety of U. Such a W is open in its closure.

By a *partition* (analytic, resp. algebraic) of a variety V we mean a locally finite family (P_α) of non-empty subsets of V such that for each α both the closure $\overline{P_\alpha}$ and the boundary $\partial P_\alpha := \overline{P_\alpha} - P_\alpha$ are closed subvarieties of V (so that P_α is a locally closed subvariety of V), such that $P_{\alpha_1} \cap P_{\alpha_2} = \emptyset$ whenever $\alpha_1 \neq \alpha_2$, and such that $V = \cup_\alpha P_\alpha$. The P_α will be referred to as the *parts* of the partition.

For example, a partition may consist of the fibers of an upper-semicontinuous function μ from V into a well-ordered set I, where "upper-semicontinuous" means that for all $i \in I$, $\{x \in V \mid \mu(x) \geq i\}$ is a closed subvariety of V.

We say that a partition $\mathfrak{P}_1$ *refines* a partition $\mathfrak{P}_2$, or write $\mathfrak{P}_1 \prec \mathfrak{P}_2$, if, $\mathfrak{P}_1$ and $\mathfrak{P}_2$ being identified with equivalence relations on V – subsets of $V \times V$ – we have $\mathfrak{P}_1 \subset \mathfrak{P}_2$. So $\mathfrak{P}_1 \prec \mathfrak{P}_2$ iff the following (equivalent) conditions hold:

(i) Every $\mathfrak{P}_1$-part is contained in a $\mathfrak{P}_2$-part.

(i)′ With $\mathfrak{P}_x$ denoting the $\mathfrak{P}$-part containing x, $\mathfrak{P}_{1,x} \subset \mathfrak{P}_{2,x}$ for all $x \in V$.

(ii) Every $\mathfrak{P}_2$-part is a union of $\mathfrak{P}_1$-parts.

Let $\mathfrak{P}$ be a partition of V, and let $W \subset V$ be an *irreducible* locally closed subvariety. For any $\mathfrak{P}$-part P, $P \cap W = \overline{P} \cap W - \partial P \cap W$ is the difference of two closed subvarieties of W, so its closure in W is a closed subvariety of W, either equal to W or of lower dimension than W; and W is the union of the locally finite family of all such closures, hence equal to one of them. Consequently there is a unique part – denoted $\mathfrak{P}_W$ – whose intersection with W is dense in W. Moreover, $W - \mathfrak{P}_W = \partial\mathfrak{P}_W \cap W$ is a proper closed subvariety of W.

2.2. Condition (ii) in the following definition of stratification is non-standard – but suits a discussion of equisingularity (and may be necessary for Proposition 2.7).

Definition. A *stratification* (analytic, resp. algebraic) of V is a decomposition into a locally finite disjoint union of non-empty, connected, locally closed subvarieties, the *strata*, satisfying:

(i) For any stratum S, with closure $\overline{S}$, the boundary $\partial S := \overline{S} - S$ is a union of strata (and hence is a closed subvariety of V).

(ii) For any $\overline{S}$ as in (i), $\mathrm{Sing}(\overline{S})$ is a union of strata.

From (i) it follows that a stratification is a partition, whose parts are the strata. Noting that a subset $Z \subset V$ is a union of strata iff Z contains every stratum which it meets, and that $\mathrm{Sing}(\overline{S})$ is nowhere dense in $\overline{S}$, we deduce:

(iii) Every stratum is smooth.

The strata of codimension i are called i-strata.

For stratifications $\mathfrak{S}_1, \mathfrak{S}_2$, we have $\mathfrak{S}_1 \prec \mathfrak{S}_2 \Leftrightarrow$ the closure of any $\mathfrak{S}_2$-stratum is the closure of a $\mathfrak{S}_1$-stratum.

2.3. One way to give a stratification is via a filtration

$$V = V_0 \supset V_1 \supset V_2 \supset \cdots$$

by closed subvarieties such that:

(1) For all $i \geq 0$, V_{i+1} contains $\mathrm{Sing}(V_i)$, but contains no component of V_i – in other words, $V_i \setminus V_{i+1}$ is a dense submanifold of V_i.

The strata are the connected components of $V_i \setminus V_{i+1}$ $(i \geq 0)$, and their closures are the irreducible components of the V_i. (Note that (1) forces the germs of the V_i

at any $x \in V$ to have strictly decreasing dimension, whence $\cap V_i$ is empty.) It follows that a closed subset Z of V is a union of strata $\Leftrightarrow$ for any component W of any V_i, if $W \not\subset Z$ then $W \cap Z \subset V_{i+1}$. So for (i) and (ii) above to hold we need:

(2) If W, W' are irreducible components of V_i, V_j respectively and $W \not\subset W'$ then $W \cap W' \subset V_{i+1}$, and if $W \not\subset \mathrm{Sing}(W')$ then $W \cap \mathrm{Sing}(W') \subset V_{i+1}$.

For any stratification, let $0 = n_0 < n_1 < n_2 < \ldots$ be the integers occurring as codimensions of strata. Redefine $V_i :=$ union of all strata of codimension $\geq n_i$, thereby obtaining a filtration which gives back the stratification and which satisfies:

(3) There is a sequence of integers $0 = n_0 < n_1 < n_2 < \ldots$ such that V_i has pure codimension n_i in V.

Thus there is a one-one correspondence between the set of stratifications and the set of filtrations which satisfy (1), (2) and (3).

In the following illustration, a tent which should be imagined to be bottomless and also to stretch out infinitely in both horizontal directions, V_1 can be taken to be the union of the ridges (labeled E, F), and V_2 to be the vertex O.

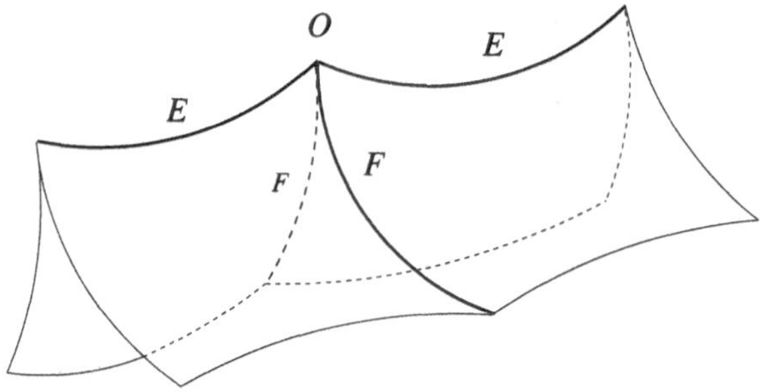

To say that an irreducible subvariety W is not contained in a closed subvariety Z is to say that $W \cap Z$ is nowhere dense in W. It follows that the conditions (1), (2) and (3) are local: they hold for a filtration if and only if they hold for the germ of that filtration at every point $x \in V$ – and indeed, they can be checked inside the complete local ring $\widehat{\mathcal{O}_{V,x}}$. Moreover if a filtration $\mathfrak{F}_x$ of a germ (V, x) satisfies these conditions, then there is such a filtration $\mathfrak{F}_N$ of an open neighborhood N of x in V, whose germ at x equals $\mathfrak{F}_x$. Thus there is a *sheaf of stratifications* whose stalk at x can be identified with the set of all filtrations $\mathfrak{F}$ of (V, x) satisfying (1), (2) and (3).

2.4. Stratifications can be determined by *local stratifying conditions*, as follows. We consider conditions $C = C(W_1, W_2, x)$ defined for all $x \in V$ and all pairs $(W_1, x) \supset (W_2, x)$ of subgerms of (V, x) with (W_1, x) *equidimensional* (all components of W_1 containing x have the same dimension) and (W_2, x) *smooth*. Such a pair can be thought of as two radical ideals $p_1 \subset p_2$ in the local ring $R := \mathcal{O}_{V,x}$, with R/p_1 equidimensional and R/p_2 regular. For example, $C(W_1, W_2, x)$ might be defined to mean that W_1 is equimultiple along W_2 at x – i.e., the local rings

R/p_1 and $(R/p_1)_{p_2}$ have the same multiplicity. Or, $C(W_1, W_2, x)$ could signify that the Whitney conditions $\mathfrak{W}(W_1, W_2)$ hold at x, these conditions being expressible within $\mathcal{O}_{V,x}$ (see §1, Example (d)).

(Recall that there is a natural equivalence between the category of local analytic $\mathbb{C}$-algebras and the category of complex analytic germs – the category of pointed spaces localized with respect to open immersions, i.e., enlarged by the adjunction of formal inverses for all open immersions. A similar equivalence holds in the algebraic context.)

For such a C and for any subvarieties W_1, W_2 of V with W_1 closed and locally equidimensional, and W_2 locally closed, set

$$C(W_1, W_2) := \{\, x \in W_2 \mid W_2 \text{ smooth at } x, \text{ and if } x \in W_1 \text{ then } \\ ((W_1, x) \supset (W_2, x) \text{ and } C(W_1, W_2, x))\,\}.$$

$$\not{C}(W_1, W_2) := W_2 - C(W_1, W_2).$$

For example, if W_1 contains no component of W_2 then

$$\not{C}(W_1, W_2) = \mathrm{Sing}(W_2) \cup (W_1 \cap W_2).$$

The condition C is called *stratifying* if for any such W_1 and W_2, $\not{C}(W_1, W_2)$ is contained in a nowhere dense closed subvariety of W_2. (It suffices that this be so whenever W_2 is smooth, connected, and contained in W_1.)

For any $S \subset V$, denote by S^{Zar} the Zariski closure of S, i.e., the smallest closed subvariety of V which contains S. If S is a union of $\mathfrak{S}$-strata for some stratification $\mathfrak{S}$, then S^{Zar} is just the topological closure of S.

A stratification $\mathfrak{S}$ is a *C-stratification* if for any strata S_1, S_2 with $S_2 \subset \overline{S_1}$ it holds that $\not{C}(\overline{S_1}, \overline{S_2})^{\mathrm{Zar}}$ is a union of strata. For *any* two strata S_1, S_2 of a C-stratification, $\not{C}(\overline{S_1}, \overline{S_2})^{\mathrm{Zar}}$ is a union of strata, since

$$\overline{S_1} \not\supset S_2 \implies \not{C}(\overline{S_1}, \overline{S_2}) = \mathrm{Sing}(\overline{S_2}) \cup (\overline{S_1} \cap \overline{S_2});$$

and if moreover C is stratifying then $C(\overline{S_1}, S_2) = S_2$.

For given stratifying C, the condition "$C(\overline{S_1}, S_2) = S_2$ for all $\mathfrak{S}$-strata S_1, S_2" might not by itself ensure that $\mathfrak{S}$ is a C-stratification. If every $\mathfrak{S}$ for which it holds *is* a C-stratification then we'll say that C is a *good* stratifying condition. (See, for instance, Example (d) below.) If C is good and if C' is any stratifying condition which implies C, then every C' stratification is a C-stratification.

We will show in 2.6 below that for each stratifying condition C there exists a C-stratification. In fact there exists a *coarsest* C-stratification – one which is refined by all others (see 2.7).

2.5. Examples. (a) If C is the empty condition (i.e., $C(W_1, W_2, x)$ holds for all pairs $(W_1, x) \supset (W_2, x)$ as in §2.4) then C is a good stratifying condition, and every stratification is a C-stratification.

(b) The logical conjunction ("and") of a finite family of stratifying conditions, $\hat{C} := \wedge_{i=1}^{n} C_i$, is again stratifying. If a stratification $\mathfrak{S}$ is a C_i-stratification for all i

then $\mathfrak{S}$ is a $\hat{C}$-stratification; but the converse does not always hold. It does hold if each C_i is good – in which case $\hat{C}$ is good too.

(c) Let $\mathfrak{P} = (P_\alpha)$ be a partition of V. For $x \in V$ let α_x be such that $x \in P_{\alpha_x}$. Define $C_{\mathfrak{P}}(W_1, W_2, x)$ to mean that $(W_2, x) \subset (P_{\alpha_x}, x)$. (In this example W_1 plays no role.) Then $C_{\mathfrak{P}}$ is stratifying: as in 2.1, for any irreducible locally closed subvariety $W_2 \subset V$ there is a unique α such that $W_2 - P_\alpha$ is a proper closed subvariety of W_2, and then for any closed locally equidimensional $W_1 \supset W_2$ we have $\not{C}_{\mathfrak{P}}(W_1, W_2) = (W_2 - P_\alpha) \cup \operatorname{Sing}(W_2)$, a nowhere dense closed subvariety of W_2. The condition $C_{\mathfrak{P}}(\overline{S_1}, S_2)$ on a pair of strata of a stratification $\mathfrak{S}$ means that the part $\mathfrak{P}_{S_2}$ of $\mathfrak{P}$ which meets S_2 in a dense subset actually contains S_2, and hence is the only part of $\mathfrak{P}$ meeting S_2. One deduces that *a $C_{\mathfrak{P}}$-stratification is just a stratification which refines $\mathfrak{P}$.*

Exercise. For two partitions $\mathfrak{P}$, $\mathfrak{P}'$, $(C_{\mathfrak{P}'} \Rightarrow C_{\mathfrak{P}}) \iff \mathfrak{P}' \prec \mathfrak{P}$.

(d) More generally, suppose given a partition $\mathfrak{P}_W$ of W for every equidimensional locally closed subvariety $W \subset V$. (For example, if (P_α) is a partition of V, one can set $\mathfrak{P}_W := (W \cap P_\alpha)$.) With $C_W := C_{\mathfrak{P}_W}$ as in Example (c), define $C(W_1, W_2, x)$ to mean $C_{W_1}(W_1, W_2, x)$. Arguing as before, one finds that C is a stratifying condition on V, and that a C-stratification is one such that, if S is any stratum then each part of $\mathfrak{P}_{\overline{S}}$ is a union of strata – a condition which holds if and only if for any two strata S_1, S_2, we have $C(\overline{S_1}, S_2) = S_2$. Thus this C is a good stratifying condition.

(e) The Whitney conditions $\mathfrak{W}(W_1, W_2, x)$ (see §2.4) are stratifying. This was first shown, of course, by Whitney, [Wh, p. 540, Lemma 19.3]. It was shown much later, by Teissier, that $\not{\mathfrak{W}}(W_1, W_2)$ itself is analytic ([T2, p. 477, Prop. 2.1]. Indeed, the main result in [T2], Theorem 1.2 on p. 455, states in part that the stratifying condition $\mathfrak{W}$ is of the type described in example (d) above, the partition $\mathfrak{P}_W$ being given by the level sets of the polar multiplicity sequence on W.

Proposition-Definition 2.6. *For any stratifying condition C, the following inductively defined filtration of V gives rise, as in 2.3, to a C-stratification $\mathfrak{S}_C$: $V_0 = V$, and for $i > 0$, V_{i+1} is the union of all the $W^* \subset V_i$ such that*

(a) *W^* is a component either of $\operatorname{Sing}(V_j)$ for some $j \le i$ or of $\not{C}(W', W)^{\mathrm{Zar}}$ for some components W', W of V_j, V_k respectively, with $j \le k \le i$; and*
(b) *W^* is not a component of V_i.*

Proof. It is clear that V_{i+1} contains every component of $\operatorname{Sing}(V_i)$, but no component of V_i. If W and W' are components of V_i and V_j respectively, and if $j > i$, then $W \cap W' \subset V_{i+1}$; while if $j \le i$ and $W \not\subset W'$ then $W \cap W' \subset \not{C}(W', W) \subset V_{i+1}$, since by the definition of stratifying condition, no component of $\not{C}(W', W)^{\mathrm{Zar}}$ is a component of V_i. So, strata being connected components of $V_i \setminus V_{i+1}$ $(i \ge 0)$, the closure of any stratum – i.e., any component of any V_i – is a union of strata (see 2.3). Now any component W^* of $\operatorname{Sing}(V_j)$ is a component of V_i for some $i > j$: otherwise, by the definition of V_i, $W^* \subset \cap V_i = \emptyset$ (see 2.3). Hence $\operatorname{Sing}(V_j)$ is a union of strata. Similarly, if $W' \supset W$ are closures of strata, and so components of V_j, V_k respectively, with $j \leq k$, then $\mathrm{Sing}(W', W)^{\mathrm{Zar}}$ is a union of strata. Thus the filtration does indeed give a C-stratification. □

Remarks.

1. If $C(W, W, x)$ holds for every smooth point x of every component W of V, then $V_1 = \mathrm{Sing}(V)$.
2. Any stratification $\mathfrak{S}$ is $\mathfrak{S}_C$ for a good C, viz. $C := C_{\mathfrak{S}}$ (Example 2.5(c)). This results at once from Proposition 2.7 below. Or, if $\mathfrak{V}$: $V = V_0^* \supset V_1^* \supset \dots$ is the filtration associated to $\mathfrak{S}$, satisfying conditions (1), (2) and (3) in 2.3, so that $V_i^* \subset V$ has pure codimension, say n_i, then one can check for irreducible components W, W' of V_i^*, V_j^* respectively, that
$$\not{C}(W', W) = \not{C}(W', W)^{\mathrm{Zar}} \subset W \cap V_{i+1}^* = \not{C}(W, W),$$
whence a component of $\not{C}(W', W)$ is a component of V_ℓ^* iff it is the closure of an n_ℓ-stratum on the boundary of W; and it follows that the filtration described in Proposition-Definition 2.6 is identical with $\mathfrak{V}$.

Proposition 2.7. *For any stratifying condition C, $\mathfrak{S}_C$ (defined in 2.6) is the coarsest C-stratification of V – every C-stratification refines $\mathfrak{S}_C$. In particular, if C is* good (§2.4) *then $\mathfrak{S}_C$ is the coarsest among all stratifications with $C(\overline{S_1}, S_2) = S_2$ for any two strata S_1, S_2.*

Proof. Let $V = V_0 \supset V_1 \supset V_2 \supset \dots$ be as in 2.6, and let $\mathfrak{S}$ be a C-stratification of V. The assertion to be proved is: *If Z is an irreducible component of V_j ($j \geq 0$), then $Z = \overline{\mathfrak{S}_Z}$, the closure of $\mathfrak{S}_Z$.* (Recall from 2.1 that $\mathfrak{S}_Z$ is the unique stratum containing a dense open subset of Z, and see the last assertion in 2.2.)

For $j = 0$, it is clear that $Z = \overline{\mathfrak{S}_Z}$. Assume the assertion for all $j \leq i$. Let Z be a component of V_{i+1}. By 2.6, Z is a component either of (a): $\mathrm{Sing}(V_j)$ ($j \leq i$) or of (b): $\not{C}(W', W)^{\mathrm{Zar}}$ with W' a component of V_j and W a component of V_k ($j \leq k \leq i$).

In case (a), the inductive hypothesis gives that every component W'' of V_j is the closure of a $\mathfrak{S}$-stratum, so both W'' and $\mathrm{Sing}(W'')$ are unions of $\mathfrak{S}$-strata (see 2.2); and it follows easily that $\mathrm{Sing}(V_j)$ is a union of $\mathfrak{S}$-strata, one of which must be $\mathfrak{S}_Z$ (because $\mathrm{Sing}(V_j)$ meets $\mathfrak{S}_Z$). Since $Z \subset \overline{\mathfrak{S}_Z} \subset \mathrm{Sing}(V_j)$ and Z is a component of $\mathrm{Sing}(V_j)$, therefore $Z = \overline{\mathfrak{S}_Z}$.

Similarly, in case (b) the inductive hypothesis gives that W' and W are both closures of $\mathfrak{S}$-strata, and so, $\mathfrak{S}$ being a C-stratification, $\not{C}(W', W)^{\mathrm{Zar}}$ is a union of $\mathfrak{S}$-strata, one of which must be $\mathfrak{S}_Z$, whence, as before $Z = \overline{\mathfrak{S}_Z}$. Thus the statement holds for $j = i + 1$, and the Proposition results, by induction. □

Corollary. *If C and C' are stratifying conditions such that C is good and $C' \Rightarrow C$ then $\mathfrak{S}_{C'} \prec \mathfrak{S}_C$.*

Corollary. (Cf. [Wh, p. 536, Thm. 18.11, and p. 540, Thm. 19.2].) *If C is a good stratifying condition and $\mathfrak{P}_1, \mathfrak{P}_2, \dots, \mathfrak{P}_n$ are partitions of V, there is a coarsest C-stratification refining all the $\mathfrak{P}_i$.*

Proof. The stratifications $C_{\mathfrak{P}_i}$ of Example 2.5(c) are good (see Example 2.5(d)), and so in view of Example 2.5(b)), Proposition 2.7 shows that $\mathfrak{S}_{C\wedge C_{\mathfrak{P}_1}\wedge\cdots\wedge C_{\mathfrak{P}_n}}$ does the job. □

2.8. For a stratifying condition C say that V is *C-equisingular* at a subgerm (W,x) if there is a C-stratification $\mathfrak{S}$ such that $x\in\mathfrak{S}_W$, the unique $\mathfrak{S}$-stratum containing a dense open subvariety of W – or equivalently, if $(W,x)\subset(\mathfrak{S}_{C,x},x)$ where $\mathfrak{S}_{C,x}$ is the unique $\mathfrak{S}_C$-stratum containing x.

As a special case of the exercise in Example 2.5(c), it holds for any two stratifying conditions C and C' on V that

$$(\mathfrak{S}_{C'}\prec\mathfrak{S}_C)\iff\text{(for all subgerms }(W,x)\text{ of }V,$$
$$V\text{ is }C'\text{-equisingular at }(W,x)\Rightarrow V\text{ is }C\text{-equisingular at }(W,x)\text{)}.$$

In particular, by the first corollary of Proposition 2.7, if C is good and $C'\Rightarrow C$ then C'-equisingularity implies C-equisingularity.

3. The Zariski stratification

In the early 1960's, Zariski developed a comprehensive theory of equisingularity in codimension 1 (see [Z3]). Let x be a point on a hypersurface V, around which the singular locus W is a smooth manifold of codimension 1. The fibers of any local retraction of V onto W form a family of plane curve germs. Zariski showed that if the singularity type of the members of one such family is constant – where singularity type is determined in the classical sense via characteristic pairs, embedded topology, multiplicity sequence, etc. – then the *Whitney conditions* hold along W at x; and conversely, the Whitney conditions imply *constancy of singularity type* in any such family. In this case, moreover, the Whitney conditions are equivalent to *topological triviality* of V along W near x. Furthermore, if these conditions hold, the singularities of V along W near x can be resolved by blowing up W and its successive strict transforms (along which the conditions continue to hold), in a way corresponding exactly to the desingularization of any of the above plane curve germs by successive blowing up of infinitely near points – and conversely. This last condition can be thought of as *simultaneous desingularization* of any family of plane curve germs arising as the fibers of a retraction.

Unfortunately, in higher codimensions no two of these characterizations of equisingularity in codimension 1 remain equivalent; and anyway there is as yet no definitive sense in which two hypersurface singularities can be said to have the same singularity type. Each of the characterizations suggests a higher-dimensional theory. (See in particular the remarks in §1, following Example (e), pointing to the work of Teissier, Gaffney, etc.) Exploration of the remaining connections among these characterizations leads to interesting open questions, a few of which are stated below.

After some tentative attempts at generalizing the notion of equisingularity to higher codimensions (see e.g., [Z1, p. 487, §4]), Zariski formulated (in essence) the following definition of the dimensionality type $\mathrm{dt}_Z(V,x)$ of a hypersurface germ (V,x) – possibly empty – of dimension d, where $d=-1$ if (V,x) is empty, and otherwise x is a $\mathbb{C}$-rational point of the algebraic or analytic variety V. (One can regard such a (V,x) algebraically as being a local $\mathbb{C}$-algebra of the form $R/(f)$ where R is a formal power-series ring in $d+1$ variables over $\mathbb{C}$ and $0\neq f\in R$.) Zariski-equisingularity is defined by local constancy of dt_Z.

The intuition which inspired Zariski's definition does not lie on the surface. It may have come out of his extensive work on ramification of algebraic functions and on fundamental groups of complements of projective hypersurfaces, or have been partly inspired by Jung's method of desingularization (which begins with a desingularization – assumed, through induction on dimension, to exist – of the discriminant of a general projection.) Anyway, here it is:

$$\begin{aligned}\mathrm{dt}_Z(V,x)&=-1 &&\text{if } (V,x) \text{ is empty,}\\ \mathrm{dt}_Z(V,x)&=1+\mathrm{dt}_Z(\Delta_\pi,0) &&\text{otherwise,}\end{aligned}$$

where $\pi\colon (V,x)\to(\mathbb{C}^d,0)$ is a general finite map germ, with discriminant $(\Delta_\pi,0)$ (the hypersurface subgerm of $(\mathbb{C}^d,0)$ consisting of points over which the fibers of π have less than maximal cardinality, i.e., at whose inverse image π is not étale). Here π is defined by its coordinate functions $\xi_1,\dots,\xi_d$, power series in $d+1$ variables, with linearly independent linear terms; and if (V,x) is then represented – via Weierstrass preparation – by an equation

$$f(Z)=Z^n+a_1(\xi_1,\dots,\xi_d)Z^{n-1}+\dots+a_n(\xi_1,\dots,\xi_d)=0 \qquad (a_i\in\mathbb{C}[[i_1,\dots,i_d]])$$

then Δ_π is given by the vanishing of the Z-discriminant of f, so that $(\Delta_\pi,0)$ is a hypersurface germ of dimension $d-1$, whose dt_Z may be assumed, by induction on dimension, already to have been defined.

A property of finite map germs π holds for *almost all* π if there is a finite set of polynomials in the (infinitely many) coefficients of $\xi_1,\dots,\xi_d$ such that the property holds for all π for whose coefficients these polynomials do not simultaneously vanish. (The coefficients depend on a choice of generators for the maximal ideal of the local $\mathbb{C}$-algebra of (V,x) – i.e., of an embedding of (V,x) into $\mathbb{C}^{d+1}$ – but the notion of "holding for almost all π" does not.) It follows from [Z2, p. 490, Proposition 5.3] that $\mathrm{dt}_Z(\Delta_\pi,0)$ has the same value for almost all π; and that enables the preceding inductive definition of dt_Z (which is a variant of the original definition in [Z2, §4]).

It is readily seen that (V,x) is smooth iff Δ_π is empty for almost all π, i.e., $\mathrm{dt}_Z(V,x)=0$. So $\mathrm{dt}_Z(V,x)=1$ means that almost all π have smooth discriminant, which amounts to Zariski's classical definition of codimension-1 equisingularity of (V,x) (see e.g., [Z3, pp. 20–21].

For another example, suppose the components $(V_i,0)$ of $(V,0)\subset(\mathbb{C}^{d+1},0)$ are smooth, with *distinct* tangent hyperplanes $\sum_{j=1}^{d+1}a_{ij}X_j=0$. Then a similar property holds for $(\Delta_\pi, 0) = (\cup_{i\neq i'}\pi(V_i \cap V_{i'}), 0)$ if π is general enough; and it follows by induction on d that $\mathrm{dt}_Z(V,0)$ is one less than the rank of the matrix (a_{ij}).

Hironaka proved, in [H2], that on a d-dimensional *algebraic* $\mathbb{C}$-variety V which is locally embeddable in $\mathbb{C}^{d+1}$, *the function* dt_Z *is upper-semicontinuous.* More precisely, it follows from the main Theorem on p. 417 of [H2] and from [Z2, p. 476, Proposition 4.2] that the set

$$V_i = \{\, x \in V \mid \mathrm{dt}_Z(V,x) \geq i \,\} \qquad (i \geq 0)$$

is a closed subvariety of V. Thus we have a partition $\mathfrak{P}_Z$ of V, by the connected components of the fibers of the function dt_Z, and correspondingly a stratifying condition $\mathfrak{Z}(W_1, W_2, x)$, defined to mean that dt_Z is constant on a neighborhood of x in W_2. This is a good stratifying condition, see Example 2.5(d). We call the stratification $\mathfrak{S}_{\mathfrak{Z}}$ the *Zariski stratification.* By Proposition 2.7, the Zariski stratification is the coarsest stratification with dt_Z constant along each stratum.

One would expect to have a similar stratification for *complex-analytic* locally-hypersurface varieties; but this is not explicitly contained in the papers of Zariski and Hironaka. To be sure, I asked Hironaka during the workshop (September, 1997) whether his key semi-continuity proof applies in the analytic case, and he said not necessarily, there are obstructions to overcome. Thus:

Problem 3.1. Investigate the upper-semicontinuity of dt_Z on analytic hypersurfaces, and the possibility of extending the Zariski stratification to this context.

Zariski proved in [Z2, §6] that with V_i as above, $V_i \setminus V_{i+1}$ is smooth, of pure codimension i in V [Z2, p. 508, Theorem 6.4]; and that the closure of a part of $\mathfrak{P}_Z$ is a union of parts [Z2, pp. 510–511]. One naturally asks then whether $\mathfrak{P}_Z = \mathfrak{S}_Z$ – what is still missing is condition (ii) of Definition 2.2, that the singular locus of the closure of a part is a union of parts. But that is indeed so, as follows from the equimultiplicity assertion in the next Theorem (which assertion implies that any part meeting the set $\mathrm{Sing}(\overline{P_1})$ of points where $\overline{P_1}$ has multiplicity > 1 is entirely contained in $\mathrm{Sing}(\overline{P_1})$).

Theorem 3.2. *Let V be a purely d-dimensional algebraic $\mathbb{C}$-variety, everywhere of embedding dimension $\leq d+1$. Then for any two parts P_1, P_2 of the partition $\mathfrak{P}_Z$ with $\overline{P_1} \supset P_2$, the Whitney conditions $\mathfrak{W}(\overline{P_1}, P_2, x)$ hold at all $x \in P_2$. So $\overline{P_1}$ is equimultiple along P_2, and $\mathfrak{P}_Z = \mathfrak{S}_Z$ is a Whitney stratification of V.*

Remarks.

(a) Let us say that V is *Zariski-equisingular* along a subvariety W at a smooth point x of W if dt_Z is constant on a neighborhood of x in W (see [Z2, p. 472]). The equality $\mathfrak{P}_Z = \mathfrak{S}_Z$ entails that Zariski-equisingularity is the same as $\mathfrak{Z}$-equisingularity (see §2.8).

(b) From Theorem 3.2 we see via Thom-Mather that Zariski-equisingularity implies topological triviality of V along W at x, a result originally due to Varchenko (who actually needed only *one* sufficiently general projection with discriminant topologically trivial along the image of W), see [Va].

(c) In connection with the idea that *analytic* equisingularity should involve something more than Whitney-equisingularity, note the example of Briançon and Speder [Z3, p. 3, (2)] showing that dt_Z need not be constant along the strata of a Whitney stratification. In fact, in [BH] Briançon and Henry show that for families of isolated singularities of surfaces in $\mathbb{C}^3$, constancy of dt_Z is equivalent to constancy of the generic polar multiplicities (i.e., Whitney-equisingularity) and of two *additional* numerical invariants of such singularities.

Here is a sketch of a *proof of Theorem 3.2.* We first need a formulation of $\mathfrak{W}(W_1, W_2, x)$ (x a smooth point of $W_1 \subset W_2$) in terms of the local $\mathbb{C}$-algebra $R := \mathcal{O}_{W_1,x}$ and the prime ideal $\mathfrak{p}$ in R corresponding to W_2 (so that $R/\mathfrak{p}$ is a regular local ring). One fairly close to the geometric definition is, in outline, as follows. We assume for simplicity that W_1 is equidimensional, of dimension d. Let m be the maximal ideal of R, and in the Zariski tangent space $T := \mathrm{Hom}_{R/m}(m/m^2, R/m)$ let T_2 be the tangent space of W_2, i.e., the subspace of maps in T vanishing on $(\mathfrak{p} + m^2)/m^2$. Set $W_1' := \mathrm{Spec}(R)$, $W_2' := \mathrm{Spec}(R/\mathfrak{p})$, let $B \to W_1'$ be the blowup of W_1' along W_2', let Ω be the universal finite differential module of $W_1'/\mathbb{C}$, and let $G \to W_1'$ be the Grassmannian which functorially parametrizes rank d locally free quotients of Ω. Then the canonical map $p\colon B \times_{W_1'} G \to W_1'$ is an isomorphism over $V_0 := W_1' - W_2' - \mathrm{Sing}(W_1')$; and we let Y be the closure of $p^{-1}V_0$ in $B \times_{W_1'} G$. (This Y is the "birational join" of B and the so-called Nash modification of W_1'.)

Any $\mathbb{C}$-rational point y of the closed fiber of $Y \to W_1'$ gives rise naturally, via standard universal properties of B and G, to a pair (E, F) of vector subspaces of T, where E contains T_2 as a codimension-1 subspace, and $\dim F = d$: E consists of all maps in T vanishing on the kernel of the surjection $(\mathfrak{p} + m^2)/m^2 \cong \mathfrak{p}/m\mathfrak{p} \twoheadrightarrow \mathfrak{q}$, where $\mathfrak{q}$ is the 1-dimensional quotient corresponding to the W_1'-homomorphism $\mathrm{Spec}(R/m) \to B \subset \mathbb{P}(\mathfrak{p})$ whose image is the projection of y, $\mathbb{P}(\mathfrak{p})$ being the projective bundle parametrizing 1-dimensional quotients of $\mathfrak{p}$; and F is dual to a similarly-obtained d-dimensional quotient of the R/m-vector space $\Omega/m\Omega \cong m/m^2$.

Then

$$\mathfrak{W}(W_1, W_2, x) \text{ holds} \iff \begin{cases} \mathfrak{W}(R, \mathfrak{p})\colon E \subset F \textit{ for all } (E, F) \\ \textit{arising from points in the} \\ \textit{closed fiber of } Y \to W_1'. \end{cases}$$

(The idea, which can be made precise with some local analytic geometry, is that for any embedding of W_1 into $\mathbb{C}^t$ ($t = \dim T$) with W_2 linear, the set of pairs (E, F) can naturally be identified with the set of limits of sequences (E_n, F_n) defined as follows: let $x_n \to x$ be a sequence in $W_1 - W_2 - \mathrm{Sing}(W_1)$, let E_n be the join of W_2 and x_n – a linear space containing W_2 as a 1-codimensional subspace – and let F_n be the tangent space to W_1 at x_n.)

It is not hard to see that $\mathfrak{W}(R, \mathfrak{p}) \iff \mathfrak{W}(\hat{R}, \mathfrak{p}\hat{R})$.

This leads to a reduction of the proof of Theorem 3.2 to that of the similar statement for the *formal* Zariski partition of $\mathrm{Spec}(\widehat{\mathcal{O}_{V,x}})$. (That partition is by constancy of formal dimensionality type, defined inductively through *generic* finite map germs – those whose coefficients (see above) are independent indeterminates.) Indeed, we have to prove $\mathfrak{W}(R,\mathfrak{p})$ whenever R is the local ring of the closure of a Zariski-stratum W_1 of V at a $\mathbb{C}$-rational point x, and $\mathfrak{p}$ is the prime ideal corresponding to the Zariski-stratum W_2 through x. (We refer – prematurely – to the parts of the Zariski partition as Zariski-strata.) The dimensionality type of the generic point of $\operatorname{Spec}(R)$ is equal to that of all the closed points of W_1, i.e., to the codimension of W_1 in V (see [H2, p. 419, Theorem] and [Z2, p. 508, Thm. 6.4]); and the analogous statement follows for the generic point of each component of $\operatorname{Spec}(\hat{R})$ (see [Z2, p. 507, bottom]) – so that each of those components is the closure of a formal Zariski stratum of $\operatorname{Spec}(\widehat{\mathcal{O}_{V,x}})$ ([Z2, p. 480, Thm. 5.1]).

To complete the reduction, observe that if $\widehat{R_1},\dots,\widehat{R_s}$ are the quotients of $\hat{R}$ by its minimal primes then $\mathfrak{W}(\hat{R},\mathfrak{p}\hat{R}) \iff \mathfrak{W}(\widehat{R_i},\mathfrak{p}\widehat{R_i})$ for all $i = 1,2,\dots,s$. (The corresponding geometric statement is clear.)

Once we are in the formal situation, where $V = \operatorname{Spec}(\mathbb{C}[[X_1,\dots,X_{d+1}]]/(f))$ with f an irreducible formal power series, [Z2, p. 490, Prop. 5.3] gives an embedding of V into $\mathbb{C}^{d+1}$ such that for almost all *linear* projections $\pi\colon \mathbb{C}^{d+1} \to \mathbb{C}^d$, if w is either the generic point of W_1 or the generic point of W_2 or the closed point of V then the dimensionality type of the point πw on the discriminant hypersurface Δ_π of $\pi|V$ is one less than that of w, while the dimension of the closure $\overline{\pi w}$ is the same as the dimension of $\overline{w}$. It follows then from [Z2, p. 491, Prop. 5.4] that $\pi(\overline{W_i})$ $(i = 1,2)$ is the closure of a Zariski stratum of Δ_π, except when $\overline{W_i} = V$.

At this point we have reached a situation which is much like the one treated by Speder in [S]. In the context of local analytic geometry, Speder shows that a certain version of equisingularity does imply the Whitney conditions. That version is not the same as the one we have called Zariski-equisingularity: while it is also based on an induction involving the discriminant, the induction is with respect to *one* sufficiently general linear projection, whereas Zariski's induction is with respect to *almost all* projections. Nevertheless there are enough similarities in the two approaches that Speder's arguments can now be adapted to give us what we want. Note that Speder asserts only that equisingularity gives the Whitney conditions when $\overline{W_1} = V$; but in view of the remark about $\pi(\overline{W_i})$ in the preceding paragraph, the remaining cases can be treated *in the present situation* by reasoning like that in section IV of [S]. Details (copious) are left to the reader.

4. Equiresolvable stratifications

A standard way to analyze singularities is to *resolve* them, and on the resulting manifold to study the inverse image of the original singularity. For example, after three point-blowups the cusp at the origin of the plane curve $Y^2 = X^3$ is replaced by a configuration of three lines crossing normally, a configuration which represents the complexity of the original singularity.

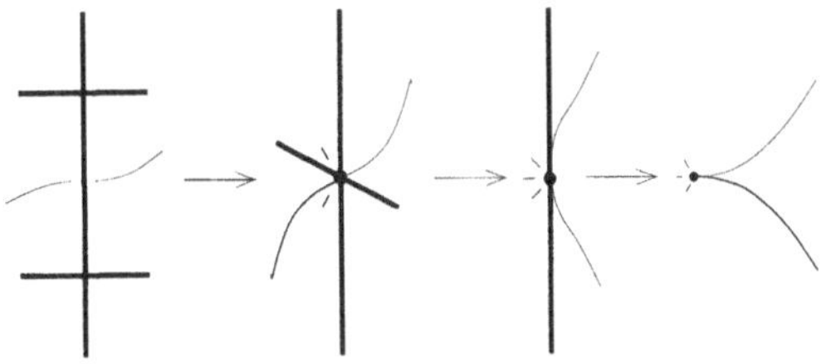

Quite generally, the embedded topological type of any plane plane curve singularity is determined by the intersection matrix of its inverse image – a collection of normally crossing projective lines – on a minimal embedded desingularization.

This suggests one sense in which a family of singularities could be regarded as being equisingular: if its members can be resolved simultaneously, in such a way that their inverse images all "look the same."

Several ways to make such an idea precise come to mind. Here is one. With notation as in section 2.4, define the *local equiresolvability* condition $ER(W_1, W_2, x)$ to mean there exists an embedding of the germ (W_1, x) into a smooth germ $(M, 0)$, a proper birational (or bimeromorphic) map of manifolds $f\colon M' \to M$ such that $f^{-1}W_1$ is a divisor with smooth components having only normal crossing intersections – so that $f^{-1}W_1$ (considered as a *reduced* variety) has an obvious stratification, by multiplicity – and such that $f^{-1}W_2$ is a union of strata, each of which f makes into a C^∞-fiber bundle, with smooth fibers, over W_2.

Proposition 4.1. (Cf. [T2, p. 401].) *The condition ER is stratifying (see §2.4). Hence, by Proposition 2.7, every variety has a coarsest equiresolvable stratification.*

Proof. Let $W_2 \subset W_1 \subset V$, with W_2 smooth and connected. The question being local, we may embed W_1 in some open $M \subset \mathbb{C}^n$. Let $f_2\colon M_2 \to M$ be the blowup of $W_2 \subset M$, so that $f_2^{-1}W_2$ is a divisor in M_2. Let $f_1\colon M' \to M_2$ be an embedded resolution of $f_2^{-1}W_1$, and set $f := f_2 f_1$, so that $f^{-1}W_1 \subset M'$ is a finite union of normally crossing smooth divisors, as is its subvariety $f^{-1}W_2$ (also a divisor in M'). After removing some proper closed subvarieties from W_2 we may assume that each intersection W of components of $f^{-1}W_2$ is mapped onto W_2 and, by the theorem of Bertini-Sard, that the restriction of f maps each such W *submersively* onto W_2. Now Thom's first isotopy lemma ([GMc, p. 41], or [Ve, p. 311, Thm. 4.14]) applied to the multiplicity stratification – clearly a Whitney stratification – of $f^{-1}W_2$, shows that each stratum is, via f, a smooth C^∞-fiber bundle over W_2. □

Following Teissier, [T1, p. 107, Définition 3.1.5], we can define a stronger equiresolvability condition ER^+ by adding to ER the requirement that the *not-necessarily-reduced* space $f^{-1}W_2$ be locally analytically trivial over each $w \in W_2$, i.e., germwise the product of W_2 with the fiber $f^{-1}w$. *Condition ER^+ is still stratifying:* with notation as in the preceding proof, set $E_2 := f_2^{-1}W_2$, and realize the desingularization f_1 as a composition of blowups $f_{i+1}\colon M_{i+1} \to M_i$ $(i \geq 2)$ of smooth $C_i \subset M_i$ such that if $E_{j+1} := f_j^{-1}(C_j \cup E_j)$ $(j \geq 2)$, then C_i has normal crossings with E_i; then note for any $z \in M_i$ at which both C_i and $(f_2 f_3 \cdots f_i)^{-1}W_2$

are locally analytically trivial, that $(f_2 f_3 \dots f_i f_{i+1})^{-1} W_2$ is locally analytically trivial everywhere along $f_{i+1}^{-1} z$; and finally argue as before, using Bertini-Sard etc.

The result that ER^+ is stratifying is similar to [T1, p. 109, Prop. 2]; but Teissier's formulation of strong simultaneous resolution refers to birational rather than embedded resolution, and so seems a priori weaker than ER^+. It would be good to know whether in fact Teissier's condition is equivalent to ER^+.

It would also be good to know whether either of the stratifying conditions ER and ER^+ is *good* (see §2.4), or in case they are not, whether there is a convincing formulation of equiresolvability which does lead to a good stratifying condition.

Teissier showed that strong simultaneous resolution along any smooth subvariety W of the hypersurface V implies the Whitney conditions for the pair $(V - \mathrm{Sing}(V), W)$, see [T1, p. 111, Prop. 4].

A converse, Whitney $\Rightarrow$ strong simultaneous resolution, was proved by Laufer (see [L]) in case V is the total space of a family of isolated two-dimensional surface singularities, for example if V is three-dimensional, $\mathrm{Sing}(V) =: W$ is a nonsingular curve, and there is a retraction $V \to W$ whose fibers are the members of the family.

The basic motivation behind this paper is our interest in possible relations between Zariski-equisingularity and simultaneous resolution. Such relations played a prominent role in Zariski's thinking – see e.g., [Z1, p. 490, F, G. H] and [Z3, p. 7].

To begin with, Zariski showed that if $\mathrm{dt_Z}(V, x) = 1$ then $\mathrm{Sing}(V)$ is smooth, of codimension 1 at x, and that for some neighborhood V' of x in V, the Zariski stratification of V' *is* equiresolvable (in a strong sense, via successive blowups of the singular locus, which stays smooth of codimension 1 until it disappears altogether) – see [Z3, p. 93, Thm. 8.1]). Zariski's result involves birational resolution, but can readily be extended to cover embedded resolution.

The *converse,* that any simultaneous resolution of a family of plane curve singularities entails equisingularity, was shown by Abhyankar [A]. This converse fails in higher dimension, for instance for the Briançon-Speder family of surface singularities referred to in Remark (c) following Theorem 3.2, a family which is Whitney-equisingular and hence – by the above-mentioned result of Laufer – strongly simultaneously resolvable, but not Zariski-equisingular.

We mention in passing a different flavor of work, by Artin, Wahl, and others, on simultaneous resolution and infinitesimal equisingular deformations of normal surface singularities, see e.g., [Wa].

At any rate, a positive answer to the next question would surely enhance the appeal of defining analytic equisingularity via $\mathrm{dt_Z}$.

Problem 4.2. Is the Zariski stratification of a hypersurface V equiresolvable in some reasonable sense? For example, is it an ER- or ER^+-stratification?

Initial discouragement is generated by an example of Luengo [Lu], a family of quasi-homogeneous two-dimensional hypersurface singularities which is equisingular in Zariski's sense, but cannot be resolved simultaneously by blowing up smooth centers. There is however another way. That is the subject of the next section.

5. Simultaneous resolution of quasi-ordinary singularities

We pursue the question "Does Zariski-equisingularity imply equiresolvability?" (A positive answer would settle Problem 4.2 if equiresolvability were expressed by a good stratifying condition, see §2.8.)

In fact we ask a little more: if $x \in V$, with V a hypersurface in $\mathbb{C}^{d+1}$, does there exist a neighborhood M of x in $\mathbb{C}^{d+1}$ and a proper birational (or bimeromorphic) map of manifolds $f\colon M' \to M$ such that $f^{-1}V$ is a divisor with smooth components having only normal crossing intersections and such that for *every* Zariski-stratum W of $V \cap M$, $f^{-1}W$ is a union of multiplicity-strata of $f^{-1}(V \cap M)$, each of which is made by f into a C^∞-fiber bundle, with smooth fibers, over W? We could further require that the not-necessarily-reduced space $f^{-1}W$ be locally analytically trivial over W.

As mentioned above, Zariski gave an affirmative answer when $\mathrm{dt}_Z\, x = 1$. We outline now an approach to the question which most likely gives an affirmative answer when $\mathrm{dt}_Z\, x = 2$. (At this writing, I have checked many, but not all, of the details.) Roughly speaking, this approach is the classical desingularization method of Jung, applied to families. The main roadblock to extending it inductively to dimensionality types ≥ 3 is indicated below (Problem 5.1).

For simplicity, assume that $\dim V = 3$ and that $\mathrm{dt}_Z\, x = 2$, so that the Zariski 2-stratum is a non-singular curve W through x. By definition we can choose a finite map germ whose discriminant Δ has the image of W as its Zariski 1-stratum; and after reimbedding V we may assume this map germ to be induced by a linear map $\pi\colon \mathbb{C}^4 \to \mathbb{C}^3$. In what follows, we need only *one* such π.

According to Zariski, there is an embedded resolution $h\colon M^3 \to \mathbb{C}^3$ of Δ such that $h^{-1}\pi(W)$ is a union of strata. (We abuse notation by writing $\mathbb{C}^3$ for a suitably small neighborhood of $\pi(x)$ in $\mathbb{C}^3$.) Let $g\colon M^4 = M^3 \times_{\mathbb{C}^3} \mathbb{C}^4 \to \mathbb{C}^4$ be the projection, a birational map of manifolds. There is a finite map from the codimension-1 subvariety $V' := g^{-1}V \subset M^4$ to M^3 whose discriminant, being contained in $h^{-1}\Delta$, has normal crossings. This means that the singularities of V' are all *quasi-ordinary*, see [L2]. Moreover, as Zariski showed (see [Z2, p. 514]), the Zariski-strata of V are all étale over the corresponding strata of Δ, so their behavior under pullback through $V' \to V$ is essentially the same as that of the strata of Δ under h, giving rise to nice fiber-bundle structures. More precisely, if $\mathfrak{S}'$ is the pullback on V' of the multiplicity stratification on the normal crossings divisor $g^{-1}\Delta$, and S is any Zariski-stratum of V, then $g^{-1}S$ is a union of $\mathfrak{S}'$-strata, each of which g makes into a C^∞-fiber bundle over S. Thus we have achieved an "equisimplification," but not yet an equiresolution – the singularities of V along W have simultaneously been made quasi-ordinary.

So now we need only deal with quasi-ordinary singularities, keeping in mind however that V' is no longer a localized object – it contains projective subvarieties in the fibers over the original singularities of V. Fortunately, the above-defined stratification $\mathfrak{S}'$ can be characterized *intrinsically* (even *topologically*) on V', see

[L3, §6.5]. So we can forget about the projection $V' \to M^3$ via which $\mathfrak{S}'$ was determined, and deal directly with this canonical stratification on V' (closely related, presumably, with the Zariski stratification, though $V' \to M^3$ may not be sufficiently general at every point of V'). The problem is thus reduced to showing that *the canonical stratification $\mathfrak{S}'$ on the quasi-ordinary space V' is equiresolvable.*

Any germ (V', y) (assumed, for simplicity, irreducible) is given by the vanishing of a polynomial

$$Z^m + a_1(X,Y,t)Z^{m-1} + \cdots + a_m(X,Y,t) \qquad (a_i \in \mathbb{C}[[X,Y,t]])$$

whose discriminant is of the form

$$\delta = X^a Y^b \epsilon(X,Y,t), \qquad \epsilon(0,0,0) \neq 0.$$

(Here t is a local parameter along W.) The roots of this polynomial are fractional power series $\zeta_i(X^{1/n}, Y^{1/n}, t)$; and since $\delta = \prod(\zeta_i - \zeta_j)$, we have, for some non-negative integers a_{ij}, b_{ij},

$$\zeta_i - \zeta_j = X^{a_{ij}/n} Y^{b_{ij}/n} \epsilon_{ij}(X^{1/n}, Y^{1/n}, t), \qquad \epsilon_{ij}(0,0,0) \neq 0.$$

Modulo some standardization, the monomials $X^{a_{ij}/n} Y^{b_{ij}/n}$ so obtained are called the *characteristic monomials* of the quasi-ordinary germ (V', y). They provide a very effective tool for studying such germs. They are higher dimensional generalizations of the characteristic pairs of plane curve singularities (which are always quasi-ordinary), and they control many basic features of (V', y), for example the number of components of the singular locus, the multiplicities of these components on V' and at the origin, etc. (See [L2], [L3] for more details.) In particular, two quasi-ordinary singularities have the same embedded topology iff they have the same characteristic monomials [G]. It is therefore natural to ask:

Problem 5.1. Do the characteristic monomials of a quasi-ordinary singularity determine a canonical embedded resolution procedure?

This question makes sense in any dimension, and would arise naturally in any attempt to extend the preceding argument inductively to higher dimensionality types. A positive answer would imply that the canonical stratification of any quasi-ordinary hypersurface is equiresolvable.

At this point, if we weren't concerned with embedded resolution (which we need to be, if there is to be any possibility of induction), we could *normalize* V' to get a family of cyclic quotient singularities whose structure is completely determined by the characteristic monomials. Simultaneously resolving such a family is old hat (see [L1, Lecture 2]), leading to the following result (for germs of families of two-dimensional hypersurface singularities): *If there is a projection with an equisingular branch locus, then we have simultaneous resolution; and if that projection is transversal (i.e., its direction does not lie in the Zariski tangent cone) then we have strong simultaneous resolution, and hence (by Teissier) Whitney-equisingularity.*

In particular, the above-mentioned example of Luengo *is* in fact strongly simultaneously resolvable – though not by one of the standard blowup methods.

For *embedded resolution,* canonical algorithms are now available [BM], [EV] (see also [O] for a two-dimensional precursor), so that Problem 5.1 is quite concrete: verify that the invariants which drive a canonical procedure, operating on a quasi-ordinary singularity parametrized as above (by the ζ_i), are completely determined by the characteristic monomials.

The idea is then that the intersections of V' with germs of smooth varieties transversal to the strata form families of quasi-ordinary singularities with the same characteristic monomials, and so any resolution of one member of the family should propagate along the entire stratum – whence the equiresolvability

A key point in dimension 2 is *monoidal stability:* any permissible blowup of a 2-dimensional quasi-ordinary singularity is again quasi-ordinary, and the characteristic monomials of the blown up singularity depend only on those of the original one and the center of blowing up – chosen according to an algorithm depending only on the characteristic monomials; explicit formulas are given in [L2, p. 170]. When it comes to total transform, there are of course some complications; but monoidal stability can still be worked out for a configuration consisting of a quasi-ordinary singularity *together with* a normal crossings divisor such as arises in the course of embedded resolution. In principle, then, one should be able to use an available canonical resolution process and see things through. Some preliminary work along these lines was reported on at the workshop by Ban and McEwan.

Unfortunately such monoidal stability fails in higher dimensions, even for such simple quasi-ordinary singularities as the origin on the threefold $W^4 = XYZ$. So we need to look into:

Problem 5.2. Find a condition on singularities weaker than quasi-ordinariness, but which is monoidally stable, and which can be substituted for quasi-ordinariness in Problem 5.1.

Careful analysis of how the above-mentioned canonical resolution procedures work on quasi-ordinary singularities could suggest an answer.

Added in proof. Villamayor has come up with an affirmative answer to problem 4.2, based on an approach involving quasi-ordinary singularities, but different from the one suggested in §5 – see [Vi].

References

[A] Abhyankar, S. S.: A criterion of equisingularity, Amer. J. Math **90** (1968), 342–345.

[BM] Bierstone, E., Milman, P.: Canonical desingularization in characteristic zero by blowing up the maximum strata of a local invariant, Invent. Math. **128** (1997), 207–302.

[BH] Briançon, J., Henry, J. P. G.: Équisingularité générique des familles de surfaces a singularité isolée, Bull. Soc. Math. France **108** (1980), 259–281.

[BS] Briançon, J., Speder, J.-P.: La trivialité topologique n'implique pas les conditions de Whitney, C. R. Acad. Sci. Paris, Sér. A-B **280** (1975), no. 6, Aiii, A365–A367.

[DM] van den Dries, L., Miller, C. : Geometric categories and o-minimal structures, Duke Math. J. **84** (1996), 497–540.

[EV] Encinas, S., Villamayor, O. : Good points and constructive resolution of singularities, Acta Math. **181** (1998), 109–158.

[G] Gau, Y.-N. : Embedded topological classification of quasi-ordinary singularities, Memoirs Amer. Math. Soc. **388**, 1988.

[GL] Gibson, G. C., Looijenga, E., du Plessis, A., Wirthmüller, K. : Topological Stability of Smooth Maps, Lecture Notes in Math. **552**, Springer-Verlag, 1976.

[GM] Gaffney, T., Massey, D. : Trends in equisingularity theory, to appear in the Proceedings of the Liverpool Conference in Honor of C.T.C. Wall.

[GMc] Goresky, M., MacPherson, R. : Stratified Morse Theory, Springer-Verlag, 1988.

[H1] Hironaka, H. : Normal cones in analytic Whitney stratifications, Publ. Math. IHES **36** (1969), 127–138.

[H2] Hironaka, H. : On Zariski dimensionality type, Amer. J. Math. **101** (1979), 384–419.

[K] Kleiman, S. L. : Equisingularity, multiplicity, and dependence, to appear in the Proceedings of a Conference in Honor of M. Fiorentini, publ. Marcel Dekker.

[L] Laufer, H: Strong simultaneous resolution for surface singularities, in "Complex Analytic Singularities," North-Holland, Amsterdam-New York, 1987, pp. 207–214.

[LM] Lê, D. T., Mebkhout, Z. : Introduction to linear differential systems, in "Singularities," Proc. Sympos. Pure Math., vol. **40**, Part 2, Amer. Math. Soc., Providence, 1983, pp. 31–63.

[LT] Lê, D. T., Teissier, B. : Cycles évanescents, Sections Planes, et conditions de Whitney. II. in "Singularities," Proc. Sympos. Pure Math., vol. **40**, Part 2, Amer. Math. Soc., Providence, 1983, pp. 65–103.

[L1] Lipman, J. : Introduction to resolution of singularities, in "Algebraic Geometry," Proc. Sympos. Pure Math., vol. **29**, Amer. Math. Soc., Providence, 1975, Lecture 3, pp. 218–228.

[L2] Lipman, J. : Quasi-ordinary singularities of surfaces in $\mathbf{C}^3$, in "Singularities," Proc. Sympos. Pure Math., vol. **40**, Part 2 , Amer. Math. Soc., Providence, 1983, pp. 161–172.

[L3] Lipman, J. : Topological invariants of quasi-ordinary singularities, Memoirs Amer. Math. Soc. **388**, 1988.

[Lu] Luengo, I. : A counterexample to a conjecture of Zariski, Math. Ann. **267**, (1984), 487–494.

[Mc] MacPherson, R. : Global questions in the topology of singular spaces, Proceedings of the International Congress of Mathematicians, Vol. 1, 2 (Warsaw, 1983), 213–236, PWN, Warsaw, 1984.

[M1] Mather, J. : Stratifications and Mappings, in Dynamical Systems, Ed. M. M. Peixoto, Academic Press, New York, 1973, pp. 195–232.

[M2] Mather, J. : How to stratify mappings and jet spaces, in "Singularités d'Applications Différentiables, Lecture Notes in Math. **535**, Springer-Verlag, 1976, pp. 128–176.

[O] Orbanz, U.: Embedded resolution of algebraic surfaces, after Abhyankar (characteristic 0), in "Resolution of Surface Singularities," Springer Lecture Notes **1101** (1984), pp. 1–50.

[P] Parusiński, A.: Lipschitz stratification of subanalytic sets, Ann. Scient. Éc. Norm. Sup. **27** (1994), 661–696.

[S] Speder, J.-P.: Equisingularité et conditions de Whitney, Amer. J. Math. **97** (1975), 571–588.

[T1] Teissier, B.: Résolution Simultanée–II, in "Séminaire sur les Singularités des Surfaces," Lecture Notes in Math. **777**, Springer-Verlag, 1980, pp. 82–146.

[T2] Teissier, B.: Variétés polaires II: multiplicités polaires, sections planes, et conditions de Whitney, in "Algebraic Geometry, Proceedings, La Rabida, 1981," Lecture Notes in Math. **961**, Springer-Verlag, 1982, pp. 314–491.

[T3] Teissier, B.: Sur la classification des singularités des espaces analytiques complexes, Proceedings of the International Congress of Mathematicians, Vol. 1, 2 (Warsaw, 1983), 763–781, PWN, Warsaw, 1984.

[Va] Varchenko, A. N.: The relation between topological and algebro-geometric equisingularities according to Zariski, Funct. Anal. Appl. **7** (1973), 87–90.

[Ve] Verdier, J.-L.: Stratifications de Whitney et théorème de Bertini-Sard, Inventiones math. **36** (1976), 295–312.

[Vi] Villamayor, O.: On equiresolution and a question of Zariski, preprint.

[Wa] Wahl, J.: Equisingular deformations of normal surface singularities, Annals of Math. **104** (1976), 325–356.

[Wh] Whitney, H.: Tangents to an analytic variety, Annals of Math. **81** (1965), 496–549.

[Z1] Zariski, O.: Some open questions in the theory of singularities, Bull. Amer. Math. Soc. **77** (1971), 481–491. (Reprinted in [Z3], pp. 238–248.)

[Z2] Zariski, O.: Foundations of a general theory of equisingularity on r-dimensional algebroid and algebraic varieties, of embedding dimension $r+1$, Amer. J. Math. **101** (1979), 453–514. (Reprinted in [Z3], pp. 573–634; and summarized in [Z3, pp. 635–651].)

[Z3] Zariski, O.: Collected Papers, vol. IV, MIT Press, Cambridge, Mass., 1979.

Dept. of Mathematics
Purdue University
West Lafayette, IN 47907, USA
lipman@math.purdue.edu
`www.math.purdue.edu/~lipman`

Progress in Mathematics, Vol. 181, © 2000 Birkhäuser Verlag Basel/Switzerland

Resolution of Weighted Homogeneous Surface Singularities

Gerd Müller

1. Weighted homogeneous singularities

The purpose of this article is to review the method of Orlik and Wagreich to resolve normal singularities on weighted homogeneous surfaces X. Moreover, we explain the description of such surfaces by automorphy factors due to Dolgachev and Pinkham.

By a weighted blowup of the singular point 0 of X one obtains a modification Λ which has the structure of a singular line bundle over the smooth compact curve $Y = (X \setminus \{0\})/\mathbb{C}^*$. The only singularities of Λ are cyclic quotient singularities corresponding to the exceptional orbits of $\mathbb{C}^*$ acting on X. Resolving these singularities one obtains a resolution of the singularity of X. The resolution graph is star shaped with center Y. Moreover, we explain a result of Dolgachev and Pinkham describing Λ and X as quotients L/H and Z/H where $L \to C$ is a suitable line bundle over a certain branched cover C of Y with group H and Z is the surface obtained by contracting the zero-section of $L \to C$. This will be used to get information about the genus and the self-intersection number of the central curve Y.

Let us fix notation and hypotheses. We consider complex affine algebraic varieties X equipped with an effective action of the multiplicative group $\mathbb{C}^*$. Such an X admits an embedding in affine space $X \subseteq \mathbb{C}^{n+1}$ which is equivariant with respect to a $\mathbb{C}^*$-action on $\mathbb{C}^{n+1}$ defined by

$$t \cdot (z_0, \ldots , z_n) = (t^{w_0} z_0, \ldots , t^{w_n} z_n)$$

with integral weights $w_0, \ldots , w_n$, see [Wag, Proposition (2.1.5)]. Throughout we suppose that

- X is not contained in any coordinate hyperplane $\{z_i = 0\}$. Then effectiveness of the action is equivalent to the property that the weights have no common divisor.
- The action is *good*, i.e., the origin 0 is contained in the closure of every orbit. This is equivalent to the property that all weights have the same sign, say, are positive.

Then X is called *weighted homogeneous* of weights $w_0, \dots, w_n$. It is defined by polynomials f which are weighted homogeneous, i.e., satisfy $f(t \cdot z) = t^d f(z)$ with some $d \in \mathbb{N}$, the degree of f. Moreover, we suppose that

- X has dimension 2 and is normal. Then X has 0 as unique singular point.

Let us look at some examples.

Example 1. Let $X = \mathbb{C}^2/G$ be a quotient with $G \subseteq \mathrm{GL}(2, \mathbb{C})$ finite, see for instance [Lo, chapter 1.C]. Then X is a normal surface and the action on $\mathbb{C}^2$ with both weights equal to 1 induces a good $\mathbb{C}^*$-action.

Example 2. Let $L \to Y$ be a negative line bundle over a smooth compact curve Y. Contracting the zero-section one obtains a normal surface X with coordinate ring $\bigoplus_{k \in \mathbb{N}} \Gamma(Y, L^{\otimes(-k)})$, see [Lo, chapter 1.D]. The action on L given by multiplication in the fibres induces a good $\mathbb{C}^*$-action on X.

Example 3. Let the normal surface $X \subseteq \mathbb{C}^{n+1}$ be defined by *homogeneous* polynomials, i.e., all weights are 1. Then X can be resolved as follows. The given polynomials define a smooth projective curve $Y \subseteq \mathbb{P}_n$. Let $\sigma : \widetilde{\mathbb{C}^{n+1}} \to \mathbb{C}^{n+1}$ be the blowup of the origin and $\pi : \widetilde{\mathbb{C}^{n+1}} \to \mathbb{P}_n$ the tautological line bundle. Restricting above Y one obtains a line bundle $\pi : L \to Y$. Then $\sigma : L \to X$ is the contraction of the zero-section of the line bundle, hence a resolution of the singularity of X with exceptional fibre Y. If $X \subseteq \mathbb{C}^3$ is a hypersurface defined by the homogeneous polynomial f of degree d then the curve $Y \subseteq \mathbb{P}_2$ has genus $g = (d-1)(d-2)/2$, hence Euler characteristic $2g - 2 = d(d-3)$ and the degree of the line bundle $L \to Y$ is $-d$.

2. Singular line bundles and resolution

In order to generalize Example 3 to the case of arbitrary weights we introduce singular line bundles. Their only singularities are cyclic quotient singularities. So let $0 \leq \beta < \alpha$ be coprime natural numbers, ζ a root of unity of order α, $C_{\alpha,\beta} \subseteq \mathrm{GL}(2, \mathbb{C})$ the cyclic group generated by $\begin{pmatrix} \zeta & 0 \\ 0 & \zeta^\beta \end{pmatrix}$ and $X_{\alpha,\beta} = \mathbb{C}^2/C_{\alpha,\beta}$ the quotient. Let $[z, t] \in X_{\alpha,\beta}$ denote the image of $(z, t) \in \mathbb{C}^2$. With the unit disk D we have a map

$$\Lambda_{\alpha,\beta} = (D \times \mathbb{C})/C_{\alpha,\beta} \to D : [z, t] \mapsto z^\alpha.$$

For $\alpha = 1, \beta = 0$ this is just the trivial line bundle over D. Consider the restriction $\dot{\Lambda}_{\alpha,\beta} = (\dot{D} \times \mathbb{C})/C_{\alpha,\beta} \to \dot{D}$ over $\dot{D} = D \backslash \{0\}$ and the trivial line bundle $\dot{D} \times \mathbb{C} \to \dot{D}$. Then

$$\phi_{\alpha,\beta} : \dot{\Lambda}_{\alpha,\beta} \to \dot{D} \times \mathbb{C} : [z, t] \mapsto (z^\alpha, z^{-\beta} t)$$

is an isomorphism over $\dot{D}$. In particular, all fibres of $\Lambda_{\alpha,\beta} \to D$ are isomorphic to the affine line $\mathbb{C}$. Let Y be a smooth compact curve. An analytic map $\pi : \Lambda \to Y$

with fibres isomorphic to $\mathbb{C}$ is called a *singular line bundle* if, for every point $p \in Y$, there are an open neighbourhood $U \subseteq Y$ with an isomorphism $U \simeq D$, coprime natural numbers $0 \leq \beta < \alpha$ and an isomorphism $\pi^{-1}(U) \simeq \Lambda_{\alpha,\beta}$ over D which is linear on fibres. If $\Lambda_0 \to Y$ is a (locally trivial) line bundle one obtains singular line bundles $\Lambda \to Y$ by removing fibres over finitely many points of Y and gluing with local models $\Lambda_{\alpha,\beta} \to D$ using the isomorphisms $\phi_{\alpha,\beta}$ from above.

Exercise 1. Every singular line bundle $\Lambda \to Y$ is obtained, by this process, from a unique line bundle $\Lambda_0 \to Y$. Hint: Use the fact that every line bundle over $Y \setminus \{\text{finitely many points}\}$ is trivial.

The crucial step in resolving weighted homogeneous singularities is

Theorem 2.1. *Let X be a weighted homogeneous normal surface.*

(i) *Then $Y = (X \setminus \{0\})/\mathbb{C}^*$ is a smooth compact curve.*

(ii) *Let Λ be the weighted blowup of 0, i.e., the closure in $X \times Y$ of the graph of the natural map $X \setminus \{0\} \to Y$. Then the first projection $\sigma : \Lambda \to X$ is a proper modification of X in 0 and the second projection $\pi : \Lambda \to Y$ is a singular line bundle. Its zero-section is the exceptional fibre of σ.*

Proof. Write $G = \mathbb{C}^*$. Fix a point p of the manifold $M = X \setminus \{0\}$. Let α be the greatest common divisor of those weights w_i for which the i-th component p_i of p does not vanish. Then, by the first hypothesis in section 1, the isotropy subgroup G_p is cyclic of order α, say generated by η. The normal space N_p to the orbit $G \cdot p \subseteq M$ is one-dimensional. The action of G_p on N_p is given by multiplication with η^ν for some $0 \leq \nu < \alpha$. Since all weights are positive the action of G on M is proper, i.e., for every compact subset $K \subseteq M$ the action map $G \times K \to M$ is proper. Hence we can use Holmann's Slice Theorem [H, Satz 4] to obtain a G-invariant open neighbourhood $W \subseteq M$ of $G \cdot p$ and a G-equivariant isomorphism $W \simeq (D \times G)/G_p$. Here D is the unit disk in N_p, G_p acts on $D \times G$ by

$$\eta \cdot (z, t) = (\eta^\nu z, t\eta^{-1}),$$

and G acts on $(D \times G)/G_p$ by

$$s \cdot [z, t] = [z, st].$$

The curve $Y = M/G$ is locally isomorphic to

$$W/G \simeq ((D \times G)/G_p)/G \simeq D/G_p.$$

The effectiveness of the action implies that α, ν are coprime. Hence also $\zeta = \eta^\nu$ is a generator of G_p. For $\alpha = 1, \nu = 0$ let $\beta = 0$. Otherwise, define $0 < \beta < \alpha$ by the congruence $\nu\beta \equiv -1 \bmod \alpha$. Then $\zeta^\beta = \eta^{-1}$ and $(D \times G)/G_p = (D \times \mathbb{C}^*)/C_{\alpha,\beta}$. The natural map $M \to Y$ is locally given by

$$\begin{array}{ccccc} (D \times G)/G_p & \to & D/G_p & \to & D \\ [z, t] & \mapsto & [z] & \mapsto & z^\alpha. \end{array}$$

This shows that $\pi : \Lambda \to Y$ is locally given by

$$(D \times \mathbb{C})/C_{\alpha,\beta} = \Lambda_{\alpha,\beta} \to D$$

and hence is a singular line bundle. Obviously, $\sigma : \Lambda \to X$ is a proper modification with exceptional fibre Y, the zero-section of π. □

The finitely many orbits $G \cdot p$ with $G_p \neq 1$ are called *exceptional.* They are contained in the coordinate hyperplanes and give rise to the singularities of Λ. The corresponding pairs (α, β) are called *orbit invariants.*

We recall from, say, [Lau, chapters IV and V] some basic facts about resolutions of normal surface singularities. Such a resolution is called *good* if its exceptional fibre is a union of smooth curves E_i intersecting transversely wherever they intersect and such that no three of them meet in one point. The intersection behaviour is encoded in the dual *resolution graph.* The vertex representing the curve E_i becomes weighted by the self-intersection number E_i^2. The *intersection matrix* $(E_i \cdot E_j)$ is negative definite. Every normal surface singularity X has a unique *minimal good resolution* $\tilde{X} \to X$. Minimality means that every good resolution factors through $\tilde{X} \to X$. A good resolution is minimal if and only if contraction of any rational (-1)-curve leads to a resolution which is not good anymore.

A cyclic quotient singularity $X_{\alpha,\beta} = \mathbb{C}^2/C_{\alpha,\beta}$ can be modified by $\widetilde{\mathbb{C}^2}/C_{\alpha,\beta}$ where $\widetilde{\mathbb{C}^2} \to \mathbb{C}^2$ is the blowup of the origin. The surface $\widetilde{\mathbb{C}^2}/C_{\alpha,\beta}$ is smooth except a cyclic quotient singularity of type (β, γ) where $\alpha = b\beta - \gamma$ with γ coprime to β and $0 \leq \gamma < \beta$. Iteration of such modifications leads to the minimal good resolution. The exceptional fibre is a string of rational curves. Their self-intersection numbers $-b_1, \ldots, -b_m$ are given by the continued fraction expansion

$$\alpha/\beta = b_1 - 1/(b_2 - 1/(\ldots - 1/b_m)\ldots).$$

A good reference for these results is [Lam, chapter IV, §§5 and 6].

Exercise 2. Show that the intersection matrix of the minimal good resolution of $X_{\alpha,\beta}$ has determinant $(-1)^m\alpha$.

Resolving the cyclic quotient singularities on the singular line bundle $\Lambda \to Y$ of Theorem 2.1 one obtains

Theorem 2.2. (Orlik, Wagreich [OW]) *Every weighted homogeneous normal surface X has a good resolution $\tilde{X} \to X$ with a star shaped resolution graph Γ. The center of Γ is the curve $Y = (X \setminus \{0\})/\mathbb{C}^*$. The branches of Γ correspond to the exceptional orbits. The vertices on the branches are rational curves and their self-intersection numbers are calculated from the orbit invariants (α, β) as above.*

There are formulas for genus and self-intersection number of the central curve in terms of the orbit invariants and data of the graded coordinate ring of X, say its Poincaré series. For details the reader is advised to consult [D3, section 3], [OW, sections 3.5 and 3.6], [O, section 3.11] and [Wag, section 2.2]. The special case of hypersurfaces will be treated in section 4 below.

In the sequel, we suppose that

- X is not a cyclic quotient.

Then, by Brieskorn's result [B, Korollar 2.12] on tautness of quotient singularities, the central curve Y can be rational only if the number of branches of Γ is at least three. In particular, the resolution of Theorem 2.2 is the minimal good resolution.

3. Automorphy factors

Next we sketch the description of weighted homogeneous surfaces by automorphy factors due to Dolgachev and Pinkham. Given a smooth compact curve C and a group H of automorphisms of C, an *automorphy factor* with respect to H is an H-line bundle $L \to C$ such that H acts freely outside the zero-section.

Let $p_1, \dots, p_r \in Y = (X \setminus \{0\})/\mathbb{C}^*$ be the exceptional orbits with orbit invariants (α_i, β_i), $i = 1, \dots, r$. As we suppose that $r \geq 3$ if Y is rational, see above, Fenchel's conjecture proven in [BN] and [F] can be applied to yield a smooth compact curve C and a finite group $H \subseteq \operatorname{Aut} C$ such that Y is the orbit space C/H and the natural map $\nu : C \to Y$ is ramified exactly over the points p_i with multiplicity α_i in every point q above p_i. Let g and g' denote the genus of Y and C. The Riemann-Hurwitz formula gives

$$2g' - 2 = \deg \nu \cdot (2g - 2 + r - \sum_{i=1}^{r} 1/\alpha_i).$$

One calls

$$\chi(X) = 2g - 2 + r - \sum_{i=1}^{r} 1/\alpha_i$$

the *virtual Euler characteristic*. By Exercise 1, the singular line bundle $\Lambda \to Y$ of Theorem 2.1 is obtained from a line bundle $\Lambda_0 \to Y$ by gluing the cyclic quotient singularities X_{α_i,β_i} into the fibres over the points p_i. Write $\Lambda_0 = \mathcal{O}(-D_0)$ with a divisor D_0 on Y. As explained in [Lam, chapter IV, §9], the smooth surface $\tilde{X}$ resolving the singularities of Λ is obtained by gluing the whole total space Λ_0 (without removing fibres) with the resolutions of the X_{α_i,β_i}. Hence the self-intersection number $-b = Y^2$ of Y in $\tilde{X}$ equals the degree of $\Lambda_0 \to Y$. Now consider the divisor

$$D = \nu^* D_0 - \sum_{i=1}^{r} \sum_{q \in \nu^{-1}(p_i)} \beta_i \cdot q$$

on C. Then the degree $-b'$ of the line bundle $L = \mathcal{O}(-D) \to C$ is

$$-b' = \deg \nu \cdot (-b + \sum_{i=1}^{r} \beta_i/\alpha_i).$$

The number

$$\mathrm{vdeg}(X) = -b + \sum_{i=1}^{r} \beta_i/\alpha_i$$

is called the *virtual degree.*

Exercise 3. Use the negative definiteness of the intersection matrix of the resolution $\tilde{X} \to X$ and Exercise 2 to show $\sum_{i=1}^{r} \beta_i/\alpha_i < b$.

Hence $L \to C$ is a negative line bundle and its zero-section can be contracted to create a normal surface Z as in Example 2.

Theorem 3.1. (Dolgachev [D3], Pinkham [Pi]) *The line bundle $L \to C$ is an automorphy factor with respect to H and the quotient $L/H \to C/H = Y$ is a singular line bundle isomorphic to $\Lambda \to Y$. The obvious actions of H and $\mathbb{C}^*$ on Z commute yielding a good $\mathbb{C}^*$-action on Z/H, and the surfaces X and Z/H are isomorphic $\mathbb{C}^*$-equivariantly.*

For the proof see [D3, section 4] and [Pi, section 3].

Example 4. We explain the description by automorphy factors for quotients $X = \mathbb{C}^2/G$ with $G \subseteq \mathrm{GL}(2,\mathbb{C})$ finite. For more details consult [Lam, §§4 and 7-9]. By [Pr, Proposition 6] one may assume without loss of generality that G contains no pseudo-reflections, i.e. no elements having only one eigenvalue different from 1. Let $C_m \subseteq G$ be the kernel of the natural map $G \to \mathrm{PGL}(2,\mathbb{C})$ and $H = G/C_m$ its image. The quotient $\mathcal{O}(-1)/C_m \to \mathbb{P}_1/C_m = \mathbb{P}_1$ is isomorphic to $\mathcal{O}(-m) \to \mathbb{P}_1$. It is an automorphy factor with respect to H because G contains no pseudo-reflections. The contraction of its zero-section is $\mathbb{C}^2/C_m$ and $X = \mathbb{C}^2/G \simeq (\mathbb{C}^2/C_m)/H$. Now specialize to Kleinian singularities $X = \mathbb{C}^2/G$ with G a non-cyclic subgroup of $\mathrm{SL}(2,\mathbb{C})$. By averaging the standard Hermitean product on $\mathbb{C}^2$ over G we may assume $G \subseteq \mathrm{SU}(2)$, hence G is the binary group corresponding to the rotation group $H \subseteq \mathrm{SO}(3)$ of a regular k-gon, tetrahedron, octahedron or icosahedron. Such a group has three exceptional orbits on $\mathbb{P}_1$, namely the mid-edge points, face centers and vertices of the polyhedron at hand. The orders $(\alpha_1, \alpha_2, \alpha_3)$ of the corresponding isotropy groups are $(2,k,2)$, (2,3,3), (2,3,4) and (2,3,5), respectively. In this case, the number m from above is 2. The line bundle $\mathcal{O}(-2) \to \mathbb{P}_1$ is the cotangent bundle and, as an H-bundle, it is the canonical one. Since the cotangent map of multiplication with ζ is multiplication with ζ^{-1} the actions of the isotropy groups near the exceptional points have parameters $(\alpha_i, \alpha_i - 1)$. As the quotient $\mathcal{O}(-2)/H \to \mathbb{P}_1/H$ is isomorphic to the singular line bundle of X one can use the equation $-2 = |H| \cdot \mathrm{vdeg}(X)$ to derive $-b = -2$. Moreover $\mathbb{P}_1/H \simeq \mathbb{P}_1$ by the Riemann-Hurwitz formula. Hence the resolution graphs are the Coxeter-Dynkin diagrams D_{k+2}, E_6, E_7, and E_8, respectively, with all vertices representing projective lines and weighted by the self-intersection number -2:

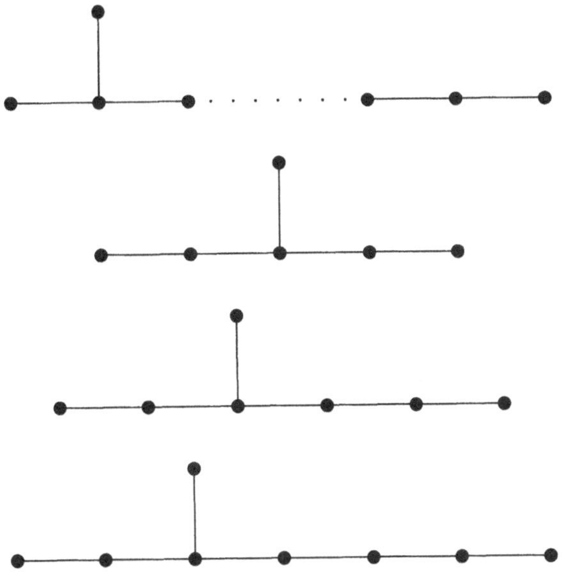

4. The hypersurface case

We shall calculate, for a hypersurface $X \subseteq \mathbb{C}^3$ defined by a weighted homogeneous polynomial f, the genus and the self-intersection number of the central curve Y in terms of the degree of f, the weights and the orbit invariants. Let us first consider, more generally, the Gorenstein case. A weighted homogeneous surface X will be called *Gorenstein* of *exponent* $\varepsilon = \varepsilon(X) \in \mathbb{Z}$ if there is a nowhere vanishing 2-form ω on $X \setminus \{0\}$ such that $t^*\omega = t^{-\varepsilon} \cdot \omega$ for all $t \in \mathbb{C}^*$.

Theorem 4.1. *Assume that X is Gorenstein of exponent $\varepsilon(X)$. Then the virtual degree and the virtual Euler characteristic are related by*

$$\varepsilon(X) \cdot \operatorname{vdeg}(X) = -\chi(X).$$

Proof. We use notation and results of section 3. Since H acts freely outside the zero-section of $L \to C$ the natural map $\nu : Z \to Z/H = X$ is unramified outside the singular point. Hence ω can be pulled back to a nowhere vanishing 2-form ω' on $Z \setminus \{0\}$. Since ν is $\mathbb{C}^*$-equivariant the surface Z is Gorenstein of the same exponent $\varepsilon = \varepsilon(X)$. This is equivalent to the property that the line bundle $L^{\otimes \varepsilon} \to C$ is isomorphic to the tangent bundle of C, see [D2, section 1] and [S2, section 5.5]. In particular,

$$-b' \cdot \varepsilon = \deg(L^{\otimes \varepsilon} \to C) = -(2g' - 2).$$

This is the claim. $\square$

Example 5. The tangent bundle of an elliptic curve is trivial. Hence simple elliptic singularities obtained by contracting the zero-section of a negative line bundle over an elliptic curve are Gorenstein of exponent 0.

Generalizing Example 3 we have

Theorem 4.2. *Let $X \subseteq \mathbb{C}^3$ be a hypersurface defined by the weighted homogeneous polynomial f of degree d and weights w_0, w_1, w_2. Then the central curve Y in the minimal good resolution has Euler characteristic*

$$2g - 2 = \frac{d(d - w_0 - w_1 - w_2)}{w_0 w_1 w_2} - r + \sum_{i=1}^{r} 1/\alpha_i$$

and self-intersection number

$$-b = -\frac{d}{w_0 w_1 w_2} - \sum_{i=1}^{r} \beta_i/\alpha_i.$$

Proof. By hypothesis, f has an isolated singularity at 0. Therefore,

$$\frac{dz_0 \wedge dz_1}{\partial_{z_2} f} = -\frac{dz_0 \wedge dz_2}{\partial_{z_1} f} = \frac{dz_1 \wedge dz_2}{\partial_{z_0} f}$$

defines a nowhere vanishing 2-form ω on $X \backslash \{0\}$. Hence X is Gorenstein of exponent $\varepsilon = d - w_0 - w_1 - w_2$. By Theorem 4.1 it is enough to prove the formula for the self-intersection number. The group $H = \mathbb{Z}/w_0 \times \mathbb{Z}/w_1 \times \mathbb{Z}/w_2$ acts on $\mathbb{C}^3$ by componentwise multiplication. The quotient map is

$$\phi : \mathbb{C}^3 \to \mathbb{C}^3/H = \mathbb{C}^3 : (z_0, z_1, z_2) \mapsto (z_0^{w_0}, z_1^{w_1}, z_2^{w_2}).$$

The preimage $X' = \phi^{-1}(X)$ is defined by the homogeneous polynomial $F = f \circ \phi$ of degree d. The curve $Y' \subseteq \mathbb{P}_2$ defined by F may have singularities. Let $\nu : C \to Y'$ be its normalization, $L' \to Y'$ the restriction of the tautological line bundle $\widetilde{\mathbb{C}^3} \to \mathbb{P}_2$ and $L \to C$ its pullback. The group H acts on C and L with quotient maps $C \to C/H = Y$ and $L \to L/H = \Lambda$, the singular line bundle of Theorem 2.1. Similarly as in the (omitted) proof of Theorem 3.1 one shows, see [OW, section 3.6], that

$$-d = \deg(L \to C) = \deg(\phi \circ \nu) \cdot \operatorname{vdeg}(X)$$

with $\deg(\phi \circ \nu) = |H| = w_0 w_1 w_2$, as claimed. □

Exercise 4. Use the equations of the Kleinian singularities as hypersurfaces in $\mathbb{C}^3$, see for instance [Lo, chapter 1.C], to determine the orbit invariants and the data of the central curve. Reprove in this way that the resolution graphs are the Coxeter-Dynkin diagrams with all vertices weighted by -2.

Example 6. Consider $f = z_0^6 + z_1^3 + z_2^2$ with weights 1, 2, 3 and $\deg f = 6$. Since the weights are pairwise coprime there are no exceptional orbits. Theorem 4.2 gives $g = 1$ and $-b = -1$. This is a simple elliptic singularity, [S1].

Example 7. Consider $f = z_0^7 + z_1^3 + z_2^2$ with weights 6, 14, 21 and $\deg f = 42$. There are exactly three exceptional orbits, the intersections of X with the coordinate hyperplanes. The orders α of the isotropy groups are 7, 3, 2. The reader may check that the three numbers β are all 1. Theorem 4.2 gives $g = 0$ and $-b = -1$. Thus, the resolution graph is

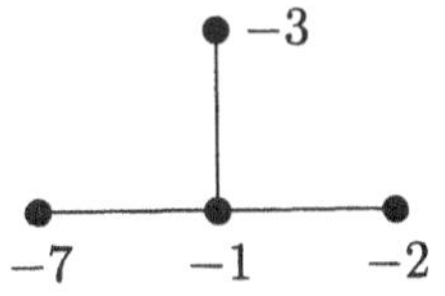

This is the weighted homogeneous member of one of the fourteen exceptional families of unimodal singularities, [AGV, part II], i.e., one of the fourteen triangle singularities which can be embedded as a hypersurface in $\mathbb{C}^3$, [D1].

5. Automorphisms

We apply Theorem 2.1 to obtain information on the group $\mathcal{A}_X$ of $\mathbb{C}^*$-equivariant automorphisms of X, i.e., automorphisms respecting the weighted homogeneous structure:

Theorem 5.1. *Suppose that X is not a cyclic quotient. Then $\mathcal{A} = \mathcal{A}_X$ is a one-dimensional algebraic group. The finite group $\mathcal{A}/\mathbb{C}^*$ embeds into the group of automorphisms of the central curve Y.*

Proof. Every $\phi \in \mathcal{A}$ induces an automorphism of $Y = (X \setminus \{0\})/\mathbb{C}^*$. Moreover, $\mathcal{A}$ acts on Λ such that the modification $\sigma : \Lambda \to X$ and the singular line bundle $\pi : \Lambda \to Y$ are $\mathcal{A}$-equivariant. Every element of the kernel of the homomorphism $\rho : \mathcal{A} \to \operatorname{Aut} Y$ acts on the fibres $\pi^{-1}(p)$ by multiplication with a constant depending analytically on p. As Y is compact the constant is the same for all fibres and the kernel is equal to $\mathbb{C}^*$. It remains to show that the image of ρ is finite. This is clear if the genus g of Y is at least two. Then, in fact, by the Hurwitz bound, $|\mathcal{A}/\mathbb{C}^*| \leq 84(g-1)$. Every element of $\mathcal{A}$ permutes the exceptional orbits $p_1, \dots, p_r \in Y$. The group $\operatorname{Aut}_0 Y$ of automorphisms of an elliptic curve Y fixing a given point $0 \in Y$ is finite cyclic. An automorphism of $\mathbb{P}_1$ with three fixed points is trivial. Therefore, the kernel of the induced map from $\operatorname{im}(\rho)$ to the symmetric group S_r is finite if $g = 1$ and $r \geq 1$ or if $g = 0$ because then, by hypothesis and as mentioned at the end of section 2, we have $r \geq 3$. Finally consider a simple elliptic X, i.e., Y is an elliptic curve and $r = 0$. Then $\Lambda \to Y$ is a locally trivial line bundle which we may assume to be $\mathcal{O}(-b \cdot 0)$ with 0 the zero element of the Abelian group Y and $-b = Y^2$. The group of automorphisms of Y is $Y \rtimes \operatorname{Aut}_0 Y$ with Y acting on itself by translations. The line bundles $\mathcal{O}(-b \cdot p)$ and $\mathcal{O}(-b \cdot 0)$ are isomorphic if and only if the divisors $-b \cdot p$ and $-b \cdot 0$ are linearly equivalent if and only if $bp = 0$ in the group Y. Using this one shows that $\operatorname{im}(\rho) \simeq (\mathbb{Z}/b \times \mathbb{Z}/b) \rtimes \operatorname{Aut}_0 Y$

where $\mathbb{Z}/b \times \mathbb{Z}/b \subseteq Y$ is the group of b-torsion points of the group Y, see [M, Theorem 3]. □

For $g = 0$ it follows from the given proof that $\mathcal{A}/\mathbb{C}^*$ embeds into the group $\mathrm{Aut}\,\Gamma$ of automorphisms of the weighted resolution graph Γ. In fact $\mathcal{A} \simeq \mathbb{C}^* \times \mathrm{Aut}\,\Gamma$ when $g = 0$ and $r = 3$, see [M, Corollary to Theorem 4]. This applies, for instance, to all non-cyclic quotients [B, section 2.4] and to Dolgachev's triangle singularities [D1], [Lo, chapter 1.D]. More details can be found in [M] and [Wah, section 3].

Acknowledgment. The author thanks Claus Hertling for some explanations concerning singular line bundles.

References

[AGV] Arnold, V. I., Gusein-Zade, S. M., Varchenko, A. N.: Singularities of differentiable maps, vol. I. Birkhäuser, Boston 1985.

[B] Brieskorn, E.: Rationale Singularitäten komplexer Flächen. Invent. Math. **4** (1968), 336–358.

[BN] Bundgaard, S., Nielsen, J.: On normal subgroups with finite index in F-groups. Matematisk Tidsskrift B (1951), 56–58.

[D1] Dolgachev, I. V.: Quotient-conical singularities on complex surfaces. Funct. Anal. Appl. **8** (1974), 160–161.

[D2] Dolgachev, I. V.: On the link space of a Gorenstein quasihomogeneous surface singularity. Math. Ann. **265** (1983), 529–540.

[D3] Dolgachev, I. V.: Automorphic forms and weighted homogeneous equations. Unpublished.

[F] Fox, R. H.: On Fenchel's conjecture about F-groups. Matematisk Tidsskrift B (1952), 61–65.

[H] Holmann, H.: Quotientenräume komplexer Mannigfaltigkeiten nach komplexen Lieschen Automorphismengruppen. Math. Ann. **139** (1960), 383–402.

[Lam] Lamotke, K.: Regular solids and isolated singularities. Vieweg, Braunschweig 1986.

[Lau] Laufer, H. B.: Normal two-dimensional singularities. Ann. of Math. Study 71. Princeton University Press, Princeton 1971.

[Lo] Looijenga, E. J. N.: Isolated singular points on complete intersections. Cambridge University Press, Cambridge 1984.

[M] Müller, G.: Symmetries of surface singularities. J. London Math. Soc., to appear.

[O] Orlik, P.: Seifert manifolds. Springer, Berlin, Heidelberg, New York 1972.

[OW] Orlik, P., Wagreich, P.: Isolated singularities of algebraic surfaces with $\mathbb{C}^*$-action. Ann. of Math. (2) **93** (1971), 205–228.

[Pi] Pinkham, H.: Normal surface singularities with $\mathbb{C}^*$-action. Math. Ann. **227** (1977), 183–193.

[Pr] Prill, D.: Local classification of quotients of complex manifolds by discontinuous groups. Duke Math. J. **34** (1967), 375–386.

[S1] Saito, K.: Einfach-elliptische Singularitäten. Invent. Math. **23** (1974), 289–325.

[S2] Saito, K.: Regular system of weights and associated singularities. In: Complex analytic singularities. Advanced Studies in Pure Mathematics 8. North-Holland, Amsterdam, New York, Oxford 1986.

[Wag] Wagreich, P.: The structure of quasihomogeneous singularities. Proc. Symp. Pure Math. **40**, part 2 (1983), 593–611.

[Wah] Wahl, J. M.: Derivations, automorphisms and deformations of quasi-homogeneous singularities. Proc. Symp. Pure Math. **40**, part 2 (1983), 613–624.

Fachbereich Mathematik
Universität Mainz
D 55099 Mainz, Germany
mueller@mat.mathematik.uni-mainz.de

Progress in Mathematics, Vol. 181,

Alterations and Birational Anabelian Geometry

Florian Pop

1. Introduction

The basic idea of Grothendieck's anabelian geometry is that under certain "anabelian hypotheses" the étale fundamental group of a scheme contains all the geometric and arithmetic information about the scheme in discussion, that is to say, the scheme is *functorially encoded* in its étale fundamental group. Such ideas are not completely new, the first assertion of this type being the celebrated result of ARTIN-SCHREIER from the middle of the Twenties, which asserts that if the absolute Galois group of some field is non-trivial and finite, then the field in discussion is real closed. This is nevertheless not an assertion about the structure/isomorphy type of the field in discussion, but rather about its elementary theory. It was the attempt to give a *p-adic analogue* of the Artin-Schreier Theorem which lead Neukirch to the question whether the isomorphy type of a number field (as a field) is *group theoretically* encoded in the isomorphy type of the absolute Galois group (as a profinite group) of the number field in discussion. After partial results by several authors, this question was positively settled by the following quite fundamental theorem, which in its final form was proved by Uchida in the middle of Seventies:

Theorem 1.1. *Let K, L be global fields, and let $\Phi : G_K \to G_L$ be an isomorphism of their absolute Galois groups G_K and G_L. Then there exists a unique field isomorphism $\phi : L^s \to K^s$ of corresponding separable closures such that $\Phi(g) = \phi^{-1}\, g\, \phi$ for all $g \in G_K$.*

In particular, if G_K and G_L are isomorphic (as profinite groups), then K and L are isomorphic as fields. Moreover, the canonical mapping

$$\mathrm{Isom}(L, K) \to \mathrm{Isom}(G_K, G_L)/\mathrm{Inn}(G_L)$$

is a bijection, where Inn *denotes the group of inner automorphisms acting in a natural way on* Isom.

By a global field here we mean a global field in the sense of number theory, i.e., either a number field or a function field of one variable over a finite field. The natural context in which the above result appears as a prominent example is

Grothendieck's anabelian geometry, see [G1], [G2]. Here is one way to formulate the hints/questions posed by Grothendieck.

Let X be a connected scheme, and let $\pi_1(X)$ denote the étale fundamental group of X (with respect to some fixed geometric point x of X). We will say that X is *anabelian* if the isomorphy type of X up to pure inseparable covers is encoded group theoretically in $\pi_1(X)$ in a functorial way, or equivalently, if there exists a *group theoretic* recipe to recover the isomorphy type of X up to pure inseparable covers from $\pi_1(X)$. Next we want to define anabelian categories of schemes. For that let first $\mathcal{G}$ be the category of all profinite groups with outer isomorphisms as morphisms. This means that for $G, H \in \mathcal{G}$ one has by definition $\mathrm{Hom}_{\mathcal{G}}(G,H) = \mathrm{Isom}(G,H)/\mathrm{Inn}(H)$, where Isom denotes isomorphisms of profinite groups, and Inn denotes the group of inner isomorphisms acting in the obvious way on Isom. Now let $\mathcal{A}$ be a category whose objects are anabelian schemes and the morphisms are scheme isomorphisms up to pure inseparable covers. Then choosing for every $X \in \mathcal{A}$ a geometric point of X one gets a covariant functor

$$\mathcal{A} \to \mathcal{G}, \quad X \mapsto \pi_1(X).$$

We will say that $\mathcal{A}$ is an anabelian category, if the above functor is fully faithful, i.e., if $\mathcal{A}$ is equivalent to a full subcategory of $\mathcal{G}$. Equivalently, for every $X, Y \in \mathcal{A}$ one has

$$\mathrm{Isom}_{\mathcal{A}}(X,Y) \cong \mathrm{Isom}\big(\pi_1(X), \pi_1(Y)\big)/\mathrm{Inn}\big(\pi_1(Y)\big)$$

in a functorial way.

Correspondingly, one can also define S-anabelian schemes X_S and S-anabelian categories $\mathcal{A}_S$, where S is a fixed, connected base scheme. Namely, an S-anabelian category $\mathcal{A}_S$ is an anabelian category in the usual sense, but the $\mathcal{A}_S$-morphism are morphisms $\phi : X \to Y$ in $\mathcal{A}_S$ viewed as an anabelian category, such that there exists an isomorphism ϕ_S (up to a pure inseparable cover of S) such that

$$\begin{array}{ccc} X & \xrightarrow{\ \phi\ } & Y \\ \downarrow & & \downarrow \\ S & \xrightarrow{\ \phi_S\ } & S \end{array}$$

is commutative. Correspondingly, one has to replace $\mathcal{G}$ by the category $\mathcal{G}_S$ of all profinite groups G with an "augmentation" morphism $G \to \pi_1(S)$, and define correspondingly the notion of a morphism in $\mathcal{G}_S$. The fundamental group functor defined above defines then a covariant functor $\mathcal{A}_S \to \mathcal{G}_S$, and $\mathcal{A}_S$ is called S-anabelian, if the above functor is fully faithful.

The idea of Grothendieck is that schemes, as well as categories of such schemes (even considered over a base scheme S which is of finite type over $\mathbf{Z}$) that one is mostly interested in arithmetic geometry are anabelian (respectively S-anabelian). If so, then these categories are equivalent to full subcategories of $\mathcal{G}$, respectively $\mathcal{G}_S$, and then one can reformulate questions in arithmetic geometry in purely group-theoretic terms. This is indeed a great vision! Thus the problem of describing anabelian schemes, respectively S-anabelian schemes is of fundamental importance. The first conjecture Grothendieck does in this respect is the so called "fundamental conjecture of the birational anabelian geometry". After defining a finitely generated field as being a field which is finitely generated over its prime field, the conjecture in discussion is the following

Conjecture 1.2 (Birational Conjecture). *The category of all finitely generated infinite fields is anabelian.*

Equivalently, if K is a finitely generated, infinite field, then up to a pure inseparable extension, K is group theoretically encoded in its absolute Galois group G_K in a functorial way. Further, if K and L are such fields then the canonical mapping

$$\mathrm{Isom}(L^{\mathrm{i}}, K^{\mathrm{i}}) \to \mathrm{Isom}(G_K, G_L)/\mathrm{Inn}(G_L)$$

is a bijection, where $(\ )^{\mathrm{i}}$ means "up to a pure inseparable cover".

Before going into further details concerning birational anabelian geometry, let us remember that one defines the Kronecker dimension $\dim(K)$ of a finitely generated field K as being its absolute transcendence degree $\mathrm{td}(K)$ if K has positive characteristic, and $\mathrm{td}(K) + 1$ else. Hence $\dim(K) = 1$ if and only if K is a global field. Further we remark that for K and L finitely generated infinite fields, if $G_K \cong G_L$ then $\dim(K) = \dim(L)$ and $\mathrm{char}(K) = \mathrm{char}(L)$. This follows by the fact that $\dim(K)$ and $\mathrm{char}(K)$ are both encoded in G_K as follows: First, there exists at most one prime number p such that $\mathrm{cd}_p(G_K) = 1$, and if such a p exists, then $\mathrm{char}(K) = p$, and second, for all other prime numbers ℓ one has $\dim(K) = \mathrm{cd}_\ell(G_K) - 1$. See [S1], Ch. II, especially §4, for details on the cohomological dimension of (finitely generated) fields.

Coming back to the Birational Conjecture, we remark that Theorem 1 above gives a positive answer to the Birational Conjecture in the case of finitely generated fields K with $\dim(K) = 1$. A first proof of the Birational Conjecture in general was given in Pop [P2].

Theorem 1.3. *Every finitely generated infinite field K is up to a pure inseparable extension group theoretically encoded in G_K in a functorial way. Further, if L is a finitely generated field, then there exists a functorial bijection*

$$\mathrm{Isom}\,(L^{\mathrm{i}}, K^{\mathrm{i}}) \longrightarrow \mathrm{Isom}\,(G_K, G_L)/\mathrm{Inn}(G_L),$$

where $(\ )^{\mathrm{i}}$ means "up to pure inseparable cover", and Inn *acts in the obvious way on* Isom.

The proof in loc.cit. is by induction on $d = \dim(K)$ the Kronecker dimension of K. The case $d = 1$ is the one treated in Theorem 1.1 above. In the proof given in loc.cit., the induction step "$d \Rightarrow d+1$" is by no means obvious. We will explain in the next section the strategy of the proof. Among other things one needs a more technical result which relies on the existence of complete, regular models for function fields. The existence of such models in characteristic zero follows from Hironaka's *Desingularization Theorem* [H]. In the positive characteristic case the existence of such models is not known in general, but for our purposes we can use the *theory of alterations* by de Jong [J].

We will describe in the following two sections the main lines of the proof, emphasizing the relevance of the theory of alterations for the proof of the Birational Conjecture.

Remark. We end up this section with the following remarks, both for the sake of completeness, but mostly because of the relevance of the facts we want to mention for the anabelian phenomena.

1. First we would like to mention that one can even give a *pro-ℓ version* of the Birational Conjecture as follows: Let ℓ be a fixed prime number. Then we can consider the full subcategory $\mathcal{G}^\ell$ of $\mathcal{G}$ introduced above whose objects are pro-ℓ groups. There exists a (canonical) functor from $\mathcal{G}$ to $\mathcal{G}^\ell$ which sends every $G \in \mathcal{G}$ to its maximal pro-ℓ quotient $G \mapsto G^\ell$. For any scheme $\mathcal{A}$ as above we can then consider the composition
$$\mathcal{A} \to \mathcal{G} \to \mathcal{G}_\ell, \quad X \mapsto \pi_1^\ell(X)$$
and formulate *ℓ-anabelian* questions as above. Naturally, one cannot expect that all schemes which are anabelian are also ℓ-anabelian. One can nevertheless prove that every finitely generated field which contains the ℓth roots of unity and has Kronecker dimension > 1 is ℓ-anabelian. Even more, the category of all such fields is ℓ-anabelian. One expects that the same is true also for global fields. We remark that one can also prove similar assertions for "truncated" quotients of the maximal pro-ℓ quotient of G_K.
2. Second, the birational anabelian geometry should be viewed (from the geometric point of view) as a 0-dimensional anabelian geometry. Grothendieck made precise conjectures about what the 1-dimensional anabelian varieties should be: These are the hyperbolic curves over finitely generated fields. Here, a hyperbolic curve over some base field κ is a geometrically integral, smooth curve over κ, whose geometric étale fundamental group is non-abelian. But it is not clear which higher dimensional schemes are/should be anabelian.
3. After some partial results by Nakamura [Na], it was Tamagawa [T1] who proved that the category of all *affine,* hyperbolic curves over finite fields is anabelian, and as a consequence, he derived the same result about the category of all affine curves over finitely generated fields of *characteristic zero.*

   Using this fact, Mochizuki [M1] proved that the category of all hyperbolic curves over finitely generated fields of characteristic zero is anabelian. Finally, with a new point of view, namely employing p-adic Hodge-Tate techniques, Mochizuki [M2], proved that the category of all hyperbolic curves over any subfield κ of a *finitely generated field extension of the p-adics* is κ-anabelian.

From this he deduced a *new proof* of the Birational Conjecture – in the relative situation over κ as above –, even of the much stronger form of it, namely of the so called Hom-form.

4. Third, the facts mentioned above shad new light on anabelian phenomena. In contrast to GROTHENDIECK's idea – where "anabelian" was related to "finite generacy", hence to strong arithmetic hypotheses –, it appears that the anabelian type phenomena are much more widespread. Indeed, as explained above, they appear in a pro-ℓ context; and over finite fields; and over p-adic fields. Moreover, answering a question by Harbater [Ha], Tamagawa [T2] shows that the positive characteristic alone has some anabelian flavour.

   For more information about the case of global fields see [N1], [N2], [N3], [Ik], [Km], [U1], [U2], [U3]. See also Speiss [Sp]. Also, for more facts about the anabelian phenomena and related questions see the corresponding articles by Ihara-Nakamura [I-N], Pop [P3] from [SL].

2. Outline of the proof of the Birational Conjecture

Let K be a finitely generated field, p its characteristic exponent, and G_K its absolute Galois group. For every integer m with $(p, m) = 1$ the Kummer Theory gives a functorial isomorphism $\delta_m : K^\times/m \to \mathrm{H}^1\big(K, \mathbf{Z}/m(1)\big)$, hence taking limits over all integers m as above we get a functorial isomorphism of m-adically complete groups

$$\hat{\delta} : \hat{K} \to \mathrm{H}^1\big(K, \hat{\mathbf{Z}}(1)\big)$$

where $\hat{K}$ and $\hat{\mathbf{Z}}$ denote the m-adic completion of the multiplicative group $K^\times$ of K, respectively of $\mathbf{Z}$ with m as above, and hence H^1 is the m-adic cohomology of K. A first remark then is: The knowledge of the cyclotomic character χ_K of K (and that of χ_M for all $M|K$ finite and separable) allows one to recover $\hat{K}$ (and further $\hat{M}$ for $M|K$ as above). Before going further we remark that the completion morphism $\hat{\jmath} : K^\times \to \hat{K}$ is injective. This is not a completely trivial fact: First, the torsion μ of $K^\times$ consists exactly of the roots of unity in K, hence it is finite. Second, $K^\times/\mu$ is a free abelian group (this follows by induction on $\dim(K)$, by using e.g. the *divisor theory* for Dedekind domains). As consequences of these two remarks we are confronted with the following questions:

1. Give a group theoretic recipe to decode χ_K from G_K.
2. Give a group theoretic recipe to detect the image $\hat{\jmath}(K^\times)$ inside $\hat{K}$.

In particular, the answer to 2) gives the *multiplicative structure* of K.

For approaching both of the above questions, an essential step is to develop a *local theory,* which roughly speaking means to describe how geometric information of local nature about K is encoded group theoretically in G_K.

2.1. Local Theory

Let K be a finitely generated infinite field. A *model* for K is by definition a separated, integral scheme of finite type over $\mathbf{Z}$ whose function field is K. One can speak about (quasi)projective or (quasi)affine models, as well as about regular or normal models, etc. In particular, if X is a model of K, then $\dim(K)$ equals $\dim(X)$ as a scheme. One has:

I) K has Kronecker dimension 1 (i.e., K is a global field in the usual terminology in number theory) if and only if every normal model X of K is an arithmetic curve, and in this case there exists a unique, normal, "maximal" model X_K of K, in the sense that any other normal model X for K is (isomorphic to) an affine open of X_K. Indeed, if K is a number field and $\mathcal{O}_K$ is its ring of integers, then $X_K := \mathrm{Spec}(\mathcal{O}_K)$; if K is a global function field, hence a function field of one variable over a finite field $\mathbf{F}$, then X_K is the unique projective, smooth curve over $\mathbf{F}$ with function field K. Further, there exists a natural bijection between the *prime Weil divisors* of X_K and the non-archimedean places of K. Back to Local Theory, it is a basic result of Neukirch [N1], Satz 4, which is used in an essential way in the proof of the Birational Conjecture and which asserts, that the non-archimedean places of K are group theoretically encoded in G_K in the following way: For every non-archimedean place $\mathfrak{p}$ of K let $Z_{\mathfrak{p}}$ be the decomposition group of some prolongation $\mathfrak{p}^{\mathrm{s}}$ of $\mathfrak{p}$ to K^{s}. Then one has: If Z is a closed subgroup of G_K and $Z \cong Z_{\mathfrak{p}}$ for some $\mathfrak{p}$, then Z is contained in some $Z_{\mathfrak{q}}$ for some place $\mathfrak{q}$ of K. On the other hand, by a theorem of F. K. Schmidt, if $Z_{\mathfrak{p}} \cap Z_{\mathfrak{q}}$ is non-trivial, for some $\mathfrak{p}, \mathfrak{q}$ places of K, then $\mathfrak{p} = \mathfrak{q}$, $\mathfrak{p}^{\mathrm{s}} = \mathfrak{q}^{\mathrm{s}}$ and $Z_{\mathfrak{p}} = Z_{\mathfrak{q}}$. Thus:

Every subgroup Z of G_K which is isomorphic to a decomposition group is contained in a maximal subgroup Z_0 of G_K with this property, and the conjugacy classes of the maximal ones are in bijection with the places of K.

This gives then the group theoretic recipe for describing the prime Weil divisors of X_K, even in a functorial way.

II) In general, i.e., if K is not necessarily a global field, there is a variety of models for K, with apparently no obvious relations among themselves and/or their prime Weil divisors. In particular, we cannot hope to obtain much information about a specific model X of K, as in general there is no priviledged model for K as in the global field case. One can nevertheless do the following: For every model X of K one can choose an isomorphic copy X_i of X such that the structure sheaf of X is a sheaf of subrings of K with inclusions as structure morphisms. On the space of all the X_i's there exists a naturally defined domination relation defined as follows: $X_j \geq X_i \Leftrightarrow$ there exists a surjective morphism $\varphi_{ji} : X_j \to X_i$ which at the structure sheaf level is defined by inclusions. Let $\mathcal{P}_K$ be the set of all projective, normal models of K. By Zariski-Samuel [Z-S], Ch.VI, especially §17, the following is well known:

- Every complete model is dominated by some $X_i \in \mathcal{P}_K$ (Chow Lemma, see e.g. [Hr], II, Ex.4.10).

- The set $\mathcal{P}_K$ is increasingly filtered with respect to $\geq$, hence it is a surjective projective system.
- Set $\mathfrak{D}_K = \varprojlim X_i$. Then the points of $\mathfrak{D}_K$ are in bijection with the space of all valuation rings of K. For $v = (x_i)_i$ in $\mathfrak{D}_K$ one has: x_i is the centre of v on X_i in the usual sense, see e.g. [A], §5.

We remark that for a point $v = (x_i)_i$ in $\mathfrak{D}_K$ as above, the following conditions are equivalent, one uses e.g. [BOU], Ch.IV, §3, or [P1], The Local Theory:

i) For i sufficiently large, x_i has codimension 1, or equivalently, x_i is the generic point of a prime Weil divisor of X_i. Hence v is the discrete valuation of K with valuation ring $\mathcal{O}_{X_i,x_i}$.

ii) The valuation v is discrete, and its residue field Kv is a finitely generated field with $\dim(Kv) = \dim(K) - 1$.

iii) $\dim(Kv) = \dim(K) - 1$.

A valuation v on K with the above equivalent properties is called a Zariski prime divisor of K. As a corollary of the observations above we have: *The space of all Zariski prime divisors of K is the union of the spaces of prime Weil divisors of all normal models of K* (if we identify every prime Weil divisor with the discrete valuation on K it defines). A Zariski prime divisor v is called geometric if K and Kv have equal characteristic (or equivalently, if v is trivial on the prime field of K), and arithmetic otherwise. We denote by $\mathfrak{D}^1_K \subset \mathfrak{D}_K$ the space of all geometric Zariski prime divisors of K.

We are now able to announce the main result of the higher dimensional Local Theory: For every Zariski prime divisor $v \in \mathfrak{D}_K$ of K let Z_v be the decomposition group of some prolongation v^{s} of v to K^{s}. Then one has, see e.g. [P1], Local Theory:

a) First, the numerical data $\mathrm{char}(K)$, $\mathrm{char}(Kv)$, and $\dim(K)$ are group theoretically encoded in Z_v, in particular, whether v is geometric or not. Further, the inertia group $T_v \subset Z_v$ of $v^{\mathrm{s}}|v$, and the canonical projection $\pi_v : Z_v \to G_{Kv}$ are also encoded group theoretically in Z_v.

b) Second, if Z is a closed subgroup of G_K which is isomorphic to Z_v for some v, then Z is contained in some Z_w for some Zariski prime divisor w of K.

On the other hand, by the result of F. K. Schmidt cited above, if $Z_v \cap Z_w$ is non-trivial, then $v = w$, $v^{\mathrm{s}} = w^{\mathrm{s}}$ and $Z_v = Z_w$. Hence, $\mathfrak{D}^1_K$ is encoded in G_K as the set of conjugacy classes of the maximal subgroups which are isomorphic to the decomposition groups of geometric Zariski prime divisors of K. In particular, if v and w are Zariski prime divisors of finitely generated fields K, resp. L with $Z_v \cong Z_w$, then K and L have the same Kronecker dimension, the same characteristic, and v and w have the same residue characteristic. Further, every isomorphism $Z_v \to Z_w$ induces in a functorial way an isomorphism $G_{Kv} \cong G_{Lw}$.

The facts above have the following consequence: Let K and L be finitely generated fields of Kronecker dimension > 1, and $\Phi : G_K \to G_L$ an isomorphism (of profinite groups). Then Φ defines a functorial bijection

$$\varphi : \mathcal{D}^1_K \to \mathcal{D}^1_L, \quad v \mapsto w$$

and Φ induces functorially "local" isomorphisms $\Phi_v : G_{Kv} \to G_{Lw}$. The mapping φ is called the *local correspondence* defined by Φ.

2.2. Consequences of the Local Theory

In the notations introduced above suppose that $\dim(K) = d > 1$. Using Theorem 1, we can suppose by induction on d, that the birational conjecture is true in Kronecker dimension $< d$. Thus, the local isomorphisms Φ_v are defined by field isomorphisms $\phi_v : (Lw)^{\mathrm{s}} \to (Kv)^{\mathrm{s}}$, i.e., $\Phi_v\big(\pi_v(g)\big) = \phi_v^{-1}\pi_v(g)\phi_v$ for all $g \in Z_v$. This provides a hint about how to proceed in order to prove the Birational Conjecture for finitely generated fields of Kronecker dimension d. Namely one should "patch together" the local data arising from the family of residue fields Kv, $v \in \mathcal{D}^1_K$, in order to get the "global" object K.

Next one shows that from the local information as mentioned above, the cyclotomic character χ_K of K is group theoretically encoded in G_K. Indeed, T_v is contained in the kernel of χ_K for all geometric Zariski prime divisors v, and the "local" cyclotomic characters χ_{Kv} are Galois theoretically encoded in G_{Kv} (as G_{Kv} "knows" everything about Kv, again by induction on the Kronecker dimension). Using the higher dimensional Chebotarev Density Theorem, as done by Serre [S2], it follows that $\ker(\chi_K)$ is generated by all the $\ker(\chi_{Kv} \circ \pi_v)$, $v \in \mathcal{D}^1_K$.

In summary, in the notations of the beginning of this section, one has a Galois theoretic description of the m-adic completion $\hat{K}$ of the multiplicative group of K via the isomorphism $\hat{\delta} : \hat{K} \to \mathrm{H}^1\big(K, \hat{\mathbf{Z}}(1)\big)$.

2.3. Global Theory

It is the task of a "global theory" to patch together (in other words, to "globalise") the local information obtained so far, and this means mainly to give a group theoretic way for describing the image of the completion functor $\hat{\jmath} : K^\times \to \hat{K}$. After this problem is solved, one gets the additive structure on $\hat{\jmath}(K) \cup \{0\}$ as follows: First, for every $v \in \mathcal{D}^1_K$, the canonical projection π_v has sections, see e.g. [K-P-R]. For every v let $G_v \subset Z_v \subset G_K$ be a complement of T_v in Z_v. The inclusion $\imath_v : G_v \to G_K$ defines via cohomology a homomorphism $\hat{K} \to (Kv)\hat{}$ which on the v-units is compatible with the residue map $\mathcal{O}_v \to Kv$, $x \mapsto xv$. For $x, y, z \in K^\times$, the equality $x + y = z$ holds in K if and only if residually $xv + yv = zv$ for "sufficiently many" $v \in \mathcal{D}^1_K$ where x, y, z are v-units. On the other hand, the residue field structure is encoded in G_{Kv}. Hence finally the addition is encoded in G_K.

3. Alterations and geometric sets of Zariski prime divisors

As we have seen at the end of the previous section, our task now is to describe the image of the completion functor $\hat{\jmath}$. This will be done by first getting more "geometry" from the absolute Galois group of a finitely generated field.

3.1. Alterations and lines on varieties

Let κ be an arbitrary field. By a κ-variety we mean a separated, integral κ-scheme of finite type $X \to \kappa$. A *line* on a κ-variety is by definition an integral κ-subvariety $\mathfrak{l}$ of X which is a curve of geometric genus equal to 0. We denote by X^{line} the union of all the lines on X. We will say that X is *non-uniruled* if the set X^{line} is not dense in X.

Remark. The following facts are more or less obvious:

1. Being *non-uniruled* is a birational notion.
2. If X is non-uniruled and $f : Y \to X$ is a generically finite morphism of κ-varieties, then Y is non-uniruled. (This is a consequence of the Lüroth Theorem.)
3. Every integral variety $X \to \kappa$ has "many" finite covers $Z \to X$ which are defined over κ such that $Z \to \kappa$ is non-uniruled. Indeed, as the question is of birational nature, we can suppose that X is a finite cover of the affine d-dimensional space $\mathbf{A}^d$, where d is the dimension of X. For every $1 \le j \le d$ we consider a finite, separable, geometrically integral cover $X_j \to \mathbf{A}^1$ which has geometric genus ≥ 1. We set $X_0 = \Pi_j X_j$, and we let $X_0 \to \mathbf{A}^d$ be the corresponding finite cover. Finally we let $Z' = X \times_{\mathbf{A}^d} X_0$ be the fibre product of X and X_0 over $\mathbf{A}^d$, and $Z' \to X$ the structural projection. Then the set of lines on Z' is empty (again, by the Lüroth Theorem), hence every irreducible component $Z \to X$ of Z' has the property we wanted.
4. Let $X \to \kappa$ be a normal, non-uniruled κ-variety. Then almost all prime Weil divisors X_1 of X are non-uniruled (when viewed as κ-varieties).

Next we want to give a birational interpretation of the facts above. A function field in $d \ge 0$ variables over κ is by definition a finitely generated field $K|\kappa$ with $\text{td}(K|\kappa) = d$. Let $K|\kappa$ be a function field in $d \ge 0$ variables. A model for $K|\kappa$ is any κ-variety $X \to \kappa$ together with a κ-identification of K with the function field $\kappa(X)$ of X. We will say that $K|\kappa$ is *non-uniruled* if $K|\kappa$ has models $X \to \kappa$ which are non-uniruled. By the remark 3) above every finitely generated field $K|\kappa$ has "many" finite extensions $M|K$ such that $M|\kappa$ is non-uniruled. A (Zariski) prime divisor of $K|\kappa$ is any valuation v of K which is trivial on κ and satisfies $\text{td}(Kv|\kappa) = \text{td}(K|\kappa) - 1$. As in Section 2, i), ii), iii), one shows that a valuation v on K is a prime divisor of $K|\kappa$ if and only if there exists a (projective) normal model X of K over κ such that v is defined by some prime Weil divisor of X. In particular, a Zariski prime divisor v of $K|\kappa$ is a discrete valuation of K and its residue field Kv is also a function field over κ. We call a Zariski prime divisor of $K|\kappa$

non-uniruled if $(Kv)|\kappa$ is non-uniruled. A prime divisor v of $K|\kappa$ is non-uniruled if and only if there exists a normal model $X \to \kappa$ of $F|\kappa$ and a non-uniruled prime Weil divisor W of X such that $v = v_W$ is the valuation of K defined by W. In general, if v is a Zariski prime divisor of $K|\kappa$ and X is a normal model of $K|\kappa$, we denote by x_v the centre of v on X, and by X_v the Zariski closure of x_v in X.

Our next aim is to give a birational variant/sharpening of the remark 4) above. This is a very essential step in the globalisation procedure we announced at the end of the last section. Its proof relies on de Jong's theory of alterations.

Theorem 3.1. *Let $K|\kappa$ be a function field. If $X \to \kappa$ is a normal model for $K|\kappa$, then X_v is a prime Weil divisor of X for* almost all *non-uniruled prime divisors v of $K|\kappa$.*

Proof. The first step in the proof of the theorem above is the following lemma, which is based on a result of Nagata [N] concerning uniruled function fields.

Lemma 3.2. *Let $K|\kappa$ be a function field, and let $X \to \kappa$ be a model for $K|\kappa$. Suppose that v is a non-uniruled prime divisor of $K|\kappa$ with x_v a regular point of X. Then X_v is a (non-uniruled) prime Weil divisor of X.*

Proof. The proof of the lemma follows immediately using a result of Zariski, see e.g. [A], Theorem 5.2. Namely by loc.cit. there exists a finite sequence of blow-ups $X^{i+1} \to X^i$, $0 \le i \le n$, of centres x^i such that: $X^0 = X$ and $x^0 = x_v$, x^i is the centre of v on X^i, and $x^{i+1} \mapsto x^i$; the codimension d_i of x^i is $d_i > 1$ for $i < n$ and $d_n = 1$. As $x^0 = x_v$ is a regular point, it follows that x^i is a regular point of X^i for all i. Further, if $\kappa(x^i)$ is the residue field of x^i, then $\kappa(x^{i+1})|\kappa(x^i)$ is a rational function field in $d_{i+1} - d_i$ variables. Now supposing that X_v is not a Weil prime divisor of X, or equivalently, that $n > 0$, then $\kappa(x^n) = Kv$ is a rational function field in $d_n - d_{n-1} > 0$ variables over $\kappa(x^{n-1})$. This is a contradiction, as $Kv = \kappa(x^n)$ is non-uniruled. □

As an application we have: Let $K|\kappa$ be a finitely generated field, and $X \to \kappa$ a complete, regular model for $K|\kappa$. Then the correspondence

$$v \mapsto X_v$$

defines a bijection between the set of all non-uniruled prime divisors of $K|\kappa$ and the set of all non-uniruled prime divisors of $X \to \kappa$.

We now conclude the proof of Theorem 3.1 as follows: By the theory of alterations of de Jong [J], there exists a finite extension $M|K$ such that M has a projective, regular model $Z \to \kappa$ (which is not necessarily geometrically connected when viewed as a variety over κ). For such a finite extension $M|K$ and a model $Z \to \kappa$ of $M|\kappa$, let $Y \to X$ be the normalisation of X in the field extension $M|K$, hence $Y \to X$ is a finite morphism. Further, Z and Y are birational over κ, and if $\psi : Z \dashrightarrow Y$ is a birational map defined over κ, then for almost all points $z_1 \in Z$ of codimension 1 in Z one has: ψ is defined at z_1, and $\psi(z_1)$ has codimension one in Y; moreover, almost all points y_1 of codimension one of Y lie in the image of

ψ. Thus for almost all prime Weil divisors W of Z one has: If w is the Zariski prime divisor of $M|\kappa$ defined by W then Y_w is a prime Weil divisor of Y. Next, if v is a non-uniruled prime divisor of K and w is a prolongation of v to M, then $(Mw)|(Kv)$ is a finite extension, hence by the remark 2) above (in a birational form), it follows that w is a non-uniruled prime divisor of $M|\kappa$. Hence by the remark following the lemma it follows that Z_w is a prime Weil divisor of Z.

Hence by the discussion above, for almost all non-uniruled prime divisors v of K and every prolongation w of v to M, the resulting Y_w is a prime Weil divisor of Y. Since $Y \to X$ is finite, hence the restriction $Y_w \to X_v$ is finite too, it follows that X_v is a prime Weil divisor of X if Y_w is a prime Weil divisor of Y. □

3.2. Geometric sets of prime divisors

We will apply the above facts in the following context: K is a finitely generated field with $\dim(K) > 1$ and κ is its prime field. As remarked in section 2, if $\mathrm{td}(K|\kappa) > 1$, one cannot expect to get information about a specific normal κ-model X of K, e.g. to describe its set of prime Weil divisors. Let us say that a set $D \subset \mathfrak{D}^1_K$ is *geometric,* if there exists a quasi-projective, normal model X of K such that D is exactly the set of prime Weil divisors of X. The point is, that using the theorem of the previous subsection one can show that the *geometric sets* of Zariski prime divisors of K are group theoretically encoded in G_K. In particular, if $\Phi : G_K \to G_L$ is an isomorphism as usual, then the local correspondence φ associated to Φ induces a functorial bijection between the geometric sets of Zariski prime divisors of K and those of L. Before doing that let us remark that for any given set D of Zariski prime divisors of K and for any finite extension $M|K$ of K we define the *prolongation* D_M of D to M as being the set of all the prolongations of all the $v \in D$ to M, n.b. all these prolongations are Zariski prime divisors of M. Further, if Z_v is the decomposition group of some prolongation of v to the separable closure of K, then there exists a functorial bijection between the set of all the prolongations of v to M and the G_M-conjugacy classes of all the subgroups $Z_v^g \cap G_M$ with $g \in G_K$. This is just general valuation theory. Further, since $\dim(K) > 1$ and $v \in \mathfrak{D}^1_K$, then by induction on $d = \dim(K)$ we can suppose that Kv is functorially encoded in G_{Kv}, hence in particular, we can see whether v is a non-uniruled Zariski prime divisor of $K|\kappa$ just by knowing Z_v.

The group theoretic recipe to detect the geometric sets of Zariski prime divisors is the following:

Theorem 3.3. *In the context above, a set $D \subset \mathfrak{D}^1_K$ is geometric if and only if for all "sufficiently large" finite extensions $M|K$, the prolongation D_M of D to M consists of exactly almost all non-uniruled prime divisors of M.*

Proof. It follows from the discussion above by induction on the Kronecker dimension (relying on the Birational Conjecture in $\dim < d$). □

3.3. Special sets of prime divisors

We finally want to give a hint on how the facts above can be used for identifying the image of the completion functor $\hat{\jmath} : K^\times \to \hat{K}$ as introduced at the end of Section 1. Let K be a finitely generated field with $\dim(K) > 1$, and let κ be the prime field of K. Hence with $K|\kappa$ we are in the context of the previous subsection. Let D be a geometric set of prime divisors of K. We will say that D is small, if there exists a regular model $X \to \kappa$ of K such that

1. The divisor class group of X is trivial.
2. The set of prime Weil divisors of X consists of exactly all the X_v, $v \in D$.

We remark that every cofinite subset of a special set of prime divisors is again a special set of prime divisors. Secondly, since the divisor class group of a regular κ-variety is finitely generated (n.b., κ is a finitely generated field), it follows that there exist "many" special sets of prime divisors of K.

Let D be a special set of prime divisors, and $X \to \kappa$ the corresponding regular model of $K|\kappa$. One has an exact sequence

$$1 \to K_D \longrightarrow K^\times \longrightarrow \oplus_v \mathbf{Z}v \to 0$$

where the sum on the right is taken over all $v \in D$, and K_D is the subgroup of $K^\times$ consisting of all the v-units ($v \in D$). Now the point is that for $v \in D$ "sufficiently general", K_D consists of v-units only, and the residue completion map $\hat{p}_v : \hat{K} \to (Kv)\hat{}$ mentioned in Section 2 maps $\hat{\jmath}(K_D)$ isomorphically onto $\hat{\jmath}_v(K_Dv)$, where K_Dv is the image of K_D in the residue field Kv, and $\hat{\jmath}_v : (Kv)^\times \to (Kv)\hat{}$ is the m-adic completion homomorphism for $(Kv)^\times$. Thus we can identify $\hat{\jmath}(K)$ as being the set of all $x \in \hat{K}$ with the property that for all small enough special sets D of Zariski prime divisors of K, and all sufficiently general $v \in D$ one has: $x \in (K_D)\hat{}$ and $\hat{p}_v(x) \in \hat{\jmath}_v(Kv)$.

References

[Ab] Abhyankar, S., Resolution of singularities of algebraic surfaces, pp. 1–12 in Algebraic Geometry, Oxford University Press, Oxford 1969.

[A] Artin, M., Néron Models, in *Arithmetic Geometry*, ed. Cornell-Silverman, Springer Verlag 1987.

[B] Bogomolov, F. A., On two conjectures in birational algebraic geometry, in Algebraic Geometry and Analytic Geometry, ICM-90 Satellite Conference Proceedings, ed. A. Fujiki et al., Springer Verlag Tokyo 1991.

[BOU] Bourbaki, N., Algèbre commutative, Hermann Paris 1964.

[Ch] Chevalley, C., *Introduction to the theory of algebraic functions of one variable*, AMS, New York 1951.

[De] Deligne, P., Le groupe fondamental de la droite projective moins trois points, Galois groups over **Q**, Math. Sci. Res. Inst. Publ. **16**, 79–297, Springer 1989.

[D] Deuring, M., Reduktion algebraischer Funktionenkörper nach Primdivisoren des Konstantenkörpers, Math. Z. **47** (1942), 643–654.

[F] Faltings, G., Endlichkeitssätze für abelsche Varietäten über Zahlkörpern, Invent. Math. **73** (1983), 349–366.

[G1] Grothendieck, A., Brief an Faltings, June 1983, see [SL] below; original p. 49–58; english translation p. 285–293.

[G2] Grothendieck, A., Esquisse d'un programme, 1984, see [SL] below; original p. 5–48; english translation p. 243–283.

[Ha] Harbater, D., Galois groups with prescribed ramification, in *Arithmetic geometry,* ed. Childress-Jones, Contemporary Math. 174, 1994.

[Hr] Hartshorne, R. Algebraic Geometry, Springer Verlag 1993.

[H] Hironaka, H., Resolution of singularities of an algebraic variety over a field of characteristic zero, Ann. of Math. **79** (1964), 109–203; 205–326.

[I] Ihara, Y., On Galois repr. arising from towers of covers of $\mathbf{P}^1 \backslash \{0, 1, \infty\}$, Invent. math. **86** (1986), 427–459.

[I-N] Ihara, Y., Nakamura, H., Some illustrative examples from anabelian geometry in high dimensions, in [SL], see below.

[Ik] Ikeda, M., Completeness of the absolute Galois group of the rational number field, J. reine angew. Math. **291** (1977), 1–22.

[J] de Jong, A. J., Smoothness, semi-stability, and alterations, Publ. Math. I.H.E.S. **83** (1996), 51-93.

[K] Koenigsmann, J., From p-rigid emelents to valuations, J. reine angew. Math. **465** (1995), 165–182.

[Km] Komatsu, K., A remark to a Neukirch's conjecture, Proc. Japan Acad. **50** (1974), 253–255.

[K-P-R] Kuhlmann, F.-V., Pank, M. and Roquette, P., Immediate and purely wild extensions of valued fields, Manuscripta Math. **55** (1986), 39–67.

[M1] Mochizuki, Sh., The local pro-p Grothendieck conjecture for hyperbolic curves, RIMS Preprint 1045, Nov 1995.

[M2] Mochizuki, Sh., The local pro-p anabelian geometry of curves, RIMS Preprint 1097, Aug 1996.

[M3] Mochizuki, Sh., A version of the Grothendieck Conjecture for p-adic local fields, The Int. J. Math. **8**, No.4 (1997), 499–506.

[N] Nagata, M., A theorem on valuation rings and its applications, Nagoya Math. Journal, **29** (1967), 85–91.

[Na] Nakamura, H., Galois rigidity of the étale fundamental groups of punctured projective lines, J. reine angew. Math. **411** (1990), 205–216.

[N1] Neukirch, J., Über eine algebraische Kennzeichnung der Henselkörper, J. reine angew. Math. **231** (1968), 75–81.

[N2] Neukirch, J., Kennzeichnung der p-adischen und endlichen algebraischen Zahlkörper, Invent. Math. **6** (1969), 269–314.

[N3] Neukirch, J., Über die absoluten Galoisgruppen algebraischer Zahlkörper, Astérisque **41–42** (1977), 67–79.

[O] Oda, T., A note on ramification of the Galois representation of the fundamental group of an algebraic curve I, J. Number Theory (1990) 225–228.

[Pa] Parshin, A. N., Finiteness Theorems and Hyperbolic Manifolds, in *The Grothendieck Festschrift* III, ed. P. Cartier et al., Progress in Math. Series Vol 88, Birkhäuser Boston Basel Berlin 1990.

[P1] Pop, F., On Grothendieck's conjecture of birational anabelian geometry, Ann. of Math. **138** (1994), 145–182.

[P2] Pop, F., On Grothendieck's conjecture of birational anabelian geometry II, Heidelberg-Mannheim Preprint Series Arithmetik II, No 16, April 1995.

[P3] Pop, F., Glimpses of Grothendieck's Anabelian Geometry, in [SL], see below.

[R] Roquette, P., Zur Theorie der Konstantenreduktion algebraischer Mannigfaltigkeiten, J. reine angew. Math. **200** (1958), 1–44.

[S1] Serre, J.-P., *Cohomologie Galoisienne,* LNM 5, Springer 1965.

[S2] Serre, J.-P., On the Chebotarev density theorem, in *Arithmetical Algebraic Geometry,* ed O. F. G. Schilling, Harper&Row, New York 1965.

[SL] Schneps, L. and Lochak, P. (editors) *Geometric Galois Actions 1,* LMS, Cambridge Univ. Press, 1997.

[Sp] Spiess, M., An arithmetic proof of Pop's Theorem concerning Galois groups of function fields over number fields, J. reine angew. Math. **478** (1966), 107–126.

[T1] Tamagawa, A., The Grothendieck conjecture for affine curves, Compositio Math. **109** (1997), 135–194.

[T2] Tamagawa, A., On the fundamental group of curves over algebraically closed fields of characteristic > 0, RIMS Preprint 1182, Jan 1998.

[U1] Uchida, K., Isomorphisms of Galois groups of algebraic function fields, Ann. of Math. **106** (1977), 589–598.

[U2] Uchida, K., Isomorphisms of Galois groups of solvably closed Galois extensions, Tôhoku Math. J. **31** (1979), 359–362.

[U3] Uchida, K., Homomorphisms of Galois groups of solvably closed Galois extensions, J. Math. Soc. Japan **33**, No.4, 1981.

[W] Ware, R., Valuation Rings and rigid Elemets in Fields, Can. J. Math. **33** (1981), 1338–1355.

[Z-S] Zariski, O. and Samuel, P., *Commutative Algebra,* Van Nostrand, Princeton 1958 and 1960.

Mathematisches Institut
Universität Bonn
53115 Bonn, Germany
pop@math.uni-bonn.de

Progress in Mathematics, Vol. 181,

The Turbulent Fifties in Resolution of Singularities

Heinrich Reitberger

Introduction

The years between 1944 when O. Zariski established the resolution of singularities for three-dimensional varieties, and 1964 when H. Hironaka completed the tour de force of the general case (char $p = 0$), seem to have passed without any activities in this field except for Abhyankar's work [Ab 1]. The present note is meant to recall that this is not the whole truth. There have been undertaken great efforts but they remained without any real succes. These false respectively incomplete proofs which were published in well renowned journals and their very thorough and examplaric reviews in the "Mathematical Reviews" are set out to be important steps and should not be forgotten completely.

I wish to thank one of the referees for valuable suggestions.

L. Derwidué's "résolution" and O. Zariski's review

L. Derwidué has studied the resolution problem since 1948, when he submitted a first incomplete paper to his teacher L. Godeaux. In 1949 Derwidué publishes an article of 139 pages in the Bulletin de l'Académie Royale de Belgique [Dw 2] wherein the problem seems to be resolved for him. In the version [Dw 3] printed in the Mathematische Annalen in 1951 and written chiefly for the purpose of convincing the unbelievers, Derwidué takes van der Waerden into responsibility by mentioning his indebtedness to him who, "in a series of conferences with the author, during a stay at Laren, has closely and critically scrutinized every single detail of the present proof".

The essential ingredients in Derwidué's work are "elementary transformations" (= Zariski's monoidal transformations), first polars and an adaptation of van der Waerden's theory of infinitely near points.

In his outstanding five-columns review on [Dw 3] O. Zariski says initially: "*In cases of doubtful points, the reviewer has made an effort to either* (a) *complete the proof himself or* (b) *find precisely what is wrong with the proof or/and* (c) *find a counterexample. With this approach, the geometric language of the author has the effect of shifting a good deal of the burden from the author to the reader, for in many cases it was at least as difficult to accomplish* (a) *or* (b) or (c) *as it was for the author to find his incomplete proofs.*"

Finally Zariski comes to the devastating statement: "*The two main pillars of the "proof"* – (a) *the theorem on first polars and* (b) *the "raisonnement fondamental" – are represented by statements of which the first is false and the second is far from having been proved.*" Zariski gives a counterexample in the simplest case of plane curves: The curve V given through $f(x,y) = y^3 - x^5$ has a triple point at the origin O and a double point at the infinitely near point O' in the direction $y = 0$. V is transformed by the quadratic transformation $T : x = x', y = x'y'$ into the curve V' give by the equation $x'^2 - y'^3 = 0$, the infinitely near point O' now being represented by the origin $x' = y' = 0$. The generic polar Φ of V is $uf'_x + vf'_y + w(xf'_x + yf'_y - 5f) = 0$, i.e. $3vy^2 - 5ux^4 - 2wy^3 = 0$. The T-transform of Φ is $3vy'^2 - 5ux'^2 - 2wx'y'^3 = 0$ and O' is now a singular point of multiplicity 2 and not 1, as is claimed in Derwidué's theorem. So this crucial theorem is "almost always" false.

But also the master himself has underestimated the power of "blowing up": Zariski says: "… *the deeper difficulties of the resolution problem can be effected by such straightforward tools (= monoidal transformations) only to a very moderate extent* … "

In contrast to this final blow for Derwidué – since 1950 member of the Société Royale de Sciences de Liège – the reviewer of [Dw 3], O.-H. Keller, in the Zentralblatt states succinctly: *"The author gives a proof with the help of a false proposition on the first polar."* In the review of [Dw 1] Keller says: *"Meanwhile the main theorem on the resolution of singularities was being proved by Deuring with valuation theory."* A bit later the reviewer of [Dw 2], W. Gröbner, uses a moderate formulation: *"The problem seems to be still open since also the proof announced by Deuring is not yet published."* And this state rests! But now Derwidué himself provides a rectification [Dw 4]: As a result of a counter-example sent to the author by B. Segre on 28th August 1951 (see also [Sg 2]), he wishes to state that the main results rest upon an unproved hypothesis.

B. Segre's "scioglimento" and J.G. Semple's remarks

The long paper of B. Segre [Sg 1] aims to investigate the possibility that any irreducible algebraic variety over the complex domain can be birationally transformed by a finite sequence of elementary transformations called "dilatations" into a nonsingular variety. A dilatation, apart from certain restrictions, is the usual blowing up.

First of all Segre demonstrates heavy faults in the work of Derwidué [Dw 3] with the previously mentioned counterexample of the plane curve $y^{12} + x^{113} = 0$ and her first polar $ay^{11} + by^{12} + cx^{112} = 0$ that does not contain a certain multiple point lying in a neighborhood (higher than the first) of a multiple point of the curve.

Then he claims to save the elementary geometrical method. The advantages aimed at, plainly, are the simplicity of conception that such a resolution would have, and its ability, by untying the singularities one-by-one and step-by-step, to give insight into their structure. In order to achieve this by a finite number of dilatations Segre defines an "associated variety": First for hypersurfaces in a projective space a sequence of polars seems to fulfil this role. Then Segre extends this concepts to arbitrary varieties V_d over **C**: He intends to define a more general class of hypersurfaces W_d associated to V_d (see also [Dw 6]): W_d has associated behavior of index 1 with V_d at a point P if the equations $f(x) = 0$ of the hypersurface "received by a suitable projection" of V_d and $g(x) = 0$ of W_d (in nonhomogeneous coordinates $x_1, \ldots, x_{d+1}$) are such that there exist polynomials $h, p, p_1, \ldots, p_{d+1}$ with the properties

(i) h does not vanish at P

and

(ii) $hg = pf + \sum_{i=1}^{d+1} p_i \partial_{x_i} f$

If P is of multiplicity m on V_d, then it has at least multiplicity $m-1$ on W_d; and the association is said to be regular if this is precise $m-1$. By an inductive procedure the associated behavior is defined for any index i $(1 \leq i \leq m-1)$.

The main part of the paper is devoted to the case of surfaces, only in the final chapter the problem for dimension $d \geq 3$ is surveyed. The dilatation may produce unwelcome effects and Segre confines himself to some general indications of possible strategies.

J.G. Semple judges: *"But the claim, in the judgement of the reviewer, is at present unjustified, since the argument at some important points is either obscure or not as yet complete"* and later " ... *outlines a bold plan, but the possibility of proof is still very much in suspense* ... "

W. Gröbner's "Auflösung" and P. Abellanas' review

W. Gröbner claims in a lecture at Bologna in 1956, published in [Gr 2]: "*We hope to give a rigorous proof of the following heuristic considerations: Any irreducible algebraic variety can be transformed birationally with a finite sequence of our type of transformations in a variety without singularities. A large part of the results is valid for arbitrary fields.*"

Let $d_x = (p_1, p_2, \ldots, p_s), p_j \in K[x_0, \ldots, x_m], j = 1, \ldots, s; s \geq 2$, be a homogeneous ideal and

$$\rho y_{ij} = x_i p_j(x), i = 0, \ldots, m; j = 1, \ldots, s \qquad (*)$$

a monoidal correspondence. Gröbner states:

a) One can assume that all the forms p_i are of the same degree.

b) For the inverse correspondence of (*) holds:

$$\rho x_i = y_{ij}, i = 0, \ldots, m; j = 1, \ldots, s. \qquad (**)$$

c) The correspondence (*) has as original variety the m-dimensional projective space S_m and $V(v_y)$ as image variety, where

$$v_y = K[y](\ldots, y_{ij}y_{kl} - y_{il}y_{kj}, \ldots, y_{ij}p_\sigma(y_{kl}) - y_{i\sigma}p_j(y_{kl}), \ldots)$$

d) The points of $V(d_x)$ have not any homologous point by (*).

e) To every point of $V(v_y)$ corresponds a well determined point of S_m.

f) The correspondence $V(v_y) \to S_m$ is one to one on the whole $V(v_y)$, except for the points of the subvariety $V(d_y)$, where $d_y = (v_y, p_1(y_{i1}), \ldots, p_s(y_{is}))$.

g) There is a one-to-one correspondence between the points of $V(d_y)$ and the tangential spaces to $V(d_x)$.

h) If $V(d_x)$ has only nonsingular points, it follows that $V(v_y)$ has no singularities.

i) If $V(d_x)$ is the singular subvariety of an irreducible variety $V(p_x)$, the correspondence $((*), (**))$ transforms $V(p_x)$ into a variety $V(p_y)$ that, generally, has no singularities.

But the reviewer P. Abellanas adds the following observations: "*The correspondence (*) is not an irreducible one, as the author implicitly assumes, - that means the bihomogeneous ideal*

$$I := K[x,y](y_{ij}x_k p_l(x) - y_{kl}x_i p_j(x), i,k = 0, \ldots, m; j,l = 1, \ldots, s)$$

corresponding to (*) and obtained by elimination of the parameter ρ among the (*) is not a prime one in the product. – In order to obtain a non-trivial irreducible correspondence one must add certain polynomials and therefore the ideal d_x is not transformed into an irrelevant ideal, in contradiction with d). *The equations (**) are true for the new correspondence, but not for (*). If d_x is a prime ideal, the proposition* a) *does not hold in every case. The proposition* e) *is wrong, since there are fundamental subvarieties on both varieties S_m and $V(v_y)$. The author doesn't prove the proposition* i) *but his reasoning purporting to justify its validity uses the false proposition* d)."

Two years earlier Gröbner was considerably more careful in [Gr 1]: "*As things stand the resources are not sufficient to characterize the behaviour of a variety in the neighbourhood of a singularity and to follow during a transformation,*" but Abellanas remarked: "... *Hence the transformation doesn't signify any progress for the resolution of the singularities ...* "

In a lecture in the fall semester 1972 Gröbner himself gives a correction of (**): "*We cannot assert that v_y is generated by the mentioned quadratic forms. This problem has produced great difficulties for us, especially for the Grassmann variety.*"

References

[Ab 1] Abhyankar, S.S.: Local uniformization on algebraic surfaces over ground fields of characteristic $p \neq 0$. Ann. of Math. **63** (1956), 491–526.

[Dw 1] Derwidué, L.: Méthode simplifiée de réduction des singularités d'une variété algébrique. Bull. Acad. Roy. Belgique, Cl. Sci.,V. **35** (1949), 880–885. **Zbl 38**, 319 (O.-H. Keller)

[Dw 2] Derwidué, L.: Le problème général de la réduction des singularités d'une variété algébrique. Mem. Soc. Roy. Sci. Liège (4) **9** (1949), fasc.2, 1–139. **Zbl 41**, 286 (W. Gröbner)

[Dw 3] Derwidué, L.: Le problème de la réduction des singularités d'une variété algébrique. Math. Ann. **123** (1951), 302–330. **MR 13**, 67 (O. Zariski)

[Dw 4] Derwidué, L.: Rectification. Math. Ann. **124** (1952), 316.

[Dw 5] Derwidué, L.: Sur la réduction des singularités d'une variété algébrique. Mem. Soc. Roy. Sci. Liège (4) **13** (1953), 1–41. **MR 15**, 551 (J.G. Semple)

[Dw 6] Derwidué, L.: Sur le comportement associé et la reduction des singularités. Bull. Soc. Roy. Sci. Liège **24** (1955), 212–238. **MR 17**, 1133 (J.G. Semple)

[Gr 1] Gröbner, W.: Die birationalen Transformationen der Polynomideale. Monatsh. Math. **58** (1954), 266–286. **MR 16**, 741, 1337 (P. Abellanas)

[Gr 2] Gröbner, W.: Sopra lo scioglimento delle singolarità delle varietà algebriche. Boll. Un. Mat. Ital. (3) **11** (1956), 319–327. **MR 18**, 513 (P. Abellanas)

[Gr 3] Gröbner, W.: Aggiunta alla Nota: Sopra lo scioglimento delle singolarità delle varietà algebriche. Boll. Un. Mat. Ital. (3) **11** (1956), 589–590.

[Sg 1] Segre, B.: Sullo scioglimento delle singolarità delle varietà algebriche. Ann. Mat. Pura Appl. (4) **33** (1952), 5–48. **MR 14**, 683 (J.G. Semple)

[Sg 2] Segre, B.: Geometria algebrica ed aritmetica. Atti IV. Congr. Un. mat. Ital. 1951 **1** (1953), 88–98. **MR 15**, 344 (J.G. Semple) **Zbl 50**, 373 (W. Gröbner)

Institut für Mathematik
Universität Innsbruck
A-6020 Austria
heinrich.reitberger@uibk.ac.at

Progress in Mathematics, Vol. 181,

Valuations

Michel Vaquié

Introduction

La notion de place d'un corps K, notion qui est plus ou moins équivalente à la notion de valuation de K, a été introduite par Dedekind et Weber en 1882. Pour étudier les courbes algébriques planes et trouver une version algébrique des constructions de Riemann qui évite en particulier toute considération topologique ou transcendante, ils utilisent l'analogie entre la théorie des nombres algébriques et la théorie des fonctions algébriques d'une variable. Ils considèrent donc un corps K, extension finie du corps des fractions rationnelles $\mathbb{C}(X)$, et pour définir les points de l'équivalent de la Surface de Riemann correspondant à K, c'est à dire la courbe projective non singulière ayant comme corps des fonctions K, ils définissent les places de K.

A la suite des travaux de Dedekind et Weber, Hensel poursuit l'analogie entre la surface de Riemann d'un corps de fonctions algébriques et les idéaux premiers d'un corps de nombres K. Il développe la théorie des nombres p-adiques qui permet d'associer à tout élément x de K une "série p-adique" de la forme $\sum \alpha_i p^{i/e}$.

En 1913 Kürschak définit la notion de valeur absolue, et en particulier de valeur absolue ultramétrique, généralisant ainsi la valeur absolue p-adique. Et c'est Krull qui définit et étudie la notion générale de valuation en 1931 (cf. la note historique de [Bo] pour un exposé plus précis).

Les premiers travaux sur les places, et par conséquent sur les valuations se trouvent ainsi dans le domaine de l'arithmétique (Ostrowski, Deuring). Pour une approche selon ce point de vue, ainsi que du point de vue des corps valués complets et de la théorie des modèles, je renvoie à l'article de Kuhlmann dans ce volume [Ku], et à sa bibliographie. En particulier l'auteur s'y intéresse au problème du prolongement d'une valuation à une extension du corps et plus précisément à la théorie de la ramification que je n'aborde pas dans cet article. (Nous pouvons la considérer comme une généralisation au cas des valuations quelconques de la théorie de la ramification pour une valuation discrète de rang un telle qu'elle est exposée dans l'étude des corps locaux (cf. [Se]).)

Les valuations ont aussi joué un rôle très important en géométrie algébrique, en particulier avec les travaux de Zariski, puis d'Abhyankar. Leur étude était motivée par le problème de la résolution des singularités. Plus récemment l'étude des valuations pour aborder de manière totalement nouvelle le problème de la résolution des singularités a été entreprise, notamment par Spivakovsky [Sp 2], [Sp 3] et Teissier [Te]. Je renvoie pour le problème de l'uniformisation locale et de la résolution des singularités suivant les idées originales de Zariski à l'article de Cossart dans ce volume [Co].

Dans cet article je donne une présentation des résultats élémentaires principaux sur les valuations avec leur démonstration. Ces résultats sont bien connus et se trouvent dans les livres d'algèbre commutative (par exemple dans [Bo], [Na], [Z-S]). Je donne aussi les résultats nécessaires sur la Variété de Riemann pour comprendre comment Zariski obtient la résolution des singularités à partir du théorème d'uniformisation locale; ainsi que le théorème d'Abhyankar sur les valuations dominant un anneau local noethérien.

Je voudrais aussi signaler le rôle des valuations dans l'étude des idéaux d'un anneau local, étude de leurs propriétés asymptotiques, de leur multiplicité, etc ... Je renvoie au livre de Rees [Re] pour cette approche et pour une bibliographie plus détaillée. Pour finir je voudrais mentionner l'approche très originale des valuations d'un corps K extension de type fini de $\mathbb{Q}$ donnée par Morgan et Shalen [M-S]. Plus précisément ils s'intéressent à une variété algebrique X_o définie sur $\mathbb{Q}$, ou sur une extension finie de $\mathbb{Q}$, dont le corps des fonctions $F(X_o)$ est égal à K, et ils considèrent la variété algébrique complexe X obtenue à partir de X_o par extension des scalaires. Ils peuvent alors décrire toutes les valuations de K en considérant des suites de points sur X, c'est à dire sans avoir besoin de regarder des variétés obtenues à partir de X_o par des éclatements.

Je remercie Bernard Teissier, Vincent Cossart et Franz-Viktor Kuhlmann pour les corrections, les remarques et les conseils qui m'ont permis d'apporter certaines précisions et de rendre plus claires plusieurs parties de ce texte.

1. Anneau de valuation

Soient A et B deux anneaux locaux d'idéaux maximaux respectifs $\max(A)$ et $\max(B)$, nous disons que B domine A si $A \subset B$ et $\max(A) = A \cap \max(B)$; si nous supposons l'inclusion $A \subset B$ alors la deuxième condition est équivalente à $\max(A) \subset \max(B)$.
La relation "B domine A", que nous notons $A \preceq B$, est une relation d'ordre sur l'ensemble des anneaux locaux. Si nous avons la relation $A \preceq B$, alors l'injection de A dans B définit un isomorphisme du corps résiduel $\kappa(A) = A/\max(A)$ sur un sous-corps du corps résiduel $\kappa(B) = B/\max(B)$.

Remarque. Si A est un anneau local noethérien le complété $\hat{A}$ de A pour la valuation $\max(A)$-adique domine A.
Soient A et B deux anneaux intègres avec $A \subset B$, alors pour tout idéal premier $\mathcal{Q}$ de B l'anneau localisé $B_{\mathcal{Q}}$ domine $A_{\mathcal{P}}$, où $\mathcal{P}$ est l'idéal premier de A défini par $\mathcal{P} = A \cap \mathcal{Q}$.

Définition. Soit V un anneau intègre contenu dans un corps K; alors V est un anneau de valuation de K si K est le corps des fractions de V et si V est un élément maximal de l'ensemble des sous-anneaux locaux de K ordonné par la relation de domination: i.e. V est un anneau local et si W est un sous-anneau local de K qui domine V, alors $W = V$.
Soit V un anneau intègre, V est un anneau de valuation si V est un anneau de valuation de son corps des fractions.

Avant de donner les propriétés caractéristiques des anneaux de valuation, nous allons rappeler le théorème suivant [Ma].

Théorème de Cohen-Seidenberg. *Soient A et B deux anneaux avec $A \subset B$ et B entier sur A, alors pour tout idéal premier $\mathcal{P}$ de A il existe un idéal premier $\mathcal{Q}$ de B au dessus de $\mathcal{P}$, c'est à dire tel que $\mathcal{P} = A \cap \mathcal{Q}$.*

Théorème 1.1. *Soit V un anneau intègre contenu dans un corps K, alors les conditions suivantes sont équivalentes:*

a) *V est un anneau de valuation de K;*
b) *soit x un élément de K, si x n'appartient pas à l'anneau V alors son inverse x^{-1} appartient à V;*
c) *K est le corps des fractions de V et l'ensemble des idéaux de V est totalement ordonné par la relation d'inclusion.*

Preuve. a) $\Rightarrow$ b): soit x un élément du corps K, $x \neq 0$, nous allons montrer que x ou x^{-1} appartient à V. Si x est entier sur V, nous considérons l'anneau $W = V[x]$. D'après le théorème de Cohen-Seidenberg, il existe un idéal premier $\mathcal{Q}$ de W au dessus de l'idéal maximal de V. L'anneau local $W_{\mathcal{Q}}$ domine alors l'anneau V, d'où $W \subset W_{\mathcal{Q}} = V$, et x appartient à V.

Si x n'est pas entier sur V, nous considérons l'anneau $W = V[x^{-1}]$. Comme x n'est pas entier sur V, x^{-1} n'est pas un élément inversible de l'anneau W, en effet toute relation de la forme $x^{-1}.w = 1$ avec $w \in W = V[x^{-1}]$, i.e. $w = \sum a_j x^{-j}$, donnerait une relation de dépendance intégrale de x sur V. Par conséquent il existe un idéal maximal $\mathcal{Q}$ de W contenant x^{-1} et soit V' le localisé $V' = W_{\mathcal{Q}}$. Comme x^{-1} appartient à l'idéal $\mathcal{Q}$, le morphisme composé $V \to W = V[x^{-1}] \to k = W/\mathcal{Q}$ est surjectif, et son noyau $V \cap \mathcal{Q}$ est l'idéal maximal de V. Nous en déduisons que V' est un sous-anneau de K qui domine V, par conséquent $V' = V$ et x^{-1} appartient à V.

b) $\Rightarrow$ c): soient $\mathcal{I}$ et $\mathcal{J}$ deux idéaux de V et nous supposons que $\mathcal{J}$ n'est pas inclus dans $\mathcal{I}$. Alors il existe un élément x de $\mathcal{J}$ n'appartenant pas à $\mathcal{I}$ et pour tout élément y appartenant à $\mathcal{I}$, $y \neq 0$, nous avons $x \notin (y)V$; par conséquent x/y est un élément de K n'appartenant pas à V et nous en déduisons que y/x appartient à V, c'est à dire $y \in (x)V$, d'où $y \in \mathcal{J}$. Nous avons ainsi montré que $\mathcal{I}$ est inclus dans $\mathcal{J}$ et il est clair aussi que K est le corps des fractions de V.

c) $\Rightarrow$ a): comme l'ensemble des idéaux de V est totalement ordonné par l'inclusion V possède un seul idéal maximal $\max(V)$. Soit W un sous-anneau local de K qui domine V et soit x appartenant à W, nous allons montrer que x appartient aussi à V; nous pouvons écrire $x = a/b$ avec $a \in V$ et $b \in V$. Si l'idéal $(a)V$ est inclus dans $(b)V$ alors x appartient à V. Si l'idéal $(b)V$ est inclus dans $(a)V$ alors x^{-1} appartient à V, nous en déduisons x et x^{-1} appartiennent tous les deux à W d'où $x^{-1} \notin \max(W)$ et $x^{-1} \notin \max(V)$ car W domine V. L'élément x^{-1} de K vérifie alors $x^{-1} \in V$ et $x^{-1} \notin \max(V)$ par conséquent, comme V est local, x appartient à V. □

Remarque. En "a) ⇒ b)" nous avons montré que tout anneau de valuation est intégralement clos.

Nous pouvons remplacer la condition c) par la condition équivalente suivante:

c′) K est le corps des fractions de V et l'ensemble des idéaux principaux de V est totalement ordonné par la relation d'inclusion. Nous en déduisons en particulier que tout idéal de type fini de V est un idéal principal.

Nous allons maintenant montrer l'existence d'anneaux de valuation.

Proposition 1.2. *Soit A un sous-anneau d'un corps K et soit h un morphisme de A dans un corps algébriquement clos L, il existe alors un anneau de valuation V de K et un morphisme h' de V dans L tel que V contienne A, h' prolonge h et* $\max(V) = h'^{-1}(0)$.

Preuve. Nous considérons l'ensemble $\mathcal{H}$ formés des couples (B, f), où B est un sous-anneau de K et f est un homomorphisme de B dans L; nous définissons sur cet ensemble la relation d'ordre $(B, f) \preceq (C, g)$ par $B \subset C$ et g prolonge f. L'ensemble $\mathcal{H}$ muni de cette relation d'ordre est un ensemble inductif, i.e. toute partie totalement ordonnée admet une borne supérieure - si nous avons la partie $\big((B_\alpha, f_\alpha)\big)$ il suffit de prendre pour borne supérieure le couple (B, f) où B est l'union des B_α et où f est défini par les restrictions f_α. D'après le lemme de Zorn nous en déduisons que $\mathcal{H}$ admet un élément maximal (W, g). Si nous appelons $\mathcal{P}$ le noyau du morphisme $g\colon W \to L$, l'anneau V cherché est le localisé $V = W_{\mathcal{P}}$. □

Corollaire. *Tout sous-anneau local A d'un corps K est dominé par au moins un anneau de valuation de K.*

Preuve. Il suffit d'appliquer la proposition précédente à $h\colon A \to L$, où L est une cloture algébrique du corps résiduel $A/\max(A)$. □

Remarque. Le plus souvent nous nous donnerons un corps de base k et nous considérerons uniquement des corps K extensions de k et les sous-anneaux A qui sont des k-algèbres. Nous trouvons comme précédemment le résultat d'existence suivant.

Soit A une sous k-algèbre de K et soit h un k-morphisme de A dans un corps algébriquement clos L, il existe alors un anneau de valuation V de K qui est une k-algèbre et un k-morphisme h' de V dans L tel que V contienne A, h' prolonge h et $\max(V) = h'^{-1}(0)$.

2. Valuation

Dans la suite Γ est un groupe commutatif totalement ordonné, en particulier Γ est un groupe sans torsion. Nous notons Γ^+ le sous-ensemble des éléments "positifs" et nous avons:

$$\Gamma = \Gamma^+ \cup \Gamma^-, \Gamma^+ \cap \Gamma^- = \{0\} \quad \text{et} \quad \alpha \geq \beta \Longleftrightarrow \alpha - \beta \in \Gamma^+ .$$

Nous adjoignons au groupe Γ un élément $+\infty$ et nous appelons Γ_∞ l'ensemble ainsi obtenu: $\Gamma_\infty = \Gamma \cup \{+\infty\}$. Nous munissons cet ensemble d'une relation d'ordre total en posant pour tout α dans Γ, $\alpha < +\infty$ et nous posons aussi:

$$\text{pour tout } \alpha \in \Gamma, (+\infty) + \alpha = (+\infty) + (+\infty) = +\infty .$$

Définition. Soit A un anneau, nous appelons valuation de A à valeurs dans Γ une application $\nu \colon A \to \Gamma_\infty$ vérifiant les conditions suivantes:

1) $\nu(x.y) = \nu(x) + \nu(y)$ pour tout $x, y \in A$,
2) $\nu(x + y) \geq \inf(\nu(x), \nu(y))$ pour tout $x, y \in A$,
3) $\nu(1) = 0$ et $\nu(0) = +\infty$.

Remarque. Si nous supposons que l'application ν vérifie les conditions 1) et 2) et ne prend pas uniquement la valeur $+\infty$, alors nous avons obligatoirement $\nu(1) = 0$. Plus généralement pour tout élément z de A vérifiant $z^n = 1$ avec $n \in \mathbb{N}^*$, nous avons encore $\nu(z) = 0$ car le groupe Γ est sans torsion, en particulier $\nu(-1) = 0$.

Définition. L'unique valuation ν de A vérifiant $\nu(x) = 0$ pour tout x appartenant à A^* est appelée valuation impropre de A.

Proposition 2.1. *Soit ν une valuation d'un anneau A, pour toute famille finie $(x_1, \dots, x_n)$ d'éléments de A nous avons l'inégalité:*

$$\nu\Big(\sum_{i=1}^{n} x_i\Big) \geq \inf_{1 \leq i \leq n} \big(\nu(x_i)\big) .$$

De plus s'il existe un indice k tel que pour tout $i \neq k$ nous ayons l'inégalité stricte $\nu(x_i) > \nu(x_k)$, alors nous avons l'égalité:

$$\nu\Big(\sum_{i=1}^{n} x_i\Big) = \inf_{1 \leq i \leq n} \big(\nu(x_i)\big) = \nu(x_k) .$$

Preuve. La première partie se démontre par récurrence sur n en utilisant l'axiome 2) de la définition d'une valuation. Pour la deuxième partie nous pouvons nous ramener grâce à ce qui précède au cas $n = 2$. Si x et y sont deux éléments de A avec $\nu(x) < \nu(y)$, nous déduisons de la définition les deux inégalités $\nu(x + y) \geq \nu(x)$ et $\nu(x) \geq \inf(\nu(x + y), \nu(-y))$, et comme nous avons $\nu(-y) = \nu(y) > \nu(x)$ nous trouvons l'égalité cherchée. □

Remarque. Si ν est une valuation de A à valeurs dans Γ et si $f \colon B \to A$ est un morphisme d'anneaux, l'application composée $\nu \circ f \colon B \to \Gamma_\infty$ définit une valuation de B à valeurs dans Γ.

Remarque. Pour toute valuation ν d'un anneau A à valeurs dans Γ, l'image réciproque $\nu^{-1}(+\infty)$ est un idéal premier $\mathcal{P}$ de A. L'application $\bar{\nu}\colon A/\mathcal{P} \to \Gamma_\infty$ déduite de ν par passage au quotient définit une valuation de l'anneau intègre $A/\mathcal{P}$ telle que l'image réciproque de $+\infty$ est réduite à 0.

Proposition 2.2. *Soient A un anneau intègre de corps des fractions K et ν une valuation de A à valeurs dans Γ telle que pour tout $x \neq 0$ nous ayons $\nu(x) \neq +\infty$. Alors il existe une valuation μ de K et une seule qui prolonge ν.*
De plus $\mu(K^*)$ est le sous-groupe de Γ engendré par $\nu(A^*)$.

Preuve. Pour tout x dans K^* il existe y et z appartenant à A^* tels que $x = y/z$, il suffit alors de poser $\mu(x) = \nu(y) - \nu(z)$. Nous vérifions immédiatement que $\mu(x)$ ne dépend pas des éléments y et z choisis et que l'application μ ainsi définie est une valuation de K qui prolonge ν, et qu'elle est unique. Par construction il est clair que $\mu(K^*)$ est le sous-groupe de Γ engendré par le semi-groupe $\nu(A^*)$. □

Nous allons montrer maintenant la relation qui existe entre les valuations d'un corps K et les anneaux de valuations de ce corps.

Proposition 2.3. *Soit ν une valuation d'un corps K à valeurs dans un groupe Γ. Alors l'ensemble A des éléments x de K vérifiant $\nu(x) \geq 0$ est un anneau de valuation de K, dont l'idéal maximal* $\max(A)$ *est l'ensemble des x vérifiant $\nu(x) > 0$.*
Réciproquement, si V est un anneau de valuation de K nous pouvons lui associer une valuation ν de K à valeurs dans un groupe Γ_V telle que l'anneau V soit l'image réciproque $\nu^{-1}(\Gamma_V^+)$.

Preuve. Nous déduisons des axiomes d'une valuation que l'ensemble A des éléments x de K vérifiant l'inégalité $\nu(x) \geq 0$ est un sous-anneau de K et nous déduisons de la condition b) du théorème 1.1 que c'est un anneau de valuation de K. De plus, commme A est local, un élément x de K vérifie $\nu(x) = 0$ si et seulement si x et x^{-1} appartiennent à A, c'est à dire si et seulement si x appartient à $A \setminus \max(A)$.

Pour la réciproque, plus généralement nous considérons un anneau intègre C de corps des fractions K; l'ensemble $U(C)$ des éléments inversibles de C est un sous-groupe du groupe multiplicatif K^* et nous notons Γ_C le groupe quotient. La relation de divisibilité dans C: $x|y \iff y \in (x)C$, définit une structure de groupe ordonné sur Γ_C. Plus précisément, si nous notons respectivement $\bar{x}$ et $\bar{y}$ les classes des éléments x et y de K^* dans le groupe quotient $\Gamma_C = K^*/U(C)$, alors la relation est définie par $\bar{x} \leq \bar{y} \iff \exists z \in C$ tel que $y = zx$. La relation "$\leq$" est bien définie sur l'espace quotient $K^*/U(C)$, en effet la relation $\bar{x} \leq \bar{y}$ ne dépend pas des représentants x et y choisis. Cette relation est une relation d'ordre sur le groupe Γ_C, compatible avec la structure de groupe, et qui correspond à la relation d'ordre définie par l'inclusion sur l'ensemble des idéaux principaux de l'anneau C. Nous déduisons alors de la remarque suivant le théorème 1.1 que le groupe Γ_C est totalement ordonné si et seulement si C est un anneau de valuation de K. Comme C est un anneau local nous avons l'égalité $U(C) = C \smallsetminus \max(C)$. L'application canonique $\nu\colon K^* \to \Gamma_C = K^*/U(C)$ est alors une valuation de K telle que l'anneau C est égal à l'anneau de valuation associé $\{x \in K \,/\, \nu(x) \geq 0\}$. □

Définition. L'anneau de valuation V de K associé à la valuation ν est appelé l'anneau de la valuation ν et le corps $\kappa(V) = V/\max(V)$ est appelé le corps résiduel de la valuation. Le sous-groupe $\nu(K^*)$ de Γ est appelé groupe des ordres ou groupe des valeurs de ν. Nous déduisons de ce qui précède qu'il est isomorphe au groupe quotient $\Gamma_V = K^*/U(V)$.

Pour toute valuation ν d'un corps K, nous noterons R_ν son anneau de valuation, κ_ν son corps résiduel et Γ_ν son groupe des ordres.

Définition. Nous disons que deux valuations ν et ν' de K sont équivalentes si elles ont même anneau.

Proposition 2.4. *Deux valuations ν et ν' d'un corps K sont équivalentes si et seulement si il existe un isomorphisme de groupes ordonnés λ de $\nu(K^*)$ dans $\nu'(K^*)$ tel que $\nu' = \lambda \circ \nu$.*

Preuve. En effet il suffit de remarquer que par définition la valuation détermine l'anneau V et réciproquement l'anneau V de la valuation détermine le groupe des ordres: $\Gamma = K^*/U(V)$, ainsi que le sous-ensemble des éléments "positifs": $\Gamma^+ = V^*/U(V)$. □

3. Hauteur d'une valuation

Nous supposons toujours que Γ est un groupe commutatif totalement ordonné.

Définition. Soit Γ un groupe totalement ordonné, une partie Δ de Γ est appelée un segment si pour tout élément α appartenant à Δ, tout élément β de Γ compris entre α et $-\alpha$, i.e. β vérifiant soit $-\alpha \leq \beta \leq \alpha$ soit $\alpha \leq \beta \leq -\alpha$, appartient à Δ. Un sous-groupe Γ' de Γ est appelé un sous-groupe isolé si Γ' est à la fois un sous-groupe propre de Γ et un segment.

Proposition 3.1. *Le noyau d'un homomorphisme croissant de Γ dans un groupe ordonné est un sous-groupe isolé de Γ.*
Réciproquement si Γ' est un sous-groupe isolé de Γ, le groupe quotient Γ/Γ' possède une structure naturelle de groupe ordonné telle que $\Gamma \to \Gamma/\Gamma'$ soit un homorphisme croissant.

Nous considérons une valuation ν d'un corps K à valeurs dans le groupe Γ, avec Γ égal au groupe des ordres, i.e. nous supposons que ν est surjective, et soit V l'anneau de valuation associé à ν. Pour toute partie A de V contenant 0 nous définissons le sous-ensemble Δ_A de Γ comme le complémentaire dans Γ_∞ de $(\nu(A)) \cup (-\nu(A))$.

Théorème 3.2. *Si $\mathcal{I}$ est un idéal propre de V le sous-ensemble $\Delta_{\mathcal{I}}$ est un segment de Γ. L'application $\mathcal{I} \mapsto \Delta_{\mathcal{I}}$ est une bijection de l'ensemble des idéaux de V sur l'ensemble des segments de Γ, et nous avons l'équivalence:*

$$\mathcal{I} \subset \mathcal{J} \Longleftrightarrow \Delta_{\mathcal{J}} \subset \Delta_{\mathcal{I}}.$$

Le segment $\Delta_{\mathcal{I}}$ est un sous-groupe isolé de Γ si et seulement si $\mathcal{I}$ est un idéal premier de V.

Preuve. Soit b un élément de Γ^+ n'appartenant pas au sous-ensemble $\Delta_{\mathcal{I}}$, il suffit de montrer que pour tout a dans Γ nous avons: $a \geq b \Longrightarrow a \notin \Delta_{\mathcal{I}}$. Par hypothèse sur b il existe un élément x de l'idéal $\mathcal{I}$ tel que $b = \nu(x)$, comme l'application ν est surjective nous déduisons de l'inégalité $a \geq b$ l'existence d'un élément y de l'anneau V tel que $\nu(y) = a - b$. Alors xy appartient à l'idéal $\mathcal{I}$ de V et $a = \nu(xy)$ n'appartient pas à $\Delta_{\mathcal{I}}$.

Réciproquement si Δ est un segment de Γ, il faut montrer que le sous-ensemble $\{x \in V \,/\, \nu(x) \notin \Delta\}$ est un idéal de V:

$$\begin{aligned}
x \in \mathcal{I} \text{ et } y \in V \quad &\Rightarrow \quad \nu(x) \notin \Delta\,,\ \nu(x) \text{ et } \nu(y) \geq 0 \\
&\Rightarrow \quad \nu(x) + \nu(y) \notin \Delta \\
&\Rightarrow \quad xy \in \mathcal{I}; \\
x \text{ et } y \in \mathcal{I} \quad &\Rightarrow \quad \nu(x) \text{ et } \nu(y) \notin \Delta \\
&\Rightarrow \quad \nu(x+y) \notin \Delta \text{ car } \nu(x+y) \geq \inf\bigl(\nu(x), \nu(y)\bigr), \\
&\Rightarrow \quad x + y \in \mathcal{I}.
\end{aligned}$$

La relation $\mathcal{I} \subset \mathcal{J} \Longleftrightarrow \Delta_{\mathcal{J}} \subset \Delta_{\mathcal{I}}$ est évidente, d'où la bijection car l'ensemble des idéaux de V et l'ensemble des segments de Γ sont totalement ordonnés par l'inclusion. L'idéal $\mathcal{I}$ de V est un idéal premier si et seulement si le complémentaire $V \setminus \mathcal{I}$ est stable par multiplication, c'est à dire si et seulement si son image $\nu\bigl(V \setminus \mathcal{I}\bigr)$ est stable par addition, ce qui est bien équivalent à la condition $\Delta_{\mathcal{I}}$ sous-groupe de Γ. □

Définition. Le rang "rang(Γ)" d'un groupe totalement ordonné Γ est égal au nombre de ses sous-groupes isolés si ceux ci sont en nombre fini, et est infini sinon. La hauteur ou le rang de la valuation ν d'un corps K est le rang du groupe des valeurs Γ, et nous le notons ht(ν) ou rang(ν).

Remarque. Au lieu de considérer les segments Δ du groupe Γ, nous pouvons considérer les sous-ensembles majeurs M, c'est à dire les sous-ensembles M de Γ vérifiant: $x \in M$ et $y \geq x \Longrightarrow y \in M$. Nous obtenons une bijection croissante entre l'ensemble des sous-ensembles majeurs M de Γ et l'ensemble des sous V-modules $\mathcal{M}$ de K, bijection définie par $M \mapsto \mathcal{M} = \bigl\{x \in K \,/\, \nu(x) \in M \cap \{+\infty\}\bigr\}$.

Pour tout élément α du groupe $\Gamma = \Gamma_\nu$, nous pouvons définir les idéaux $\mathcal{P}_\alpha(R_\nu)$ et $\mathcal{P}_{\alpha+}(R_\nu)$ de l'anneau de valuation $V = R_\nu$ par:

$$\mathcal{P}_\alpha(R_\nu) = \bigl\{x \in R_\nu \,/\, \nu(x) \geq \alpha\bigr\} \quad \text{et} \quad \mathcal{P}_{\alpha+}(R_\nu) = \bigl\{x \in R_\nu \,/\, \nu(x) > \alpha\bigr\}.$$

Nous pouvons définir aussi une algèbre de Rees associée par:

$$\mathcal{A}_\nu(R_\nu) = \bigoplus_{\alpha\in\Gamma} \mathcal{P}_\alpha(R_\nu) v^{-\alpha} \subset R_\nu[v^\Gamma].$$

Ces idéaux et cette algèbre graduée jouent un rôle important dans l'étude de la valuation ν (cf. [Te]).

Remarque. Si l'anneau de valuation V n'est pas noethérien, il peut avoir d'autres idéaux que les idéaux $\mathcal{P}_\alpha(V)$ et $\mathcal{P}_{\alpha+}(V)$. Nous allons donner deux exemples d'un anneau de valuation V et d'un idéal $\mathcal{P}$ de V qui n'est pas de cette forme.

Exemple 1. Si le groupe des ordres de la valuation est égal à $\mathbb{Q}$, pour tout nombre réel $\beta > 0$ appartenant à $\mathbb{R} \setminus \mathbb{Q}$ l'ensemble $\mathcal{P} = \{x \in V \,/\, \nu(x) \geq \beta\}$, qui est égal aussi à $\{x \in V \,/\, \nu(x) > \beta\}$ car $\beta \notin \Gamma$, est un idéal de l'anneau de valuation V. Mais il n'existe pas d'élément α appartenant à $\Gamma = \mathbb{Q}$ tel que cet idéal soit égal à $\mathcal{P}_\alpha(V)$ ou à $\mathcal{P}_{\alpha+}(V)$. Si la valuation ν est de hauteur supérieure ou égale à 2 et si $\mathcal{P}$ est un idéal premier de l'anneau de valuation V, distinct de (0) et de l'idéal maximal, il n'existe pas d'élément α appartenant au groupe des ordres Γ de ν tel que l'idéal $\mathcal{P}$ soit égal à $\mathcal{P}_\alpha(V)$ ou à $\mathcal{P}_{\alpha+}(V)$.

Corollaire. *La hauteur de la valuation ν est égale à la dimension de l'anneau de valuation V associé à ν.*

Preuve. En effet la hauteur de la valuation ν est égale au nombre de sous-groupes isolés de Γ, donc au nombre d'idéaux premiers propres de l'anneau V. Comme l'ensemble de ces idéaux est totalement ordonné par l'inclusion ce nombre, s'il est fini, est la dimension de l'anneau V. □

Proposition 3.3. *Soient K un corps et V un anneau de valuation de K.*

a) *Tout anneau local R vérifiant $V \subset R \subset K$ est un anneau de valuation de K. L'idéal maximal* $\max(R)$ *de R est contenu dans l'anneau V et est un idéal premier de V.*
b) *L'application $\mathcal{P} \mapsto V_\mathcal{P}$ est une bijection décroissante de l'ensemble des idéaux premiers $\mathcal{P}$ de V dans l'ensemble des anneaux locaux R tels que $V \subset R \subset K$. La bijection réciproque est définie par $R \mapsto$* $\max(R)$.

Preuve. De la condition b) du théorème 1.1 nous déduisons que l'anneau R est un anneau de valuation et que son idéal maximal $\max(R)$ est inclus dans V. Comme $\max(R)$ est un idéal premier de R, c'est aussi un idéal premier de V.
Pour tout idéal premier $\mathcal{P}$ de V, l'anneau localisé $V_\mathcal{P}$ vérifie bien $V \subset V_\mathcal{P} \subset K$, et l'application $\mathcal{P} \mapsto V_\mathcal{P}$ est strictement décroissante. De plus nous vérifions que l'idéal maximal $\mathcal{P}V_\mathcal{P}$ du localisé est égal à l'idéal premier $\mathcal{P}$ de V. □

Nous voyons ainsi que l'étude des idéaux premiers $\mathcal{P}$ de V, c'est à dire l'étude des sous-groupes isolés du groupe des ordres Γ, se ramène à l'étude des anneaux R vérifiant $V \subset R \subset K$.

Soit V l'anneau de valuation associé à une valuation ν de K de groupe des valeurs Γ, et nous supposons que ν est de rang fini r. Nous notons $\mathcal{P}_i$, V_i et Δ_i, $0 \leq i \leq r$, respectivement les idéaux premiers de V, les sous-anneaux de K contenant V et les sous-groupes isolés de Γ, avec les relations: $\mathcal{P}_i = \max(V_i)$ et $V_i = V_{\mathcal{P}_i}$; $\Delta_i = \Delta_{\mathcal{P}_i}$, c'est à dire le complémentaire de $\big(\nu(\mathcal{P}_i)\big) \cup \big(-\nu(\mathcal{P}_i)\big)$ dans Γ. Nous notons Γ_i le groupe quotient Γ/Δ_i qui est un groupe totalement ordonné. Nous avons alors les inclusions:

$$\begin{aligned}
&(0) = \mathcal{P}_0 \subset \mathcal{P}_1 \subset \cdots \subset \mathcal{P}_{r-1} \subset \mathcal{P}_r = \max(V)\\
&V = V_r \subset V_{r-1} \subset \cdots \subset V_1 \subset V_0 = K\\
&(0) = \Delta_r \subset \Delta_{r-1} \subset \cdots \subset \Delta_1 \subset \Delta_0 = \Gamma.
\end{aligned}$$

Le groupe des ordres de la valuation ν_i associée à l'anneau de valuation $V_i = V_{\mathcal{P}_i}$ est alors le groupe quotient $\Gamma_i = \Gamma/\Delta_i$ et l'application $\nu_i\colon K^* \to \Gamma_i$ est la composée de l'application $\nu\colon K^* \to \Gamma$ et de l'application canonique $\Gamma \to \Gamma_i$. Nous vérifions aussi que l'inclusion de $U(V) = V \setminus \mathcal{P}_r$ dans $U(V_i) = V_i \setminus \mathcal{P}_i$ définit l'application canonique du groupe $\Gamma = K^*/U(V)$ dans le groupe quotient $\Gamma_i = K^*/U(V_i)$ (cf. Proposition 4.1).

Exemple 2. La valuation impropre de K, c'est à dire la valuation ν définie par $\nu(x) = 0$ pour tout $x \in K^*$, est l'unique valuation de hauteur nulle.

Exemple 3. La valuation ν de K est de hauteur 1 si et seulement si le groupe des ordres Γ de ν est isomorphe à un sous-groupe de $(\mathbb{R}, +)$. C'est équivalent à dire que le groupe Γ est archimédien, c'est à dire que pour tout α et β dans Γ avec $\beta > 0$, il existe un entier n tel que $n\beta \geq \alpha$. L'anneau de valuation V associé à ν est de dimension 1, et nous déduisons de la proposition précédente que l'anneau V est maximal parmi les sous-anneaux propres de K.

Exemple 4. La valuation ν est une valuation discrète de K si son groupe des ordres Γ est un groupe discret de rang fini, i.e. isomorphe à un sous-groupe de $\mathbb{Z}^n$.
En particulier nous disons que la valuation ν est discrète de rang 1 si son groupe des ordres est isomorphe à un sous-groupe de $\mathbb{Z}$, et nous pouvons toujours supposer qu'il est égal à $\mathbb{Z}$; nous disons que l'anneau associé V est un anneau de valuation discrète de rang 1.

Proposition 3.4. *Soit A un anneau local intègre distinct de son corps des fractions K. Alors les conditions suivantes sont équivalentes:*

a) *A est un anneau de valuation discrète de rang 1;*
b) *A est un anneau principal;*
c) *l'idéal maximal* $\max(A)$ *est principal et l'anneau A est noethérien;*
d) *A est un anneau de valuation noethérien.*

Preuve. a) $\Rightarrow$ b): par hypothèse le groupe des ordres est isomorphe à $\mathbb{Z}$, les seuls segments sont alors de la forme $[-n, n]$, pour $n \in \mathbb{N}$. Par conséquent tout idéal $\mathcal{I}$ de A est un idéal du type $\mathcal{P}_n$ et est engendré par tout élément x de l'anneau A vérifiant $\nu(x) = n$ où $n = \nu(\mathcal{I}) = \inf\{\nu(y) \,/\, y \in \mathcal{I}\}$.

b) $\Rightarrow$ c): évident.

c) $\Rightarrow$ d): nous allons définir une valuation ν sur A appelée la valuation $\mathcal{M}$-adique, où $\mathcal{M}$ est l'idéal maximal $\max(A)$ de A. Comme A est noethérien nous avons $\bigcap_{n\geq 0}\mathcal{M}^n = (0)$, par conséquent pour tout élément non nul x de A nous pouvons définir $\nu(x)$ comme le plus grand entier n tel que x appartienne à $\mathcal{M}^n$, c'est à dire $\nu(x) \geq n \Longleftrightarrow x \in \mathcal{M}^n$. Si nous appelons u un générateur de l'idéal maximal $\mathcal{M}$ de A, tout élément x de A s'écrit sous la forme $x = yu^n$ avec $n = \nu(x)$ et où y est un élément inversible de A. Tout élément z de K s'écrit alors $z = yu^n$ avec $n \in \mathbb{Z}$ et y élément inversible de A, nous en déduisons que ν est bien une valuation discrète de rang 1 de K et que A est l'anneau associé.

d) $\Rightarrow$ a): si A est noethérien, toute suite croissante d'idéaux de A est stationnaire, par conséquent toute suite décroissante d'éléments de Γ^+ doit aussi être stationnaire. Alors le groupe Γ est isomorphe à $\mathbb{Z}$. □

Remarque. Si ν est une valuation discrète de rang 1, nous supposons que son groupe des ordres Γ est égal à $\mathbb{Z}$, c'est à dire que la valuation ν est bien la valuation $\mathcal{M}$-adique définie précédemment, où $\mathcal{M}$ est l'idéal maximal de l'anneau de valuation A. Alors tout élément u de K vérifiant $\nu(u) = 1$ est un générateur de l'idéal maximal $\mathcal{M}$ de A. Un tel élément u est appelé une uniformisante. De plus les seuls idéaux de A sont les idéaux $(u^n)A$.

Définition. Le rang rationnel d'un groupe commutatif Γ est la dimension du $\mathbb{Q}$-espace vectoriel $\Gamma \otimes_{\mathbb{Z}} \mathbb{Q}$. Le rang rationnel d'une valuation ν d'un corps K est le rang rationnel de son groupe des ordres Γ, nous le notons rang rat.(ν) = rang rat.$(\Gamma) = \dim_{\mathbb{Q}}(\Gamma \otimes_{\mathbb{Z}} \mathbb{Q})$.

Le rang rationnel d'un groupe Γ est nul si et seulement si Γ est un groupe de torsion. Le rang rationnel d'un groupe Γ est égal au plus grand entier r tel qu'il existe r éléments de Γ linéairement indépendants sur $\mathbb{Z}$, si le rang rationnel est fini.

Proposition 3.5. *Soient Γ un groupe commutatif et Γ' un sous-groupe de Γ. Alors nous avons l'égalité:*

$$\textit{rang rat.}(\Gamma) = \textit{rang rat.}(\Gamma') + \textit{rang rat.}(\Gamma/\Gamma')\,.$$

Si le groupe Γ est totalement ordonné nous avons l'inégalité:

$$\textit{rang}(\Gamma) \leq \textit{rang}(\Gamma') + \textit{rang rat.}(\Gamma/\Gamma')\,.$$

En particulier nous avons toujours l'inégalité rang$(\Gamma) \leq$ rang rat.(Γ), d'où pour toute valuation ν l'inégalité:

$$ht(\nu) \leq \textit{rang rat.}(\nu)\,.$$

Preuve. L'égalité rang rat.(Γ) = rang rat.(Γ')+ rang rat.(Γ/Γ') est immédiate par définition du rang rationnel (et car $\mathbb{Q}$ est plat sur $\mathbb{Z}$).

Pour démontrer l'inégalité nous allons montrer par récurrence sur n que si nous avons une suite strictement croissante de longueur n, $\Gamma_0 \subset \Gamma_1 \subset \cdots \subset \Gamma_n$, de sous-groupes isolés de Γ, alors nous avons: $n \leq \text{rang}(\Gamma') + \text{rang rat.}(\Gamma/\Gamma')$.

C'est évident pour $n = 0$. Supposons que c'est vrai à l'ordre $n-1$, par hypothèse de récurrence appliquée à Γ_{n-1} nous avons:

$$n - 1 \leq \text{rang}(\Gamma' \cap \Gamma_{n-1}) + \text{rang rat.}\big(\Gamma_{n-1}/\Gamma' \cap \Gamma_{n-1}\big)\,.$$

Si $\Gamma' \cap \Gamma_{n-1}$ est égal à Γ', c'est à dire si Γ' est inclus dans Γ_{n-1}, nous avons:

$$n - 1 \leq \text{rang}(\Gamma') + \text{rang rat.}\big(\Gamma_{n-1}/\Gamma'\big)\,;$$

comme le groupe Γ/Γ_{n-1} est totalement ordonné, il est sans torsion et vérifie rang rat.$\big(\Gamma/\Gamma_{n-1}\big) \geq 1$. Nous déduisons alors de la première partie de la proposition rang rat.$\big(\Gamma_{n-1}/\Gamma'\big) \leq$ rang rat.$\big(\Gamma/\Gamma'\big) - 1$, d'où l'inégalité cherchée. Si $\Gamma' \cap \Gamma_{n-1}$ n'est pas égal à Γ', c'est un sous-groupe isolé propre de Γ', d'où:

$$\text{rang}(\Gamma') \geq \text{rang}(\Gamma' \cap \Gamma_{n-1}) + 1$$

et nous déduisons de rang rat.$\big(\Gamma/\Gamma'\big) \geq$ rang rat.$\big(\Gamma_{n-1}/\Gamma' \cap \Gamma_{n-1}\big)$, l'inégalité cherchée. □

Définition. Un groupe Γ est divisible s'il vérifie la propriété suivante:

$$\text{pour tout } x \in \Gamma\,,\ \text{ pour tout } n \in \mathbb{N}^*\,, \exists x' \in \Gamma \quad \text{tel que} \quad nx' = x\,.$$

Pour tout groupe sans torsion Γ il existe un plus petit groupe divisible le contenant, c'est à dire un groupe Γ^* divisible vérifiant $\Gamma \subset \Gamma^*$ et tout groupe divisible Γ" contenant Γ contient aussi Γ^*. Nous appelons le groupe Γ^* la fermeture divisible de Γ, c'est le groupe $\Gamma \otimes_{\mathbb{Z}} \mathbb{Q}$. Il peut aussi être défini comme l'ensemble quotient de $\big\{(\gamma, m, n) \in \Gamma \times \mathbb{Z} \times \mathbb{N}^*\big\}$ par la relation d'équivalence $(\gamma, m, n) \simeq (\gamma', m', n') \iff n'm\gamma = nm'\gamma'$.

Remarque. Tout groupe divisible totalement ordonné Γ de rang fini h, est isomorphe à un groupe ordonné Γ' de la forme

$$\Gamma' = \big(\Gamma_1 \oplus \cdots \oplus \Gamma_h\big)\,,$$

où chaque Γ_i est un groupe divisible archimédien, i.e. est un sous-groupe de $(\mathbb{R}, +)$, et où l'ordre sur Γ' est l'ordre lexicographique [Ab 2]. Nous déduisons de ce résultat et de l'existence de la fermeture divisible que tout groupe totalement ordonné Γ de rang fini h est isomorphe à un sous-groupe ordonné de $\big(\mathbb{R}^h, +\big)_{lex}$.

Exemple 5. Il existe des groupes ordonnés qui ne sont isomorphes à aucun groupe de la forme $\Gamma = \big(\Gamma_1 \oplus \cdots \oplus \Gamma_h\big)_{lex}$. Soit Γ le sous-groupe ordonné de $G = \big(\mathbb{R}^2, +\big)_{lex}$ engendré par les éléments $\alpha_k = \big(1/2^k, 0\big)$, pour $k \geq 0$, et l'élément $\beta = \big(1/3, 1/3\big)$. Le groupe Γ est un sous-groupe non divisible de rang rationnel deux. Tout sous-groupe isolé $\bar{\Gamma}$ de Γ est de la forme $\Gamma \cap \bar{G}$, où $\bar{G}$ est un sous-groupe isolé de $\big(\mathbb{R}^2, +\big)_{lex}$. Nous vérifions que le sous-groupe $\bar{\Gamma}$ défini par $\bar{\Gamma} = \Gamma \cap \big(\{0\} \times \mathbb{R}\big)$ est égal à $\big(\{0\} \times \mathbb{Z}\big)$ et nous en déduisons que Γ est un groupe ordonné de rang deux. Si Γ pouvait s'écrire sous la forme $\big(\Gamma_1 \oplus \Gamma_2\big)_{lex}$, alors Γ_1 et Γ_2 seraient deux sous-groupes de Γ vérifiant que Γ_1 est isomorphe au groupe quotient $\Gamma' = \Gamma/\bar{\Gamma}$ et que Γ_2 est égal à $\bar{\Gamma}$. Le groupe Γ' est un sous-groupe de $\mathbb{R}$ et l'application canonique de Γ dans Γ' est la projection $\gamma = (x, y) \longmapsto x$. Alors comme Γ' est engendré par les éléments $1/2^k$, $k \geq 0$ et $1/3$, il devrait exister des éléments γ_k et δ dans le sous-groupe Γ_1 de la forme $\gamma_k = (1/2^k, y_k)$ et $\delta = (1/3, z)$. Par construction les éléments $\gamma_k - \alpha_k$, $k \geq 0$, et $\delta - \beta$ sont dans le sous-groupe $\bar{\Gamma}$, par conséquent les y_k et $z - 1/3$ sont des entiers. De plus pour tout $k \geq 0$, $\gamma_0 - 2^k\gamma_k$ appartient à l'intersection $\Gamma_1 \cap \Gamma_2$, donc est nul et nous en déduisons $y_0 = 0$. De même $\gamma_0 - 3\delta$ est nul et nous en déduisons $z = 0$, ce qui contredit la condition $z - 1/3 \in \mathbb{Z}$.

Nous pouvons remarquer que si nous munissons ce groupe Γ de l'ordre lexicographique inverse, ou ce qui revient au même si nous considérons le sous-groupe $\tilde{\Gamma}$ de $(\mathbb{R}^2, +)_{lex}$ engendré par les éléments $\tilde{\alpha}_k = (0, 1/2^k)$, pour $k \geq 0$, et l'élément $\beta = (1/3, 1/3)$, nous pouvons trouver une décomposition $\tilde{\Gamma} = (\tilde{\Gamma}_1 \oplus \tilde{\Gamma}_2)_{lex}$, où $\tilde{\Gamma}_1$ est le sous-groupe isomorphe à $\mathbb{Z}$ engendré par β et où $\tilde{\Gamma}_2$ est le sous-groupe isolé engendré par les $\tilde{\alpha}_k$, $k \geq 0$.

4. Valuations composées

Soit ν une valuation d'un corps K, d'anneau de valuation V et de groupe des ordres Γ. Nous supposons que le rang de ν est strictement plus grand que 1, par conséquent il existe une valuation ν' de K dont l'anneau de valuation V' contient l'anneau V. Nous appelons $\mathcal{P}$ et $\mathcal{P}'$ les idéaux maximaux respectifs de V et V', alors $\mathcal{P}'$ est un idéal premier de V et V' est égal au localisé $V_{\mathcal{P}'}$. Nous appelons Γ' le groupe des ordres de la valuation ν' et nous appelons $\bar{\Gamma}$ le sous-groupe isolé de Γ correspondant à l'idéal premier $\mathcal{P}'$ de V.

Proposition 4.1.

a) *Le groupe des ordres* Γ' *est naturellement isomorphe au groupe quotient* $\Gamma/\bar{\Gamma}$.

b) *L'anneau quotient* $\bar{V} = V/\mathcal{P}'$ *est un anneau de valuation du corps résiduel* $\bar{K} = \kappa(V') = V'/\mathcal{P}'$, *et la valuation associée* $\bar{\nu}$ *admet pour groupe des ordres le groupe* $\bar{\Gamma}$.

Preuve. a) Les valuations ν et ν' peuvent être définies comme les applications naturelles $\nu \colon K^* \to K^*/U(V) = \Gamma$ et $\nu' \colon K^* \to K^*/U(V') = \Gamma'$. Nous déduisons de $V \subset V'$ et $\mathcal{P} \subset \mathcal{P}'$ que $U(V) = V \setminus \mathcal{P}$ est inclus dans $V' \setminus \mathcal{P}' = U(V')$, d'où un morphisme surjectif de groupes ordonnés $\phi \colon \Gamma \to \Gamma'$ tel que $\nu' = \phi \circ \nu$. Nous voulons montrer que le noyau de ϕ est le sous-groupe $\bar{\Gamma}$, pour cela il suffit de montrer qu'un élément α de Γ^+ appartient à Kerϕ si et seulement s'il appartient à $\bar{\Gamma}^+$. Soit $\alpha = \nu(x)$, c'est à dire α est l'image de $x \in V$ dans $\Gamma^+ = V^*/U(V)$; alors α appartient à Kerϕ si et seulement si x appartient à $U(V') = V \setminus \mathcal{P}'$. Or d'après le théorème 3.2, le sous-groupe isolé $\bar{\Gamma}$ associé à l'idéal premier $\mathcal{P}'$ de V est défini par $\bar{\Gamma} \cap \Gamma^+ = \bar{\Gamma}^+ = \nu(V \setminus \mathcal{P}')$, d'où Ker$\phi \cap \Gamma^+ = \bar{\Gamma}^+$.

b) Comme V est un anneau de valuation de K, l'anneau $\bar{V} = V/\mathcal{P}'$ est un anneau de valuation de $\bar{K} = V'/\mathcal{P}'$, il suffit par exemple de considérer la condition b) du théorème 1.1. Le groupe des ordres de la valuation $\bar{\nu}$ de associée à $\bar{V}$ est isomorphe au groupe $\bar{K}^*/U(\bar{V})$ et il suffit alors de trouver une suite exacte:

$$0 \longrightarrow \bar{K}^*/U(\bar{V}) \longrightarrow K^*/U(V) \longrightarrow K^*/U(V') \longrightarrow 0\,.$$

Nous définissons un morphisme $\bar{K}^* \to K^*/U(V)$ de la manière suivante: soit $\bar{x} \in \bar{K}$ correspondant à la classe d'un élément x de V' et comme $\bar{x} \neq 0$, x n'appartient pas à l'idéal $\mathcal{P}'$, nous associons alors à $\bar{x}$ l'élément $\hat{x}$ classe de x dans $K^*/U(V)$. Montrons que ce morphisme est bien défini: si $\bar{x}$ est égal à $\bar{y}$, $x-y$ appartient à l'idéal $\mathcal{P}'$; nous avons alors $\nu(x-y) \notin \bar{\Gamma}$. Par hypothèse x et y n'appartiennent pas à $\mathcal{P}'$, c'est à dire $\nu(x)$ et $\nu(y)$ sont dans $\bar{\Gamma}$ et comme $\bar{\Gamma}$ est un sous-groupe isolé de Γ nous avons forcément $\nu(x) = \nu(y)$, c'est à dire $\hat{x} = \hat{y}$ dans $K^*/U(V)$. Il est ensuite facile de vérifier que nous avons une suite exacte de groupes qui respecte la structure de groupes ordonnés. □

Définition. La valuation ν est appelée la valuation composée avec les valuations ν' et $\bar{\nu}$ et nous écrivons $\nu = \nu' \circ \bar{\nu}$.

Corollaire. *Si ν est la valuation composée avec les valuations ν' et $\bar{\nu}$ nous avons les égalités:*

$$\mathit{rang}(\nu) = \mathit{rang}(\nu') \; + \; \mathit{rang}(\bar{\nu})$$
$$\mathit{rang\ rat.}(\nu) = \mathit{rang\ rat.}(\nu') \; + \; \mathit{rang\ rat.}(\bar{\nu})\,.$$

Preuve. Nous appelons Γ, Γ' et $\bar{\Gamma}$ les groupes des ordres respectifs des valuations ν, ν' et $\bar{\nu}$. Nous avons alors un isomorphisme de groupes ordonnés entre Γ' et le groupe quotient $\Gamma/\bar{\Gamma}$. La deuxième égalité est exactement l'égalité de la proposition 3.5. Pour démontrer la première égalité il suffit de vérifier que si $\bar{\Gamma}$ est un sous-groupe isolé d'un groupe totalement ordonné Γ nous avons:

$$\mathrm{rang}(\Gamma) = \; \mathrm{rang}(\bar{\Gamma}) \; + \; \mathrm{rang}(\Gamma/\bar{\Gamma})\,. \qquad \square$$

Réciproquement il est toujours possible de définir la valuation composée ν d'une valuation ν' de K et d'une valuation $\bar{\nu}$ du corps résiduel $\kappa_{\nu'}$.

Proposition 4.2. *La valuation composée $\nu = \nu' \circ \bar{\nu}$ de la valuation ν' de K et de la valuation $\bar{\nu}$ du corps résiduel $\kappa_{\nu'}$ est la valuation de K associée au sous-anneau V de l'anneau $R_{\nu'}$ défini par $V = \{x \in R_{\nu'} \,/\, \bar{\nu}(\bar{x}) \geq 0\}$.*

Preuve. Comme V est l'image inverse de l'anneau $R_{\bar{\nu}}$ associé à la valuation $\bar{\nu}$ du corps résiduel $\kappa_{\nu'}$ par l'application canonique $\iota\colon R_{\nu'} \longrightarrow \kappa_{\nu'}$, c'est un anneau de valuation de K, associé à une valuation ν. Nous pouvons alors vérifier que la valuation ν ainsi définie est bien la valuation composée $\nu' \circ \bar{\nu}$. □

En particulier la donnée des deux valuations ν' et $\bar{\nu}$ respectivement sur les corps K et $\kappa_{\nu'}$ nous donne une extension du groupe des ordres Γ' de ν' par le groupe des ordres $\bar{\Gamma}$, c'est à dire une suite exacte:

$$0 \longrightarrow \bar{\Gamma} \longrightarrow \Gamma \longrightarrow \Gamma' \longrightarrow 0\,.$$

Nous allons donner une description du groupe ordonné Γ et de la valuation composée $\nu = \nu' \circ \bar{\nu}$ à partir de cette suite exacte. Nous choisissons une section $\underline{x}$ de la valuation ν', c'est à dire pour tout élément γ du groupe Γ', nous choisissons x_γ dans K tel que $\nu'(x_\gamma) = \gamma$; et nous supposons en plus que les éléments x_γ vérifient $x_{-\gamma} = x_\gamma^{-1}$, et $x_0 = 1$. Alors pour tout couple (γ, γ') de $\Gamma' \times \Gamma'$, nous appelons $z_{\gamma,\gamma'}$ l'image de $\dfrac{x_\gamma . x_{\gamma'}}{x_{\gamma+\gamma'}}$ dans $\kappa^*_{\nu'}$, et nous définissons une application $F_{\underline{x}}$ de $\Gamma' \times \Gamma'$ dans $\bar{\Gamma}$ par: $F_{\underline{x}}(\gamma, \gamma') = \bar{\nu}(z_{\gamma,\gamma'})$.

Proposition 4.3. *L'application $F_{\underline{x}}$ définit un élément de* $\mathrm{Ext}^1(\Gamma', \bar{\Gamma})$ *qui correspond à la suite exacte*

$$0 \longrightarrow \bar{\Gamma} \longrightarrow \Gamma \longrightarrow \Gamma' \longrightarrow 0$$

associée à la composition des valuations $\nu = \nu' \circ \bar{\nu}$, où Γ est le groupe des ordres de ν. Plus précisément le groupe Γ est l'ensemble produit $\Gamma' \times \bar{\Gamma}$ muni de la loi interne:

$$(\gamma_1, \bar{\delta}_1) + (\gamma_2, \bar{\delta}_2) = \big(\gamma_1 + \gamma_2, \bar{\delta}_1 + \bar{\delta}_2 + F_{\underline{x}}(\gamma_1, \gamma_2)\big)\,,$$

et muni de l'ordre lexicographique induit par les ordres sur Γ' et $\bar{\Gamma}$. La valuation composée ν à valeurs dans Γ est définie en posant pour tout y appartenant à K^*, $\nu(y) = (\gamma, \bar{\delta})$ dans $\Gamma' \times \bar{\Gamma}$, avec $\gamma = \nu'(y)$ et $\bar{\delta} = \bar{\nu}(z)$ où z est l'image de (y/x_γ) dans $\kappa^*_{\nu'}$ par l'application ι.

Preuve. Nous vérifions que nous définissons ainsi une loi de groupe sur $\Gamma' \times \bar{\Gamma}$, pour cela il suffit de remarquer que nous avons les égalités:

$$F_{\underline{x}}(\gamma, -\gamma) = 0 \text{ pour tout } \gamma \in \Gamma'\,,$$
$$F_{\underline{x}}(\gamma_1, \gamma_2) + F_{\underline{x}}(\gamma_1 + \gamma_2, \gamma_3) = F_{\underline{x}}(\gamma_1, \gamma_2 + \gamma_3) + F_{\underline{x}}(\gamma_2, \gamma_3) \text{ pour tout } \gamma_1, \gamma_2, \gamma_3\,.$$

Pour montrer que l'ordre lexicographique munit $\Gamma' \times \bar{\Gamma}$ d'une structure de groupe ordonné il faut vérifier la relation suivante:

$$(\gamma_1, \bar{\delta}_1) \geq (\gamma_2, \bar{\delta}_2) \Longrightarrow \big(\gamma_1 + \gamma, \bar{\delta}_1 + \bar{\delta} + F_{\underline{x}}(\gamma_1, \gamma)\big) \geq \big(\gamma_2 + \gamma, \bar{\delta}_2 + \bar{\delta} + F_{\underline{x}}(\gamma_2, \gamma)\big)\,;$$

où la relation d'ordre $\geq$ est définie par $(\gamma_1, \bar{\delta}_1) \geq (\gamma_2, \bar{\delta}_2)$ si et seulement si $\gamma_1 > \gamma_2$ ou $\gamma_1 = \gamma_2$ et $\bar{\delta}_1 \geq \bar{\delta}_2$. Vérifions que l'application ν définie par $\nu(y) = (\gamma, \bar{\delta})$, avec $\gamma = \nu'(y)$ et $\bar{\delta} = \bar{\nu}\big(\iota(y/x_\gamma)\big)$ est une valuation. Soient y' et y'' deux éléments de K^* dont les valuations respectives sont $\nu(y') = (\gamma', \bar{\delta}')$ et $\nu(y'') = (\gamma'', \bar{\delta}'')$.

1) $\nu(y'.y'') = \nu(y') + \nu(y'')$, il suffit de remarquer que nous avons les égalités suivantes:

$$\gamma' + \gamma'' = \nu'(y') + \nu'(y'') = \nu'(y'.y'') = \gamma\,,$$
$$\begin{aligned}\bar{\delta}' + \bar{\delta}'' + F_{\underline{x}}(\gamma', \gamma'') &= \bar{\nu}\big(\iota(y'/x_{\gamma'})\big) + \bar{\nu}\big(\iota(y''/x_{\gamma''})\big) + \bar{\nu}\big(\iota(x_{\gamma'}.x_{\gamma''}/x_{\gamma'+\gamma''})\big)\\ &= \bar{\nu}\big(\iota(y'/x_{\gamma'}).\iota(y''/x_{\gamma''}).\iota(x_{\gamma'}.x_{\gamma''}/x_\gamma)\big)\\ &= \bar{\nu}\big(\iota(y'.y''/x_\gamma)\big) = \bar{\delta}\,.\end{aligned}$$

2) $\nu(y' + y'') \geq \inf\big(\nu(y'), \nu(y'')\big)$, nous allons considérer les trois cas suivants:

si $\nu'(y'') > \nu'(y')$, alors $\nu'(y' + y'') = \gamma'$ et $\iota(y' + y''/x_{\gamma'}) = \iota(y'/x_{\gamma'})$, ce qui implique $\bar{\nu}\big(\iota(y' + y''/x_{\gamma'})\big) = \bar{\delta}'$, c'est à dire $\nu(y' + y'') = \nu(y')$;

si $\nu'(y'') = \nu'(y') = \gamma'$ et $\bar{\nu}\big(\iota(y''/x_{\gamma'})\big) > \bar{\nu}\big(\iota(y'/x_{\gamma'})\big) = \bar{\delta}'$, alors nous avons encore $\nu'(y' + y'') = \gamma'$, (en effet nous déduisons de $\bar{\delta}'' > \bar{\delta}'$ que $\iota(y' + y''/x_{\gamma'}) =$

$\iota(y'/x_{\gamma'}) + \iota(y''/x_{\gamma'})$ est non nul, d'où $\nu'(y'+y'') = \gamma'$) et nous avons aussi $\bar{\nu}\big(\iota(y'+y''/x_{\gamma'})\big) = \bar{\nu}\big(\iota(y'/x_{\gamma'})\big) = \bar{\delta}'$, c'est à dire $\nu(y'+y'') = \nu(y')$;

si $\nu'(y'') = \nu'(y') = \gamma'$ et $\bar{\nu}\big(\iota(y''/x_{\gamma'})\big) = \bar{\nu}\big(\iota(y'/x_{\gamma'})\big) = \bar{\delta}'$, soit $\nu'(y'+y'') > \gamma'$ et nous avons alors $\nu(y'+y'') > \inf\big(\nu(y'), \nu(y'')\big)$, soit $\nu'(y'+y'') = \gamma'$ et nous avons $\bar{\nu}\big(\iota(y'+y''/x_{\gamma'})\big) \geq \inf(\bar{\delta}', \bar{\delta}'')$, d'où $\nu(y'+y'') \geq \inf\big(\nu(y'), \nu(y'')\big)$.

3) $\nu(1) = 0$ car nous avons choisi $x_0 = 1$.

Pour montrer que la valuation ν est égale à la valuation composée $\nu' \circ \bar{\nu}$, il faut montrer que l'anneau de valuation associé R_ν est le sous-anneau de $R_{\nu'}$ défini par $R_\nu = \iota^{-1}\big(R_{\bar{\nu}}\big)$, où ι est l'application canonique de $R_{\nu'}$ dans $\kappa_{\nu'}$, c'est à dire il faut montrer que pour tout y dans K^* nous avons:

$$\nu(y) \geq 0 \Longleftrightarrow \nu'(y) \geq 0 \text{ et } \bar{\nu}(\iota(y)) \geq 0 .$$

Par définition de l'ordre sur $\Gamma = \Gamma' \times \bar{\Gamma}$, nous avons $\nu(y) = (\gamma, \bar{\delta}) \geq 0$ si et seulement si $\gamma > 0$ ou $\gamma = 0$ et $\bar{\delta} \geq 0$: dans le cas $\gamma > 0$, nous avons $\nu'(y) > 0$, d'où $\iota(y) = 0$ dans $\kappa_{\nu'}$ et $\bar{\delta} = \bar{\nu}\big(\iota(y)\big) = +\infty$, dans le cas $\gamma = 0$, nous avons $\nu'(y) = 0$ et $\bar{\delta} = \bar{\nu}(\iota(y))$ car $x_0 = 1$. □

Remarque. Si la suite exacte $0 \longrightarrow \bar{\Gamma} \longrightarrow \Gamma \longrightarrow \Gamma' \longrightarrow 0$ est scindée, le groupe des ordres Γ de la valuation composée $\nu = \nu' \circ \bar{\nu}$ est isomorphe au groupe produit $(\Gamma' \times \bar{\Gamma})$ muni de l'ordre lexicographique. En particulier si la valuation ν' est une valuation discrète de rang un, c'est à dire pour $\Gamma' \simeq \mathbb{Z}$, la suite exacte est toujours scindée et nous pouvons décrire la valuation composée $\nu = \nu' \circ \bar{\nu}$, de la manière suivante. L'idéal maximal de l'anneau de valuation $R_{\nu'}$ est engendré par un élément u et à tout élément non nul x de K nous pouvons associer l'élément non nul $\bar{y}$ du corps résiduel $\kappa_{\nu'}$ obtenu comme classe de $y = x.u^{-\nu'(x)}$. La valuation composée ν est alors définie par $\nu(x) = \big(\nu'(x), \bar{\nu}(\bar{y})\big)$.

5. Prolongement d'une valuation

Soient K un corps et L une extension de K. Si μ est une valuation de L, la restriction de μ à K est une valuation ν de K dont le groupe des ordres Γ_ν est un sous-groupe du groupe des ordres Γ_μ de μ. De plus l'anneau de valuation V de ν est égal à $W \cap K$, où W est l'anneau de valuation de μ, et W domine V.

Définition. Dans la situation précédente nous disons que la valuation μ de L prolonge la valuation ν de K ou que la valuation μ est un prolongement de ν.

Remarque. Si V et W sont deux anneaux de valuation respectivement de K et de L, où L est une extension de K, W domine V si et seulement si $V = W \cap K$. En effet si W domine V, V est inclus dans $W \cap K$; et si x est un élément de K n'appartenant pas à V, son inverse x^{-1} appartient à $\max(V)$, donc à $\max(W)$, par conséquent x n'appartient pas à W. Réciproquement si V est égal à $W \cap K$, V est inclus dans W; et si x est dans $\max(V)$, son inverse x^{-1} n'appartient pas à V, par conséquent n'appartient pas à W et x est dans $\max(W)$.

Remarque. Pour toute valuation ν de K, il existe au moins une valuation μ de L qui prolonge ν. En effet l'anneau de valuation V de ν est un sous-anneau local de L, et d'après le corollaire à la proposition 1.2 il existe un anneau de valuation W de L qui domine V. Grâce à la remarque précédente nous en déduisons que la valuation μ associée à W prolonge la valuation ν.

Nous nous proposons d'étudier les extensions Γ_μ et κ_μ respectivement du groupe des valeurs Γ_ν et du corps résiduel κ_ν d'une valuation ν de K, correspondant à un prolongement μ de ν à une extension L de K donnée.

Nous allons d'abord étudier le cas où L est une extension algébrique de K. Nous notons respectivement Γ_ν et Γ_μ, et V et W, les groupes des ordres, et les anneaux de valuations, associés à ν et μ. Nous notons κ_ν et κ_μ les corps résiduels respectifs des valuations ν et μ, c'est à dire les corps définis par $\kappa_\nu = V/\max(V)$ et $\kappa_\mu = W/\max(W)$.

Définition. L'indice de ramification de μ par rapport à ν est égal à l'indice du groupe des ordres Γ_ν dans Γ_μ:

$$e(\mu/\nu) = \left[\Gamma_\mu : \Gamma_\nu\right].$$

Le degré résiduel de μ par rapport à ν est égal au degré de l'extension du corps résiduels κ_ν dans le corps κ_μ:

$$f(\mu/\nu) = \left[\kappa_\mu : \kappa_\nu\right].$$

L'indice de ramification $e(\mu/\nu)$ et le degré résiduel $f(\mu/\nu)$ sont des éléments de $\bar{\mathbb{N}} = \mathbb{N} \cup \{+\infty\}$.

Remarque. Si L' est une extension de L et si μ' est une valuation de L' qui prolonge μ, alors μ' prolonge ν et nous avons les égalités:

$$e(\mu'/\nu) = e(\mu'/\mu)e(\mu/\nu) \text{ et } f(\mu'/\nu) = f(\mu'/\mu)f(\mu/\nu).$$

En particulier $e(\mu'/\nu)$, (resp. $f(\mu'/\nu)$,) est fini si et seulement si $e(\mu'/\mu)$ et $e(\mu/\nu)$, (resp. $f(\mu'/\mu)$ et $f(\mu/\nu)$,) sont finis.

Proposition 5.1. *Si L est une extension finie de K de degré n nous avons l'égalité:*

$$e(\mu/\nu)f(\mu/\nu) \leq n.$$

En particulier l'indice de ramification $e(\mu/\nu) = \left[\Gamma_\mu : \Gamma_\nu\right]$ et le degré résiduel $f(\mu/\nu) = \left[\kappa_\mu : \kappa_\nu\right]$ sont finis.

Preuve. Soient r et s deux entiers avec $r \leq e(\mu/\nu)$ et $s \leq f(\mu/\nu)$; il suffit de montrer que nous avons $rs \leq n$. Par hypothèse il existe r éléments $x_1, x_2, \ldots, x_r$ de L tels que pour tout couple (i,j), avec $i \neq j$, $\mu(x_i) \not\equiv \mu(x_j) \operatorname{mod} \Gamma_\nu$. De même il existe s éléments $y_1, y_2, \ldots, y_s$ de W dont les images $\bar{y}_1, \bar{y}_2, \ldots, \bar{y}_s$ dans κ_μ sont linéairement indépendants sur κ_ν. Il suffit de montrer que les rs éléments $x_i y_k$, $1 \leq i \leq r$ et $1 \leq k \leq s$, sont indépendants sur K. Supposons que ce n'est pas le cas et qu'il existe une relation linéaire non triviale entre eux: $(*) \sum a_{i,k} x_i y_k = 0$, avec $a_{i,k} \in K$.

Nous choisissons un indice (j, m) tel que pour tout (i, k) nous avons l'inégalité $\mu(a_{j,m}x_jy_m) \leq \mu(a_{i,k}x_iy_k)$; en particulier $a_{j,m} \neq 0$. Pour $i \neq j$, nous avons l'inégalité $\mu(a_{j,m}x_jy_m) \neq \mu(a_{i,k}x_iy_k)$; en effet si nous avions égalité, $\mu(x_j) - \mu(x_i)$ serait égal à $\nu(a_{i,k}) - \nu(a_{j,m})$ car $\mu(y_k)$ est nul pour tout y_k vérifiant $\bar{y}_k \neq 0$ et car μ prolonge ν, et nous aurions alors la relation $\mu(x_j) \equiv \mu(x_i) \operatorname{mod}\Gamma_\nu$, ce qui est impossible pour $i \neq j$. En multipliant la relation $(*)$ par $(a_{j,m}x_j)^{-1}$, nous obtenons alors une relation: $\sum b_ky_k + z = 0$, avec $b_k = a_{j,k}/a_{j,m} \in W \cap K$ et $z \in \max(W)$. Nous obtenons ainsi dans le corps $\kappa_\mu = W/\max(W)$ la relation $\sum \bar{b}_k\bar{y}_k = 0$, avec $\bar{b}_m = 1$. C'est une relation non triviale de dépendance linéaire sur κ_ν des $\bar{y}_k$, ce qui est impossible par hypothèse sur les y_k. □

Proposition 5.2. *Si L est une extension algébrique de K, alors le groupe quotient Γ_μ/Γ_ν est un groupe de torsion et κ_μ est une extension algébrique de κ_ν.*

Preuve. Nous pouvons écrire $L = \varinjlim L_\alpha$, avec L_α extensions finies de K. Alors, si nous posons $\Gamma_\alpha = \mu(L_\alpha^*)$, le groupe Γ_μ est réunion de la famille filtrante des sous-groupes Γ_α. D'après la proposition précédente, le groupe Γ_ν est un sous-groupe d'indice fini dans chacun des groupes Γ_α, par conséquent le groupe Γ_μ/Γ_ν est de torsion. De même, si nous notons κ_α le corps résiduel associé à la valuation μ sur L_α, le corps κ_μ est égal à $\varinjlim \kappa_\alpha$, où les κ_α sont des extensions finies de κ_ν, par conséquent est une extension algébrique de κ_ν. □

Nous pouvons aussi montrer directement que pour tout élément δ de Γ_μ, il existe un entier $k \neq 0$ tel que $k.\delta$ appartienne à Γ_ν. Pour cela nous utilisons la méthode du polygone de Newton.

Soit Γ un groupe ordonné et nous définissons la droite D de l'espace $\mathbb{Z} \times \Gamma$ comme le sous-ensemble $D = \{(n, \gamma) \in \mathbb{Z} \times \Gamma \,/\, q\gamma + \alpha n + \beta = 0\}$, où $q \in \mathbb{Z}$, et α et $\beta \in \Gamma$. La pente $p(D)$ de la droite D d'équation $q\gamma + \alpha n + \beta = 0$ est l'élément $p(D) = -\alpha/q$ appartenant à $\Gamma \otimes_\mathbb{Z} \mathbb{Q}$.

Chaque droite D définit deux demi-espaces $H^D_\geq$ et $H^D_\leq$ de $\mathbb{Z} \times \Gamma$ par:

$$H^D_\geq = \{(n, \gamma) \in \mathbb{Z} \times \Gamma \,/\, q\gamma + \alpha n + \beta \geq 0\}$$

$$H^D_\leq = \{(n, \gamma) \in \mathbb{Z} \times \Gamma \,/\, q\gamma + \alpha n + \beta \leq 0\}.$$

Pour tout sous-ensemble A de $\mathbb{Z} \times \Gamma$ nous définissons son enveloppe convexe $Conv(A)$ par $Conv(A) = \bigcap H$, où H parcourt l'ensemble des demi-espaces de $\mathbb{Z} \times \Gamma$ contenant A. Une face F de $Conv(A)$ est un sous-ensemble F de $Conv(A)$ défini par $F = Conv(A) \cap D$, où D est une droite de $\mathbb{Z} \times \Gamma$ vérifiant:

$Conv(A)$ est contenu dans l'un des demi-espaces $H^D_\geq$ ou $H^D_\leq$ définis par D,
$F = Conv(A) \cap D$ contient au moins deux points distincts.

Nous définissons la pente $p(F)$ de la face F comme la pente de la droite D qui définit F. Soit ν une valuation d'un corps K à valeurs dans Γ_ν, alors pour tout polynôme $P(X) = a_0X^d + a_1X^{d-1} + \cdots + a_d$ de $K[X]$, nous définissons dans $\mathbb{Z} \times \Gamma_\nu$ le polygone de Newton $N(P)$ de $P(X)$ comme l'enveloppe convexe $Conv(\mathrm{Supp}(P))$ du support de $P(X)$, avec $\mathrm{Supp}(P) = \{(d - j, \nu(a_j)) \,/\, a_j \neq 0, 0 \leq j \leq d\}$. Soit L

une extension algébrique de K et soit z appartenant à L, nous appelons polygone de Newton de z sur K et nous notons $N_K(z)$ le polygone de Newton du polynôme minimal de z sur K.

Proposition 5.3. *Pour toute valuation μ de L qui prolonge ν il existe une face F du polygone de Newton $N_K(z)$ de z sur K dont la pente $p(F)$ est égale à $-\mu(z)$. En particulier il existe un entier k non nul tel que $k.\mu(z)$ appartienne à Γ_ν.*

Preuve. Soit $P(X) = X^d + a_1X^{d-1} + \cdots + a_d$ le polynôme minimal de z sur K; comme $P(z)$ est nul, pour toute valuation μ de L la valeur minimale des $\mu\big(a_jz^{d-j}\big)$, $0 \leq j \leq d$, est atteinte pour au moins deux termes distincts, c'est à dire qu'il existe $i > j$ tels que pour tout k nous ayons $\mu\big(a_kz^{d-k}\big) \geq \mu\big(a_iz^{d-i}\big) = \mu\big(a_jz^{d-j}\big)$. Si la valuation μ est un prolongement de ν et si nous posons $\delta = \mu(z)$, nous trouvons $\nu(a_k) + (d-k)\delta \leq \nu(a_i) + (d-i)\delta = \nu(a_j) + (d-j)\delta$.

Soit D la droite de $\mathbb{Z}\times\Gamma_\nu$ qui passe les points $\big(d-i, \nu(a_i)\big)$ et $\big(d-j, \nu(a-j)\big)$ appartenant au polygone de Newton $N_K(z)$ de z, c'est à dire la droite D d'équation $q\gamma+\alpha n+\beta = 0$, avec $q = i-j$, $\alpha = \nu(a_i)-\nu(a_j)$ et $\beta = \big((d-i)\nu(a_j)-(d-j)\nu(a_i)\big)$. L'élément $\delta = \mu(z)$ de Γ_μ vérifie $(i-j)\delta = \nu(a_i) - \nu(a_j)$, c'est à dire que $-\delta$ est égal à la pente $p(D)$ de la droite D et en particulier δ appartient à $\Gamma \otimes_{\mathbb{Z}} \mathbb{Q}$. Pour tout k, $0 \leq k \leq d$, nous avons $\nu(a_k) + (d-k)\delta \leq \nu(a_i) + (d-i)\delta$, c'est à dire que tout point $\big(d-k, \nu(a_k)\big)$ de $N_K(z)$ appartient au demi-espace $H^D_{\geq}$ d'équation $q\gamma + \alpha n + \beta \geq 0$. Comme $F = D \cap N_K(z)$ contient les deux points distincts $\big(d-i, \nu(a_i)\big)$ et $\big(d-j, \nu(a-j)\big)$ nous en déduisons que F est une face de $N_K(z)$ dont la pente $p(F)$ est égale à $-\mu(z)$. □

Remarque. Nous déduisons du raisonnement précédent que si L est une extension algébrique finie de degré n de K, alors pour tout élément α du groupe Γ_μ nous avons $N\alpha$ qui appartient à Γ_ν, avec $N = n!$.

Corollaire. *Si L est une extension algébrique de K, le rang de la valuation μ de L est égal au rang de la valuation ν de K. De plus si L est une extension finie de K, la valuation μ est discrète si et seulement si la valuation ν l'est aussi.*

Preuve. La première partie du corollaire est une conséquence du résultat suivant. Si H est un sous-groupe d'un groupe totalement ordonné G, l'application définie par $F \mapsto F \cap H$ est une surjection de l'ensemble des sous-groupes isolés de G dans l'ensemble des sous-groupes isolés de H. De plus cette application est bijective si le groupe quotient G/H est un groupe de torsion.

Si L est une extension finie de K, le sous-groupe Γ_ν est d'indice fini dans le groupe Γ_μ, et d'après ce qui précède l'application $\Delta'_i \mapsto \Delta_i = \Delta_i \cap \Gamma_\nu$ définit une bijection de l'ensemble des sous-groupes isolés de Γ_μ dans l'ensemble des sous-groupes isolés de Γ_ν. Soient $\Delta_i \subset \Delta_{i+1}$ deux sous groupes-isolés successifs d'un groupe totalement ordonné Γ, le groupe quotient est de hauteur 1, c'est à dire est isomorphe à un sous-groupe de $\mathbb{R}$. Alors si le groupe Γ est discret le groupe quotient Δ_{i+1}/Δ_i est un sous-groupe discret de $\mathbb{R}$, c'est à dire est isomorphe à $\mathbb{Z}$. Réciproquement si tous les groupes quotients Δ_{i+1}/Δ_i sont isomorphes à $\mathbb{Z}$, le groupe Γ est discret. Il suffit donc de montrer l'équivalence:

$$\Delta'_{i+1}/\Delta'_i \simeq \mathbb{Z} \iff \Delta_{i+1}/\Delta_i \simeq \mathbb{Z}\,.$$

Comme Δ_{i+1}/Δ_i est un sous-groupe non trivial du groupe Δ'_{i+1}/Δ'_i, nous avons $\Delta'_{i+1}/\Delta'_i \simeq \mathbb{Z} \Longrightarrow \Delta_{i+1}/\Delta_i \simeq \mathbb{Z}$; Réciproquement comme Δ_{i+1}/Δ_i est d'indice fini dans Δ'_{i+1}/Δ'_i, nous avons $\Delta_{i+1}/\Delta_i \simeq \mathbb{Z} \Longrightarrow \Delta'_{i+1}/\Delta'_i \simeq \mathbb{Z}$. □

Remarque. Dans le cas d'une extension algébrique finie L de K nous pouvons aussi étudier l'ensemble de toutes les valuations μ_i, définies à équivalence près, qui prolongent une valuation ν de K. Nous pouvons aussi nous intéresser au complété $\hat{K}$ de K pour la valuation ν ainsi qu'aux complétés respectifs $\hat{L}_i$ de L pour les valuations μ_i [Bo]. Comme nous n'avons pas besoin de ces résultats dans la suite de l'exposé, nous n'aborderons pas ces problèmes ici.

Nous allons étudier maintenant le cas d'une extension transcendante de K. Nous considérons une valuation ν d'un corps K, de groupe des ordres Γ, et un prolongement ν' de ν à un corps K', extension transcendante de K, de groupe des ordres Γ'. Nous appelons respectivement V et V' et $\kappa = V/\max(V)$ et $\kappa' = V'/\max(V')$ les anneaux de valuation et les corps résiduels de ν et ν'. Nous voulons alors étudier les grandeurs suivantes:

$\text{dim.alg.}_K K'$ = degré de transcendance de l'extension K'/K,

$\text{dim.alg.}_\kappa \kappa'$ = degré de transcendance de l'extension κ'/κ,

rang rat.(Γ'/Γ) = rang rationnel du groupe quotient Γ'/Γ,

qui est égal d'après la proposition 3.5 à rang rat.$(\nu')-$ rang rat.(ν) et est toujours supérieur ou égal à rang$(\nu')-$ rang(ν) = dim $V'-$ dim V.

Nous commençons par étudier le cas d'une extension transcendante monogène $K' = K(X)$.

Proposition 5.4. *Soit ν une valuation d'un corps K, de groupe des ordres Γ et de corps résiduel κ.*

a) *Si Γ'' est un groupe totalement ordonné contenant Γ et si ξ est un élément de Γ'' vérifiant la condition $n.\xi \in \Gamma \Longrightarrow n = 0$, il existe une valuation ν' et une seule du corps $K' = K(X)$ prolongeant ν, à valeurs dans Γ'' et telle que $\nu'(X) = \xi$. Dans ce cas le groupe des ordres Γ' de la valuation ν' est égal au sous-groupe $\Gamma + \mathbb{Z}.\xi$ de Γ'' et le corps résiduel κ' de ν' est égal à κ.*

b) *Il existe une valuation ν' et une seule du corps $K' = K(X)$ qui prolonge ν telle que $\nu'(X) = 0$ et telle que l'image t de X dans le corps résiduel κ' soit transcendante sur κ. Dans ce cas le groupe des ordres Γ' de la valuation ν' est égal à Γ et le corps résiduel κ' de ν' est égal à $\kappa(t)$.*

Remarque. Il peut exister d'autres prolongements ν' de la valuation ν à l'extension monogène $K' = K(X)$, tels que le groupe quotient Γ'/Γ est un groupe de torsion non trivial et tels que le corps résiduel κ' est une extension algébrique non triviale de κ.

Remarque. Dans deux articles [McL 1] et [McL 2] MacLane donne une description explicite du prolongement d'une valuation ν de K à une extension transcendante $K(x)$ ou à une extension algébrique séparable $K[x]$, dans le cas où ν est une valuation discrète de rang un. Il construit une suite finie ou infinie (ν_k) de valuations qui sont des approximations successives du prolongement cherché. Pour cela il définit par récurrence la valuation ν_k à partir de la valuation ν_{k-1} et d'un polynôme $\phi_k(x)$ de $K[x]$, appelé "polynôme clé".

Preuve de la proposition. Nous vérifions d'abord que pour tout élément ξ d'un groupe ordonné Γ'' contenant Γ, l'application ν' de $K' = K(X)$ dans Γ'' définie pour tout $P = \sum a_j X^j$ appartenant à $K[X]$ par l'égalité $\nu'(P) = \inf_j\big(\nu(a_j)+j.\xi\big)$ est un valuation de K' qui prolonge ν.

Dans le premier cas, si la valuation μ est un prolongement de ν qui vérifie $\mu(X) = \xi$, nous avons toujours l'inégalité $\mu(\sum a_j X^j) \geq \inf_j\big(\nu(a_j) + j.\xi\big)$. Par hypothèse sur ξ, tous les éléments $\nu(a_j)+j.\xi$ de Γ'' sont distincts, par conséquent nous avons en fait l'égalité $\mu(\sum a_j X^j) = \inf_j\big(\nu(a_j) + j.\xi\big)$ (cf. Proposition 2.1). La valuation μ est donc la valuation ν' définie précédemment et a pour groupe des ordres $\Gamma + \mathbb{Z}.\xi$. Tout élément x de $K' = K(X)$ peut s'écrire sous la forme $x = X^n b(1+u)$, avec $n \in \mathbb{Z}$, $b \in K^*$ et $u \in K'$, $\nu'(u) > 0$; nous en déduisons que si $\nu'(x)$ est nul, alors nous avons $n.\xi + \nu(b) = 0$, c'est à dire $n = 0$ et $\nu(b) = 0$, par conséquent le corps résiduel κ' est égal à κ.

Dans le deuxième cas, quitte à multiplier l'élément x de K' par un élément de K^*, nous pouvons toujours supposer que x s'écrit sous la forme $x = \sum a_j X^j$ avec $\nu(a_j) \geq 0$ pour tout j, i.e. $a_j \in V$, et que l'un des $\nu(a_j)$ est nul. Alors, comme précédemment, si la valuation μ est un prolongement de ν qui vérifie $\mu(X) = 0$, nous avons l'inégalité $\mu(x) \geq \inf_j\big(\nu(a_j)\big) = 0$.

L'image $\bar{x}$ de x dans le corps résiduel κ_μ est égale à $\sum \bar{a}_j t^j$, où nous notons $\bar{a}_j$ l'image de a_j dans κ, et $\bar{x}$ est nul si et seulement si nous avons l'inégalité stricte $\mu(x) > 0$. Par hypothèse sur la valuation μ, l'élément t est transcendant sur κ et comme l'un des $\bar{a}_j$ est non nul, $\bar{x}$ ne peut pas être nul et nous avons $\mu(x) = 0$, c'est à dire que la valuation μ est la valuation ν' définie par $\nu'\big(\sum a_j X^j\big) = \inf_j\big(\nu(a_j)\big)$. Réciproquement la valuation ν' de K' définie par $\nu'\big(\sum a_j X^j\big) = \inf_j\big(\nu(a_j)\big)$ a pour groupe des ordres Γ, et le même raisonnement que précédemment permet de vérifier que l'image t de X est transcendante sur κ. Il reste alors à montrer que le corps résiduel κ' est égal à $\kappa(t)$, ce qui se déduit directement du fait que K' est égal à $K(X)$. □

Nous revenons au cas général, K' est une extension de K de degré de transcendance $\text{dim.alg.}_K K'$ et ν' est un prolongement à K' d'une valuation ν de K.

Théorème 5.5. *Soient $x_1, x_2, \dots, x_s$ des éléments de l'anneau de valuation V' dont les images canoniques $\bar{x}_1, \bar{x}_2, \dots, \bar{x}_s$ dans κ' sont algébriquement indépendantes sur κ, et soient $y_1, y_2, \dots, y_r$ des éléments de K' tels que les images canoniques de $\nu'(y_1), \nu'(y_2), \dots, \nu'(y_r)$ dans le groupe quotient Γ'/Γ sont linéairement indépendantes sur $\mathbb{Z}$. Alors les $r + s$ éléments $x_1, \dots, x_s, y_1, \dots, y_r$ de K' sont* *algébriquement indépendants sur K et la restriction ν'' de ν' au sous-corps $K'' = K(x_1, \ldots, x_s, y_1, \ldots, y_r)$ de K' admet $\Gamma'' = \Gamma + \mathbb{Z}.\nu'(y_1) + \cdots + \mathbb{Z}.\nu'(y_r)$ pour groupe des ordres et $\kappa'' = \kappa(\bar{x}_1, \ldots, \bar{x}_s)$ pour corps résiduel.*

Pour un polynôme $f = \sum a_{(\underline{\beta},\underline{\gamma})} x^{\underline{\beta}} y^{\underline{\gamma}}$ de $K[x_1, \ldots, x_s, y_1, \ldots, y_r]$, la valuation de f est égale au minimum des valuations des monômes de f:

$$\nu''(f) = \inf_{(\underline{\beta},\underline{\gamma})} \nu''\left(a_{(\underline{\beta},\underline{\gamma})} x^{\underline{\beta}} y^{\underline{\gamma}}\right) = \inf_{(\underline{\beta},\underline{\gamma})} \left(\nu\left(a_{(\underline{\beta},\underline{\gamma})}\right) + \sum_{1 \leq j \leq r} \gamma_j \nu'(y_j)\right).$$

Preuve. Nous faisons une démonstration par récurrence sur $r + s$. Le théorème est évident pour $r + s = 0$. Nous supposons que le résultat est démontré pour r' et s' avec $r' \leq r$, $s' \leq s$ et $r' + s' \leq r + s - 1$; alors quitte à remplacer K par $K(x_1, \ldots, x_{s'}, y_1, \ldots, y_{r'})$ et Γ par $\Gamma + \mathbb{Z}.\nu'(y_1) + \cdots + \mathbb{Z}.\nu'(y_{r'})$, nous pouvons nous ramener aux deux cas suivants:

1) $r = 1$ et $s = 0$, c'est à dire il existe x dans V' dont l'image $\bar{x}$ dans κ' est transcendante sur κ;

2) $r = 0$ et $s = 1$, c'est à dire il existe y dans K' dont l'image $\xi = \nu'(y)$ dans Γ' vérifie $n.\xi \in \Gamma \Longrightarrow n = 0$.

De la proposition 5.2 sur le prolongement de ν à une extension algébrique de K, nous déduisons dans le cas 1) que x est transcendant sur K et dans le cas 2) que y est transcendant sur K. Le théorème se déduit alors de la proposition précédente. □

Corollaire.

a) *Nous avons l'inégalité:*

$$\textit{rang rat.}\left(\Gamma'/\Gamma\right) \; + \; \textit{dim.alg.}_{\kappa}\kappa' \; \leq \; \textit{dim.alg.}_K K'.$$

De plus, si nous avons égalité et si K' est une extension de type fini de K le groupe Γ'/Γ est un $\mathbb{Z}$-module de type fini et κ' est une extension de type fini de κ.

b) *Nous avons l'inégalité:*

$$\textit{rang}(\nu') \; + \; \textit{dim.alg.}_{\kappa}\kappa' \; \leq \; \textit{rang}(\nu) \; + \; \textit{dim.alg.}_K K'.$$

De plus, si nous avons égalité, si K' est une extension de type fini de K et si Γ est le groupe $\mathbb{Z}^h$ muni de l'ordre lexicographique, Γ' est isomorphe au groupe $\mathbb{Z}^{h'}$ muni de l'ordre lexicographique et κ' est une extension de type fini de κ.

Preuve. a) L'inégalité est une conséquence directe du théorème. Si K' est une extension de type fini de K alors $d = \text{dim.alg.}_K K'$ est fini, par conséquent $r = \text{rang}\left(\Gamma'/\Gamma\right)$ et $s = \text{dim.alg.}_{\kappa}\kappa'$ sont aussi finis. Nous supposons que nous avons l'égalité $r + s = d$.

Soit ν'' le prolongement de ν au sous-corps $K'' = K(x_1, \ldots, x_s, y_1, \ldots, y_r)$ de K' obtenue par restriction de ν'. Son groupe des ordres Γ'' est isomorphe au sous-groupe $\Gamma + \mathbb{Z}.\nu'(y_1) + \cdots + \mathbb{Z}.\nu'(y_r)$ de Γ' et son corps résiduel κ'' est isomorphe à $\kappa(\bar{x}_1, \ldots, \bar{x}_s)$. Par hypothèse K' est une extension algébrique finie de K'' et nous déduisons de la proposition 5.1 que Γ'/Γ'' est un groupe fini et que κ' est une extension algébrique finie de κ'', par conséquent Γ'/Γ est un $\mathbb{Z}$-module de type fini et κ' une extension de type fini de κ.

b) L'inégalité est une conséquence de l'inégalité précédente et de l'inégalité suivante démontrée à la proposition 3.5: $\text{rang}(\Gamma') \leq \text{rang}(\Gamma) + \text{ rang rat.}(\Gamma'/\Gamma)$. Si nous avons égalité en b) alors nous avons égalité dans a), par conséquent Γ'/Γ est un $\mathbb{Z}$-module de type fini et κ' une extension de type fini de κ. De plus le rang rationnel du groupe Γ'/Γ est égal à son rang, par conséquent si Γ est isomorphe à $\mathbb{Z}^h$ muni de l'ordre lexicographique alors Γ' est aussi isomorphe à $\mathbb{Z}^{h'}$ muni de l'ordre lexicographique avec $h' = h + \text{rang}(\Gamma'/\Gamma)$. □

Remarque. Par récurrence sur le degré de transcendance de l'extension K'/K, $d = \text{dim.alg.}_K K'$, nous déduisons de la remarque suivant la proposition 5.4 que pour tout triplet (h, r, s) vérifiant $1 \leq h \leq r \leq d - s \leq d$ ou $0 = h = r \leq d - s \leq d$, et pour toute valuation ν de K, il existe un prolongement ν' de ν à K' tel que nous ayons rang $(\Gamma'/\Gamma) = h$, rang rat.$(\Gamma'/\Gamma) = r$ et $\text{dim.alg.}_K K' = s$. Mais dans le cas où le groupe Γ est soit réduit à $\{0\}$, c'est à dire si la valuation ν de K est la valuation impropre, soit de la forme $(\mathbb{R}^h, +)_{lex}$, il est impossible de trouver une valuation ν' qui corresponde aux valeurs $0 = h < r$.

Exemple 6. Soit un corps K muni d'une valuation ν, alors d'après le théorème 5.5, il existe une unique valuation ν' de l'extension transcendante pure $K' = K(x_1, \ldots, x_s)$ de K telle que $\nu'(x_i) = 0$ pour tout x_i et telle que les images canoniques $\bar{x}_1, \bar{x}_2, \ldots, \bar{x}_s$ dans le corps résiduel $\kappa_{\nu'}$ sont algébriquement indépendantes sur κ_ν, c'est à dire que ν' est la valuation définie par:

$$\nu'\left(\sum a_{\underline{\beta}} x^{\underline{\beta}}\right) = \inf_{\underline{\beta}} \left(\nu(a_{\underline{\beta}})\right).$$

La valuation ν' est appelée valuation de Gauss.

Toute valuation μ d'une extension L de K prolongeant une valuation ν telle que $\text{dim.alg.}_{\kappa_\nu} \kappa_\mu = \text{dim.alg.}_K L = s$, est le prolongement d'une valuation de Gauss ν' d'une extension transcendante pure $K' = K(x_1, \ldots, x_s)$ de K telle que L est algébrique sur K'.

Définition. Soit ν une valuation d'un corps K et soit μ un prolongement de ν à une extension L de K tel que les groupes des ordres Γ_ν et Γ_μ et les corps résiduels κ_ν et κ_μ soient égaux, c'est à dire que nous supposons que les inclusions naturelles de Γ_ν dans Γ_μ et de κ_ν dans κ_μ sont des isomorphismes, c'est à dire que nous avons $e(\mu/\nu) = 0$ et $f(\mu/\nu) = 0$. Nous disons alors que (L, μ) est un extension immédiate de (K, ν). Par exemple si $\hat{K}$ est le complété de K pour la valuation ν et si $\hat{\nu}$ est l'unique valuation de $\hat{K}$ qui prolonge ν, $(\hat{K}, \hat{\nu})$ est une extension immédiate de (K, ν). Un corps K muni d'une valuation ν est dit maximal s'il n'existe aucune extension propre immédiate de (K, ν). Les propriétes de ces corps ont été en particulier étudiées par Kaplansky [Ka 1], [Ka 2].

6. Centre d'une valuation

Soient K un corps et ν une valuation de K, nous appelons R_ν l'anneau de valuation de ν et $\mathcal{M}_\nu$ son idéal maximal.

Définition. Soit A un sous-anneau de K sur lequel la valuation ν est positive, c'est à dire un anneau A inclus dans R_ν, alors le centre de la valuation ν dans A est l'idéal $\mathcal{P}$ de A défini par $\mathcal{P} = A \cap \mathcal{M}_\nu$.

Remarque. Le centre $\mathcal{P}$ de la valuation est un idéal premier de A. Si A est un anneau local , le centre de la valuation ν dans A est l'idéal maximal $\mathcal{M}$ de A si et seulement si l'anneau de valuation R_ν domine A. En particulier, le centre $\mathcal{P}$ de la valuation ν dans l'anneau A est l'unique idéal premier $\mathcal{Q}$ de A tel que l'anneau R_ν de la valuation domine le localisé $A_\mathcal{Q}$.

Nous avons vu dans la démonstration du théorème 1.1, que tout anneau de valuation est intégralement clos. Nous allons montrer que la cloture intégrale du localisé d'un anneau A en un de ses idéaux premiers $\mathcal{P}$ peut être définie grâce aux valuations de A ayant $\mathcal{P}$ pour centre.

Soient K un corps et A un sous-anneau de K, en particulier le corps K est une extension du corps des fractions L de A. Nous considérons $\mathcal{P}$ un idéal premier de A et nous appelons $N(\mathcal{P})$ l'ensemble des valuations ν de K, positives sur A et dont le centre dans A est l'idéal $\mathcal{P}$.

Théorème 6.1. *La fermeture intégrale $\big(A_\mathcal{P}\big)'$ du localisé $A_\mathcal{P}$ dans le corps K est égale à l'intersection des anneaux de valuations R_ν associés aux valuation de $N(\mathcal{P})$:*

$$\big(A_\mathcal{P}\big)' = \bigcap_{\nu \in N(\mathcal{P})} R_\nu \,.$$

Preuve. Quitte à remplacer l'anneau A par son localisé $A_\mathcal{P}$, nous pouvons supposer que l'anneau A est local, d'idéal maximal $\mathcal{P}$. Si la valuation ν de K appartient à l'ensemble $N(\mathcal{P})$, alors l'anneau A est inclus dans l'anneau de valuation R_ν. Comme cet anneau est intégralement clos la fermeture intégrale de A de K est aussi incluse dans R_ν, d'où $(A)' \subset \bigcap_{\nu \in N(\mathcal{P})} R_\nu$.

Réciproquement soit y un élément de K qui n'est pas entier sur A, alors y^{-1} est un élément non inversible de l'anneau $A\big[y^{-1}\big]$ et il existe un idéal maximal $\mathcal{M}$ de cet anneau contenant y^{-1}. Nous appelons B le localisé de $A\big[y^{-1}\big]$ en $\mathcal{M}$, et d'après le corollaire de la proposition 1.2 il existe un anneau de valuation R_ν de K qui domine B. Comme l'élément y^{-1} appartient à $\mathcal{M}$, il appartient aussi à l'idéal maximal $\mathcal{M}_\nu$ de R_ν. Et comme R_ν domine B, nous avons $A \subset B \subset R_\nu$ et il reste à montrer que l'idéal $\mathcal{M}_\nu \cap A = \mathcal{M} \cap A$ est maximal. Pour cela nous vérifions que si un élément x de A admet un inverse x^{-1} dans B alors x^{-1} appartient à A, c'est encore une conséquence du fait que y n'est pas entier sur A. □

Remarque. Si nous supposons que le corps K est une extension de degré fini du corps des fractions L de A et si nous supposons que l'anneau A est noethérien, ou plus généralement que A est un anneau de Krull, nous pouvons remplacer l'ensemble $N(\mathcal{P})$ par un sous-ensemble N' vérifiant:

- pour tout $\nu \in N'$, ν est une valuation discrète de rang un;
- pour tout $x \in A_{\mathcal{P}}$, il n'existe qu'un nombre fini de valuations $\nu_1, \nu_2, \dots, \nu_r$ appartenant à N' telles que $\nu_i(x) > 0$.

Nous nous intéressons maintenant à une variété algébrique X et nous voulons définir le centre dans X d'une valuation ν d'un corps K, extension du corps des fonctions $F(X)$ de X.

Considérons d'abord le cas d'un schéma affine $X = \operatorname{Spec} A$, où A est un sous-anneau du corps K, et une valuation ν de K. Si A est inclus dans l'anneau R_ν de la valuation ν, alors le centre de ν dans A est l'idéal premier $\mathcal{P}$ de A défini par $\mathcal{P} = A \cap \mathcal{M}_\nu$. Cet idéal définit un fermé Z intègre, i.e. irréductible et réduit, du schéma X. Si A n'est pas inclus dans l'anneau R_ν de la valuation nous prenons pour fermé Z l'ensemble vide.

Nous voulons généraliser la notion de centre d'une valuation ν pour tout schéma X, nous supposerons que X est irréductible et réduit de corps des fonctions $F(X)$ et nous considérons une valuation ν d'un corps K, extension de $F(X)$.

Proposition 6.2. *Le sous-ensemble Z de X formé des points x dont l'anneau local $\mathcal{O}_{X,x}$ est inclus dans l'anneau R_ν, $Z = \{x \in X \,/\, \mathcal{O}_{X,x} \subset R_\nu\}$, est un fermé irréductible du schéma X, qui peut être vide.*

Définition. Le centre de la valuation ν dans X est le point générique du sous-schéma intègre Z de X obtenu en mettant la structure de schéma réduit sur le fermé Z défini précédemment, quand Z est non vide. Plus précisément, le centre de la valuation ν dans X est l'unique point x de X, quand il existe, dont l'anneau local $\mathcal{O}_{X,x}$ est dominé par l'anneau R_ν de la valuation. Nous appellerons parfois centre de la valuation le fermé Z. Ainsi par centre nous pouvons considérer soit un sous-schéma fermé intègre Z, soit le point générique x de ce schéma, soit dans le cas affine l'idéal premier $\mathcal{P}$ qui définit ce fermé.

Preuve de la proposition. Il suffit de vérifier que si X est un schéma affine $X = \operatorname{Spec} A$, le fermé Z de X défini par le centre $\mathcal{P}$ de la valuation ν dans l'anneau A, a bien pour support l'ensemble des points x de X tels que $\mathcal{O}_{X,x} \subset R_\nu$. Il s'agit de montrer que pour tout idéal premier $\mathcal{Q}$ de l'anneau A, nous avons l'équivalence suivante: $\mathcal{P} \subset \mathcal{Q} \iff A_{\mathcal{Q}} \subset R_\nu$.

Supposons d'abord que $\mathcal{P}$ est inclus $\mathcal{Q}$. Tout élément x de $A_{\mathcal{Q}}$ peut s'écrire sous la forme $x = u/v$, avec $u \in A$ et $v \in A \smallsetminus \mathcal{Q}$, nous avons alors $\nu(u) \geq 0$ et $\nu(v) = 0$, d'où $\nu(x) \geq 0$. Réciproquement, si v appartient à $A \smallsetminus \mathcal{Q}$, l'élément $x = 1/v$ appartient à $A_{\mathcal{Q}}$; nous avons donc l'inégalité $\nu(x) \geq 0$, d'où $\nu(v) \leq 0$ et v n'appartient pas à $\mathcal{P}$. □

Si X est un schéma intègre quelconque, il peut arriver que la valuation ν n'ait pas de centre dans X. Par exemple si X est un schéma affine $X = \operatorname{Spec} A$, la valuation ν a un centre dans X si et seulement si l'anneau A est inclus dans l'anneau R_ν. Mais dans certains cas nous pouvons affirmer l'existence d'un centre dans X pour toute valuation ν.

Avant de donner une condition sur la variété X pour que toute valuation ν d'une extension K du corps des fonctions $F(X)$ ait un centre dans X, rappelons le rôle que jouent les anneaux de valuation dans l'étude locale des morphismes de schémas. Plus précisément nous pouvons définir des critères valuatifs de séparation et de propreté pour un morphisme de schémas $f\colon X \longrightarrow Y$, en considérant les diagrammes commutatifs de la forme:

$$\begin{array}{ccc} U = \operatorname{Spec} K & \xrightarrow{g} & X \\ \downarrow{\scriptstyle i} & & \downarrow{\scriptstyle f} \\ T = \operatorname{Spec} V & \xrightarrow{h} & Y \end{array}$$

où V est un anneau de valuation d'un corps K et où i est le morphisme naturel de $U = \operatorname{Spec} K$ dans $T = \operatorname{Spec} V$.

Nous avons ainsi les deux critères valuatifs suivants [EGA], Proposition 7.2.3 et Théorème 7.3.8; [Ha], chapitre II, Theorem 4.3 et Theorem 4.7:

Critère valuatif de séparation. *Soit $f\colon X \longrightarrow Y$ un morphisme de schémas, avec X noethérien. Le morphisme f est séparé si et seulement si pour tout anneau de valuation V, pour tout couple de morphismes (g,h) de $\big(U \xrightarrow{i} T\big)$ dans $\big(X \xrightarrow{f} Y\big)$, il existe au plus un morphisme $\bar{h}$ de T dans X tel que $f \circ \bar{h} = h$.*

Critère valuatif de propreté. *Soit $f\colon X \longrightarrow Y$ un morphisme de type fini de schémas, avec X noethérien. Le morphisme f est propre si et seulement si pour tout anneau de valuation V, pour tout couple de morphismes (g,h) de $\big(U \xrightarrow{i} T\big)$ dans $\big(X \xrightarrow{f} Y\big)$, il existe un et un seul morphisme $\bar{h}$ de T dans X tel que $f \circ \bar{h} = h$.*

Le critère valuatif de séparation signifie que pour tout diagramme commutatif de la forme:

$$\begin{array}{ccc} U = \operatorname{Spec} K & \xrightarrow{g} & X \\ \downarrow{\scriptstyle i} & & \downarrow{\scriptstyle f} \\ T = \operatorname{Spec} V & \xrightarrow{h} & Y \end{array}$$

il existe au plus un morphisme $\bar{h}$ de T dans X tel que le nouveau diagramme soit encore commutatif, et le critère valuatif de propreté assure en plus l'existence d'un tel morphisme $\bar{h}$.

Dans le cas d'un anneau de valuation discrète V, nous pouvons aussi considérer le schéma T comme un germe de courbe régulière et U comme l'ouvert constitué par le point générique η de T. Alors le critère valuatif de séparation signifie que pour tout morphisme h de T dans Y, il y a unicité du relèvement $\bar{h}$ de h de T dans X une fois que l'image du point générique η de T a été fixée, et le critère valuatif de propreté signifie qu'il y a existence et unicité du relèvement.

Soit k un corps et soit X une variété définie sur k, c'est à dire un schéma intègre de type fini sur k, nous avons un morphisme naturel $f\colon X \to \operatorname{Spec} k$ et par définition, la variété X est complète si et seulement si le morphisme f est propre.

Nous considérons un corps K, extension du corps des fonctions $F(X)$ de la variété X, et une valuation ν de K telle que l'anneau R_ν associé est une k-algèbre. En particulier, nous en déduisons que la restriction de ν à k est la valuation impropre, c'est à dire la valuation ν vérifie $\nu(x) = 0$ pour tout élément x de k^*.

Proposition 6.3. *Si la variété X est complète, toute valuation ν vérifiant la condition précédente a un centre dans X.*

Preuve. Par hypothèse, nous avons le diagramme commutatif suivant:

$$\begin{array}{ccc} \operatorname{Spec} K & \xrightarrow{\bar{g}} & X \\ \downarrow{\scriptstyle i} & & \downarrow{\scriptstyle f} \\ \operatorname{Spec} R_\nu & \xrightarrow{g} & \operatorname{Spec} k \end{array}$$

D'après le critère valuatif de propreté, il existe alors un unique morphisme h de $\operatorname{Spec} R_\nu$ dans X qui rend le diagramme commutatif, c'est à dire tel que $h \circ i = \bar{g}$ et $f \circ h = g$. L'image du point fermé $\mathcal{M}_\nu$ de $\operatorname{Spec} R_\nu$ par le morphisme h est le centre de la valuation ν dans X. □

Proposition 6.4. *Soit X un k-schéma intègre excellent complet de corps des fonctions $F(X) = K$. Pour toute valuation ν de K, impropre sur k, la dimension du centre de ν sur X est inférieure ou égale à dim.alg.$_k\kappa_\nu$. De plus il existe un Y éclaté de X tel que le centre de ν sur Y est de dimension égale à dim.alg.$_k\kappa_\nu$.*

Preuve. Soit x le centre de la valuation ν sur X et soit $Z = \overline{\{x\}}$ le fermé engendré. Nous notons A l'anneau local $\mathcal{O}_{X,x}$ et $\mathcal{M}$ son idéal maximal et nous déduisons de l'inclusion de $F(Z) = A/\mathcal{M}$ dans $\kappa_\nu = R_\nu/\mathcal{M}_\nu$ l'inégalité $\dim Z \leq$ dim.alg.$_k\kappa_\nu$.

Si l'inégalité est stricte, nous prenons une base de transcendance $\bar{x}_1, \ldots, \bar{x}_n$ de l'extension $F(Z) \subset \kappa_\nu$, avec $\bar{x}_i$ dans κ_ν image de $x_i = \dfrac{p_i}{q} \in R_\nu$, $1 \leq i \leq n$, et $q, p_1, \ldots, p_n$ appartenant à A. Nous considérons alors un idéal $\mathcal{I}$ de $\mathcal{O}_X$ localement engendré par $q, p_1, \ldots, p_n$ et Y l'éclaté de X le long de $\mathcal{I}$. Le centre Z' de ν sur Y vérifie $F(Z)(\bar{x}_1, \ldots, \bar{x}_n) \subset F(Z')$, par conséquent Z' est de dimension dim.alg.$_k\kappa_\nu$. □

7. Variété de Riemann

Dans ce paragraphe nous suivons la présentation donnée par Zariski et Samuel dans leur livre [Z-S], Chapitre 7, Paragraphe 17. Nous nous proposons d'étudier l'ensemble S de toutes les valuations d'un corps K donné, mais comme cet ensemble est trop gros nous allons faire la restriction suivante. Nous nous donnons un sous-anneau k du corps K, le plus souvent k sera un sous-corps de K, et nous nous restreignons aux valuations ν de K dont la restriction à k est impropre, c'est à dire aux valuations ν de K qui vérifient $\nu(x) = 0$ pour tout x dans k^*.

Définition. La Variété de Riemann, ou la Variété de Riemann-Zariski, de K relativement à k, ou Variété de Riemann de K/k, est l'ensemble $S = S(K/k)$ de toutes les valuations ν de K dont la restriction à k est impropre.

Remarque. Si k' est la fermeture intégrale de k dans K, alors les Variétés de Riemann $S(K/k)$ et $S(K/k')$ peuvent être identifiées; il existe une bijection canonique entre ces deux ensembles.

Remarque. Si k est un corps et si K est une extension algébrique de k, alors la seule valuation ν de K qui prolonge la valuation impropre de k est la valuation impropre de K d'après le corollaire de la proposition 5.2. Dans ce cas la Variété de Riemann $S(K/k)$ est réduite à un seul élément. Nous donnerons parfois une définition un peu différente de la Variété de Riemann et nous la noterons $S^*(K/k)$: c'est l'ensemble des valuations propres ν de K dont la restriction à k est impropre. Dans le cas d'une extension algébrique K de k, nous trouvons alors que la Variété de Riemann $S^*(K/k)$ est l'ensemble vide.

Dans la suite nous étudions le cas où le corps K est une extension de type fini, ou de degré de transcendance fini, d'un sous-corps k. Les valuations ν appartenant à $S(K/k)$ sont les valuations de K qui prolongent la valuation impropre de k. Nous pouvons munir les espaces $S = S(K/k)$ et $S^*(K/k)$ d'une topologie.

Définition. Pour tout sous-anneau A du corps K contenant le sous-anneau k, nous appelons $E(A)$ l'ensemble des valuations ν de S qui sont positives sur A, c'est à dire l'ensemble défini par $E(A) = \{\nu \in S(K/k) \,/\, A \subset R_\nu\}$, où nous notons R_ν l'anneau de la valuation ν. Nous définissons la topologie de la Variété de Riemann S en prenant comme base d'ouverts la famille E constituée des ensembles $E(A)$, où A parcourt l'ensemble des sous k-algèbres de type fini de K, c'est à dire où A est de la forme $k[x_1, \ldots, x_n]$, avec les x_j appartenant à K. Nous appelons cette topologie, la topologie de Zariski sur $S(K/k)$.

Il faut vérifier que nous définissons bien ainsi une base d'ouverts, pour cela il suffit de vérifier que l'intersection de deux éléments de E est encore un élément de E. Si A et A' sont deux sous-anneaux de K, nous notons $[A, A']$ le sous-anneau de K engendré par A et A'. Nous avons alors l'égalité $E(A) \cap E(A') = E([A, A'])$, et si A et A' sont de type fini sur k, $[A, A']$ est aussi de type fini sur k. Nous avons aussi la relation: $A \subset A' \Longrightarrow E(A') \subset E(A)$.

Théorème 7.1. *Pour toute valuation ν de la Variété de Riemann S, l'adhérence de $\{\nu\}$ est formée des valuations ν' qui sont composées avec ν:*

$$\overline{\{\nu\}} = \{\nu' \in S \;/\; \nu' \text{ est composée avec } \nu\}.$$

Plus précisément l'adhérence $\overline{\{\nu\}}$ est isomorphe à la Variété de Riemann $S(\kappa_\nu/k)$ du corps résiduel κ_ν de la valuation ν.

Preuve. Par définition si ν' et ν sont deux valuations d'un même corps K, la valuation ν' est composée avec ν si et seulement si son anneau de valuation $R_{\nu'}$ est inclus dans l'anneau R_ν de ν. Si ν' appartient à l'adhérence de $\{\nu\}$, pour toute sous k-algèbre A de K de type fini telle que ν' appartient à l'ouvert $E(A)$, ν appartient aussi à $E(A)$. Par conséquent pour toute sous k-algèbre A de K de type fini, nous avons $A \subset R_{\nu'} \Longrightarrow A \subset R_\nu$, d'où $R_{\nu'}$ est inclus dans R_ν.

Réciproquement, si ν' n'appartient pas à l'adhérence de $\{\nu\}$, il existe une sous k-algèbre A de K incluse dans l'anneau $R_{\nu'}$ de la valuation ν' et qui n'est pas incluse dans R_ν. Il existe alors un élément x de A appartenant à $R_{\nu'}$ et n'appartenant pas à R_ν.

D'après la proposition 4.2, l'application Φ de $\overline{\{\nu\}}$ dans $S(\kappa_\nu/k)$ qui à toute valuation μ composée avec ν associe la valuation $\bar{\mu}$ de κ_ν telle que $\mu = \nu \circ \bar{\mu}$, est une bijection. Il est facile de vérifier que cette bijection est un homéomorphisme de $\overline{\{\nu\}}$ dans $S(\kappa_\nu/k)$. En effet, par construction de l'anneau de valuation R_μ à partir de l'anneau $R_{\bar{\mu}}$, nous avons pour tout élément x de R_ν $\mu(x) \geq 0$ si et seulement si $\bar{\mu}(\bar{x}) \geq 0$, où $\bar{x}$ est l'image de x dans κ_ν. □

Remarque. Si nous appelons ν_0 la valuation impropre de K, toute valuation ν de K est composée avec ν_0, en effet l'anneau associé à la valuation impropre ν_0 est le corps K tout entier. Nous vérifions ainsi que ν_0 appartient à tout ouvert de la Variété de Riemann. Mais même si nous considérons la Variété de Riemann S^* obtenue en enlevant la valuation impropre ν_0, nous voyons grâce au théorème précédent que S^* n'est presque jamais un espace séparé. En effet nous déduisons du théorème qu'une condition nécessaire pour que tout point ν de la Variété de Riemann $S^*(K/k)$ soit fermé est que K soit une extension de degré de transcendance un de k, ou une extension algébrique du corps des fractions de k si k n'est pas un corps.

Théorème 7.2. *La Variété de Riemann $S(K/k)$ est un espace quasi-compact.*

D'après la remarque précédente, la valuation impropre ν_0 appartient à tout ouvert de la Variété de Riemann S. Par conséquent la Variété de Riemann S est quasi-compacte si et seulement si S^*, obtenue à partir de S en enlevant ν_0, l'est aussi.

Preuve du théorème. Nous donnons le démonstration de Chevalley telle qu'elle est exposée dans [Z-S]. Toute valuation ν de K est complètement déterminée par l'anneau de valuation R_ν associé, c'est à dire si nous savons pour quels éléments x de K, $\nu(x)$ est strictement positif, strictement négatif ou nul. Par conséquent nous pouvons considérer $S = S(K/k)$ comme un sous-ensemble de l'ensemble Z^K des applications de K dans $Z = \{+, 0, -\}$.

Nous définissons une topologie sur l'ensemble Z telle que les ouverts sont $\emptyset$, $\{0, +\}$ et Z, et nous en déduisons naturellement une topologie sur Z^K, considéré comme espace produit. La topologie induite sur le sous-ensemble S de Z^K est alors définie de la manière suivante: pour tout sous-ensemble fini $\{x_1, \dots, x_n\}$ de K, l'ensemble des valuations ν de S telles que pour tout x_j de cet ensemble, $\nu(x_j)$ appartienne à $\{0, +\}$, est un ouvert E de S, et les sous-ensembles E ainsi obtenus forment une base d'ouverts de la topologie de S. Nous retrouvons bien ainsi la topologie sur S définie précédemment, l'ouvert E correspond en effet à l'ouvert $E(A)$ pour la k-algèbre $A = k[x_1, \dots, x_n]$.

Nous considérons maintenant sur Z une topologie plus forte que celle définie précédemment, telle que Z soit un espace séparé et telle que S soit un sous-espace fermé de Z^K. Alors si Z muni de cette topologie est un espace compact, les espaces Z^K muni de la topologie produit et S comme fermé de Z^K le sont aussi. Nous en déduisons que S muni de la topologie de Zariski qui est moins forte, est un espace quasi-compact. La topologie que nous allons considérer sur Z est la topologie discrète, c'est à dire telle que tout sous-ensemble de Z est un ouvert. Alors Z est un espace séparé et comme Z est un ensemble fini, c'est un espace compact. Il reste à montrer que la Variété de Riemann est un fermé de Z^K pour cette topologie. Une application $f\colon K \to Z$ appartient au sous-ensemble S de Z^K si et seulement si elle vérifie les trois conditions suivantes:

a) l'ensemble $V = \{x \in K \,/\, f(x) \in \{0,+\}\}$ est stable par addition et par multiplication;

b) l'anneau k est inclus dans V;

c) si l'élément x de K^* n'appartient pas à V, son inverse x^{-1} appartient à V.

Pour tout élément x de K, nous appelons pr_x l'application continue de Z^K dans Z définie par $f \mapsto f(x)$. Alors pour tout couple (x,y) de K^2 nous définissons le fermé $F_{x,y}$ de Z^K par

$$F_{x,y} = A_{x,y} \cap M_{x,y}\,,$$

avec

$$A_{x,y} = \big(pr_x\big)^{-1}(\{-\}) \cup \big(pr_y\big)^{-1}(\{-\}) \cup \big(pr_{x+y}\big)^{-1}(\{0,+\})$$

et

$$M_{x,y} = \big(pr_x\big)^{-1}(\{-\}) \cup \big(pr_y\big)^{-1}(\{-\}) \cup \big(pr_{x.y}\big)^{-1}(\{0,+\})\,.$$

Alors l'ensemble V est stable par addition (resp. par multiplication) si et seulement si l'application f vérifie:

$$\text{pour tout } (x,y) \in K^2,\ f(x) \geq 0 \text{ et } f(y) \geq 0 \Longrightarrow f(x+y) \geq 0 \ (\text{resp.} f(x.y) \geq 0)\,,$$

c'est à dire si et seulement si f appartient à $A_{x,y}$ (resp. à $M_{x,y}$).

Par conséquent la condition a) est équivalente à la condition fermée suivante:

a′) pour tout couple (x,y) de K^2, f appartient à $F_{x,y}$.

De même les conditions b) et c) sont équivalentes aux conditions fermées suivantes:

b′) pour tout x de k, f appartient à $\big(pr_x\big)^{-1}(\{0,+\})$;

c′) pour tout x de K^*, f appartient à $\big(pr_x\big)^{-1}(\{0,+\}) \cup \big(pr_{x^{-1}}\big)^{-1}(\{0,+\})$.

Nous avons ainsi montré que S est un sous-espace fermé de Z^K, donc que S est un espace compact. □

Nous allons montrer que la Variété de Riemann $S = S(K/k)$ peut être considérée comme limite projective de schémas: $S = \varprojlim X_\alpha$. Nous considérons un corps k et une extension K de k. Nous allons nous intéresser aux modèles M de K sur k, c'est à dire aux schémas sur k intègres dont le corps des fonctions est K, et nous considérerons surtout les modèles complets.

Définition. Nous appelons L l'ensemble des k-algèbres locales P contenant k et contenues dans K, et pour toute algèbre P de L nous notons $m(P)$ son idéal maximal: $L = \{P\, k\text{-algèbre locale} \,/\, k \subset P \subset K\}$. Pour toute k-algèbre A contenant k et contenue dans K, non nécessairement locale, nous appelons $L(A)$ le sous-ensemble de L formé des k-algèbres P contenant A: $L(A) = \{P \in L \,/\, A \subset P\}$. Alors nous définissons une topologie sur L telle que l'ensemble des $L(A)$, pour A k-algèbre de type fini, est une base d'ouverts.

Si A est une k-algèbre de type fini, nous pouvons considérer le schéma affine $\operatorname{Spec} A$, et pour A avec $k \subset A \subset K$, nous pouvons définir une application naturelle $f_A\colon L(A) \to \operatorname{Spec} A$ par $f_A(P) = m(P) \cap A = \mathcal{P}$. Nous rappelons que $\operatorname{Spec} A$ est muni d'une topologie, la topologie de Zariski, telle que les fermés sont les sous-ensembles $V(\mathcal{I})$ de $\operatorname{Spec} A$ définis par $V(\mathcal{I}) = \{\mathcal{P} \in \operatorname{Spec} A \,/\, \mathcal{I} \subset \mathcal{P}\}$, pour tout idéal $\mathcal{I}$ de A. De plus $V(\mathcal{I})$ est isomorphe au schéma affine $\operatorname{Spec} A/\mathcal{I}$ associé à l'anneau quotient $A/\mathcal{I}$.

Proposition 7.3.

a) *L'application f_A est continue de $L(A)$ dans* $\operatorname{Spec} A$.

b) *De plus f_A est un homéomorphisme de $V(A)$ dans* $\operatorname{Spec} A$, *où $V(A)$ est le sous-ensemble de $L(A)$ défini par $V(A) = \{A_{\mathcal{P}} \,/\, \mathcal{P} \in \operatorname{Spec} A\}$.*

Preuve. a) Il suffit de montrer que pour tout ouvert O de $\operatorname{Spec} A$, l'image inverse ${f_A}^{-1}(O)$ est un ouvert de $L(A)$, c'est à dire un ouvert de L. Nous pouvons nous restreindre aux ouverts $D(x) = \{\mathcal{P} \in \operatorname{Spec} A \,/\, x \notin \mathcal{P}\}$, et nous avons $D(x) \simeq \operatorname{Spec} A_x$, où A_x est l'anneau $S^{-1}A$ pour la partie multiplicative S engendrée par x. Nous pouvons alors vérifier que l'image inverse ${f_A}^{-1}\big(D(x)\big)$ est l'ouvert $L\big(A_x\big)$. En effet l'anneau local P appartient à ${f_A}^{-1}\big(D(x)\big)$ si et seulement si l'idéal premier $\mathcal{P} = m(P) \cap A$ de A ne contient pas x, c'est à dire x n'appartient pas à l'idéal maximal $m(P)$. Comme x appartient à A qui est inclus dans P, c'est équivalent à demander que l'inverse x^{-1} appartienne à l'anneau local P, donc à ce que A_x soit inclus dans P.

b) Par définition, l'application f_A est une bijection du sous-ensemble $V(A)$ dans le schéma affine $\operatorname{Spec} A$, et il faut montrer que f_A permet d'identifier la topologie sur $V(A)$ induite par celle sur L, et la topologie de Zariski sur $\operatorname{Spec} A$. Par définition de la topologie sur L, tout ouvert de $V(A)$ est intersection d'un nombre fini d'ouverts $O(x)$ de la forme $O(x) = \{P \in V(A) \,/\, x \in P\}$, pour x un élément non nul du corps des fractions de l'anneau A. Il suffit alors de montrer que le sous-ensemble $f_A\big(O(x)\big) = \{\mathcal{P} \in \operatorname{Spec} A \,/\, x \in A_{\mathcal{P}}\}$ est un ouvert de $\operatorname{Spec} A$.

Appelons $\mathcal{I}$ l'idéal de A défini par $\mathcal{I} = (A : x) = \{c \in A \,/\, cx \in A\}$. Un élément x du corps des fractions de A appartient à l'anneau localisé $A_{\mathcal{P}}$ si et seulement si x peut s'écrire sous la forme $x = a/b$ avec a et b dans A et b n'appartenant pas à $\mathcal{P}$, c'est à dire si et seulement s'il existe un élément b de $A \smallsetminus \mathcal{P}$ tel que bx appartienne à A, ce qui est équivalent à demander l'existence d'un élément b de l'idéal $\mathcal{I}$ n'appartenant pas à $\mathcal{P}$. Nous avons ainsi montré que $f\big(O(x)\big)$ est le complémentaire dans $\operatorname{Spec} A$ du fermé $V(\mathcal{I})$. □

Remarque. Si nous associons à chaque valuation ν appartenant à la Variété de Riemann $S = S(K/k)$ son anneau de valuation R_ν, nous pouvons considérer S comme un sous-ensemble de L. Pour toute k-algèbre $A \subset K$, nous avons l'égalité $E(A) = L(A) \cap S$, par conséquent la topologie sur L définie précédemment induit sur S la topologie de Zariski.

Nous considérons une k-algèbre A intègre de type fini, de corps des fractions K. L'ensemble des valuations ν appartenant à la Variété de Riemann $S = S(K/k)$ qui admettent un centre sur le schéma affine $X = \operatorname{Spec} A$ est égal à l'ouvert $E(A)$ de S. L'application g_A de $E(A)$ dans X qui à la valuation ν associe son centre x_ν sur X est la restriction à $E(A)$ de l'application f_A de $L(A)$ dans $\operatorname{Spec} A$ définie précédemment. En effet par définition, le centre x_ν de la valuation ν sur X correspond à l'idéal premier $\mathcal{P}_\nu$ de A défini par $\mathcal{P}_\nu = \max(R_\nu) \cap A$. Plus généralement, nous avons le résultat suivant.

Proposition 7.4. *Soit X est un k-schéma intègre, noethérien, de corps des fonctions K. L'ensemble des valuations ν de $S(K/k)$ qui admettent un centre sur X est un ouvert $U(X)$ de la Variété de Riemann $S(K/k)$ et l'application g_X qui à toute valuation ν de $U(X)$ associe son centre x_ν sur X est une application continue. De plus si le schéma X est complet sur k, l'ouvert $U(X)$ est égal à S tout entier, et nous avons une application continue $g_X : S \to X$.*

Preuve. Comme par définition le schéma X est séparé, le critère valuatif de séparation nous permet d'affirmer que toute valuation ν appartenant à S admet au plus un seul centre x_ν sur X. De même le critère valuatif de propreté nous donne l'existence du centre x_ν de la valuation ν sur X quand le schéma X est complet, d'où l'égalité $U(X) = S$ dans ce cas. Nous pouvons écrire le schéma X comme réunion d'un nombre fini d'ouverts affines $X_i = \operatorname{Spec} A_i$, $1 \leq i \leq n$, où les A_i sont des k-algèbres intègres de type fini, de corps des fractions K.

Le sous-ensemble $U(X)$ de S cherché est alors la réunion des ouverts $E(A_i)$, et la restriction de l'application g_X à l'ouvert A_i est l'application g_{A_i} définie précédemment. Nous en déduisons que $U(X)$ est un ouvert de S. Les applications g_{A_i} sont obtenues comme restrictions des applications f_{A_i} qui sont continues d'après la proposition 7.3, par conséquent elles sont continues elles aussi, ainsi que l'application g_X. □

Rappelons que si la valuation ν admet un centre x_ν sur X, c'est à dire si ν appartient à l'ouvert $U(X)$ de S, il existe un morphisme birationnel s_ν du schéma affine $\operatorname{Spec} R_\nu$ dans X, tel que l'image du point fermé soit le centre x_ν de la valuation sur X. En effet, le centre x_ν est l'unique point x de X dont l'anneau local $\mathcal{O}_{X,x}$ est dominé par l'anneau de valuation R_ν, et le morphisme s_ν est défini par l'inclusion de $\mathcal{O}_{X,x}$ dans R_ν.

Nous considérons deux k-schémas intègres X et X' de même corps des fonctions K et un morphisme birationnel h de X' dans X, si la valuation ν appartient à l'ouvert $U(X')$, alors nous avons un morphisme s'_ν de $\operatorname{Spec} R_\nu$ dans X'

tel que l'image du point fermé est le centre x'_ν de ν sur X'. Le morphisme composé $s_\nu = h \circ s'_\nu$ permet de définir le centre $x_\nu = h(x'_\nu)$ de la valuation ν sur X. En particulier l'ouvert $U(X')$ est inclus dans l'ouvert $U(X)$ et la restriction de l'applications g_X à $U(X')$ vérifie $g_X = h \circ g_{X'}$. De plus, si nous supposons que le morphisme h est propre, le critère valuatif de propreté appliqué au diagramme suivant:

$$\begin{array}{ccc} \operatorname{Spec} K & \xrightarrow{j} & X' \\ \big\downarrow{\scriptstyle i} & & \big\downarrow{\scriptstyle h} \\ \operatorname{Spec} R_\nu & \xrightarrow{s_\nu} & X \end{array}$$

permet d'affirmer l'existence d'un unique morphisme $s'_\nu \colon \operatorname{Spec} R_\nu \to X'$ qui le rend commutatif. Nous en déduisons que toute valuation ν qui admet un centre sur X, admet aussi un centre sur X', c'est à dire que les ouverts $U(X)$ et $U(X')$ sont égaux.

Nous nous fixons un k-schéma intègre noethérien X de corps des fonctions K, et nous notons U l'ouvert $U(X)$ de la Variété de Riemann $S = S(K/k)$ formé des valuations ν admettant un centre sur X. Pour tout schéma Y tel qu'il existe un morphisme $h_Y \colon Y \to X$, propre birationnel, l'ouvert $U(Y)$ est égal à U et les applications continues $g_X \colon U \to X$ et $g_Y \colon U \to Y$ vérifient $g_X = h_Y \circ g_Y$. Si nous avons deux tels schémas Y et Y' propres au dessus de X et un morphisme $h_{Y',Y}$ de Y' dans Y vérifiant $h_{Y'} = h_Y \circ h_{Y',Y}$, nous notons $Y \prec Y'$. Nous appelons $\mathcal{C}$ le système projectif ainsi obtenu, et nous pouvons définir la limite projective

$$\mathcal{X} = \varprojlim_{\mathcal{C}} Y\,.$$

Cette limite projective est le sous-ensemble de l'espace produit $\prod_{\mathcal{C}} Y$ formé des éléments $\bar{x} = (x_Y)$ vérifiant $h_{Y',Y}(x_{Y'}) = x_Y$ pour tout couple (Y', Y) tel que $Y \prec Y'$. Nous le munissons de la topologie induite par la topologie produit sur $\prod_{\mathcal{C}} Y$, et les applications naturelles $t_Y \colon \mathcal{X} \to Y$ sont des applications continues.

Théorème 7.5. *Il existe un homéomorphisme naturel* $t \colon U \to \mathcal{X}$ *de l'ouvert* $U = U(X)$ *de la Variété de Riemann dans la limite projective* $\mathcal{X} = \varprojlim_{\mathcal{C}} Y$. *En particulier, la Variété de Riemann* $S = S(K/k)$ *peut être obtenue comme limite projective d'un système* $\mathcal{C}$ *de* k*-schémas* Z *intègres complets, ayant* K *comme corps des fonctions:* $S = \varprojlim_{\mathcal{C}} Z$.

Remarque. Au lieu de considérer un système de schémas complets Z, nous pouvons grâce au lemme de Chow nous restreindre à un sous-système de schémas projectifs, cf. [EGA], Théorème 5.6.1; [Ha], Chapitre II, Exercice 4.10.

Remarque. Les schémas complets sont des espaces topologiques quasi-compacts, nous pouvions donc en déduire la quasi-compacité de la Variété de Riemann S comme limite projective d'espaces quasi-compacts. Mais dans la démonstration suivante du Théorème 7.5 nous utilisons la quasi-compacité de S.

Remarque. La “topologie de Zariski” a été introduite par Zariski dans son article sur la Variété de Riemann [Za]; il définit cette topologie pour une variété algébrique X en donnant comme base d'ouverts les complémentaires des sous-variétés algébriques de X. Il définit cette topologie sur les variétés algébriques et la topologie sur la variété de Riemann $S(K/k)$ de manière à ce que la bijection t de $S(K/k)$ dans la limite projective $\mathcal{X} = \varprojlim_{\mathcal{C}} Y$ soit un homéomorphisme.

Preuve du théorème. Pour tout couple (Y, Y') du système $\mathcal{C}$ vérifiant $Y \prec Y'$, les applications $g_Y \colon U \to Y$ et $g_{Y'} \colon U \to Y'$ sont continues et vérifient $g_Y = h_{Y',Y} \circ g_{Y'}$. Par conséquent nous obtenons une application continue t de l'ouvert U de S dans la limite projective $\mathcal{X}$. Nous allons d'abord montrer que cette application $t \colon U \to \mathcal{X}$ est surjective. Nous considérons un point $\bar{x} = (x_Y)$ de $\mathcal{X}$, et nous notons R_Y l'anneau local du schéma Y au point $x_Y = t_Y(\bar{x})$, $R_Y = \mathcal{O}_{Y,x_Y}$. En particulier si nous avons $Y \prec Y'$, l'anneau $R_{Y'}$ domine l'anneau R_Y. Alors l'anneau $R = \cup_{Y \in \mathcal{C}} R_Y$ est un sous-anneau local de K, d'idéal maximal $\max(R) = \cup_{Y \in \mathcal{C}} \max(R_Y)$. Il existe un anneau de valuation V de K qui domine R et la valuation ν associée appartient à la Variété de Riemann $S = S(K/k)$. L'anneau de valuation V domine alors chacun des anneaux locaux R_Y, nous en déduisons que sur chaque schéma Y de $\mathcal{C}$, la valuation ν a pour centre le point x_Y, c'est à dire que l'image de ν par l'application t est le point $\bar{x}$, et que la valuation ν appartient à l'ouvert U.

Pour montrer l'injectivité de l'application $t \colon U \to \mathcal{X}$, nous allons montrer que l'anneau $\cup_{Y \in \mathcal{C}} R_Y$ défini précédemment à partir d'un point $\bar{x} = (x_Y)$ est un anneau de valuation de K. L'anneau R est un anneau local, de corps des fractions K, et nous allons montrer que pour tout élément non nul w de K, nous avons soit w, soit son inverse w^{-1} dans R. Nous pouvons écrire w sous la forme $w = u/v$, avec u et v appartenant à l'anneau R, c'est à dire qu'il existe Y' et Y'' dans $\mathcal{C}$ tels que $u \in R_{Y'}$ et $v \in R_{Y''}$. Pour tout couple (Y', Y'') de $\mathcal{C}$, il existe Y dans $\mathcal{C}$ tel que $Y' \prec Y$ et $Y'' \prec Y$, par conséquent nous pouvons supposer que les éléments u et v appartiennent au même anneau R_Y. Nous pouvons trouver un faisceau d'idéaux $\mathcal{I}$ sur le schéma Y tel que $\mathcal{I}_{Y,x_Y}$ soit égal à l'idéal (u, v) de l'anneau local $R_Y = \mathcal{O}_{Y,x_Y}$. Nous appelons $r \colon Z \to Y$ le morphisme propre birationnel obtenu en faisant l'éclatement de centre $\mathcal{I}$ dans Y. Alors Z appartient au système projectif $\mathcal{C}$, l'anneau local $\mathcal{O}_{Z,x_Z}$ au point x_Z défini par $x_Z = t_Z(\bar{x})$, est R_Z et l'idéal $\mathcal{I}R_Z$ est engendré par l'un des deux éléments u ou v. Si $\mathcal{I}R_Z$ est engendré par u, alors $w^{-1} = v/u$ appartient à l'anneau R_Z, donc à R, et si $\mathcal{I}R_Z$ est engendré par v, alors w appartient à R. Si ν est une valuation de U dont l'image $t(\nu)$ dans $\mathcal{X}$ est le point $\bar{x}$, son anneau associé R_ν domine l'anneau local R. Par conséquent nous avons montré qu'il existe une unique valuation ν au dessus de $\bar{x}$, c'est la valuation associée à l'anneau R.

Il reste à montrer que l'application $t \colon U \to \mathcal{X}$ est une application fermée. Nous déduisons du b) de la proposition 7.3 que les applications $t_Y \colon U \to Y$ sont fermées, par conséquent l'application $t \colon U \to \mathcal{X}$ l'est aussi. En effet l'image par t d'un fermé F de U est le fermé de $\mathcal{X}$ défini par $t(F) = \mathcal{X} \cap \prod_{Y \in \mathcal{C}} t_Y(F)$. □

Proposition 7.6. *Tout fermé irréductible F de la Variété de Riemann $S(K/k)$ admet un point générique, c'est à dire est de la forme $F = \overline{\{\nu\}}$, où ν est une valuation de S.*

Preuve. Soit F un fermé irréductible de $S = S(K/k)$, alors pour tout schéma complet X dans $\mathcal{C}$, l'image de F par l'application $g_X\colon S \to X$ est un fermé irréductible $F(X) = g_X(F)$ de X. Nous appelons $x_{F(X)}$ le point générique de $F(X)$, et $\bar{x} = (x_{F(X)})$ est un élément de la limite projective $\mathcal{X} = \varprojlim_{\mathcal{C}} X$. En effet pour tout morphisme $h\colon Y \to X$ propre birationnel, l'image de $F(Y)$ par h est $F(X)$, d'où $h(x_{F(Y)}) = x_{F(X)}$. Soit ν la valuation de S définie par $\bar{x}$, c'est à dire la valuation dont le centre sur chaque schéma complet X de $\mathcal{C}$ est le point $x_{F(X)}$. Comme ν appartient à ${g_X}^{-1}(g_X(F))$ pour tout X, ν appartient au fermé F de S, et nous allons montrer que ν est le point générique de F.

Supposons qu'il existe un valuation μ de F qui n'appartienne pas à $\overline{\{\nu\}}$. Soit A une sous k-algèbre de type fini de K, avec Frac $(A) = K$, telle que μ soit dans l'ouvert $E(A)$ de S et telle que ν n'y appartienne pas, et soit X un schéma complet dans $\mathcal{C}$ tel que $O = \operatorname{Spec} A$ soit un ouvert affine de X. Alors le centre $g_X(\mu)$ de μ sur X appartient à $F(X) = g_X(F)$ et à l'ouvert O, par conséquent le point générique $x_{F(X)}$ de $F(X)$ appartient aussi à O, ce qui est impossible car $x_{F(X)}$ est le centre de ν sur X et ν n'appartient pas à $E(A)$. □

Nous rappelons que la dimension d'un espace topologique T est définie comme la longueur maximale d'une suite strictement croissante de fermés irréductibles de T, c'est à dire: dim T est le plus grand entier d tel qu'il existe une suite $\emptyset \subsetneq F_0 \subsetneq F_1 \subsetneq \cdots \subsetneq F_d$, où les F_i sont des fermés irréductibles. De même la codimension $\operatorname{codim}_T x$ d'un point x de T est définie comme le plus grand entier c tel qu'il existe une suite $x \in F_0 \subsetneq F_1 \subsetneq \cdots \subsetneq F_c$, où les F_i sont des fermés irréductibles.

Proposition 7.7. *Pour toute valuation ν de $S = S(K/k)$, il existe une suite finie unique de valuations $\{\nu_i\}_{0 \leq i \leq h}$ de longueur h maximale vérifiant $\nu = \nu_h$ et $\nu_i \in \overline{\{\nu_{i-1}\}}$ pour tout i, $1 \leq i \leq h$. De plus la longueur h de cette suite est égale au rang de la valuation ν.*

Preuve. D'après le théorème 7.1, une valuation ν_i appartient à une telle suite si ν est composée avec elle, c'est à dire si l'anneau de valuation R_ν est inclus dans R_{ν_i}. Nous déduisons alors de la proposition 3.3 que la suite cherchée $\{\nu_i\}$ correspond à la suite des anneaux de valuations contenant R_ν, ou ce qui revient au même à la suite des idéaux premiers de R_ν. □

En particulier la valuation ν_0 est la valuation impropre correspondant à l'anneau K ou à l'idéal premier (0).

Remarque. Comme les seuls fermés irréductibles de la Variété de Riemann $S = S(K/k)$ sont de la forme $F = \overline{\{\nu\}}$ nous déduisons de cette proposition que la codimension de ν dans S est égale au rang de ν.

Remarque. Nous en déduisons aussi que le point générique ν de tout fermé irréductible F de la Variété de Riemann S est unique. De plus deux fermés irréductibles sont soit disjoints, soit l'un inclus dans l'autre.

Proposition 7.8. *La dimension topologique de la Variété de Riemann $S(K/k)$ est égale au degré de transcendance de l'extension $k \subset K$.*

Preuve. Nous déduisons du corollaire du théorème 5.5 appliqué au cas où la valuation ν est considérée comme un prolongement à K de la valuation triviale de k, que nous avons pour toute valuation ν de K l'inégalité:

$$\text{rang}\nu + \text{dim.alg.}_k\kappa_\nu \leq \text{dim.alg.}_k K\,.$$

Nous avons donc $\text{codim}_S\nu \leq \text{dim.alg.}_k K$ pour toute valuation ν et il est facile de trouver une valuation de K dont le rang soit égal au degré de transcendance n de K sur k. Nous pouvons par exemple construire une "valuation divisorielle composée" $\nu\colon K \to \left(\mathbb{Z}^n, +\right)_{lex}$ (cf. l'exemple 9 du paragraphe 10). □

Corollaire. *Une valuation ν de la Variété de Riemann $S(K/k)$ est un point fermé si et seulement si le corps résiduel κ_ν est une extension algébrique de k.*

Si K est le corps des fonctions d'un k-schéma intègre excellent complet X, les points fermés de $S(K/k)$ sont les valuations dont les centres sur tous les schémas Y du système projectif $\mathcal{C}$ sont des points fermés.

Pour toute valuation ν non fermée et pour tout entier d, $1 \leq d \leq dim.alg._k\kappa_\nu$, il existe au moins un point fermé μ adhérent à ν tel que $\dim(R_\mu) = d + \dim(R_\nu)$.

Preuve. La valuation ν est un point fermé de S si et seulement si $\overline{\{\nu\}}$ est réduit à un point, c'est à dire d'après le théorème 7.1 et la proposition 7.8 si et seulement si le degré de transcendance de l'extension $k \subset K$ est nul. Soit ν une valuation dont le centre x_Y sur tout schéma Y de $\mathcal{C}$ est un point fermé. Alors l'ensemble $\{\nu\}$ est égal à $\bigcap t_Y{}^{-1}(x_Y)$ où t_Y est l'application naturelle de χ dans Y. Par conséquent $\{\nu\}$ est intersection de fermés, donc fermé. Réciproquement d'après la proposition 6.4 la dimension du centre x_Y de la valuation ν sur un schéma Y de $\mathcal{C}$ est toujours inférieure ou égale au degré de transcendance de $k \subset K$, par conséquent le centre x_Y est fermé pour tout Y. Comme l'adhérence $\overline{\{\nu\}}$ est isomorphe à la Variété de Riemann $S(\kappa_\nu/k)$, la dernière assertion du corollaire est une conséquence de la remarque suivant le corollaire du théorème 5.5. □

Remarque. Il existe un analogue de la Variété de Riemann en géométrie analytique: "la voûte étoilée" définie par Hironaka [Hi]. Au lieu de considérer la variété $S = \varprojlim_{\mathcal{C}} X'$, où $\mathcal{C}$ est le système projectif constitué par tous les morphismes propres birationnels au dessus de la variété algébrique X, la voûte étoilée $p_Y\colon \mathcal{E}_Y \to Y$ d'un espace analytique complexe Y est définie comme l'espace de certaines limites inverses de composés finis d'éclatements locaux $h_{Y'}\colon Y' \to Y$.

8. Uniformisation et résolution des singularités

Soit X un schéma excellent, un point non-singulier x de X est un point tel que l'anneau local $\mathcal{O}_{X,x}$ est un anneau régulier. Si nous supposons que le schéma X est réduit, l'ensemble des points non-singuliers x de X est un ouvert dense X_{reg} de X. Nous disons que le schéma X est non-singulier si tous les points x de X sont non-singuliers, c'est à dire si l'ouvert X_{reg} est égal à X tout entier. En particulier toutes les composantes connexes du schéma X sont irréductibles.

Par définition une résolution des singularités du schéma réduit X est un morphisme propre birationnel $\pi\colon \tilde{X} \to X$ d'un schéma non-singulier $\tilde{X}$ dans X, nous pouvons demander de plus que le morphisme π induise un isomorphisme au dessus de l'ouvert X_{reg}. Pour résoudre le problème de l'existence d'une résolution des singularités Zariski a démontré un résultat concernant les valuations centrées sur X ou plus précisément concernant la variété de Riemann associée à X, en supposant que X est irréductible, cas auquel il est toujours possible de se ramener.

Dans la suite nous considérons une variété X définie sur un corps k et de corps des fonctions K. Soit ν une valuation appartenant à l'ouvert $U(X)$ de la variété de Riemann $S = S(K/k)$, nous disons que le centre de ν sur X est non-singulier si le point $x = g_X(\nu)$ est un point non-singulier de X, c'est à dire appartenant à X_{reg}.

Théorème d'Uniformisation. *Soit X une variété définie sur un corps k de caractéristique nulle, de corps des fonctions K. Alors pour toute valuation ν appartenant à l'ouvert $U(X)$ de la variété de Riemann $S(K/k)$ il existe un morphisme propre birationnel $\varphi\colon X' \to X$ tel que le centre de la valuation ν sur X' est non-singulier.*

Nous allons montrer comment ce théorème est un pas important dans la démonstration de l'existence d'un résolution des singularités. En particulier, c'est par cette méthode que Zariski a résolu le problème pour les variétés de dimension 2 et 3, cf. [Co].

Pour simplifier nous supposons que la variété X est propre, c'est à dire que l'ouvert $U(X)$ est la variété de Riemann $S = S(K/k)$. Pour toute valuation ν appartenant à S, il existe d'après le théorème d'uniformisation un morphisme propre birationnel $\varphi_\nu\colon X'(\nu) \to X$ tel que le centre x'_ν de la valuation ν est un point non-singulier de la variété $X'(\nu)$, c'est à dire appartient à l'ouvert dense $X'(\nu)_{\text{reg}}$. L'image inverse de $X'(\nu)_{\text{reg}}$ par l'application $g_{X'(\nu)}\colon S(K/k) \to X'(\nu)$ est aussi un ouvert non vide $W(\nu)$ de la variété de Riemann S.

Ainsi pour toute valuation ν de S, nous trouvons un morphisme propre birationnel $\varphi_\nu\colon X'(\nu) \to X$ et un ouvert $W(\nu)$ de S contenant ν et tel que toute valuation μ appartenant à cet ouvert a un centre non-singulier sur la variété $X'(\nu)$. Comme la variété de Riemann S est quasi-compacte, nous en déduisons l'existence d'une famille finie d'ouverts $W_1, W_2, \dots, W_k$ recouvrant S et de morphismes propres birationnels $\varphi_i\colon X'_i \to X$, $1 \le i \le k$, telle que pour tout i le centre sur X'_i de toute valuation appartenant à l'ouvert W_i est non-singulier. En particulier, pour tout point x de la variété X et pour toute valuation ν de S de centre x sur X, il existe un voisinage ouvert U de x dans X, un indice i et un voisinage ouvert non-singulier U'_i de ν dans la variété X'_i, tels que U soit contenu dans l'image de U'_i par le morphisme φ_i.

Le problème de l'existence d'une résolution des singularités se ramène alors au problème de recollement suivant.

Problème de recollement. *Soit X une variété définie sur un corps k et soient $\varphi_i\colon X'_i \to X$, $1 \leq i \leq 2$, deux morphismes propres birationnels. Existe-t-il une variété Y et des morphismes propres birationnels $\psi_1 : Y \to X'_1$ et $\psi_2 : Y \to X'_2$ tels que les morphismes composés $\varphi_1 \circ \psi_1$ et $\varphi_2 \circ \psi_2$ de Y dans X soient égaux et tels que l'ouvert Y_{reg} vérifie $\psi_1^{-1}(X'_{1\text{reg}}) \cup \psi_2^{-1}(X'_{2\text{reg}}) \subset Y_{\text{reg}}$?*

Si nous pouvons trouver une telle variété X', nous trouvons par récurrence sur le nombre k d'ouverts W_i recouvrant la variété de Riemann S une variété $\tilde{X}$ et un morphisme propre birationnel $\psi\colon \tilde{X} \to X$ vérifiant: le centre x sur $\tilde{X}$ de toute valuation ν de la surface de Riemann S est non-singulier. Nous obtenons ainsi une résolution $\psi\colon \tilde{X} \to X$ des singularités de X.

9. Valuation sur un anneau noethérien

Soit A un anneau local noethérien de corps des fractions K, d'idéal maximal $\max(A) = \mathcal{M}$ et de corps résiduel k. Nous considérons une valuation ν de K dont l'anneau de valuation R_ν domine l'anneau A, c'est à dire vérifiant $\nu(x) \geq 0$ pour tout x appartenant à A et $\nu(x) > 0$ pour tout x appartenant à $\mathcal{M}$. Nous disons que la valuation ν domine l'anneau A ou est centrée en A.

Nous notons R_ν l'anneau de la valuation et κ_ν son corps résiduel, comme l'anneau R_ν domine l'anneau A nous avons une inclusion naturelle entre les corps résiduels: $k \subset \kappa_\nu$, et nous notons $\text{deg.tr.}_k\nu = \text{dim.alg.}_k\kappa_\nu$ le degré de transcendance de cette extension. Nous appelons Γ le groupe de la valuation ν. Par hypothèse l'image de A^* par ν est un semi-groupe Φ de Γ, inclus dans Γ^+ car $A \subset R_\nu$, et qui engendre le groupe Γ car K est le corps des fractions de A.

Proposition 9.1. *Si l'anneau A est noethérien, le semi-groupe Φ est un ensemble bien ordonné.*

Preuve. Il faut montrer que tout sous-ensemble non vide E de Φ admet un plus petit élément. Considérons l'image inverse de E dans l'anneau A par la valuation ν: $S = \nu^{-1}(E) \cap A = \{x \in A \,/\, \nu(x) \in E\}$ et l'idéal $\mathcal{I}$ de A engendré par S. L'idéal $\mathcal{I}$ possède un nombre fini de générateurs $x_1, \dots, x_l$ que nous pouvons choisir appartenant à S. Nous appelons α le plus petit élément de l'ensemble $\{\nu(x_i) \,/\, 1 \leq i \leq l\}$, et nous allons vérifier que c'est bien le plus petit élément de l'ensemble E. Comme les générateurs de l'idéal $\mathcal{I}$ ont été choisis dans S nous avons bien $\alpha \in E$ et par construction, pour tout x dans $\mathcal{I}$, nous avons l'inégalité $\nu(x) \geq \alpha$, d'où pour tout β de E, l'inégalité $\beta \geq \alpha$. □

Nous pouvons faire pour l'anneau A une construction analogue à celle que nous avons faite pour l'anneau de valuation R_ν (cf. la remarque suivant le théorème 3.2). Pour tout élément α du semi-groupe Φ, nous pouvons définir les idéaux $\mathcal{P}_\alpha(A)$ et $\mathcal{P}_{\alpha+}(A)$ par:

$$\mathcal{P}_\alpha(A) = \{x \in A \,/\, \nu(x) \geq \alpha\} \quad \text{et} \quad \mathcal{P}_{\alpha+}(A) = \{x \in A \,/\, \nu(x) > \alpha\}.$$

Les idéaux de l'anneau A qui sont des contractions d'idéaux de l'anneau de valuation R_ν, c'est à dire les idéaux $\mathcal{I}$ de la forme $\mathcal{I} = \mathcal{J} \cap A$ avec $\mathcal{J}$ idéal de R_ν, sont appelés les ν-idéaux de A. Les idéaux $\mathcal{P}_\alpha(A)$ et $\mathcal{P}_{\alpha+}(A)$ sont des ν-idéaux et comme l'anneau A est noethérien tout ν-idéal $\mathcal{I}$ de A est l'un des $\mathcal{P}_\alpha(A)$. Nous pouvons aussi considérer les deux algèbres graduées associées à la valuation ν sur l'anneau A:

$$\mathrm{Gr}_\nu A := \bigoplus_{\alpha \in \Gamma} \mathcal{P}_\alpha(A) \quad \text{et} \quad \mathrm{gr}_\nu A := \bigoplus_{\alpha \in \Phi} \mathcal{P}_\alpha(A)/\mathcal{P}_{\alpha+}(A).$$

Et comme précédemment l'étude des idéaux $\mathcal{P}_\alpha(A)$ et des algèbres $\mathrm{Gr}_\nu A$ et $\mathrm{gr}_\nu A$ joue un rôle important dans l'étude de la valuation ν sur l'anneau A.

Définition. Le semi-groupe Φ est archimédien si pour tout α et β appartenant à Φ, avec $\alpha \neq 0$, il existe un entier r tel que $r\alpha > \beta$. C'est équivalent à dire que tout sous-groupe isolé Δ de Γ vérifie $\Phi \cap \Delta = \{0\}$. C'est aussi équivalent à dire que tout ν-idéal $\mathcal{P}_\beta$ de A est $\mathcal{M}$-primaire. En effet si nous posons $\alpha = \nu(\mathcal{M}) := \inf\{\nu(x) \,/\, x \in \mathcal{M}\}$, nous avons $\alpha > 0$. Ainsi pour tout β appartenant à Φ il existe un entier r tel que $\mathcal{M}^r \subset \mathcal{P}_\beta$.

Remarque. Cette propriété ne dépend pas uniquement de la valuation ν, mais aussi de l'anneau A. En particulier c'est plus faible que demander que la valuation ν soit réelle, c'est à dire de rang 1.

Considérons par exemple un anneau local régulier A de dimension deux, de corps des fractions K et la valuation ν de K à valeurs dans $(\mathbb{Z}^2)_{lex}$ définie par $\nu(u) = (1,1)$ et $\nu(v) = (1,0)$, où (u,v) est un système régulier de paramètres de l'anneau A. Alors le semi-groupe $\Phi = \nu(A^*)$ est égal à $\{(n,m) \in \mathbb{Z}^2 \,/\, n \geq m \geq 0\}$ et le seul sous-groupe isolé $\Delta = \{(0,m) \,/\, m \in \mathbb{Z}\}$ de Γ non réduit à $\{0\}$, vérifie $\Phi \cap \Delta = \{0\}$. Nous en déduisons que le semi-groupe Φ est archimédien alors que le groupe qu'il engendre est de rang 2.

Nous pouvons maintenant énoncer le résultat principal sur les valuations sur un anneau noethérien: le théorème d'Abhyankar.

Théorème 9.2. *Soit A un anneau local noethérien de corps des fractions K, d'idéal maximal $\mathcal{M}$ et de corps résiduel k et soit ν une valuation de K qui domine l'anneau A.*

a) *Nous avons l'inégalité:*

$$\mathit{rang\ rat.}\nu + \mathit{deg.tr.}_k\nu \leq \dim A.$$

b) *Si nous avons l'égalité alors le groupe de la valuation* Γ *est isomorphe à* $\mathbb{Z}^d$ *et le corps résiduel* κ_ν *est une extension de type fini du corps* k.

c) *Si de plus nous avons l'égalité rang* $\nu +$ *deg.tr.*$_k\nu = \dim A$ *alors la valuation* ν *est discrète, c'est à dire le groupe de la valuation* Γ *est isomorphe à* $\mathbb{Z}^d$ *muni de l'ordre lexicographique.*

La démonstration originale d'Abhyankar utilisait une récurrence sur le rang de la valuation ν et dans le cas où le rang est 1, un théorème de comparaison entre les valuations centrées sur l'anneau A et sur son complété $\hat{A}$. Nous allons donner ici la démonstration de Spivakovsky [Sp 2].

Preuve du théorème. Nous allons d'abord démontrer l'inégalité:

$$\text{rang rat.}\nu + \text{deg.tr.}_k\nu \leq \dim A\,.$$

Par récurrence sur la dimension de l'anneau A nous allons nous ramener au cas où le degré de transcendance $\text{deg.tr.}_k\nu$ est nul et où le semi-groupe Φ est archimédien.

Réduction 1. Nous pouvons supposer que le degré de transcendance $\text{deg.tr.}_k\nu$ est nul. Supposons $\text{deg.tr.}_k\nu > 0$ et soit x un élément de l'anneau R_ν dont l'image $\bar{x}$ dans le corps résiduel κ_ν est transcendante sur k. Nous considérons l'idéal $\mathcal{M}_1$ de $A[x]$ défini par $\mathcal{M}_1 = \max(R_\nu) \cap A[x]$ et le localisé $B = A[x]_{\mathcal{M}_1}$. Par hypothèse nous avons $\text{dim.alg.}_k k_1 = 1$, où k_1 est le corps résiduel de l'anneau local B, d'où l'égalité

$$\text{deg. tr.}_k\nu = \text{deg.tr.}_{k_1}\nu + 1\,.$$

Pour conclure il suffit de montrer que nous avons $\dim B \leq \dim A - 1$. En effet nous pouvons alors faire une récurrence sur la dimension de l'anneau A et nous trouvons $\dim A \geq \dim B + 1 \geq \text{rang rat.}\nu + \text{deg.tr.}_{k_1}\nu + 1$. Cette inégalité est une conséquence directe du théorème suivant [Ma]:

Théorème de la dimension. *Soient* A *un anneau noethérien intègre,* C *une* A*-algèbre intègre de type fini, soient* $\mathcal{Q}$ *un idéal premier de* C *et* $\mathcal{P}$ *l'idéal premier de* A *défini par* $\mathcal{P} = \mathcal{Q} \cap A$. *Nous avons alors l'inégalité:*

$$ht(\mathcal{Q}) \leq ht\mathcal{P} + dim.alg._A C - dim.alg._{\kappa(\mathcal{P})}\kappa(\mathcal{Q})\,,$$

où $dim.alg._A C$ *et* $dim.alg._{\kappa(\mathcal{P})}\kappa(\mathcal{Q})$ *sont respectivement le degré de transcendance du corps des fractions de* C *sur celui de* A, *et le degré de transcendance du corps des fractions de* $C/\mathcal{Q}$ *sur celui de* $A/\mathcal{P}$.

Il suffit d'appliquer ce théorème à l'idéal $\mathcal{Q} = \mathcal{M}_1$ de l'anneau $C = A[x]$ pour trouver le résultat cherché. Mais dans le cas qui nous intéresse, nous allons donner une démonstration de cette inégalité. Nous allons d'abord montrer $\dim A[x] \leq \dim A$. (Plus généralement, si A et A' sont deux anneaux intègres ayant même corps des fractions K et tels que $A \subset A'$, alors $\dim A' \leq \dim A$.)

Nous avons un morphisme surjectif de l'anneau de polynômes $A[X]$ dans l'anneau $A[x]$ et nous appelons $\mathcal{P}$ son noyau. Comme l'anneau A est noethérien, nous avons $\dim A[X] = \dim A + 1$. Comme x appartient au corps K, il peut s'écrire sous la forme $x = a/b$ avec a et b appartenant à A, et l'élément $bX - a$

appartient à l'idéal premier $\mathcal{P}$ de $A[X]$; par conséquent $\mathcal{P}$ n'est pas réduit à l'idéal $\{0\}$ et nous trouvons l'inégalité:

$$\dim A[x] \leq \dim A[X] - 1 = \dim A .$$

Nous allons maintenant montrer $\dim B < \dim A[x]$. Il suffit de montrer que l'idéal premier $\mathcal{M}_1$ de $A[x]$ défini par $\mathcal{M}_1 = \max(R_\nu) \cap A[x]$ n'est pas un idéal maximal de l'anneau $A[x]$. Si $\mathcal{M}_1$ était un idéal maximal de $A[x]$, l'élément x qui n'appartient pas à cet idéal aurait un inverse y modulo $\mathcal{M}_1$. Plus précisément, il existerait un élément y dans $A[x]$ tel que $1 - xy \in \mathcal{M}_1$, ce qui contredirait l'hypothèse de transcendance de $\bar{x}$ sur le corps k. Nous avons donc:

$$\dim B \leq \dim A[x] - 1 .$$

Réduction 2. Nous pouvons supposer que le semi-groupe Φ est archimédien. Supposons que Φ n'est pas archimédien, il existe un sous-groupe isolé Δ de Γ tel que $\Delta \cap \Phi \neq \{0\}$ et nous appelons $\mathcal{Q}$ l'idéal premier de l'anneau de valuation R_ν associé au sous-groupe Δ:

$$\mathcal{Q} = \{x \in R_\nu \,/\, \nu(x) \notin \Delta\} .$$

Alors la valuation ν de K est la valuation composée avec la valuation ν_0 de K associée à l'anneau de valuation $(R_\nu)_{\mathcal{Q}}$ et la valuation $\tilde{\nu}$ associée à l'anneau de valuation $R_\nu/\mathcal{Q}$: $\nu = \nu_0 \circ \tilde{\nu}$. Si nous appelons $\mathcal{P}$ l'idéal premier de l'anneau A défini par $\mathcal{P} = \mathcal{Q} \cap A$, l'anneau local $(R_\nu)_{\mathcal{Q}}$ domine l'anneau local $A_{\mathcal{P}}$ et l'anneau quotient $A/\mathcal{P}$ est inclus dans l'anneau quotient $R_\nu/\mathcal{Q}$. Nous avons les inclusions strictes:

$$(0) \subsetneq \mathcal{P} \subsetneq \mathcal{M} .$$

En effet comme l'idéal $\mathcal{Q}$ de R_ν n'est pas réduit à (0) nous avons aussi $\mathcal{P} \neq (0)$. Et par hypothèse il existe un élément α non nul du semi-groupe Φ appartenant au sous-groupe isolé Δ, c'est à dire il existe un élément x de l'idéal maximal $\mathcal{M}$ de A n'appartenant pas à $\mathcal{P}$. Nous en déduisons que l'anneau quotient $A/\mathcal{P}$ et l'anneau localisé $A_{\mathcal{P}}$ sont de dimension strictement inférieure à la dimension de l'anneau A. En particulier nous pouvons supposer par récurrence sur la dimension de l'anneau A que le théorème d'Abhyankar est vérifié pour toute valuation centrée en $A_{\mathcal{P}}$ et toute valuation centrée en $A/\mathcal{P}$.

Nous appelons respectivement $\tilde{K}$ et $\bar{K}$ les corps des fractions des anneaux quotients $R_\nu/\mathcal{Q}$ et $A/\mathcal{P}$, $\bar{K}$ est un sous-corps de $\tilde{K}$. Le corps $\tilde{K}$ est aussi le corps résiduel de l'anneau localisé $(R_\nu)_{\mathcal{Q}}$, c'est à dire de l'anneau de valuation R_{ν_0} associé à la valuation ν_0: $\tilde{K} = \kappa_{\nu_0}$. De même le corps $\bar{K}$ est le corps résiduel de l'anneau localisé $A_{\mathcal{P}}$.

Par récurrence sur la dimension de l'anneau, le théorème est vérifié pour la valuation ν_0 du corps K centrée en l'anneau $A_{\mathcal{P}}$ et nous avons l'inégalité: $\text{deg.tr.}_{\bar{K}}\nu_0 + \text{rang rat.}\nu_0 \leq \dim A_{\mathcal{P}}$, où $\text{deg.tr.}_{\bar{K}}\nu_0$ est le degré de transcendance de l'extension $\tilde{K}/\bar{K}$ et où rang rat.ν_0 est le rang rationnel du groupe des ordres de la valuation ν_0, c'est à dire du groupe quotient Γ/Δ d'après la proposition 4.1. Nous obtenons ainsi:

$$\text{rang rat.}\Gamma/\Delta + \text{dim.alg.}_{\bar{K}}\tilde{K} \leq \text{dim}A_{\mathcal{P}}. \tag{1}$$

Nous appelons $\bar{\nu}$ la valuation de $\bar{K}$ obtenue comme restriction de la valuation $\tilde{\nu}$ et $\bar{\Gamma}$ son groupe des ordres, qui est un sous-groupe du groupe des ordres $\tilde{\Gamma}$ de $\tilde{\nu}$, isomorphe au sous-groupe Δ de Γ d'après la proposition 4.1. L'anneau associé à la valuation $\tilde{\nu}$ du corps $\tilde{K}$ est l'anneau quotient $R_\nu/\mathcal{Q}$; cet anneau domine l'anneau quotient $A/\mathcal{P}$, par conséquent l'anneau $R_{\bar{\nu}}$, associé à la valuation $\bar{\nu}$ du corps $\bar{K}$, domine aussi $A/\mathcal{P}$. Par récurrence sur la dimension de l'anneau, le théorème est vérifié pour la valuation $\bar{\nu}$ et nous avons l'inégalité: deg.tr.$_k\bar{\nu}+$ rang rat.$\bar{\nu} \leq$ dim $A/\mathcal{P}$, où deg.tr.$_k\bar{\nu}$ est le degré de transcendance de l'extension $\kappa_{\bar{\nu}}/k$. Nous obtenons:

$$\text{rang rat.}\bar{\Gamma} + \text{dim.alg.}_k\kappa_{\bar{\nu}} \leq \text{dim}A/\mathcal{P}. \tag{2}$$

Nous pouvons maintenant considérer l'inégalité a) du corollaire du théorème 5.5 pour la valuation $\tilde{\nu}$ de $\tilde{K}$ considérée comme extension de la valuation $\bar{\nu}$ de $\bar{K}$, et nous trouvons: rang rat.$\big(\tilde{\Gamma}/\bar{\Gamma}\big)$ + dim.alg.$_{\kappa_{\bar{\nu}}}\kappa_{\tilde{\nu}} \leq$ dim.alg.$_{\bar{K}}\tilde{K}$, où le corps résiduel $\kappa_{\tilde{\nu}}$ de la valuation $\tilde{\nu}$ est égal au corps résiduel de la valuation ν et où le groupe $\tilde{\Gamma}$ est isomorphe au sous-groupe Δ. Nous trouvons alors:

$$\text{rang rat.}\big(\Delta/\bar{\Gamma}\big) + \text{dim.alg.}_{\kappa_{\bar{\nu}}}\kappa_\nu \leq \text{dim.alg.}_{\bar{K}}\tilde{K}. \tag{3}$$

Nous déduisons de la proposition 3.5 et de dim $A/\mathcal{P}+$ dim $A_{\mathcal{P}} \leq$ dim A, l'inégalité cherchée pour la valuation ν centrée en A:

$$\text{rang rat.}\Gamma + \text{dim.alg.}_k\kappa_\nu \leq \text{dim}\,A\,.$$

Grâce aux deux réductions précédentes, il suffit pour montrer la partie a) du théorème de montrer l'inégalité: rang rat.$\nu \leq$ dim A, où ν est une valuation centrée en A, de degré de transcendance nul et telle que le semi-groupe $\Phi = \nu(A^*)$ est archimédien.

Soit n le rang rationnel de la valuation ν, c'est à dire le rang rationnel du groupe Γ. Nous pouvons alors choisir des éléments $x_1, \dots, x_n$ appartenant à l'anneau A tels que les éléments $\nu(x_1), \dots, \nu(x_n)$ de Φ forment une base de $\Gamma \otimes_\mathbb{Z} \mathbb{Q}$, en particulier ces éléments sont linéairement indépendants sur $\mathbb{Z}$ et nous en déduisons que tous les monômes $\underline{x}^{\underline{l}} = x_1^{l_1} \dots x_n^{l_n}$, avec $\underline{l} = (l_1, \dots, l_n) \in \mathbb{N}^n$, prennent des valeurs distinctes par la valuation ν. Comme le semi-groupe Φ est archimédien, il existe un entier $r \in \mathbb{N}$ tel que $\nu(x_i) \leq r\alpha$, pour tout i, $1 \leq i \leq n$, où α est l'élément de Γ défini par $\alpha = \nu\big(\mathcal{M}\big)$. Par conséquent, pour tout $m \in \mathbb{N}$ et tout $\underline{l} \in \mathbb{N}^n$ vérifiant $|\underline{l}| = l_1 + \dots + l_n \leq m$, les monômes $\underline{x}^{\underline{l}}$ n'appartiennent pas à l'idéal $\mathcal{P}_{rm\alpha} = \big\{\, y \,/\, \nu(y) \geq rm\alpha \,\big\}$. Nous déduisons même de l'indépendance linéaire des $\nu(x_i)$ que l' ensemble des monômes $\big\{\, \underline{x}^{\underline{l}} \,/\, |\underline{l}| \leq m \,\big\}$ forment une famille libre de $A/\mathcal{P}_{rm\alpha}$. Et nous trouvons l'inégalité:

$$\text{long.}\big(A/\mathcal{M}^{rm}\big) \geq \text{long.}\big(A/\mathcal{P}_{rm\alpha}\big) \geq \sharp\big\{\, \underline{l} \in \mathbb{N}^n \,/\, |\underline{l}| \leq m \,\big\} = \binom{m+n}{n}.$$

Comme la fonction $H(m) = \text{long.}\big(A/\mathcal{M}^{rm}\big)$ est minorée par $\binom{m+n}{n}$, c'est à dire par un polynôme en m de degré n, nous trouvons l'inégalité: $\dim A \geq n$.

Supposons maintenant que la valuation ν centrée sur l'anneau noethérien A vérifie l'égalité:

$$\text{rang rat.}\nu + \text{deg.tr.}_k\nu = \dim A\,. \qquad (*)$$

Nous allons d'abord montrer que nous pouvons encore supposer que le degré de transcendance deg tr.$_k\nu$ est nul et que le semi-groupe Φ est archimédien. Nous gardons les mêmes notations que précédemment.

Réduction 1'. Si la valuation ν centrée sur l'anneau A vérifie l'égalité $(*)$ la valuation ν centrée sur l'anneau B défini par $B = A[x]_{\mathcal{M}_1}$, vérifie aussi l'égalité:

$$\text{rang rat.}\nu + \text{deg.tr.}_{k_1}\nu = \dim B\,. \qquad (*)$$

Comme le corps k_1 est une extension de type fini du corps k, si le théorème est vérifié pour la valuation ν centrée en B par hypothèse de récurrence sur la dimension de l'anneau, il est aussi vérifié pour la valuation ν centrée en A.

Réduction 2'. Si la valuation ν centrée sur l'anneau A vérifie l'égalité $(*)$ les trois inégalités (1), (2) et (3) apparaissant précédemment sont en fait des égalités. En particulier les valuations ν_0 centrée en $A_{\mathcal{P}}$ et $\bar{\nu}$ centrée en $A/\mathcal{P}$ vérifient aussi l'égalité $(*)$. Nous en déduisons par hypothèse de récurrence sur la dimension de l'anneau, que les groupes Γ/Δ et $\bar{\Gamma}$ sont des $\mathbb{Z}$-modules libres de rang fini et que les extensions $\tilde{K}/\bar{K}$ et $\kappa_{\bar{\nu}}/k$ sont de type fini. Nous avons aussi l'égalité:

$$\text{dim.alg.}_{\kappa_{\bar{\nu}}}\kappa_\nu + \text{rang rat.}\big(\Delta/\bar{\Gamma}\big) = \text{dim.alg.}_{\bar{K}}\tilde{K}\,.$$

Comme l'extension $\tilde{K}/\bar{K}$ est de type fini, nous déduisons du a) du corollaire du théorème 5.5 que l'extension $\kappa_\nu/\kappa_{\bar{\nu}}$ est de type fini et que le groupe $\Delta/\bar{\Gamma}$ est un $\mathbb{Z}$-module de type fini. Il reste à montrer que le groupe Γ est un $\mathbb{Z}$-module libre. C'est une conséquence des propriétés des groupes $\bar{\Gamma}$, $\Delta/\bar{\Gamma}$ et Γ/Δ et du fait que le groupe Γ est ordonné donc sans torsion.

Nous supposons donc que nous avons une valuation ν centrée sur un anneau noethérien A, telle que le degré de transcendance deg.tr.$_k\nu$ est nul et telle que le semi-groupe $\Phi = \nu(A^*)$ est archimédien, vérifiant l'égalité rang rat.$\nu = \dim A$. Nous reprenons les notations de la démonstration de la partie a) du théorème.

Par hypothèse sur la dimension de A, la fonction $H(m) = \text{long.}\big(A/\mathcal{M}^{rm}\big)$ est majorée par un polynôme en m de degré n, la fonction $H'(m) = \text{long.}\big(A/\mathcal{P}_{rm\alpha}\big)$ est aussi majorée par un polynôme en m de degré n. Nous avons l'égalité:

$$H'(m) = \text{long.}\big(A/\mathcal{P}_{rm\alpha}\big) = \sum_{\substack{\beta\in\Phi\\ \beta\leq rm\alpha}} \text{long.}\big(\mathcal{P}_\beta/\mathcal{P}_{\beta+}\big)\,.$$

Pour tout élément β appartenant au semi-groupe Φ, il existe un élément x de l'anneau A tel que $\nu(x) = \beta$, d'où l'inclusion stricte $\mathcal{P}_{\beta+} \subsetneq \mathcal{P}_\beta$, et l'inégalité $\text{long.}\big(\mathcal{P}_\beta/\mathcal{P}_{\beta+}\big) \geq 1$. Par conséquent le nombre $N(m) = \sharp\big\{\beta \in \Phi\,/\,\beta \leq rm\alpha\big\}$

vérifie $N(m) \leq H(m)$ et la fonction $N(m)$ est majorée par un polynôme en m de degré n.

Nous appelons Γ' le sous-groupe de Γ engendré par $\nu(x_1), \dots, \nu(x_n)$. Par hypothèse sur la famille $x_1, \dots, x_n$, c'est un groupe isomorphe à $\mathbb{Z}^n$ et le semi-groupe $\Gamma'^+ = \bigoplus_{i=1}^n \mathbb{N}\,\nu(x_i)$ est inclus dans Φ.

Pour tout entier m, nous avons alors l'inégalité

$$N(m) \geq \binom{m+n}{n}$$

car les $\underline{x}^{\underline{l}}$, pour $|\underline{l}| \leq m$ prennent des valeurs différentes par la valuation ν. Nous pouvons alors écrire:

$$\binom{m+n}{n} \leq N(m) \leq c\binom{m+n}{n}$$

pour tout entier m, où c est une constante dans $\mathbb{N}^*$, par conséquent Γ' est un sous-groupe d'indice fini du groupe Γ, d'où Γ est isomorphe à $\mathbb{Z}^n$.

Montrons maintenant que pour tout β appartenant à Φ, nous avons l'inégalité $\text{long.}\big(\mathcal{P}_\beta/\mathcal{P}_{\beta+}\big) \leq c$. En effet, s'il existe $\gamma \in \Phi$ tel que $\text{long.}\big(\mathcal{P}_\gamma/\mathcal{P}_{\gamma+}\big) > c$, en multipliant $\mathcal{P}_\gamma$ par le monôme $\underline{x}^{\underline{l}}$ nous avons aussi $\text{long.}\big(\mathcal{P}_\beta/\mathcal{P}_{\beta+}\big) > c$, où $\beta = \gamma + \sum l_i \nu(x_i)$. Nous trouvons alors l'inégalité $\text{long.}\big(\mathcal{P}_\beta/\mathcal{P}_{\beta+}\big) > c$ pour tout $\beta \in \alpha + \bigoplus_{i=1}^n \mathbb{N}\nu(x_i)$, d'où l'inégalité:

$$N(m) > c\binom{m+n}{n}$$

quand m devient suffisamment grand, ce qui est impossible. Nous choisissons s éléments $y_1, \dots, y_s$ de l'anneau R_ν tels que leurs images respectives $\bar{y}_1, \dots, \bar{y}_s$ dans le corps résiduel $\kappa_\nu = R_\nu/\max(R_\nu)$ sont linéairement indépendantes sur le corps k. Nous pouvons écrire chaque élément y_i sous la forme $y_i = a_i/b_i$ avec a_i et b_i appartenant à l'anneau A. Nous posons $\gamma = \nu\big(\prod_{j=1}^s b_j\big) = \sum_{j=1}^s \nu(b_j)$, les éléments $z_i = y_i \prod_{j=1}^s b_j$ appartiennent à l'idéal $\mathcal{P}_\gamma$, et leurs images z_i dans l'espace quotient $\mathcal{P}_\gamma/\mathcal{P}_{\gamma+}$ sont linéairement indépendants sur k. Supposons en effet que nous ayons une relation de la forme $\nu\big(\sum_{i=1}^s u_i z_i\big) > \gamma$, avec $u_i \in k$, $1 \leq i \leq s$. Nous en déduisons $\nu\big(\sum_{i=1}^s u_i y_i\big) > 0$, d'où une relation $\sum_{i=1}^s u_i \bar{y}_i = 0$, ce qui est impossible car nous avons choisi les $\bar{y}_i$ linéairement indépendants sur k. Nous avons ainsi montré l'inégalité:

$$s \leq \text{long.}\big(\mathcal{P}_\gamma/\mathcal{P}_{\gamma+}\big) \leq c\,.$$

Par conséquent le corps résiduel κ_ν est une extension finie du corps k, ce qui termine la démonstration de la partie b) du théorème.

Pour démontrer la partie c), rappelons que nous avons toujours l'inégalité $\text{rang}\ \nu \leq \text{rang rat.}\nu$. Par conséquent si la valuation ν centrée sur l'anneau A vérifie l'égalité:

$$\text{rang}\,\nu + \text{deg.tr.}_k\nu = \dim A\,,$$

elle vérifie aussi l'égalité rang rat.ν+ deg.tr.$_k\nu$ = dim A. Nous déduisons de la partie b) du théorème que son groupe des ordres Γ est isomorphe à $\mathbb{Z}^d$, et comme le rang de Γ est égal à son rang rationnel, l'ordre sur $\Gamma \simeq \mathbb{Z}^d$ est l'ordre lexicographique. □

10. Exemples de valuation

Les premiers exemples que nous allons donner sont peut être parmi les plus importants de la géométrie algébrique.

Exemple 7. Nous allons d'abord nous intéresser à une famille très importante de valuations qui vérifient la deuxième égalité du théorème d'Abhyankar.
Plus précisémment, nous considérons un anneau local noethérien intègre A de dimension n, de corps des fractions K.

Définition. Une valuation ν du corps K centrée en A est divisorielle si elle vérifie:

$$\operatorname{rang} \nu = 1 \quad \text{et} \quad \text{deg.tr.}_k \nu = n - 1\,.$$

D'après le théorème 9.2, la valuation ν est alors une valuation discrète de rang un, en particulier son groupe des ordres Γ est le groupe $\mathbb{Z}$ et l'anneau de valuation associé R_ν est un anneau noethérien.

Remarque. Parmi les valuations divisorielles, certaines proviennent de la situation géométrique suivante. Soit $\pi\colon Y \to X = \operatorname{Spec} A$ un morphisme propre birationnel et soit D une composante irréductible de l'image inverse $\pi^{-1}(o)$ du point fermé o de $\operatorname{Spec} A$ telle que l'anneau local $\mathcal{O}_{Y,D}$ soit un anneau de valuation discrète. Alors nous considérons la valuation ν associée à cet anneau, c'est la valuation divisorielle géométrique ν_D définie précédemment sur le corps $K = F(Y)$. Il est facile de vérifier qu'une telle valuation $\nu = \nu_D$ est une valuation divisorielle centrée sur l'anneau A. En effet c'est une valuation discrète de rang un et son corps résiduel κ_ν est isomorphe au corps $F(D)$ des fonctions sur le diviseur D, par conséquent vérifie dim.alg.$_k\kappa_\nu$ =deg.tr.$_k\nu = n - 1$.

Définition. Nous disons qu'une valuation ν de K centrée sur l'anneau A obtenue comme précédemment par une valuation divisorielle géométrique ν_D est une valuation divisorielle géométrique de l'anneau A.

Nous allons rappeler la définition d'un anneau de Nagata.

Définition. Un anneau A est un anneau de Nagata si A est noethérien et si pour tout idéal premier $\mathfrak{p}$ de A l'anneau intègre quotient $A/\mathfrak{p}$ vérifie la propriété suivante: pour toute extension finie L du corps des fractions $\kappa(\mathfrak{p}) = Fr(A/\mathfrak{p})$, la fermeture intégrale de $A/\mathfrak{p}$ dans L est un $A/\mathfrak{p}$-module fini.

Tout anneau B essentiellement de type fini sur un anneau de Nagata A, c'est à dire obtenu comme localisé d'une A-algèbre de type fini, est encore un anneau de Nagata. En particulier les anneaux de la géométrie algébriques, qui sont essentiellement de type fini sur un corps k, sont des anneaux de Nagata.

Proposition 10.1. *Si l'anneau A est un anneau de Nagata toute valuation divisorielle centrée en A est une valuation divisorielle géométrique.*

Spivakovsky démontre que si l'anneau A est un anneau de Nagata ou est régulier de dimension deux, l'anneau R_ν de toute valuation divisorielle ν centrée en A est essentiellement de type fini sur A, c'est équivalent à dire que la valuation ν est une valuation de la forme $\nu = \nu_D$, obtenue comme précédemment en demandant que le morphisme $\pi\colon Y \to X$ est birationnel de type fini, [Sp 1], [Sp 2]. Dans le cas d'un anneau de Nagata, il peut ensuite se ramener au cas d'une valuation divisorielle géométrique [Sp 2].

Preuve de la proposition. Soit A un anneau local de Nagata de dimension n, de corps résiduel k, de corps des fractions K et soit ν une valuation divisorielle de K centrée en A, c'est à dire telle que son anneau de valuation R_ν domine A, est de dimension dim $R_\nu =$ rang $\nu = 1$ et vérifie $\text{dim.alg.}_k\kappa_\nu = \text{deg.tr.}_k\nu = n-1$.

Montrons d'abord que l'anneau R_ν est essentiellement de type fini sur l'anneau A. Dans la démonstration du théorème 9.2, par la **Réduction 1** nous pouvons trouver un anneau local B, de corps résiduel l, essentiellement de type fini sur A, en particulier B est aussi un anneau de Nagata, de corps des fractions K, tel que $\text{deg.tr.}_k\nu = 0$, et d'après la **Réduction 1'**, comme l'anneau A vérifie le cas d'égalité c) du théorème 9.2, l'anneau B vérifie aussi l'égalité c), c'est à dire que nous avons rang $\nu+ \text{deg.tr.}_k\nu = \dim B$.

Nous en déduisons que B est un anneau local de dimension 1, dominé par l'anneau de valuation discrète R_ν et ayant même corps de fractions. Par conséquent R_ν est égal à un localisé $(\bar{B})_{\mathfrak{q}}$ du normalisé $\bar{B}$ de B. Comme B est un anneau de Nagata, $\bar{B}$ est un B-module fini, par conséquent R_ν est essentiellement de type fini sur B, donc sur A.

La suite de la démonstration est semblable à la démonstration de la proposition 6.4. Comme $\text{dim.alg.}_k\kappa_\nu$ est égal à $n-1$, nous pouvons trouver des générateurs $(x_1, x_2, \dots, x_N)$, $n \leq N$, de l'anneau A tel que $\left(x_2^{v_1} / x_1^{v_2}, \dots, x_n^{v_1} / x_1^{v_n}\right)$ soit une base de transcendance de κ_ν sur k, où nous avons posé $v_i = \nu(x_i)$, pour $1 \leq i \leq N$. Soit I l'idéal de A engendré par les $\left(x_i^{n_i}\right)$, $1 \leq i \leq N$, avec $n_i = \sum_{1\leq j\leq N,\, j\neq i} v_j$, et soit $\phi\colon Z \to X = \operatorname{Spec} A$, l'éclatement de centre I dans $X = \operatorname{Spec} A$.

Alors le centre de la valuation ν sur Z est de codimension 1. Comme A est un anneau de Nagata la normalisation $n\colon Y \to Z$ est un morphisme fini et le morphisme composé $\pi = \phi \circ n\colon Y \to X$ est propre birationnel. Le centre de la valuation ν sur Y est une composante irréductible D du diviseur exceptionnel, et comme Y est une variété normale l'anneau $\mathcal{O}_{Y,D}$ est un anneau de valuation discrète, dominé par R_ν, donc qui lui est égal. □

Exemple 8. Plus généralement, soit X un schéma intègre de type fini sur un corps k, de corps des fonctions K, soit D un sous-schéma intègre de X tel que l'anneau local $\mathcal{O}_{X,D}$ est un anneau régulier. Nous pouvons alors définir une valuation ν_D associée à D, $\nu_D\colon K^* \to \mathbb{Z}$, par $\nu_D(f) = \text{ord}_D(f)$ pour toute fonction non nulle f du corps K, où $\text{ord}_D(f)$ est l'ordre du zéro ou du pôle de f au point générique de D. Plus précisémment, si $\underline{m}$ est l'idéal maximal de l'anneau local $\mathcal{O}_{X,D}$, pour toute fonction f appartenant à $\mathcal{O}_{X,D}$ nous posons $\nu_D(f) = \sup\{ n / f \in \underline{m}^n \}$.

Comme toute fonction f de K peut s'écrire sous la forme $f = \dfrac{g}{h}$, avec g et h appartenant à $\mathcal{O}_{X,D}$, nous pouvons définir $\nu_D(f)$ par $\nu_D(f) = \nu_D(g) - \nu_D(h)$, et comme nous avons supposé l'anneau local $\mathcal{O}_{X,D}$ régulier, nous définissons bien ainsi une valuation du corps K. En fait nous pouvons définir une valuation ν_D de K pour tout sous-schéma intègre D de X tel que la fonction ord_D est additive sur l'anneau local $\mathcal{O}_{X,D}$, ce qui est équivalent à demander que l'anneau gradué $\bigoplus_{n=0}^{\infty} \underline{m}^n / \underline{m}^{n+1}$ est un anneau intègre. Une valuation ν_D définie de la manière précédente est encore appelée une valuation divisorielle géométrique. Si le sous-schéma D de X n'est pas un diviseur, il suffit de considérer un morphisme $\pi\colon X' \to X$ qui se factorise par l'éclatement $\pi_D\colon E_D \to X$ de centre D pour trouver un diviseur $D' = \pi^{-1}(D)$ sur X' qui définit la même valuation du corps K, et si nous supposons que le schéma X' est normal l'anneau local $\mathcal{O}_{X,D}$ est un anneau de valuation discrète associé à ν_D.

Exemple 9. Nous considérons encore un schéma intègre X de type fini sur k, de corps des fonctions $K = F(X)$. Grâce à la notion de valuations composées introduite au chapitre 4, nous pouvons définir une valuation ν du corps K à partir de la situation géométrique suivante.

Soient D et E deux sous-schémas fermés fermés intègres de X avec $E \subset D$, tels que les anneaux locaux $\mathcal{O}_{X,D}$ et $\mathcal{O}_{D,E}$ sont réguliers, ou plus généralement tels que les fonctions ord_D et ord_E respectivement sur les anneaux locaux $\mathcal{O}_{X,D}$ et $\mathcal{O}_{D,E}$ sont additives. Nous supposons aussi pour simplifier que le sous-schéma D est un diviseur du schéma X. Grâce à ce qui précède, nous pouvond définir les valuations divisorielles géométriques ν_D et ν_E respectivement sur les corps $K = F(X)$ et $L = F(D)$. Ce sont des valuations discrètes de rang un. Alors comme le corps $L = F(D)$ est isomorphe au corps résiduel de la valuation ν_D, nous pouvons définir la valuation composée $\nu = \nu_D \circ \nu_E$ sur le corps K à valeur dans le groupe $\mathbb{Z}^2$ muni de l'ordre lexicographique.

Nous pouvons définir directement cette valuation $\nu\colon K \to \mathbb{Z}^2$ de la manière suivante (cf. la remarque qui suit la proposition 4.3). Comme l'anneau $\mathcal{O}_{X,D}$ est un anneau de valuation discrète, nous choisissons un générateur u de son idéal maximal $m_{X,D} = \max(\mathcal{O}_{X,D})$. Pour tout élément non nul f de K, nous appelons $\bar{f}$ la classe de l'élément $f.u^{-\nu_D(f)}$ dans le corps $L \simeq \mathcal{O}_{X,D}/m_{X,D}$, et nous avons $\nu(f) = \big(\nu_D(f), \nu_E(\bar{f})\big)$. Plus généralement, nous considérons un drapeau $D_n \subset \cdots \subset D_1 \subset D_0 = X$ de sous-schémas intègres de X tel que pour tout i, $1 \leq i \leq n$, la fonction ord_{D_i} est additive sur l'anneau local $\mathcal{O}_{D_{i-1},D_i}$. Alors il existe une unique valuation $\nu\colon K^* \to \mathbb{Z}^n$, avec $\mathbb{Z}^n$ muni de l'ordre lexicographique, définie à partir des valuations divisorielles ν_{D_i} sur le corps $F(D_{i-1})$.

Remarque. D'après le b) du corollaire du théorème 5.5 les valuations ν de K, impropres sur k, dont le rang est maximal, c'est à dire vérifiant $\mathrm{rang}(\nu) = \mathrm{dim.alg.}_k K = h$, ont pour corps résiduel κ une extension algébrique de k et pour groupe des ordres le groupe $\mathbb{Z}^h$ muni de l'ordre lexicographique. Dans l'exemple précédent, si le drapeau $D_h \subset \cdots \subset D_1 \subset D_0 = X$ est de longueur h égale à la dimension de X, c'est à dire si chaque D_i est un diviseur de D_{i-1}, nous obtenons une valuation de rang maximal, comme composée de h valuations divisorielles géométriques.

Dans la suite, nous allons étudier des valuations centrées sur un anneau local régulier de dimension deux. Nous considérons un corps k et une k-algèbre de type fini A; nous supposons que A est un anneau local régulier de dimension deux, de corps résiduel k et de corps des fractions K isomorphe à l'extension transcendante pure $k(u,v)$. Nous pouvons prendre par exemple pour anneau A, l'anneau localisé $k[u,v]_{(u,v)}$. Nous considérons une valuation ν de K, impropre sur k, centrée sur A et de groupe des ordres Γ.

Exemple 10. Nous supposons que le groupe des ordres Γ de la valuation ν est le groupe $\mathbb{Z}^2$ muni de l'ordre lexicographique, c'est à dire le groupe libre engendré par deux éléments $\alpha = (1,0)$ et $\beta = (0,1)$. Nous définissons la valuation $\nu\colon K^* \to \Gamma \simeq \mathbb{Z}^2$ par $\nu(u) = \alpha$ et $\nu(v) = \beta$. Comme les éléments α et β de Γ sont linéairement indépendants sur $\mathbb{Q}$, la valuation est uniquement déterminée par la donnée de $\nu(u)$ et de $\nu(v)$. En particulier, pour tout polynôme $P(u,v) = \sum c_{r,s} u^r v^s$ considéré comme élément de A, la valeur de la valuation $\nu(P)$ est égale à:

$$\nu(P) = \mathrm{Inf}\left\{ \nu(u^r v^s) \,/\, c_{r,s} \neq 0 \right\} = \mathrm{Inf}\left\{ (r,s) \,/\, c_{r,s} \neq 0 \right\}.$$

L'anneau de valuation R_ν associé à la valuation ν est le sous-anneau de $K = k(u,v)$ engendré par les monômes $u^r v^s$ vérifiant l'inégalité $r \geq 0$ et les monômes v^s vérifiant l'inégalité $s \geq 0$, son idéal maximal $\max(R_\nu)$ est engendré par les monômes $u^r v^s$ vérifiant $r > 0$ et les monômes v^s vérifiant $s > 0$. Le corps résiduel de la valuation κ_ν est isomorphe au corps k.

L'anneau de valuation R_ν est un anneau de valuation discrète de dimension deux, mais ce n'est pas un anneau noethérien. Si nous considérons la valuation ν comme une valuation centrée sur l'anneau A, nous vérifions que la seconde égalité du théorème d'Abhyankar est vérifiée. En effet nous avons:

$$\text{rang rat.}\, \nu = \text{rang}\, \nu = 2 \quad \text{et} \quad \deg \text{tr.}_k \nu = 0\,.$$

Exemple 11. Nous supposons que le groupe des ordres de la valuation ν est le sous-groupe libre Γ de $\mathbb{R}$ engendré par deux réels α et β linéairement indépendants sur $\mathbb{Q}$, par exemple $\alpha = 1$ et $\beta = \sqrt{2}$. Nous définissons la valuation $\nu\colon K^* \to \Gamma \simeq \mathbb{Z}^2$ par $\nu(u) = \alpha$ et $\nu(v) = \beta$. Comme dans l'exemple précédent les éléments α et β de Γ sont linéairement indépendants sur $\mathbb{Q}$ et la valuation est uniquement déterminée par la donnée de $\nu(u)$ et de $\nu(v)$. En particulier, pour tout polynôme $P(u,v) = \sum c_{r,s} u^r v^s$ considéré comme élément de A, la valeur de la valuation $\nu(P)$ est égale à:

$$\nu(P) = \mathrm{Inf}\left\{ \nu(u^r v^s) \,/\, c_{r,s} \neq 0 \right\} = \mathrm{Inf}\left\{ r\alpha + s\beta \,/\, c_{r,s} \neq 0 \right\}.$$

L'anneau de valuation R_ν associé à la valuation ν est le sous-anneau de $K = k(u,v)$ engendré par les monômes $u^r v^s$ vérifiant l'inégalité $r\alpha + s\beta \geq 0$ et son idéal maximal $\max(R_\nu)$ est engendré par les monômes $u^r v^s$ vérifiant l'inégalité stricte $r\alpha + s\beta > 0$. Comme α et β sont linéairement indépendants sur $\mathbb{Q}$, nous avons l'égalité $r\alpha + s\beta = 0$ uniquement pour le couple $(u, v) = (0, 0)$. Nous en déduisons que le corps résiduel de la valuation κ_ν est isomorphe au corps k.

L'anneau de valuation R_ν est un anneau de dimension un, mais ce n'est pas un anneau de valuation discrète, en particulier il n'est pas noethérien. Si nous considérons la valuation ν comme une valuation centrée sur l'anneau A, nous vérifions que la première égalité du théorème d'Abhyankar est vérifiée, mais pas la seconde. En effet nous avons:

$$\text{rang rat.}\,\nu = 2\,, \ \text{rang}\,\nu = 1 \quad \text{et} \quad \text{deg.tr.}_k\nu = 0\,.$$

Exemple 12. Nous supposons que le groupe des ordres Γ de la valuation ν est le groupe $\mathbb{Q}$ et nous définissons la valuation de la manière suivante. Nous posons:

$$\nu(v) = 1 \quad \text{et} \quad \nu(u) = 1 + \frac{1}{2} = \frac{3}{2}\,.$$

Le "premier" élément w_2 de l'anneau A, écrit sous la forme d'un polynôme $w_2 = P(u, v)$ de $k[u, v]$, pour lequel la valeur de la valuation $\nu(w_2)$ n'est pas déterminée par les valeurs de ν en $w_0 = v$ et en $w_1 = u$ est l'élément $w_2 = u^2 + v^3$. Nous avons toujours $\nu(w_2) \geq 3$ et nous posons:

$$\nu(w_2) = 3 + \frac{1}{3} = \frac{10}{3}\,.$$

Nous allons définir une suite (w_n) d'éléments de l'anneau A telle que pour tout entier $n \geq 2$, la valeur de la valuation $\nu(w_n)$ n'est pas déterminée par les valeurs de la valuation $\nu(w_r)$ pour $r < n$. Pour tout entier $n \geq 1$, nous choisissons pour la valeur de la valuation $\nu(w_n)$ un nombre rationnel γ_n qui s'écrit sous la forme $\gamma_n = \dfrac{p_n}{n+1}$, où p_n est un entier positif premier avec $n+1$. La suite (w_n) est construite par récurrence de la manière suivante. Nous supposons que nous avons trouvé les éléments w_r pour $1 \leq r \leq n$, avec $n \geq 1$, et que nous avons les valeurs de la valuation $\nu(w_r) = \gamma_r$ pour ces éléments. Nous posons alors:

$$w_{n+1} = {w_n}^{n+1} + v^{p_n}\,,$$

avec comme valeur de la valuation:

$$\nu(w_{n+1}) = \gamma_{n+1} = p_n + \frac{1}{n+2}\,,$$

c'est à dire avec $p_{n+1} = p_n(n+2) + 1$. Nous pouvons alors déterminer la valuation ν par la suite $(w_n)_{n\in\mathbb{N}}$ d'éléments appartenant à l'idéal maximal de l'anneau A et par les valeurs de la valuation en ces éléments $\gamma_n = \nu(w_n)$.

Par construction l'anneau de valuation R_ν associé est un anneau de valuation de dimension un, non noethérien, de corps résiduel κ_ν isomorphe au corps k. Il est possible de montrer que tout élément non nul x appartenant à l'anneau de valuation R_ν peut s'écrire sous la forme $x = \lambda \prod_{i=0}^{N} {w_i}^{n_i} + x'$, où N est un entier positif, où les n_i appartiennent à $\mathbb{N}$, où λ est un élément non nul du corps k et où x' est un élément de R_ν vérifiant $\nu(x') > \nu(x)$. En particulier, nous avons $\nu(x) = \nu(\prod_{i=0}^{N} w_i{}^{n_i}) = \sum_{i=0}^{N} n_i.\gamma_i$. La valuation ν considérée comme centrée sur l'anneau A ne vérifie aucune des égalités du théorème d'Abhyankar. En effet nous avons:

$$\text{rang rat.}\,\nu = \text{rang}\,\nu = 1 \quad \text{et} \quad \text{deg.tr.}_k\nu = 0\,.$$

Exemple 13. Nous considérons l'anneau $\hat{A}$, complété de l'anneau A, et son corps des fractions $\hat{K}$. En particulier nous pouvons écrire l'anneau $\hat{A}$ sous la forme d'un anneau de séries formelles $\hat{A} = k[[u,v]]$, et nous considérons un élément t du complété $\hat{A}$ n'appartenant pas au corps K et nous supposons t de la forme $t = u + \sum_{i=1}^{\infty} c_i v^i$, avec pour tout $i \geq 1$, $c_i \in k^*$.

Nous pouvons définir une valuation $\hat{\nu}$ du corps $\hat{K}$ à valeurs dans le groupe $\hat{\Gamma} = \mathbb{Z}^2$ muni de l'ordre lexicographique, en posant:

$$\hat{\nu}(v) = (0,1) \quad \text{et} \quad \hat{\nu}(t) = (1,0)\,.$$

Pour toute série formelle $\hat{x} = \sum_{i=0}^{\infty} d_i v^i$ nous avons l'inégalité $\hat{\nu}(\hat{x}) \geq 0$. Par conséquent, comme nous avons supposé que tous les coefficients c_i sont non nuls, pour tout entier $N \geq 1$ nous avons:

$$\hat{\nu}\big(\sum_{i=N}^{\infty} c_i v^i\big) = \hat{\nu}(v^N) = (0,N)\,.$$

Par hypothèse, pour tout entier $N \geq 1$, la valeur de la valuation:

$$\hat{\nu}(t) = \hat{\nu}\big((u + \sum_{i=1}^{N-1} c_i v^i) + (\sum_{i=N}^{\infty} c_i v^i)\big) = (1,0)$$

est strictement supérieure aux valeurs de la valuation $\hat{\nu}$ pour chacun des deux éléments $u + \sum_{i=1}^{N-1} c_i v^i$ et $\sum_{i=N}^{\infty} c_i v^i$. Nous en déduisons que pour tout $N \geq 1$ nous avons:

$$\hat{\nu}\big(u + \sum_{i=1}^{N-1} c_i v^i\big) = (0,N)\,.$$

Nous considérons maintenant la valuation ν du corps K obtenue comme restriction de la valuation $\hat{\nu}$ de $\hat{K}$. Alors le groupe des ordres Γ de la valuation ν est le sous-groupe de $\hat{\Gamma}$ engendré par $\hat{\nu}(v) = \hat{\nu}(u) = (0,1)$, c'est à dire le sous-groupe isolé non trivial de $\hat{\Gamma}$. En particulier le groupe des ordres Γ est isomorphe à $\mathbb{Z}$, le rang et le rang rationnel de la valuation ν sont égaux à un.

L'anneau de valuation R_ν associé est un anneau de valuation discrète de rang un, c'est un anneau noethérien. De plus nous avons l'égalité $R_\nu = R_{\hat{\nu}} \cap K$, où $R_{\hat{\nu}}$ est l'anneau de valuation associé à la valuation $\hat{\nu}$ de $\hat{K}$, c'est à dire le sous-anneau de $\hat{K}$ engendré par les monômes $t^a v^b$ avec $(a,b) \geq (0,0)$ dans $\mathbb{Z}^2$ muni de l'ordre lexicographique. Comme la valuation $\hat{\nu}$ du corps $\hat{K}$ est une valuation discrète de rang deux centrée sur l'anneau $\hat{A}$, nous avons vu que son corps résiduel $\kappa_{\hat{\nu}}$ est isomorphe au corps k. Nous en déduisons que le corps résiduel κ_ν de la valuation ν est aussi isomorphe à k. Nous avons ainsi obtenu une valuation discrète ν de rang un, centrée sur l'anneau A, ne vérifiant aucune des des deux égalités d'Abhyankar. En effet nous avons:

$$\operatorname{rang\,rat.} \nu = \operatorname{rang} \nu = 1 \quad \text{et} \quad \operatorname{deg.tr.}_k \nu = 0 .$$

Géométriquement cette valuation correspond à l'arc formel défini sur le germe de surface régulière $S = \operatorname{Spec} A$, par $(t = 0)$. Nous pouvons aussi définir la valuation ν de la manière suivante. Nous avons un morphisme injectif de $A = k[u,v]_{(u,v)}$ dans $k[[v]]$ défini par $u \mapsto -\sum_{i=1}^{\infty} c_i v^i$ et $v \mapsto v$, et la valuation ν est la restriction à A de la valuation v-adique sur l'anneau $k[[v]]$.

Exemple 14. Nous voulons donner un exemple d'une valuation ν sur un corps de fonctions K ayant pour groupe des ordres le groupe Γ de l'exemple 5 du paragraphe 3.

Soit $K = k(u,v,w)$ et ν la valuation de K à valeurs dans $(\mathbb{R}^2,+)_{lex}$ définie par $\nu(u) = \beta = (1/3, 1/3$, $\nu(v) = \alpha_0 = (1,0)$ et pour tout $k \geq 1$ nous définissons par récurrence $w_k \in K$ et $\gamma_k = \nu(w_k) \in \mathbb{R}^2$ de la manière suivante:

$$\begin{aligned} w_1 &= w \text{ et } \nu(w_1) = \gamma_1 = (1/2, 0) = (p_1/2, 0), \\ w_2 &= v + w_1^2 \text{ et } \nu(w_2) = \gamma_2 = (1 + 1/4, 0) = (p_2/4, 0) \text{ avec } p_2 = 5, \\ w_k &= v^{p_{k-1}} + w_{k-1}^{2^{k-1}} \text{ et } \nu(w_k) = \gamma_k = (p_{k-1} + 1/2^k, 0) = (p_k/2^k, 0) \\ &\quad \text{avec } p_k = 2^k p_{k-1} + 1. \end{aligned}$$

En particulier nous vérifions que le groupe des ordres de la valuation ν est le sous-groupe de $(\mathbb{R}^2, +)$ engendré par β et les γ_k, c'est à dire est égal au groupe Γ. Au sous-groupe isolé $\bar{\Gamma} = (\{0\} \times \mathbb{Z})$ de Γ correspond la d'ecomposition $\nu = \nu' \circ \bar{\nu}$ où ν' est la valuation de K définie par $\nu'(u) = 1/3$, $\nu'(v) = 1$ et $\nu'(w_k) = p_k/2^k$, et $\bar{\nu}$ est une valuation discrète de rang 1 de $\kappa_{\nu'} = k((u^3)^a, (w_k^{2^k})^c, v^{-a-p_k c})$.

Références

[Ab 1] S.S. Abhyankar: On the valuations centered in a local domain. Amer. J. Math. **78** (1956), 321–348.

[Ab 2] S.S. Abhyankar: *Ramification Theoretic Methods in Algebraic Geometry*, Princeton University Press, 1959.

[Bo] N. Bourbaki: *Algèbre Commutative, Chapitres 5 à 7*, Masson, 1985.

[Co] V. Cossart: Uniformisation et désingularisation des surfaces d'après Zariski, dans ce volume.

[EGA] A. Grothendieck, J. Dieudonné: *Eléments de Géométrie Algébrique II*, Publ. Math. IHES **8**, 1961.

[En] O. Endler: *Valuation Theory*, Springer Verlag, 1972.

[Ha] R. Hartshorne: *Algebraic Geometry*, Graduate Texts in Math. 52, Springer Verlag, 1977.

[Hi] H. Hironaka: La voûte étoilée, dans *Singularités à Cargèse*, Astérisque 7 et 8, 1973.

[Ka 1] I. Kaplansky: Maximal fields with valuations I. Duke Math. Journ. **9** (1942), 303–321.

[Ka 2] I. Kaplansky: Maximal fields with valuations II. Duke Math. Journ. **12** (1945), 243–248.

[Ku] F.-V. Kuhlmann: Valuation theoric and model theoric aspects of local uniformization, dans ce volume.

[Ma] H. Matsumura: *Commutative Algebra*, Benjamin/Cummings Publ. Co., 1980.

[McL 1] S. MacLane: A construction for absolute values in polynomial rings. Trans. Amer. Math. Soc. **40** (1936), 363–395.

[McL 2] S. MacLane: A construction for prime ideals as absolute values of an algebraic field. Duke Math. J. **2** (1936), 492–510.

[M-S] J. Morgan, P. Shalen: Valuations, trees, and degenerations of hyperbolic structures, I. Annals of Math. **120** (1984), 401–476.

[Na] M. Nagata: *Local Rings*, Krieger Publ. Co., 1975.

[Re] D. Rees: *Lectures on the asymptotic theory of ideals*, London Math. Society, Lecture Notes Series 113, Cambridge University Press, 1988.

[Ri] P. Ribenboim: *Théorie des valuations*, Les Presses de l'Université de Montréal, 1964.

[Se] J.P. Serre: *Corps locaux*, Hermann, 1968.

[Sp 1] M. Spivakovsky: Valuations in function fields of surfaces. Amer. J. Math. **112** (1990), 107–156.

[Sp 2] M. Spivakovsky: On the structure of valuations centered in a local domain. Prepublication 1993.

[Sp 3] M. Spivakovsky: Resolution of singularities I: local uniformization. Prepublication 1997.

[Te] B. Teissier: Valuations, deformations and toric geometry, á paraître.

[Za] O. Zariski: The compactness of the Riemann manifold of an abstract field of algebraic functions. Bull. Amer. Math. Soc. **45** (1944), 683–691.

[Z-S] O. Zariski, P. Samuel: *Commutative Algebra II*, Van Nostrand, 1958.

Laboratoire de Mathématiques
Ecole Normale Supérieure
45 rue d'Ulm
75230 Paris, France
vaquie@dmi.ens.fr

Index